INTERNATIONAL SERIES OF MONOGRAPHS IN

NATURAL PHILOSOPHY

GENERAL EDITOR: D. TER HAAR

VOLUME 45

STATISTICAL MECHANICS

OTHER TITLES IN THE SERIES IN NATURAL PHILOSOPHY

Vol. 1. DAVIDOV—Quantum Mechanics
Vol. 2. FOKKER—Time and Space, Weight and Inertia
Vol. 3. KAPLAN—Interstellar Gas Dynamics
Vol. 4. ABRIKOSOV, GOR'KOV and DZYALOSHINSKII—Quantum Field Theoretical Methods in Statistical Physics
Vol. 5. OKUN'—Weak Interaction of Elementary Particles
Vol. 6. SHKLOVSKII—Physics of the Solar Corona
Vol. 7. AKHIEZER *et al.*—Collective Oscillations in a Plasma
Vol. 8. KIRZHNITS—Field Theoretical Methods in Many-body Systems
Vol. 9. KLIMONTOVICH—The Statistical Theory of Non-equilibrium Processes in a Plasma
Vol. 10. KURTH—Introduction to Stellar Statistics
Vol. 11. CHALMERS—Atmospheric Electricity (2nd Edition)
Vol. 12. RENNER—Current Algebras and their Applications
Vol. 13. FAIN and KHANIN—Quantum Electronics, Volume 1 – Basic Theory
Vol. 14. FAIN and KHANIN—Quantum Electronics, Volume 2 – Maser Amplifiers and Oscillators
Vol. 15. MARCH—Liquid Metals
Vol. 16. HORI—Spectral Properties of Disordered Chains and Lattices
Vol. 17. SAINT JAMES, THOMAS and SARMA—Type II Superconductivity
Vol. 18. MARGENAU and KESTNER—Theory of Intermolecular Forces
Vol. 19. JANCEL—Foundations of Classical and Quantum Statistical Mechanics
Vol. 20. TAKAHASHI—An Introduction to Field Quantization
Vol. 21. YVON—Correlations and Entropy in Classical Statistical Mechanics
Vol. 22. PENROSE—Foundations of Statistical Mechanics
Vol. 23. VISCONTI—Quantum Field Theory, Volume 1
Vol. 24. FURTH—Fundamental Principles of Theoretical Physics
Vol. 25. ZHELEZNYAKOV—Radioemission of the Sun and Planets
Vol. 26. GRINDLAY—An Introduction to the Phenomenological Theory of Ferroelectricity
Vol. 27. UNGER—Introduction to Quantum Electronics
Vol. 28. KOGA—Introduction to Kinetic Theory Stochastic Processes in Gaseous Systems
Vol. 29. GALASIEWICZ—Superconductivity and Quantum Fluids
Vol. 30. CONSTANTINESCU and MAGYARI—Problems in Quantum Mechanics
Vol. 31. KOTKIN and SERBO—Collection of Problems in Classical Mechanics
Vol. 32. PANCHEV—Random Functions and Turbulence
Vol. 33. TALPE—Theory of Experiments in Paramagnetic Resonance
Vol. 34. TER HAAR—Elements of Hamiltonian Mechanics (2nd Edition)
Vol. 35. CLARK and GRAINGER—Polarized Light and Optical Measurement
Vol. 36. HAUG—Theoretical Solid State Physics, Volume 1
Vol. 37. JORDAN and BEER—The Expanding Earth
Vol. 38. TODOROV—Analytic Properties of Feynman Diagrams in Quantum Field Theory
Vol. 39. SITENKO—Lectures in Scattering Theory
Vol. 40. SOBEL'MAN—Introduction to the Theory of Atomic Spectra
Vol. 41. ARMSTRONG and NICHOLLS—Emission, Absorption and Transfer of Radiation in Heated Atmospheres
Vol. 42. BRUSH—Kinetic Theory, Volume 3
Vol. 43. BOGOLYUBOV—A Method for Studying Model Hamiltonians
Vol. 44. TSYTOVICH—An Introduction to the Theory of Plasma Turbulence

STATISTICAL MECHANICS

by

R. K. PATHRIA

Department of Physics,
University of Waterloo, Ontario, Canada

PERGAMON PRESS

Oxford · New York · Toronto
Sydney · Braunschweig

Pergamon Press Ltd., Headington Hill Hall, Oxford
Pergamon Press Inc., Maxwell House, Fairview Park, Elmsford, New York 10523
Pergamon of Canada Ltd., 207 Queen's Quay West, Toronto 1
Pergamon Press (Aust.) Pty. Ltd., 19a Boundary Street, Rushcutters Bay, N.S.W. 2011, Australia
Vieweg & Sohn GmbH, Burgplatz 1, Braunschweig

First edition 1972
Library of Congress Catalog Card No. 73-181690

Printed in Hungary

08 016747 0

TO

PROFESSORS F. C. AULUCK AND D. S. KOTHARI

who initiated me into the study of this subject

CONTENTS

PREFACE

This book has arisen out of the notes of lectures that I have given to the graduate students at the McMaster University (1964–5), the University of Alberta (1965–7), the University of Waterloo (1969–71) and the University of Windsor (1970–1). While the subject matter, in its finer details, has changed considerably during the preparation of the manuscript, the style of presentation remains the same as followed in these lectures.

Statistical mechanics is an indispensable tool for studying physical properties of matter "in bulk" on the basis of the dynamical behavior of its "microscopic" constituents. Founded on the well-laid principles of *mathematical statistics* on one hand and *hamiltonian mechanics* on the other, the formalism of statistical mechanics has proved to be of immense value to the physics of the last 100 years. In view of the universality of its appeal, a basic knowledge of this subject is considered essential for every student of physics, irrespective of the area(s) in which he may be planning to specialize. To provide this knowledge, in a manner that brings out the essence of the subject with due rigor but without undue pain, is the main purpose of this work.

The fact that *the dynamics of a physical system is represented by a set of quantum states* and the assertion that *the thermodynamics of the system is determined by the multiplicity of these states* constitute the basis of our treatment. The fundamental connection between the microscopic and the macroscopic descriptions of a system is uncovered by investigating the conditions for equilibrium between two physical systems in thermodynamic contact. This is best accomplished by working in the spirit of the quantum theory right from the beginning; the entropy and other thermodynamic variables of the system then follow in a most natural manner. After the formalism is developed, one may (if the situation permits) go over to the limit of the classical statistics. This message may not be new, but here I have tried to follow it as far as is reasonably possible in a textbook. In doing so, an attempt has been made to keep the level of presentation fairly uniform so that the reader does not encounter fluctuations of too wild a character.

The text is confined to the study of the *equilibrium states* of physical systems and is intended to be used for a *graduate course* in statistical mechanics. Within these bounds, the coverage is fairly wide and provides enough material for tailoring a good two-semester course. The final choice always rests with the individual instructor; I, for one, regard Chapters 1–9 (*minus* a few sections from these chapters *plus* a few sections from Chapter 13) as the "essential part" of such a course. The contents of Chapters 10–12 are relatively advanced (not necessarily difficult); the choice of material out of these chapters will depend entirely on the taste of the instructor. To facilitate the understanding of the subject, the text has been illustrated with a large number of graphs; to assess the understanding, a large number of problems have been included. I hope these features are found useful.

I feel that one of the most essential aspects of teaching is to arouse the curiosity of the students in their subject, and one of the most effective ways of doing this is to discuss with them (in a reasonable measure, of course) the circumstances that led to the emergence of the subject. One would, therefore, like to stop occasionally to reflect upon the manner in which the various developments really came about; at the same time, one may not like the flow of the text to be hampered by the discontinuities arising from an intermittent addition of historical material. Accordingly, I decided to include in this account an Historical Introduction to the subject which stands separate from the main text. I trust the readers, especially the instructors, will find it of interest.

For those who wish to continue their study of statistical mechanics beyond the confines of this book, a fairly extensive bibliography is included. It contains a variety of references —old as well as new, experimental as well as theoretical, technical as well as pedagogical. Hopefully, this will make the book useful for a wider readership.

Waterloo, Ontario, Canada R. K. P.

ACKNOWLEDGMENTS

THE completion of this task has left me indebted to many. Like most authors, I owe considerable debt to those who have written on the subject before. The bibliography at the end of the book is the most obvious tribute to them; nevertheless, I would like to mention, in particular, the works of the Ehrenfests, Fowler, Guggenheim, Schrödinger, Rushbrooke, ter Haar, Hill, Landau and Lifshitz, Huang and Kubo, which have been my constant reference for several years and which have influenced my understanding of the subject in a variety of ways. As for the preparation of the text, I am indebted to Robert Teshima who drew most of the graphs and checked most of the problems, to Ravindar Bansal, Vishwa Mittar and Surjit Singh who went through the entire manuscript and made several suggestions that helped me unkink the exposition at a number of points, to Mary Annetts who typed the manuscript with exceptional patience, diligence and care, and to Fred Hetzel, Jim Briante and Larry Kry who provided technical help during the preparation of the final version.

As this work progressed I felt increasingly gratified towards Professors F. C. Auluck and D. S. Kothari of the University of Delhi with whom I started my career and who initiated me into the study of this subject, and towards Professor R. C. Majumdar who took keen interest in my work on this and every other project that I have undertaken from time to time. I am grateful to Dr. D. ter Haar of the University of Oxford who, as the general editor of this series, gave valuable advice on various aspects of the preparation of the manuscript and made several useful suggestions towards the improvement of the text. I am thankful to Professors J. W. Leech, J. Grindlay and A. D. Singh Nagi of the University of Waterloo for their interest and hospitality that went a long way in making this task a pleasant one.

The final tribute must go to my wife whose cooperation and understanding, at all stages of this project and against all odds, have been simply overwhelming.

Waterloo, Ontario, Canada R. K. P.

HISTORICAL INTRODUCTION

STATISTICAL mechanics is a formalism which aims at explaining the physical properties of matter *in bulk* on the basis of the dynamical behavior of its *microscopic* constituents. The scope of the formalism is almost as unlimited as the very range of the natural phenomena, for in principle it is applicable to matter in any state whatsoever. It has, in fact, been applied, with considerable success, to the study of matter in the solid state, the liquid state or the gaseous state, matter composed of several phases and/or several components, matter under extreme conditions of density and temperature, matter in equilibrium with radiation (as, for example, in astrophysics), matter in the form of a biological specimen, etc. Furthermore, the formalism of statistical mechanics enables us to investigate the *nonequilibrium* states of matter as well as the *equilibrium* states; indeed, these investigations help us to understand the manner in which a physical system that happens to be "out of equilibrium" at a given time t approaches a "state of equilibrium" as time passes.

In contrast with the present status of its development, the success of its applications and the breadth of its scope, the beginnings of statistical mechanics were rather modest. Barring certain primitive references, such as those of Gassendi, Hooke, etc., the real work started with the contemplations of Bernoulli (1738), Herapath (1821) and Joule (1851) who, in their own ways, attempted to lay foundation for the so-called *kinetic theory of gases*—a discipline that finally turned out to be the forerunner of statistical mechanics. The pioneering work of these investigators established the fact that the pressure of a gas arose from the motion of its molecules and could be computed by considering the dynamical influence of the molecular bombardment on the walls of the container. Thus, Bernoulli and Herapath could show that, if the temperature remained constant, the pressure P of an ordinary gas was inversely proportional to the volume V of the container (Boyle's law), and that it was essentially independent of the shape of the container. This, of course, involved the explicit assumption that, *at a given temperature*, the (mean) speed of the molecules is independent of both pressure and volume. Bernoulli even attempted to determine the (first-order) correction to this law, arising from the *finite* size of the molecules, and showed that the volume V appearing in the statement of the law should be replaced by $(V-b)$, where b is the "actual" volume of the molecules.* Joule was the first to show that the pressure P is directly proportional to the square of the molecular speed c, which he had assumed to be the same for all the molecules. Krönig (1856) went a step further. Introducing the "quasi-statistical" assumption that, *at any time* t, one-sixth of the gas molecules could be assumed to be flying in

* As is well known, this "correction" was correctly evaluated, much later, by van der Waals (1873) who showed that, for large V, b is equal to *four times* the "actual" volume of the molecules; see Problem 1.4.

each of the six "independent" directions, namely $+x$, $-x$, $+y$, $-y$, $+z$ and $-z$, he derived the equation

$$P = \tfrac{1}{3}nmc^2, \tag{1}$$

where n is the number density of the molecules and m the molecular mass. Krönig, too, assumed the molecular speed c to be the same for all the molecules; of course, from (1), he concluded that the kinetic energy of the molecules should be directly proportional to the absolute temperature of the gas.

Krönig justified his method in these words: "The path of each molecule must be so irregular that it will defy all attempts at calculation. However, according to the laws of probability, one could assume a completely regular motion in place of a completely irregular one!" It must, however, be noted that it is only because of the special form of the summations appearing in the calculation of the pressure that Krönig's model leads to the same result as the one following from more refined models. In other problems, such as diffusion, viscosity or heat conduction, this is no longer the case.

It was at this stage that Clausius entered into the field. First of all, in 1857, he derived the ideal-gas law under assumptions far less stringent than Krönig's. He discarded both of the leading assumptions of Krönig and showed that eqn. (1) was still true; of course, c^2 now became the *mean square speed* of the molecules. In a later paper (1859), Clausius introduced the concept of the *mean free path* and thus became the first to analyze the transport phenomena. It was in these studies that he introduced the famous "Stosszahlansatz"—the hypothesis on the number of collisions (among the molecules)—which had to play, later on, a prominent role in the monumental work of Boltzmann.* With Clausius, the introduction of the microscopic and statistical points of view into physical theory was definitive, rather than speculative. Accordingly, Maxwell, in a popular article entitled "Molecules", written for the *Encyclopedia Britannica*, has referred to him as the "principal founder of the kinetic theory of gases", while Gibbs, in his Clausius obituary notice, has called him the "father of statistical mechanics".†

The work of Clausius attracted Maxwell to the field. He made his first appearance with the great memoir "Illustrations in the dynamical theory of gases" (1860), in which he went much ahead of his predecessors by deriving his famous law of "distribution of molecular speeds". This derivation was based on the elementary principles of probability and was clearly inspired by the Gaussian law of "distribution of random errors". A derivation based on the requirement that "the *equilibrium* distribution of molecular speeds, once acquired, should remain invariant under molecular collisions" appeared in 1867. This led Maxwell to establish what is known as *Maxwell's transport equation* which, if skilfully used, leads to the same results as the ones following from the more fundamental equation due to Boltzmann.‡

* For an excellent review of this and related topics, see P. and T. Ehrenfest (1912).

† For further details, refer to Montroll (1963) where an account is also given of the pioneering work of Waterston (1846, 1892).

‡ This has been demonstrated in Guggenheim (1960) where the coefficients of viscosity, thermal conductivity and diffusion of a gas of hard spheres have been calculated on the basis of Maxwell's transport equation.

Maxwell's contributions to the subject diminished considerably after his appointment, in 1871, as the Cavendish Professor at Cambridge. By that time Boltzmann had already made his first strides. In the period 1868–71 he generalized Maxwell's distribution law to polyatomic gases, also taking into account the presence of external forces, if any; this gave rise to the famous *Boltzmann factor* $\exp(-\beta\varepsilon)$, where ε denotes the *total* energy of a molecule. These investigations also led to the *equipartition theorem*. Boltzmann further showed that, just like the original distribution of Maxwell, the generalized distribution (which we now call the *Maxwell–Boltzmann distribution*) is stationary with respect to molecular collisions. In 1872 came the celebrated *H-theorem* which provided a molecular basis for the natural tendency of physical systems to approach, and stay in, a state of equilibrium. This established the fundamental connection between the microscopic approach (which characterizes statistical mechanics) and the phenomenological approach (which characterized thermodynamics) much more transparently than ever before; it also provided a direct method of computing the entropy of a given physical system from a purely microscopic standpoint. As a corollary to the H-theorem, Boltzmann showed that the Maxwell–Boltzmann distribution is the *only* distribution that stays invariant under molecular collisions and that any other distribution, under the influence of molecular collisions, ultimately goes over into a Maxwell–Boltzmann distribution. In 1876 Boltzmann derived his famous transport equation which, in the hands of Chapman and Enskog (1916–17), has proved to be an extremely powerful tool for investigating the macroscopic properties of systems in nonsteady states.

Things, however, proved quite harsh for Boltzmann. His H-theorem, and the consequent *irreversible* character of physical systems, came under heavy attack, mainly from Loschmidt (1876–7) and Zermelo (1896). While Loschmidt wondered how the consequences of this theorem could be reconciled with the reversible character of the basic equations of motion of the molecules, Zermelo wondered how these consequences could be made to fit with the *quasi-periodic* behavior of closed systems (which arose in view of the so-called Poincaré cycles). Boltzmann defended himself against these attacks with all his might but could not convince his opponents of the correctness of his work. At the same time, the energeticists, led by Mach and Ostwald, were criticizing the very (molecular) basis of the kinetic theory,* while Lord Kelvin was emphasizing the "nineteenth-century clouds hovering over the dynamical theory of light and heat".†

All this left Boltzmann in a state of despair and induced in him a persecution complex.‡ He wrote in the introduction to the second volume of his treatise *Vorlesungen über Gastheorie* (1898):§

> I am convinced that the attacks (on the kinetic theory) rest on misunderstandings and that the role of the kinetic theory is not yet played out. In my opinion it would be a blow to science if the contemporary opposition were to cause the kinetic theory to sink into the oblivion which was the fate suffered by

---

* These critics were silenced by Einstein whose work on the Brownian motion (1905b) established atomic theory *once and for all*.

† The first of these clouds was concerned with the mysteries of the "aether", and was dispelled by the theory of relativity. The second was concerned with the inadequacy of the "equipartition theorem", and was dispelled by the quantum theory.

‡ Some people attribute Boltzmann's suicide on September 5, 1906 to this cause.

§ Quotation from Montroll (1963).

the wave theory of light through the authority of Newton. I am aware of the weakness of an individual against the prevailing currents of opinion. In order to insure that not too much will have to be rediscovered when people return to the study of the kinetic theory I will present the most difficult and misunderstood parts of the subject in as clear a manner as I can.

We shall not dwell any further on the kinetic theory; we would rather consider the development of the more sophisticated approach known as the *ensemble theory*, which may in fact be regarded as the statistical mechanics proper.* In this approach, the dynamical state of a given system, as characterized by the generalized coordinates q_i and the generalized momenta p_i of the system, is represented by a *phase point* $G(q_i, p_i)$ in a *phase space* of appropriate dimensionality. The evolution of the dynamical state in time is depicted by the *trajectory* of the G-point in the phase space, the "geometry" of the trajectory being governed by the equations of motion of the system and by the nature of the physical constraints acting on the system. To evolve an appropriate formalism, one considers the given system along with an infinitely large number of "mental copies" thereof, i.e. an *ensemble* of systems under identical physical constraints (though, at any time t, the various systems in the ensemble would differ widely in respect of their dynamical states). In the phase space, then, one has a swarm of infinitely many G-points (which, at any time t, are widely dispersed in this space and, with time, move along their respective trajectories). The fiction of a host of infinitely many, identical but independent, systems allows one to replace certain dubious assumptions of the kinetic theory of gases by readily acceptable statements of statistical mechanics. The explicit formulation of these statements was first given by Maxwell (1879) who on this occasion used the word "statistico-mechanical" to describe the study of ensembles (of gaseous systems). However, eight years earlier, Boltzmann (1871) had already worked with essentially the same kind of ensembles.

The most important quantity in the ensemble theory is the *density function*, $\varrho(q, p; t)$, of the G-points in the phase space; a stationary distribution ($\partial\varrho/\partial t = 0$) characterizes a *stationary ensemble*, which in turn represents a system *in equilibrium*. Maxwell and Boltzmann confined their study to those ensembles for which the function ϱ depended solely on the energy E of the system. This included the special case of the *ergodic* systems which were so defined that "the undisturbed motion of such a system, if pursued for an unlimited time, would ultimately traverse (the neighborhood of) each and every phase point compatible with the *fixed* value E of the energy". Consequently, the *ensemble average*, $\langle f \rangle$, of a physical quantity f, taken at *any* time t, would be the same as the *long-time average*, $\bar{f}$, pertaining to *any* member of the ensemble. Now, $\bar{f}$ is the value we expect to obtain for the quantity in question when we make an appropriate measurement on the system; the result of this measurement should, therefore, agree with the theoretical estimate $\langle f \rangle$. We thus acquire a recipe which enables us to bring about a direct contact between the theory and the experiment. At the same time, we lay down a rational basis for a microscopic theory of matter as an alternative to the empirical approach of thermodynamics!

A significant advance in this direction was made by Gibbs who, by his *Elementary Principles of Statistical Mechanics* (1902), turned the ensemble theory into a most efficient tool for the theorist. He emphasized the use of "generalized" ensembles and developed

* For a review of the historical development of kinetic theory leading to statistical mechanics, see Brush (1957, 1958, 1961, 1965–6).

schemes which, in principle, enabled one to compute the complete set of thermodynamic quantities of a given system from purely mechanical properties of its microscopic constituents.* In its methods and results, the work of Gibbs turned out to be much more general than any preceding treatment of the subject; it applied to any physical system that met the simple-minded requirements that (i) it was mechanical in structure and (ii) it obeyed Lagrange's and Hamilton's equations of motion. In this respect, Gibbs's work may be considered to have accomplished as much for thermodynamics as Maxwell's had for electrodynamics.

These developments almost coincided with the great revolution that Planck's work of 1900 brought into physics. As is well known, Planck's *quantum hypothesis* successfully resolved the essential mysteries of the black-body radiation—a subject where the three best-established disciplines of the nineteenth century, namely mechanics, electrodynamics and thermodynamics, were all focused. At the same time, it disclosed all the strengths and weaknesses of these disciplines. It would have been strange if statistical mechanics, which links thermodynamics with mechanics, could have escaped the repercussions of this revolution.

The subsequent work of Einstein (1905a) on the photoelectric effect and of Compton (1923) on the scattering of x-rays established, so to say, the "existence" of the *quantum of radiation*, or the *photon* as we now call it.† It was then natural that someone tried to derive Planck's radiation formula by treating the black-body radiation as a *gas of photons* in much the same way as Maxwell had derived his law of distribution (of molecular speeds) for a gas of conventional molecules. But, then, does a gas of photons differ so radically from a gas of conventional molecules that the two laws of distribution should be so different from one another?

The answer to this question was provided by the manner in which Planck's formula was derived by Bose. In his historic paper of 1924, Bose treated the black-body radiation as a gas of photons; however, instead of considering the allocation of the "individual" photons to the various energy states of the system, he fixed his attention on the number of states that contained "a particular number" of photons. Einstein, who seems to have translated this paper into German from an English manuscript sent to him by Bose, at once recognized the importance of this approach and added the following note to his translation: "Bose's derivation of Planck's formula is in my opinion an important step forward. The method employed here would also yield the quantum theory of an ideal gas, which I propose to demonstrate elsewhere."

Implicit in Bose's approach was the fact that in the case of photons what really mattered was "the set of numbers (of photons) in various energy states of the system" and not the specification as to "which photon was in which state"; in other words, the photons

* In much the same way as Gibbs, but quite independently of him, Einstein (1902, 1903) also developed the theory of ensembles.

† Strictly speaking, it might be somewhat misleading to cite Einstein's work on the photoelectric effect as a proof of the "existence" of photons. In fact, many of the effects (including the photoelectric effect), for which it seems necessary to invoke photons, can be explained away on the basis of a wave theory of radiation. The only phenomena for which photons seem indispensable are the ones involving *fluctuations*, such as the Hanbury Brown–Twiss effect or the Lamb shift. For the relevance of fluctuations to the problem of radiation, see ter Haar (1967, 1968).

were *mutually indistinguishable*. Einstein argued that what Bose had implied for photons should be true for material particles as well (for the property of indistinguishability arose essentially from the wave character of these entities and, according to de Broglie, material particles also possessed that character).* In two papers, which appeared soon after, Einstein (1924, 1925) applied this method to the study of an ideal gas and thereby developed what we now call *Bose–Einstein statistics*. In the second of these papers, the fundamental difference between the new statistics and the classical *Maxwell–Boltzmann* statistics comes out transparently in terms of the indistinguishability of the molecules.† In the same paper Einstein discovered the famous phenomenon of *Bose–Einstein condensation* which, thirteen years later, was adopted by London (1938) as the basis for a microscopic understanding of the curious properties of liquid helium at low temperatures.

Following the enunciation of Pauli's exclusion principle (1925), Fermi (1926) showed that certain physical systems would obey a different kind of statistics, viz. the *Fermi–Dirac statistics*, in which not more than one particle could occupy the same energy state ($n_i = 0, 1$). It seems important to mention here that Bose's method leads to the Fermi–Dirac distribution as well, provided that one limits the occupancy of an energy state to *at most* one particle!‡

Soon after its appearance, the Fermi–Dirac statistics were applied, by Fowler (1926), to discuss the equilibrium states of the white dwarf stars and, by Pauli (1927), to explain the weak, temperature-independent paramagnetism of the alkali metals; in each case, one had to deal with a "highly degenerate" gas of electrons. In the wake of this, Sommerfeld produced his monumental work of 1928 which not only put the electron theory of metals on a physically secure foundation but also gave it a fresh start in the right direction. Thus, Sommerfeld could explain practically all the major properties of metals that arose from conduction electrons and, in each case, obtained results which showed much better agreement with experiment than the ones following from the classical theories of Riecke (1898), Drude (1900) and Lorentz (1904–5). Around the same time, Thomas (1927) and Fermi (1928) investigated the electron distribution in heavier atoms and obtained theoretical estimates for the relevant binding energies; these investigations led to the development of the so-called *Thomas–Fermi model* of the atom, which has been considerably extended so as to be applicable to molecules, solids and nuclei as well.§

Thus, the whole structure of statistical mechanics was overhauled by the introduction of the concept of indistinguishability of (identical) particles.|| The statistical aspect of the problem, which was already there in view of the large number of particles present, was

* Of course, in the case of material particles, the total number N (of the particles) would also have to be conserved; this had not to be done in the case of photons. For details, see Sec. 6.1.

† It is here that one encounters the *correct* method of counting "the number of distinct ways in which g_i energy states can accommodate n_i particles", depending upon whether the particles are (i) distinguishable or (ii) indistinguishable. The occupancy of the individual states was, in each case, *unrestricted*, i.e. $n_i = 0, 1, 2, \ldots$.

‡ Dirac, who was the first to investigate the connection between statistics and wave mechanics, showed, in 1926, that the wave functions describing a system of (identical) particles obeying Bose–Einstein (or Fermi–Dirac) statistics must be symmetric (or antisymmetric) with respect to an interchange of two particles.

§ For an excellent review of this model, see N. H. March (1957).

|| Of course, in many a situation, especially at high temperatures and low densities, where the wave nature of the particles is not very important the classical statistics are still applicable.

augmented by another statistical aspect that arose from the probabilistic nature of the wave mechanical description. One had, therefore, to carry out a *two-fold* averaging of the dynamical variables over the states of the given system in order to obtain the relevant expectation values. That sort of a situation was bound to necessitate a reformulation of the ensemble theory itself. This was carried out step by step. First of all, Landau (1927) and von Neumann (1927) introduced the so-called *density matrix*, which was meant to be the quantum-mechanical analogue of the *density function* of the classical phase space; this was discussed, both from statistical and quantum-mechanical points of view, by Dirac (1929–31). Guided by the classical ensemble theory, these authors considered both *microcanonical* and *canonical* ensembles; the introduction of *grand canonical* ensembles in quantum statistics was made by Pauli (1927).*

The important question as to which particles would obey Bose–Einstein statistics and which Fermi–Dirac remained theoretically unsettled until Belinfante (1939) and Pauli (1940) discovered the vital connection between spin and statistics.† It turns out that those particles whose spin is an integral multiple of $\hbar$ obey Bose–Einstein statistics while those whose spin is a half-odd integral multiple of $\hbar$ obey Fermi–Dirac statistics. To date, no third category of particles has been discovered.

Apart from the foregoing milestones, several notable contributions towards the development of statistical mechanics have been made from time to time; however, most of these contributions are concerned with the development or perfection of mathematical techniques which make the application of the basic formalism to actual physical problems more fruitful. A review of these developments is clearly out of place here; we better discuss them at their appropriate place in the text.

* A detailed treatment of this development has been given by Kramers (1938).

† See also Lüders and Zumino (1958).

CHAPTER 1

THE STATISTICAL BASIS OF THERMODYNAMICS

IN THE annals of thermal physics, the fifties of the last century mark a very definite epoch. By that time the science of thermodynamics, which grew essentially out of an experimental study of the macroscopic behavior of physical systems, had become, through the work of Carnot, Joule, Clausius and Kelvin, a secure and stable discipline of physics. The theoretical conclusions following from the first two laws of thermodynamics were found to be in very good agreement with the corresponding experimental results.* At the same time, the kinetic theory of gases, which aimed at explaining the macroscopic behavior of gaseous systems in terms of the motion of the molecules and had so far developed more on speculation than on calculation, began to emerge as a real, mathematical theory. Its initial successes were indeed glaring; however, a real contact with thermodynamics could not be made until about 1872 when Boltzmann developed his H-theorem and thereby established a direct connection between entropy on the one hand and the dynamics of the molecules on the other. Almost simultaneously, the conventional (kinetic) theory began giving way to its more sophisticated successor—the ensemble theory. The power of the techniques that finally emerged reduced thermodynamics to the status of an "essential" consequence of the get-together of the *statistics* and the *mechanics* of the molecules constituting a given physical system. It was then natural to give the resulting formalism the name *Statistical Mechanics*.

As a preparation towards the development of the formal theory, we start with a few general considerations regarding the statistical nature of a macroscopic system. These considerations should, in some measure, provide ground for a statistical interpretation of thermodynamics. It may be mentioned here that, unless a statement is made to the contrary, the system under study is supposed to be in one of its equilibrium states.

* The third law, which is also known as *Nernst's heat theorem*, did not arrive until about 1906. For a general discussion of this law, see Simon (1930) and Wilks (1961). These references also provide an extensive bibliography on this subject.

1.1. The macroscopic and the microscopic states

We consider a physical system composed of N identical particles confined to a space of volume V. In a typical case, N would be an extremely large number—generally, of the order of 10^{23}. In view of this, it is customary to carry out analysis in the so-called *thermodynamic limit*, viz. $N \to \infty$, $V \to \infty$ (such that the ratio N/V, which is generally denoted by the symbol n and is referred to as the *particle density*, stays fixed at a preassigned value). In this limit, the *extensive* properties of the system become directly proportional to the size of the system (i.e. proportional to N or to V), while the *intensive* properties become independent thereof; the particle density, of course, remains an important parameter for all physical properties of the system.

Next we consider the total energy of the system. If the particles comprising the system could be regarded as noninteracting, the total energy E of the system would be equal to the sum of the energies ε_i of the individual particles:

$$E = \sum_i n_i \varepsilon_i, \tag{1}$$

where n_i denotes the number of particles with energy ε_i. Clearly,

$$N = \sum_i n_i. \tag{2}$$

According to quantum mechanics, the possible values of the single-particle energies ε_i are discrete and their magnitude depends crucially on the volume V to which the particles are confined. Accordingly, the possible values of the total energy E are also discrete. However, for large V, the spacings of the different energy values are so small in comparison with the total energy of the system that the parameter E might be regarded as almost a *continuous* variable. This would be true even if the particles were mutually interacting; however, in that case the total energy E cannot be written in the form (1).

The specification of the actual values of the parameters N, V and E then defines a particular *macrostate* of the given system.

At the molecular level, however, a large number of possibilities still exist, because at that level there will *in general* be a large number of different ways in which the macrostate (N, V, E) of the given system can be realized. In the case of a noninteracting system, since the total energy E consists of a simple sum of the N single-particle energies ε_i, there will obviously be a large number of different ways in which the individual ε_i's can be chosen so as to make the total energy equal to E. In other words, there will be a large number of different ways in which the total energy E of the system can be distributed among the N particles constituting it. Each of these (different) ways specifies a particular *microstate, or complexion*, of the given system. In general, the various microstates, or complexions, of a given system could be identified with the independent solutions $\psi(\boldsymbol{r}_1, \ldots, \boldsymbol{r}_N)$ of the Schrödinger equation of the system, corresponding to the eigenvalue E of the relevant Hamiltonian. In any case, to a given macrostate of the system there does in general correspond a large number of microstates, and it seems natural to assume that at any time t the system is *equally likely* to be in any one of these microstates. This assumption forms the backbone of our formalism and is generally referred to as the postulate of "equal *a priori* probabilities" for all microstates consistent with a given macrostate.

The actual number of all possible microstates will, of course, be a function of N, V and E and may be denoted by the symbol $\Omega(N, V, E)$; the dependence on V comes in because the possible values ε_i of the single-particle energy ε are a function of this parameter.* Curiously enough, it is from the magnitude of the number Ω, and from the nature of its dependence on the parameters N, V and E, that the complete thermodynamics of the given system can be derived!

We shall not stop here to discuss the ways in which the number $\Omega(N, V, E)$ can be computed; we shall do that only when we have developed our considerations sufficiently so that we can carry out further derivations from it. First we have to discover the manner in which this number is related to any of the leading thermodynamic quantities. To do this, we consider here the problem of the "thermal contact" between two physical systems, in the hope that this consideration will bring out the true nature of the number Ω.

1.2. Contact between statistics and thermodynamics: physical significance of $\Omega(N, V, E)$

We consider two physical systems, A_1 and A_2, which are separately in equilibrium; see Fig. 1.1. Let the macrostate of A_1 be represented by the parameters N_1, V_1 and E_1, so that it has $\Omega_1(N_1, V_1, E_1)$ possible complexions, and the macrostate of A_2 be represented

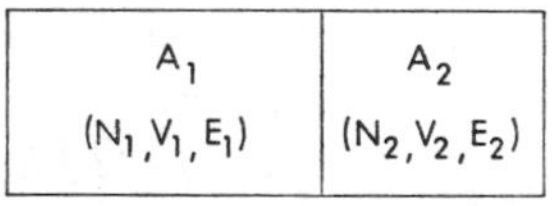

FIG. 1.1. Two physical systems, being brought into thermal contact.

by the parameters N_2, V_2 and E_2, so that it has $\Omega_2(N_2, V_2, E_2)$ possible complexions. The mathematical form of the function Ω_1 need not be the same as that of the function Ω_2, because that ultimately depends upon the nature of the system. We, of course, believe that all thermodynamic properties of the systems A_1 and A_2 can be completely derived from the respective functions $\Omega_1(N_1, V_1, E_1)$ and $\Omega_2(N_2, V_2, E_2)$.

We now bring the two systems into thermal contact with each other, thus allowing the possibility of an exchange of energy between the two; this can be done by sliding in a conducting wall and removing the impervious one. For simplicity, the two systems are still separated by a rigid, impenetrable wall, so that the respective volumes V_1 and V_2 and the respective particle numbers N_1 and N_2 remain unchanged. The energies E_1 and E_2, however, become variable and the only condition that restricts their variation is

$$E^{(0)} = E_1 + E_2 = \text{const.} \tag{1}$$

* It may be noted that the manner in which the ε_i's depend on V is itself determined by the nature of the system. For instance, it is not the same for relativistic systems as it is for nonrelativistic ones; compare, for instance, the cases dealt with in Sec. 1.4 and in Problem 1.8. We should also note that, *in principle*, the dependence of Ω on V arises from the fact that it is the *physical dimensions* of the container that enter into the boundary conditions imposed on the wave functions of the system.

Here, $E^{(0)}$ denoted the energy of the composite system $A^{(0)}$ ($\equiv A_1 + A_2$); the energy of interaction between A_1 and A_2, if any, is being neglected. Now, at any time t, the sub-system A_1 is equally likely to be in any one of the $\Omega_1(E_1)$ microstates while the sub-system A_2 is equally likely to be in any one of the $\Omega_2(E_2)$ microstates; therefore, the composite system $A^{(0)}$ is equally likely to be in any one of the $\Omega^{(0)}(E_1, E_2)$ microstates, where

$$\Omega^{(0)}(E_1, E_2) = \Omega_1(E_1)\,\Omega_2(E_2) = \Omega_1(E_1)\,\Omega_2(E^{(0)} - E_1) = \Omega^{(0)}(E^{(0)}, E_1), \text{ say.} \qquad (2)^*$$

Clearly, the number $\Omega^{(0)}$ itself varies with E_1. The question then arises: at what value of the variable E_1 will the composite system be in equilibrium? Or, in other words, at what stage of the energy exchange will A_1 and A_2 be in mutual equilibrium?

We assert that this will happen at that value of E_1 which *maximizes* the number $\Omega^{(0)}(E^{(0)}, E_1)$. The philosophy behind this assertion is that a physical system, left to itself, naturally proceeds in a direction that enables it to assume an ever-increasing number of complexions until it finally settles down in a state that affords the *largest possible* number of complexions. Statistically speaking, we regard a state with a larger number of complexions as a more probable state, and the one with the largest number of complexions as the most probable. Detailed studies show that for a typical system the number of complexions pertaining to any state which departs even slightly from the most probable state is "orders of magnitude" smaller than the number pertaining to the most probable state. Thus, the most probable state of a system is *the* state in which the system spends an "overwhelmingly" large fraction of its time. It is then natural to identify this state with the *equilibrium* state of the system.

Denoting the equilibrium value of E_1 by $\bar{E}_1$ (and that of E_2 by $\bar{E}_2$), we obtain, on maximizing $\Omega^{(0)}$,

$$\left(\frac{\partial \Omega_1(E_1)}{\partial E_1}\right)_{E_1=\bar{E}_1} \Omega_2(\bar{E}_2) + \Omega_1(\bar{E}_1)\left(\frac{\partial \Omega_2(E_2)}{\partial E_2}\right)_{E_2=\bar{E}_2} \cdot \frac{\partial E_2}{\partial E_1} = 0.$$

Since $\partial E_2/\partial E_1 = -1$, see eqn. (1), the foregoing condition can be written in the form

$$\left(\frac{\partial \ln \Omega_1(E_1)}{\partial E_1}\right)_{E_1=\bar{E}_1} = \left(\frac{\partial \ln \Omega_2(E_2)}{\partial E_2}\right)_{E_2=\bar{E}_2}.$$

Thus, our condition for equilibrium reduces to the equality of the respective values of the parameter $\{\partial \ln \Omega(N, V, E)/\partial E\}_{N,V}$ for the sub-systems A_1 and A_2. We denote this parameter by the symbol β:

$$\beta \equiv \left(\frac{\partial \ln \Omega(N, V, E)}{\partial E}\right)_{N,V}, \qquad (3)$$

whence the condition for equilibrium becomes

$$\beta_1 = \beta_2. \qquad (4)$$

We thus find that when two physical systems are brought into thermal contact, which enables an exchange of energy to take place between them, this exchange continues *until*

* It should be obvious that the macrostate of the composite system $A^{(0)}$ has to be defined by two energies, viz. E_1 and E_2 (or else $E^{(0)}$ and E_1).

the equilibrium values $\bar{E}_1$ and $\bar{E}_2$ of the variables E_1 and E_2 are reached. Once these values are reached, there is no more exchange of energy between the two systems; the systems are then said to have attained a state of mutual equilibrium. According to our analysis, this happens only when the respective values of the parameter β, namely β_1 and β_2, become equal.* It is then natural that the parameter β be regarded as the statistical analogue of the *thermodynamic temperature* T. To determine the precise relationship between β and T, we consider the same problem from the conventional, thermodynamic point of view.

When the sub-systems A_1 and A_2 were isolated, their individual entropies $S_1(N_1, V_1, E_1)$ and $S_2(N_2, V_2, E_2)$ were determined by their respective macrostates. When they came into thermal contact, an exchange of energy set in. Let us consider, during this period of exchange, the transfer of an energy ΔE from the sub-system A_1 to the sub-system A_2. The corresponding change in the entropy of the composite system $A^{(0)}$ would be

$$\Delta S^{(0)} = (\Delta S)_{A_1} + (\Delta S)_{A_2} = \left(\frac{\partial S_1}{\partial E_1}\right)_{N_1, V_1}(-\Delta E) + \left(\frac{\partial S_2}{\partial E_2}\right)_{N_2, V_2}(+\Delta E)$$

$$= \Delta E\left[\left(\frac{\partial S_2}{\partial E_2}\right)_{N_2, V_2} - \left(\frac{\partial S_1}{\partial E_1}\right)_{N_1, V_1}\right]. \tag{5}$$

According to the second law of thermodynamics, the total entropy $S^{(0)}$ must increase, or at best stay constant, in the transfer process. Hence, for the "natural" direction of energy flow to be from A_1 to A_2, i.e. for ΔE to be positive, we must have

$$\left(\frac{\partial S_2}{\partial E_2}\right)_{N_2, V_2} > \left(\frac{\partial S_1}{\partial E_1}\right)_{N_1, V_1}. \tag{6}$$

The foregoing condition is essentially the same as the empirical one, namely

$$T_1 > T_2, \tag{7}$$

provided we recall the thermodynamic formula

$$\left(\frac{\partial S}{\partial E}\right)_{N, V} = \frac{1}{T}. \tag{8}$$

It is clear that the state of equilibrium for the composite system $A^{(0)}$ will obtain when $\Delta S^{(0)} = 0$, i.e. when

$$T_1 = T_2. \tag{9}^{\dagger}$$

Comparing the respective conditions of equilibrium, viz. (4) and (9), we conclude that the results of the statistical approach are essentially equivalent to those of the conventional thermodynamic approach. A further comparison of the formulae (3) and (8) prompts us to draw the desired correspondence between the statistical and the thermodynamic para-

* This result may be compared with the so-called "zeroth law of thermodynamics", which stipulates the existence of a *common* parameter T for two or more physical systems in equilibrium.

† It may be pointed out here that the onset of equilibrium does not necessarily require: $\Delta E = 0$. In fact, when $T_1 = T_2$, the process of energy transfer becomes *reversible* and ΔE is then as likely to be positive as negative; of course, we must have: $\overline{\Delta E} = 0$.

meters, viz.

$$\frac{\Delta S}{\Delta(\ln \Omega)} = \frac{1}{\beta T} = \text{const.} \tag{10}$$

This correspondence was first established by Boltzmann who also believed that since the relationship between the two approaches seems to be of a fundamental character the constant appearing in (10) must be a *universal constant*. It was Planck who first wrote the explicit formula

$$S = k \ln \Omega, \tag{11}$$

without any additive constant S_0. As it stands, formula (11) determines the *absolute* value of the entropy of a given physical system in terms of the total number of microstates accessible to it in conformity with the given macrostate. The zero of the entropy then corresponds to the special state for which only one microstate is accessible ($\Omega = 1$)—the so-called "unique configuration"; the statistical approach thus provides a theoretical basis for the third law of thermodynamics as well!

The formula (11) is of fundamental importance in physics. It provides a bridge between the microscopic and the macroscopic. Now, in the study of the second law of thermodynamics we are told that the law of increase of entropy is related to the fact that the energy content of the universe, in its natural course, is becoming less and less available for conversion into work; accordingly, the entropy of a system may be supposed to be a measure of the so-called disorder prevailing in the system. Formula (11) tells us how disorder arises microscopically. Clearly, this disorder is a manifestation of the largeness of the number of complexions the system can have. The larger the choice of complexions, the lesser the degree of predictability or the degree of order in the system. Complete order prevails when and only when the system has no other choice but to be in a *unique* state ($\Omega = 1$); this, in turn, corresponds to a state of vanishing entropy.

By (10) and (11), we also have

$$\beta = \frac{1}{kT}. \tag{12}$$

The universal constant k is usually referred to as the Boltzmann constant. We shall discover in Sec. 1.4 how k is related to the gas constant R and the Avogadro number N_A; see eqn. (1.4.3).*

1.3. Further contact between statistics and thermodynamics

In continuation of the preceding considerations, we now examine a more elaborate exchange between the sub-systems A_1 and A_2. If we assume that the wall separating the two sub-systems is movable as well as conducting, the respective volumes V_1 and V_2 (of the sub-systems A_1 and A_2) also become variable; indeed, the total volume $V^{(0)}$ ($\equiv V_1 + V_2$) remains constant, so that effectively we have only one more independent variable. Of course, the wall is still assumed to be impenetrable to particles, so the numbers N_1 and N_2

* We follow the notation whereby eqn. (1.4.3) means eqn. (3) of Sec. 1.4. However, while referring to an equation in its own section, we will omit the mention of the section number.

remain fixed. Arguing as before, the state of equilibrium for the composite system $A^{(0)}$ will obtain when the number $\Omega^{(0)}(V^{(0)}, E^{(0)}; V_1, E_1)$ attains its largest value, i.e. when not only

$$\left(\frac{\partial \ln \Omega_1}{\partial E_1}\right)_{N_1, V_1; E_1=\bar{E}_1} = \left(\frac{\partial \ln \Omega_2}{\partial E_2}\right)_{N_2, V_2; E_2=\bar{E}_2}, \tag{1a}$$

but also

$$\left(\frac{\partial \ln \Omega_1}{\partial V_1}\right)_{N_1, E_1; V_1=\bar{V}_1} = \left(\frac{\partial \ln \Omega_2}{\partial V_2}\right)_{N_2, E_2; V_2=\bar{V}_2}. \tag{1b}$$

The conditions of equilibrium now take the form of an equality between the pair of parameters (β_1, η_1) of the sub-system A_1 and the parameters (β_2, η_2) of the sub-system A_2 where, by definition,

$$\eta \equiv \left(\frac{\partial \ln \Omega(N, V, E)}{\partial V}\right)_{N, E}. \tag{2}$$

Similarly, if A_1 and A_2 came into contact through a wall which allowed an exchange of particles as well, the conditions of equilibrium would be further augmented by the equality of the parameter ζ_1 of the sub-system A_1 and the parameter ζ_2 of the sub-system A_2 where, by definition,

$$\zeta \equiv \left(\frac{\partial \ln \Omega(N, V, E)}{\partial N}\right)_{V, E}. \tag{3}$$

To determine the physical meaning of the parameters η and ζ, we note that, for an infinitesimal variation in the macrostate of a system in equilibrium,

$$d(\ln \Omega) = \left(\frac{\partial \ln \Omega}{\partial E}\right)_{N, V} dE + \left(\frac{\partial \ln \Omega}{\partial V}\right)_{N, E} dV + \left(\frac{\partial \ln \Omega}{\partial N}\right)_{V, E} dN, \tag{4}$$

which, in view of our definitions, may be written as

$$d(\ln \Omega) = \beta\, dE + \eta\, dV + \zeta\, dN.$$

Making use of eqns. (1.2.11) and (1.2.12), this becomes

$$dE = T\, dS - (\eta kT)\, dV - (\zeta kT)\, dN,$$

which is identical with the fundamental formula of thermodynamics, viz.

$$dE = T\, dS - P\, dV + \mu\, dN, \tag{5}$$

provided that we invoke the correspondence:

$$\eta = \frac{P}{kT} \quad \text{and} \quad \zeta = -\frac{\mu}{kT}; \tag{6}$$

here, P and μ are the *thermodynamic pressure* and the *chemical potential* of the system. From the physical point of view, these results are completely satisfactory because, thermodynamically too, the conditions of equilibrium between two systems A_1 and A_2, if the wall separating them is both conducting and movable (thus making their respective energies and volumes variable), are indeed the same as the ones contained in eqn. (1), namely

$$T_1 = T_2 \quad \text{and} \quad P_1 = P_2. \tag{7}$$

On the other hand, if the two systems can exchange particles as well as energy but have their volumes fixed, the conditions of equilibrium, obtained thermodynamically, are indeed

$$T_1 = T_2 \quad \text{and} \quad \mu_1 = \mu_2. \tag{8}$$

Finally, if the exchange is such that all the three (macroscopic) parameters become variable, then the conditions of equilibrium become

$$T_1 = T_2, \quad P_1 = P_2 \quad \text{and} \quad \mu_1 = \mu_2. \tag{9}^*$$

It is gratifying that these conclusions are identical with the ones following from statistical considerations.

Combining the results of the foregoing discussion, we arrive at the following recipe for deriving thermodynamics from a statistical beginning: determine, for the macrostate (N, V, E) of the given system, the number of all possible complexions accessible to the system; call this number $\Omega(N, V, E)$. Then, the entropy of the system in that state follows from the fundamental formula

$$S(N, V, E) = k \ln \Omega(N, V, E), \tag{10}$$

while the leading *intensive* parameters, viz. temperature, pressure and chemical potential, are given by

$$\left(\frac{\partial S}{\partial E}\right)_{N,V} = \frac{1}{T}; \quad \left(\frac{\partial S}{\partial V}\right)_{N,E} = \frac{P}{T}; \quad \left(\frac{\partial S}{\partial N}\right)_{V,E} = -\frac{\mu}{T}. \tag{11}$$

Alternatively, we can write

$$P = \left(\frac{\partial S}{\partial V}\right)_{N,E} \Big/ \left(\frac{\partial S}{\partial E}\right)_{N,V} \equiv -\left(\frac{\partial E}{\partial V}\right)_{N,S} \tag{12a}^\dagger$$

and

$$\mu = -\left(\frac{\partial S}{\partial N}\right)_{V,E} \Big/ \left(\frac{\partial S}{\partial E}\right)_{N,V} \equiv \left(\frac{\partial E}{\partial N}\right)_{V,S}, \tag{12b}^\dagger$$

while

$$T = \left(\frac{\partial E}{\partial S}\right)_{N,V}. \tag{12c}$$

Formulae (12) are just the ones that follow from the basic law of thermodynamics, as stated in (5). The evaluation of P, μ and T from these formulae indeed requires that the

* It may be noted that the same would be true for *any* two parts of a single thermodynamic system; consequently, in equilibrium, the parameters μ, P and T must be constant *throughout* the system.

† In writing these formulae, we have made use of the well-known relationship in partial differential calculus, namely that "if three variables x, y and z are mutually related, then

$$\left(\frac{\partial x}{\partial y}\right)_z \left(\frac{\partial y}{\partial z}\right)_x \left(\frac{\partial z}{\partial x}\right)_y = -1\text{''}.$$

energy E be expressed as a function of the quantities N, V and S; this should, in principle, be possible once S is known as a function of N, V and E; see (10).

The rest of the thermodynamics follows straightforwardly. For instance, the Helmholtz free energy A, the Gibbs free energy G and the enthalpy H are given by

$$A = E-TS, \tag{13a}$$

$$G = A+PV = E-TS+PV$$
$$= \mu N \tag{13b}^*$$

and

$$H = E+PV = G+TS. \tag{13c}$$

The specific heat at constant volume, C_V, and the one at constant pressure, C_P, would be given by

$$C_V \equiv T\left(\frac{\partial S}{\partial T}\right)_{N,V} = \left(\frac{\partial E}{\partial T}\right)_{N,V} \tag{14a}$$

and

$$C_P \equiv T\left(\frac{\partial S}{\partial T}\right)_{N,P} = \left(\frac{\partial(E+PV)}{\partial T}\right)_{N,P} = \left(\frac{\partial H}{\partial T}\right)_{N,P}. \tag{14b}$$

1.4. The classical ideal gas

To illustrate the approach developed in the preceding sections, we derive here the various thermodynamic properties of a classical ideal gas composed of monatomic molecules. The main reason why we choose to consider this specialized system is that it affords an explicit, though asymptotic, evaluation of the number $\Omega(N, V, E)$. The example becomes all the more instructive when we find that its study enables us, in a most straightforward manner, to identify the Boltzmann constant k in terms of other physical constants. Moreover, the behavior of this system serves as a useful basic reference with which the behavior of other physical systems, especially real gases (with or without the inclusion of quantum effects), can be compared. And, indeed, in the limit of very high temperatures and very low densities the ideal-gas behavior becomes typical of most of the real systems.

Before undertaking a detailed study of this case it appears worth while to make a remark which applies to all *classical* systems composed of *noninteracting* particles, irrespective of the internal structure of the particles. The remark is related to the explicit dependence of the number $\Omega(N, V, E)$ on V and hence to the *equation of state* of these systems. Now, if there does not exist any spatial correlation among the particles, i.e. if the probability of any one of them being found in a particular region of the available space is completely independent of the location of the other particles,† then the total number of ways in which the N particles

* The relation $E-TS+PV = \mu N$ follows directly from (5). For this, we have to regard the given system as having grown to its present size in a gradual manner, such that the intensive parameters T, P and μ stayed constant throughout the process while the extensive parameters N, V and E (and hence S) increased *proportionately* with one another.

† This will be true if (i) the mutual interaction between the particles is negligible and (ii) the wave packets of the individual particles do not significantly overlap (or, in other words, the quantum effects are also negligible).

can be spatially distributed in the system will be simply equal to the product of the numbers of ways in which the individual particles can be independently accommodated in the available space. With N and E fixed, each of these numbers will be directly proportional to V, the volume of the container; accordingly, the total number of ways would be directly proportional to the Nth power of V:

$$\Omega(N, E, V) \propto V^N. \tag{1}$$

Combined with eqns. (1.3.10) and (1.3.11), this gives

$$\frac{P}{T} \equiv k\left(\frac{\partial \ln \Omega(N, E, V)}{\partial V}\right)_{N,E} = k\frac{N}{V}. \tag{2}$$

If the system contains n moles of the gas, then $N = nN_A$, where N_A is the *Avogadro number*. Equation (2) then becomes

$$PV = nRT \qquad (R = kN_A), \tag{3}$$

which is the famous *ideal-gas* law, R being the *gas constant*. Thus, for any classical system composed of noninteracting particles the ideal-gas law holds.

For deriving other thermodynamic properties of this system, we require a detailed knowledge of the way Ω depends on the parameters N, V and E. The problem essentially reduces to determining the total number of ways in which eqns. (1.1.1) and (1.1.2) can be mutually satisfied. In other words, we must determine the total number of (independent) ways of satisfying the equation

$$\sum_{r=1}^{3N} \varepsilon_r = E, \tag{4}$$

where ε_r's denote the energies associated with the various degrees of freedom of the N particles. The reason why this number should depend upon the parameters N and E is quite obvious. Nevertheless, this number would also depend upon the "spectrum of values" which the variables ε_r can assume; it is through the nature of this spectrum that the dependence on V comes in. Now, the energy eigenvalues for a *free, nonrelativistic* particle confined to a cubical box of side $L(V = L^3)$, under the condition that the wave function $\psi(\boldsymbol{r})$ vanishes everywhere on the boundary, are given by

$$\varepsilon(n_x, n_y, n_z) = \frac{h^2}{8mL^2}(n_x^2+n_y^2+n_z^2); \qquad n_x, n_y, n_z = 1, 2, 3, \ldots, \tag{5}$$

where h is Planck's constant and m the mass of the particle. The number of distinct eigenfunctions (or microstates) for a particle of energy ε would, therefore, be equal to the number of independent, positive-integral solutions of the equation

$$(n_x^2+n_y^2+n_z^2) = \frac{8mV^{2/3}\varepsilon}{h^2} = \varepsilon^*, \quad \text{say}. \tag{6}$$

We may denote this number by $\Omega(1, \varepsilon, V)$. It then follows that the desired number

$\Omega(N, E, V)$ would be equal to the number of independent, positive-integral solutions of the equation

$$\sum_{r=1}^{3N} n_r^2 = \frac{8mV^{2/3}E}{h^2} = E^*, \quad \text{say.} \tag{7}$$

An important result follows straightforwardly from (7), even if the number $\Omega(N, E, V)$ is not explicitly evaluated. From the nature of the expression appearing on the right-hand side of this equation we conclude that the volume V and the energy E of the system enter into the expression for Ω in the form of the combination $(V^{2/3}E)$. Consequently, we can write

$$S(N, V, E) \equiv S(N, V^{2/3}E). \tag{8}$$

Hence, for the constancy of S and N, which defines an *adiabatic* process, we must have:

$$V^{2/3}E = \text{const.} \tag{9}$$

Equation (1.3.12a) then gives

$$P = -\left(\frac{\partial E}{\partial V}\right)_{N,S} = \frac{2}{3}\frac{E}{V}, \tag{10}$$

that is, the pressure of a system of nonrelativistic, noninteracting particles is precisely equal to two-thirds of its energy density.† It must be noted here that since an explicit computation of the number Ω has not yet been done, the results (9) and (10) hold for *quantum* as well as *classical* statistics; equally general is the result obtained by combining these, viz.

$$PV^{5/3} = \text{const.}, \tag{11}$$

which tells us how P varies with V during an *adiabatic* process.

We shall now attempt to evaluate the number Ω. In this evaluation we shall explicitly assume the particles to be *distinguishable*, so that if a particle in the state i gets interchanged with a particle in the state j the resulting complexion is counted as distinct. Consequently, the number $\Omega(N, V, E)$, or better $\Omega_N(E^*)$ {see eqn. (7)}, is equal to the number of positive-integral lattice points lying on the surface of a $3N$-dimensional sphere of radius $\sqrt{E^*}$.‡ Clearly, this number will be an extremely irregular function of E^*, i.e. for two given values of E^* which may be very close to one another the values of this number could be very very different. In contrast, the number $\Sigma_N(E^*)$, which is defined as the number of positive-integral lattice points lying *on or within* the surface of a $3N$-dimensional sphere of radiust $\sqrt{E^*}$, would be much less irregular. In terms of our physical problem, this would correspond to the number, $\Sigma(N, V, E)$, of the various complexions of the given system consisten with *all* macrostates characterized by the specified values of the parameters N and V but having energy of any amount *less than or equal to E*, i.e.

$$\Sigma(N, V, E) = \sum_{E' \leqslant E} \Omega(N, V, E') \tag{12}$$

† Combining (10) with (2), we obtain for the classical ideal gas: $E = \frac{3}{2}NkT$. Accordingly, eqn. (9) reduces to the well-known thermodynamic relationship: $V^{\gamma-1}T = \text{const.}$, which holds during an *adiabatic* process, with $\gamma = \frac{5}{3}$.

‡ If the particles are regarded as *indistinguishable*, the evaluation of the number Ω becomes rather intricate. The problem is then solved by having recourse to the theory of "partitions of numbers"; see Auluck and Kothari (1946, 1947).

or

$$\Sigma_N(E^*) = \sum_{E^{*\prime} \leqslant E^*} \Omega_N(E^{*\prime}). \tag{13}$$

Of course, the number Σ will also be somewhat irregular; however, we expect that its asymptotic behavior, as $E^* \to \infty$, would be a lot more reasonable than that of Ω. We shall see in the sequel that the thermodynamics of the system follows as well from the number Σ as from the number Ω.

To appreciate the point made here, let us digress a little to examine the behavior of the numbers $\Omega_1(\varepsilon^*)$ and $\Sigma_1(\varepsilon^*)$, which correspond to the case of a single particle confined to the given volume V. The *exact* values of these numbers, for $\varepsilon^* \leqslant 10{,}000$, can be extracted from a table compiled by Gupta (1947). The wild irregularities of the number $\Omega_1(\varepsilon^*)$ can hardly be missed. The number $\Sigma_1(\varepsilon^*)$, on the other hand, exhibits a much neater asymptotic behavior. From the geometry of the problem, we note that, *asymptotically*, $\Sigma_1(\varepsilon^*)$ should be equal to the volume of an octant of a sphere of radius $\sqrt{\varepsilon^*}$, i.e.

$$\operatorname*{Lim}_{\varepsilon^* \to \infty} \frac{\Sigma_1(\varepsilon^*)}{(\pi/6)\varepsilon^{*3/2}} = 1. \tag{14}$$

A more detailed analysis shows (see Pathria, 1966b) that, to the next approximation,

$$\Sigma_1(\varepsilon^*) \simeq \frac{\pi}{6}\varepsilon^{*3/2} - \frac{3\pi}{8}\varepsilon^*; \tag{15}$$

the correction term arises from the fact that the volume of an octant somewhat overestimates the number of desired lattice points, for it includes, partly though, some points with one or more coordinates equal to zero. Figure 1.2 shows a histogram of the actual values of $\Sigma_1(\varepsilon^*)$, for ε^* lying between 200 and 300; the theoretical estimate (15) is also shown. In the figure, we have also included a histogram of the actual values of the corresponding number of complexions, $\Sigma_1'(\varepsilon^*)$, when the quantum numbers n_x, n_y and n_z can assume the value zero as well. In the latter case, the volume of an octant somewhat underestimates the number of desired lattice points; we now have

$$\Sigma_1'(\varepsilon^*) \simeq \frac{\pi}{6}\varepsilon^{*3/2} + \frac{3\pi}{8}\varepsilon^*. \tag{16}$$

Asymptotically, however, the number $\Sigma_1'(\varepsilon^*)$ also satisfies eqn. (14).

Returning to the N-particle system, the number $\Sigma_N(E^*)$ should be *asymptotically* equal to the "volume" of the "positive compartment" of a $3N$-dimensional sphere of radius $\sqrt{E^*}$. Referring to eqn. (7) of Appendix C, we obtain

$$\Sigma_N(E^*) \simeq \left(\frac{1}{2}\right)^{3N} \left\{\frac{\pi^{3N/2}}{(3N/2)!} E^{*3N/2}\right\},$$

which, on substitution for E^*, gives

$$\Sigma(N, V, E) \simeq \left(\frac{V}{h^3}\right)^N \frac{(2\pi mE)^{3N/2}}{(3N/2)!}. \tag{17}$$

Taking logarithms and applying Stirling's formula (B.38), viz.

$$\ln(n!) \simeq n \ln n - n \qquad (n \gg 1), \tag{18}$$

we get

$$\ln \Sigma(N, V, E) \simeq N \ln \left[\frac{V}{h^3}\left(\frac{4\pi m E}{3N}\right)^{3/2}\right] + \frac{3}{2} N. \tag{19}$$

For deriving the thermodynamic properties of the given system we must, first of all, fix the precise value of, or limits for, the energy of the system. In view of the extremely irregular

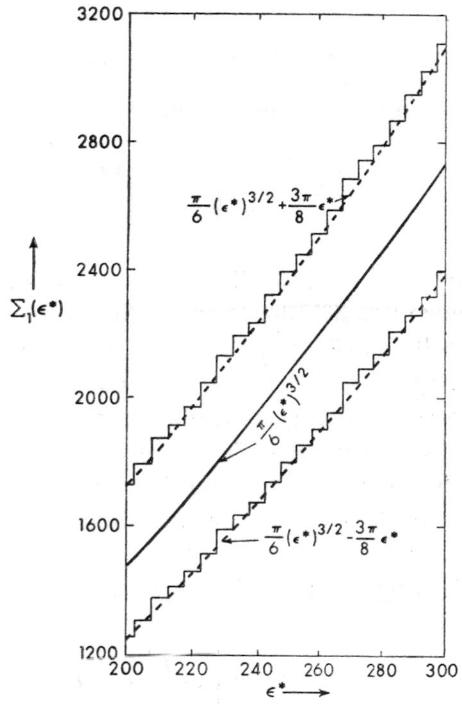

FIG. 1.2. Histograms showing the *actual* number of complexions available to a particle in a cubical enclosure; the lower histogram corresponds to the so-called Dirichlet boundary conditions, while the upper one corresponds to the Neumann boundary conditions (see Appendix A). The corresponding theoretical estimates, (15) and (16), are shown by the dashed lines; the customary estimate, as embodied in (14), is shown by the solid line.

nature of the function $\Omega(N, V, E)$, the specification of a precise value for the energy of the system cannot be justified on physical grounds, for that would fail to yield well-behaved expressions for the thermodynamic functions of the system. From a practical point of view, too, an absolutely isolated system is too much of an idealization. In the real world, almost every system has some contact with the surroundings, however little it may be; as a result, its energy cannot be sharply defined.* Of course, the effective width of the range over which the energy may be varying would, in general, be small in comparison with the mean value of

* Actually, the very act of making measurements on a given physical system brings about, inevitably, a contact between the system and the surroundings.

the energy. Let us specify this range by the limits $\left(E-\frac{1}{2}\Delta\right)$ and $\left(E+\frac{1}{2}\Delta\right)$ where, by assumption, $\Delta \ll E$. The corresponding number of microstates, $\Gamma(N, V, E; \Delta)$, is then given by

$$\Gamma(N, V, E; \Delta) \simeq \Delta \frac{\partial \Sigma(N, V, E)}{\partial E} \simeq \frac{3N}{2} \frac{\Delta}{E} \Sigma(N, V, E), \tag{17a}$$

whence we get

$$\ln \Gamma(N, V, E; \Delta) \simeq N \ln \left[\frac{V}{h^3}\left(\frac{4\pi m E}{3N}\right)^{3/2}\right] + \frac{3}{2} N + \left\{\ln\left(\frac{3N}{2}\right) + \ln\left(\frac{\Delta}{E}\right)\right\}. \tag{19a}$$

Now, for $N \gg 1$, the first term in the curly bracket is negligible in comparison with any of the terms outside the bracket, for $\lim_{N \to \infty} (\ln N)/N = 0$. Further, for any reasonable value of Δ/E, the same is true for the second term in this bracket.* Hence, for all practical purposes,

$$\ln \Gamma \simeq \ln \Sigma \simeq N \ln \left[\frac{V}{h^3}\left(\frac{4\pi m E}{3N}\right)^{3/2}\right] + \frac{3}{2} N. \tag{20}$$

Thus, we arrive at the baffling result that, for all practical purposes, the actual width of the range allowed for the energy of the system does not make much difference—the energy could lie between the values $\left(E-\frac{1}{2}\Delta\right)$ and $\left(E+\frac{1}{2}\Delta\right)$ or equally well between the values 0 and E. The physical reason underlying this situation is that the rate at which the number of complexions of the system increases with energy is so fantastic that even if we allow *all* values of energy between zero and a particular value E it is only the "immediate neighborhood" of E that makes an overwhelmingly dominant contribution to this number! And since we are finally concerned only with the logarithm of this number, even the "width" of that neighborhood is inconsequential!

The stage is now set for deriving the thermodynamics of our system. First of all, we have

$$S(N, V, E) = k \ln \Gamma = Nk \ln \left[\frac{V}{h^3}\left(\frac{4\pi m E}{3N}\right)^{3/2}\right] + \frac{3}{2} Nk, \tag{21†}$$

whence it follows that

$$E(S, V, N) = \frac{3h^2 N}{4\pi m V^{2/3}} \exp\left(\frac{2S}{3Nk} - 1\right). \tag{22}$$

The temperature of the gas then follows with the help of the formula (1.3.12c):

$$T = \left(\frac{\partial E}{\partial S}\right)_{N, V} = \frac{2}{3} \frac{E}{Nk}, \tag{23}$$

* It should be clearly understood that while Δ/E must be much smaller than 1, it must not tend to 0, for that would make $\Gamma \to 0$ and $\ln \Gamma \to -\infty$. A situation of this kind would be purely artificial and would have nothing to do with physical reality. Actually, in most physical systems, $\Delta/E = 0(E^{-1/2})$, whereby $\ln(\Delta/E)$ becomes of the order of $\ln E$, which again is negligible in comparison with the terms outside the curly bracket.

† Henceforth, we shall replace the sign $\simeq$, which characterizes the *asymptotic* character of a relationship, by the sign of equality because for most physical systems the asymptotic results are as good as exact.

whence we obtain the energy–temperature relationship

$$E = N(\tfrac{3}{2}kT) = n(\tfrac{3}{2}RT); \tag{24}$$

here, n is the number of moles of the gas and R the gas constant. The specific heat at constant volume now follows with the help of the formula (1.3.14a):

$$C_V = \left(\frac{\partial E}{\partial T}\right)_{N,V} = \frac{3}{2}Nk = \frac{3}{2}nR. \tag{25}$$

For the equation of state, we obtain from (1.3.12a) and (22):

$$P = \left(\frac{\partial E}{\partial V}\right)_{N,S} = \frac{2}{3}\frac{E}{V}, \tag{26}$$

which agrees with our earlier result (10). Further, with the help of (24), the foregoing result becomes

$$P = \frac{NkT}{V} \quad \text{or} \quad PV = nRT, \tag{27}$$

which is the same as (3). The specific heat at constant pressure would be, see (1.3.14b),

$$C_P = \left(\frac{\partial(E+PV)}{\partial T}\right)_{N,P} = \frac{5}{2}nR, \tag{28}$$

so that, for the ratio of the two specific heats, we have

$$\gamma = C_P/C_V = \tfrac{5}{3}. \tag{29}$$

Now, suppose that the gas undergoes an *isothermal* change of state ($T =$ const. and $N =$ const.); then, according to (24), the total energy of the gas would also remain constant while, according to (27), its pressure would vary inversely with volume (Boyle's law). The change in the entropy of the gas, between the initial state i and the final state f, would be, see eqn. (21),

$$S_f - S_i = Nk \ln (V_f/V_i); \tag{30}$$

the same result follows from the thermodynamic formula (1.3.5) and the equation of state (27), whence

$$(dS)_{\text{isoth}} = \frac{P}{T}dV = \frac{Nk}{V}dV \tag{31}$$

and hence

$$S_f - S_i = \int_i^f dS = Nk \ln (V_f/V_i).$$

On the other hand, if the gas undergoes an *adiabatic* change of state ($S =$ const. and $N =$ const.), then, according to (22) and (23), both E and T would vary as $V^{-2/3}$; moreover, according to (26) or (27), P would vary as $V^{-5/3}$. These results are in complete agreement with the conventional thermodynamic ones, namely

$$PV^{\gamma} = \text{const.} \quad \text{and} \quad TV^{\gamma-1} = \text{const.}, \tag{32}$$

with $\gamma = \frac{5}{3}$. It may be noted that, thermodynamically, the change in E during an adiabatic process arises solely from the external work done by the gas on the surroundings or vice versa:

$$(dE)_{\text{adiab}} = -P\,dV = -\frac{2E}{3V}\,dV; \tag{33}$$

see eqns. (1.3.5) and (26). The precise dependence of E on V follows readily from this relationship.

The considerations of this section have clearly demonstrated the manner in which the complete thermodynamics of a macroscopic system can be deduced from the multiplicity of its microstates (as represented by the number Ω or Γ or Σ). The whole problem then hinges upon an asymptotic enumeration of these numbers, which unfortunately is tractable only in a few idealized cases, such as the one considered in this section; see also Problems 1.8 and 1.9. Even in an idealized case like this, there remains an inadequacy which could not be detected in the derivations made so far; this relates to the *explicit* dependence of S on N. The discussion of the next section is intended not only to bring out this inadequacy but also to provide the necessary remedy for that.

1.5. The entropy of mixing and the Gibbs paradox

One thing we readily observe from expression (1.4.21) is that, contrary to what is logically desired, the entropy of an ideal gas, as given by this expression, is *not* an extensive property of the system! That is, if we increase the size of the system by a factor α, keeping the intensive variables unchanged,* then the value of the entropy, which should just increase

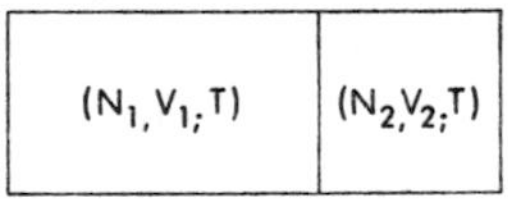

FIG. 1.3. The mixing together of two ideal gases 1 and 2.

by the same factor α, does not do so; actually, the presence of the ln V term in the expression affects the result adversely. This in a way means that the entropy of the system is different from the sum of the entropies of its various parts, which is simply absurd! A more common way of looking at this problem is to consider the so-called *Gibbs paradox.*

Gibbs visualized the mixing of two ideal gases 1 and 2, both being initially at the same temperature T; see Fig. 1.3. Clearly, the temperature of the mixture would also be the same. Now, before the mixing took place, the respective entropies of the two gases were, see eqns.

* This means an increase of the parameters N, V and E to αN, αV and αE, so that the energy per particle and the volume per particle remain unchanged.

(1.4.21) and (1.4.23),

$$S_i = N_i k \ln V_i + \frac{3}{2} N_i k \left\{1 + \ln \left(\frac{2\pi m_i kT}{h^2}\right)\right\}; \qquad i = 1, 2. \tag{1}$$

After the mixing has taken place, the total entropy would be

$$S_T = \sum_{i=1}^{2} \left[N_i k \ln V + \frac{3}{2} N_i k \left\{1 + \ln \left(\frac{2\pi m_i kT}{h^2}\right)\right\}\right], \tag{2}$$

where $V = V_1 + V_2$. Thus, the net increase in the value of S, which may be called the *entropy of mixing*, is given by

$$(\Delta S) = S_T - \sum_{i=1}^{2} S_i = k\left[N_1 \ln \frac{V_1+V_2}{V_1} + N_2 \ln \frac{V_1+V_2}{V_2}\right]; \tag{3}$$

the quantity ΔS is indeed positive, as it must be for an *irreversible* process like mixing. Now, in the particular case, when the initial particle densities of the two gases (and, hence, the particle density of the mixture) are the same, eqn. (3) becomes

$$(\Delta S)^* = k\left[N_1 \ln \frac{N_1+N_2}{N_1} + N_2 \ln \frac{N_1+N_2}{N_2}\right], \tag{4}$$

which is again positive definite.

So far, it is all right. However, a paradoxical situation arises when we consider the mixing of two samples of the same gas. Once again, the entropies of the individual samples will be given by (1); of course, now $m_1 = m_2 = m$, say. And the entropy after mixing will be given by

$$S_T = Nk \ln V + \frac{3}{2} Nk \left\{1 + \ln \left(\frac{2\pi mkT}{h^2}\right)\right\}, \tag{2a}$$

where $N = N_1 + N_2$; however, this expression is numerically the same as expression (2), with $m_i = m$. Therefore, the entropy of mixing in this case will also be given by expression (3) or, if $N_1/V_1 = N_2/V_2 = (N_1+N_2)/(V_1+V_2)$, by expression (4). The last conclusion, however, is totally unacceptable because the mixing of two samples of the same gas, with a common initial temperature T and a common initial particle density n, is clearly a *reversible* process, for we can simply reinsert the partitioning wall into the system and obtain a situation which is in no way different from the one we had before mixing. Of course, we imply that in dealing with (a system of) identical particles we cannot track them down individually; all we can reckon with is their numbers. When two dissimilar gases, even with a common initial temperature and a common initial particle density, mixed together the process *was* irreversible, for by reinserting the partitioning wall into the system one would only obtain two samples of the mixture and not the two gases that were originally present; to that case, expression (4) would indeed apply. However, in the present case, the corresponding result should be

$$(\Delta S)^*_{1\equiv 2} = 0. \tag{4a}$$ †

† In view of this, we expect that expression (3) may also be inapplicable to this case!

The foregoing result would also be consistent with the requirement that the entropy of a given system be simply equal to the sum of the entropies of its various parts. Of course, we had already noticed that this is not ensured by expression (1.4.21). Thus, we are once again led to believe that there is something basically wrong with that expression.

To see how the above paradoxical situation can be avoided, we recall that, for the entropy of mixing of two samples of the same gas, with a common T and a common n, we were led to the result (4), which can also be written as

$$(\Delta S)^* = S_T-(S_1+S_2) \simeq k\left[\ln\{(N_1+N_2)!\}-\ln(N_1!)-\ln(N_2!)\right], \tag{4}$$

instead of the logical result (4a). A closer look at this expression shows that we would obtain the correct result if our original expression for S were diminished by an *ad hoc* term, namely $k\ln(N!)$, for that would diminish S_1 by $k\ln(N_1!)$, S_2 by $k\ln(N_2!)$ and S_T by $k\ln\{(N_1+N_2)!\}$, with the result that $(\Delta S)^*$ would turn out to be zero instead of the expression in (4), as it actually should. Clearly, this would amount to an *ad hoc* reduction of the statistical numbers Γ and Σ by a factor $N!$. This is indeed the remedy proposed by Gibbs to avoid the paradox in question.

If we agree to the foregoing suggestion, the modified expression for the entropy of a classical ideal gas would be

$$S(N, V, E) = Nk\ln\left[\frac{V}{Nh^3}\left(\frac{4\pi mE}{3N}\right)^{3/2}\right]+\frac{5}{2}Nk \tag{1.4.21a}$$

$$= Nk\ln\left(\frac{V}{N}\right)+\frac{3}{2}Nk\left\{\frac{5}{3}+\ln\left(\frac{2\pi mkT}{h^2}\right)\right\}. \tag{1a}$$

Now, if we mix two samples of the same gas at a common initial temperature T, the entropy of mixing turns out to be

$$(\Delta S)_{1\equiv 2} = k\left[(N_1+N_2)\ln\left(\frac{V_1+V_2}{N_1+N_2}\right)-N_1\ln\left(\frac{V_1}{N_1}\right)-N_2\ln\left(\frac{V_2}{N_2}\right)\right]; \tag{3a}$$

and, if the initial particle densities of the samples were also equal, the result would indeed be

$$(\Delta S)^*_{1\equiv 2} = 0. \tag{4a}$$

It may be noted that for the mixing of two dissimilar gases, the expressions (3) and (4) continue to hold even when (1.4.21) is replaced by (1.4.21a).† The paradox of Gibbs is thereby resolved.

Equation (1a) is generally referred to as the *Sackur–Tetrode* equation. We note that, by this equation, the entropy does indeed become a truly extensive property of the system! Thus the very root of the trouble has been eliminated by the recipe of Gibbs. We shall discuss the physical implications of this recipe in Sec. 1.6; here, let us jot down some of its immediate consequences.

First of all, we note that the expression for the energy E of the gas, written as a function of

† Because, in this case, the entropy S_T of the mixture would be diminished by the term $k\ln(N_1!\,N_2!)$ rather than by the term $k\ln\{(N_1+N_2)!\}$.

S, V and N, is also modified. We now have

$$E(S, V, N) = \frac{3h^2N^{5/3}}{4\pi mV^{2/3}} \exp\left(\frac{2S}{3Nk} - \frac{5}{3}\right), \tag{1.4.22a}$$

which, unlike its predecessor (1.4.22), does make the energy of the system a truly extensive quantity! Of course, the thermodynamic results (1.4.23) through (1.4.33), derived in the previous section, remain unchanged. However, there are some which were intentionally left out, for they would come out correctly only from the modified expression for $S(N, V, E)$ or $E(S, V, N)$. The most important of these is the chemical potential of the gas, for which we obtain from (1.3.12b) and (1.4.22a):

$$\mu \equiv \left(\frac{\partial E}{\partial N}\right)_{V,S} = E\left[\frac{5}{3N} - \frac{2S}{3N^2k}\right]. \tag{5}$$

In view of (1.4.23) and (1.4.26), it becomes

$$\mu = \frac{1}{N}[E+PV-TS] \equiv \frac{G}{N}, \tag{6}$$

where G is the Gibbs free energy of the system. In terms of the variables N, V and T, the chemical potential takes the form

$$\mu(N, V, T) = kT \ln\left\{\frac{N}{V}\left(\frac{h^2}{2\pi mkT}\right)^{3/2}\right\}. \tag{7}$$

Another quantity of importance is the Helmholtz free energy:

$$A = E-TS = G-PV = NkT\left[\ln\left\{\frac{N}{V}\left(\frac{h^2}{2\pi mkT}\right)^{3/2}\right\}-1\right]. \tag{8}$$

It will be noted that A is an extensive property of the system while μ is an intensive one.

1.6. The "correct" enumeration of the microstates

We have seen in the preceding section that an *ad hoc* diminution in the entropy of an N-particle system by an amount $k \ln (N!)$, which implies an *ad hoc* reduction in the number of complexions accessible to the system by a factor $(N!)$, was able to correct the unphysical features of some of our former expressions. It is now natural to enquire: why, *in principle*, should the number of complexions, computed in Sec. 1.4, be suppressed in this manner? The physical reason for doing this is that the particles constituting the given system are not only identical but also *indistinguishable*; accordingly, it is unphysical to label them as No. 1, No. 2, No. 3, etc., and to speak of their being *individually* in the various single-particle states ε_i. All we can sensibly speak of is their distribution over the states ε_i *by numbers*, e.g. n_1 particles being in the state ε_1, n_2 in the state ε_2, and so on. Thus, the correct way of specifying a complexion is through the distribution numbers $\{n_i\}$, and not through the statement as to "which particle is where". To elaborate the point, we may say that if we consider a complexion which differs from another one merely in an interchange of two particles in different energy states, then according to our original mode of counting we would regard the two complexions as distinct; in view of the indistinguishability of the particles, however, these complexions are no longer distinct (for, physically, there exists no way of distinguishing between them).*

Now, the total number of permutations that can be effected among the N particles, distributed according to the set $\{n_i\}$, is

$$\frac{N!}{n_1!\,n_2!\,\ldots}, \tag{1}$$

where the n_i's must be consistent with the basic constraints, viz. (1.1.1) and (1.1.2).† If the particles were distinguishable, then all these permutations would lead to "distinct" complexions. However, in view of the indistinguishability of the particles, these permutations must be regarded as leading to one and the same thing; consequently, for *any* distribution set $\{n_i\}$, we have one, and only one, distinct complexion. As a result of this, the total number of distinct complexions accessible to the system, consistent with a given macrostate (N, V, E), would be severely cut down. However, since the factor (1) itself depends upon the numbers n_i constituting a particular distribution set and for a given macrostate there will be many such sets, there is no straightforward way to "correct down" the number of complexions computed on the basis of the classical concept of "distinguishability" of the particles.

The recipe of Gibbs obviously amounted to disregarding the details of the numbers n_i and slashing down the whole sequence of complexions by a *common* factor $N!$; this is indeed correct for situations in which all the N particles happen to be in different energy states, but is certainly wrong for other situations. We must keep in mind that by adopting this recipe we are still using a spurious permutation factor or *weight factor*, viz.

$$w\{n_i\} = \frac{1}{n_1!\,n_2!\,\ldots}, \tag{2}$$

for the distribution set $\{n_i\}$, whereas in principle we should use a factor of *unity*, irrespective of the values of the numbers n_i.‡ Nonetheless, the recipe of Gibbs does correct the situation in a gross manner, though in matters of detail it is still inadequate. In fact, it is by taking $w\{n_i\}$ to be equal to unity (or zero) that we obtain the so-called *quantum statistics*!

We have seen that the recipe of Gibbs corrects the enumeration of the microstates, as necessitated by the indistinguishability of the particles, only in a gross manner. Numerically, this would approach closer and closer to reality as the probability of the n_i's being greater than unity becomes less and less. This in turn happens when the given system is at a sufficiently high temperature (so that many more energy states become accessible) and has a sufficiently low density (so that there are not as many particles to accommodate). Thus, the "corrected" classical statistics represents truth more closely if the expectation values of the *occupation*

* Of course, if an interchange took place between particles in the same energy state, then even our original mode of counting did not regard the two complexions as distinct.

† The presence of the factors $(n_i!)$ in the denominator is related to the comment made in the preceding footnote.

‡ Or a factor of *zero* if the distribution set $\{n_i\}$ is disallowed on certain physical grounds, such as the Pauli exclusion principle.

numbers n_i are much less than unity:

$$\langle n_i \rangle \ll 1, \tag{3}$$

i.e. if the numbers n_i are generally equal to 0, occasionally equal to 1, and hardly ever $\geqslant 2$. The condition (3) in a way defines the *classical limit*. It must, however, be noted that it is because of the application of the correction factor $1/N!$, which replaces (1) by (2), that our results agree with reality *at least* in the classical limit. Without this correction, we encounter, even in the classical limit, serious physical discrepancies, one of which was discussed in Sec. 1.5.

In Sec. 5.5 we shall demonstrate, in an independent manner, that the factor by which the number of microstates, as computed for the "labeled" molecules, be reduced so that the formalism of classical statistical mechanics does become a true limit of the formalism of the quantum statistical mechanics is indeed $N!$.

Problems

1.1. (a) Show that, for two *large* systems in thermal contact, the number $\Omega^{(0)}(E^{(0)}, E_1)$ of Sec. 1.2 can be expressed as a Gaussian function in the variable E_1. Determine the expression for the root-mean-square deviation of E_1 from the mean value $\overline{E}_1$ in terms of other quantities pertaining to the problem.

(b) Make an explicit evaluation of the root-mean-square deviation of E_1 in the special case when the systems A_1 and A_2 are ideal gases.

1.2. Assuming that the entropy S and the statistical number Ω of a physical system are related through an arbitrary functional form

$$S = f(\Omega),$$

show that the additive character of S and the multiplicative character of Ω *necessarily* require the function $f(\Omega)$ to be of the form (1.2.11).

1.3. Two systems A and B, of identical composition, are brought together and allowed to exchange both energy and particles, the volumes V_A and V_B remaining constant. Show that the minimum value of the quantity (dE_A/dN_A) is given by

$$\frac{\mu_A T_B - \mu_B T_A}{T_B - T_A},$$

where μ's and T's are the respective chemical potentials and temperatures.

1.4. In a classical gas of hard spheres (of diameter σ), the spatial distribution of the particles is no longer uncorrelated. Roughly speaking, the presence of N' particles in the system leaves only a volume $(V - N'v_0)$ available for the $(N'+1)$th particle; clearly, v_0 would be proportional to σ^3. Assuming that $Nv_0 \ll V$, determine the dependence of $\Omega(N, V, E)$ on V {cf. (1.4.1)} and show that, as a result of this, V in the gas law (1.4.3) gets replaced by $(V - b)$, where b is equal to four times the actual space occupied by the particles.

1.5. Establish the formulae (1.4.15) and (1.4.16). Estimate the importance of the linear term in these formulae, relative to the main term $(\pi/6)\varepsilon^{*3/2}$, for an oxygen molecule confined to a cube of side 10 cm; take $\varepsilon = 0.05$ eV.

1.6. Prove that the "volume" $V_n(R)$ of an n-dimensional sphere of radius R can be written as

$$V_n(R) = (2R)^n \prod_{s=2}^{n} \left[\int_0^{\pi/2} \cos^s \theta \, d\theta \right].$$

Next, prove that

$$\prod_{s=2}^{s=n} \left[\int_0^{\pi/2} \cos^s \theta \, d\theta \right] = \frac{(\pi/4)^{n/2}}{(n/2)!},$$

whence it follows that

$$V_n(R) = \frac{\pi^{n/2}}{(n/2)!} R^n.$$

1.7. A cylindrical vessel 1 m long and 0.1 m in diameter is filled with a monatomic gas at $P = 1$ atm and $T = 300°K$. The gas is heated by an electrical discharge, along the axis of the vessel, which releases an energy of 10^4 joules. What will be the temperature of the gas immediately after the discharge?

1.8. Study the statistical mechanics of an extreme relativistic gas, which is characterized by the single-particle energy states

$$\varepsilon(n_x, n_y, n_z) = \frac{hc}{2L}(n_x^2+n_y^2+n_z^2)^{1/2},$$

instead of (1.4.5), along the lines of the nonrelativistic study carried out in Sec. 1.4. Show that the ratio C_P/C_V is now equal to 4/3, and not 5/3.

1.9. Consider a system of microscopic entities whose energy eigenvalues are given by

$$\varepsilon(n) = nh\nu; \qquad n = 0, 1, 2, \ldots .$$

Obtain an asymptotic expression for the number Ω of this system, for a given number N of the microscopic entities and a given total energy E. Determine the temperature T of the system as a function of E/N and $h\nu$. Finally, discuss the situation in the limit $E/(Nh\nu) \to \infty$.

1.10. Making use of the fact that the entropy $S(N, V, E)$ of a thermodynamic system is an extensive quantity, prove that

$$N\left(\frac{\partial S}{\partial N}\right)_{V,E} + V\left(\frac{\partial S}{\partial V}\right)_{N,E} + E\left(\frac{\partial S}{\partial E}\right)_{N,V} = S.$$

Note that this result implies: $(-N\mu+PV+E)/T = S$, i.e. $N\mu = E+PV-TS$, a relationship well known in thermodynamics.

1.11. A mole of argon and a mole of helium are contained in vessels of equal volume. If argon is at 300°K, what should be the temperature of helium so that the two have the same entropy?

1.12. Four moles of nitrogen and one mole of oxygen at $P = 1$ atm and $T = 300°K$ are mixed together to form air at the same pressure and temperature. Calculate the entropy of mixing per mole of the air formed.

1.13. Show that the various expressions for the entropy of mixing, as derived in Sec. 1.5, satisfy the following relations:

(a) For all N_1, V_1, N_2 and V_2, $(\Delta S)_{1\equiv 2} \geqslant 0$, the equality holding when $N_1/V_1 = N_2/V_2$.
(b) For all N_1, V_1, N_2 and V_2,

$$(\Delta S) - (\Delta S)_{1\equiv 2} = (\Delta S)^* \geq 0,$$

the last equality holding only if either N_1 or N_2 is zero.
(c) For a given value of (N_1+N_2), $(\Delta S)^* \leqslant (N_1+N_2)\, k \ln 2$, the equality holding when $N_1 = N_2$.

1.14. If the two gases considered in the mixing process of Sec. 1.5 were initially at different temperatures, say T_1 and T_2, what would be the entropy of mixing in that case? Does the contribution arising from this cause depend upon whether the two gases were different or identical?

1.15. Prove that for an ideal gas composed of monatomic molecules the entropy change, between any two temperatures, when the pressure is kept constant is 5/3 times the corresponding entropy change when the volume is kept constant. Verify this result *numerically* by calculating the actual values of $(\Delta S)_P$ and $(\Delta S)_V$ per mole of an ideal gas whose temperature is raised from 300°K to 400°K.

1.16. We have seen that the P–V relationship characterizing an adiabatic process in an ideal gas is governed by the exponent γ, i.e.

$$PV^\gamma = \text{const.}$$

Consider a mixture of two ideal gases, with mole fractions f_1 and f_2 and the respective exponents γ_1 and γ_2.

Show that the corresponding exponent γ for the mixture is given by

$$\frac{1}{\gamma-1} = \frac{f_1}{\gamma_1-1}+\frac{f_2}{\gamma_2-1}.$$

1.17. Establish thermodynamically the formulae

$$V\left(\frac{\partial P}{\partial T}\right)_\mu = S, \quad V\left(\frac{\partial P}{\partial \mu}\right)_T = N$$

and verify them in the case of a classical ideal gas.

CHAPTER 2

ELEMENTS OF ENSEMBLE THEORY

IN THE preceding chapter we noted that, for a given *macrostate* (N, V, E), a statistical system, at any time t, is equally likely to be in any one of an extremely large number of distinct *microstates*. As time passes, the system continually switches over from one microstate to another, with the result that over a reasonable span of time all one observes of the system is a behavior "averaged" over the variety of microstates. It may, therefore, make sense if we consider, at a *single* instant of time, a large number of systems—all being some sort of "mental copies" of the given system—which are characterized by the same macrostate as that of the original system but are, naturally enough, in all sorts of possible microstates. Then, under ordinary circumstances, we may expect that the average behavior of any system in this collection, which we call an *ensemble*, would be identical with the time-averaged behavior of the given system. It is on the basis of this expectation that we proceed to develop the so-called *ensemble theory*.

For classical systems, the most appropriate workshop for developing theoretical formalism is the *phase space*. Accordingly, we begin our study of the various ensembles with an analysis of the basic features of this space.

2.1. Phase space of a classical system

The microstate of a given classical system, at any time t, can be defined by specifying the *instantaneous* positions and momenta of all the particles constituting the system. Thus, if N is the number of particles in the system, the definition of a microstate requires the specification of $3N$ position coordinates $q_1, q_2, \ldots, q_{3N}$ and $3N$ momentum coordinates $p_1, p_2, \ldots, p_{3N}$. Geometrically, the set of coordinates (q_i, p_i), where $i = 1, 2, \ldots, 3N$, may be regarded as a point, representing a particular microstate of the given system, in a space of $6N$ dimensions. We refer to this space as the *phase space*, and the phase point (q_i, p_i) as a *representative point*, for the system. If the system is gaseous in character, then the phase space is also referred to as the Γ-space and the phase points as the G-points.

Of course, the coordinates q_i and p_i are functions of the time t; the precise manner in which they change with t is determined by the canonical equations of motion, viz.

$$\left.\begin{aligned}\dot{q}_i &= \frac{\partial H(q_i, p_i)}{\partial p_i}\\ \dot{p}_i &= -\frac{\partial H(q_i, p_i)}{\partial q_i}\end{aligned}\right\} \quad i = 1, 2, \ldots, 3N, \tag{1}$$

where $H(q_i, p_i)$ is the *Hamiltonian* of the system. Now, as time passes the set of coordinates (q_i, p_i), which also defines the microstate of the system, undergoes a continual change. Correspondingly, the representative point G in the phase space carves out a *trajectory* whose direction, at any time t, is determined by the *velocity vector* $\boldsymbol{v} \equiv (\dot{q}_i, \dot{p}_i)$, which in turn is given by the equations of motion (1). It is not difficult to see that the trajectory of the representative point must remain within a limited region of the phase space; this is so because a finite volume V directly limits the values of the coordinates q_i, while a finite energy E limits the values of both q's and p's [through the functional form of the Hamiltonian $H(q_i, p_i)$]. In particular, if the total energy of the system is known to have a *precise* value, say E, the corresponding trajectory in the phase space will be restricted to the "hypersurface"

$$H(q_i, p_i) = E; \tag{2}$$

on the other hand, if the total energy may lie anywhere in the range $\left(E-\frac{1}{2}\Delta,\, E+\frac{1}{2}\Delta\right)$, the corresponding trajectory will be restricted to the "hypershell" defined by these limits.

Now, if we consider an ensemble of systems (i.e. the given system along with a large number of mental copies of it) then, at any time t, the various members of the ensemble will be expected to be in all sorts of possible microstates; indeed, each one of these microstates must be consistent with the given macrostate, which is supposed to be common to all the members of the ensemble. In the phase space, the corresponding picture will consist of a swarm of representative points, one for each member of the ensemble, all lying within the "allowed" region of the space. As time passes, every member of the ensemble undergoes a continual change of microstates; correspondingly, the representative points constituting the swarm continually move along their respective trajectories. The overall picture of this movement possesses certain interesting features which are best appreciated in terms of what we call a *density function* $\varrho(q, p; t)$. This function is defined in such a way that, at any time t, the number of representative points in the "volume element" $(d^{3N}q\, d^{3N}p)$ around the point (q, p) of the phase space is given by the product $\varrho(q, p; t)\, d^{3N}q\, d^{3N}p$.* Clearly, the density function $\varrho(q, p; t)$ symbolizes the manner in which the members of the ensemble are distributed over various possible microstates at various instants of time. Accordingly, the *ensemble average* $\langle f \rangle$ of a given physical quantity $f(q, p)$, which may be different for systems in different microstates, would be given by

$$\langle f \rangle = \frac{\int f(q, p)\, \varrho(q, p; t)\, d^{3N}q\, d^{3N}p}{\int \varrho(q, p; t)\, d^{3N}q\, d^{3N}p}. \tag{3}$$

The integrations in (3) extend over the whole of the phase space; however, it is only the populated regions of the phase space ($\varrho \neq 0$) that actually contribute to the integrals. We note that, in general, the ensemble average $\langle f \rangle$ may itself be a function of time.

* Note that (q, p) is a further abbreviation of $(q_i, p_i) \equiv (q_1, \ldots, q_{3N}, p_1, \ldots, p_{3N})$.

An ensemble is said to be *stationary* if ϱ does not depend explicitly on time, i.e. at all times

$$\frac{\partial \varrho}{\partial t} = 0. \tag{4}$$

Clearly, for such an ensemble the average value $\langle f \rangle$ of *any* physical quantity $f(q, p)$ will be independent of time. Naturally, then, a stationary ensemble qualifies to represent a system *in equilibrium*. To determine the circumstances under which eqn. (4) can hold, we have to make a rather detailed study of the movement of the representative points in the phase space.

2.2. Liouville's theorem and its consequences

Consider an arbitrary "volume" ω in the relevant region of the phase space and let the "surface" enclosing this volume be denoted by σ; see Fig. 2.1. Then, the rate at which the number of representative points in this volume increases with time is given by

$$\frac{\partial}{\partial t} \int_{\omega} \varrho \, d\omega, \tag{1}$$

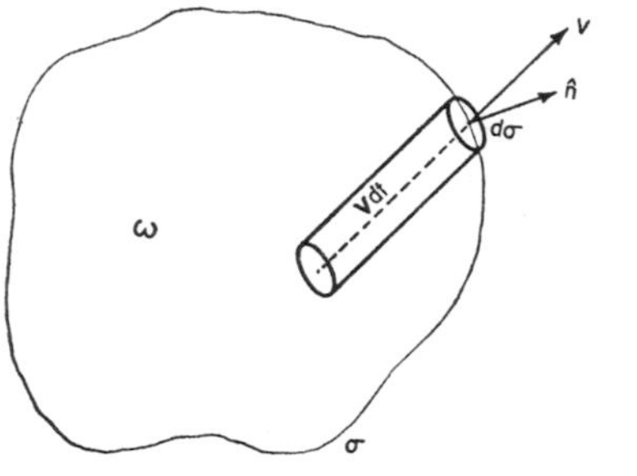

FIG. 2.1. The "hydrodynamics" of the representative points in the phase space.

where $d\omega \equiv (d^{3N}q \, d^{3N}p)$. On the other hand, the *net* rate at which the representative points "flow" out of ω (across the bounding surface σ) is given by

$$\int_{\sigma} \varrho(\boldsymbol{v} \cdot \hat{\boldsymbol{n}}) \, d\sigma; \tag{2}$$

here $\boldsymbol{v}$ is the velocity vector of the representative points in the region of the surface element $d\sigma$ while $\hat{\boldsymbol{n}}$ is the (outward) unit vector normal to this element. By the divergence theorem, (2) can be written as

$$\int_{\omega} \operatorname{div} (\varrho \boldsymbol{v}) \, d\omega; \tag{3}$$

of course, the operation of divergence here means the following:

$$\operatorname{div} (\varrho \boldsymbol{v}) \equiv \sum_{i=1}^{3N} \left\{ \frac{\partial}{\partial q_i} (\varrho \dot{q}_i) + \frac{\partial}{\partial p_i} (\varrho \dot{p}_i) \right\}. \tag{4}$$

In view of the fact that there are no "sources" or "sinks" in the phase space and hence the total number of representative points must be conserved,* we have, by (1) and (3),

$$\int_{\omega} \operatorname{div}(\varrho \boldsymbol{v})\, d\omega = -\frac{\partial}{\partial t}\int_{\omega} \varrho\, d\omega \tag{5}$$

or

$$\int_{\omega} \left\{\frac{\partial \varrho}{\partial t} + \operatorname{div}(\varrho \boldsymbol{v})\right\} d\omega = 0. \tag{6}$$

Now, the necessary and sufficient condition that the volume integral (6) vanish for all arbitrary volumes ω is that the integrand must vanish *everywhere* in the relevant region of the phase space. Thus, we must have

$$\frac{\partial \varrho}{\partial t} + \operatorname{div}(\varrho \boldsymbol{v}) = 0, \tag{7}$$

which is the *equation of continuity* for the swarm of the representative points.

Combining (4) and (7), we obtain

$$\frac{\partial \varrho}{\partial t} + \sum_{i=1}^{3N}\left(\frac{\partial \varrho}{\partial q_i}\dot{q}_i + \frac{\partial \varrho}{\partial p_i}\dot{p}_i\right) + \varrho \sum_{i=1}^{3N}\left(\frac{\partial \dot{q}_i}{\partial q_i} + \frac{\partial \dot{p}_i}{\partial p_i}\right) = 0. \tag{8}$$

The last group of terms vanishes identically because, by the equations of motion, we have, for all i,

$$\frac{\partial \dot{q}_i}{\partial q_i} = \frac{\partial^2 H(q_i, p_i)}{\partial q_i\, \partial p_i} \equiv \frac{\partial^2 H(q_i, p_i)}{\partial p_i\, \partial q_i} = -\frac{\partial \dot{p}_i}{\partial p_i}. \tag{9}$$

Further, since $\varrho \equiv \varrho(q_i, p_i; t)$, the remaining terms in (8) may be combined to give the "total" time derivative of ϱ.† Thus, we finally have

$$\frac{d\varrho}{dt} \equiv \frac{\partial \varrho}{\partial t} + [\varrho, H] = 0. \tag{10‡}$$

Equation (10) embodies the so-called *Liouville's theorem* (1838). According to this theorem, the "local" density of the representative points, *as viewed by an observer moving with a representative point*, stays constant in time. Thus, the swarm of the representative points moves in the phase space in essentially the same manner as an incompressible fluid moves in the physical space!

* This means that in the ensemble under consideration neither are any new members being admitted, nor are any old ones being expelled.

† It may be noted that the derivative d/dt is also referred to as the "substantial" time derivative or the "mobile" time derivative.

‡ We recall that the *Poisson bracket* $[\varrho, H]$ stands for the sum

$$\sum_{i=1}^{3N}\left(\frac{\partial \varrho}{\partial q_i}\frac{\partial H}{\partial p_i} - \frac{\partial \varrho}{\partial p_i}\frac{\partial H}{\partial q_i}\right),$$

which is identical with the group of terms in the middle of (8).

A distinction must be made, however, between eqn. (10) on the one hand and eqn. (2.1.4) on the other. While the former derives from the basic mechanics of the particles and is therefore *quite generally* true, the latter is only a requirement which, in a given case, may or may not be satisfied. The condition which ensures simultaneous validity of the two equations can be obtained by setting their difference equal to zero, i.e.

$$[\varrho, H] \equiv \sum_{i=1}^{3N} \left(\frac{\partial \varrho}{\partial q_i} \dot{q}_i + \frac{\partial \varrho}{\partial p_i} \dot{p}_i \right) = 0. \tag{11}$$

Now, one possible way of satisfying (11) is to assume that ϱ, which is already assumed to have no explicit dependence on time, is *independent* of the coordinates (q, p) as well, i.e.

$$\varrho(q, p) = \text{const.}; \tag{12}$$

this means that we are dealing with a swarm of representative points *uniformly distributed* over the relevant region of the phase space.* Physically, the choice (12) corresponds to an ensemble of systems which at *all* times are *uniformly* distributed over all possible microstates. The ensemble average (2.1.3) thereby reduces to

$$\langle f \rangle = \frac{1}{\omega} \int_{\omega} f(q, p)\, d\omega, \tag{13}$$

where ω denotes the total "volume" of the accessible region of the phase space. Clearly, in this case, *any* member of the ensemble is equally likely to be in *any* one of the various possible microstates, inasmuch as *any* representative point in the swarm is equally likely to be in the neighborhood of *any* phase point in the allowed region of the phase space. This statement is usually referred to as the "postulate of equal *a priori* probabilities" for the various possible microstates (or for the various volume elements in the allowed region of the phase space); the corresponding ensemble is referred to as the *microcanonical ensemble*.

A more general way of satisfying (11) is to assume that the dependence of the function ϱ on the coordinates (q, p) comes only through an explicit dependence on the Hamiltonian $H(q, p)$, i.e.

$$\varrho(q, p) = \varrho[H(q, p)]; \tag{14}$$

condition (11) is then identically satisfied. The functional form (14) provides a whole class of density functions for which the corresponding ensembles are stationary. In Chapter 3 we shall see that the most natural choice in this class of ensembles is the one for which

$$\varrho(q, p) \propto \exp\left[-H(q, p)/kT\right]. \tag{15}$$

The ensemble defined by (15) is referred to as the *canonical ensemble*.

* Remember that, outside the relevant region, ϱ is identically zero.

the energy of the system does not make much difference

2.3. The microcanonical ensemble

In this ensemble the macrostate of a system is defined by the number of molecules N, the volume V and the energy E. However, in view of the considerations developed in Sec. 1.4, we may prefer to specify a range of energy values, say from $\left(E-\frac{1}{2}\Delta\right)$ to $\left(E+\frac{1}{2}\Delta\right)$, rather than a sharply defined value E. With a specified macrostate, a choice still remains for the systems of the ensemble to be in *any one* of a large number of possible microstates. In the phase space, correspondingly, the representative points of the ensemble have a choice to lie *anywhere* within a "hypershell" defined by the condition

$$\left(E-\tfrac{1}{2}\Delta\right) \leqslant H(q,\, p) \leqslant \left(E+\tfrac{1}{2}\Delta\right). \tag{1}$$

The volume of the phase space enclosed within this shell is given by

$$\omega = \int' d\omega \equiv \int' (d^{3N}q \; d^{3N}p), \tag{2}$$

where the primed integration extends only over that part of the phase space which conforms to the condition (1). It is clear that ω will be a function of the parameters N, V, E and Δ.

Now, the microcanonical ensemble is a collection of systems for which the density function ϱ is, at all times, given by

$$\begin{cases} \varrho(q,\, p) = \text{const.} & \text{if} \quad \left(E-\tfrac{1}{2}\Delta\right) \leqslant H(q,\, p) \leqslant \left(E+\tfrac{1}{2}\Delta\right) \\ \qquad\quad = 0 & \text{otherwise.} \end{cases} \tag{3}$$

Accordingly, the expectation value of the number of representative points in a volume element $d\omega$ in the relevant hypershell of the phase space is simply proportional to $d\omega$. In other words, the *a priori* probability of finding a representative point in a given volume element $d\omega$ is the same as that of finding a representative point in an equivalent volume element located *anywhere* in the hypershell. In our original parlance, this implies an equal *a priori* probability for a given member of the ensemble to be in *any one* of the various possible microstates. In view of these considerations, the ensemble average $\langle f \rangle$, as given by eqn. (2.2.13), acquires a simple physical meaning. To see this, we proceed as follows.

Since the ensemble under study is a stationary one, the ensemble average of any physical quantity f must be independent of time; accordingly, taking a time average thereof will not produce any new result. Thus

$$\langle f \rangle \equiv \text{the ensemble average of } f$$
$$= \text{the time average of (the ensemble average of } f).$$

Now, the processes of time averaging and ensemble averaging are completely independent processes, so the order in which they are performed may be reversed without causing any change in the value of $\langle f \rangle$. Thus

$$\langle f \rangle = \text{the ensemble average of (the time average of } f).$$

Now, the time average of any physical quantity whatsoever, taken over a reasonably long interval of time, must be the same for *every* member of the ensemble, for after all, we are dealing with only the *mental copies* of a given system.* Therefore, taking an ensemble average thereof should be inconsequential and we may write

$$\langle f \rangle = \text{the long-time average of } f,$$

where the latter may be taken over *any* member of the ensemble. We further observe that the long-time average of a physical quantity is all one obtains by making a measurement of that quantity on a given system; therefore, it should be identified with the value one expects to obtain through experiment. Thus, we finally have

$$\langle f \rangle = f_{\text{exp}}. \tag{4}$$

This brings us to the most important result: *the ensemble average of any physical quantity f is identical with the value one expects to obtain on making an appropriate measurement on the given system.*

The next thing we look for is the establishment of a connection between the mechanics of the microcanonical ensemble and the thermodynamics of the member systems. To do this, we observe that there exists a direct correspondence between the various microstates of the given system and the various locations in the phase space. The volume ω (of the allowed region of the phase space) is, therefore, a direct measure of the multiplicity Γ of the microstates obtaining in the system. To establish a numerical correspondence between ω and Γ, we must discover a *fundamental volume* ω_0 which could be regarded as "equivalent to one microstate". Once this is done, we can right away conclude that, asymptotically,

$$\Gamma = \omega/\omega_0. \tag{5}$$

The thermodynamics of the system would then follow in the same way as in Sec. 1.3, namely through the relationship

$$S(N, V, E) = k \ln \Gamma = k \ln (\omega/\omega_0), \quad \text{etc.} \tag{6}$$

The basic problem then consists in determining ω_0. From dimensional considerations, see (2), ω_0 must be in the nature of an "angular momentum raised to the power $3N$".

* To provide a *rigorous* justification for this assertion is not a trivial job. One can readily see that if, for any particular member of the ensemble, the quantity f is averaged only over a *short* span of time, the result is bound to depend upon the relevant "subset of microstates" through which the system passes during that time. In the phase space, this will mean an averaging over only a "part of the allowed region". However, if we employ instead a reasonably long interval of time, the system may be expected to pass through *almost* all possible microstates "without fear or favor"; consequently, the result of the averaging process would depend only upon the macrostate of the system, and not upon the variety of microstates. Correspondingly, the averaging in the phase space would go over *practically* all parts of the allowed region, again "without fear or favor". In other words, the representative point of our system will have traversed each and every part of the allowed region *almost* uniformly. This statement embodies the so-called *ergodic theorem* or *ergodic hypothesis*, which was first introduced by Boltzmann (1871). According to this hypothesis, the trajectory of a representative point passes, in the course of time, through *each and every* point of the relevant region of the phase space. A little reflection, however, shows that the statement as such cannot be strictly true; we better replace it by the so-called *quasi-ergodic hypothesis*, according to which the trajectory of a representative point traverses, in the course of time, *any neighborhood of any point* of the relevant region. For further details, see ter Haar (1954, 1955), Farquhar (1964).

Now, when we consider an ensemble of systems, the foregoing statement should hold for every member of the ensemble; thus, *irrespective of the initial (and final) states* of the various systems, the long-time average of any physical quantity f should be the same for every member system.

To determine it exactly, we consider in the sequel certain simplified systems, both from the point of view of the phase space and from the point of view of the distribution of quantum states. We find that

$$\omega_0 = (h)^{3N}, \tag{7}$$

where h is the Planck constant.*

2.4. Examples

We consider, first of all, the problem of a classical, ideal gas composed of monatomic particles; see Sec. 1.4. In the microcanonical ensemble, the volume ω of the phase space accessible to the representative points of the (member) systems is given by

$$\omega = \int' \dots \int' (d^{3N}q\, d^{3N}p), \tag{1}$$

the integrations being restricted by the conditions that (i) the particles of the system are confined to a physical space of volume V and (ii) the total energy of the system lies within the limits $(E-\frac{1}{2}\Delta)$ and $(E+\frac{1}{2}\Delta)$. Since the Hamiltonian in this case is a function of the p's alone, the integrations over the q's can be carried out straightforwardly: this gives a factor of V^N. The remaining integral is

$$\underset{(E-\frac{1}{2}\Delta)\leq \sum_{i=1}^{3N}(p_i^2/2m)\leq(E+\frac{1}{2}\Delta)}{\int \dots \int} d^{3N}p = \underset{2m(E-\frac{1}{2}\Delta)\leq \sum_{i=1}^{3N} y_i^2 \leq 2m(E+\frac{1}{2}\Delta)}{\int \dots \int} d^{3N}y,$$

which is equal to the volume of a $3N$-dimensional hypershell, bounded by two hyperspheres of radii

$$\sqrt{[2m(E+\tfrac{1}{2}\Delta)]} \quad \text{and} \quad \sqrt{[2m(E-\tfrac{1}{2}\Delta)]}.$$

For $\Delta \ll E$, this is very nearly equal to the thickness of the shell, $\simeq \Delta(m/2E)^{1/2}$, *times* the surface area of a $3N$-dimensional hypersphere of radius $\sqrt{(2mE)}$. By eqn. (7) of Appendix C, we obtain for this integral

$$\Delta\left(\frac{m}{2E}\right)^{1/2}\left\{\frac{2\pi^{3N/2}}{[(3N/2)-1]!}(2mE)^{(3N-1)/2}\right\},$$

whence it follows that

$$\omega \simeq \frac{\Delta}{E} V^N \frac{(2\pi mE)^{3N/2}}{[(3N/2)-1]!}. \tag{2}$$

Comparing (2) with (1.4.17, 17a), we obtain the desired correspondence, viz.

$$(\omega/\Gamma)_{\text{asymp}} \equiv \omega_0 = h^{3N};$$

see also Problem 2.10. Quite generally, if the system under study has $\mathcal{N}$ degrees of freedom, the desired conversion factor is

$$\boxed{\omega_0 = h^{\mathcal{N}};} \tag{3}$$

remember, as well, the footnote on this page.

* The expression (7) has to be multiplied by the factor $(N!)$ if we wish to take into account (in a gross manner, of course) the indistinguishability of the particles constituting the system; cf. Sec. 1.6.

In the case of a single particle, $\mathcal{N} = 3$; accordingly, the number of available microstates would be asymptotically equal to the volume of the allowed region of the phase space divided by h^3. Let $\Sigma(P)$ denote the number of microstates available to a free particle confined to volume V of the physical space, its momentum p being less than or equal to a specified value P. We then have

$$\Sigma(P) \simeq \frac{1}{h^3} \int \cdots \int_{p \leq P} (d^3q\, d^3p) = \frac{V}{h^3} \frac{4\pi}{3} P^3, \tag{4}$$

whence we obtain for the number of microstates with momentum lying between p and $p+dp$

$$g(p)\, dp = \frac{d\Sigma(p)}{dp}\, dp \simeq \frac{V}{h^3} 4\pi p^2\, dp. \tag{5}$$

Expressed in terms of the particle energy $E = P^2/2m$, these expressions assume the form

$$\Sigma(E) \simeq \frac{V}{h^3} \frac{4\pi}{3} (2mE)^{3/2} \tag{6}$$

and

$$a(\varepsilon)\, d\varepsilon = \frac{d\Sigma(\varepsilon)}{d\varepsilon}\, d\varepsilon \simeq \frac{V}{h^3} 2\pi (2m)^{3/2} \varepsilon^{1/2}\, d\varepsilon. \tag{7}$$

Of course, the foregoing results are only asymptotic in character; they hold only if the physical dimensions of the enclosure are "infinitely" large. To be applicable to systems of finite extent, they must be suitably corrected. We find that, *in the main*, these corrections are proportional to S, the surface area of the enclosure. For instance, the corrected expressions for the functions $\Sigma(P)$ and $g(p)$ are, see Appendix A,

$$\Sigma(P) \simeq \frac{V}{h^3} \frac{4\pi}{3} P^3 + \theta \frac{S}{h^2} \frac{\pi}{4} P^2 \tag{4'}$$

and

$$g(p)\, dp \simeq \frac{V}{h^3} 4\pi p^2\, dp + \theta \frac{S}{h^2} \frac{\pi}{2} p\, dp, \tag{5'}$$

where θ is a number that depends upon the nature of the boundary conditions imposed on the wave functions of the particle: $-1 \leq \theta \leq 1$. In terms of the particle energy, we have

$$\Sigma(E) \simeq \frac{V}{h^3} \frac{4\pi}{3} (2mE)^{3/2} + \theta \frac{S}{h^2} \frac{\pi m E}{2}, \tag{6'}$$

and

$$a(\varepsilon)\, d\varepsilon \simeq \frac{V}{h^3} 2\pi (2m)^{3/2} \varepsilon^{1/2}\, d\varepsilon + \theta \frac{S}{h^2} \frac{\pi m}{2}\, d\varepsilon. \tag{7'}$$

The formula (6′) may be compared with the formulae (1.4.15) and (1.4.16), with $\varepsilon^* = (8mL^2E)/h^2$, $V = L^3$ and $S = 6L^2$.

The next case we would like to consider here is that of a one-dimensional *harmonic oscillator*. The classical expression for the Hamiltonian of this system is

$$H(q, p) = \frac{1}{2}kq^2 + \frac{1}{2m}p^2, \tag{8}$$

where k is the spring constant and m the mass of the oscillating particle. The space coordinate q and the momentum coordinate p of the oscillator are given by

$$q = A\cos(\omega t+\varphi), \quad p = -m\omega A\sin(\omega t+\varphi), \tag{9}$$

A being the amplitude and ω the (angular) frequency of vibration:

$$\omega = \sqrt{(k/m)}. \tag{10}$$

The energy of the oscillator is a constant of the motion, and is given by

$$E = \tfrac{1}{2}m\omega^2 A^2. \tag{11}$$

The phase-space trajectory of the representative point (q, p) of this system is determined by eliminating t between the expressions (9) for $q(t)$ and $p(t)$; we obtain

$$\frac{q^2}{(2E/m\omega^2)} + \frac{p^2}{(2mE)} = 1, \tag{12}$$

which is an ellipse, with axes proportional to $\sqrt{E}$ and area proportional to E; to be precise, the area is equal to $2\pi E/\omega$. Now, if we restrict the oscillator energy to the interval $(E-\frac{1}{2}\Delta, E+\frac{1}{2}\Delta)$, its representative point will be confined to a limited region of the phase space, viz. the region bounded by two elliptical trajectories corresponding to the energy values $(E+\frac{1}{2}\Delta)$ and $(E-\frac{1}{2}\Delta)$. The "volume" (in this case, the area) of this region will be

$$\iint\limits_{(E-\frac{1}{2}\Delta)\leq H(q,p)\leq(E+\frac{1}{2}\Delta)} (dq\,dp) = \frac{2\pi(E+\frac{1}{2}\Delta)}{\omega} - \frac{2\pi(E-\frac{1}{2}\Delta)}{\omega} = \frac{2\pi}{\omega}\Delta. \tag{13}$$

According to quantum mechanics, the energy eigenvalues of the harmonic oscillator are given by

$$E_n = (n+\tfrac{1}{2})\hbar\omega; \qquad n = 0, 1, 2, \ldots. \tag{14}$$

In the *classical* picture of the phase space, one could say that the representative point of the system must move along one of the "chosen" trajectories, as shown in Fig. 2.2; the area of the phase space enclosed between two consecutive trajectories, for which $\Delta = \hbar\omega$, is simply $2\pi\hbar$.* For arbitrary values of E and Δ,† the number of eigenstates within the allowed energy interval is very nearly equal to $\Delta/\hbar\omega$. Hence, the area of the phase space per eigenstate is, asymptotically, given by

$$\omega_0 = (2\pi\Delta/\omega)/(\Delta/\hbar\omega) = 2\pi\hbar = h. \tag{15}$$

* Strictly speaking, the very concept of phase space is invalid in quantum mechanics, because then it is wrong in principle to assign to a particle the coordinates q and p *simultaneously*. Nevertheless, the ideas discussed here are tenable in the correspondence limit!

† Of course, with $E \gg \Delta \gg \hbar\omega$.

Next, if we consider a system of N harmonic oscillators along the same lines as above, we arrive at the result: $\omega_0 = h^N$ (see Problem 2.7). Thus, our findings in these cases are consistent with our earlier result (3).

Another system that may be considered here is the one consisting of one or more *rigid rotators* (in two or more dimensions); once again we arrive at the same conversion factor as before (see Problem 2.4).

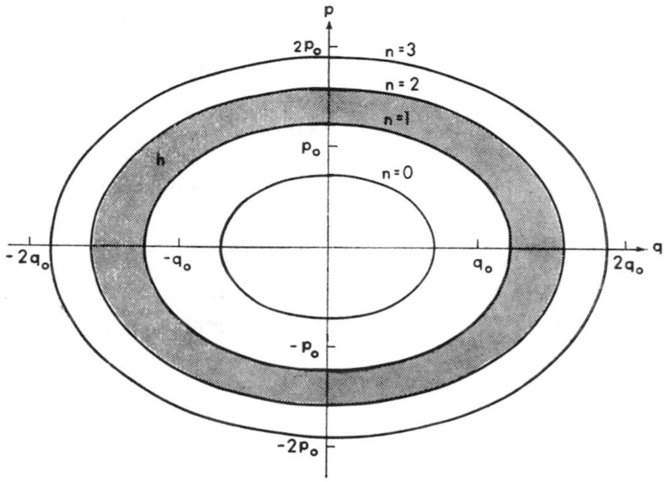

FIG. 2.2. Eigenstates of a linear harmonic oscillator, shown in relation to its phase space.

2.5. Quantum states and the phase space

At this stage we would like to say a few words on the central role played here by the Planck constant h. The best way to appreciate this role is to recall the implications of the Heisenberg uncertainty principle, according to which it is impossible to specify *simultaneously* both the position and the momentum of a particle accurately. An element of uncertainty is inherently present and can be expressed as follows: assuming that all conceivable uncertainties of measurement are eliminated, even then, by the very nature of things, the product of the uncertainties Δq and Δp in the *simultaneous* measurement of the canonically conjugate coordinates q and p would be of the order of $\hbar$:

$$(\Delta q\ \Delta p)_{\min} \approx \hbar. \tag{1}$$

Thus, it is impossible to define the position of a representative point in the phase space of the given system more accurately than is allowed by the condition (1). In other words, around any point (q, p) in the (two-dimensional) phase space there exists an area of the order of $\hbar$ within which the position of the representative point cannot be pin-pointed. In a phase space of $2\mathcal{N}$ dimensions, the corresponding "volume of uncertainty" around any point would be of the order of $\hbar^{\mathcal{N}}$. Therefore, it seems reasonable to regard the phase space as made up of elementary cells, of volume $\approx \hbar^{\mathcal{N}}$, and to consider the various positions within a cell as non-distinct. These cells could then be put into one-to-one correspondence with the quantum-mechanical states of the system.

It is, however, clear that considerations of uncertainty alone cannot give us the *exact* value of the conversion factor ω_0; see (2.3.5). This could only be done by an *actual* counting of states on the one hand and a computation of volume of the phase space on the other, as was done in the examples of the previous section. Clearly, a procedure along these lines could not be possible until after the work of Schrödinger and others. Historically, however, the first to establish the result (2.4.3) was Tetrode (1912) who, in this well-known work on the chemical constant and the entropy of a monatomic gas, assumed that

$$\omega_0 = (zh)^{\mathcal{N}},$$

where z was supposed to be an unknown numerical factor. On comparing theoretical results with the experimental data for mercury, Tetrode found that z was very nearly equal to unity; from this he concluded that "it seems rather plausible that z is *exactly* equal to unity, as has already been taken by O. Sackur (1911)".*

In the extreme relativistic limit, the same result was established by Bose (1924). In his famous treatment of the photon gas, Bose made use of Einstein's relationship between the momentum of a photon and the frequency of the associated vibration, namely

$$p = \frac{h\nu}{c}, \tag{2}$$

and observed that for a photon, confined to a three-dimensional cavity of volume V, the relevant "volume" of the phase space, which is

$$\int' (dq\, dp) = V 4\pi p^2\, dp = V(4\pi h^3 \nu^2/c^3)\, d\nu, \tag{3}$$

would correspond exactly to the Rayleigh expression

$$V(4\pi\nu^2/c^3)\, d\nu, \tag{4}$$

for the number of normal modes of a radiation oscillator, *provided that* we divide the phase space into elementary cells of volume h^3 and put these cells into one-to-one correspondence with the vibrational modes of Rayleigh. It may be added here that a two-fold multiplicity of these states ($g = 2$) arises from the spin orientations of the photon (or from the states of polarization of the vibrational modes); this requires a multiplication of the expressions (3) and (4) by a factor of 2, leaving the conversion factor h^3 unchanged.

2.6. Two important theorems—the "equipartition" and the "virial"

We consider once again a microcanonical ensemble, in which the macrostate of the (member) systems is defined by the parameters N, V, E and $\Delta(\Delta \ll E)$, and evaluate the ensemble average of the quantity $x_i(\partial H/\partial x_j)$, where $H(q_i, p_i)$ is the Hamiltonian of the system(s) while x_i and x_j are any of the $6N$ coordinates (q_i, p_i); note that the coordinate

* For proof, see Sec. 5.5, especially eqn. (5.5.22).

x_i *may* or *may not* be the same as the coordinate x_j. Then by (2.2.13),

$$\left\langle x_i \frac{\partial H}{\partial x_j} \right\rangle = \frac{\displaystyle\int_{(E-\frac{1}{2}\Delta)\leq H\leq(E+\frac{1}{2}\Delta)} \cdots \int \left(x_i \frac{\partial H}{\partial x_j}\right) d\omega}{\displaystyle\int_{(E-\frac{1}{2}\Delta)\leq H\leq(E+\frac{1}{2}\Delta)} \cdots \int d\omega}$$

$$= \frac{\int dq_i dp_i x_i \frac{\partial H}{\partial x_j} e^{-\beta H}}{\int dq_i dp_i e^{-\beta H}}$$

$$\simeq \frac{\Delta \cdot \dfrac{\partial}{\partial E}\left[\displaystyle\int_{0<H\leq E} \cdots \int \left(x_i \frac{\partial H}{\partial x_j}\right) d\omega\right]}{\Delta \cdot \dfrac{\partial}{\partial E}\left[\displaystyle\int_{0<H\leq E} \cdots \int d\omega\right]}. \tag{1}$$

Noting that $\partial E/\partial x_j = 0$, we write the integral in the numerator as

$$\int_{0<H\leq E} \cdots \int \left(x_i \frac{\partial H}{\partial x_j}\right) d\omega = \int_{0<H\leq E} \cdots \int \left\{x_i \frac{\partial(H-E)}{\partial x_j}\right\} d\omega$$

and carry out integration over x_j by parts. The integrated portion contains

$$x_i(H-E)\Big|_{(x_j)_1}^{(x_j)_2}, \tag{2}$$

where $(x_j)_1$ and $(x_j)_2$ are the "extreme" values of the coordinate x_j. Now, it is obvious that the representative point, when any of its coordinates assumes an "extreme" value, necessarily lies on the energy surface $H(q_i, p_i) = E$; accordingly, the expression (2) vanishes at both the limits. The remaining integral would be

$$-\int_{0<H\leq E} \cdots \int (H-E) \frac{\partial x_i}{\partial x_j} d\omega = \delta_{ij} \int_{0<H\leq E} \cdots \int (E-H)\, d\omega, \tag{3}$$

for $(\partial x_i/\partial x_j)$ is equal to 1 if $x_i = x_j$ and 0 if $x_i \neq x_j$. Equation (1) then becomes

$$\left\langle x_i \frac{\partial H}{\partial x_j} \right\rangle = \delta_{ij} \frac{\dfrac{\partial}{\partial E} \displaystyle\int_{0<H\leq E} \cdots \int (E-H)\, d\omega}{\dfrac{\partial}{\partial E} \displaystyle\int_{0<H\leq E} \cdots \int d\omega}. \tag{4}$$

Finally, noting that

$$\frac{\partial}{\partial \alpha} \int_{x=f(\alpha)}^{x=g(\alpha)} F(\alpha, x)\, dx = \int_{x=f(\alpha)}^{x=g(\alpha)} \frac{\partial F(\alpha, x)}{\partial \alpha} dx + \left\{\frac{\partial g(\alpha)}{\partial \alpha} F[\alpha, g(\alpha)] - \frac{\partial f(\alpha)}{\partial \alpha} F[\alpha, f(\alpha)]\right\},$$

we obtain

$$-\frac{1}{\beta}\int dq_i\, dp_i\; x_i \frac{\partial}{\partial x_j}\left(e^{-\beta H}\right)\qquad \left\langle x_i \frac{\partial H}{\partial x_j}\right\rangle = \delta_{ij}\frac{\displaystyle\int \cdots \int_{0<H\le E} d\omega}{\displaystyle\frac{\partial}{\partial E}\int \cdots \int_{0<H\le E} d\omega}$$

$$= \delta_{ij}\frac{1}{\dfrac{\partial}{\partial E}\ln\left[\displaystyle\int\cdots\int_{0<H\le E} d\omega\right]} = \delta_{ij}\frac{k}{\left(\dfrac{\partial S}{\partial E}\right)_{N,\,V}}$$

$$= \delta_{ij}kT, \tag{5}$$

which is the desired result.*

In the special case $x_i = x_j = p_i$, eqn. (5) takes the form

$$\left\langle p_i \frac{\partial H}{\partial p_i}\right\rangle \equiv \langle p_i \dot{q}_i\rangle = kT, \tag{6}$$

while $x_i = x_j = q_i$, it becomes

$$\left\langle q_i \frac{\partial H}{\partial q_i}\right\rangle \equiv -\langle q_i \dot{p}_i\rangle = kT. \tag{7}$$

Adding over all i, from $i = 1$ to $i = 3N$, we obtain

$$\left\langle \sum_i p_i \frac{\partial H}{\partial p_i}\right\rangle \equiv \left\langle \sum_i p_i \dot{q}_i\right\rangle = 3NkT \tag{8}$$

and

$$\left\langle \sum_i q_i \frac{\partial H}{\partial q_i}\right\rangle \equiv -\left\langle \sum_i q_i \dot{p}_i\right\rangle = 3NkT. \tag{9}$$

Now, in several physical situations the Hamiltonian of the system happens to be a *quadratic* function of its coordinates; so, through a canonical transformation, it can be brought into the form

$$H = \sum_j A_j P_j^2 + \sum_j B_j Q_j^2, \quad \Rightarrow \text{equipartition function} \tag{10}$$

where P_j and Q_j are the transformed, canonically conjugate, coordinates while A_j and B are certain constants of the problem. For such a system, we clearly have[†]

$$\sum_j \left(P_j \frac{\partial H}{\partial P_j} + Q_j \frac{\partial H}{\partial Q_j}\right) = 2H; \tag{11}$$

accordingly, we have from (5)

$$\langle H \rangle = \tfrac{1}{2}fkT, \tag{12}$$

---

* For a simpler proof of (5), see Sec. 3.6.

† Otherwise too, in view of the *quadratic* nature of the function $H(x_i)$, we have, by *Euler's theorem*,
$$\sum_i \left(x_i \frac{\partial H}{\partial x_i}\right) = 2H.$$

where f is the number of nonvanishing coefficients in the expression (10). We, therefore, conclude that each harmonic term in the (transformed) Hamiltonian makes a contribution of $\frac{1}{2}kT$ towards the internal energy of the system and, hence, a contribution of $\frac{1}{2}k$ towards the specific heat C_v. This result embodies the classical theorem of *equipartition of energy* (among the various degrees of freedom of the system). It may be mentioned that, for the distribution of kinetic energy alone, the equipartition theorem was first stated by Boltzmann (1871).

In our subsequent study we shall find that the equipartition theorem as stated here cannot always be valid; it applies only when the relevant degrees of freedom can be *freely* excited. At a given temperature T, there may be certain degrees of freedom which, due to the insufficiency of the energy available, are more or less "frozen". Such degrees of freedom can hardly make a contribution towards the internal energy of the system or towards its specific heat; see, for example, Secs. 6.6, 7.3 and 8.3. Of course, the higher the temperature of the system the better the validity of the theorem.

We now consider the implications of the formula (9). First of all, we note that this formula embodies the so-called *virial theorem* of Clausius (1870), for the quantity $\langle \sum_i q_i \dot{p}_i \rangle$, which is the expectation value of the sum of the products of the coordinates of the various particles and the respective forces acting on them, is by definition the *virial* of the system (and is generally denoted by the symbol $\mathcal{V}$). The virial theorem then states:

$$\mathcal{V} = -3NkT. \tag{13}$$

The relationship between the virial and other physical quantities of the system is best understood by investigating the simplest system we have, viz. a classical gas of noninteracting particles. In this case, the only forces that come into play are the ones arising from the walls of the container; these can be designated by an external pressure P acting upon the system by virtue of the fact that it is bounded by the walls of the container. Consequently, we have here a force $-P\,d\boldsymbol{S}$ associated with an element of area $d\boldsymbol{S}$ of the walls; the negative sign appears because the force is directed *inward* while the vector $d\boldsymbol{S}$ is directed outward. The virial of the gas is then given by

$$\mathcal{V}_0 = \left\langle \sum_i q_i F_i \right\rangle_0 = -P \oint_S \boldsymbol{r} \cdot d\boldsymbol{S}, \tag{14}^*$$

where $\boldsymbol{r}$ is the position vector of a particle which happens to be in the (close) vicinity of the surface element $d\boldsymbol{S}$; accordingly, $\boldsymbol{r}$ may be considered to be the position vector of the surface element itself. By the divergence theorem, eqn. (14) becomes

$$\mathcal{V}_0 = -P \int_V (\text{div } \boldsymbol{r})\, dV = -3PV. \tag{15}$$

Comparing (15) with (13), we obtain the well-known result:

$$PV = NkT. \tag{16}$$

* It may be noted that the summation over the various particles of the system, which appears in the definition of the virial, has been replaced by an integration over the surface of the container, for the simple reason that no contributions arise from the interior of the container.

The internal energy of the gas, which in this case is wholly kinetic, follows from the equipartition theorem (12) and is equal to $\frac{3}{2}NkT$, $3N$ being the number of degrees of freedom. Comparing this result with (13), we obtain the classical relationship

$$\mathcal{V} = -2K, \tag{17}$$

where K denotes the average kinetic energy of the system.

It is straightforward to carry out an extension of this study to a system of particles interacting through a two-body potential $u(\boldsymbol{r}_j - \boldsymbol{r}_i)$. The virial $\mathcal{V}$ then draws a contribution from the interior as well. Assuming the interparticle potential to be central and denoting it by the symbol $u(r)$, where $r = |\boldsymbol{r}_j - \boldsymbol{r}_i|$, the relevant contribution arising from the pair of particles i and j, with position vectors $\boldsymbol{r}_i$ and $\boldsymbol{r}_j$, is given by

$$\boldsymbol{r}_i \cdot \left(-\frac{\partial u(r)}{\partial \boldsymbol{r}_i}\right) + \boldsymbol{r}_j \cdot \left(-\frac{\partial u(r)}{\partial \boldsymbol{r}_j}\right) = -\frac{\partial u(r)}{\partial r^2}\left\{\boldsymbol{r}_i \cdot \frac{\partial |\boldsymbol{r}_j - \boldsymbol{r}_i|^2}{\partial \boldsymbol{r}_i} + \boldsymbol{r}_j \cdot \frac{\partial |\boldsymbol{r}_j - \boldsymbol{r}_i|^2}{\partial \boldsymbol{r}_j}\right\} = -r\frac{\partial u(r)}{\partial r}. \tag{18}$$

The net contribution, arising from all the $N(N-1)/2$ pairs of particles, would then be, for $N \gg 1$,

$$\frac{1}{2}N^2\left\langle -r\frac{\partial u(r)}{\partial r}\right\rangle = -\frac{N^2}{2}\iint\left\{r\frac{\partial u(r)}{\partial r}\right\} g(\boldsymbol{r}_2 - \boldsymbol{r}_1)\frac{d\boldsymbol{r}_1\, d\boldsymbol{r}_2}{V^2}$$
$$= -\frac{N^2}{2V}\int_0^\infty \left\{r\frac{\partial u(r)}{\partial r}\right\} g(r)\,(4\pi r^2\, dr), \tag{19}$$

where $g(r)$, the *pair distribution function* of the particles, is a measure of the probability of finding a pair of particles separated by a distance r; as used here, $g(r) \to 1$ as $r \to \infty$. Combining (19) with (15) and comparing the sum with (13), we obtain for a classical, interacting system

$$PV = NkT\left[1 - \frac{2\pi n}{3kT}\int_0^\infty \frac{\partial u(r)}{\partial r} g(r)\, r^3\, dr\right], \tag{20}$$

where n is the particle density in the system. The internal energy of the system can also be expressed in terms of the functions $u(r)$ and $g(r)$. Noting that the average kinetic energy is still given by the expression $\frac{3}{2}NkT$, we have for the total energy

$$E = \frac{3}{2}NkT + \frac{1}{2}N^2\iint u(r)\, g(r)\frac{d\boldsymbol{r}_1\, d\boldsymbol{r}_2}{V^2}$$
$$= \frac{3}{2}NkT\left[1 + \frac{4\pi n}{3kT}\int_0^\infty u(r)\, g(r) r^2\, dr\right]; \tag{21}$$

the second term here denotes the average potential energy of the system. Clearly, a knowledge of the functions $u(r)$ and $g(r)$, the latter itself depending crucially on the nature of the former, is essential before we can make use of eqns. (20) and (21). For further study along these lines, refer to Hill (1956), chap. 6; see also Problem 2.16.

Problems

2.1. Show that the volume element

$$d\omega = \prod_{i=1}^{3N} (dq_i\, dp_i)$$

of the phase space remains *invariant* under a canonical transformation of the set of (generalized) coordinates (q, p) to any other set of (generalized) coordinates (Q, P). [*Hint.* Before considering the most general transformation of this kind, which is referred to as a *contact* transformation, it may be helpful to consider a *point* transformation—one in which the new coordinates Q_i and the old coordinates q_i transform only among themselves.]

2.2. (a) Verify *explicitly* the invariance of the element $d\omega$ of the phase space of a single particle under transformation from the Cartesian coordinates (x, y, z, p_x, p_y, p_z) to the spherical polar coordinates $(r, \theta, \varphi, p_r, p_\theta, p_\varphi)$.

(b) The foregoing result seems to contradict the intuitive notion of "equal weights for equal solid angles", because the factor $\sin\theta$ is invisible in the expression for $d\omega$. Show that if we average out any physical quantity, whose dependence on p_θ and p_φ comes only through the kinetic energy of the particle, then as a result of integration over these variables we do indeed recover the factor $\sin\theta$ to appear with the sub-element $(d\theta\, d\varphi)$.

2.3. Starting with the line of zero energy and working in the (two-dimensional) phase space of a classical rotator, draw lines of constant energy which divide the phase space into cells of "volume" h. Calculate the energies of these states and compare them with the energy eigenvalues of the corresponding quantum-mechanical rotator.

2.4. By evaluating the "volume" of the relevant region of its phase space, show that the number of microstates available to a three-dimensional rigid rotator, with angular momentum $\leq M$, is $(M/\hbar)^2$. Hence determine the number of microstates that may be associated with the quantized angular momentum $M_j = \sqrt{(j(j+1))}\,\hbar$, where $j = 0, 1, 2, \ldots$ or $\frac{1}{2}, \frac{3}{2}, \frac{5}{2}, \ldots$. Interpret the result physically.

2.5. Consider a particle of energy E moving in a one-dimensional potential well $V(q)$, such that

$$m\hbar \left| \frac{dV}{dq} \right| \ll \{m(E-V)\}^{3/2}.$$

Show that the allowed values of the momentum p of the particle satisfy the condition

$$\oint p\, dq = (n+\tfrac{1}{2})h,$$

where n is an integer. Note that this result compares "favorably" with the Wilson–Sommerfeld quantization condition.

2.6. The generalized coordinates of a simple pendulum are the angular displacement θ and the angular momentum $ml^2\dot{\theta}$. Study, both mathematically and graphically, the nature of the corresponding trajectories in the phase space and show that the area A enclosed by a trajectory is precisely equal to the product of the total energy E and the time period T of the pendulum.

2.7. Derive (i) an *asymptotic* expression for the number of ways in which a given energy E can be distributed among a set of N one-dimensional harmonic oscillators, the energy eigenvalues of the oscillators being $(n+\frac{1}{2})\,\hbar\omega$: $n = 0, 1, 2, \ldots$, and (ii) the corresponding expression for the "volume" of the relevant region of the phase space of this system. Establish the correspondence between the two results and show that the conversion factor ω_0 is precisely equal to h^N.

2.8. The "volume" of a $3N$-dimensional sphere of radius R is formally given by the integral

$$V_{3N} = \underset{0 \leq \sum_{i=1}^{3N} x_i^2 \leq R^2}{\int \cdots \int} (dx_1\, dx_2\, dx_3)(dx_4\, dx_5\, dx_6) \ldots$$

and is equal to $C_{3N}R^{3N}$, where $C_n = \pi^{n/2}/(n/2)!$; see eqn. (7) of Appendix C. Derive the same result by writing the integral as

$$V_{3N} = \int\limits_{0 \leqslant \sum_{i=1}^{N} r_i^2 \leqslant R^2} \cdots \int (4\pi r_1^2\, dr_1)(4\pi r_2^2\, dr_2)\ \ldots .$$

[*Hint*. It is better to first establish a recurrence relationship between V_{3N} and $V_{3(N-1)}$.]

2.9. In a manner similar, prove that

$$V'_{3N} = \int\limits_{0 \leqslant \sum_{i=1}^{N} r_i \leqslant R} \cdots \int \prod_{i=1}^{N} (4\pi r_i^2\, dr_i) = (8\pi R^3)^N/(3N)!.$$

Using this formula, compute the "volume" of the relevant region of the phase space of an extreme relativistic gas ($\varepsilon = pc$) of N particles moving in three dimensions. Hence, derive expressions for the various thermodynamic properties of this system. Compare your results with those of Problem 1.8.

2.10. (a) Solve the integral

$$\int\limits_{0 \leqslant \sum_{i=1}^{3N} |x_i| \leqslant R} \cdots \int (dx_1 \ldots dx_{3N})$$

and use it to determine the "volume" of the relevant region of the phase space of an extreme relativistic gas ($\varepsilon = pc$) of $3N$ particles moving in one dimension. Determine, as well, the number of ways of distributing a given energy E among this system of particles and establish that, asymptotically, $\omega_0 = h^{3N}$.

(b) Compare the thermodynamics of this system with the thermodynamics of the system considered in Problem 2.9.

2.11. Consider the long-time averaged behavior of the quantity dG/dt, where

$$G = \sum_i q_i p_i,$$

and show that the validity of eqn. (2.6.8) implies the validity of eqn. (2.6.9), and vice versa.*

2.12. Show that, for a statistical system in which the interparticle potential energy $u(\mathbf{r})$ is a homogeneous function (of degree n) of the particle coordinates, we have for the *virial* $\mathcal{V}$

$$\mathcal{V} = -3PV - nU$$

and, hence, for the *mean kinetic energy* K

$$K = -\frac{1}{2}\mathcal{V} = \frac{1}{2}(3PV + nU) = \frac{1}{(n+2)}(3PV + nE);$$

here, U denotes the *mean potential energy* of the system and $E = K + U$. Note that the foregoing result holds not only for a classical system but also for a quantum-mechanical one.

2.13. (a) Calculate the time-averaged kinetic energy and potential energy of a one-dimensional harmonic oscillator, both classically and quantum-mechanically, and show that the results obtained are consistent with the theorem established in the preceding problem (with $n = +2$).

(b) Consider, similarly, the case of the hydrogen atom ($n = -1$) on the basis of (i) the Bohr–Sommerfeld model and (ii) the Schrödinger model.

(c) Finally, consider the case of a planet moving in (i) a circular orbit or (ii) an elliptic orbit.

2.14. The restoring force of an anharmonic oscillator is proportional to the cube of the displacement. Show that the mean kinetic energy of the oscillator is *twice* its mean potential energy.

2.15. Examine in detail the relativistic version of Problem 2.12.

---

* In this connection, it would be of interest to note that, in the case of an *ideal* gas, the left-hand sides of both these equations turn out to be equal to $3PV$; see eqns. (6.4.3) and (2.6.15), respectively.

2.16. (a) For a dilute gas, the pair distribution function $g(r)$ may be approximated as

$$g(r) \simeq \exp\{-u(r)/kT\}.$$

Show that, under this approximation, eqn. (2.6.20) takes the form

$$\frac{PV}{NkT} \simeq 1 - 2\pi n \int_0^\infty f(r)\, r^2\, dr,$$

where $f(r)$ [$= \exp\{-u(r)/kT\} - 1$] is the so-called *Mayer function*; see eqn. (9.1.6).

(b) What form will this result take for *a gas of hard spheres*, for which

$$\begin{aligned} u(r) &= \infty \text{ for } r \leqslant \sigma \\ &= 0 \text{ otherwise?} \end{aligned}$$

Compare your result with that of Problem 1.4.

CHAPTER 3

THE CANONICAL ENSEMBLE

IN THE preceding chapter we established the basis of ensemble theory and made a somewhat detailed study of the microcanonical ensemble. In that ensemble the macrostate of the systems was defined through a fixed number of particles N, a fixed volume V and a fixed energy E [or, preferably, a fixed energy range $(E-\frac{1}{2}\Delta, E+\frac{1}{2}\Delta)$]. The basic problem then consisted in determining the total number $\Omega(N, V, E)$, or $\Gamma(N, V, E; \Delta)$, of *distinct* microstates accessible to the given system. From the asymptotic expressions for these numbers, complete thermodynamics of the system could be derived in a straight-forward manner. However, for most physical systems, the mathematical problem of determining these numbers is quite formidable. For this reason alone, a search for an alternative approach within the framework of the ensemble theory seems rather necessary.

Physically, too, the concept of a fixed energy (or even an energy range) for a system belonging to the real world does not appear satisfactory. For one thing, the total energy E of a system is hardly ever measured; for another, it is hardly possible to keep its value under strict physical control. A far better alternative appears to be to speak of a fixed temperature T of the system—a parameter which is not only directly observable (by placing a "thermometer" in contact with the system) but also controllable (by keeping the system in contact with an appropriate "heat reservoir"). For most purposes, the precise nature of the reservoir is not very relevant; all one requires is that it should have an infinitely large heat capacity, so that, irrespective of the energy exchange between the system and the reservoir, an overall constant temperature can be maintained. Now, if the reservoir simply consists of an infinitely large number of mental copies of the given system we have once again an ensemble of systems—this time, however, it is an ensemble in which the macrostate of the systems is defined through the parameters N, V and T. Such an ensemble is referred to as a *canonical* ensemble.

In a canonical ensemble, the energy E of a system is necessarily variable; in principle, it can take up values anywhere between zero and infinity. The question then arises: what is the probability that, at any time t, a system is found to be in one of the states characterized by the energy value E_r?* We denote this probability by the symbol P_r. Clearly, there are two

* In what follows, the energy levels E_r appear as purely *mechanical* concepts—independent of the temperature of the system. For a treatment involving "temperature-dependent energy levels", see Elcock and Landsberg (1957).

ways in which the dependence of P_r on E_r can be determined. One consists in regarding the system as in equilibrium with a heat reservoir at a *common* temperature T and studying the statistics of the energy exchange between the two. The other consists in regarding the system as a member of a canonical ensemble (N, V, T), in which an energy $\mathcal{E}$ is being shared by $\mathcal{N}$ identical systems constituting the ensemble, and studying the statistics of this sharing process. We expect that in either case the final result would be the same. Once P_r is determined, the rest follows without any difficulty.

3.1. Equilibrium between a system and a heat reservoir

Let us consider the given system A, immersed in a very large heat reservoir A'; see Fig. 3.1. On attaining a state of mutual equilibrium, the system and the reservoir would have a *common* temperature, T say. Their energies, however, would be variable and, in principle, could

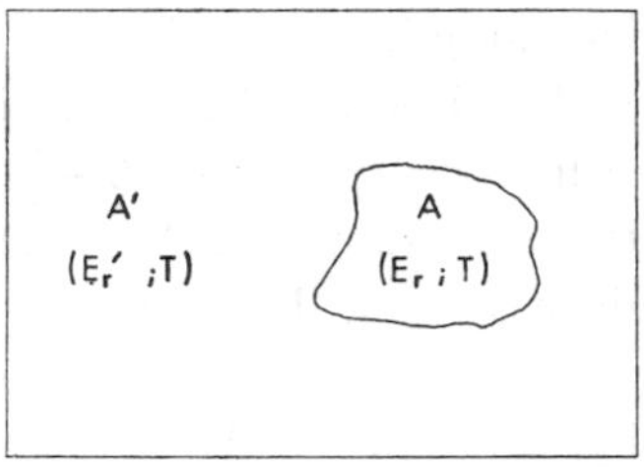

FIG. 3.1. A given system A immersed in a heat reservoir A'; in equilibrium, they have a common temperature T.

have, at any time t, values lying anywhere between 0 and $E^{(0)}$, where $E^{(0)}$ denotes the energy of the composite system $A^{(0)}$ ($\equiv A+A'$). If, at any particular instant of time, the system A happens to be in a state characterized by the energy value E_r, then the reservoir would have an energy E_r' such that

$$E_r+E_r' = E^{(0)} = \text{const.} \tag{1}$$

Of course, since the reservoir is supposed to be much larger than the given system, any *practical* value of E_r would be a very small fraction of $E^{(0)}$; therefore, for all practical purposes,

$$\frac{E_r}{E^{(0)}} = \left(1-\frac{E_r'}{E^{(0)}}\right) \ll 1. \tag{2}$$

Now, the state of the system A having been specified, the reservoir A' can still be in *any one* of a large number of states which are compatible with the energy value E_r'. Let the number of these states be denoted by $\Omega'(E_r')$. The prime on the symbol Ω emphasizes the fact that its functional form may depend upon the physical nature of the reservoir; of course, the details of this dependence are not going to be of any particular relevance to our discussion. Now, the larger the number of states available to the reservoir, the larger the probability of the reservoir assuming that particular energy E_r' (and, hence, of the system A assuming the corresponding energy value E_r). Moreover, since the various possible states (with a given energy value) are *equally likely* to occur, the relevant probability should be directly proportional to this number; thus,

$$P_r \propto \Omega'(E_r') \equiv \Omega'(E^{(0)}-E_r). \tag{3}$$

In view of (2), we may carry out an expansion of (3) around the value $E_r' = E^{(0)}$, i.e. around $E_r = 0$. However, since Ω' is an extremely large number, it is preferable to effect the expansion of its logarithm instead:

$$\begin{aligned}\ln \Omega'(E_r') &= \ln \Omega'(E^{(0)}) + \left(\frac{\partial \ln \Omega'}{\partial E'}\right)_{E'=E^{(0)}} (-E_r) + \dots \\ &\simeq \ln \Omega'(E^{(0)}) - \beta E_r,\end{aligned} \tag{4}$$

where use has been made of the formula (1.2.3), namely

$$\left(\frac{\partial \ln \Omega}{\partial E}\right)_{N,V} \equiv \beta = \frac{1}{kT}. \tag{5}$$

From (3) and (4), we obtain the desired result:

$$P_r \propto \exp(-\beta E_r). \tag{6}$$

Normalizing (6), we get

$$P_r = \frac{\exp(-\beta E_r)}{\sum_r \exp(-\beta E_r)}, \tag{7}$$

where the summation in the denominator goes over *all* the states accessible to the system A. We note that formula (7) bears no relation whatsoever to the physical nature of the reservoir A'.

We now examine the same problem from the ensemble point of view.

3.2. A system in the canonical ensemble

We consider an ensemble of $\mathcal{N}$ identical systems (which can certainly be labelled as $1, 2, \dots, \mathcal{N}$), sharing a total energy $\mathcal{E}$; let E_r $(r = 0, 1, 2 \dots)$ denote the energy eigenvalues of the systems. If n_r denotes the number of systems which, at any time t, have the energy value E_r, then the set of numbers $\{n_r\}$ must satisfy the obvious conditions

$$\left.\begin{aligned}\sum_r n_r &= \mathcal{N} \\ \text{and} \qquad \sum_r n_r E_r &= \mathcal{E} = \mathcal{N}U,\end{aligned}\right\} \tag{1}$$

where U stands for the average energy per system. Any set $\{n_r\}$, of the numbers n_r, which satisfies the restrictive conditions (1) represents a possible mode of distribution of the total energy $\mathcal{E}$ among the $\mathcal{N}$ members of the ensemble. Furthermore, any such mode can be realized in a number of ways, for we may effect a reshuffle among those members of the ensemble for which the energy values are different and thereby obtain a state of the ensemble which is distinct from the original state. Denoting the number of different ways of doing so by the symbol $W\{n_r\}$, we have

$$W\{n_r\} = \frac{\mathcal{N}!}{n_0!\, n_1!\, n_2!\, \ldots}. \tag{2}$$

In view of the fact that all possible states of the ensemble, which are compatible with the conditions (1), are *equally likely* to occur, the frequency with which the distribution set $\{n_r\}$ may appear will be directly proportional to the number $W\{n_r\}$. Accordingly, the "most probable" mode of distribution will be the one for which the number W is a maximum. We denote the corresponding distribution set by $\{n_r^*\}$; clearly, the set $\{n_r^*\}$ must also satisfy the conditions (1). As will be seen in the sequel, the probability of appearance of other modes of distribution, however little they may be differing from the most probable mode of distribution, is extremely low! Therefore, for all practical purposes, it is only the *most probable distribution set* $\{n_r^*\}$ one has anything to do with.

However, unless this is mathematically demonstrated, we must take into account *all* possible modes of distribution, as characterized by the various distribution sets $\{n_r\}$, along with their respective weight factors $W\{n_r\}$. Accordingly, the *expectation values*, or the *mean values*, $\langle n_r\rangle$ of the numbers n_r would be given by

$$\langle n_r\rangle = \frac{\sum_{\{n_r\}}' n_r W\{n_r\}}{\sum_{\{n_r\}}' W\{n_r\}}, \tag{3}$$

where the primed summations go over all the distribution sets that conform to the conditions (1). Physically, the mean value $\langle n_r\rangle$, as a fraction of the total number $\mathcal{N}$, should be a natural analogue of the probability P_r evaluated in the preceding section. In practice, however, the fraction $n_r^*/\mathcal{N}$ is also the same.

We now proceed to derive expressions for the numbers n_r^* and $\langle n_r\rangle$ and to show that, in the limit $\mathcal{N} \to \infty$, they become identical. These derivations are quite instructive, for the mathematical techniques involved occupy an important place in the methodology of our subject.

(i) *The method of most probable values.* Here we determine that distribution set which, while satisfying the restrictive conditions (1), maximizes the weight factor (2). Again, in view of the largeness of the number W, we prefer to work with its logarithm, viz.

$$\ln W = \ln(\mathcal{N}!) - \sum_r \ln(n_r!). \tag{4}$$

Since in the end we propose to resort to the limit $\mathcal{N} \to \infty$, the values of n_r (which are going to be of any practical significance) would also, in that limit, tend to infinity. It is, therefore, justified to apply the Stirling formula $\ln(n!) \simeq n \ln n - n$ to (4) and write

$$\ln W = \mathcal{N} \ln \mathcal{N} - \sum_r n_r \ln n_r. \tag{5}$$

If we change over from the set $\{n_r\}$ to a slightly different set $\{n_r + \delta n_r\}$, the expression (5) would change by

$$\delta(\ln W) = -\sum_r (\ln n_r + 1)\, \delta n_r. \tag{6}$$

Now, if the set $\{n_r\}$ were maximal, the variation $\delta(\ln W)$ must vanish. At the same time, in view of the restrictive conditions (1), the variations δn_r cannot be completely arbitrary; they must satisfy the conditions

$$\left.\begin{aligned} &\sum_r \delta n_r = 0 \\ \text{and} \qquad &\sum_r E_r\, \delta n_r = 0. \end{aligned}\right\} \tag{7}$$

The desired set $\{n_r^*\}$ is then determined by the method of *Lagrangian multipliers*,† by which the condition determining this set becomes

$$\sum_r \{-(\ln n_r^*+1)-(\alpha+\beta E_r)\}\, \delta n_r = 0, \tag{8}$$

where α and β are the Lagrangian undetermined multipliers that take care of the restrictive conditions (1). In (8), the variations δn_r become completely arbitrary. Accordingly, the only way to satisfy this condition is that all its coefficients must vanish identically; hence, *for all r*,

$$\ln n_r^* = -(\alpha+1)-\beta E_r,$$

that is,

$$n_r^* = C \exp(-\beta E_r), \tag{9}$$

where C is again an undetermined parameter. To determine C and β, we subject (9) to the conditions (1), whence it follows that

$$\frac{n_r^*}{\mathcal{N}} = \frac{\exp(-\beta E_r)}{\sum_r \exp(-\beta E_r)}, \tag{10}$$

the parameter β being a solution of the equation

$$\frac{\mathcal{E}}{\mathcal{N}} = U = \frac{\sum_r E_r \exp(-\beta E_r)}{\sum_r \exp(-\beta E_r)}. \tag{11}$$

Combining mechanical considerations with thermodynamical ones, see Sec. 3.3, we can show that the parameter β is exactly the same as the one occurring in Sec. 3.1, viz. $\beta = 1/kT$.

(ii) *The method of mean values.* Here, we attempt to evaluate expression (3) for $\langle n_r \rangle$, taking into account the weight factors (2) and the restrictive conditions (1). To do this, we prefer to replace (2) by

$$\tilde{W}\{n_r\} = \frac{\mathcal{N}!\, \omega_0^{n_0}\omega_1^{n_1}\omega_2^{n_2} \ldots}{n_0!\, n_1!\, n_2! \ldots}, \tag{12}$$

with the express understanding that in the end all the ω's will be set equal to unity. We also introduce a function

$$\Gamma(\mathcal{N}, U) = \sum_{\{n_r\}}{}' \tilde{W}\{n_r\}, \tag{13}$$

† For the method of Lagrangian multipliers, see ter Haar and Wergeland (1966), Appendix C.1.

where the primed summation, as before, goes over all the distribution sets that conform to the conditions (1). The expression (3) can then be written as

$$\langle n_r \rangle = \omega_r \frac{\partial}{\partial \omega_r} (\ln \Gamma) \Bigg|_{\text{all } \omega\text{'s}=1} . \tag{14}$$

Thus, all we need to know is the dependence of the quantity $\ln \Gamma$ on the parameters ω_r. Now

$$\Gamma(\mathcal{N}, U) = \mathcal{N}! \sum_{\{n_r\}}{}' \left(\frac{\omega_0^{n_0}}{n_0!} \cdot \frac{\omega_1^{n_1}}{n_1!} \cdot \frac{\omega_2^{n_2}}{n_2!} \cdots \right), \tag{15}$$

but this summation cannot be evaluated explicitly because of the fact that it is restricted to those sets alone which conform to the pair of conditions (1). If our distribution sets were restricted by the condition $\Sigma_r n_r = \mathcal{N}$ alone, then the evaluation of (15) would have been practically trivial; in fact, by the multinomial theorem, $\Gamma(\mathcal{N})$ would have been simply $(\omega_0 + \omega_1 + \ldots)^{\mathcal{N}}$. The added restriction $\Sigma_r n_r E_r = \mathcal{N} U$, however, permits the inclusion of only a "limited" number of terms in the sum—and that constitutes the real difficulty of the problem. Nevertheless, we can still hope to make some progress because, from a physical point of view, we do not require anything more than an *asymptotic* result, viz. a result that holds in the limit $\mathcal{N} \to \infty$. The method used for this purpose is the one developed by Darwin and Fowler (1922, 1923), which itself makes use of the so-called *saddle-point method* of integration or the *method of steepest descents*.

We construct a *generating function* $G(\mathcal{N}, z)$ for the quantity $\Gamma(\mathcal{N}, U)$:

$$G(\mathcal{N}, z) = \sum_{U=0}^{\infty} z^{\mathcal{N} U} \Gamma(\mathcal{N}, U), \tag{16}$$

which, in view of eqn. (15) and the second of the restrictive conditions (1), becomes

$$G(\mathcal{N}, z) = \sum_{U=0}^{\infty} \left[\sum_{\{n_r\}}{}' \frac{\mathcal{N}!}{n_0!\, n_1! \ldots} (\omega_0 z^{E_0})^{n_0} (\omega_1 z^{E_1})^{n_1} \ldots \right]. \tag{17}$$

It is easy to see that the summation over *doubly* restricted sets $\{n_r\}$, followed by a summation over all possible values of U, is equivalent to a summation over *singly* restricted sets $\{n_r\}$, namely the ones which satisfy only one condition: $\Sigma_r n_r = \mathcal{N}$. The expression (17) can, therefore, be evaluated with the help of the multinomial theorem, with the result

$$\begin{aligned} G(\mathcal{N}, z) &= (\omega_0 z^{E_0} + \omega_1 z^{E_1} + \ldots)^{\mathcal{N}} \\ &= [f(z)]^{\mathcal{N}}, \quad \text{say.} \end{aligned} \tag{18}$$

Now, if we suppose that the E_r's (and hence the total energy values $\mathcal{E} = \mathcal{N} U$) are all integers, then, by (16), the quantity $\Gamma(\mathcal{N}, U)$ is simply the coefficient of $z^{\mathcal{N} U}$ in the expansion of the function $G(\mathcal{N}, z)$ as a power series in z. It can, therefore, be computed by the method of residues in the complex z-plane.

To make this plan work, we assume to have chosen, *right at the outset*, a unit of energy so small that, to any desired degree of accuracy, we can regard the energies E_r (and the prescribed total energy $\mathcal{N}U$) as integral multiples of this unit. In terms of this unit, any energy value we come across must be an integral number. We further assume, without loss of generality, that the sequence $E_0, E_1 \ldots$ is a *nondecreasing* sequence, *with no common divisor.** Also, for the sake of simplicity, we assume that $E_0 = 0$.†

The solution is now obvious:

$$\Gamma(\mathcal{N}, U) = \frac{1}{2\pi i} \oint \frac{[f(z)]^{\mathcal{N}}}{z^{\mathcal{N}U+1}}\, dz, \tag{19}$$

where the integration is to be conducted along any closed contour around the origin; of course, we better remain *within* the circle of convergence of the function $f(z)$, so that the need of an analytic continuation does not arise.

Let us first of all examine the behavior of the integrand as we proceed from the origin along the real positive axis, remembering that all our ω's are virtually equal to unity and that $0 = E_0 \leqslant E_1 \leqslant E_2 \ldots$. We find that the factor $[f(z)]^{\mathcal{N}}$ starts from the value 1 at $z = 0$, increases monotonically and tends to infinity as z approaches the circle of convergence of $f(z)$, wherever that may be. The factor $z^{-(\mathcal{N}U+1)}$, on the other hand, starts from an infinite, positive value at $z = 0$ and decreases monotonically as z increases. Moreover, the relative rate of increase of the factor $[f(z)]^{\mathcal{N}}$ itself increases monotonically while the relative rate of decrease of the factor $z^{-(\mathcal{N}U+1)}$ decreases monotonically. Under these circumstances, the integrand must exhibit a minimum (and no other extremum) at some value of z, say x_0, *within* the circle of convergence. And, in view of the largeness of the numbers $\mathcal{N}$ and $\mathcal{N}U$, this minimum is expected to be rather steep!

Thus, at $z = x_0$ the first derivative of the integrand must vanish, while the second derivative must be positive and, hopefully, very large. Clearly, then, if we proceed through the point $z = x_0$ in a direction orthogonal to the real axis, the integrand must exhibit an equally steep maximum.‡ Thus, in the complex z-plane, as we move along the real axis our integrand shows a minimum at $z = x_0$, whereas if we move along a path parallel to the imaginary axis but passing through the point $z = x_0$, the integrand shows a maximum there. It is thus natural to call the point x_0 a *saddle point*; see Fig. 3.2. For the contour of integration we take a circle, with center at $z = 0$ and radius equal to x_0, hoping that on integrating along this contour only the immediate neighborhood of the sharp maximum at the point x_0 will make the most dominant contribution to the value of the integral.§

* Actually, this is no restriction at all, for a common divisor, if any, can be removed by selecting the unit of energy correspondingly larger.

† This too is not serious, for by doing so we are merely shifting the zero of the energy; the mean energy U then becomes $U - E_0$, but we can agree to call it U again.

‡ This can be seen by noting that (i) an analytic function must possess a *unique* derivative everywhere (so, in our case, it must be zero, irrespective of the direction in which we pass through the point x_0), and (ii) by the Cauchy–Riemann conditions of analyticity, the second derivative of the function with respect to y must be equal and opposite to the second derivative with respect to x.

§ It is indeed true that, for large $\mathcal{N}$, the contribution from the rest of the circle is negligible. The intuitive reason for this is that the single terms $(\omega_r z^{E_r})$, which constitute $f(z)$, "reinforce" one another *only* at the point $z = x_0$; elsewhere, there must be a disagreement between their phases, so that at *all* other points along the circle, $|f(z)| < f(x_0)$. Now, the factor that actually governs the relative contributions is $[|f(z)|/f(x_0)]^{\mathcal{N}}$; for $\mathcal{N} \gg 1$, this is clearly negligible. For a rigorous demonstration of this point, see Schrödinger (1960), pp. 31–33.

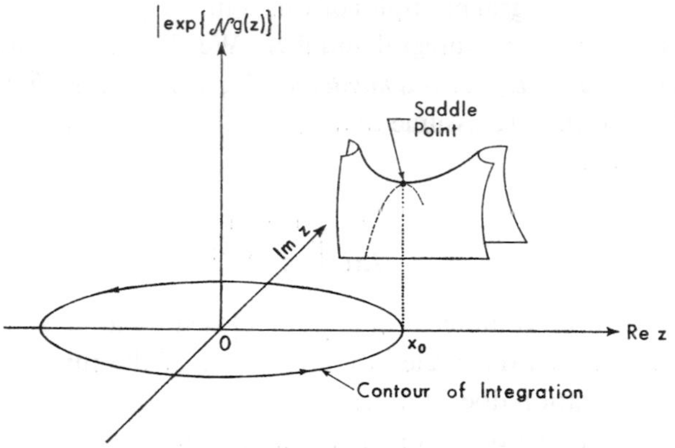

FIG. 3.2. The saddle point.

To carry out the integration we must first of all locate the point x_0. For this we write the integrand as

$$\frac{[f(z)]^{\mathcal{N}}}{z^{\mathcal{N}U+1}} = \exp\,[\mathcal{N}g(z)], \tag{20}$$

where

$$g(z) = \ln f(z) - \left(U + \frac{1}{\mathcal{N}}\right) \ln z \tag{21}$$

while

$$f(z) = \sum_r \omega_r z^{E_r}. \tag{22}$$

The number x_0 is then determined by the condition

$$g'(x_0) = \frac{f'(x_0)}{f(x_0)} - \frac{\mathcal{N}U+1}{\mathcal{N}x_0} = 0, \tag{23}$$

which, in view of the fact that $1 \ll \mathcal{N}U$, can be written as

$$U \simeq x_0 \frac{f'(x_0)}{f(x_0)} = \frac{\sum_r \omega_r E_r x_0^{E_r}}{\sum_r \omega_r x_0^{E_r}}. \tag{24}$$

We further have

$$\begin{aligned} g''(x_0) &= \left(\frac{f''(x_0)}{f(x_0)} - \frac{[f'(x_0)]^2}{[f(x_0)]^2}\right) + \frac{\mathcal{N}U+1}{\mathcal{N}x_0^2} \\ &\simeq \frac{f''(x_0)}{f(x_0)} - \frac{U^2 - U}{x_0^2}. \end{aligned} \tag{25}$$

It may be noted here that, in the limit $\mathcal{N} \to \infty$ and $\mathcal{E}(\equiv \mathcal{N}U) \to \infty$, *with U staying constant*, the number x_0 and the quantity $g''(x_0)$ become independent of $\mathcal{N}$.

Expanding $g(z)$ about the value $z = x_0$ but along the direction of integration, i.e. along the line $z = x_0 + iy$, we have

$$g(z) = g(x_0) - \tfrac{1}{2}g''(x_0)y^2 + \ldots;$$

accordingly, the integrand (20) might be approximated as

$$\frac{[f(x_0)]^{\mathcal{N}}}{x_0^{\mathcal{N}U+1}} \exp\left[-\frac{\mathcal{N}}{2} g''(x_0) y^2\right]. \tag{26}$$

Equation (19) then gives

$$\Gamma(\mathcal{N}, U) \simeq \frac{1}{2\pi i} \frac{[f(x_0)]^{\mathcal{N}}}{x_0^{\mathcal{N}U+1}} \int_{-\infty}^{\infty} \exp\left[-\frac{\mathcal{N}}{2} g''(x_0) y^2\right] i\, dy$$

$$\simeq \frac{[f(x_0)]^{\mathcal{N}}}{x_0^{\mathcal{N}U+1}} \cdot \frac{1}{\{2\pi \mathcal{N} g''(x_0)\}^{1/2}}, \tag{27}$$

whence it follows that

$$\frac{1}{\mathcal{N}} \ln \Gamma(\mathcal{N}, U) = \{\ln f(x_0) - U \ln x_0\} - \frac{1}{\mathcal{N}} \ln x_0 - \frac{1}{2\mathcal{N}} \ln \{2\pi \mathcal{N} g''(x_0)\}. \tag{28}$$

In the limit $\mathcal{N} \to \infty$ (with U staying constant), the last two terms in this expression tend to zero; hence, in this limit, we obtain

$$\frac{1}{\mathcal{N}} \ln \Gamma(\mathcal{N}, U) = \ln f(x_0) - U \ln x_0 . \tag{29}$$

Substituting for $f(x_0)$ and introducing a new variable β, defined by the relationship

$$x_0 \equiv \exp(-\beta), \tag{30}$$

we get

$$\frac{1}{\mathcal{N}} \ln \Gamma(\mathcal{N}, U) = \ln \left\{\sum_r \omega_r \exp(-\beta E_r)\right\} + \beta U. \tag{31}$$ *

The expectation value of the number n_r then follows from (14) and (31):

$$\frac{\langle n_r \rangle}{\mathcal{N}} = \left[\frac{\omega_r \exp(-\beta E_r)}{\sum_r \omega_r \exp(-\beta E_r)} + \left\{-\frac{\sum_r \omega_r E_r \exp(-\beta E_r)}{\sum_r \omega_r \exp(-\beta E_r)} + U\right\} \omega_r \frac{\partial \beta}{\partial \omega_r}\right]\Bigg|_{\text{all } \omega\text{'s}=1} . \tag{32}$$ *

The term inside the curly brackets vanishes identically because of (24) and (30). It has, however, been included to emphasize the fact that, for a fixed value of U, the number $\beta(\equiv -\ln x_0)$ must depend upon the choice of the ω's; see (24). We will appreciate the importance of this fact when we evaluate the mean square fluctuation in the value of the number n_r; in the calculation of the expectation value of n_r, however, it does not matter. We therefore obtain

$$\frac{\langle n_r \rangle}{\mathcal{N}} = \frac{\exp(-\beta E_r)}{\sum_r \exp(-\beta E_r)}, \tag{33}$$

which is identical with the expression (10) for $n_r^*/\mathcal{N}$. The physical significance of the parameter β is also the same as in that expression, for it is determined by eqn. (24), with all

ω's = 1, i.e. by the equation

$$U = \frac{\sum_r E_r \exp(-\beta E_r)}{\sum_r \exp(-\beta E_r)}, \tag{34}$$

which is again identical with its counterpart (11). It may be mentioned here that eqn. (34) fits very naturally with eqn. (33) because U is nothing but the ensemble average of the variable E_r, so we must have

$$U = \sum_r E_r P_r = \frac{1}{\mathcal{N}} \sum_r E_r \langle n_r \rangle. \tag{35}$$

Finally, we compute the statistical fluctuations in the values of the numbers n_r. We have, first of all,

$$\langle n_r^2 \rangle \equiv \frac{\sum'_{\{n_r\}} n_r^2 W\{n_r\}}{\sum'_{\{n_r\}} W\{n_r\}} = \frac{1}{\Gamma} \left(\omega_r \frac{\partial}{\partial \omega_r}\right) \left(\omega_r \frac{\partial}{\partial \omega_r}\right) \Gamma \Bigg|_{\text{all } \omega\text{'s}=1}; \tag{36}$$

see eqns. (12)–(14). It then follows that

$$\langle (\Delta n_r)^2 \rangle \equiv \langle \{n_r - \langle n_r \rangle\}^2 \rangle = \langle n_r^2 \rangle - \langle n_r \rangle^2 = \left(\omega_r \frac{\partial}{\partial \omega_r}\right) \left(\omega_r \frac{\partial}{\partial \omega_r}\right) (\ln \Gamma) \Bigg|_{\text{all } \omega\text{'s}=1}. \tag{37}$$

Substituting from (31) and making use of (32), we get

$$\frac{\langle (\Delta n_r)^2 \rangle}{\mathcal{N}} = \omega_r \frac{\partial}{\partial \omega_r} \left[\frac{\omega_r \exp(-\beta E_r)}{\sum_r \omega_r \exp(-\beta E_r)} + \left\{ -\frac{\sum_r \omega_r E_r \exp(-\beta E_r)}{\sum_r \omega_r \exp(-\beta E_r)} + U \right\} \omega_r \frac{\partial \beta}{\partial \omega_r} \right]_{\text{all } \omega\text{'s}=1}. \tag{38}$$

We note that the term in the curly brackets would not make any contribution because it is identically equal to zero, *whatever the choice of the ω's*. However, in the differentiation of the first term, we must not forget to take into account the *implicit* dependence of β on the ω's, which arises from the fact that unless ω's are set equal to unity the relation determining β does contain ω's; see eqns. (24) and (30) whereby

$$U = \frac{\sum_r \omega_r E_r \exp(-\beta E_r)}{\sum_r \omega_r \exp(-\beta E_r)} \Bigg|_{\text{all } \omega\text{'s}=1}. \tag{39}$$

A straightforward calculation gives

$$\left(\frac{\partial \beta}{\partial \omega_r}\right)_U \Bigg|_{\text{all } \omega\text{'s}=1} = \frac{E_r - U}{\langle E_r^2 \rangle - U^2} \frac{\langle n_r \rangle}{\mathcal{N}}. \tag{40}$$

We can now evaluate the relevant term on the right-hand side of (38), with the result

$$\begin{aligned} \frac{\langle (\Delta n_r)^2 \rangle}{\mathcal{N}} &= \frac{\langle n_r \rangle}{\mathcal{N}} - \left(\frac{\langle n_r \rangle}{\mathcal{N}}\right)^2 + \frac{\langle n_r \rangle}{\mathcal{N}} (U - E_r) \left(\frac{\partial \beta}{\partial \omega_r}\right)_U \Bigg|_{\text{all } \omega\text{'s}=1} \\ &= \frac{\langle n_r \rangle}{\mathcal{N}} \left[1 - \frac{\langle n_r \rangle}{\mathcal{N}} - \frac{\langle n_r \rangle}{\mathcal{N}} \frac{(E_r - U)^2}{\langle (E_r - U)^2 \rangle} \right]. \end{aligned} \tag{41}$$

Statistically speaking, the dispersion of the variable n_r would be *normal* if the last term in (41) were zero. We, however, find that the dispersion is *infra-normal*.

For the relative fluctuation in the value of n_r, we have

$$\left\langle\left(\frac{\Delta n_r}{\langle n_r\rangle}\right)^2\right\rangle = \frac{1}{\langle n_r\rangle} - \frac{1}{\mathcal{N}}\left\{1+\frac{(E_r-U)^2}{\langle(E_r-U)^2\rangle}\right\}. \tag{42}$$

As $\mathcal{N} \to \infty$, all $\langle n_r\rangle$'s $\to \infty$, with the result that all relative fluctuations tend to zero; accordingly, the (canonical) distribution becomes infinitely sharp. In this limit, the mean values, the most probable values—in fact, any values that occur with a nonvanishing probability—become identical! And that is the reason why two wildly different methods of calculating the canonical distribution, which were adopted in this section, have led to identical results.

3.3. Physical significance of the various statistical quantities

Having derived the *canonical distribution*, namely

$$P_r = \frac{\langle n_r\rangle}{\mathcal{N}} \simeq \frac{n_r^*}{\mathcal{N}} = \frac{\exp(-\beta E_r)}{\sum_r \exp(-\beta E_r)} \equiv -\frac{1}{\beta}\frac{\partial}{\partial E_r}\ln\left\{\sum_r \exp(-\beta E_r)\right\}, \tag{1}$$

where β is given by the equation

$$U = \frac{\sum_r E_r \exp(-\beta E_r)}{\sum_r \exp(-\beta E_r)} \equiv -\frac{\partial}{\partial\beta}\ln\left\{\sum_r \exp(-\beta E_r)\right\}, \tag{2}$$

we must now look for a recipe to extract information about the various macroscopic properties of the given system on the basis of the foregoing statistical results.

We begin by introducing a quantity A, defined by the relationship

$$A \equiv -\frac{1}{\beta}\ln\left\{\sum_r \exp(-\beta E_r)\right\}; \tag{3}$$

naturally, A is a function of the parameter β and of the spectrum of the eigenvalues E_r. Taking the differential of the quantity βA and making use of the formulae (1) and (2), we get

$$d(\beta A) = U\,d\beta + \frac{\beta}{\mathcal{N}}\sum_r \langle n_r\rangle\,dE_r, \tag{4}$$

so that

$$d\{\beta(U-A)\} = \beta\left(dU - \frac{1}{\mathcal{N}}\sum_r \langle n_r\rangle\,dE_r\right). \tag{5}$$

To interpret (5), we consider the following physical process.

We assume that the systems constituting the ensemble are fitted with identical mechanisms whereby their nature (in particular, their energy levels) can be altered; of course, when we do so we must ensure that the energy levels of all the systems are affected *alike*, for otherwise the very notion (of $\mathcal{N}$ *identical* systems) on which all our reasoning rests will break down. We also assume that our ensemble is coupled with an infinitely large heat bath, so that the temperature of the ensemble (and hence that of all the systems constituting it) can be changed. Viewed in this light, the various terms appearing in (5) acquire the following meaning:

(i) The quantity $\Sigma_r \langle n_r \rangle dE_r$ denotes the mechanical work done by an external agency in "lifting" the systems of the ensemble from the old energy levels E_r to the altered energy levels $(E_r + dE_r)$. Hence, $-\Sigma_r \langle n_r \rangle dE_r$ represents the mechanical work done on the external agency by the systems of the ensemble, and one-$\mathcal{N}$th of it stands for the *average* work done by any one system.

(ii) The quantity dU is obviously the *average* energy increase per system of the ensemble.

(iii) Accordingly, the quantity within the parentheses represents the *average* amount of heat, per system, transferred from the heat bath into the ensemble. The parameter β is, clearly, the "integrating factor" for the heat transferred!

Recalling the corresponding situation in thermodynamics, we readily infer that, firstly, the parameter β must be equivalent to the "reciprocal of the absolute temperature":

$$\beta \propto \frac{1}{T} = \frac{1}{kT}, \quad \text{say.} \tag{6}$$

The constant k is yet undetermined; however, we expect that, for our reasoning to be of universal applicability, k is a universal constant. We shall later see that k is nothing but the familiar *Boltzmann constant*. Secondly, the quantity $\beta(U-A)$, whose differential appears on the left-hand side of (5), must be directly associated with the entropy of the system:

$$\beta(U-A) = S/k, \tag{7}$$

where k is the same constant as in (6). Combining (6) and (7), we obtain for A

$$A = U - TS. \tag{8}$$

Thus, the quantity A, which was introduced by definition, see eqn. (3), gets identified as the *Helmholtz free energy* of the given system. This identification provides the most straightforward link between the thermodynamics of a physical system and the statistics of the canonical ensemble (of which this system is *supposedly* a member). We finally write

$$A(N, V, T) \equiv -kT \ln Q_N(V, T), \tag{9}$$

where

$$Q_N(V, T) = \sum_r \exp(-\beta E_r). \tag{10}$$

The quantity $Q_N(V, T)$ is referred to as the *partition function* of the system; sometimes, it is also called the "sum-over-states" (German: *Zustandssumme*). The dependence of Q on T is quite obvious. The dependence on N and V comes through the energy values E_r; in fact, any other parameters which might govern the values E_r should also appear as the arguments of Q. Moreover, for the quantity $A(N, V, T)$ to be an extensive property of the system, $\ln Q$ must itself be an extensive quantity.

Equations (9) and (10) constitute the basic result of the canonical ensemble theory. From here onward, the other thermodynamic quantities of the system follow straightforwardly. For instance (since $dA = -S\,dT - P\,dV + \mu\,dN$),

$$S = -\left(\frac{\partial A}{\partial T}\right)_{N,V} = k\left(\frac{\partial}{\partial T}\{T \ln Q\}\right)_{N,V}, \tag{11}$$

$$P = -\left(\frac{\partial A}{\partial V}\right)_{N,T} = kT\left(\frac{\partial}{\partial V} \ln Q\right)_{N,T}, \tag{12}$$

$$\mu = +\left(\frac{\partial A}{\partial N}\right)_{V,T} = -kT\left(\frac{\partial}{\partial N} \ln Q\right)_{V,T}. \tag{13}$$

Combining (9) and (11), we obtain for the *internal energy* of the system

$$U = A + TS = kT^2 \frac{\partial}{\partial T}(\ln Q) = -\frac{\partial}{\partial \beta}(\ln Q), \tag{14}$$

which is identical with our previous result (2). In fact, if we had kept in mind the thermodynamic relationship

$$U = A - T\left(\frac{\partial A}{\partial T}\right) = -T^2\left\{\frac{\partial}{\partial T}(A/T)\right\} = \frac{\partial}{\partial \beta}(A/kT), \tag{15}$$

eqn. (2) would have immediately suggested that

$$A \equiv -kT \ln\left\{\sum_r \exp(-\beta E_r)\right\}.$$

From (14) and (15), we obtain for the specific heat *at constant volume*

$$C_v = T\left(\frac{\partial S}{\partial T}\right)_{N,V} \equiv \left(\frac{\partial U}{\partial T}\right)_{N,V} = -T\left(\frac{\partial^2 A}{\partial T^2}\right)_{N,V} = k\beta^2\left\{\frac{\partial^2}{\partial \beta^2}(\ln Q)\right\}_{N,V}. \tag{16}$$

Finally, combining (9) and (12), we obtain for the *Gibbs free energy* of the system

$$G = A + PV = A - V\left(\frac{\partial A}{\partial V}\right)_{N,T} = N\left(\frac{\partial A}{\partial N}\right)_{V,T} = N\mu; \tag{17}$$

see Problem 3.5.

At this stage it appears worth while to make a few remarks on these results. First of all, we note from eqns. (10) and (12) that the pressure

$$P = -\frac{\sum_r \frac{\partial E_r}{\partial V} \exp(-\beta E_r)}{\sum_r \exp(-\beta E_r)}, \tag{18}$$

whence it follows that

$$P\,dV = -\sum_r \frac{\langle n_r \rangle}{\mathcal{N}}\,dE_r. \tag{19}$$

The quantity on the right-hand side has already been identified as the average work done by a system (of the ensemble) during a process that alters the energy levels E_r. Equation (19) now tells us that the volume change of a system (and hence of all the systems in the ensemble) provides an example of such a process, pressure P being the "force" accompanying this process. The quantity P, which was introduced here through the thermodynamic relationship (12), thereby acquires a mechanical meaning as well.

Next, about the entropy. We obtain from eqns. (1), (8) and (9)

$$\frac{S}{k} = \frac{U-A}{kT} = \sum_r P_r(\beta E_r + \ln Q) = -\sum_r P_r \ln P_r, \tag{20}$$

which is an extremely interesting relationship, for it shows that the entropy of a system is *solely* determined by the probability values P_r (of the system being in different dynamical states accessible to it)!

From the very look of it, eqn. (20) appears to be of fundamental importance; indeed, it admits of a number of interesting conclusions. One of these relates to a system in its ground state ($T = 0°$K). If the ground state is unique, then the system is sure to be found in this particular state and in no other; consequently, P_r is equal to 1 for this state and 0 for all others. Equation (20) then tells us that the entropy of the system is precisely zero, which is essentially the content of the *Nernst heat theorem* or the *third law of thermodynamics.** We also infer that vanishing entropy and perfect statistical order (which implies a complete predictability about the system) go together. As the number of accessible states increases, we have more and more of the P's becoming nonzero; the entropy of the system thereby increases. As the number of states becomes exceedingly large, most of the P-values become exceedingly small (and their logarithms assume large, negative values); the net result is that the entropy itself becomes exceedingly large. Thus, the largeness of entropy and the high degree of statistical disorder (or unpredictability) in the system also go together.

It is because of this fundamental connection between entropy on the one hand and lack of information on the other that formula (20) became the starting point of the interesting work of Shannon (1948, 1949) in the theory of communication.

It may be pointed out that formula (20) applies in the microcanonical ensemble as well. There, we have, for each member system of the ensemble, a group of Ω states, all of which are *equally likely* to occur. The value of P_r is, then, $1/\Omega$ for each of these states and 0 for all others. Consequently, we have for the entropy of the system

$$S = -k \sum_{r=1}^{\Omega} \left\{\frac{1}{\Omega} \ln\left(\frac{1}{\Omega}\right)\right\} = k \ln \Omega, \tag{21}$$

which is the same as the central formula of the microcanonical ensemble theory; see eqn. (1.2.11) or (2.3.6).

* Of course, if the ground state of the system is *degenerate* (with a multiplicity Ω_0), then its ground-state entropy is nonzero and is given by the expression $k \ln \Omega_0$; see eqn. (21).

3.4. Alternative expressions for the partition function

In most physical cases the energy levels accessible to a system are *degenerate*, i.e. one has a group of states, g_r in number, all belonging to the same energy value E_r. In such a case it would be more appropriate to write the partition function (3.3.10) as

$$Q_N(V, T) = \sum_r g_r \exp(-\beta E_r); \tag{1}$$

the corresponding expression for P_r, the probability that the system be in *any* of the states with energy E_r, would be

$$P_r = \frac{g_r \exp(-\beta E_r)}{\sum_r g_r \exp(-\beta E_r)}. \tag{2}$$

Clearly, the g_r states with a common energy E_r are all equally likely to occur. As a result, the probability of a system having energy E_r becomes directly proportional to the multiplicity g_r of this level; g_r thus plays the role of a "weight factor" for the level E_r. The actual probability is then determined by both the weight factor g_r and the Boltzmann factor $\exp(-\beta E_r)$ of the level, as we indeed have in (2). The basic formulae established in the preceding section, of course, remain unaffected.

Now, in view of the largeness of the number of particles constituting a given system and the largeness of the volume to which these particles are confined, the consecutive energy values E_r of the system must be extremely close to one another. Accordingly, there lie, within any reasonable interval of energy $(E, E+dE)$, a very large number of energy levels. One may then regard E as a *continuous* variable and write $P(E)\,dE$ for the probability that the given system, as a member of the canonical ensemble, may have its energy in the specified range. Clearly, this will be given by the product of the relevant single-state probability and the number of energy states lying in the specified range. Denoting the latter by $g(E)\,dE$, where $g(E)$ stands for the *density of states* of the system around the energy value E, we have

$$P(E)\,dE \propto \exp(-\beta E)\,g(E)\,dE, \tag{3}$$

which, on normalization, becomes

$$P(E)\,dE = \frac{\exp(-\beta E)\,g(E)\,dE}{\int_0^\infty \exp(-\beta E)\,g(E)\,dE}. \tag{4}$$

The denominator is clearly another expression for the partition function of the system:

$$Q_N(V, T) = \int_0^\infty e^{-\beta E} g(E)\,dE. \tag{5}$$

The expression for $\langle f \rangle$, the expectation value of a physical quantity f, may then be written as

$$\langle f \rangle = \frac{\sum_r f(E_r)\,g_r e^{-\beta E_r}}{\sum_r g_r e^{-\beta E_r}} \to \frac{\int_0^\infty f(E)\,e^{-\beta E} g(E)\,dE}{\int_0^\infty e^{-\beta E} g(E)\,dE}. \tag{6}$$

Before proceeding further, let us take a closer look at eqn. (5). If we regard β as a complex variable, then the partition function $Q(\beta)$ is just the Laplace transform of the density of states $g(E)$. The integral is, of course, convergent over the positive half plane of β (because $g(E) \geqslant 0$ for all E and $\lim_{E\to\infty} g(E) \exp(-\beta E) = 0$ for all $\beta > 0$). We can, therefore, write $g(E)$ as the inverse Laplace transform of $Q(\beta)$:

$$g(E) = \frac{1}{2\pi i} \int_{\beta'-i\infty}^{\beta'+i\infty} e^{\beta E} Q(\beta)\, d\beta \qquad (\beta' > 0) \tag{7}$$

$$= \frac{1}{2\pi} \int_{-\infty}^{\infty} e^{(\beta'+i\beta'')E} Q(\beta'+i\beta'')\, d\beta''; \tag{8}$$

the path of integration runs parallel to, and to the right of, the imaginary axis, i.e. along the straight line $\operatorname{Re}\beta = \beta' > 0$. Of course, the path may be continuously deformed so long as the integral converges.

3.5. The classical systems

The theory developed in the preceding sections is of very general applicability. It applies to systems in which quantum-mechanical effects are important as well as to systems which can be treated classically. In the latter case, the formalism may be written in the language of the phase space; as a result, the summations over the quantum states get replaced by integrations over the phase space.

Let us recall the concepts developed in Secs. 2.1 and 2.2, especially the formula (2.1.3) for the ensemble average $\langle f \rangle$ of a physical quantity $f(q, p)$, namely

$$\langle f \rangle = \frac{\int f(q, p)\, \varrho(q, p)\, d^{3N}q\, d^{3N}p}{\int \varrho(q, p)\, d^{3N}q\, d^{3N}p}, \tag{1}$$

where the function $\varrho(q, p)$ denotes the density of the representative points in the phase space; we have omitted here the explicit dependence of the density function ϱ on time t because we are interested in the study of equilibrium situations alone. Evidently, the function $\varrho(q, p)$ is a measure of the probability of finding a representative point in the vicinity of the phase point (q, p) which must in turn depend upon the corresponding value $H(q, p)$ of the Hamiltonian of the system. In the canonical ensemble,

$$\varrho(q, p) \propto \exp\{-\beta H(q, p)\}; \tag{2}$$

cf. eqn. (3.1.6). The expression for $\langle f \rangle$ then takes the form

$$\langle f \rangle = \frac{\int f(q, p) \exp(-\beta H)\, d\omega}{\int \exp(-\beta H)\, d\omega}, \tag{3}$$

where $d\omega (\equiv d^{3N}q\, d^{3N}p)$ denotes an element of the phase space. The denominator of the foregoing expression is directly related to the partition function of the system. However, to write the precise expression for the latter, we must take into account the precise relationship between a volume element in the phase space and the corresponding number of distinct quantum states of the system. This relationship was established in Secs. 2.4 and 2.5, whereby an element of volume $d\omega$ in the phase space corresponds to

$$\frac{d\omega}{N!\,h^{3N}} \tag{4}$$

distinct quantum states of the system.* The precise expression for the partition function would, therefore, be

$$Q_N(V, T) = \frac{1}{N!\,h^{3N}} \int e^{-\beta H(q, p)}\, d\omega; \tag{5}$$

it is understood that the integration here goes over the whole of the phase space.

As an application of this formulation, let us consider the example of an ideal gas. Here, we have a system of identical molecules, assumed to be monatomic (so that there are no internal degrees of motion to be considered), confined to a space of volume V and being in equilibrium at temperature T. Since there are no intermolecular interactions to be taken into account, the energy of the system is wholly kinetic:

$$H(q, p) = \sum_{i=1}^{N} (p_i^2/2m). \tag{6}$$

The partition function would then be

$$Q_N(V, T) = \frac{1}{N!\,h^{3N}} \int e^{-\frac{\beta}{2m}\sum_i p_i^2} \prod_{i=1}^{N} (d^3q_i\, dp_i). \tag{7}$$

Integrations over the space coordinates are rather trivial; we obtain a factor V for each of these N integrations. Integrations over the momentum coordinates are also quite easy, once we note that the integral (7) is simply a product of N identical integrals. Thus, we get

$$Q_N(V, T) = \frac{V^N}{N!\,h^{3N}} \left[\int_0^\infty e^{-p^2/2mkT}(4\pi p^2\, dp) \right]^N \tag{8}$$

$$Q_N(V,T) = \frac{1}{N!} \left[\frac{V}{h^3} (2\pi mkT)^{3/2} \right]^N. \tag{9}$$

The Helmholtz free energy is then given by

$$A(N, V, T) \equiv -kT \ln Q_N(V, T) \simeq NkT\left[\ln\left\{\frac{N}{V}\left(\frac{h^2}{2\pi mkT}\right)^{3/2}\right\} - 1\right]; \tag{10}$$

here, use has been made of the Stirling approximation: $\ln N! \simeq N \ln N - N$. The foregoing

* Ample justification has been given for the factor h^{3N}. The factor $N!$, on the other hand, comes from the considerations of Secs. 1.5 and 1.6; it arises essentially from the fact that the particles constituting the given system are *indistinguishable*. For a complete proof, see Sec. 5.5.

result is identical with eqn. (1.5.8), which was obtained by following a very different procedure. The simplicity of the present approach is, however, quite striking. Needless to say, the complete thermodynamics of the ideal gas can be derived from eqn. (10) in a straightforward manner. For instance, we have

$$\mu \equiv \left(\frac{\partial A}{\partial N}\right)_{V,T} = kT \ln\left\{\frac{N}{V}\left(\frac{h^2}{2\pi mkT}\right)^{3/2}\right\}, \tag{11}$$

$$P \equiv -\left(\frac{\partial A}{\partial V}\right)_{N,T} = \frac{NkT}{V} \tag{12}$$

and

$$S \equiv -\left(\frac{\partial A}{\partial T}\right)_{N,V} = Nk\left[\ln\left\{\frac{V}{N}\left(\frac{2\pi mkT}{h^2}\right)^{3/2}\right\}+\frac{5}{2}\right]. \tag{13}$$

These results are identical with the ones derived previously, namely (1.5.7), (1.4.2) and (1.5.1a), respectively. In fact, the identification of formula (12) with the ideal-gas law, $PV = nRT$, establishes the identity of the (hitherto undetermined) constant k as the *Boltzmann constant*; see eqn. (3.3.6). We further obtain

$$U \equiv -\left[\frac{\partial}{\partial \beta}(\ln Q)\right]_{N,V} \equiv -T^2\left[\frac{\partial}{\partial T}\left(\frac{A}{T}\right)\right]_{N,V} \equiv A+TS = \frac{3}{2}NkT, \tag{14}$$

and so on.

At this stage we have an important remark to make. Looking at the form of eqn. (8) and the manner in which it came about, we may write

$$Q_N(V, T) = \frac{1}{N!}[Q_1(V, T)]^N, \tag{15}$$

where $Q_1(V, T)$ may be regarded as the partition function of a *single* molecule in the system. A little reflection will show that the foregoing result obtains essentially from the fact that the basic constituents of our system are noninteracting (and hence the total energy of the system is simply the sum of their individual energies). Clearly, the situation will not be altered even if the molecules had internal degrees of motion as well. What is essentially required for eqn. (15) to be valid is the absence of interactions among the basic constituents of the system (and, of course, the absence of quantum-mechanical correlations).

Going back to the ideal gas, we could as well have started with the density of states $g(E)$. From eqn. (1.4.17), and in view of the Gibbs correction factor, we have

$$g(E) = \frac{\partial \Sigma}{\partial E} = \frac{1}{N!}\left(\frac{V}{h^3}\right)^N \frac{(2\pi m)^{3N/2}}{\{(3N/2)-1\}!}E^{(3N/2)-1}. \tag{16}$$

Substituting this into (3.4.5), and noting that the integral involved is equal to $\{(3N/2)-1\}!/\beta^{3N/2}$, we readily obtain

$$Q_N(\beta) = \frac{1}{N!}\left(\frac{V}{h^3}\right)^N\left(\frac{2\pi m}{\beta}\right)^{3N/2}, \tag{17}$$

which is identical with (9). It may as well be noted that if one starts with the single-particle density of states (2.4.7), namely

$$a(\varepsilon) = \frac{2\pi V}{h^3}(2m)^{3/2}\,\varepsilon^{1/2}, \tag{18}$$

and computes the single-particle partition function,

$$Q_1(\beta) = \int_0^\infty e^{-\beta\varepsilon}a(\varepsilon)\,d\varepsilon = \frac{V}{h^3}\left(\frac{2\pi m}{\beta}\right)^{3/2}, \tag{19}$$

and then makes use of the formula (15), one arrives at the same result for $Q_N(V, T)$.

Lastly, we consider the question of determining the density of states, $g(E)$, from the expression for the partition function, $Q(\beta)$, assuming that the latter is already known; indeed, expression (9) for $Q(\beta)$ was derived without making use of any knowledge regarding the function $g(E)$. According to eqns. (3.4.7) and (9), we then have

$$g(E) = \frac{V^N}{N!}\left(\frac{2\pi m}{h^2}\right)^{3N/2}\frac{1}{2\pi i}\int_{\beta'-i\infty}^{\beta'+i\infty}\frac{e^{\beta E}}{\beta^{3N/2}}\,d\beta \qquad (\beta' > 0). \tag{20}$$

The evaluation of the integral appearing here makes quite an interesting exercise. We, however, prefer to skip over it and merely note that, for all positive n,

$$\frac{1}{2\pi i}\int_{s'-i\infty}^{s'+i\infty}\frac{e^{sx}}{s^{n+1}}\,ds = \begin{cases} \dfrac{x^n}{n!} & \text{for} \quad x \geqslant 0, \\ 0 & \text{for} \quad x \leqslant 0. \end{cases} \tag{21*}$$

Using this formula, eqn. (20) becomes

$$g(E) = \begin{cases} \dfrac{V^N}{N!}\left(\dfrac{2\pi m}{h^2}\right)^{3N/2}\dfrac{E^{(3N/2)-1}}{\{(3N/2)-1\}!} & \text{for} \quad E \geqslant 0, \\ 0 & \text{for} \quad E \leqslant 0 \end{cases} \tag{22}$$

which is indeed the correct result for the density of states of an ideal gas; cf. eqn. (16). The foregoing derivation may not appear particularly valuable because in the present case we already knew the expression for $g(E)$. However, cases do arise where the evaluation of the partition function of a given system and the consequent evaluation of its density of states turn out to be quite simple, while the direct evaluation of the density of states from first principles happens to be rather involved. In such cases, the method given here can indeed be useful; see, for example, Problem 3.15 in comparison with Problems 1.8 and 2.9.

3.6. Energy fluctuations in the canonical ensemble: correspondence with the microcanonical ensemble

In the canonical ensemble, a system can have energy anywhere between zero and infinity. On the other hand, the energy of a system in the microcanonical ensemble is restricted to a very narrow range. How, then, can we assert that the thermodynamic properties of a system

* The interested reader can find complete details of this evaluation in Kubo (1965), pp. 165–8.

derived through the formalism of the canonical ensemble would be the same as the ones derived through the formalism of the microcanonical ensemble? Of course, we do expect that the two formalisms yield identical results, for otherwise our whole scheme would be marred by inner inconsistency; and, indeed, in the case of an ideal classical gas the results obtained by following one approach were precisely the same as the ones obtained by following the other approach. What is the underlying reason for this equivalence?

The answer to this question is obtained by examining the *actual* extent of the range over which the energies of the systems (in the canonical ensemble) have a significant probability to spread; that will tell us the extent to which the canonical ensemble *really* differs from the microcanonical one. To do this, we write down the expression for the mean energy,

$$U \equiv \langle E \rangle = \frac{\sum_r E_r g_r \exp(-\beta E_r)}{\sum_r g_r \exp(-\beta E_r)}, \tag{1}$$

and differentiate it with respect to the parameter β, holding the energy values E_r constant; we obtain

$$\frac{\partial U}{\partial \beta} = -\frac{\sum_r E_r^2 g_r \exp(-\beta E_r)}{\sum_r g_r \exp(-\beta E_r)} + \frac{\left[\sum_r E_r g_r \exp(-\beta E_r)\right]^2}{\left[\sum_r g_r \exp(-\beta E_r)\right]^2}$$

$$= -\{\langle E^2 \rangle - \langle E \rangle^2\}, \tag{2}$$

whence it follows that

$$\langle(\Delta E)^2\rangle \equiv \langle E^2 \rangle - \langle E \rangle^2 = -\left(\frac{\partial U}{\partial \beta}\right) = kT^2\left(\frac{\partial U}{\partial T}\right) = kT^2 C_V. \tag{3}$$

Note that we have here the specific heat *at constant volume*, because the partial differentiation in (2) was carried out with the E_r's kept constant! For the relative root-mean-square fluctuation in E, eqn. (3) gives

$$\frac{\sqrt{[\langle(\Delta E)^2\rangle]}}{\langle E \rangle} = \frac{\sqrt{(kT^2 C_V)}}{U}, \tag{4}$$

which is $0(N^{-1/2})$, N being the number of particles in the system. Consequently, for large N (which is true for every statistical system) the relative r.m.s. fluctuation in the values of E is quite negligible! Thus, for all practical purposes, a system in the canonical ensemble has an energy equal to, or almost equal to, the mean energy U; the situation in this ensemble is, therefore, practically the same as in the microcanonical ensemble. That explains why the two ensembles should lead to practically identical results.

For further understanding of the situation, we consider the manner in which energy is distributed among the various members of the (canonical) ensemble. To do this, we treat E as a continuous variable and start our investigation from the expression (3.4.3), namely

$$P(E)\, dE \propto \exp(-\beta E) g(E)\, dE. \tag{3.4.3}$$

The probability density $P(E)$ is given by the product of two factors: (i) the Boltzmann factor, which monotonically decreases with E, and (ii) the density of states, which monotonically increases with E. The product, therefore, passes through an extremum at some value of E, say E^*.† The value E^* will be determined by the condition

$$\frac{\partial}{\partial E}\{e^{-\beta E} g(E)\}\bigg|_{E=E^*} = 0,$$

that is, by

$$\frac{\partial \ln g(E)}{\partial E}\bigg|_{E=E^*} = \beta. \tag{5}$$

Recalling that

$$S = k \ln g \quad \text{and} \quad \left(\frac{\partial S(E)}{\partial E}\right)_{E=U} = \frac{1}{T} = k\beta,$$

the foregoing condition implies that

$$E^* = U. \tag{6}$$

This is a very interesting result, for it shows that, irrespective of the physical nature of the given system, the most probable value of its energy is identical with its mean value. Accordingly, if it is advantageous, we may use one instead of the other.

We now expand the logarithm of the probability density $P(E)$ around the value $E^* \equiv U$; we get

$$\ln [e^{-\beta E} g(E)] = \left(-\beta U+\frac{S}{k}\right)+\frac{1}{2}\frac{\partial^2}{\partial E^2}\ln\{e^{-\beta E} g(E)\}\bigg|_{E=U}(E-U)^2+\ \ldots$$

$$= -\beta(U-TS)-\frac{1}{2kT^2C_V}(E-U)^2+\ \ldots, \tag{7}$$

whence

$$P(E) = e^{-\beta E} g(E) \simeq e^{-\beta(U-TS)} \exp\left\{-\frac{(E-U)^2}{2kT^2C_V}\right\}. \tag{8}$$

Thus, the distribution of energy among systems in the canonical ensemble is of the *Gaussian* type, with mean value U and dispersion $\sqrt{(kT^2C_V)}$; cf. (3). In terms of the reduced variable E/U, the distribution is again Gaussian, with mean value unity and dispersion $\sqrt{(kT^2C_V)}/U$ {which is $0(N^{-1/2})$}; thus, for $N \gg 1$, we have an extremely sharp distribution and, as $N \to \infty$, the distribution approaches a delta-function!

It would be instructive to consider here a special case, viz. that of a classical ideal gas. Here, $g(E)$ is proportional to $E^{(3N/2-1)}$ and hence increases very fast with E; the factor $e^{-\beta E}$, however, decreases with E. The product $g(E)\exp(-\beta E)$ exhibits a maximum at $E^* = (3N/2-1)\beta^{-1}$, which is practically the same as the mean value $U = (3N/2)\beta^{-1}$. At $E = E^*$, the factor $g(E)$ rises as fast as the factor $\exp(-\beta E)$ falls. For values of E significantly different from E^*, the product essentially vanishes (for smaller values of E, due to the relative paucity of the available energy states; for larger values of E, due to the rela-

† Subsequently we shall see that this extremum is actually a maximum—and that too an extremely sharp one.

tive depletion caused by the Boltzmann factor). The overall picture is shown in Fig. 3.3 where we have depicted the actual behavior of these functions in the special case: $N = 10$. The most probable value of E is now $\frac{14}{15}$ of the mean value; so, the distribution is somewhat asymmetrical. The effective width Δ can be readily calculated from (3) and is equal to $(2/3N)^{1/2}U$ which, for $N = 10$, is about a quarter of U. We can imagine that, as N becomes large, both E^* and U increase (essentially linearly with N), the ratio E^*/U approaches unity and the distribution tends to become symmetrical about E^*. At the same time, the width Δ increases (but only as $N^{1/2}$); considered in the relative sense, it tends to vanish (as $N^{-1/2}$).

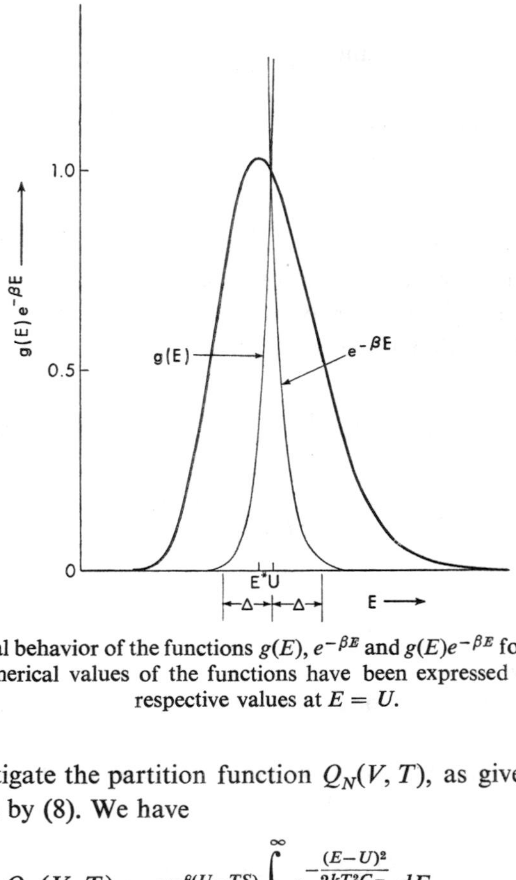

FIG. 3.3. The actual behavior of the functions $g(E)$, $e^{-\beta E}$ and $g(E)e^{-\beta E}$ for an ideal gas, with $N = 10$. The numerical values of the functions have been expressed as fractions of their respective values at $E = U$.

Let us finally investigate the partition function $Q_N(V, T)$, as given by eqn. (3.4.5), with its integrand replaced by (8). We have

$$Q_N(V, T) \simeq e^{-\beta(U-TS)} \int_0^\infty e^{-\frac{(E-U)^2}{2kT^2C_V}} dE$$

$$\simeq e^{-\beta(U-TS)} \sqrt{(2kT^2C_V)} \int_{-\infty}^{\infty} e^{-x^2} dx$$

$$= e^{-\beta(U-TS)} \sqrt{(2\pi kT^2C_V)},$$

so that

$$-kT \ln Q_N(V, T) \equiv A \simeq (U-TS) - \tfrac{1}{2}kT \ln (2\pi kT^2C_V). \tag{9}$$

The last term, being O(ln N), is negligible in comparison with the other terms, which are all O(N). Hence, in essence,

$$A = U - TS. \tag{10}$$

Note that the quantity A in this formula has come through the formalism of the canonical ensemble, while the quantity S has come through a definition belonging to the microcanonical ensemble. The fact that we finally end up with a consistent thermodynamic relationship establishes beyond doubt that the two approaches are, for all practical purposes, identical.

As another check on the identity of the results obtained by following these two approaches, we derive here, using the formalism of the canonical ensemble, the theorem (5) of Sec. 2.6. The problem here consists in determining the expectation value of the quantity $x_i(\partial H/\partial x_j)$, where $H(q, p)$ is the Hamiltonian of the system (assumed classical) while x_i and x_j are any of the generalized coordinates (q, p). In the canonical ensemble, the desired result would be given by

$$\left\langle x_i \frac{\partial H}{\partial x_j} \right\rangle = \frac{\int \left(x_i \frac{\partial H}{\partial x_j} \right) e^{-\beta H} \, d\omega}{\int e^{-\beta H} \, d\omega}. \tag{11}$$

Let us consider the integral in the numerator. Integrating over x_j *by parts*, it becomes

$$\int \left[-\frac{1}{\beta} x_i e^{-\beta H} \Big|_{(x_j)_1}^{(x_j)_2} + \frac{1}{\beta} \int \left(\frac{\partial x_i}{\partial x_j} \right) e^{-\beta H} \, dx_j \right] d\omega_{(j)};$$

here, $(x_j)_1$ and $(x_j)_2$ are the "extreme" values of the coordinate x_j, while $d\omega_{(j)}$ denotes "$d\omega$ devoid of dx_j". The integrated portion vanishes identically because when any one of the coordinates takes an "extreme" value the Hamiltonian of the system becomes infinite.* In the integral that remains, the factor $\partial x_i/\partial x_j$, being equal to δ_{ij}, comes out of the integral sign and we are left with

$$\frac{1}{\beta} \delta_{ij} \int e^{-\beta H} \, d\omega.$$

Substituting this into (11), we arrive at the desired result:

$$\left\langle x_i \frac{\partial H}{\partial x_j} \right\rangle = \delta_{ij} kT. \tag{2.6.5}$$

The major implications of this theorem were discussed in Sec. 2.6. Nevertheless, we make here an important remark on the *equipartition theorem* for a noninteracting system. This follows from eqn. (2.6.5) by taking $x_i = x_j = p_i$ and summing over the index i; we get

$$\left\langle \sum_{i=1}^{3N} p_i \frac{\partial H}{\partial p_i} \right\rangle \equiv \left\langle \sum_{i=1}^{3N} p_i \dot{q}_i \right\rangle = 3NkT. \tag{2.6.8}$$

* We observe that if x_j is a space coordinate, then its extreme values will correspond to "locations at the walls of the container"; accordingly, the potential energy of the system would become infinite. If, on the other hand, x_j is a momentum coordinate, then its extreme values will themselves be $\pm\infty$, in which case the kinetic energy of the system would become infinite.

In the nonrelativistic case

$$H = \sum_{i=1}^{3N} p_i^2/2m; \qquad \frac{\partial H}{\partial p_i} \equiv \dot{q}_i = p_i/m,$$

and the theorem yields the well-known result

$$\langle H \rangle \equiv U = \tfrac{3}{2}NkT.$$

However, in the extreme relativistic case

$$H = \sum_{i=1}^{N} (p_{i1}^2+p_{i2}^2+p_{i3}^2)^{1/2}\, c,$$

so that

$$\frac{\partial H}{\partial p_{i\alpha}} \equiv \dot{q}_{i\alpha} = (p_{i1}^2+p_{i2}^2+p_{i3}^2)^{-1/2}\, p_{i\alpha}\, c \qquad (\alpha = 1, 2, 3)$$

and hence

$$\sum_{\alpha=1}^{3} p_{i\alpha}\dot{q}_{i\alpha} = (p_{i1}^2+p_{i2}^2+p_{i3}^2)^{1/2}\, c.$$

Consequently, our theorem now yields

$$\langle H \rangle \equiv U = 3NkT.$$

Thus, in the extreme relativistic case the mean energy per degree of freedom is kT, and not $\frac{1}{2}kT$. One must, however, note that *in either case* the equation of state is: $PV = NkT$; see Problem 3.15.

3.7. A system of harmonic oscillators

We shall now study, as an example, a system of N, practically independent, harmonic oscillators. This study will not only provide an interesting illustration of the canonical ensemble formulation but will also serve as a formal basis for some of our subsequent studies in this text. Two important problems in this line are (i) the theory of the black-body radiation (or the "statistical mechanics of photons") and (ii) the theory of lattice vibrations (or the "statistical mechanics of phonons"); see Secs. 7.2 and 7.3.

We start with the specialized situation when the oscillators can be treated *classically*. The Hamiltonian of any one of them (assumed to be one-dimensional) may then be written as

$$H(q_i, p_i) = \frac{1}{2} m\omega^2 q_i^2 + \frac{1}{2m} p_i^2; \tag{1}$$

of course, the index i will run from 1 to N. For the single-oscillator partition function, we readily obtain

$$Q_1(\beta) = \frac{1}{h} \int_{-\infty}^{\infty} \int_{-\infty}^{\infty} \exp\left\{-\beta\left(\frac{1}{2} m\omega^2 q^2 + \frac{1}{2m} p^2\right)\right\} dq\, dp$$

$$= \frac{1}{h}\left(\frac{2\pi}{\beta m\omega^2}\right)^{1/2}\left(\frac{2\pi m}{\beta}\right)^{1/2} = \frac{1}{\beta\hbar\omega}, \tag{2}$$

where $\hbar = h/2\pi$. The partition function of the N-oscillator system would then be

$$Q_N(\beta) = [Q_1(\beta)]^N = (\beta\hbar\omega)^{-N}; \tag{3}$$

note that in writing (3) we have assumed the oscillators to be *distinguishable*. This is so because, as we shall see later, these oscillators are only a representation of the energy levels available in the system; they are not particles (or even "quasi-particles"). It is actually photons in one case and phonons in the other, which distribute themselves over the various oscillator levels, that are *indistinguishable*!

The Helmholtz free energy of the system is now given by

$$A \equiv -kT \ln Q_N = NkT \ln \left(\frac{\hbar\omega}{kT}\right), \tag{4}$$

whence we obtain for other thermodynamic quantities

$$\mu = kT \ln \left(\frac{\hbar\omega}{kT}\right), \tag{5}$$

$$P = 0, \tag{6}$$

$$S = Nk\left[\ln \left(\frac{kT}{\hbar\omega}\right) + 1\right], \tag{7}$$

$$U = NkT \tag{8}$$

and

$$C_P = C_V = Nk. \tag{9}$$

We note that the mean energy per oscillator is in complete accord with the equipartition theorem, namely $2 \times \frac{1}{2}kT$, for we have here two *independent* quadratic terms in the single-oscillator Hamiltonian; see (1).

We may determine the density of states, $g(E)$, of this system from the expression (3) for its partition function. We have, in view of (3.4.7),

$$g(E) = \frac{1}{(\hbar\omega)^N} \cdot \frac{1}{2\pi i} \int_{\beta'-i\infty}^{\beta'+i\infty} \frac{e^{\beta E}}{\beta^N} d\beta \qquad (\beta' > 0),$$

that is,

$$g(E) = \begin{cases} \dfrac{1}{(\hbar\omega)^N} \cdot \dfrac{E^{N-1}}{(N-1)!} & \text{for} \quad E \geqslant 0, \\ 0 & \text{for} \quad E \leqslant 0. \end{cases} \tag{10}$$

To test its correctness, we may calculate the entropy of the system with the help of this formula. Taking $N \gg 1$ and making use of the Stirling approximation, we get

$$S(N, E) = k \ln g(E) \simeq Nk\left[\ln \left(\frac{E}{N\hbar\omega}\right) + 1\right], \tag{11}$$

which yields for the temperature of the system

$$T = \left(\frac{\partial S}{\partial E}\right)_N^{-1} = \frac{E}{Nk}. \tag{12}$$

Eliminating E between these two relations, we obtain precisely our earlier result for the function $S(N, T)$. This indeed assures us of the inner consistency of our approach; more so, it gives us confidence to accept (10) as the correct expression for the density of states of this system.

We now take up the quantum-mechanical situation, according to which the energy eigenvalues of a one-dimensional harmonic oscillator are given by

$$\varepsilon_n = (n+\tfrac{1}{2})\hbar\omega; \qquad n = 0, 1, 2, \ldots. \tag{13}$$

Accordingly, we have for the single-oscillator partition function

$$Q_1(\beta) = \sum_{n=0}^{\infty} e^{-\beta(n+\frac{1}{2})\hbar\omega} = \frac{\exp(-\frac{1}{2}\beta\hbar\omega)}{1-\exp(-\beta\hbar\omega)}$$

$$= \left\{2\sinh\left(\frac{1}{2}\beta\hbar\omega\right)\right\}^{-1}. \tag{14}$$

The N-oscillator partition function is then given by

$$Q_N(\beta) = [Q_1(\beta)]^N = [2\sinh(\tfrac{1}{2}\beta\hbar\omega)]^{-N}$$

$$= e^{-(N/2)\beta\hbar\omega}\{1-e^{-\beta\hbar\omega}\}^{-N}. \tag{15}$$

For the Helmholtz free energy of the system, we have

$$A = NkT\ln[2\sinh(\tfrac{1}{2}\beta\hbar\omega)] = N[\tfrac{1}{2}\hbar\omega + kT\ln\{1-e^{-\beta\hbar\omega}\}], \tag{16}$$

whence we obtain for other thermodynamic quantities

$$\mu = A/N, \tag{17}$$

$$P = 0 \tag{18}$$

$$S = Nk\left[\frac{1}{2}\beta\hbar\omega\coth\left(\frac{1}{2}\beta\hbar\omega\right) - \ln\left\{2\sinh\left(\frac{1}{2}\beta\hbar\omega\right)\right\}\right]$$

$$= Nk\left[\frac{\beta\hbar\omega}{e^{\beta\hbar\omega}-1} - \ln\{1-e^{-\beta\hbar\omega}\}\right], \tag{19}$$

$$U = \frac{1}{2}N\hbar\omega\coth\left(\frac{1}{2}\beta\hbar\omega\right) = N\left[\frac{1}{2}\hbar\omega + \frac{\hbar\omega}{e^{\beta\hbar\omega}-1}\right] \tag{20}$$

and

$$C_p = C_V = Nk\left(\frac{1}{2}\beta\hbar\omega\right)^2\operatorname{cosech}^2\left(\frac{1}{2}\beta\hbar\omega\right)$$

$$= Nk(\beta\hbar\omega)^2\frac{e^{\beta\hbar\omega}}{(e^{\beta\hbar\omega}-1)^2}. \tag{21}$$

The result (20) is extremely interesting in that it brings out the fact that the quantum-mechanical harmonic oscillators do not obey the equipartition law. The mean energy per oscillator is different from the equipartition value kT; actually, it is always greater than kT; see curve 2 in Fig. 3.4. Only in the limit of high temperatures, where the thermal energy kT

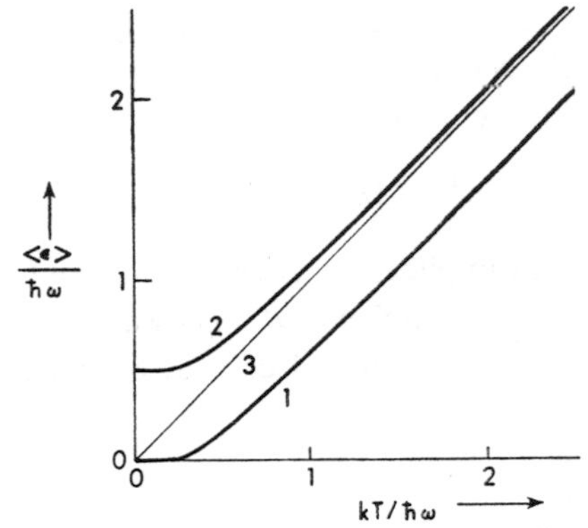

FIG. 3.4. The mean energy $\langle \varepsilon \rangle$ of a harmonic oscillator as a function of temperature. 1—the Planck oscillator; 2—the Schrödinger oscillator; 3—the classical oscillator.

is much larger than the energy quantum $\hbar\omega$, does the mean energy per oscillator tend to the equipartition value. It must be noted here that if the zero-point energy $\frac{1}{2}\hbar\omega$ were not there, the limiting value of the mean energy would be $\left(kT-\frac{1}{2}\hbar\omega\right)$, and not kT—we may call such an oscillator the *Planck oscillator*; see curve 1 in the figure. In passing, we observe that the specific heat (21), which is the same for the Planck oscillator as for the Schrödinger oscillator, is temperature-dependent; moreover, it is always less than, and at high temperatures tends to, the classical value (9).

Indeed, for $kT \gg \hbar\omega$, formulae (14)–(21) go over to their classical counterparts, namely (2)–(9), respectively.

We shall now determine the density of states $g(E)$ of the N-oscillator quantum-mechanical system from its partition function (15). Carrying out the binomial expansion of this expression, we have

$$Q_N(\beta) = \sum_{R=0}^{\infty} \binom{N+R-1}{R} e^{-\beta\left(\frac{1}{2}N\hbar\omega + R\hbar\omega\right)}. \tag{22}$$

Comparing this with the formula

$$Q_N(\beta) = \int_0^{\infty} g(E)\, e^{-\beta E}\, dE,$$

we conclude that

$$g(E) = \sum_{R=0}^{\infty} \binom{N+R-1}{R} \delta\left(E - \left\{R + \tfrac{1}{2}N\right\} \hbar\omega\right), \tag{23}$$

where $\delta(x)$ stands for the Dirac delta function. Equation (23) implies that there are $(N+R-1)!/R!(N-1)!$ states available to the system when its energy E has the discrete value $\left(R+\frac{1}{2}N\right)\hbar\omega$, where $R = 0, 1, 2, \ldots$, and that there is no state available for other values of E.

This is hardly surprising; nonetheless, it is instructive to look at this result from a slightly different point of view.

We consider the following problem: given an energy E for distribution among a set of N harmonic oscillators, each of which can be in any one of the eigenstates (13), what is the total number of *distinct* ways in which the process of distribution can be carried out? Now, in view of the form of the eigenvalues ε_n, we better give away, right in the beginning, the zero-point energy $\frac{1}{2}\hbar\omega$ to each of the N oscillators and convert the rest of it into quanta (of energy $\hbar\omega$). Let R be the number of these quanta; then

$$R = (E - \tfrac{1}{2}N\hbar\omega)/\hbar\omega. \tag{24}$$

Clearly, R must be an integer; by implication, E must be of the form $(R+\frac{1}{2}N)\hbar\omega$. The problem then reduces to determining the number of *distinct* ways of allotting R quanta to N oscillators, such that an oscillator may have 0 or 1 or 2... quanta; in other words, we have to

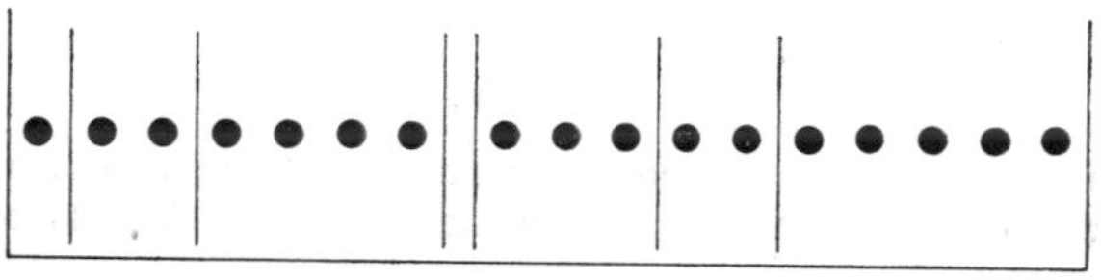

FIG. 3.5. Distributing seventeen *indistinguishable* balls among seven *distinguishable* boxes. The arrangement shown here represents *one* of the 23!/17!6! distinct ways of carrying out the distribution.

determine the number of *distinct* ways of putting R *indistinguishable* balls into N *distinguishable* boxes, such that a box may receive 0 or 1 or 2 ... balls. A little reflection will show that this is just the number of permutations that can be obtained by shuffling R balls, placed along a row, with $(N-1)$ partitioning lines;* see Fig. 3.5. The answer is now straightforward, viz.

$$\frac{(R+N-1)!}{R!\,(N-1)!}, \tag{25}$$

which agrees with (23).

We can determine the entropy of the system from the number (25). Since $N \gg 1$, we have

$$\begin{aligned} S &\simeq k\{\ln (R+N)! - \ln R! - \ln N!\} \\ &\simeq k\{(R+N)\ln (R+N) - R\ln R - N\ln N\}; \end{aligned} \tag{26}$$

the number R is, of course, a measure of the energy of the system, see (24). For the temperature of the system, we obtain

$$\frac{1}{T} = \left(\frac{\partial S}{\partial E}\right)_N = \left(\frac{\partial S}{\partial R}\right)_N \frac{1}{\hbar\omega} = \frac{k}{\hbar\omega}\ln\left(\frac{R+N}{R}\right) = \frac{k}{\hbar\omega}\ln\left(\frac{E+\frac{1}{2}N\hbar\omega}{E-\frac{1}{2}N\hbar\omega}\right), \tag{27}$$

whence it follows that

$$\frac{E}{N} = \frac{1}{2}\hbar\omega\,\frac{\exp(\hbar\omega/kT)+1}{\exp(\hbar\omega/kT)-1}, \tag{28}$$

* These lines obviously divide the space into N boxes.

which is identical with (20). It can be further checked that, by eliminating R between (26) and (27), we obtain precisely the formula (19) for $S(N, T)$. Thus, once again we find that the results obtained by following the microcanonical and canonical approaches are identically the same.

Finally, we may consider the classical limit when E/N, the mean energy per oscillator, is much larger than the energy quantum $\hbar\omega$, i.e. when $R \gg N$. The expression (25) may, in that case, be replaced by

$$\frac{(R+N-1)(R+N-2)\dots(R+1)}{(N-1)!} \simeq \frac{R^{N-1}}{(N-1)!}, \tag{25a}$$

with

$$R \simeq E/\hbar\omega.$$

The corresponding expression for the entropy would be

$$S \simeq k\{N \ln (R/N)+N\} \simeq Nk\left\{\ln\left(\frac{E}{N\hbar\omega}\right)+1\right\}, \tag{26a}$$

whence it follows that

$$\frac{1}{T} = \left(\frac{\partial S}{\partial E}\right)_N = \frac{Nk}{E}, \tag{27a}$$

so that

$$\frac{E}{N} = kT. \tag{28a}$$

These results are identical with eqns. (10)–(12) derived earlier in the section.

3.8. The statistics of paramagnetism

Next, we study a system of N magnetic dipoles, each having a magnetic moment μ. In the presence of an external magnetic field $\boldsymbol{H}$, the dipoles will experience a torque tending to align them in the direction of the field. If there were nothing else to check this tendency, the dipoles would align themselves precisely in this direction; we would thereby achieve a complete magnetization of the system. In reality, however, thermal agitation in the system offers a resistance to this tendency and, in equilibrium, we obtain only a *partial* magnetization. Clearly, as $T \to 0$°K, the thermal agitation becomes ineffective and the system exhibits a complete orientation of the dipole moments, whatever the strength of the applied field; at the other end, as $T \to \infty$, we approach a state of complete randomization of the dipole moments, which implies a vanishing magnetization. At intermediate temperatures, the situation is governed by the quantity $(\mu H/kT)$.

The model adopted for this study consists of N identical, localized (and, hence, distinguishable), practically static, mutually noninteracting and freely orientable dipoles. It is obvious that the only energy we need to consider here is the potential energy of the dipoles which arises from the presence of the external field $\boldsymbol{H}$ and depends upon the orientations of the dipoles with respect to the direction of the field:

$$E = -\sum_{i=1}^{N} \boldsymbol{\mu}_i \cdot \boldsymbol{H} = -\mu H \sum_{i=1}^{N} \cos\theta_i; \tag{1}$$

it may be noted that we are neglecting here the effect of the *induced* magnetic field. The partition function of the system is given by

$$Q_N(\beta) = \sum_{\{\theta_i\}} \exp\left\{\beta\mu H \sum_{i=1}^{N} \cos\theta_i\right\}, \tag{2}$$

where the summation goes over all sets of orientations the system can have. Since the total energy of the system is simply a sum of the individual energies of the dipoles, we can write

$$Q_N(\beta) = \sum_{\{\theta_i\}} \prod_{i=1}^{N} \exp(\beta\mu H \cos\theta_i) = \prod_{i=1}^{N}\left[\sum_{\theta_i} \exp(\beta\mu H\cos\theta_i)\right] = [Q_1(\beta)]^N, \tag{3}$$

where

$$Q_1(\beta) = \sum_{\theta} \exp(\beta\mu H\cos\theta). \tag{4}$$

The mean magnetic moment $\boldsymbol{M}$ of the system will indeed be in the direction of the field $\boldsymbol{H}$; for its magnitude we have

$$M_z = \left\langle \sum_{i=1}^{N} \mu\cos\theta_i \right\rangle_{\{\theta_i\}} = \frac{1}{\beta}\frac{\partial}{\partial H}\ln Q_N(\beta) \tag{5}$$

$$= \frac{N}{\beta}\frac{\partial}{\partial H}\ln Q_1(\beta) = N\langle\mu\cos\theta\rangle. \tag{6*}$$

Thus, to determine the degree of magnetization in the system all we have to do is evaluate the single-dipole partition function (4).

First, we do it classically (after Langevin, 1905). Using ($\sin\theta\, d\theta\, d\varphi$) for the elemental solid angle representing a small range of orientations of the dipole, we get

$$Q_1(\beta) = \int_0^{2\pi}\int_0^{\pi} e^{\beta\mu H\cos\theta}\sin\theta\, d\theta\, d\varphi = 4\pi\frac{\sinh(\beta\mu H)}{\beta\mu H}, \tag{7}$$

whence we obtain for the mean magnetic moment per dipole

$$\bar{\mu}_z \equiv \frac{M_z}{N} = \mu\left\{\coth(\beta\mu H) - \frac{1}{\beta\mu H}\right\} = \mu L(x), \tag{8}$$

where $L(x)$ is the so-called *Langevin function*,

$$L(x) = \coth x - \frac{1}{x}, \tag{9}$$

* Note that, by definition as well,

$$\langle\mu\cos\theta\rangle = \sum_{\theta} P(\theta)\mu\cos\theta,$$

where $P(\theta)$ is the relevant probability factor, viz.

$$P(\theta) = \frac{\exp\{-\beta E(\theta)\}}{\sum_{\theta}\exp\{-\beta E(\theta)\}}.$$

Hence the formula (6).

with

$$x = \beta\mu H; \tag{10}$$

we note that the parameter x determines the strength of the (magnetic) potential energy μH against the (thermal) kinetic energy kT. A plot of the Langevin function is shown in Fig. 3.6.

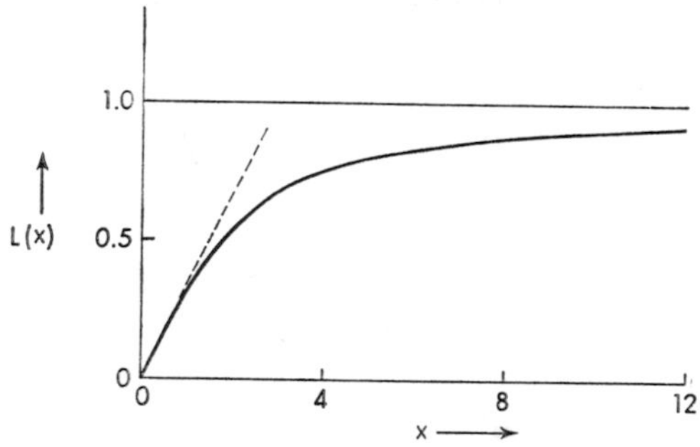

FIG. 3.6. The Langevin function $L(x)$.

If we have N_0 dipoles per unit volume in the system, then the *magnetization* of the system, viz. the mean magnetic moment per unit volume, will be given by

$$M_{z0} = N_0\bar{\mu}_z = N_0\mu L(x). \tag{11}$$

For magnetic fields so strong (or temperatures so low) that the parameter $x \gg 1$, the function $L(x)$ is almost equal to 1; the system then acquires a state of magnetic saturation:

$$\bar{\mu}_z \simeq \mu \quad \text{and} \quad M_{z0} \simeq N_0\mu. \tag{12}$$

For temperatures so high (or magnetic fields so weak) that the parameter $x \ll 1$, the function $L(x)$ may be written as

$$\frac{x}{3} - \frac{x^3}{45} + \dots \tag{13}$$

which, in the lowest approximation, gives

$$M_{z0} \simeq \frac{N_0\mu^2}{3kT} H. \tag{14}$$

The high-temperature *susceptibility* of the system is, therefore, given by

$$\chi = \lim_{H\to 0}\left(\frac{\partial M_{z0}}{\partial H}\right) \simeq \frac{N_0\mu^2}{3kT} = \frac{C}{T}, \quad \text{say.} \tag{15}$$

Equation (15) is the *Curie law* of paramagnetism, the parameter C being the *Curie constant* of the system. Figure 3.7 shows a plot of the susceptibility of a powdered sample of copper–potassium sulphate hexahydrate as a function of T^{-1}; the fact that the graph is linear and passes almost through the origin vindicates the Curie law for this particular salt.

In passing, we mention that the magnetic induction B and the magnetic permeability μ' of a medium are related by the equation $B = \mu' H = H + 4\pi I$, where $I = \chi H$; hence the formula

$$\mu' \simeq 1 + \frac{4\pi N_0 \mu^2}{3kT}, \tag{16}$$

which gives the magnetic permeability of a medium at high temperatures.

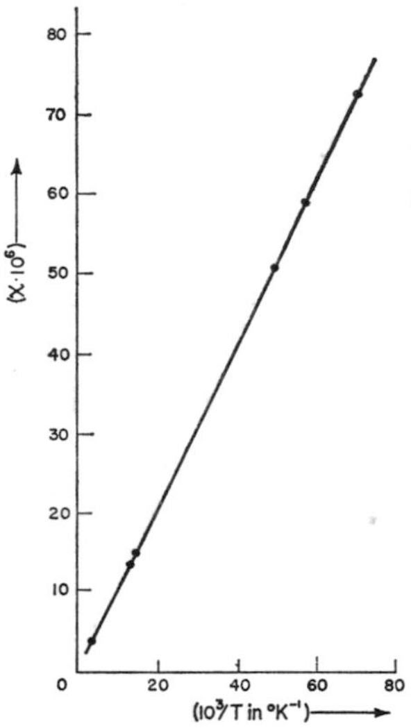

FIG. 3.7. χ vs. $1/T$ plot for a powdered sample of copper–potassium sulphate hexahydrate (after Hupse, 1942).

We shall now treat the problem of paramagnetism quantum-mechanically. The major modification here arises from the fact that the magnetic dipole moment $\boldsymbol{\mu}$ and its component μ_z in the direction of the applied field cannot have *arbitrary* values. Quite generally, we have a direct relationship between the magnetic moment $\boldsymbol{\mu}$ of a given dipole and its angular momentum $\boldsymbol{M}$;

$$\boldsymbol{\mu} = \left(g\frac{e}{2mc}\right)\boldsymbol{M}, \tag{17}$$

while

$$M = J\hbar; \quad J = \tfrac{1}{2}, \tfrac{3}{2}, \tfrac{5}{2}, \ldots \quad \text{or} \quad 0, 1, 2, \ldots. \tag{18}$$

The quantity $g(e/2mc)$ is referred to as the *gyromagnetic ratio* of the dipole while the number g is known as *Lande's g-factor*. If the net angular momentum of the dipole is due only to electron spins, then $g = 2$; on the other hand, if it is due only to orbital motions, then $g = 1$. In general, however, its origin is mixed; g is then given by the formula

$$g = \frac{3}{2} + \frac{S(S+1) - L(L+1)}{2J(J+1)}, \tag{19}$$

S and L being, respectively, the spin and the orbital quantum numbers of the dipole. Note that there is no upper or lower bound for the values that g can have!

Combining (17) and (18), we can write

$$\mu = g\mu_B J, \tag{20}$$

where $\mu_B(= e\hbar/2mc)$ is the *Bohr magneton.* The component μ_z of the magnetic moment in the direction of the applied field is, accordingly, given by

$$\mu_z = g\mu_B m, \tag{21}$$

where the magnetic quantum number m, in accordance with the rules of space quantization, can take no other values except

$$m = -J,\ -J+1,\ \ldots,\ J-1,\ J. \tag{22}$$

Thus, a dipole whose magnetic moment is given by (20) can have no other orientations with respect to the applied field except the ones conforming to the values (22) of the magnetic quantum number m; obviously, the number of allowed orientations, for a given value of J, is $(2J+1)$. In view of this, the single-dipole partition function $Q_1(\beta)$ is given by, see (4),

$$Q_1(\beta) = \sum_{m=-J}^{J} \exp\,(\beta g\mu_B Hm). \tag{23}$$

Introducing a parameter x, defined by

$$x = \beta(g\mu_B J)H, \tag{24}$$

eqn. (23) becomes

$$\begin{aligned} Q_1(\beta) &= \sum_{m=-J}^{J} e^{mx/J} = \frac{e^{-x}\{e^{(2J+1)x/J}-1\}}{e^{x/J}-1} \\ &= \frac{e^{(2J+1)x/2J}-e^{-(2J+1)x/2J}}{e^{x/2J}-e^{-x/2J}} \\ &= \sinh\left\{\left(1+\frac{1}{2J}\right)x\right\}\Big/\sinh\left\{\frac{1}{2J}x\right\}. \end{aligned} \tag{25}$$

The mean magnetic moment of the system is then given by, see (6),

$$\begin{aligned} M_z = N\bar{\mu}_z &= \frac{N}{\beta}\frac{\partial}{\partial H}\ln Q_1(\beta) \\ &= N(g\mu_B J)\left[\left(1+\frac{1}{2J}\right)\coth\left\{\left(1+\frac{1}{2J}\right)x\right\}-\frac{1}{2J}\coth\left\{\frac{1}{2J}x\right\}\right]. \end{aligned} \tag{26}$$

We thus have for the mean magnetic moment per dipole

$$\bar{\mu}_z = \mu B_J(x), \tag{27}$$

where $B_J(x)$ is the *Brillouin function* (of order J):

$$B_J(x) = \left(1+\frac{1}{2J}\right)\coth\left\{\left(1+\frac{1}{2J}\right)x\right\}-\frac{1}{2J}\coth\left\{\frac{1}{2J}x\right\}. \tag{28}$$

In Fig. 3.8 we have plotted the function $B_J(x)$ for some typical values of the quantum number J.

Let us now consider a few special cases. First of all, we note that for strong fields and low temperatures ($x \gg 1$), the function $B_J(x) \simeq 1$ *for all* J; this corresponds to a state of magnetic saturation. On the other hand, for high temperatures and weak fields ($x \ll 1$), the function $B_J(x)$ may be written as

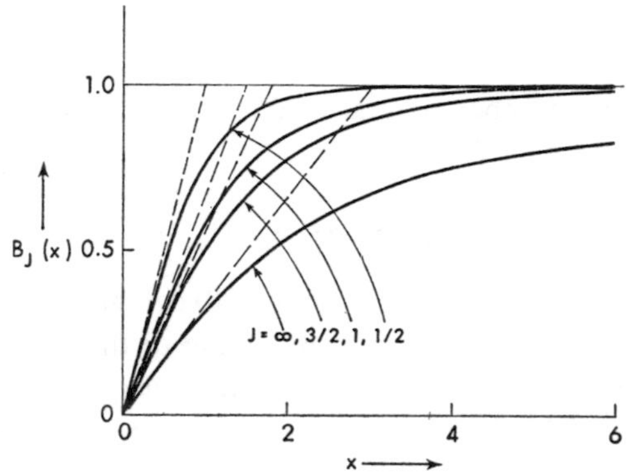

FIG. 3.8. The Brillouin function $B_J(x)$ for various values of J.

$$\tfrac{1}{3}(1+1/J)x+ \;\ldots, \tag{29}$$

so that

$$\bar{\mu}_z \simeq \frac{\mu^2}{3kT}\left(1+\frac{1}{J}\right)H = \frac{g^2\mu_B^2 J(J+1)}{3kT}H. \tag{30}$$

The Curie law, $\chi \propto 1/T$, is again obeyed; however, the Curie constant is now given by

$$C_J = \frac{N_0\mu^2}{3k}\left(1+\frac{1}{J}\right) = \frac{N_0 g^2\mu_B^2 J(J+1)}{3k}. \tag{31}$$

It is interesting to note that the high-temperature results (30) and (31) directly involve the eigenvalues of the operator $\boldsymbol{\mu}^2$; cf. the corresponding classical results (14) and (15) where we have simply μ^2.

Let us now look a little more closely at the dependence of the foregoing results on the quantum number J. First of all, we consider the extreme case, viz. $J \to \infty$; we do this with the understanding that simultaneously $g \to 0$, in such a way that the product gJ, and hence the value of μ, stays constant. From (28), we readily observe that, in this limit, the Brillouin function $B_J(x)$ tends to become (i) independent of J and (ii) identical with the Langevin function $L(x)$. This is not surprising because, in this limit, the number of allowed orientations for a magnetic dipole becomes infinitely large, with the result that the problem essentially reduces to its classical counterpart (where one allows *all* conceivable orientations). At the other extreme, we have the case: $J = \frac{1}{2}$, which allows only *two* orientations. The results in this case are naturally very different from the ones for $J \gg 1$. We now have

$$\bar{\mu}_z = \mu B_{1/2}(x) = \mu \tanh x. \tag{32}$$

For $x \gg 1$, $\bar{\mu}_z$ is, of course, very nearly equal to μ. For $x \ll 1$, however, $\bar{\mu}_z \simeq \mu x$, which corresponds to the Curie constant

$$C_{1/2} = \frac{N_0 \mu^2}{k}. \tag{33}$$

We note that the constant $C_{1/2}$ is exactly three times the classical Curie constant C_∞; cf. eqn. (15). Of course, both $C_{1/2}$ and C_∞ are in conformity with the general result (31).

In Fig. 3.9 we have reproduced the experimental values of $\bar{\mu}_z$ (in terms of μ_B) as a function of the quantity H/T, for three paramagnetic salts; the corresponding theoretical plots, viz.

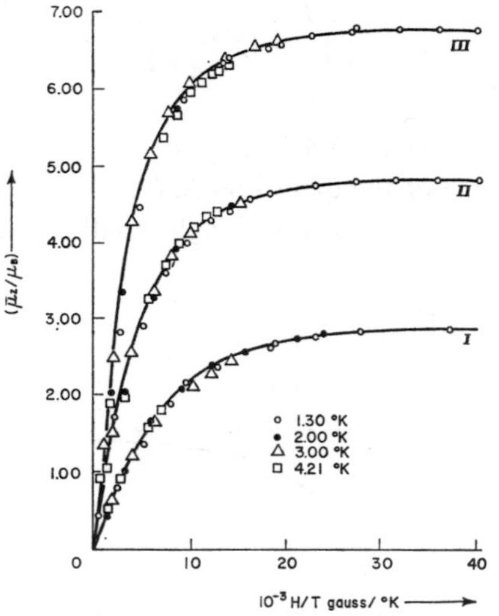

FIG. 3.9. Plots of $\bar{\mu}_z/\mu_B$ as a function of H/T. The solid curves represent the theoretical results, while the points mark the experimental findings of Henry (1952). Curve I is for potassium chromium alum ($J = \frac{3}{2}$, $g = 2$), curve II for iron ammonia alum ($J = \frac{5}{2}$, $g = 2$) and curve III for gadolinium sulphate octahydrate ($J = \frac{7}{2}$, $g = 2$).

the curves $gJB_J(x)$, are also included in the figure. The agreement between theory and experiment is indeed good. In passing, we note that, at a temperature of 1.3°K, a field of about 50,000 gauss is sufficient to produce over 99 per cent of saturation in these salts.

3.9. Thermodynamics of magnetic systems: negative temperatures

For the purpose of this section, it will suffice to consider a system of dipoles with $J = \frac{1}{2}$. Each dipole then has a choice of two orientations, the corresponding energies being $-\mu H$ and $+\mu H$; let us call these energies $-\varepsilon$ and $+\varepsilon$, respectively. The partition function of the system is then given by

$$Q_N(\beta) = (e^{\beta\varepsilon} + e^{-\beta\varepsilon})^N = \{2 \cosh (\beta\varepsilon)\}^N, \tag{1}$$

cf. the general expression (3.8.25). Accordingly, the Helmholtz free energy is given by

$$A = -NkT \ln \{2 \cosh (\varepsilon/kT)\}, \tag{2}$$

whence we obtain

$$S = -\left(\frac{\partial A}{\partial T}\right)_H = Nk\left[\ln\left\{2\cosh\left(\frac{\varepsilon}{kT}\right)\right\} - \frac{\varepsilon}{kT}\tanh\left(\frac{\varepsilon}{kT}\right)\right], \tag{3}$$

$$U = A+TS = -N\varepsilon\tanh\left(\frac{\varepsilon}{kT}\right), \tag{4}$$

$$M = -\left(\frac{\partial A}{\partial H}\right)_T = N\mu\tanh\left(\frac{\varepsilon}{kT}\right) \tag{5}$$

and, finally,

$$C = \left(\frac{\partial U}{\partial T}\right)_H = Nk\left(\frac{\varepsilon}{kT}\right)^2 \text{sech}^2\left(\frac{\varepsilon}{kT}\right). \tag{6}$$

Equation (5) is essentially the same as (3.8.32); moreover, as expected, $U = -MH$.

The temperature dependence of the quantities S, U, M and C is shown in Figs. 3.10–13. We note that the entropy of the system is vanishingly small for $kT \ll \varepsilon$; it rises rapidly when

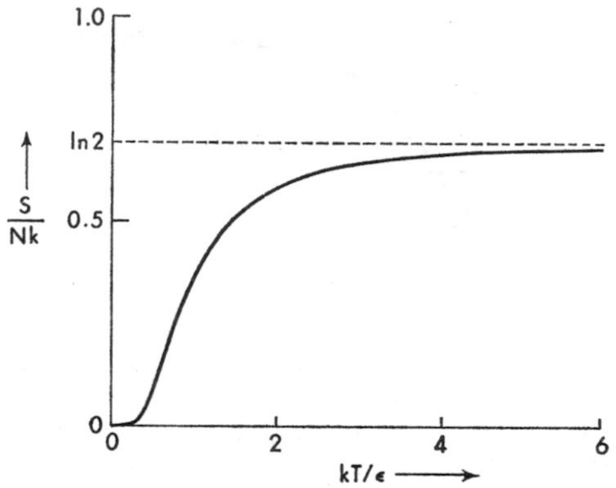

FIG. 3.10. The entropy of a system of magnetic dipoles (with $J = \frac{1}{2}$) as a function of temperature.

kT is of the order of ε and approaches the limiting value $Nk \ln 2$ for $kT \gg \varepsilon$. (This limiting value of S corresponds to the fact that at high temperatures the orientation of the dipoles assumes a completely random character, with the result that we now have 2^N *equally likely* complexions available to the system.) The energy of the system attains its lowest value, $-N\varepsilon$, as $T \to 0°K$; this clearly corresponds to a state of magnetic saturation and, hence, to a state of perfect order in the system. Towards high temperatures, the energy tends to vanish;* this implies a purely random orientation of the dipoles and hence a complete loss of magnetic order. These features are re-emphasized in Fig. 3.12, which depicts the temperature dependence of the magnetization M. The specific heat of the system is vanishingly small at low temperatures but, in view of the fact that the energy of the system tends to a constant value as $T \to \infty$, the specific heat vanishes at high temperatures as well. Somewhere around

* Note that in the present study we are completely disregarding the kinetic energy of the dipoles.

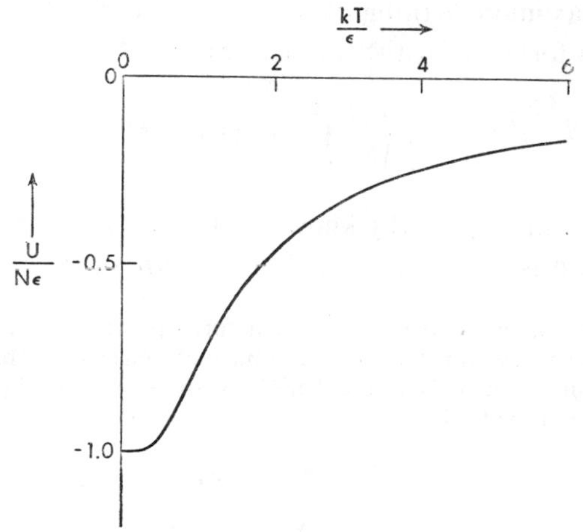

FIG. 3.11. The energy of a system of magnetic dipoles (with $J = \frac{1}{2}$) as a function of temperature.

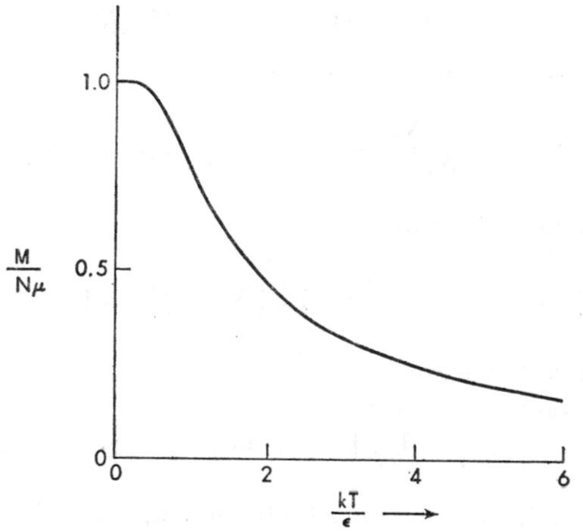

FIG. 3.12. The magnetization of a system of magnetic dipoles (with $J = \frac{1}{2}$) as a function of temperature.

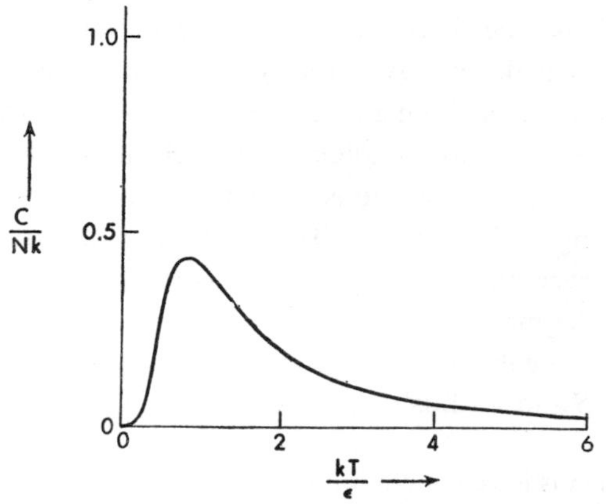

FIG. 3.13. The specific heat of a system of magnetic dipoles (with $J = \frac{1}{2}$) as a function of temperature.

$T = \varepsilon/k$, it displays a maximum. Writing Δ for the energy difference between the two allowed states of the dipole, the formula for the specific heat can be written as

$$C = Nk\left(\frac{\Delta}{kT}\right)^2 e^{\Delta/kT}(1+e^{\Delta/kT})^{-2}. \tag{7}$$

A specific heat of this form is generally known as the *Schottky specific heat*; characterized by an anomalous peak, it is observed in all systems with an excitation gap Δ.

Let us digress to consider our system from the combinatorial point of view. The question then arises: in how many different ways can our system attain a state of magnetic energy E? This can be tackled in precisely the same way as the problem of the *random walk*. Let N_+ be the number of dipoles aligned opposite to the field and N_- in the direction of the field; then

$$E = (N_+ - N_-)\varepsilon,$$

where

$$N_+ + N_- = N.$$

Solving for N_+ and N_-, we obtain

$$N_+ = \tfrac{1}{2}(N+E/\varepsilon); \qquad N_- = \tfrac{1}{2}(N-E/\varepsilon). \tag{8}$$

The desired number of ways is then given by the expression

$$\Omega(N, E) = \frac{N!}{N_+!\,N_-!} = \frac{N!}{\{\frac{1}{2}(N+E/\varepsilon)\}!\,\{\frac{1}{2}(N-E/\varepsilon)\}!}, \tag{9}$$

whence we obtain for the entropy of the system

$$S(N, E) = k \ln \Omega \simeq k\left[N \ln N - \frac{1}{2}\left(N+\frac{E}{\varepsilon}\right)\ln\left\{\frac{1}{2}\left(N+\frac{E}{\varepsilon}\right)\right\} - \frac{1}{2}\left(N-\frac{E}{\varepsilon}\right)\ln\left\{\frac{1}{2}\left(N-\frac{E}{\varepsilon}\right)\right\}\right]. \tag{10*}$$

The temperature of the system is then given by

$$\frac{1}{T} = \left(\frac{\partial S}{\partial E}\right)_N = \frac{k}{2\varepsilon}\ln\left\{\frac{N-E/\varepsilon}{N+E/\varepsilon}\right\}. \tag{11}$$

Expressing E as a function of T, we recover formula (4). Further, eliminating E between (10) and (11), we recover formula (3).

Now, throughout our study so far we have considered only those cases for which $T > 0$. For *normal* systems, this is indeed true, for otherwise we have to contend with canonical distributions that blow up as the energy of the system is indefinitely increased! If, however, the energy of a system is bounded from above, then there is no compelling reason to exclude the possibility of negative temperatures. Such specialized situations can indeed arise, and the system of magnetic dipoles being considered here provides an example thereof. From eqn. (11) we note that, so long as $E < 0$, $T > 0$—and that is the only range we covered in Figs. 3.10 to 3.13. However, the same equation tells us that if $E > 0$, then $T < 0$, which prompts us to examine the matter a little more closely. For this purpose we may consider as well the variation of the entropy S with the energy E; see Figs. 3.14 and 3.15. We note that for $E = -N\varepsilon$, both S and T vanish. As E increases, they too increase, until we reach the

* Note that this expression is in conformity with the formula (3.3.20), whereby $S = -Nk\Sigma_{i=1}^{2} p_i \ln p_i$, $p_1(= N_+/N)$ and $p_2(= N_-/N)$ being the probabilities of a dipole being in one or the other of the two possible states.

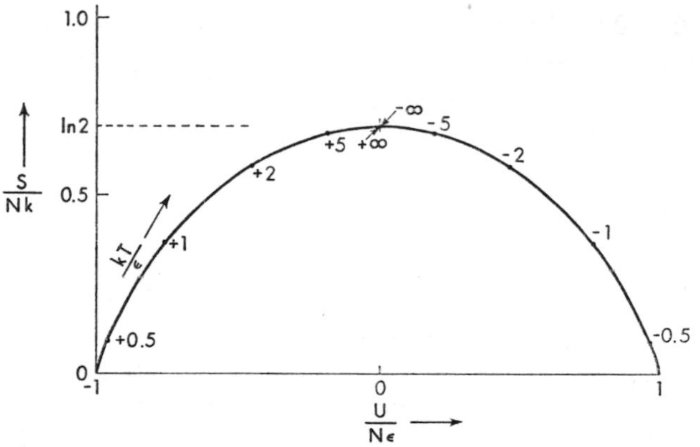

FIG. 3.14. The entropy of a system of magnetic dipoles (with $J = \frac{1}{2}$) as a function of energy. Some values of the parameter kT/ε are also shown in the figure.

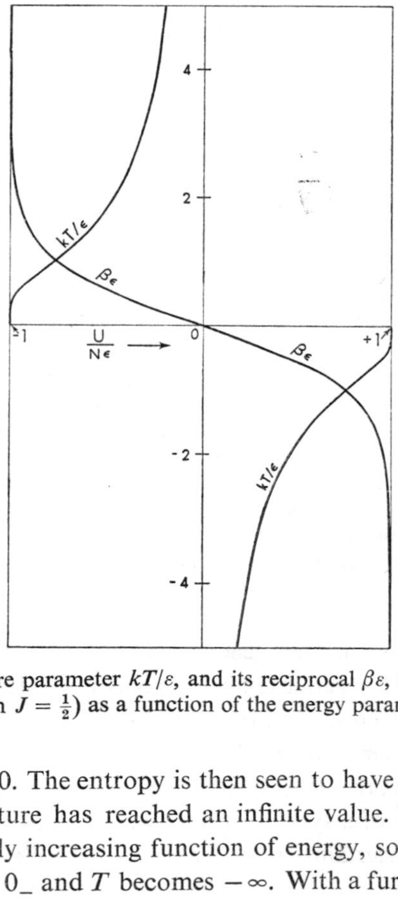

FIG. 3.15. The temperature parameter kT/ε, and its reciprocal $\beta\varepsilon$, for a system of magnetic dipoles (with $J = \frac{1}{2}$) as a function of the energy parameter $U/N\varepsilon$.

special situation where $E = 0$. The entropy is then seen to have attained its maximum value $Nk \ln 2$, while the temperature has reached an infinite value. Throughout this range, the entropy was a monotonically increasing function of energy, so T was positive. Now, as E equals 0_+, (dS/dE) becomes 0_- and T becomes $-\infty$. With a further increase in the value of E, the entropy monotonically decreases; as a result, the temperature continues to be negative, though its magnitude steadily decreases. Finally, we reach the largest value of E, namely $+N\varepsilon$, where the entropy is once again zero and $T = -0$.

The region of $E > 0$ (and hence $T < 0$) is indeed abnormal because it corresponds to a magnetization *opposite* in direction to that of the applied field! Nevertheless, it can be realized experimentally in the system of nuclear moments of a crystal in which the relaxation time t_1 for mutual interaction of the nuclear spins is very small in comparison with the relaxation time t_2 for interaction between the spins and the lattice. Let such a crystal be magnetized in a strong magnetic field, and then the field reversed so quickly that the spins are unable to follow the switch-over. This will leave the system in a state of nonequilibrium, with energy higher than the new equilibrium value U. During a period of the order of t_1 the sub-system of the nuclear spins should be able to attain a state of internal equilibrium; this state will have a negative magnetization and will, therefore, correspond to a negative temperature. The sub-system of the lattice, which involves energy parameters that are in principle unbounded, will indeed be at a positive temperature. During a period of the order of t_2, the two sub-systems would attain a state of mutual equilibrium, which again has a positive temperature.* An experiment of this kind was successfully performed by Purcell and Pound (1951) with a crystal of LiF; in this case, t_1 was of the order of 10^{-5} sec while t_2 was of the order of 5 min. A state of negative temperature for the sub-system of spins was indeed attained and was found to persist for a period of several minutes; see Fig. 3.16.

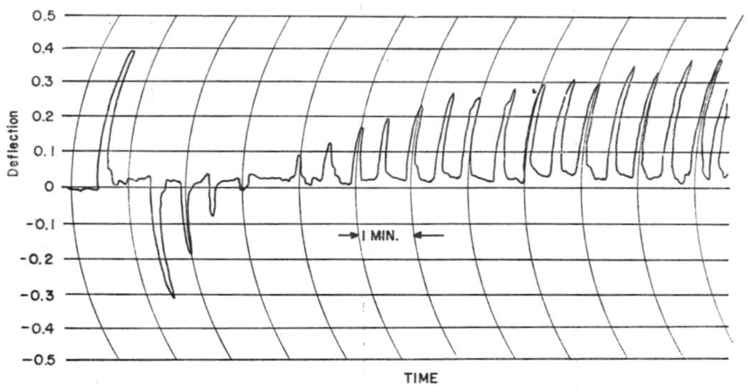

FIG. 3.16. A typical record of the reversed nuclear magnetization (after Purcell and Pound, 1951). On the left we have a deflection corresponding to normal, equilibrium magnetization ($T \approx 300°K$); it is followed by the reversed deflection (corresponding to $T \approx -350°K$), which decays through zero deflection (corresponding to a passage from $T = -\infty$ to $T = +\infty$) towards the new equilibrium state.

Before we close this discussion, a few general remarks seem in order. First of all, we should clearly note that the onset of negative temperatures is possible only if there exists an upper limit for the energy of the given system. In most physical systems this is not the case, simply because most physical systems possess kinetic energy of motion which is obviously unbounded. By the same token, the onset of positive temperatures is related to the existence of a lower limit for the energy of a system; this, however, does not present any problems because, if nothing else, the uncertainty principle alone is sufficient to set such a limit for *every*

* The reader should convince himself that in the latter process, during which the spins realign themselves (now more favorably in the *new* direction of the field), the energy will flow from the sub-system of the spins to that of the lattice, and not vice versa. This is in agreement with the fact that negative temperatures are *hotter* than positive ones; see the subsequent discussion in the text.

physical system. Thus, it is quite normal for a system to be at a positive temperature whereas it is very unusual for one to be at a negative temperature. Now, suppose that we are concerned with a system whose energy *cannot* assume unlimitedly high values. Then, we can surely visualize a temperature T such that the quantity NkT is much larger than any admissible value E_r of the energy. At such a high temperature, the mutual interaction of the microscopic entities constituting the system could be regarded as negligible; accordingly, one may write for the partition function

$$Q_N(\beta) \simeq \left[\sum_n e^{-\beta\varepsilon_n}\right]^N. \tag{12}$$

Since, by assumption, *all* $\beta\varepsilon_n \ll 1$, we have

$$Q_N(\beta) \simeq \left[\sum_n \{1-\beta\varepsilon_n+\tfrac{1}{2}\beta^2\varepsilon_n^2\}\right]^N. \tag{13}$$

Let g denote the number of possible orientations of a microscopic constituent of the system with respect to the direction of the external field; then the quantities $\Sigma_n\varepsilon_n^\alpha$ ($\alpha = 0, 1, 2$) may be replaced by the quantities $g\overline{\varepsilon^\alpha}$. So, we get

$$\begin{aligned}\ln Q_N(\beta) &\simeq N[\ln g+\ln(1-\beta\bar{\varepsilon}+\tfrac{1}{2}\beta^2\overline{\varepsilon^2})]\\ &\simeq N[\ln g-\beta\bar{\varepsilon}+\tfrac{1}{2}\beta^2\overline{(\varepsilon-\bar{\varepsilon})^2}].\end{aligned} \tag{14}$$

The Helmholtz free energy of the system is then given by

$$A(N, \beta) \simeq -\frac{N}{\beta}\ln g+N\bar{\varepsilon}-\frac{N}{2}\beta\overline{(\varepsilon-\bar{\varepsilon})^2}, \tag{15}$$

whence we obtain

$$S(N, \beta) \simeq Nk\ln g-\frac{Nk}{2}\beta^2\overline{(\varepsilon-\bar{\varepsilon})^2}, \tag{16}$$

$$U(N, \beta) \simeq N\bar{\varepsilon}-N\beta\overline{(\varepsilon-\bar{\varepsilon})^2}, \tag{17}$$

and

$$C(N, \beta) \simeq Nk\beta^2\overline{(\varepsilon-\bar{\varepsilon})^2}. \tag{18*}$$

Formulae (14)–(18) determine the thermodynamic properties of the system for $\beta \simeq 0$. The important thing to note here is that they do so not only for $\beta \gtrsim 0$ but also for $\beta \lesssim 0$! In fact, these formulae hold in the vicinity of, and on *both* sides of, the maximum in the S–U curve; see Fig. 3.14. Quite expectedly, the maximum value of S is given by $Nk \ln g$, and it occurs at $\beta = \pm 0$; S decreases both ways, whether U decreases ($\beta > 0$) or increases ($\beta < 0$). It may be noted that the specific heat of the system is positive in either case.

It is not difficult to show that if two systems, characterized by the temperature parameters β_1 and β_2, are brought into thermal contact, then energy will flow from the system with the smaller value of β to the system with the larger value of β; this will continue until the two systems acquire a common value of this parameter. What is more important to note is that

* Compare this result with eqn. (3.6.3).

this result remains *literally* true even if one or both of the β's are negative. Thus, if β_1 is $-$ve while β_2 is $+$ve, then energy will flow from system 1 to system 2, i.e. from the system at negative temperature to the system at positive temperature. In this sense, systems at negative temperatures are ***hotter*** than systems at positive temperatures; indeed, all negative temperatures are above $+\infty$, not below zero!

For further discussion of this topic, reference may be made to a paper by Ramsey (1956).

Problems

3.1. (a) Derive formula (3.2.37) from eqns. (3.2.14) and (3.2.36).
(b) Derive formulae (3.2.40) and (3.2.41) from eqns. (3.2.38) and (3.2.39).

3.2. Prove that the quantity $g''(x_0)$, see eqn. (3.2.25), is equal to $\langle (E-U)^2 \rangle \exp(2\beta)$. Hence show that eqn. (3.2.28) is physically equivalent to eqn. (3.6.9).

3.3. Using the fact that $(1/n!)$ is the coefficient of x^n in the power expansion of the function $\exp(x)$, derive an asymptotic formula for this coefficient by the method of saddle-point integration. Compare your result with the Stirling formula for $(n!)$.

3.4. Verify that the quantity $(k/\mathcal{N}) \ln \Gamma$, where

$$\Gamma(\mathcal{N}, U) = \sum_{\{n_r\}}{}' W\{n_r\},$$

is equal to the (mean) entropy of the system. Prove that we obtain essentially the same value of $\ln \Gamma$ if we take, in the foregoing summation, *only* the largest term of the sum, viz. the term $W\{n_r^*\}$ which corresponds to the *most probable* distribution set.

[Surprised? Well, note the following example:

For all N,
$$\sum_{r=0}^{N} {}^N C_r = 2^N;$$

therefore,
$$\ln \left\{ \sum_{r=0}^{N} {}^N C_r \right\} = N \ln 2. \tag{a}$$

Now, the largest term in the sum corresponds to $r \simeq N/2$; so, for large N, the logarithm of the largest term is very nearly equal to

$$\ln \{N!\} - 2 \ln \{(N/2)!\}$$
$$\simeq N \ln N - 2\frac{N}{2} \ln \frac{N}{2} = N \ln 2, \tag{b}$$

which agrees with (a).]

3.5. Making use of the fact that the Helmholtz free energy $A(N, V, T)$ of a thermodynamic system must be an *extensive* property of the system, prove that

$$N\left(\frac{\partial A}{\partial N}\right)_{V,T} + V\left(\frac{\partial A}{\partial V}\right)_{N,T} = A.$$

[Note that this result implies the well-known relationship: $N\mu = A + PV (\equiv G)$.]

3.6. (a) Assuming that the total number of states accessible to a given statistical system is Ω, show that the entropy of the system, as given by eqn. (3.3.20), is maximum when all the Ω states are equally likely to occur.
(b) If, on the other hand, we have an ensemble of systems sharing energy (with mean value $\bar{E}$), then show that the entropy, as given by the same formal expression, is maximum when $P_r \propto \exp(-\beta E_r)$, β being a constant to be determined by the given value of $\bar{E}$.
(c) Further, if we have an ensemble of systems sharing energy (with mean value $\bar{E}$) and also sharing particles (with mean value $\bar{N}$), then show that the entropy, given by a similar expression, is maximum when $P_{r,s} \propto \exp(-\alpha N_r - \beta E_s)$, α and β being constants to be determined by the given values of $\bar{N}$ and $\bar{E}$.

3.7. Prove that, quite generally,

$$C_P - C_V = -k \frac{\left[\frac{\partial}{\partial T}\left\{T\left(\frac{\partial \ln Q}{\partial V}\right)_T\right\}\right]_V^2}{\left(\frac{\partial^2 \ln Q}{\partial V^2}\right)_T} > 0.$$

Verify that the value of this quantity for an ideal gas is Nk.

3.8. Show that, for an ideal gas,

$$\frac{S}{Nk} = \ln\left(\frac{Q_1}{N}\right) + T\left(\frac{\partial \ln Q_1}{\partial T}\right)_P.$$

3.9. If a perfect monatomic gas is expanded *adiabatically* to twice its initial volume, what will be the ratio of the final pressure to the initial pressure? If during the process some heat is added to the system, will the final pressure be higher or lower than in the preceding case? Support your answer by deriving the relevant formula for the ratio P_f/P_i.

3.10. (a) The volume of a sample of helium gas is increased by withdrawing the piston of the containing cylinder. The final pressure P_f is found to be equal to the initial pressure P_i times $(V_i/V_f)^{1.2}$, V_i and V_f being the initial and final volumes. Assuming that the product PV is always equal to $\frac{2}{3}U$, do (i) the energy and (ii) the entropy of the gas increase, remain constant or decrease during the process?

(b) If the process were reversible, how much would be the work done and the heat added in doubling the volume of the gas? Take $P_i = 1$ atm and $V_i = 1\ \text{m}^3$.

3.11. Determine the work done on a gas and the amount of heat absorbed by it during a compression from volume V_1 to volume V_2, following the law $PV^n = \text{const.}$

3.12. If the "free volume" $\bar{V}$ of a classical system is defined by the equation

$$\bar{V}^N = \int e^{\{\bar{U} - U(q_i)\}/kT} \prod_{i=1}^{N} d^3q_i,$$

where $\bar{U}$ is the average potential energy of the system and $U(q_i)$ the actual potential energy as a function of the molecular configuration, then show that

$$S = Nk\left[\ln\left\{\frac{\bar{V}}{N}\left(\frac{2\pi mkT}{h^2}\right)^{3/2}\right\} + \frac{5}{2}\right].$$

In what sense is it justified to refer to the quantity $\bar{V}$ as the "free volume" of the system? Substantiate your answer by considering a particular case, e.g. the case of a hard sphere gas.

3.13. (a) Evaluate the partition function and the major thermodynamic properties of an ideal gas consisting of N_1 molecules of mass m_1 and N_2 molecules of mass m_2, confined to a space of volume V at temperature T. Assume that the molecules of a given kind are mutually indistinguishable, while those of one kind are distinguishable from those of the other kind.

(b) Compare your results with the ones pertaining to an ideal gas consisting of (N_1+N_2) molecules, *all of one kind*, of mass m, such that $m(N_1+N_2) = m_1N_1+m_2N_2$.

3.14. Consider an ideal-gas sample of atoms A, atoms B and molecules AB, undergoing the reaction $AB \rightleftarrows A+B$. If n_A, n_B and n_{AB} denote their respective concentrations (that is, the number densities of respective atoms/molecules), then show that, in equilibrium,

$$\frac{n_{AB}}{n_A n_B} = V\frac{f_{AB}}{f_A f_B} = K(T) \qquad \text{(the law of } \textit{mass action}\text{)}.$$

Here, V is the volume of the sample while f's are the respective single-particle partition functions; the quantity $K(T)$ is generally referred to as the *equilibrium constant* of the reaction.

3.15. Show that the partition function $Q_N(V, T)$ of an *extreme* relativistic gas consisting of N monatomic molecules with energy–momentum relationship $\varepsilon = pc$, c being the speed of light, is given by

$$Q_N(V, T) = \frac{1}{N!}\left\{8\pi V\left(\frac{kT}{hc}\right)^3\right\}^N.$$

Study the thermodynamics of this system, checking in particular that

$$PV = \tfrac{1}{3}U, \quad U/N = 3kT \quad \text{and} \quad \gamma = \tfrac{4}{3}.$$

Next, using the inversion formula (3.4.7), derive an expression for the density of states $g(E)$ of this system.

3.16. Consider a system similar to the one in the preceding problem, but consisting of $3N$ particles moving in *one* dimension. Show that the partition function in this case is given by

$$Q_{3N}(L, T) = \frac{1}{(3N)!}\left[2L\left(\frac{kT}{hc}\right)\right]^{3N},$$

L being the "length" of the space available. Compare the thermodynamics and the density of states of this system with the preceding one.

3.17. If we take $f(q, p)$ in (3.5.3) as equal to $U - H(q, p)$, then clearly $\langle f\rangle = 0$; formally, this would mean

$$\int [U - H(q, p)]e^{-\beta H(q, p)}\, d\omega = 0.$$

Obtain from this equation the expression (3.6.3) for the fluctuation in the energy of a system embedded in canonical ensemble.

3.18. Prove that for a system in the canonical ensemble

$$\langle(\Delta E)^3\rangle = k^2\left\{T^4\left(\frac{\partial C_V}{\partial T}\right)_V + 2T^3 C_V\right\};$$

in particular, for an ideal gas

$$\left\langle\left(\frac{\Delta E}{U}\right)^2\right\rangle = \frac{2}{3N} \quad \text{and} \quad \left\langle\left(\frac{\Delta E}{U}\right)^3\right\rangle = \frac{8}{9N^2}.$$

3.19. Prove that in the *general* case, when the given ideal-gas system is neither nonrelativistic nor extreme relativistic, the equipartition theorem reduces to

$$\langle m_0 u^2(1 - u^2/c^2)^{-1/2}\rangle = 3kT,$$

where m_0 is the rest mass of a particle and u its speed.

3.20. Develop a *kinetic* argument to show that the average value of the quantity $\Sigma_i p_i \dot{q}_i$ is precisely equal to $3PV$. Hence show that, irrespective of the relativistic considerations, $PV = NkT$.

3.21. The energy eigenvalues of an s-dimensional harmonic oscillator can be written as

$$\varepsilon_j = (j + s/2)\hbar\omega; \qquad j = 0, 1, 2, \ldots.$$

Show that the jth energy level has a multiplicity $(j+s-1)!/j!\,(s-1)!$. Evaluate the partition function and the major thermodynamic properties of a system of N such oscillators, assuming that the oscillators are independent and distinguishable. Compare your results with a corresponding system of sN one-dimensional oscillators.

3.22. Obtain an asymptotic expression for the quantity $\ln g(E)$ for a system of N quantum-mechanical harmonic oscillators by using the inversion formula (3.4.7) and the partition function (3.7.15). Hence show that

$$\frac{S}{Nk} = \left(\frac{E}{N\hbar\omega} + \frac{1}{2}\right)\ln\left(\frac{E}{N\hbar\omega} + \frac{1}{2}\right) - \left(\frac{E}{N\hbar\omega} - \frac{1}{2}\right)\ln\left(\frac{E}{N\hbar\omega} - \frac{1}{2}\right).$$

[*Hint.* Employ the Darwin–Fowler method.]

3.23. (a) When a system of N oscillators with total energy E is in thermal equilibrium, what is the probability p_n that a given oscillator among them is in the quantum state n? [*Hint.* Use expression (3.7.25).]
Verify that $\Sigma_{n=0}^{\infty} p_n = 1$.

(b) When an ideal gas of N monatomic molecules with total energy E is in thermal equilibrium, show that the probability of a given molecule having an energy in the neighborhood of ε is proportional to $\exp(-\beta\varepsilon)$. [*Hint.* Use expression (3.5.16).]

3.24. The potential energy of a classical, one-dimensional, *anharmonic* oscillator may be written as

$$V(q) = cq^2 - gq^3 - fq^4,$$

where c, g and f are positive constants; of course, g and f may be assumed to be very small in value. Show that the first-order contribution of anharmonic terms to the heat capacity of the system is given by

$$\frac{3}{2}k^2\left(\frac{f}{c^2}+\frac{5}{4}\frac{g^2}{c^3}\right)T,$$

and, to the same order, the mean value of the position coordinate q is given by

$$\frac{3}{4}\frac{gkT}{c^2}.$$

3.25. The energy levels of a quantum-mechanical, one-dimensional, *anharmonic* oscillator can be approximated as

$$\varepsilon_n = (n+\tfrac{1}{2})\hbar\omega - x(n+\tfrac{1}{2})^2\hbar\omega; \qquad n = 0, 1, 2, \ldots.$$

The parameter x, usually $\ll 1$, represents the degree of anharmonicity. Show that, to the first order in x and fourth order in $u(\equiv \hbar\omega/kT)$, the specific heat of a system of N such oscillators is given by

$$C = Nk\left[\left(1-\frac{1}{12}u^2+\frac{1}{240}u^4\right)+4x\left(\frac{1}{u}+\frac{1}{80}u^3\right)\right].$$

Note that the correction term *increases* with temperature.

3.26. Study, along the lines of Sec. 3.7, the statistical mechanics of a system of N "Fermi oscillators" which are characterized by only two eigenvalues, namely 0 and ε.

3.27. The quantum states available to a given physical system are (i) a group of g_1 *equally likely* states, with a common energy value ε_1 and (ii) a group of g_2 *equally likely* states, with a common energy value ε_2. Show that the entropy of the system is given by

$$S = -k[p_1 \ln(p_1/g_1)+p_2 \ln(p_2/g_2)],$$

where p_1 and p_2 are, respectively, the probabilities of the system being in a state belonging to group 1 or to group 2: $p_1+p_2 = 1$.

(a) Assuming that the p's are given by a canonical distribution, show that

$$S = k\left[\ln g_1+\ln\{1+(g_2/g_1)e^{-x}\}+\frac{x}{1+(g_1/g_2)e^x}\right],$$

where $x = (\varepsilon_2-\varepsilon_1)/kT$, assumed positive. Compare the special case $g_1 = g_2 = 1$ with that of the Fermi oscillator of the preceding problem.

(b) Verify the foregoing expression for S by deriving it from the partition function of the system.

(c) Check that as $T \to 0$, $S \to k \ln g_1$. Interpret this result physically.

3.28. Gadolinium sulphate obeys Langevin's theory of paramagnetism down to temperatures of the order of a few degrees Kelvin. Its molecular magnetic moment is 7.2×10^{-23} amp-m^2. Determine the degree of magnetic saturation in this salt at a temperature of 2°K in a field of flux density 2 weber/m^2.

3.29. Oxygen is a paramagnetic gas obeying Langevin's theory of paramagnetism. Its susceptibility per unit volume, at 293°K and at atmospheric pressure, is 1.80×10^{-6} mks units. Determine its molecular magnetic moment and compare it with the Bohr magneton (which is very nearly equal to 9.27×10^{-24} amp-m^2).

3.30. (a) Consider a gaseous system of N noninteracting, diatomic molecules, each having an electric dipole moment μ, placed in an external electric field of strength E. The energy of a molecule will be given by the kinetic energy of rotation as well as of translation *plus* the potential energy of orientation in the applied field:

$$\varepsilon = \frac{p^2}{2m} + \left\{\frac{p_\theta^2}{2I} + \frac{p_\varphi^2}{2I \sin^2\theta}\right\} - \mu E \cos\theta,$$

where I is the moment of inertia of the molecule. Study the thermodynamics of this system, including the electric polarization and the dielectric constant.

Assume that (i) the system is a classical one and (ii) $|\mu E| \ll kT$.*

(b) The molecule H_2O has an electric dipole moment of 1.85×10^{-18} e.s.u. Calculate, on the basis of the preceding theory, the dielectric constant of steam at 100°C and at atmospheric pressure.

3.31. The partition function of the system studied in Problem 3.30(a) can be written as

$$Q_N(\beta) = \int_0^{N\mu} \exp(\beta E M_z)\, g(M_z)\, dM_z,$$

where M_z is the net electric moment of the system (in the direction of E) while $g(M_z)$ is the density of states of the system around a particular value of the variable M_z.

(a) Show that, for $N \gg 1$,

$$\ln g(M_z) \simeq N\left[\text{const} + \ln\frac{\sinh x}{x} - xL(x)\right],$$

where $x = L^{-1}(M_z/N\mu)$, L^{-1} being the *inverse Langevin function.*

(b) Next, show that, for $M_z \ll N\mu$,

$$g(M_z) \propto \exp\left\{-\frac{3M_z^2}{2N\mu^2}\right\}.$$

Note that this implies: $(\Delta M_z)_{\text{r.m.s.}} = (N\mu^2/3)^{1/2}$ and, since in this limit

$$\langle M_z\rangle = \frac{N\mu^2 E}{3kT},$$

$$\frac{(\Delta M_z)_{\text{r.m.s.}}}{\langle M_z\rangle} = \left(\frac{3}{N}\right)^{1/2}\frac{kT}{\mu E}.$$

Compare this result with the formula (3.6.4) for the fluctuation of energy in a canonical ensemble.

3.32. Consider a pair of electric dipoles $\boldsymbol{\mu}$ and $\boldsymbol{\mu}'$, oriented in the directions (θ, φ) and (θ', φ'), respectively; the distance R between their centers is assumed to be fixed. The potential energy in this orientation is given by

$$-\frac{\mu\mu'}{R^3}\{2\cos\theta\cos\theta' - \sin\theta\sin\theta'\cos(\varphi-\varphi')\}.$$

Now, consider the pair to be in thermal equilibrium, their orientations being governed by the canonical distribution. Show that the mean force between the dipoles, at high temperatures, will be

$$-2\frac{\mu\mu'}{kT}\frac{\hat{R}}{R^7},$$

$\hat{R}$ being a unit vector in the direction of the line of centers.

3.33. Evaluating the high-temperature approximation of the partition function $Q_N(\beta)$ for a system of magnetic dipoles, show that, quite generally, the Curie constant C_J is given by

$$C_J = \frac{N_0\mu^2}{k}\overline{\cos^2\theta}.$$

Hence derive the formula (3.8.31).

3.34. Replacing the sum in (3.8.23) by an integral, evaluate $Q_1(\beta)$ and study the thermodynamics following from it. Compare these results with the ones following from the Langevin theory.

---

* In general, the electric dipole moments of molecules are of the order of 10^{-18} e.s.u. (or a *Debye unit*). In a field of 1 e.s.u. (= 300 volts/cm) and at a temperature of 300°K, the parameter $\beta\mu E = 0(10^{-4})$.

3.35. Atoms of silver vapor, each having a magnetic moment μ_B ($g = 2$, $J = \frac{1}{2}$), align themselves either parallel or antiparallel to the direction of an applied magnetic field. Determine the respective fractions of atoms aligned parallel and antiparallel to a field of flux density 0.1 weber/m^2 at a temperature of 1000°K.

3.36. (a) Show that, for any magnetizable material, the heat capacities at constant field H and at constant mean moment M are connected by the relation

$$C_H - C_M = -T\left(\frac{\partial H}{\partial T}\right)_M \left(\frac{\partial M}{\partial T}\right)_H.$$

(b) Show that for a paramagnetic material obeying Curie's law

$$C_H - C_M = CH^2/T^2,$$

where C on the right-hand side of this equation denotes the *Curie constant* of the given sample.

3.37. A system of N spins at a negative temperature ($E > 0$) is brought into contact with an ideal-gas thermometer consisting of N' molecules. What will be the nature of their state of mutual equilibrium? Will their common temperature be negative or positive, and in what manner will it be affected by the ratio N'/N?

3.38. Consider a system of charged particles (not dipoles), obeying classical mechanics and classical statistics. Show that the magnetic susceptibility of this system is identically zero (Bohr–van Leeuwen theorem).

[Note that the Hamiltonian of this system in the presence of a magnetic field $\boldsymbol{H}(= \nabla \times \boldsymbol{A})$ will be a function of the quantities $\boldsymbol{p}_j + (e_j/c)\boldsymbol{A}(\boldsymbol{r}_j)$, and not of the $\boldsymbol{p}_j$ as such. One has now to prove that the partition function of the system is independent of the applied field.]

CHAPTER 4

THE GRAND CANONICAL ENSEMBLE

IN THE preceding chapter we developed the formalism of the canonical ensemble and established a scheme of operations for deriving the various thermodynamic properties of a given physical system. The effectiveness of that approach became clear from the examples discussed there; it will become more vivid in the subsequent studies carried out in this text. However, for a number of problems, both physical and chemical, the usefulness of the canonical ensemble formalism turns out to be severely limited and it appears that a further generalization of the formalism needs to be evolved. The motivation that brings about this generalization is physically of the same nature as the one that led us from the microcanonical to the canonical ensemble—it is only the next natural step from there. It comes from the realization that not only the energy of a system but the number of particles as well is hardly ever measured in a "direct" manner; we only estimate this number through an *indirect* probing into the system. Conceptually, therefore, we may regard both N and E as *variables* of the system and identify their expectation values, $\langle N \rangle$ and $\langle E \rangle$, with the corresponding thermodynamic quantities.

The procedure for studying the statistics of the variables N and E is self-evident. We may *either* (i) consider the given system A to be immersed in a large reservoir A' with which it can exchange both energy and particles *or* (ii) regard it as a member of what we may call a *grand canonical ensemble*, which consists of the given system A and a large number of (mental) copies thereof, the members of the ensemble carrying out a mutual exchange of energy and particles. In either case, we have to fix two intensive parameters, say α and β, such that in case (i) they have the same values for the given system as well as for the reservoir, while in case (ii) they have the same values for every member of the ensemble. These parameters will define the equilibrium state and, consequently, govern the quantities $\langle N \rangle$ and $\langle E \rangle$.* Subsequently we shall find that

$$\alpha = -\mu/kT \quad \text{and} \quad \beta = 1/kT \tag{0}$$

where μ denotes the chemical potential of the system and T, as usual, the absolute temperature.

* In the spirit of the ensemble theory, one would rather say that the parameters α and β will be determined by the *preassigned* values of the numbers $\langle N \rangle$ and $\langle E \rangle$.

4.1. Equilibrium between a system and a particle-energy reservoir

We consider the given system A as immersed in a large reservoir A', with which it can exchange both energy and particles; see Fig. 4.1. After some time has elapsed, the system and the reservoir are supposed to attain a state of mutual equilibrium. Then, according to Sec. 1.3, the system and the reservoir will have a common temperature T and a common chemical potential μ. The fraction of the total number of particles $N^{(0)}$ and the fraction of the total energy $E^{(0)}$ which the system A can have at any time t are, however, variables (whose values, in principle, can lie anywhere between zero and unity). If, at a particular

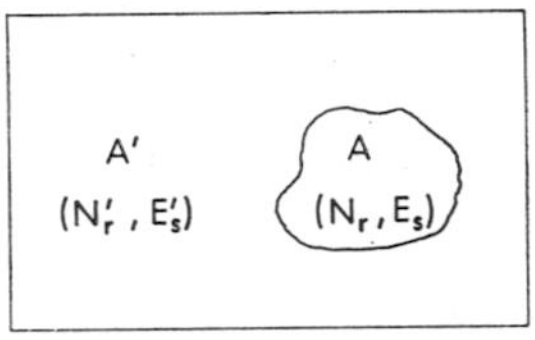

FIG. 4.1. A statistical system immersed in a particle-energy reservoir.

instant of time, the system A happens to be in *one* of its states characterized by the number N_r of particles and the amount E_s of energy, then the number of particles in the reservoir would necessarily be N_r' and its energy necessarily E_s', such that

$$N_r + N_r' = N^{(0)} = \text{const.} \tag{1}$$

and

$$E_s + E_s' = E^{(0)} = \text{const.} \tag{2}$$

Again, since the reservoir is supposed to be much larger than the given system, the values of N_r and E_s, which are going to be of practical importance, must be very small fractions of the total magnitudes $N^{(0)}$ and $E^{(0)}$, respectively; therefore, for all practical purposes, we can write

$$\frac{N_r}{N^{(0)}} = \left(1 - \frac{N_r'}{N^{(0)}}\right) \ll 1 \tag{3}$$

and

$$\frac{E_s}{E^{(0)}} = \left(1 - \frac{E_s'}{E^{(0)}}\right) \ll 1. \tag{4}$$

Now, in the manner of Sec. 3.1, the probability $P_{r,s}$ that, at any time t, the system A is found to be in an (N_r, E_s)-state would be directly proportional to the number of microstates $\Omega'(N_r', E_s')$ which the reservoir can have for the corresponding macrostate (N_r', E_s'). Thus

$$P_{r,s} \propto \Omega'(N^{(0)} - N_r, E^{(0)} - E_s). \tag{5}$$

Again, in view of (3) and (4), we can write

$$\begin{aligned} \ln \Omega'(N^{(0)} - N_r, E^{(0)} - E_s) &= \ln \Omega'(N^{(0)}, E^{(0)}) \\ &\quad + \left(\frac{\partial \ln \Omega'}{\partial N'}\right)_{N'=N^{(0)}} (-N_r) + \left(\frac{\partial \ln \Omega'}{\partial E'}\right)_{E'=E^{(0)}} (-E_s) + \dots \\ &\simeq \ln \Omega'(N^{(0)}, E^{(0)}) - \alpha N_r - \beta E_s. \end{aligned} \tag{6}$$

Recalling eqns. (1.2.3), (1.2.12), (1.3.3) and (1.3.6), we note that the parameters α and β here are directly related to the chemical potential μ and the temperature T of the reservoir (and hence of the given system too) through the formulae (0). From (5) and (6), then, we obtain the desired result:

$$P_{r,s} \propto \exp(-\alpha N_r - \beta E_s). \tag{7}$$

On normalization, it becomes

$$P_{r,s} = \frac{\exp(-\alpha N_r - \beta E_s)}{\sum_{r,s} \exp(-\alpha N_r - \beta E_s)}; \tag{8}$$

the summation in the denominator goes over *all* the (N_r, E_s)-states accessible to the system A.

We shall now examine the same problem from the ensemble point of view.

4.2. A system in the grand canonical ensemble

We now visualize an ensemble of $\mathcal{N}$ identical systems (which, of course, can be labelled as $1, 2, \ldots, \mathcal{N}$) mutually sharing a total number of particles $\mathcal{N}\bar{N}$ and a total energy $\mathcal{N}\bar{E}$.† Let $n_{r,s}$ denote the number of systems that have, at any time t, the number N_r of particles and the amount E_s of energy ($r, s = 0, 1, 2, \ldots$); then, we must have

$$\left.\begin{aligned} \sum_{r,s} n_{r,s} &= \mathcal{N}, \\ \sum_{r,s} n_{r,s} N_r &= \mathcal{N}\bar{N} \\ \text{and} \qquad \sum_{r,s} n_{r,s} E_s &= \mathcal{N}\bar{E}. \end{aligned}\right\} \tag{1}$$

Any set $\{n_{r,s}\}$, of the numbers $n_{r,s}$, which satisfies the restrictive conditions (1), represents one of the possible modes of distribution of particles and energy among the members of our ensemble. Furthermore, any such mode of distribution can be realized in $W\{n_{r,s}\}$ different ways, where

$$W\{n_{r,s}\} = \frac{\mathcal{N}!}{\prod_{r,s} (n_{r,s}!)}. \tag{2}$$

We may then define the *most probable* mode of distribution, $\{n^*_{r,s}\}$, as the one that maximizes expression (2), satisfying at the same time the restrictive conditions (1). Going through the relevant derivation, we obtain for a large ensemble

$$\frac{n^*_{r,s}}{\mathcal{N}} = \frac{\exp(-\alpha N_r - \beta E_s)}{\sum_{r,s} \exp(-\alpha N_r - \beta E_s)}; \tag{3}$$

cf. the corresponding eqn. (3.1.6) for the canonical ensemble as well as the formula (4.1.8) for $P_{r,s}$. The parameters α and β, so far undetermined, are eventually determined by the

† For simplicity, we shall henceforth use the symbols $\bar{N}$ and $\bar{E}$ instead of $\langle N \rangle$ and $\langle E \rangle$.

equations

$$\bar{N} = \frac{\sum_{r,s} N_r \exp(-\alpha N_r - \beta E_s)}{\sum_{r,s} \exp(-\alpha N_r - \beta E_s)} \equiv -\frac{\partial}{\partial \alpha}\left\{\ln \sum_{r,s} \exp(-\alpha N_r - \beta E_s)\right\} \tag{4}$$

and

$$\bar{E} = \frac{\sum_{r,s} E_s \exp(-\alpha N_r - \beta E_s)}{\sum_{r,s} \exp(-\alpha N_r - \beta E_s)} \equiv -\frac{\partial}{\partial \beta}\left\{\ln \sum_{r,s} \exp(-\alpha N_r - \beta E_s)\right\}, \tag{5}$$

where the values $\bar{N}$ and $\bar{E}$ are supposed to be preassigned.

Alternatively, we may define the *expectation* (or *mean*) values of the numbers $n_{r,s}$, namely

$$\langle n_{r,s} \rangle = \frac{\sum'_{\{n_{r,s}\}} n_{r,s} W\{n_{r,s}\}}{\sum'_{\{n_{r,s}\}} W\{n_{r,s}\}}, \tag{6}$$

where the primed summations go over all the distribution sets that conform to the conditions (1). An asymptotic expression for $\langle n_{r,s} \rangle$ can be derived by using the method of Darwin and Fowler—the only difference from the corresponding derivation in Sec. 3.2 being that in the present case we have to work with functions of more than one (complex) variable. The derivation, however, runs along similar lines, with the result

$$\lim_{\mathcal{N}\to\infty} \frac{\langle n_{r,s} \rangle}{\mathcal{N}} \simeq \frac{n^*_{r,s}}{\mathcal{N}} = \frac{\exp(-\alpha N_r - \beta E_s)}{\sum_{r,s} \exp(-\alpha N_r - \beta E_s)}. \tag{7}$$

4.3. Physical significance of the statistical quantities

In order to establish a connection between the statistics of the grand canonical ensemble and the thermodynamics of the system under study, we introduce a quantity q, as defined by

$$q = \ln\left\{\sum_{r,s} \exp(-\alpha N_r - \beta E_s)\right\}; \tag{1}$$

the quantity q is a function of the parameters α and β and of all the E's.† Taking the differential of q and making use of the formulae (4.2.4), (4.2.5) and (4.2.7), we get

$$dq = -\bar{N}\, d\alpha - \bar{E}\, d\beta - \frac{\beta}{\mathcal{N}} \sum_{r,s} \langle n_{r,s} \rangle\, dE_s, \tag{2}$$

so that

$$d(q + \alpha\bar{N} + \beta\bar{E}) = \beta\left(\frac{\alpha}{\beta}\, d\bar{N} + d\bar{E} - \frac{1}{\mathcal{N}} \sum_{r,s} \langle n_{r,s} \rangle\, dE_s\right). \tag{3}$$

To interpret the terms appearing on the right-hand side of this equation, we compare the expression enclosed within the parentheses with the statement of the first law of thermo-

† This quantity was first introduced by Kramers (1944–5), who called it the *q-potential*.

dynamics, viz.

$$\delta Q = d\bar{E} + \delta W - \mu \, d\bar{N}, \tag{4}$$

where the various symbols have their usual meanings. The following correspondence seems inevitable:

$$\delta W = -\frac{1}{\mathcal{N}} \sum_{r,s} \langle n_{r,s} \rangle \, dE_s; \qquad \mu = -\alpha/\beta, \tag{5}$$

whence we obtain

$$d(q + \alpha \bar{N} + \beta \bar{E}) = \beta \, \delta Q. \tag{6}$$

The parameter β, being the integrating factor for the heat δQ, must be equivalent to the reciprocal of the absolute temperature; so, we may write

$$\beta = 1/kT \tag{7}$$

and, hence,

$$\alpha = -\mu/kT. \tag{8}$$

The quantity $(q + \alpha \bar{N} + \beta \bar{E})$ must, therefore, be identified with the thermodynamic variable S/k. Accordingly, we obtain for the q-potential

$$q = \frac{S}{k} - \alpha \bar{N} - \beta \bar{E} = \frac{TS + \mu \bar{N} - \bar{E}}{kT}. \tag{9}$$

However, $\mu \bar{N}$ is identically equal to G, the Gibbs free energy of the system, and hence must be equal to $(\bar{E} - TS + PV)$. So, we finally obtain

$$q \equiv \ln \left\{ \sum_{r,s} \exp(-\alpha N_r - \beta E_s) \right\} = \frac{PV}{kT}. \tag{10}$$

Equation (10) represents the essential link between the thermodynamics of the given system and the statistics of the corresponding grand canonical ensemble. It is, therefore, a relationship of central importance in the formalism developed in this chapter.

To derive further thermodynamics, we prefer to introduce a parameter z, defined by the relationship

$$z \equiv e^{-\alpha} = e^{\mu/kT}; \tag{11}$$

the parameter z is generally referred to as the *fugacity* of the system. In terms of z, the q-potential takes the form

$$q \equiv \ln \left\{ \sum_{r,s} z^{N_r} e^{-\beta E_s} \right\} \tag{12}$$

$$= \ln \left\{ \sum_{N_r=0}^{\infty} z^{N_r} Q_{N_r}(V, T) \right\} \qquad (\text{with } Q_0 \equiv 1), \tag{13}$$

so that we may write

$$q(z, V, T) \equiv \ln \mathcal{Q}(z, V, T), \tag{14}$$

where

$$\mathcal{Q}(z, V, T) \equiv \sum_{N_r=0}^{\infty} z^{N_r} Q_{N_r}(V, T) \qquad (\text{with } Q_0 \equiv 1). \tag{15}$$

Note that in going from (12) to (13), we have (mentally) carried out the summation over the energy values E_s, with N_r kept fixed, thus giving rise to the partition function $Q_{N_r}(V, T)$; of course, the dependence of Q_{N_r} on V comes from the dependence of the E's on V. In going from (13) to (14), we have (again mentally) carried out the summation over all the numbers N_r, viz. 0, 1, 2, ... ∞, thus giving rise to the *grand partition function* $\mathcal{Q}(z, V, T)$ of the system. The q-potential, which we have already identified with PV/kT, is, therefore, given by the logarithm of the grand partition function.

It appears that in order to evaluate the grand partition function $\mathcal{Q}(z, V, T)$ we have to pass through the routine of the partition function $Q(N, V, T)$. In principle, this is indeed true. In practice, however, we find that on many occasions an explicit evaluation of the partition function is extremely hard to carry out but at the same time considerable progress can be made in the evaluation of the grand partition function. This is particularly true when we deal with systems in which the influence of quantum statistics and/or interparticle interactions is important. The formalism of the grand canonical ensemble then proves to be of considerable value. Moreover, if the partition function of a system can be evaluated explicitly, then there may not be much point in pursuing the formalism of the grand partition function.

We are now in a position to write down the full recipe for deriving the leading thermodynamic quantities of a given system from its q-potential. We have first of all, for the pressure of the system,

$$P(z, V, T) = \frac{kT}{V} q(z, V, T) \equiv \frac{kT}{V} \ln \mathcal{Q}(z, V, T). \tag{16}$$

Next, writing N for $\bar{N}$ and U for $\bar{E}$, we obtain with the help of (4.2.4), (4.2.5) and (11)

$$N(z, V, T) = z\left[\frac{\partial}{\partial z} q(z, V, T)\right]_{V,T} = kT\left[\frac{\partial}{\partial \mu} q(\mu, V, T)\right]_{V,T} \tag{17}$$

and

$$U(z, V, T) = -\left[\frac{\partial}{\partial \beta} q(z, V, T)\right]_{z,V} = kT^2\left[\frac{\partial}{\partial T} q(z, V, T)\right]_{z,V}. \tag{18}$$

Eliminating z between eqns. (16) and (17), one obtains the equation of state, viz. the (P, V, T)-relationship, of the system. On the other hand, eliminating z between eqns. (17) and (18), one obtains U as a function of N, V and T, whence one readily computes the specific heat at constant volume as $(\partial U/\partial T)_{N,V}$. The Helmholtz free energy is given by the formula

$$\begin{aligned} A &= N\mu - PV = NkT \ln z - kTq(z, V, T) \\ &= -kT \ln \frac{\mathcal{Q}(z, V, T)}{z^N}, \end{aligned} \tag{19}$$

which may be compared with the canonical ensemble formula: $A = -kT \ln Q(N, V, T)$; see also Problem 4.2. Finally, we have for the entropy of the system

$$\begin{aligned} S &= \frac{U-A}{T} = kT\left(\frac{\partial q}{\partial T}\right)_{z,V} - Nk \ln z + kq \\ &= k\left[\frac{\partial}{\partial T}(Tq)\right]_{\mu,V}, \end{aligned} \tag{20}$$

for

$$\left[\frac{\partial}{\partial T}(Tq)\right]_{\mu, V} = q+T\left\{\left(\frac{\partial q}{\partial T}\right)_{z, V}+\left(\frac{\partial q}{\partial \ln z}\right)_{T, V}\left(\frac{\partial \ln z}{\partial T}\right)_{\mu, V}\right\}$$
$$= q+T\left(\frac{\partial q}{\partial T}\right)_{z, V} - N \ln z; \tag{21}$$

here, use has been made of the formulae (11) and (17). It appears worth while to point out here that formula (20) is directly related to the thermodynamic formula

$$d(PV) = P\, dV + N\, d\mu + S\, dT, \tag{22}$$

whence it follows that

$$S = \left[\frac{\partial(PV)}{\partial T}\right]_{\mu, V} = k\left[\frac{\partial(T \ln \mathcal{Q})}{\partial T}\right]_{\mu, V}. \tag{23}$$

4.4. Examples

We shall now study a couple of simple problems, with the explicit purpose of demonstrating how the method of the q-potential works. This is not intended to be a demonstration of the power of this method, for we shall consider here only those problems which can be solved equally well by the method of the preceding chapter. The real power of the new method will become apparent only when we study problems involving quantum-statistical effects and effects arising from the interparticle interactions; many such problems will appear in Chapters 6–12.

The first problem we propose to consider here is that of the classical ideal gas. In Sec. 3.5 we showed that the partition function $Q_N(V, T)$ of this system could be written as

$$Q_N(V, T) = \frac{[Q_1(V, T)]^N}{N!}, \tag{1}$$

where $Q_1(V, T)$ may be regarded as the partition function of a single particle in the system. First of all, we should note that eqn. (1) does not imply any restrictions on the particles having *internal* degrees of motion; these degrees of motion, if present, would affect the results only through Q_1. Secondly, we should recall that the factor $N!$ in the denominator arises from the fact that the particles constituting the gas are, in fact, *indistinguishable*. Closely related to the indistinguishability of the particles is the fact that they are *non-localized*, for otherwise we could distinguish them through their very sites; recall, for example, the system of harmonic oscillators, which was studied in Sec. 3.7. Now, since our particles are nonlocalized they can be *anywhere* in the space available to them; consequently, the function Q_1 will be directly proportional to V:

$$Q_1(V, T) = Vf(T), \tag{2}$$

where $f(T)$ is a function of temperature alone. We thus obtain for the grand partition function of the gas

$$\mathcal{Q}(z, V, T) = \sum_{N_r=0}^{\infty} z^{N_r} Q_{N_r}(V, T) = \sum_{N_r=0}^{\infty} \frac{\{zVf(T)\}^{N_r}}{N_r!}$$
$$= \exp\{zVf(T)\}, \tag{3}$$

whence it follows that

$$q(z, V, T) = zVf(T). \tag{4}$$

Formulae (4.3.16)–(4.3.20) then lead to the following results:

$$P = zkTf(T), \tag{5}$$

$$N = zVf(T), \tag{6}$$

$$U = zVkT^2f'(T), \tag{7}$$

$$A = NkT \ln z - zVkTf(T) \tag{8}$$

and

$$S = -Nk \ln z + zVk\{Tf'(T) + f(T)\}. \tag{9}$$

Eliminating z between (5) and (6), we obtain the equation of state of the system:

$$PV = NkT. \tag{10}$$

We note that eqn. (10) holds irrespective of the form of the function $f(T)$. Next, eliminating z between (6) and (7), we obtain

$$U = NkT^2f'(T)/f(T), \tag{11}$$

whence it follows that

$$C_V = Nk \frac{2Tf(T)f'(T) + T^2\{f(T)f''(T) - [f'(T)]^2\}}{[f(T)]^2}. \tag{12}$$

In simple cases, the function $f(T)$ turns out to be directly proportional to a certain power of T. Supposing that $f(T) \propto T^n$, eqns. (11) and (12) become

$$U = n(NkT) \tag{11$'$}$$

and

$$C_V = n(Nk). \tag{12$'$}$$

Accordingly, the pressure in such cases is directly proportional to the energy density of the gas, the constant of proportionality being $1/n$. The reader will recall that the case $n = 3/2$ corresponds to a nonrelativistic gas while the case $n = 3$ corresponds to an extreme relativistic one.

Finally, eliminating z between eqn. (6) and eqns. (8, 9), we obtain expressions for A and S as functions of N, V and T. This essentially completes our study of the classical, ideal gas.

The next problem to be considered here is that of a system of independent, *localized* particles—a model which, in some respects, approximates a solid. Mathematically, the problem is similar to that of a system of harmonic oscillators. In either case, the microscopic entities constituting the system are mutually *distinguishable*. The partition function $Q_N(V, T)$ of such a system can be written as

$$Q_N(V, T) = [Q_1(V, T)]^N. \tag{13}$$

Next, in view of the localized nature of the particles, the single-particle partition function $Q_1(V, T)$ is essentially independent of the volume occupied by the system. Consequently, we may write

$$Q_1(V, T) = \varphi(T), \tag{14}$$

where $\varphi(T)$ is a function of temperature alone. We then obtain for the grand partition function of the system

$$\mathcal{Q}(z, V, T) = \sum_{N_r=0}^{\infty} [z\varphi(T)]^{N_r} = [1 - z\varphi(T)]^{-1}. \tag{15}$$

Clearly, the quantity $z\varphi(T)$ must stay below unity, so that the summation over N_r remains convergent.

The thermodynamics of the system follows straightforwardly from eqn. (15). We have, to begin with,

$$P \equiv \frac{kT}{V} q(z, T) = -\frac{kT}{V} \ln \{1 - z\varphi(T)\}. \tag{16}$$

Since both z and T are intensive variables, the right-hand side of (16) vanishes as $V \to \infty$. Hence, in the thermodynamic limit, $P = 0$.* For other quantities of interest, we obtain, with the help of eqns. (4.3.17)–(4.3.20),

$$N = \frac{z\varphi(T)}{1 - z\varphi(T)}, \tag{17}$$

$$U = \frac{zkT^2\varphi'(T)}{1 - z\varphi(T)}, \tag{18}$$

$$A = NkT \ln z + kT \ln \{1 - z\varphi(T)\} \tag{19}$$

and

$$S = -Nk \ln z - k \ln \{1 - z\phi(T)\} + \frac{zkT\varphi'(T)}{1 - z\varphi(T)}. \tag{20}$$

From (17), we readily obtain

$$z\varphi(T) = \frac{N}{N+1} \simeq 1 - \frac{1}{N}, \tag{21}$$

because N is necessarily a large number. It also follows that

$$1 - z\varphi(T) = \frac{1}{N+1} \simeq \frac{1}{N}. \tag{22}$$

Substituting (21) and (22) into eqns. (18)–(20), we get

$$U/N = kT^2\varphi'(T)/\varphi(T), \tag{18'}$$

$$A/N = -kT \ln \varphi(T) + 0\left(\frac{\ln N}{N}\right) \tag{19'}$$

* It will be seen in the sequel that P actually vanishes like $\{(\ln V)/V\}$.

and

$$S/Nk = \ln \varphi(T) + T\varphi'(T)/\varphi(T) + 0\left(\frac{\ln N}{N}\right). \tag{20'}$$

Substituting

$$\varphi(T) = [2 \sinh (\hbar\omega/2kT)]^{-1} \tag{23}$$

into the foregoing set of formulae, we obtain results pertaining to a system of *quantum-mechanical*, one-dimensional harmonic oscillators. The substitution

$$\varphi(T) = kT/\hbar\omega, \tag{24}$$

on the other hand, leads to results pertaining to a system of *classical*, one-dimensional harmonic oscillators.

As a corollary, we may study here the problem of the *solid–vapor equilibrium*. Consider a single-component system, having two phases—solid and vapor—in equilibrium, contained in a closed vessel of volume V at temperature T. Since the phases are free to exchange particles, a state of mutual equilibrium would require that their chemical potentials be equal; this, in turn, requires that they have a common fugacity as well. Now, eqn. (6) gives for the fugacity z_g of the gaseous phase

$$z_g = \frac{N_g}{V_g f(T)}, \tag{25}$$

where N_g is the number of particles in the gaseous phase and V_g the volume occupied by it; in a typical case, $V_g \simeq V$. The fugacity z_s of the solid phase, on the other hand, is given by eqn. (21); we have

$$z_s \simeq \frac{1}{\varphi(T)}. \tag{26}$$

Equating (25) and (26), we obtain for the *equilibrium particle density* in the vapor phase

$$N_g/V_g = f(T)/\varphi(T). \tag{27}$$

Now, if the density in the vapor phase is sufficiently low and the temperature of the system sufficiently high, the vapor pressure P would be given by

$$P = \frac{N_g}{V_g} kT = kT \frac{f(T)}{\varphi(T)}. \tag{28}$$

To be specific, we may assume the vapor to be monatomic; the function $f(T)$ is then given by

$$f(T) = (2\pi mkT)^{3/2}/h^3. \tag{29}$$

Further, let the solid phase be approximated by a set of three-dimensional harmonic oscillators characterized by a single frequency ω (the *Einstein* model); the function $\varphi(T)$ is then given by

$$\varphi(T) = [2 \sinh (\hbar\omega/2kT)]^{-3}. \tag{30}$$

However, there is one important difference here. An atom in a solid is energetically more stabilized than an atom which is free—that is why a certain threshold energy is always needed to transform a solid into separate atoms. Let ε denote the value of this energy per atom; this in a way implies that the zeros of the energy spectra ε_g and ε_s, which led to the functions (29) and (30) respectively, are displaced with respect to one another by an amount ε. A *true* comparison between the functions $f(T)$ and $\varphi(T)$ must take this difference into account. As a result, we obtain for the vapor pressure

$$P = kT\left(\frac{2\pi mkT}{h^2}\right)^{3/2} [2 \sinh (\hbar\omega/2kT)]^3 e^{-\varepsilon/kT}. \tag{31}$$

In passing, we note that eqn. (27) also gives us the condition necessary for the formation of the solid phase. The condition clearly is:

$$N \geqslant V\frac{f(T)}{\varphi(T)}, \tag{32}$$

where N is the total number of particles in the system. Alternatively, we require:

$$T \leqslant T_c, \tag{33}$$

where T_c is a *characteristic* temperature, given by the implicit relationship

$$\frac{f(T_c)}{\varphi(T_c)} = \frac{N}{V}. \tag{34}$$

Once the two phases appear, the number $N_g(T)$ will have a value determined by eqn. (27) while the remaining particles, viz. $N-N_g$, will constitute the solid phase.

4.5. Density and energy fluctuations in the grand canonical ensemble: correspondence with other ensembles

In a grand canonical ensemble, the variables N and E, for any member of the ensemble, can lie anywhere between zero and infinity. Therefore, at the face of it, the grand canonical ensemble appears to be very different from its predecessors—the canonical ensemble and the microcanonical ensemble. However, as far as thermodynamics is concerned, the results obtained from this ensemble turn out to be identical with the ones obtained from the other two ensembles. Thus, in spite of strong facial differences, the overall behavior of a given physical system is practically the same whether it belongs to one kind of ensemble or to another. The basic reason for this is related to the fact that the "relative fluctuations" in the values of the quantities that vary from member to member in an ensemble are practically negligible. Therefore, in spite of the different surroundings which different ensembles provide to a given physical system, the overall behavior of the system is not significantly affected.

To appreciate the point, we evaluate here the relative fluctuations in the density n and the energy E of a given physical system in the grand canonical ensemble. Recalling that

$$P_{r,s} = \frac{z^{N_r}e^{-\beta E_s}}{\mathcal{Q}(z, V, T)}, \tag{1}$$

where

$$\mathcal{Q}(z, V, T) = \sum_{r,s} z^{N_r} e^{-\beta E_s}, \tag{2}$$

we have

$$\bar{N} = \sum_{r,s} N_r P_{r,s} = \frac{1}{\mathcal{Q}}\left(z\frac{\partial}{\partial z}\right)_{V,T} \mathcal{Q} = z\left(\frac{\partial \ln \mathcal{Q}}{\partial z}\right)_{V,T} \tag{3}$$

and

$$\overline{N^2} = \sum_{r,s} N_r^2 P_{r,s} = \frac{1}{\mathcal{Q}}\left(z\frac{\partial}{\partial z}\right)^2_{V,T} \mathcal{Q}$$
$$= \frac{1}{\mathcal{Q}}\left\{z\frac{\partial}{\partial z}+z^2\frac{\partial^2}{\partial z^2}\right\}_{V,T} \mathcal{Q}; \tag{4}$$

it then follows that

$$\overline{(\Delta N)^2} \equiv \overline{N^2} - \bar{N}^2 = \left(z\frac{\partial}{\partial z}\right)^2_{V,T} \ln \mathcal{Q}$$
$$= \left(z\frac{\partial}{\partial z}\bar{N}\right)_{V,T} = kT\left(\frac{\partial \bar{N}}{\partial \mu}\right)_{V,T}. \tag{5}$$

From (5), we obtain for the relative mean-square fluctuation in the particle density $n(\equiv N/V)$

$$\frac{\overline{(\Delta n)^2}}{\bar{n}^2} = \frac{\overline{(\Delta N)^2}}{\bar{N}^2} = \frac{kT}{N^2}\left(\frac{\partial N}{\partial \mu}\right)_{V,T}; \tag{6}$$

for simplicity, we have replaced, in the last expression, the symbol $\bar{N}$ by its thermodynamic counterpart N.

To put expression (6) in a more familiar form, we proceed as follows. From the thermodynamic relationship (4.3.22), we have

$$d\mu = \frac{V}{N}\,dP - \frac{S}{N}\,dT, \tag{7}$$

whence it follows that

$$\left(\frac{\partial \mu}{\partial v}\right)_T = v\left(\frac{\partial P}{\partial v}\right)_T, \tag{8}$$

where $v = V/N$. From (8), we obtain

$$-\frac{N^2}{V}\left(\frac{\partial \mu}{\partial N}\right)_{V,T} = V\left(\frac{\partial P}{\partial V}\right)_{N,T}. \tag{9}$$

Substituting (9) into (6), we finally obtain

$$\frac{\overline{(\Delta n)^2}}{\bar{n}^2} = -\frac{kT}{V^2}\left(\frac{\partial V}{\partial P}\right)_{N,T} = \frac{kT}{V}\varkappa_T, \tag{10}$$

where $\varkappa_T$ denotes the *isothermal compressibility* of the system.

Thus, the relative root-mean-square fluctuation in the particle density of the given system is *ordinarily* $0(N^{-1/2})$ and, hence, negligible. However, there can be exceptions, like the ones met with in situations accompanying *phase transitions*. In these situations, the compressibility of a given system can become excessively large, as is evidenced by an almost "flattening" of the isotherms. Under these circumstances, the derivative $(\partial v/\partial P)_T$, and hence the quantity $\varkappa_T$, can very well be $0(N)$; consequently, the relative root-mean-square fluctuation in the particle density n can be $0(1)$. Thus, in the regions of phase transitions, especially at the critical points, we expect to encounter unusually large fluctuations in the particle density of the system. Such fluctuations indeed exist and account for phenomena like the *critical opalescence*. It is clear that under these circumstances the formalism of the grand canonical ensemble could, in principle, lead to results which are not necessarily identical with the ones following from the formalism of the corresponding canonical ensemble. In such cases, it is the formalism of the grand canonical ensemble which will have to be preferred because only this one will provide a correct picture of the actual physical situation.

We shall now examine the degree of fluctuations in the energy of the system. From eqns. (1) and (2), we obtain

$$\bar{E} = \sum_{r,s} E_s P_{r,s} = -\frac{1}{\mathscr{Q}}\left(\frac{\partial \mathscr{Q}}{\partial \beta}\right)_{z,V} = -\left(\frac{\partial \ln \mathscr{Q}}{\partial \beta}\right)_{z,V} \tag{11}$$

and

$$\overline{E^2} = \sum_{r,s} E_s^2 P_{r,s} = \frac{1}{\mathscr{Q}}\left(\frac{\partial^2 \mathscr{Q}}{\partial \beta^2}\right)_{z,V}, \tag{12}$$

whence it follows that

$$\overline{(\Delta E)^2} \equiv \overline{E^2} - \bar{E}^2 = \left[\frac{\partial^2}{\partial \beta^2}\ln \mathscr{Q}\right]_{z,V} = -\left(\frac{\partial \bar{E}}{\partial \beta}\right)_{z,V}. \tag{13}$$

From (13), we obtain for the relative mean-square fluctuation in E

$$\frac{\overline{(\Delta E)^2}}{\bar{E}^2} = \frac{kT^2}{U^2}\left(\frac{\partial U}{\partial T}\right)_{z,V}; \tag{14}$$

in the last expression, we have replaced the symbol $\bar{E}$ by its thermodynamic counterpart U.

To put expression (14) in a more comprehensible form, we proceed as follows. We first note that

$$\left(\frac{\partial U}{\partial T}\right)_{z,V} = \left(\frac{\partial U}{\partial T}\right)_{N,V} + \left(\frac{\partial U}{\partial N}\right)_{T,V}\left(\frac{\partial N}{\partial T}\right)_{z,V}. \tag{15}$$

The quantity $(\partial U/\partial T)_{N,V}$ is the familiar one, namely C_V. To understand the rest of it, we first observe that*

$$\left(\frac{\partial U}{\partial N}\right)_{T,V} = \mu + T\left(\frac{\partial S}{\partial N}\right)_{T,V} = \mu - T\left(\frac{\partial \mu}{\partial T}\right)_{N,V}; \tag{16}$$

* Here, use has been made of the fact that

(i) $dU = T\,dS - P\,dV + \mu\,dN$ and (ii) $(\partial S/\partial N)_{T,V} = -(\partial \mu/\partial T)_{N,V}$;

the latter result follows from the formulae

$$S = -\left(\frac{\partial A}{\partial T}\right)_{N,V}; \qquad \mu = +\left(\frac{\partial A}{\partial N}\right)_{T,V}.$$

next, we have

$$\left(\frac{\partial N}{\partial T}\right)_{z,V} = \left(\frac{\partial N}{\partial T}\right)_{\mu,V} + \left(\frac{\partial N}{\partial \mu}\right)_{T,V}\left(\frac{\partial \mu}{\partial T}\right)_{z,V}$$

$$= -\left(\frac{\partial N}{\partial \mu}\right)_{T,V}\left(\frac{\partial \mu}{\partial T}\right)_{N,V} + \left(\frac{\partial N}{\partial \mu}\right)_{T,V}\frac{\mu}{T} \qquad (\because \mu = kT \ln z)$$

$$= \frac{1}{T}\left(\frac{\partial N}{\partial \mu}\right)_{T,V}\left\{\mu - T\left(\frac{\partial \mu}{\partial T}\right)_{N,V}\right\}. \tag{17}$$

Making use of the formulae (15), (16) and (17), we finally obtain from eqn. (14)

$$\overline{(\Delta E)^2} = kT^2C_V + kT\left(\frac{\partial N}{\partial \mu}\right)_{T,V}\left\{\left(\frac{\partial U}{\partial N}\right)_{T,V}\right\}^2$$

$$= \langle(\Delta E)^2\rangle_{\text{can}} + \overline{(\Delta N)^2}\left\{\left(\frac{\partial U}{\partial N}\right)_{T,V}\right\}^2. \tag{18}$$

The formula (18) is highly instructive; it tells us that the mean-square fluctuation in the energy E of a given system in the grand canonical ensemble is equal to the value it would have in the canonical ensemble, see eqn. (3.6.3), *plus* a contribution arising from the fact that now the particle number N is also fluctuating. Again, under ordinary circumstances, the relative root-mean-square fluctuation in the energy density of the system would be practically negligible. However, in the regions of phase transitions, unusually large fluctuations in the values of this variable can arise by virtue of the second term in the formula.

Problems

4.1. Show that the entropy of a system in the grand canonical ensemble can be written as

$$S = -k \sum_{r,s} P_{r,s} \ln P_{r,s},$$

where $P_{r,s}$ is given by eqn. (4.1.8).

4.2. "In the thermodynamic limit (when the extensive properties of the system become infinitely large, while the intensive ones remain constant), the q-potential of the system may be calculated by taking only the largest term in the sum

$$\sum_{N_r=0}^{\infty} z^{N_r} Q_{N_r}(V, T)."$$

Verify this statement and interpret the result physically.

4.3. A vessel of volume $V^{(0)}$ contains $N^{(0)}$ molecules. Assuming that there is no correlation whatsoever between the locations of the various molecules, calculate the probability $p(N, V)$ that a region of volume V (located anywhere in the vessel) contains exactly N molecules.

(i) Show that $\bar{N} = N^{(0)}p$ and $(\Delta N)_{\text{r.m.s.}} = \{N^{(0)}p(1-p)\}^{1/2}$, where $p = V/V^{(0)}$.
(ii) Show that if both $N^{(0)}$ and N are large numbers, the function $p(N, V)$ assumes a Gaussian form.
(iii) Further, if $p \ll 1$ (the numbers $N^{(0)}$ and N still being large), the distribution function $p(N, V)$ assumes the form of the Poisson distribution:

$$p(N) = e^{-\bar{N}}\frac{\bar{N}^N}{N!}.$$

4.4. "The probability that a system in the grand canonical ensemble has *exactly* N particles is given by

$$p(N) = \frac{z^N Q_N(V, T)}{\mathcal{Q}(z, V, T)}."$$

Verify this statement and show that in the case of a classical, ideal gas the distribution of particles among the members of a grand canonical ensemble is identically a Poisson distribution. Calculate the root-mean-square value of (ΔN) for this system both from the general formula (4.5.6) and from the Poisson distribution, and show that the two results are identical.

4.5. Define the function $Y(N, \gamma, T)$ as

$$Y(N, \gamma, T) = \int_0^\infty Q(N, V, T)e^{\gamma V}\, dV.$$

What physical meaning must be given to the parameter γ so that the function Y assumes a direct relevance to thermodynamics? Set up a scheme to derive the various thermodynamic quantities of a given physical system from the function Y and illustrate it by considering a few simple examples.

4.6. Consider a classical system of noninteracting, diatomic molecules enclosed in a box of volume V at temperature T. The Hamiltonian of a single molecule is given by

(i) $$H(\boldsymbol{r}_1, \boldsymbol{r}_2, \boldsymbol{p}_1, \boldsymbol{p}_2) = \frac{1}{2m}(p_1^2+p_2^2) + \frac{1}{2}K\,|\boldsymbol{r}_1-\boldsymbol{r}_2|^2 \quad \text{or}$$

(ii) $$H(\boldsymbol{r}_1, \boldsymbol{r}_2, \boldsymbol{p}_1, \boldsymbol{p}_2) = \frac{1}{2m}(p_1^2+p_2^2)+\varepsilon\,|r_{12}-r_0|.$$

Study the thermodynamics of this system, including the dependence of the quantity $\langle r_{12}^2\rangle$ on T.

4.7. Determine the grand partition function of a gaseous system of "magnetic" atoms (with $J = \frac{1}{2}$ and $g = 2$) which can have, in addition to the kinetic energy, a magnetic potential energy equal to $\mu_B H$ or $-\mu_B H$, depending upon their orientation with respect to an applied magnetic field H. Derive an expression for the magnetization of the system and calculate how much heat will be given off by the system when the magnetic field is reduced from H to zero at constant volume and constant temperature.

4.8. Study the problem of the solid–vapor equilibrium (Sec. 4.4) by setting up the grand partition function of the system.

4.9. A surface with N_0 adsorption centers has $N(\leqslant N_0)$ gas molecules adsorbed on it. Show that the chemical potential of the adsorbed molecules is given by

$$\mu = kT \ln \frac{N}{(N_0-N)\,a(T)},$$

where $a(T)$ is the partition function of a single adsorbed molecule. Solve the problem by constructing the grand partition function as well as the partition function of the system.

[Neglect the intermolecular interaction between the adsorbed molecules.]

4.10. Study the state of equilibrium between a gaseous phase and an adsorbed phase in a single-component system. Show that the pressure in the gaseous phase is given by the Langmuir equation

$$P_g = \frac{\theta}{1-\theta} \times \text{(a certain function of temperature)},$$

where θ is the equilibrium fraction of the adsorption sites which *are* occupied by the adsorbed molecules.

4.11. Prove that for a system in the grand canonical ensemble

$$\{\overline{(NE)}-\bar{N}\bar{E}\} = \overline{(\Delta N)^2}\left(\frac{\partial U}{\partial N}\right)_{T,V}.$$

4.12. Define a quantity J as

$$J = E-N\mu = TS-PV.$$

Show that for a system in the grand canonical ensemble

$$\overline{(\Delta J)^2} = kT^2C_V+\overline{(\Delta N)^2}\left\{\left(\frac{\partial U}{\partial N}\right)_{T,V}-\mu\right\}^2.$$

CHAPTER 5

FORMULATION OF QUANTUM STATISTICS

THE scope of the ensemble theory developed in Chapters 2–4 is extremely general, though the applications considered so far were confined either to classical systems or to quantum-mechanical systems composed of *distinguishable* entities. When it comes to quantum-mechanical systems composed of *indistinguishable* entities, as most physical systems are, the considerations of the preceding chapters have to be applied with considerable care. One finds that in this case it is advisable to rewrite the ensemble theory in a language that is more natural to a quantum-mechanical treatment, namely the language of the operators and the wave functions. Insofar as statistics are concerned, this rewriting of the theory may not seem to introduce any new physical ideas as such; nonetheless, it provides us with a tool which is highly suited for studying typical quantum systems. And once we set out to study these systems in detail, we encounter a stream of new, and altogether different, physical concepts. In particular, we find that the behavior of even a noninteracting system, such as the ideal gas, departs considerably from the pattern set by the so-called classical treatments. In the presence of interactions, the pattern becomes still more complicated. Of course, in the limit of high temperatures and low densities, the behavior of all physical systems tends *asymptotically* to what we expect on classical grounds. In the process of demonstrating this point, we automatically obtain a criterion which tells us whether a given physical system may or may not be treated classically. At the same time, we obtain rigorous evidence in support of the procedure, employed in the previous chapters, for computing the number, Γ, of microstates (corresponding to a given macrostate) of the system under study from the volume, ω, of the relevant region of its phase space, viz. $\Gamma \simeq \omega/h^f$, where f is the number of "degrees of freedom".

5.1. Quantum-mechanical ensemble theory: the density matrix

We consider an ensemble of $\mathcal{N}$ identical systems, where $\mathcal{N} \gg 1$. These systems are characterized by a (common) Hamiltonian, which may be denoted by the operator $\hat{H}$. At time t, the physical states of the various systems will be characterized by the wave functions $\psi(\boldsymbol{r}_i, t)$, where the $\boldsymbol{r}_i$ denote the position coordinates relevant to the system(s) under study. Let $\psi^k(\boldsymbol{r}_i, t)$ denote the (normalized) wave function characterizing the physical state in which the kth system of the ensemble happens to be at time t; naturally, $k = 1, 2, \ldots, \mathcal{N}$. The time variation of the function $\psi^k(t)$ will be determined by the Schrödinger equation†

$$\hat{H}\psi^k(t) = i\hbar\dot{\psi}^k(t). \tag{1}$$

Introducing a complete set of orthonormal functions φ_n, the wave functions $\psi^k(t)$ may be written as

$$\psi^k(t) = \sum_n a_n^k(t)\varphi_n, \tag{2}$$

where

$$a_n^k(t) = \int \varphi_n^* \psi^k(t)\, d\tau; \tag{3}$$

here, φ_n^* denotes the complex conjugate of φ_n while $d\tau$ denotes the volume element of the coordinate space of the given system.

Obviously enough, the physical state of the kth system can be described equally well in terms of the coefficients $a_n^k(t)$. The time variation of these coefficients will be given by

$$\begin{aligned} i\hbar\dot{a}_n^k(t) &= i\hbar \int \varphi_n^* \dot{\psi}^k(t)\, d\tau = \int \varphi_n^* \hat{H}\psi^k(t)\, d\tau \\ &= \int \varphi_n^* \hat{H} \left\{ \sum_m a_m^k(t)\varphi_m \right\} d\tau \\ &= \sum_m H_{nm} a_m^k(t), \end{aligned} \tag{4}$$

where

$$H_{nm} = \int \varphi_n^* \hat{H} \varphi_m\, d\tau. \tag{5}$$

The physical significance of the coefficients $a_n^k(t)$ is evident from eqn. (2). They are the *probability amplitudes* for the kth system of the ensemble to be in the respective states φ_n; to be practical, the number $|a_n^k(t)|^2$ represents the probability that a measurement at time t finds the kth system of the ensemble to be in the particular state φ_n. Clearly, we must have

$$\sum_n |a_n^k(t)|^2 = 1 \qquad \text{(for all } k\text{)}. \tag{6}$$

We now introduce the *density operator* $\hat{\varrho}(t)$, as defined by the matrix elements

$$\varrho_{mn}(t) = \frac{1}{\mathcal{N}} \sum_{k=1}^{\mathcal{N}} \{a_m^k(t)\, a_n^{k*}(t)\}; \tag{7}$$

clearly, the matrix element $\varrho_{mn}(t)$ is the ensemble average of the quantity $a_m(t)\, a_n^*(t)$ which, as a rule, varies from member to member in the ensemble. In particular, the diagonal element $\varrho_{nn}(t)$ is the ensemble average of the probability $|a_n(t)|^2$, the latter itself being a (quantum-mechanical) average. Thus, we are concerned here with a double averaging process—once due to the probabilistic aspect of the wave functions and again due to the statistical aspect of the ensemble! The quantity $\varrho_{nn}(t)$ now represents the probability that a system, chosen *at random* from the ensemble, at time t, is found to be in the state φ_n. In view

† For simplicity of notation, we suppress the coordinates $\mathbf{r}_i$ in the argument of the wave function ψ^k.

of (6) and (7), we have

$$\sum_n \varrho_{nn} = 1. \tag{8}$$

We shall now determine the equation of motion for the density matrix $\varrho_{mn}(t)$. We obtain, with the help of the foregoing equations,

$$\begin{aligned} i\hbar\dot{\varrho}_{mn}(t) &= \frac{1}{\mathcal{N}}\sum_{k=1}^{\mathcal{N}}\left[i\hbar\{\dot{a}_m^k(t)\,a_n^{k*}(t)+a_m^k(t)\,\dot{a}_n^{k*}(t)\}\right] \\ &= \frac{1}{\mathcal{N}}\sum_{k=1}^{\mathcal{N}}\left[\left\{\sum_l H_{ml}a_l^k(t)\right\}a_n^{k*}(t)-a_m^k(t)\left\{\sum_l H_{nl}^*a_l^{k*}(t)\right\}\right] \\ &= \sum_l\{H_{ml}\varrho_{ln}(t)-\varrho_{ml}(t)\,H_{ln}\} \\ &= (\hat{H}\hat{\varrho}-\hat{\varrho}\hat{H})_{mn}; \end{aligned} \tag{9}$$

here, use has been made of the fact that, in view of the Hermitian character of the operator $\hat{H}$, $H_{nl}^* = H_{ln}$. Using the commutator notation, eqn. (9) may be written as

$$i\hbar\dot{\hat{\varrho}} = [\hat{H}, \hat{\varrho}]_-. \tag{10}$$

Equation (10) is the quantum-mechanical analogue of the classical equation of Liouville; cf. (2.2.10). As expected in going from a classical equation of motion to its quantum-mechanical counterpart, the Poisson bracket $[\varrho, H]$ has given place to the commutator $(\hat{\varrho}\hat{H}-\hat{H}\hat{\varrho})/i\hbar$.

If the given system is known to be in a state of equilibrium, the corresponding ensemble must naturally be *stationary*, i.e. $\dot{\varrho}_{mn} = 0$. Equations (9) and (10) then tell us that, for this to be the case, (i) the density operator $\hat{\varrho}$ must be an explicit function of the Hamiltonian operator $\hat{H}$ (because the two operators will then necessarily commute) and (ii) the Hamiltonian must not depend explicitly on time; in symbols, we must have (i) $\hat{\varrho} = \hat{\varrho}(\hat{H})$ and (ii) $\dot{\hat{H}} = 0$. Now, if the basis functions φ_n were the eigenfunctions of the Hamiltonian itself, then the matrices H and ϱ would be diagonal:

$$\varrho_{mn} = \varrho_n\delta_{mn}. \tag{11}^\dagger$$

The diagonal element ϱ_n, being a measure of the probability that a system, chosen *at random* (and at *any* time) from the ensemble, is found to be in the eigenstate φ_n, will naturally depend upon the corresponding eigenvalue E_n of the Hamiltonian; the precise nature of this dependence is, however, determined by the "kind" of ensemble we wish to construct.

In any other representation, the density matrix may or may not be diagonal. However, quite generally, it will be symmetric:

$$\varrho_{mn} = \varrho_{nm}. \tag{13}$$

† It may be noted that in this (so-called energy) representation the density operator $\hat{\varrho}$ may be written as

$$\hat{\varrho} = \sum_n |\varphi_n\rangle\,\varrho_n\langle\varphi_n|, \tag{12}$$

for then

$$\varrho_{kl} = \sum_n \langle\varphi_k|\varphi_n\rangle\,\varrho_n\langle\varphi_n|\varphi_l\rangle = \sum_n \delta_{kn}\varrho_n\delta_{nl} = \varrho_k\delta_{kl}.$$

The physical reason underlying this symmetry is that, under statistical equilibrium, the tendency of a physical system to switch over from one state (in the new representation) to another must be counterbalanced by an equally strong tendency to switch between the same states in the reverse direction. This condition of *detailed balancing* is essential for the maintenance of an equilibrium distribution within the ensemble.

Finally, we consider the expectation value of a physical quantity G, which is dynamically represented by an operator $\hat{G}$. This will naturally be determined by the double averaging process, viz.

$$\langle G\rangle = \frac{1}{\mathcal{N}}\sum_{k=1}^{\mathcal{N}}\int \psi^{k*}\hat{G}\psi^k\, d\tau. \tag{14}$$

In terms of the coefficients a_n^k, this becomes

$$\langle G\rangle = \frac{1}{\mathcal{N}}\sum_{k=1}^{\mathcal{N}}\left[\sum_{m,n} a_n^{k*}a_m^k G_{nm}\right] \tag{15}$$

where

$$G_{nm} = \int \varphi_n^*\hat{G}\varphi_m\, d\tau. \tag{16}$$

Introducing the density matrix ϱ, eqn. (15) takes a particularly neat form:

$$\langle G\rangle = \sum_{m,n}\varrho_{mn}G_{nm} = \sum_m(\hat{\varrho}\hat{G})_{mm} = \mathrm{Tr}\,(\hat{\varrho}\hat{G}). \tag{17}$$

Taking $\hat{G} = \hat{1}$, where $\hat{1}$ is the unit operator, we have

$$\mathrm{Tr}\,(\hat{\varrho}) = 1, \tag{18}$$

which is identical with (8). It must be noted here that if the original wave functions ψ^k were not normalized, then the expectation value $\langle G\rangle$ would be given by the formula

$$\langle G\rangle = \frac{\mathrm{Tr}\,(\hat{\varrho}\hat{G})}{\mathrm{Tr}\,(\hat{\varrho})} \tag{19}$$

instead. In view of the mathematical structure of the formulae (17) and (19), the expectation value of any physical quantity G is *manifestly* independent of the choice of the basis $\{\varphi_n\}$. Indeed, the expectation value of a physical quantity is expected to possess this property anyway!

5.2. Statistics of the various ensembles

(i) *The microcanonical ensemble*

The construction of the microcanonical ensemble is based on the premise that the systems constituting the ensemble are characterized by a fixed number of particles N, a fixed volume V and an energy lying within the interval $(E-\frac{1}{2}\Delta, E+\frac{1}{2}\Delta)$, where $\Delta \ll E$. The total number of distinct microstates accessible to a system is then denoted by the symbol $\Gamma(N, V, E;\Delta)$ and, by assumption, any one of these microstates is just as likely to occur as any other. This assumption enters into our theory in the nature of a postulate, and is often referred to as the postulate of *equal a priori probabilities* for the various accessible states. Accordingly, the density matrix ϱ_{mn} (which, in the energy representation, must be a diagonal matrix) will be of the form

$$\varrho_{mn} = \varrho_n \delta_{mn}, \tag{1}$$

with

$$\varrho_n = \begin{cases} 1/\Gamma & \text{for each of the accessible states,} \\ 0 & \text{for all other states;} \end{cases} \tag{2}$$

the normalization condition (5.1.18) is clearly satisfied. As we already know, the thermodynamics of the system is completely determined by the expression for its entropy which, in turn, is given by

$$S = k \ln \Gamma. \tag{3}$$

Since Γ, the total number of distinct, accessible states, is supposed to be computed quantum-mechanically, taking into account the indistinguishability of the particles right from the beginning, no paradox, such as Gibbs', is now expected to arise. Moreover, if the quantum state of the system turns out to be unique ($\Gamma = 1$), the entropy of the system would identically vanish. This provides us with a sound theoretical basis for the hitherto empirical theorem of Nernst (also known as the *third law of thermodynamics*).

The situation corresponding to the case $\Gamma = 1$ is usually referred to as a *pure* case. In such a case, the construction of an ensemble is practically superfluous, because every system in the ensemble has got to be in one and the same state. Accordingly, there is only one diagonal element ϱ_{nn} which is nonzero (actually equal to unity), while all others are equal to zero. The density matrix, therefore, satisfies the condition

$$\varrho^2 = \varrho. \tag{4}$$

In a different representation, the pure case will correspond to

$$\varrho_{mn} = \frac{1}{\mathcal{N}} \sum_{k=1}^{\mathcal{N}} a_m^k a_n^{k*} \equiv a_m a_n^*, \tag{5}$$

because all values of k are now literally equivalent. We then have

$$\begin{aligned} \varrho^2_{mn} &= \sum_l \varrho_{ml}\varrho_{ln} = \sum_l a_m a_l^* a_l a_n^* \\ &= a_m a_n^* \qquad \left(\text{because } \sum_l a_l^* a_l = 1\right) \\ &= \varrho_{mn}. \end{aligned} \tag{6}$$

Thus, the condition (4) holds in *all* representations.

A situation in which $\Gamma > 1$ is usually referred to as a *mixed* case. The density matrix, in the energy representation, is then given by eqns. (1) and (2). If we now change over to any other representation, the general form of the density matrix will remain the same, namely (i) the off-diagonal elements will continue to be zero, while (ii) the diagonal elements (over the allowed range) will continue to be equal to one another. Now, had we constructed our ensemble on a representation other than the energy representation, how could we have possibly anticipated *ab initio* the property (i) of the density matrix, though property (ii) could have been easily invoked through a postulate of *equal a priori probabilities*? To ensure that the property (i), as well as property (ii), holds in *every* representation, we invoke yet another postulate—which is referred to as the postulate of *random a priori phases* for the probability amplitudes a_n^k, which in turn implies that the wave function ψ^k, for all k, is an *incoherent* superposition of the basis $\{\varphi_n\}$. As a consequence of the postulate of random *a priori* phases, coupled with the postulate of equal a *priori* probabilities, we would have in *any* representation

$$\begin{aligned}\varrho_{mn} &\equiv \frac{1}{\mathcal{N}}\sum_{k=1}^{\mathcal{N}} a_m^k a_n^{k*} = \frac{1}{\mathcal{N}}\sum_{k=1}^{\mathcal{N}} |a|^2 e^{i(\theta_m^k-\theta_n^k)}\\ &= c\langle e^{i(\theta_m^k-\theta_n^k)}\rangle\\ &= c\delta_{mn},\end{aligned} \tag{7}$$

as it should be for a microcanonical ensemble.

Thus, contrary to what might have been expected on customary thought, to secure the physical situation corresponding to a microcanonical ensemble, we require in general two postulates instead of one! The second postulate arises solely from quantum-mechanics and is intended to ensure noninterference (and hence a complete absence of correlations) among the member systems; this, in turn, enables us to form a mental picture of each system, one at a time, completely disentangled from other systems in the ensemble.

(ii) *The canonical ensemble*

In this ensemble the macrostate of a member system is defined through the parameters N, V and T; the energy E now becomes a variable quantity. The probability that a system, chosen *at random* from the ensemble, possesses an energy E_r is determined by the Boltzmann factor $\exp(-\beta E_r)$, where $\beta = 1/kT$; see Secs. 3.1 and 3.2. The density matrix in the energy representation is therefore taken as

$$\varrho_{mn} = \varrho_n \delta_{mn}, \tag{8}$$

with

$$\varrho_n = C\exp(-\beta E_n); \qquad n = 0, 1, 2, \ldots. \tag{9}$$

The constant C is determined by the normalization condition (5.1.18), whence it follows that

$$C = \frac{1}{\sum_n \exp(-\beta E_n)} = \frac{1}{Q_N(\beta)}, \tag{10}$$

where $Q_N(\beta)$ is the *partition function* of the system. In view of (5.1.12), the density operator in the canonical ensemble may be written as

$$\begin{aligned}\hat{\varrho} &= \sum_n |\varphi_n\rangle \frac{1}{Q_N(\beta)} e^{-\beta E_n}\langle\varphi_n|\\ &= \frac{1}{Q_N(\beta)} e^{-\beta\hat{H}}\sum_n |\varphi_n\rangle\langle\varphi_n|\\ &= \frac{1}{Q_N(\beta)} e^{-\beta\hat{H}} = \frac{e^{-\beta\hat{H}}}{\mathrm{Tr}\,(e^{-\beta\hat{H}})},\end{aligned} \tag{11}$$

?

because the operator $\Sigma_n |\varphi_n\rangle\langle\varphi_n|$ is identically the unit operator. It is understood that the operator $\exp\left(-\beta\hat{H}\right)$ in (11) stands for the sum

$$\sum_{j=0}^{\infty}(-1)^j\frac{(\beta\hat{H})^j}{j!}. \tag{12}$$

The expectation value $\langle G\rangle_N$ of a physical quantity G, which is represented by an operator $\hat{G}$, is now given by

$$\langle G\rangle_N = \mathrm{Tr}\,(\hat{\varrho}\hat{G}) = \frac{1}{Q_N(\beta)}\,\mathrm{Tr}\,(\hat{G}e^{-\beta\hat{H}}) = \frac{\mathrm{Tr}\,(\hat{G}e^{-\beta\hat{H}})}{\mathrm{Tr}\,(e^{-\beta\hat{H}})}; \tag{13}$$

the suffix N here emphasizes the fact that the averaging has been done over an ensemble *with N fixed.*

(iii) *The grand canonical ensemble*

In this ensemble the density operator $\hat{\varrho}$ operates on a Hilbert space with an indefinite number of particles. The density operator must therefore commute not only with the Hamiltonian operator $\hat{H}$ but also with a number operator $\hat{n}$ whose eigenvalues are 0, 1, 2, ... The precise form of the relevant density operator can now be obtained by a straightforward generalization of the preceding case, with the result

$$\hat{\varrho} = \frac{1}{\mathscr{Q}(\mu, V, T)}\,e^{-\beta(\hat{H}-\mu\hat{n})}, \tag{14}$$

where

$$\mathscr{Q}(\mu, V, T) = \sum_{r,\,s} e^{-\beta(E_r-\mu N_s)} = \mathrm{Tr}\left\{e^{-\beta(\hat{H}-\mu\hat{n})}\right\}. \tag{15}$$

The ensemble average $\langle G\rangle$ is now given by

$$\langle G\rangle = \frac{1}{\mathscr{Q}(\mu, V, T)}\,\mathrm{Tr}\,(\hat{G}e^{-\beta\hat{H}}e^{\beta\mu\hat{n}}) = \frac{\sum_{N=0}^{\infty} z^N\langle G\rangle_N\, Q_N(\beta)}{\sum_{N=0}^{\infty} z^N Q_N(\beta)}, \tag{16}$$

where $z(\equiv e^{\beta\mu})$ is the *fugacity* of the system while $\langle G\rangle_N$ is the canonical-ensemble average, as given by eqn. (13). The quantity $\mathscr{Q}(\mu, V, T)$ appearing in these formulae is, clearly, the *grand partition function* of the system.

5.3. Examples

(i) *An electron in a magnetic field*

Let us consider, for illustration, the case of a single electron which possesses an intrinsic spin $\frac{1}{2}\hbar\hat{\sigma}$ and a magnetic moment of magnitude μ_B, where $\hat{\sigma}$ is the Pauli spin operator and $\mu_B = e\hbar/2mc$. The spin of the electron can have any of the two possible orientations,

↑ or ↓, with respect to an applied magnetic field $\boldsymbol{B}$. If the applied field is taken to be in the direction of the z-axis, the configurational Hamiltonian of the electron spin takes the form

$$\hat{H} = -\mu_B(\hat{\boldsymbol{\sigma}} \cdot \boldsymbol{B}) = -\mu_B B \hat{\sigma}_z. \tag{1}$$

In the representation which makes $\hat{\sigma}_z$ diagonal, namely

$$\sigma_x = \begin{pmatrix} 0 & 1 \\ 1 & 0 \end{pmatrix}, \quad \sigma_y = \begin{pmatrix} 0 & -i \\ i & 0 \end{pmatrix}, \quad \sigma_z = \begin{pmatrix} 1 & 0 \\ 0 & -1 \end{pmatrix}, \tag{2}$$

the density matrix in the canonical ensemble would be

$$\begin{aligned} (\hat{\varrho}) &= \frac{(e^{-\beta \hat{H}})}{\mathrm{Tr}\,(e^{-\beta \hat{H}})} \\ &= \frac{1}{e^{\beta\mu_B B} + e^{-\beta\mu_B B}} \begin{pmatrix} e^{\beta\mu_B B} & 0 \\ 0 & e^{-\beta\mu_B B} \end{pmatrix}. \end{aligned} \tag{3}$$

We readily obtain for the expectation value of σ_z:

$$\langle \sigma_z \rangle = \mathrm{Tr}\,(\hat{\varrho}\hat{\sigma}_z) = \frac{e^{\beta\mu_B B} - e^{-\beta\mu_B B}}{e^{\beta\mu_B B} + e^{-\beta\mu_B B}} = \tanh\,(\beta\mu_B B), \tag{4}$$

in agreement with the findings of Secs. 3.8 and 3.9.

Alternatively, we could proceed as follows: since $\sigma_z^2 = 1$,

$$\begin{aligned} e^{-\beta\hat{H}} &= e^{\beta\mu_B B \hat{\sigma}_z} = \sum_{j=0}^{\infty} \frac{1}{j!} (\beta\mu_B B)^j \hat{\sigma}_z^j \\ &= \sum_{j\ \text{even}} \frac{1}{j!} (\beta\mu_B B)^j + \sum_{j\ \text{odd}} \frac{1}{j!} (\beta\mu_B B)^j \hat{\sigma}_z \\ &= \cosh\,(\beta\mu_B B) + \hat{\sigma}_z \sinh\,(\beta\mu_B B). \end{aligned} \tag{5}$$

Now, $\mathrm{Tr}\ \hat{1} = 2$, $\mathrm{Tr}\ \hat{\sigma}_z = 0$ and $\mathrm{Tr}\ \hat{\sigma}_z^2 = 2$; therefore,

$$\begin{aligned} \langle \sigma_z \rangle &= \mathrm{Tr}\,(\hat{\sigma}_z e^{-\beta\hat{H}}) / \mathrm{Tr}\,(e^{-\beta\hat{H}}) \\ &= 2 \sinh\,(\beta\mu_B B)/2 \cosh\,(\beta\mu_B B) \\ &= \tanh\,(\beta\mu_B B), \end{aligned} \tag{6}$$

which is the same as (4). Thus, it is sometimes possible to carry out a desired calculation without making an explicit evaluation of the density matrix ϱ_{mn}.

(ii) *A free particle in a box*

We now consider the case of a free particle, of mass m, in a cubical box of side L. The Hamiltonian of the particle is given by

$$\hat{H} = -\frac{\hbar^2}{2m}\nabla^2 \equiv -\frac{\hbar^2}{2m}\left(\frac{\partial^2}{\partial x^2} + \frac{\partial^2}{\partial y^2} + \frac{\partial^2}{\partial z^2}\right), \tag{7}$$

while the eigenfunctions of the Hamiltonian, which satisfy periodic boundary conditions

$$\varphi(x+L, y, z) = \varphi(x, y+L, z) = \varphi(x, y, z+L) \\ = \varphi(x, y, z), \tag{8}$$

are given by

$$\varphi_E(\boldsymbol{r}) = \frac{1}{L^{3/2}} \exp(i\boldsymbol{k}\cdot\boldsymbol{r}), \tag{9}$$

the corresponding eigenvalues E being

$$E = \frac{\hbar^2 k^2}{2m}, \tag{10}$$

with

$$\boldsymbol{k} \equiv (k_x, k_y, k_z) = \frac{2\pi}{L}(n_x, n_y, n_z); \tag{11}$$

the quantum numbers n_x, n_y and n_z must be integers (positive, negative or zero). Symbolically, we may write for the wave vector $\boldsymbol{k}$

$$\boldsymbol{k} = \frac{2\pi}{L}\boldsymbol{n}, \tag{12}$$

where $\boldsymbol{n}$ is a vector *with integral components* $0, \pm 1, \pm 2, \ldots$.

We now proceed to evaluate the canonical density matrix ($\hat{\varrho}$) of this system; we shall do this in the *coordinate* representation. In view of eqn. (5.2.11), we have

$$\langle \boldsymbol{r} | e^{-\beta\hat{H}} | \boldsymbol{r}' \rangle = \sum_E \langle \boldsymbol{r} | E \rangle e^{-\beta E} \langle E | \boldsymbol{r}' \rangle \\ = \sum_E e^{-\beta E} \varphi_E(\boldsymbol{r}) \varphi_E^*(\boldsymbol{r}'). \tag{13}$$

Substituting from (9) and making use of (10) and (12), we obtain

$$\langle \boldsymbol{r} | e^{-\beta\hat{H}} | \boldsymbol{r}' \rangle = \frac{1}{L^3} \sum_{\boldsymbol{k}} \exp\left[-\frac{\beta\hbar^2}{2m} k^2 + i\boldsymbol{k}\cdot(\boldsymbol{r}-\boldsymbol{r}') \right] \\ \simeq \frac{1}{(2\pi)^3} \int \exp\left[-\frac{\beta\hbar^2}{2m} k^2 + i\boldsymbol{k}\cdot(\boldsymbol{r}-\boldsymbol{r}') \right] d^3k \\ = \left(\frac{m}{2\pi\beta\hbar^2} \right)^{3/2} \exp\left[-\frac{m}{2\beta\hbar^2} (\boldsymbol{r}-\boldsymbol{r}')^2 \right], \tag{14}$$

whence it follows that

$$\mathrm{Tr}\,(e^{-\beta\hat{H}}) = \int \langle \boldsymbol{r} | e^{-\beta\hat{H}} | \boldsymbol{r} \rangle \, d^3r \\ = V \left(\frac{m}{2\pi\beta\hbar^2} \right)^{3/2}. \tag{15}$$

The expression (15) is indeed the partition function, $Q(\beta)$, of a single particle confined to a box of volume V; cf. eqn. (3.5.19). Combining (14) and (15), we obtain for the density matrix in the coordinate representation

$$\langle \boldsymbol{r} | \hat{\varrho} | \boldsymbol{r}' \rangle = \frac{1}{V} \exp\left[-\frac{m}{2\beta\hbar^2} (\boldsymbol{r}-\boldsymbol{r}')^2 \right]. \tag{16}$$

As expected, the matrix $\varrho_{r,r'}$ is symmetric between the states $\boldsymbol{r}$ and $\boldsymbol{r}'$. Moreover, the diagonal element $\langle \boldsymbol{r} | \varrho | \boldsymbol{r} \rangle$, which represents the *probability density* for the particle to be in the neighborhood of the point $\boldsymbol{r}$, is independent of $\boldsymbol{r}$; this implies that, in the case of a single free particle, *all* positions within the box are equally likely to obtain. A nondiagonal element $\langle \boldsymbol{r} | \varrho | \boldsymbol{r}' \rangle$, on the other hand, is a measure of the probability of "spontaneous transition" between the position coordinates $\boldsymbol{r}$ and $\boldsymbol{r}'$ and is therefore a measure of the relative "intensity" of the wave packet (associated with the particle) at a distance $|\boldsymbol{r}-\boldsymbol{r}'|$ from the centre of the packet. The spatial extent of the wave packet, which in turn is a measure of the uncertainty involved in locating the position of the particle, is clearly of the order of $\hbar/(mkT)^{1/2}$; the latter is also a measure of the *mean thermal wavelength* of the particle. The spatial spread encountered here is a pure quantum-mechanical effect; quite expectedly, it tends to vanish at high temperatures. Actually, as $\beta \to 0$, the matrix element (14) approaches a delta function which implies a return to the classical picture.

Finally, we determine the expectation value of the Hamiltonian itself. From eqns. (7) and (16), we obtain

$$\begin{aligned}\langle H \rangle = \mathrm{Tr}\,(\hat{\varrho}\hat{H}) &= -\frac{\hbar^2}{2mV} \int \left\{ \nabla_r^2 \exp\left[-\frac{m}{2\beta\hbar^2} (\boldsymbol{r}-\boldsymbol{r}')^2 \right] \right\}_{\boldsymbol{r}=\boldsymbol{r}'} d^3r \\ &= \frac{1}{2\beta V} \int \left\{ \left[3 - \frac{m}{\beta\hbar^2} (\boldsymbol{r}-\boldsymbol{r}')^2 \right] \exp\left[-\frac{m}{2\beta\hbar^2} (\boldsymbol{r}-\boldsymbol{r}')^2 \right] \right\}_{\boldsymbol{r}=\boldsymbol{r}'} d^3r \\ &= \frac{3}{2\beta} = \frac{3}{2} kT, \end{aligned} \tag{17}$$

which was indeed expected. Otherwise too,

$$\langle H \rangle = \frac{\mathrm{Tr}\,(\hat{H} e^{-\beta \hat{H}})}{\mathrm{Tr}\,(e^{-\beta \hat{H}})} = -\frac{\partial}{\partial \beta} \ln \mathrm{Tr}\,(e^{-\beta \hat{H}}), \tag{18}$$

which, on combination with (15), leads to the same result.

(iii) *A linear harmonic oscillator*

Next, we consider the case of a linear harmonic oscillator, whose Hamiltonian is given by

$$\hat{H} = -\frac{\hbar^2}{2m} \frac{\partial^2}{\partial q^2} + \frac{1}{2} m\omega^2 q^2. \tag{19}$$

The eigenvalues of this Hamiltonian are well known, namely

$$E_n = (n+\tfrac{1}{2})\hbar\omega; \qquad n = 0, 1, 2, \ldots, \tag{20}$$

the corresponding eigenfunctions being

$$\varphi_n(q) = \left(\frac{m\omega}{\pi\hbar}\right)^{1/4} \frac{H_n(\xi)}{(2^n n!)^{1/2}} e^{-(1/2)\xi^2}, \tag{21}$$

where

$$\xi = \left(\frac{m\omega}{\hbar}\right)^{1/2} q \tag{22}$$

and

$$H_n(\xi) = (-1)^n e^{\xi^2}\left(\frac{d}{d\xi}\right)^n e^{-\xi^2}. \tag{23}$$

The matrix elements of the operator $\exp(-\beta\hat{H})$ in the q-representation are given by

$$\begin{aligned}\langle q|e^{-\beta\hat{H}}|q'\rangle &= \sum_{n=0}^{\infty} e^{-\beta E_n}\varphi_n(q)\,\varphi_n(q') \\ &= \left(\frac{m\omega}{\pi\hbar}\right)^{1/2} e^{-(1/2)(\xi^2+\xi'^2)} \sum_{n=0}^{\infty}\left\{e^{-(n+1/2)\beta\hbar\omega}\frac{H_n(\xi)\,H_n(\xi')}{(2^n n!)}\right\}.\end{aligned} \tag{24}$$

The summation over n is somewhat difficult to evaluate; nevertheless, the final result is*

$$\begin{aligned}&\langle q|e^{-\beta\hat{H}}|q'\rangle \\ &= \left[\frac{m\omega}{2\pi\hbar\sinh(\beta\hbar\omega)}\right]^{1/2}\exp\left[-\frac{m\omega}{4\hbar}\left\{(q+q')^2\tanh\left(\frac{\beta\hbar\omega}{2}\right)+(q-q')^2\coth\left(\frac{\beta\hbar\omega}{2}\right)\right\}\right].\end{aligned} \tag{25}$$

From here, we readily obtain

$$\begin{aligned}\mathrm{Tr}\left(e^{-\beta\hat{H}}\right) &= \int_{-\infty}^{\infty}\langle q|e^{-\beta\hat{H}}|q\rangle\,dq \\ &= \left[\frac{m\omega}{2\pi\hbar\sinh(\beta\hbar\omega)}\right]^{1/2}\int_{-\infty}^{\infty}\exp\left[-\frac{m\omega q^2}{\hbar}\tanh\left(\frac{\beta\hbar\omega}{2}\right)\right]dq \\ &= \frac{1}{2\sinh(\frac{1}{2}\beta\hbar\omega)} = \frac{e^{-(1/2)\beta\hbar\omega}}{1-e^{-\beta\hbar\omega}},\end{aligned} \tag{26}$$

which is indeed the *partition function* of a linear harmonic oscillator; see eqn. (3.7.14). At the same time, we find that the *probability density* for the oscillator coordinate to be in the vicinity of a particular value q is given by

$$\langle q|\hat{\varrho}|q\rangle = \left[\frac{m\omega\tanh(\frac{1}{2}\beta\hbar\omega)}{\pi\hbar}\right]^{1/2}\exp\left[-\frac{m\omega q^2}{\hbar}\tanh\left(\frac{\beta\hbar\omega}{2}\right)\right]; \tag{27}$$

we note that this is a Gaussian distribution in the variable q, with mean value zero and the root-mean-square deviation

$$q_{\text{r.m.s.}} = \left[\frac{\hbar}{2m\omega\tanh(\frac{1}{2}\beta\hbar\omega)}\right]^{1/2}. \tag{28}$$

* The mathematical details of this derivation can be found in R. Kubo (1965), pp. 175–7.

The probability distribution (27) was first derived by Bloch in 1932. In the classical limit ($\beta\hbar\omega \ll 1$), the distribution becomes *purely thermal*—free from quantum effects:

$$\langle q \,|\, \hat{\varrho} \,|\, q\rangle \simeq \left(\frac{m\omega^2}{2\pi kT}\right)^{1/2} \exp\left[-\frac{m\omega^2 q^2}{2kT}\right], \tag{29}$$

with dispersion $(kT/m\omega^2)^{1/2}$. We note that the limiting distribution (29) is precisely the one expected on the basis of classical statistics. At the other extreme ($\beta\hbar\omega \gg 1$), the distribution becomes *purely quantum-mechanical*—free from thermal effects:

$$\langle q \,|\, \hat{\varrho} \,|\, q\rangle \simeq \left(\frac{m\omega}{\pi\hbar}\right)^{1/2} \exp\left[-\frac{m\omega q^2}{\hbar}\right], \tag{30}$$

with dispersion $(\hbar/2m\omega)^{1/2}$. Again we note that the limiting distribution (30) is precisely the one expected for an oscillator in its ground state ($n = 0$), viz. a distribution with probability density $\varphi_0^2(q)$; see eqns. (21)–(23).

In view of the fact that the mean energy of the oscillator is given by

$$\langle H\rangle = -\frac{\partial}{\partial\beta} \ln \operatorname{Tr}\left(e^{-\beta\hat{H}}\right) = \frac{1}{2}\hbar\omega \coth\left(\frac{1}{2}\beta\hbar\omega\right), \tag{31}$$

we observe that the temperature dependence of the distribution (27) is *solely* determined through the expectation value $\langle H\rangle$. Actually, we can write

$$\langle q \,|\, \hat{\varrho} \,|\, q\rangle = \left(\frac{m\omega^2}{2\pi\langle H\rangle}\right)^{1/2} \exp\left[-\frac{m\omega^2 q^2}{2\langle H\rangle}\right], \tag{27a}$$

with

$$q_{\text{r.m.s.}} = \left(\frac{\langle H\rangle}{m\omega^2}\right)^{1/2}. \tag{28a}$$

From (27a) we obtain for the mean value of the potential energy ($\frac{1}{2}m\omega^2 q^2$) of the oscillator:

$$\left\langle \frac{1}{2} m\omega^2 q^2 \right\rangle = \left(\frac{m\omega^2}{2\pi\langle H\rangle}\right)^{1/2} \int_{-\infty}^{\infty} \exp\left[-\frac{m\omega q^2}{2\langle H\rangle}\right]\left(\frac{1}{2} m\omega^2 q^2\right) dq$$

$$= \frac{1}{2} m\omega^2 q^2_{\text{r.m.s.}} = \frac{1}{2}\langle H\rangle. \tag{32}$$

Naturally, then, the mean value of the kinetic energy ($p^2/2m$) must also be equal to $\frac{1}{2}\langle H\rangle$:

$$\left\langle -\frac{\hbar^2}{2m}\frac{\partial^2}{\partial q^2}\right\rangle = \frac{1}{2}\langle H\rangle. \tag{33}$$

The foregoing results are in complete accord with the *virial theorem*; see Problem 2.13a.

Finally, we would like to make some remarks on the formulae (25) and (26). For ($\beta\hbar\omega$) $\ll 1$, we obtain

$$\langle q \,|\, e^{-\beta\hat{H}} \,|\, q'\rangle \simeq \left(\frac{m}{2\pi\hbar^2\beta}\right)^{1/2} \exp\left[-\frac{\beta m\omega^2}{8}(q+q')^2 - \frac{m}{2\hbar^2\beta}(q-q')^2\right] \tag{25a}$$

and

$$\mathrm{Tr}\,(e^{-\beta\hat{H}}) \simeq (\beta\hbar\omega)^{-1}. \tag{26a}$$

Equation (25a), with $\omega = 0$, corresponds exactly to eqn. (14) for the free particle. This extreme situation, however, does not fit with the expression (26a); actually, the corresponding result for the free particle has to be obtained directly from the reduced expression for $\langle q|e^{-\beta\hat{H}}|q'\rangle$. Formally, however, one notices that this amounts to replacing the parameter ω of this problem by the parameter $L^{-1}(2\pi/m\beta)^{1/2}$ of that one; here, L is the length of the one-dimensional box in which the free particle is contained. We also note that the partition function (26a) yields for the mean energy of the oscillator: $\langle H\rangle = kT$. At the other extreme ($\beta\hbar\omega \gg 1$), the formulae (25) and (26) become

$$\langle q\,|\,e^{-\beta\hat{H}}\,|\,q'\rangle \simeq \left(\frac{m\omega}{\pi\hbar}\right)^{1/2} e^{-(1/2)\beta\hbar\omega} \exp\left[-\frac{m\omega}{2\hbar}(q^2+q'^2)\right] \tag{25b}$$

and

$$\mathrm{Tr}\left(e^{-\beta\hat{H}}\right) \simeq e^{-(1/2)\beta\hbar\omega}\,; \tag{26b}$$

as expected, the matrix element (25b) is now directly related to the product of the ground state wave functions $\varphi_0(q)$ and $\varphi_0(q')$. Moreover, the partition function (26b) now yields a value of $\frac{1}{2}\hbar\omega$ for the mean energy of the oscillator.

5.4. Systems composed of indistinguishable particles

We shall now formulate the quantum-mechanical description of a system of N identical particles. To fix ideas, it seems advisable to study a system which is physically simple-minded, such as a gas of *noninteracting* particles. The findings of this study will be of relevance to other systems as well.

Now, the Hamiltonian of a system of N noninteracting particles is simply a sum of the individual single-particle Hamiltonians:

$$\hat{H}(\boldsymbol{q},\boldsymbol{p}) = \sum_{i=1}^{N} \hat{H}_i(q_i, p_i); \tag{1}$$

here, (q_i, p_i) are the coordinates and momenta of the ith particle while $\hat{H}_i$ is its Hamiltonian.* Since the particles are identical, the Hamiltonians $\hat{H}_i$ $(i = 1, 2, \ldots, N)$ are *formally* the same; they only differ in the values of their arguments. The time-independent Schrödinger equation for the system is

$$\hat{H}\psi_E(\boldsymbol{q}) = E\psi_E(\boldsymbol{q}), \tag{2}$$

where E is an eigenvalue of the Hamiltonian of the system and $\psi_E(\boldsymbol{q})$ the corresponding eigenfunction. In view of (1), we can write a straightforward solution of the Schrödinger equation, namely

$$\psi_E(\boldsymbol{q}) = \prod_{i=1}^{N} u_{\varepsilon_i}(q_i), \tag{3}$$

* We are studying here a single-component system composed of "spinless" particles. Generalizations to a system composed of particles with spin and to a system composed of two or more components is quite straightforward.

with

$$E = \sum_{i=1}^{N} \varepsilon_i; \tag{4}$$

the factor $u_{\varepsilon_i}(q_i)$ in (3) is an eigenfunction of the single-particle Hamiltonian $\hat{H}_i(q_i, p_i)$, with eigenvalue ε_i:

$$\hat{H}_i u_{\varepsilon_i}(q_i) = \varepsilon_i u_{\varepsilon_i}(q_i). \tag{5}$$

Thus, a stationary state of the given system may be described in terms of the single-particle states of the constituent particles. In general, we may do so by specifying the distribution set $\{n_i\}$ to represent a particular state of the system; this would imply that there are n_i particles in the eigenstate characterized by the energy value ε_i $(i = 0, 1, 2, \ldots)$. Clearly, the distribution set $\{n_i\}$ must conform to the conditions:

$$\sum_i n_i = N \tag{6}$$

and

$$\sum_i n_i \varepsilon_i = E. \tag{7}$$

Accordingly, the wave function of this state may be written as

$$\psi_E(\boldsymbol{q}) = \prod_{m=1}^{n_1} u_1(m) \prod_{m=n_1+1}^{n_1+n_2} u_2(m) \ldots, \tag{8}$$

where the symbol $u_i(m)$ stands for the single-particle wave function $u_{\varepsilon_i}(q_m)$.

Suppose we effect a permutation among the N coordinates appearing on the right-hand side of (8); as a result, the numbers $(1, 2, \ldots, N)$ get replaced by the numbers $(P1, P2, \ldots, PN)$, say. The resulting wave function, which we shall call $P\psi_E(\boldsymbol{q})$, will be

$$P\psi_E(\boldsymbol{q}) = \prod_{m=1}^{n_1} u_1(Pm) \prod_{m=n_1+1}^{n_1+n_2} u_2(Pm) \ldots. \tag{9}$$

In classical physics, where the particles of a given system, though identical, are regarded as mutually *distinguishable*, any permutation that brings about an interchange of particles in two *different* single-particle states is recognized to have led to a *new, physically distinct* microstate of the system. For example, classical physics regards a microstate in which the so-called 5th particle is in the state u_i and the so-called 7th particle in the state u_j $(j \neq i)$ as *distinct* from a microstate in which the 7th particle is in the state u_i and the 5th in the state u_j. This leads to

$$\frac{N!}{n_1!\, n_2! \ldots} \tag{10}$$

(supposedly distinct) microstates of the system, corresponding to a given mode of distribution $\{n_i\}$. The number (10) would then be ascribed as a "statistical weight factor" to the distribution set $\{n_i\}$. Of course, one must not forget that a "correction" applied by Gibbs, which has been discussed in Secs. 1.5 and 1.6, reduces this weight factor to

$$W_c\{n_i\} = \frac{1}{n_1!\, n_2! \ldots}. \tag{11}$$

And the only way to understand the physical basis of this "correction" is in terms of the inherent *indistinguishability* of the particles.

According to quantum physics, however, the situation remains unsatisfactory even after the Gibbs correction has been incorporated. According to a genuine quantum-mechanical point of view, an interchange among identical particles, *even if* they are in different single-particle states, should not lead to any new microstate of the system! Thus, if we want to take into account the indistinguishability of the particles properly, we must not regard a microstate in which the "5th" particle is in the state u_i and the "7th" in the state u_j as distinct from a microstate in which the "7th" particle is in the state u_i and the "5th" in the state u_j (*even if* $i \neq j$), for the labeling of the particles as No. 1, No. 2, etc. (which one often resorts to) is at most a matter of convenience—it is hardly a matter of reality. In other words, all that matters in the description of a particular state of the given system is the set of numbers n_i which tell us "*how many* particles there are in the various single-particle states u_i"; the question "*which* particle is in *which* single-particle state" makes no relevance at all. Accordingly, the microstates resulting from any permutation P among the N particles (so long as the numbers n_i remain the same) must be regarded as *one and the same* microstate. For the same reason, the weight factor associated with a distribution set $\{n_i\}$, provided that the set is not disallowed on other physical grounds, should be identically equal to unity, whatever the values of the numbers n_i;

$$W_q\{n_i\} \equiv 1. \qquad (12)^*$$

Indeed, if for some other reason the set $\{n_i\}$ is physically disallowed, then the weight factor W_q should be identically equal to zero; see eqn. (19).

At the same time, a wave function of the type (8), which we may call *Boltzmannian* and denote by the symbol $\psi_{\text{Boltz}}(\boldsymbol{q})$, is inappropriate for describing the state of a system composed of indistinguishable particles, because an interchange of arguments among the factors u_i and u_j, where $i \neq j$, would lead to a wave function which is physically different from the one we started with. Now, since a mere interchange of the particle coordinates must not lead to a new microstate of the system, the wave function $\psi_E(\boldsymbol{q})$ must be constructed in such a way that, for all practical purposes, it is insensitive to any interchange among the arguments. The simplest way to do this is to set up a linear combination of all the $N!$ functions of the type (9) which obtain from (8) by all possible permutations among its arguments; of course, the combination must be such that if a permutation of coordinates is carried out in it, then the wave functions ψ and $P\psi$ must satisfy the property

$$|P\psi|^2 = |\psi|^2. \qquad (13)$$

This leads to the following possibilities:

(i) $$P\psi = \psi \quad \text{for all } P, \qquad (14)$$

* It may be mentioned here that as early as in 1905 Ehrenfest pointed out that to obtain Planck's formula for the black-body radiation one must assign equal *a priori* probabilities to the various states $\{n_i\}$.

which means that the wave function is *symmetric* in all its arguments, or

(ii)
$$P\psi = \begin{cases} +\psi & \text{if} \quad P \text{ is an } \textit{even} \text{ permutation,} \\ -\psi & \text{if} \quad P \text{ is an } \textit{odd} \text{ permutation,} \end{cases} \tag{15*}$$

which means that the wave function is *antisymmetric* in its arguments. We call these wave functions ψ_S and ψ_A, respectively; their mathematical structure is given by

$$\psi_S(q) = \text{const.} \sum_P P\psi_{\text{Boltz}}(q) \tag{16}$$

and

$$\psi_A(q) = \text{const.} \sum_P \delta_P P\psi_{\text{Boltz}}(q), \tag{17}$$

where δ_P in the expression for ψ_A is equal to $+1$ or -1 according as the permutation P is even or odd.

We note that the function $\psi_A(q)$ can be written in the form of a *Slater determinant*:

$$\psi_A(q) = \text{const.} \begin{vmatrix} u_i(1) & u_i(2) & \dots & u_i(N) \\ u_j(1) & u_j(2) & \dots & u_j(N) \\ \cdot & \cdot & \dots & \cdot \\ \cdot & \cdot & \dots & \cdot \\ \cdot & \cdot & \dots & \cdot \\ u_l(1) & u_l(2) & \dots & u_l(N) \end{vmatrix}, \tag{18}$$

where the leading diagonal is precisely the Boltzmannian wave function while the other terms of the expansion are the various possible permutations thereof; positive and negative signs in the combination (17) appear automatically as we expand the determinant. On interchanging a pair of arguments (which amounts to interchanging the corresponding columns of the Slater determinant), the wave function ψ_A merely changes its sign, as it indeed should. However, if two or more particles happen to be in the same single-particle state, the corresponding rows of the determinant become identical and the wave function identically vanishes.† Such a state is physically impossible to realize. We conclude that if a system composed of indistinguishable particles is characterized by an antisymmetric wave function, then the particles of the system must all be in different single-particles states—a result equivalent to *Pauli's exclusion principle* for electrons. Conversely, a statistical system composed of particles obeying an exclusion principle must be described by a wave function which is antisymmetric in all its arguments. The statistics governing the behavior of this class of particles is called *Fermi–Dirac*, or simply *Fermi*; statistics and the particles themselves are referred to as *fermions*. The statistical weight

* An even (odd) permutation is one which can be arrived at from the original order by an even (odd) number of "pair interchanges" among the arguments. For example, of the six permutations

(1, 2, 3), (2, 3, 1), (3, 1, 2), (1, 3, 2), (3, 2, 1), and (2, 1, 3),

of the arguments 1, 2 and 3, the first three are *even* permutations while the last three are *odd*. A single interchange, among any two arguments, is clearly an *odd* permutation.

† This is directly related to the fact that if we effect an interchange of two particles in the *same* single-particle state, then $P\psi_A$ must obviously be identical with ψ_A. At the same time, if we also have: $P\psi_A = -\psi_A$, then ψ_A must be identically zero.

factor $W_{\text{F.D.}}\{n_i\}$ for such a system is unity so long as the n_i's in the distribution set are either 0 or 1; otherwise, the weight factor is zero:

$$W_{\text{F.D.}}\{n_i\} = \begin{cases} 1 & \text{if} \quad \sum_i n_i^2 = N, \\ 0 & \text{if} \quad \sum_i n_i^2 > N. \end{cases} \tag{19}$$

(19)*

No such problems arise in the case of systems characterized by symmetric wave functions; in particular, we have no restriction whatsoever on the values of the numbers n_i. The statistics governing the behavior of this class of particles is called *Bose–Einstein*, or simply *Bose*, statistics and the particles themselves are referred to as *bosons*. The weight factor $W_{\text{B.E.}}\{n_i\}$ in this case is identically equal to unity, whatever the values of the numbers n_i:

$$W_{\text{B.E.}}\{n_i\} = 1; \qquad n_i = 0, 1, 2, \ldots . \tag{20}$$

It should be pointed out here that there exists an intimate connection between the statistics governing a particular species of particles and the intrinsic spin of the particles. For instance, particles with an integral spin (in units of $\hbar$, of course) obey Bose–Einstein statistics, while particles with a half-odd integral spin obey Fermi–Dirac statistics. Examples in the first category are photons, phonons, π-mesons, gravitons, He^4-atoms, etc., while those in the second category are electrons, nucleons (protons and neutrons), μ-mesons, neutrinos, He^3-atoms, etc.

Finally, it must be emphasized that although we have derived our conclusions here on the basis of a study of noninteracting systems, the basic results hold for interacting systems as well. In general, the desired wave function $\psi(\boldsymbol{q})$ may not be obtainable in terms of single-particle wave functions; nonetheless, it must be either of the kind $\psi_S(\boldsymbol{q})$, satisfying eqn. (14), or of the kind $\psi_A(\boldsymbol{q})$, satisfying eqn. (15).

5.5. The density matrix and the partition function of a system of free particles†

Suppose that the given system, which is composed of N indistinguishable, noninteracting particles confined to a cubical box of volume V, is a member of a canonical ensemble characterized by the temperature parameter β. The *density matrix* of the system in the coordinate representation will be given by

$$\langle \boldsymbol{r}_1, \ldots, \boldsymbol{r}_N | \hat{\varrho} | \boldsymbol{r}_1', \ldots, \boldsymbol{r}_N' \rangle = \frac{1}{Q_N(\beta)} \langle \boldsymbol{r}_1, \ldots, \boldsymbol{r}_N | e^{-\beta \hat{H}} | \boldsymbol{r}_1', \ldots, \boldsymbol{r}_N' \rangle, \tag{1}$$

where $Q_N(\beta)$ is the *partition function* of the system:

$$Q_N(\beta) = \text{Tr}\,(e^{-\beta\hat{H}}) = \int \langle \boldsymbol{r}_1, \ldots, \boldsymbol{r}_N | e^{-\beta\hat{H}} | \boldsymbol{r}_1, \ldots, \boldsymbol{r}_N \rangle \, d^{3N}r. \tag{2}$$

For brevity, we denote the vector $\boldsymbol{r}_i$ by the letter i and the primed vector $\boldsymbol{r}_i'$ by i'. Further, let $\psi_E(1, \ldots, N)$ denote the eigenfunctions of the Hamiltonian, the suffix E representing

* We note that the condition $\Sigma_i n_i^2 = N$ necessarily implies that all the n_i's are either 0 or 1. On the other hand, if any of the n_i's is greater than 1, the sum $\Sigma_i n_i^2$ is necessarily greater than N.

† For a general survey of the density matrix and its applications, see Husimi (1940) and ter Haar (1961).

the corresponding eigenvalues. We then have

$$\langle 1, \ldots, N \left| e^{-\beta \hat{H}} \right| 1', \ldots, N' \rangle = \sum_E e^{-\beta E}[\psi_E(1, \ldots, N)\psi_E^*(1', \ldots, N')], \tag{3}$$

where the summation goes over all possible values of E.

Since the particles constituting the given system are noninteracting, we may express the eigenfunctions $\psi_E(1, \ldots, N)$ and the eigenvalues E in terms of the single-particle wave functions $u_i(m)$ and the single-particle energies ε_i. Moreover, we find it advisable to work with the wave vectors $\boldsymbol{k}_i$ rather than with the energies ε_i; so we write

$$E = \frac{\hbar^2 K^2}{2m} = \frac{\hbar^2}{2m}(k_1^2 + k_2^2 + \ldots + k_N^2), \tag{4}$$

where the k's on the right-hand side are the wave vectors of the individual particles. Imposing periodic boundary conditions, we obtain for the *normalized* single-particle wave functions

$$u_k(\boldsymbol{r}) = V^{-1/2} \exp\{i(\boldsymbol{k} \cdot \boldsymbol{r})\} \tag{5}$$

with

$$\boldsymbol{k} = 2\pi V^{-1/3}\boldsymbol{n}; \tag{6}$$

here, $\boldsymbol{n}$ stands for a three-dimensional vector whose components can have values 0, ± 1, ± 2, ... The wave function ψ of the total system would then be, see eqns. (5.4.3), (5.4.16) and (5.4.17),

$$\psi_K(1, \ldots, N) = (N!)^{-1/2} \sum_P \delta_P P\{u_{k_1}(1) \ldots u_{k_N}(N)\}, \tag{7}$$

where the magnitudes of the individual $\boldsymbol{k}$'s are such that

$$(k_1^2 + \ldots + k_N^2) = K^2. \tag{8}$$

The number δ_P in the expression for ψ_K is identically equal to $+1$ if the particles are bosons. For fermions, on the other hand, it is equal to $+1$ or -1 according as the permutation P is even or odd. Thus, quite generally,

$$\delta_P = (\pm 1)^{[P]}, \tag{9}$$

where $[P]$ denotes the order of the permutation; note that the upper sign in this equation holds for bosons while the lower sign holds for fermions. The factor $(N!)^{-1/2}$ has been introduced here to secure the normalization of the total wave function.

Now, it makes no difference to the wave function (7) whether the permutations P are carried out on the coordinates $1, \ldots, N$ or on the wave vectors $\boldsymbol{k}_1, \ldots, \boldsymbol{k}_N$, because after all we are going to sum over *all* the $N!$ permutations. Denoting the permuted coordinates by $P1, \ldots, PN$ and the permuted wave vectors by $P\boldsymbol{k}_1, \ldots, P\boldsymbol{k}_N$, eqn. (7) can be written as

$$\psi_K(1, \ldots, N) = (N!)^{-1/2} \sum_P \delta_P\{u_{k_1}(P1) \ldots u_{k_N}(PN)\} \tag{10a}$$

$$= (N!)^{-1/2} \sum_P \delta_P\{u_{Pk_1}(1) \ldots u_{Pk_N}(N)\}. \tag{10b}$$

Equations (10) may now be substituted into (3). We get

$$\langle 1, \ldots, N | e^{-\beta \hat{H}} | 1', \ldots, N' \rangle$$
$$= (N!)^{-1} \sum_K e^{-\beta\hbar^2 K^2/2m} \left[\sum_P \delta_P \{u_{k_1}(P1) \ldots u_{k_N}(PN)\} \sum_{\tilde{P}} \delta_{\tilde{P}} \{u^*_{\tilde{P}k_1}(1') \ldots u^*_{\tilde{P}k_N}(N')\} \right], \tag{11}$$

where P and $\tilde{P}$ are any of the $N!$ possible permutations. Now, since a permutation among the $\boldsymbol{k}$'s changes the wave function ψ at most by a sign, the quantity $[\psi\psi^*]$ in (11) is insensitive to such a permutation; the same holds for the exponential factor as well. The summation over K is, therefore, equivalent to $(1/N!)$ times a summation over all the vectors $\boldsymbol{k}_1, \ldots, \boldsymbol{k}_N$ *independently of one another*. Next, in view of the N-fold summation over the $\boldsymbol{k}$'s, all the permutations $\tilde{P}$ will make equal contributions towards the sum (because they differ from one another only in the ordering of the $\boldsymbol{k}$'s). Therefore, we may consider only one of these permutations, say the one for which $\tilde{P}\boldsymbol{k}_1 = \boldsymbol{k}_1, \ldots, \tilde{P}\boldsymbol{k}_N = \boldsymbol{k}_N$ (and hence $\delta_{\tilde{P}} = 1$ for both the statistics), and include a factor of $(N!)$ along. This factor will just cancel the previous factor $(1/N!)$, with the result

$$\langle 1, \ldots, N | e^{-\beta \hat{H}} | 1', \ldots, N' \rangle$$
$$= (N!)^{-1} \sum_{k_1, \ldots, k_N} e^{-\beta\hbar^2(k_1^2 + \ldots + k_N^2)/2m} \left[\sum_P \delta_P \{u_{k_1}(P1)\, u^*_{k_1}(1')\} \ldots \{u_{k_N}(PN)\, u^*_{k_N}(N')\} \right]. \tag{12}$$

Substituting from (5) and noting that, in view of the largeness of V, the summations over $\boldsymbol{k}$'s may be replaced by corresponding integrations, according to the rule

$$\sum_k \to \frac{V}{(2\pi)^3} \int d^3k, \tag{13}$$

eqn. (12) becomes

$$\langle 1, \ldots, N | e^{-\beta \hat{H}} | 1', \ldots, N' \rangle$$
$$= \frac{1}{N!\,(2\pi)^{3N}} \sum_P \delta_P \left[\int e^{-\beta\hbar^2 k_1^2/2m + i k_1 \cdot (P1 - 1')}\, d^3k_1 \ldots \int e^{-\beta\hbar^2 k_N^2/2m + i k_N \cdot (PN - N')}\, d^3k_N \right]$$
$$= \frac{1}{N!} \left(\frac{m}{2\pi\beta\hbar^2} \right)^{3N/2} \sum_P \delta_P [f(P1 - 1') \ldots f(PN - N')], \tag{14}$$

where

$$f(\xi) = \exp\left(-\frac{m}{2\beta\hbar^2} \xi^2 \right). \tag{15}$$

Here, use has been made of the mathematical result (5.3.14), which is also a special case of the present formula, viz. for $N = 1$.

Introducing the mean thermal wavelength

$$\lambda = \frac{h}{(2\pi mkT)^{1/2}} = \hbar \left(\frac{2\pi\beta}{m} \right)^{1/2}, \tag{16}$$

and rewriting our coordinates as $\boldsymbol{r}_1, \ldots, \boldsymbol{r}_N$, the diagonal elements among (14) take the form

$$\langle \boldsymbol{r}_1, \ldots, \boldsymbol{r}_N | e^{-\beta\hat{H}} | \boldsymbol{r}_1, \ldots, \boldsymbol{r}_N \rangle = \frac{1}{N!\,\lambda^{3N}} \sum_P \delta_P [f(P\boldsymbol{r}_1 - \boldsymbol{r}_1) \ldots f(P\boldsymbol{r}_N - \boldsymbol{r}_N)], \quad (17)$$

where

$$f(\boldsymbol{r}) = \exp(-\pi r^2/\lambda^2). \quad (18)$$

To obtain the partition function of the system, we must integrate (17) over all the space coordinates involved. However, before we do that, we would like to make some observations on the summation Σ_P. First of all, we note that the leading term in this summation, namely the one for which $P\boldsymbol{r}_i = \boldsymbol{r}_i$, is identically equal to unity (because $f(0) = 1$). This is followed by a group of terms in which a *single* pair interchange (among the coordinates) has taken place; a typical term in this group will be $f(\boldsymbol{r}_j - \boldsymbol{r}_i) f(\boldsymbol{r}_i - \boldsymbol{r}_j)$ where $i \neq j$. This group of terms is followed by other groups of terms in which *more than one* pair interchanges (among the coordinates) have taken place. Thus, we can write

$$\sum_P = 1 \pm \sum_{i<j} f_{ij} f_{ji} + \sum_{i<j<k} f_{ij} f_{jk} f_{ki} \pm \ldots, \quad (19)$$

where $f_{ij} \equiv f(\boldsymbol{r}_i - \boldsymbol{r}_j)$; again note that the upper (lower) signs in this expansion correspond to a system of bosons (fermions). Now, the function f_{ij} vanishes rapidly as the distance r_{ij} becomes much larger than the mean thermal wavelength λ. It then follows that if the mean interparticle distance $(V/N)^{1/3}$ in the system is much larger than the mean thermal wavelength, i.e. if

$$\frac{nh^3}{(2\pi mkT)^{3/2}} \ll 1, \quad (20)$$

where n is the particle density in the system, then the sum Σ_P might be approximated by unity. Accordingly, the partition function of the system becomes, see eqn. (17),

$$Q_N(V, T) \equiv \mathrm{Tr}\,(e^{-\beta\hat{H}}) \simeq \frac{1}{N!\,\lambda^{3N}} \int 1\,(d^{3N}r) = \frac{1}{N!}\left(\frac{V}{\lambda^3}\right)^N. \quad (21)$$

This is precisely the result obtained earlier for the ideal *classical* gas; see eqn. (3.5.9). Thus, we have obtained from our quantum-mechanical treatment the precise classical limit for the partition function $Q_N(V, T)$. Incidentally, we have achieved something more. Firstly, we have automatically recovered here the Gibbs correction factor $(1/N!)$ which was introduced into the classical treatment on an *ad hoc*, semi-empirical basis. We, of course, tried to understand its origin in terms of the inherent indistinguishability of the particles. Here, on the other hand, we have seen it coming in a very natural manner and its source indeed lies in the symmetrization of the wave functions of the system (which is ultimately related to the indistinguishability of the particles); cf. Problem 5.5. Secondly, we find here a formal, rigorous justification for computing the number of microstates of a system corresponding to a given region of its phase space by dividing the volume of that region into cells of a "suitable" size and then counting instead the number of these cells. This correspondence becomes all the more transparent if we note that formula (21) is exactly equivalent to the classical expression

$$Q_N(V, T) = \frac{1}{N!}\int e^{-\beta(p_1^2+\ldots+p_N^2)/2m}\left(\frac{d^{3N}q\, d^{3N}p}{\omega_0}\right), \tag{22}$$

where $\omega_0 = h^{3N}$. Thirdly, in deriving the classical limit we have also evolved a criterion which enables us to determine whether a given physical system can be treated classically; mathematically, this criterion is given by the condition (20). Now, in statistical mechanical studies, a system which cannot be treated classically is said to be *degenerate*; the quantity $(n\lambda^3)$ is, therefore, referred to as the *degeneracy discriminant*. Accordingly, the condition that classical considerations may be applicable to a given physical system is that "the value of the degeneracy discriminant of the system must be much less than unity".

Next, we note that, in the classical limit, the diagonal elements of the density matrix are given by

$$\langle \boldsymbol{r}_1, \ldots, \boldsymbol{r}_N \,|\hat{\varrho}|\, \boldsymbol{r}_1, \ldots, \boldsymbol{r}_N\rangle \simeq \left(\frac{1}{V}\right)^N, \tag{23}$$

which is simply a product of N factors, each equal to $(1/V)$. Recalling that for a particle enclosed in a box of volume V, $\langle \boldsymbol{r}\,|\hat{\varrho}|\,\boldsymbol{r}\rangle = (1/V)$, see eqn. (5.3.16), we conclude that in the classical limit there is no spatial correlation between the various particles of the system. In general, however, spatial correlations exist even if the particles are supposedly noninteracting; they arise because of the symmetry properties of the wave functions and their magnitude is quite significant if the interparticle distances in the system are comparable with the mean thermal wavelength of the particles. To see it more clearly, we may consider the simplest relevant case, namely the one with $N = 2$. In this case, the sum Σ_P is exactly equal to $1 \pm [f(r_{12})]^2$. Accordingly,

$$\langle \boldsymbol{r}_1, \boldsymbol{r}_2 \left| e^{-\beta\hat{H}} \right| \boldsymbol{r}_1, \boldsymbol{r}_2\rangle = \frac{1}{2\lambda^6}\,[1 \pm \exp(-2\pi r_{12}^2/\lambda^2)] \tag{24}$$

and hence

$$\begin{aligned} Q_2(V, T) &= \frac{1}{2\lambda^6}\iint [1 \pm \exp(-2\pi r_{12}^2/\lambda^2)]\, d^3r_1\, d^3r_2 \\ &= \frac{1}{2}\left(\frac{V}{\lambda^3}\right)^2\left[1 \pm \frac{1}{V}\int_0^\infty \exp(-2\pi r^2/\lambda^2)\, 4\pi r^2\, dr\right] \\ &= \frac{1}{2}\left(\frac{V}{\lambda^3}\right)^2\left[1 \pm \frac{1}{2^{3/2}}\left(\frac{\lambda^3}{V}\right)\right]. \end{aligned} \tag{25}$$

The term in (λ^3/V) is generally negligible, so we may write

$$Q_2(V, T) \simeq \frac{1}{2}\left(\frac{V}{\lambda^3}\right)^2. \tag{26}$$

Combining (24) and (26), we obtain

$$\langle \boldsymbol{r}_1, \boldsymbol{r}_2 \,|\hat{\varrho}|\, \boldsymbol{r}_1, \boldsymbol{r}_2\rangle = \frac{1}{V^2}\,[1 \pm \exp(-2\pi r_{12}^2/\lambda^2)]. \tag{27}$$

Thus, if r_{12} is comparable to λ, the probability density (27) may differ considerably from the classical value $(1/V)^2$. In particular, the probability density for a pair of *bosons* to be a distance r apart is larger than the classical, r-independent value by a factor of $[1+\exp(-2\pi r^2/\lambda^2)]$ which becomes as high as 2 as $r \to 0$. The corresponding result for a pair of *fermions* is smaller than the classical value by a factor of $[1-\exp(-2\pi r^2/\lambda^2)]$ which becomes as low as 0 as $r \to 0$. Thus, we obtain a *positive* spatial correlation among particles obeying Bose statistics and a *negative* spatial correlation among particles obeying Fermi statistics; see also Sec. 6.3.

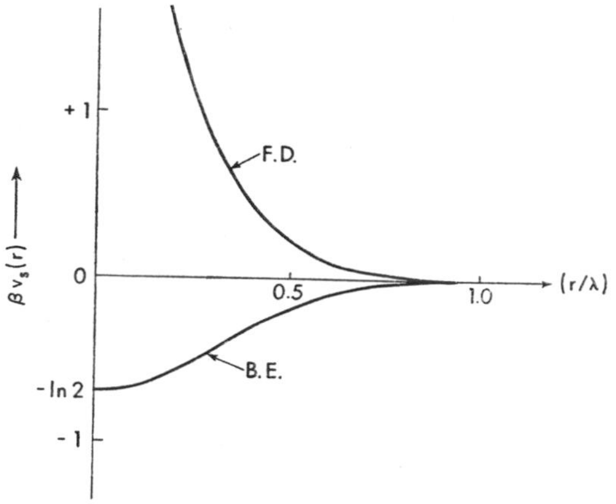

FIG. 5.1. The statistical potential $v_s(r)$ between a pair of particles obeying Bose–Einstein statistics or Fermi–Dirac statistics.

Another way of expressing correlations (among otherwise noninteracting particles) is by introducing a *statistical* interparticle potential $v_s(r)$ and then treating the particles classically (Uhlenbeck and Gropper, 1932). The potential $v_s(r)$ must be such that the Boltzmann factor $\exp(-\beta v_s)$ is precisely equal to the correlation factor $[\ldots]$ in (27), i.e.

$$v_s(r) = -kT\ln[1\pm\exp(-2\pi r^2/\lambda^2)]. \tag{28}$$

Figure 5.1 shows a plot of the statistical potential $v_s(r)$ for a pair of particles obeying Bose–Einstein statistics or Fermi–Dirac statistics. In the Bose case, the potential is throughout attractive, thus giving rise to a "statistical attraction" among bosons; in the Fermi case, it is throughout repulsive, giving rise to a "statistical repulsion" among fermions. In either case, the potential vanishes rapidly as r becomes larger than λ; accordingly, its influence becomes less important as the temperature of the system increases.

Problems

5.1. Evaluate the density matrix ϱ_{mn} of an electron spin in the representation which makes $\hat{\sigma}_x$ diagonal. Next, show that the value of $\langle\sigma_z\rangle$, resulting from this representation, is precisely the same as the one obtained in Sec. 5.3.

Hint. The representation needed here follows from the one used in Sec. 5.3 by carrying out a transformation with the help of the unitary operator

$$U = \begin{pmatrix} 1/\sqrt{2} & 1/\sqrt{2} \\ -1/\sqrt{2} & 1/\sqrt{2} \end{pmatrix}.$$

5.2. Verify eqn. (5.3.33) by making an explicit evaluation of the matrix elements of the operator on the left-hand side.

5.3. Prove that

$$\langle q \,|\, e^{-\beta \hat{H}} \,|\, q' \rangle \equiv \exp\left[-\beta \hat{H}\left(-i\hbar \frac{\partial}{\partial q}, q\right)\right] \delta(q-q'),$$

where $\hat{H}(-i\hbar\, \partial/\partial q, q)$ is the Hamiltonian of the system, in the q-representation, which formally operates upon the Dirac delta function $\delta(q-q')$. Writing δ-function in a suitable form, apply this result to (i) a free particle and (ii) a linear harmonic oscillator.

5.4. Derive the density matrix ϱ for (i) a free particle and (ii) a linear harmonic oscillator in the *momentum* representation and study its main properties along the lines of Sec. 5.3.

5.5. Study the density matrix and the partition function of a system of free particles, using the *unsymmetrized* wave function (5.4.3) instead of the *symmetrized* wave function (5.5.7). Show that, following this procedure, one obtains neither the Gibbs' correction factor ($1/N!$) nor a spatial correlation among the particles.

5.6. Show that in the *first* approximation the partition function of a system of N noninteracting, indistinguishable particles is given by

$$Q_N(V, T) = \frac{1}{N!\, \lambda^{3N}} Z_N(V, T),$$

where

$$Z_N(V, T) = \int \exp\left\{-\beta \sum_{i<j} v_s(r_{ij})\right\} d^{3N}r,$$

$v_s(r)$ being the statistical potential (5.5.28). Hence evaluate the *first-order* correction to the equation of state of this system.

5.7. Determine the values of the degeneracy discriminant ($n\lambda^3$) for hydrogen, helium and oxygen at N.T.P. Make an estimate of the respective temperature ranges where the magnitude of this quantity becomes comparable to unity and hence quantum effects become important.

5.8. Prove that the quantum-mechanical partition function for a system of N *interacting* particles approaches the classical form

$$Q_N(V, T) = \frac{1}{N!\, h^{3N}} \int e^{-\beta E(q, p)} (d^{3N}q\, d^{3N}p)$$

as the mean thermal wavelength λ becomes much smaller than (i) the mean interparticle distance $(V/N)^{1/3}$ and (ii) a characteristic length r_0 of the interparticle potential.*

5.9. Prove the following theorem due to Peierls.†

"If $\hat{H}$ is the hermitian Hamiltonian operator of a given physical system and $\{\varphi_n\}$ an arbitrary orthonormal set of wave functions satisfying the symmetry requirements and the boundary conditions of the problem, then the partition function of the system satisfies the following inequality:

$$Q(\beta) \geqslant \sum_n \exp\left\{-\beta \langle \varphi_n \,|\, \hat{H} \,|\, \varphi_n \rangle\right\};$$

the equality holds when $\{\varphi_n\}$ is the complete orthonormal set of eigenfunctions of the Hamiltonian itself."

* See K. Huang (1963), Sec. 10.2.
† R. E. Peierls (1938), *Phys. Rev.* **54**, 918. See also K. Huang (1963), Sec. 10.3.

CHAPTER 6

THE THEORY OF SIMPLE GASES

WE ARE now fully equipped with the formalism required for determining the macroscopic properties of a large variety of physical systems. In most cases, however, the derivations run into serious mathematical difficulties, with the result that one is forced to restrict one's analysis either to simpler kinds of systems or to simplified models of actual systems. In practice, even these restricted studies are carried out in a series of stages, the first stage of the process being highly "idealized". The best example of such an idealization is the familiar *ideal* gas, a study of which is not only helpful in acquiring facility with the mathematical procedures but also throws considerable light on the physical behavior of gases actually met with in nature. In fact, it also serves as a formal base on which the theory of real gases is ultimately founded; see Chapter 9.

In this chapter we propose to derive, and at some length discuss, the most basic properties of simple gaseous systems obeying quantum statistics; the discussion will include some of the essential features of diatomic and polyatomic gases as well.

6.1. An ideal gas in a quantum-mechanical microcanonical ensemble

We consider a gaseous system of N noninteracting, *indistinguishable* particles, confined to a space of volume V and sharing a given energy E. The statistical quantity of interest in this case is $\Omega(N, V, E)$ which, by definition, denotes the number of *distinct* microstates accessible to the given system under the macrostate (N, V, E). While determining this number for the present case, we have to be very careful because a failure to take into account the indistinguishability of the particles in a proper manner could lead to results which, except in the classical limit, may not be acceptable. With this in mind, we proceed as follows.

Since, for large V, the single-particle energy levels in the system must be extremely close to one another, we may divide the energy spectrum into a large number of "groups of levels", which may be referred to as the *energy cells*; see Fig. 6.1. Let ε_i denote the average energy of a level, and g_i the (arbitrary) number of levels, in the ith cell; we assume that all $g_i \gg 1$. In a particular situation, we may have n_1 particles in the first cell, n_2 particles in the second cell, and so on. Clearly, the distribution set $\{n_i\}$ must conform to the conditions

$$\sum_i n_i = N \tag{1}$$

and

$$\sum_i n_i \varepsilon_i = E. \tag{2}$$

Then

$$\Omega(N, V, E) = \sum_{\{n_i\}}{}' W\{n_i\}, \tag{3}$$

where $W\{n_i\}$ is the number of *distinct* microstates associated with the distribution set $\{n_i\}$

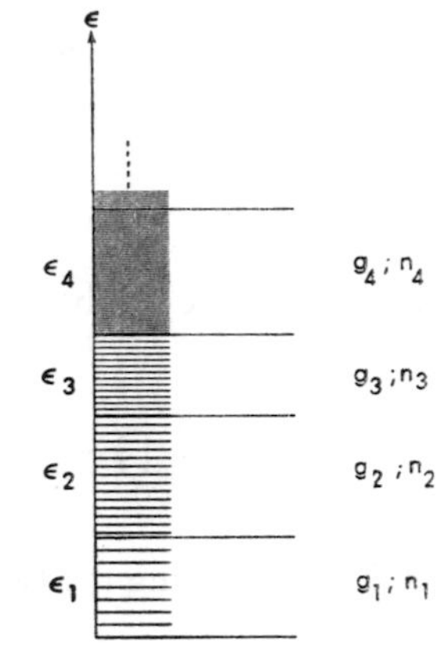

FIG. 6.1. The grouping of single-particle energy levels into "cells".

while the primed summation goes over all the distribution sets that conform to the conditions (1) and (2). Next,

$$W\{n_i\} = \prod_i w(i), \tag{4}$$

where $w(i)$ is the number of *distinct* microstates associated with the ith cell of the spectrum (the cell that contains n_i particles, to be accommodated among g_i levels) while the product goes over all the cells in the spectrum. Clearly, $w(i)$ is the number of *distinct* ways in which the n_i identical, and *indistinguishable*, particles can be distributed among the g_i levels of the ith cell. This number in the Bose–Einstein case is given by, see eqn. (3.7.25),

$$w_{\text{B.E.}}(i) = \frac{(n_i+g_i-1)!}{n_i!\,(g_i-1)!}, \tag{5}$$

with the result that

$$W_{\text{B.E.}}\{n_i\} = \prod_i \frac{(n_i+g_i-1)!}{n_i!\,(g_i-1)!}. \tag{6}$$

In the Fermi–Dirac case no single level can accommodate more than one particle; accordingly, the number n_i must not exceed g_i. The number $w(i)$ is then given by the "number of ways in which the g_i levels can be divided into two sub-groups—one consisting of n_i levels (which will be occupied by one particle each) and the other consisting of (g_i-n_i) levels (which will be unoccupied)". This number is given by

$$w_{\text{F.D.}}(i) = \frac{g_i!}{n_i!\,(g_i-n_i)!}, \tag{7}$$

with the result that

$$W_{\text{F.D.}}\{n_i\} = \prod_i \frac{g_i!}{n_i!\,(g_i-n_i)!}. \tag{8}$$

For completeness, we may include the classical—or what is generally known as the Maxwell–Boltzmann—case as well. In that case the particles are regarded as *distinguishable*. Therefore, any of the n_i particles may be put into any of the g_i levels, independently of one another, and the resulting complexions may all be counted as distinct; their number is clearly $(g_i)^{n_i}$. Moreover, the distribution set $\{n_i\}$ is itself regarded as obtainable in

$$\frac{N!}{n_1!\,n_2!\,\ldots} \tag{9}$$

different ways which, on the introduction of the Gibbs correction factor, lead to a "weight factor" of

$$\frac{1}{n_1!\,n_2!\,\ldots} = \prod_i \frac{1}{n_i!}; \tag{10}$$

see also Sec. 1.6, especially eqn. (1.6.2). Combining the two results, we obtain

$$W_{\text{M.B.}}\{n_i\} = \prod_i \frac{(g_i)^{n_i}}{n_i!}. \tag{11}$$

Now, the entropy of the system would be given by

$$S(N, V, E) = k \ln \Omega(N, V, E) = k \ln \left[\sum_{\{n_i\}}{}' W\{n_i\}\right]. \tag{12}$$

We, however, expect that, in view of the largeness of the numbers involved, the logarithm of the sum on the right-hand side of (12) can be approximated by the logarithm of the largest term in the sum; cf. Problem 3.4. We may, therefore, replace (12) by

$$S(N, V, E) \simeq k \ln W\{n_i^*\}, \tag{13}$$

where $\{n_i^*\}$ is the distribution set that maximizes the number $W\{n_i\}$; the numbers n_i^* are clearly the *most probable values* of the distribution numbers n_i. The maximization, however, is to be carried out under the restrictions that the quantities N and E remain constant. This can be done by the method of Lagrange's undetermined multipliers; cf. the corresponding treatment in Sec. 3.2. Our condition for the determination of the most probable distribution set $\{n_i^*\}$ then turns out to be, see eqns. (1), (2) and (13),

$$\delta \ln W\{n_i\} - \left[\alpha \sum_i \delta n_i + \beta \sum_i \varepsilon_i\, \delta n_i\right] = 0. \tag{14}$$

For $\ln W\{n_i\}$, we obtain from eqns. (6), (8) and (11), assuming that not only all g_i but also all $n_i \gg 1$ (so that the Stirling formula $\ln(x!) \simeq x \ln x - x$ can be applied to all the factorials that appear),

$$\ln W\{n_i\} = \sum_i \ln w(i)$$

$$\simeq \sum_i \left[n_i \ln \left(\frac{g_i}{n_i} - a \right) - \frac{g_i}{a} \ln \left(1 - a \frac{n_i}{g_i} \right) \right], \tag{15}$$

where $a = -1$ for the B.E. case, $+1$ for the F.D. case and 0 for the M.B. case. Equation (14) then becomes

$$\sum_i \left[\ln \left(\frac{g_i}{n_i} - a \right) - \alpha - \beta \varepsilon_i \right]_{n_i = n_i^*} \delta n_i = 0. \tag{16}$$

In view of the arbitrariness of the increments δn_i in (16), we must have (for all i)

$$\ln \left(\frac{g_i}{n_i^*} - a \right) - \alpha - \beta \varepsilon_i = 0, \tag{17}$$

whence it follows that†

$$n_i^* = \frac{g_i}{e^{\alpha + \beta \varepsilon_i} + a}. \tag{18}$$

The fact that n_i^* turns out to be *directly* proportional to g_i prompts us to interpret the quantity

$$\frac{n_i^*}{g_i} = \frac{1}{e^{\alpha + \beta \varepsilon_i} + a}, \tag{18a}$$

which is actually the most probable number of particles *per* energy level in the ith cell, as the most probable number of particles in a *single* level of energy ε_i. Incidentally, the final result (18a) is entirely independent of the manner in which the energy levels of the particles are grouped into cells, provided that the number of levels in each cell is sufficiently large. As will be seen in the next section, formula (18a) can also be derived without grouping energy levels into cells; in fact, it is only then that this result becomes fully acceptable.

Substituting (18) into (15) we obtain for the entropy of the gas

$$\frac{S}{k} \simeq \ln W\{n_i^*\} = \sum_i \left[n_i^* \ln \left(\frac{g_i}{n_i^*} - a \right) - \frac{g_i}{a} \ln \left(1 - a \frac{n_i^*}{g_i} \right) \right]$$

$$= \sum_i \left[n_i^* (\alpha + \beta \varepsilon_i) + \frac{g_i}{a} \ln \{1 + a e^{-\alpha - \beta \varepsilon_i}\} \right]. \tag{19}$$

The first sum on the right-hand side of (19) is identically equal to αN while the second sum is identically equal to βE. For the third sum, therefore, we have

$$\frac{1}{a} \sum_i g_i \ln \{1 + a e^{-\alpha - \beta \varepsilon_i}\} = \frac{S}{k} - \alpha N - \beta E. \tag{20}$$

Now, the physical interpretation of the parameters α and β here is precisely the same as in

† For a critique of this derivation, see Landsberg (1954a, 1961).

Sec. 4.3, namely

$$\alpha = -\frac{\mu}{kT}, \quad \beta = \frac{1}{kT}; \tag{21}$$

see also Sec. 6.2. The right-hand side of eqn. (20) is, therefore, equal to

$$\frac{S}{k}+\frac{\mu N}{kT}-\frac{E}{kT} = \frac{G-(E-TS)}{kT} = \frac{PV}{kT}. \tag{22}$$

The thermodynamic pressure of the system is, therefore, given by

$$PV = \frac{kT}{a}\sum_i [g_i \ln \{1+ae^{-\alpha-\beta\varepsilon_i}\}]. \tag{23}$$

In the Maxwell–Boltzmann case ($a \to 0$), eqn. (23) takes the form

$$PV = kT\sum_i g_i e^{-\alpha-\beta\varepsilon_i} = kT\sum_i n_i^*,$$

that is,

$$PV = NkT, \tag{24}$$

which is the familiar equation of state of the classical ideal gas. It should be noted here that eqn. (24) for the Maxwell–Boltzmann case holds irrespective of the details of the energy spectrum ε_i.

It will be recognized that the expression $a^{-1}\Sigma_i[\;\;]$ in eqn. (23), being equal to the thermodynamic quantity (PV/kT), must be identical with the logarithm of the grand partition function of the ideal gas. One may, therefore, expect to obtain from this expression all the macroscopic properties of this system. However, we prefer to first develop the formal theory of an ideal gas in the canonical and grand canonical ensembles.

6.2. An ideal gas in other quantum-mechanical ensembles

In the canonical ensemble the thermodynamics of the system is obtained from its partition function:

$$Q_N(V, T) = \sum_E e^{-\beta E}, \tag{1}$$

where E's denote the energy eigenvalues of the system while $\beta = 1/kT$. Now, an energy value E can be expressed in terms of the single-particle energies ε; for instance,

$$E = \sum_\varepsilon n_\varepsilon \varepsilon, \tag{2}$$

where n_ε is the number of particles in the single-particle energy state ε. The values of the numbers n_ε must satisfy the condition:

$$\sum_\varepsilon n_\varepsilon = N. \tag{3}$$

The expression for the partition function may then be written as

$$Q_N(V,T) = \sum_{\{n_\varepsilon\}}{}' g\{n_\varepsilon\} e^{-\beta \sum_\varepsilon n_\varepsilon \varepsilon}, \tag{4}$$

where $g\{n_\varepsilon\}$ is the *statistical weight factor* for the distribution set $\{n_\varepsilon\}$ and the summation Σ' goes over all the sets that conform to the restrictive condition (3). The statistical weight factor in the various cases is given by

$$g_{\text{B.E.}}\{n_\varepsilon\} = 1, \tag{5}$$

$$g_{\text{F.D.}}\{n_\varepsilon\} = \begin{cases} 1 & \text{if all } n_\varepsilon = 0 \text{ or } 1 \\ 0 & \text{otherwise} \end{cases} \tag{6}$$

and

$$g_{\text{M.B.}}\{n_\varepsilon\} = \prod_\varepsilon \frac{1}{n_\varepsilon!}. \tag{7}$$

Note that in the present treatment we are dealing with the single-particle states as *individual* states, without requiring them to be grouped into cells; indeed, the weight factors (5), (6) and (7) follow straightforwardly from their respective predecessors (6.1.6), (6.1.8) and (6.1.11) by putting all $g_i = 1$.

First of all, we work out the Maxwell–Boltzmann case. Substituting (7) into (4), we obtain

$$\begin{aligned} Q_N(V,T) &= \sum_{\{n_\varepsilon\}}{}' \left[\left(\prod_\varepsilon \frac{1}{n_\varepsilon!}\right) \prod_\varepsilon (e^{-\beta\varepsilon})^{n_\varepsilon}\right] \\ &= \frac{1}{N!} \sum_{\{n_\varepsilon\}}{}' \left[\frac{N!}{\prod_\varepsilon n_\varepsilon!} \prod_\varepsilon (e^{-\beta\varepsilon})^{n_\varepsilon}\right]. \end{aligned} \tag{8}$$

Since the summation here is governed by condition (3), it can be evaluated with the help of the multinomial theorem; thus

$$\begin{aligned} Q_N(V,T) &= \frac{1}{N!}\left[\sum_\varepsilon e^{-\beta\varepsilon}\right]^N \\ &= \frac{1}{N!}[Q_1(V,T)]^N, \end{aligned} \tag{9}$$

which is identical with the result we already have in eqn. (3.5.15). The evaluation of Q_1 is, of course, straightforward: one obtains, using the asymptotic formula (2.4.7) for the number of single-particle states with energies between ε and $\varepsilon + d\varepsilon$,

$$\begin{aligned} Q_1(V,T) \equiv \sum_\varepsilon e^{-\beta\varepsilon} &\xrightarrow[V\to\infty]{} \frac{2\pi V}{h^3}(2m)^{3/2} \int_0^\infty e^{-\beta\varepsilon}\varepsilon^{1/2}\, d\varepsilon \\ &= V/\lambda^3, \end{aligned} \tag{10}$$

where $\lambda[= h/(2\pi mkT)^{1/2}]$ is the mean thermal wavelength of the particles. Hence

$$Q_N(V,T) = \frac{V^N}{N!\,\lambda^{3N}}, \tag{11}$$

from which the complete thermodynamics of this system can be derived; see, for example, Sec. 3.5. Further, we obtain for the grand partition function of this system

$$\mathcal{Q}(z, V, T) = \sum_{N=0}^{\infty} z^N Q_N(V, T) = \exp(zV/\lambda^3); \tag{12}$$

cf. eqn. (4.4.3). We know that the thermodynamics of the system follows equally well from the expression for $\mathcal{Q}$.

In the Bose–Einstein and Fermi–Dirac cases we obtain, by substituting (5) and (6) into (4),

$$\mathcal{Q}_N(V, T) = \sum_{\{n_\varepsilon\}}{}' \left(e^{-\beta \sum_\varepsilon n_\varepsilon \varepsilon} \right); \tag{13}$$

the difference between the two quantum cases, namely B.E. and F.D., arises from the values the numbers n_ε can take. Now, in view of restriction (3) on the summation Σ', an explicit evaluation of the partition function Q_N is rather cumbersome. On the other hand, the grand partition function $\mathcal{Q}$ turns out to be more easily tractable; we have

$$\begin{aligned}\mathcal{Q}(z, V, T) &= \sum_{N=0}^{\infty} \left[z^N \sum_{\{n_\varepsilon\}}{}' e^{-\beta \sum_\varepsilon n_\varepsilon \varepsilon} \right] \\ &= \sum_{N=0}^{\infty} \left[\sum_{\{n_\varepsilon\}}{}' \prod_\varepsilon (ze^{-\beta\varepsilon})^{n_\varepsilon} \right].\end{aligned} \tag{14}$$

The double summation in (14) (first over the numbers n_ε under the constraint of a *fixed* value of the total number N, and then over *all* possible values of N) is equivalent to a summation over all possible values of the numbers n_ε, *independently of one another*. Hence, we can write

$$\begin{aligned}\mathcal{Q}(z, V, T) &= \sum_{n_0, n_1, \ldots} \left[(ze^{-\beta\varepsilon_0})^{n_0} (ze^{-\beta\varepsilon_1})^{n_1} \ldots \right] \\ &= \left[\sum_{n_0} (ze^{-\beta\varepsilon_0})^{n_0} \right] \left[\sum_{n_1} (ze^{-\beta\varepsilon_1})^{n_1} \right] \ldots .\end{aligned} \tag{15}$$

Now, in the Bose–Einstein case, the n's can be either 0 or 1 or 2 or . . ., while in the Fermi–Dirac case we can have only 0 or 1. Therefore, eqn. (15) becomes

$$\mathcal{Q}(z, V, T) = \begin{cases} \prod_\varepsilon \dfrac{1}{(1 - ze^{-\beta\varepsilon})} & \text{in the B.E. case,} \\ \prod_\varepsilon (1 + ze^{-\beta\varepsilon}) & \text{in the F.D. case.} \end{cases} \tag{16}$$

The grand potential q is then given by

$$\begin{aligned}q(z, V, T) \equiv \frac{PV}{kT} &\equiv \ln \mathcal{Q}(z, V, T) \\ &= \mp \sum_\varepsilon \ln (1 \mp ze^{-\beta\varepsilon});\end{aligned} \tag{17}$$

cf. eqn. (6.1.23), with all $g_i = 1$. The identification of the fugacity z with the quantity $e^{-\alpha}$ of eqn. (6.1.23) is quite natural; incidentally, this supports our anticipation that $\alpha = -\mu/kT$. As usual, the upper (lower) signs in eqn. (17) correspond to the Bose (Fermi) case. In the end, we may write our result for q in a form which is applicable to all the three cases:

$$q(z, V, T) \equiv \frac{PV}{kT} = \frac{1}{a} \sum_{\varepsilon} \ln (1 + aze^{-\beta\varepsilon}), \tag{18}$$

where $a = -1, +1$ or 0, depending upon the statistics governing the system. In particular, we note that in the classical case, viz. $a \to 0$, eqn. (18) gives

$$q_{\text{M.B.}} = z \sum_{\varepsilon} e^{-\beta\varepsilon} = zQ_1, \tag{19}$$

which is consistent with eqn. (4.4.4).

From (18) it follows that

$$N \equiv z\left(\frac{\partial q}{\partial z}\right)_{V,T} = \sum_{\varepsilon} \frac{1}{z^{-1}e^{\beta\varepsilon}+a} \tag{20}$$

and

$$U \equiv -\left(\frac{\partial q}{\partial \beta}\right)_{z,V} = \sum_{\varepsilon} \frac{\varepsilon}{z^{-1}e^{\beta\varepsilon}+a}. \tag{21}$$

Moreover, for $\langle n_\varepsilon \rangle$, the *mean occupation number* of the level ε, we obtain

$$\begin{aligned} \langle n_\varepsilon \rangle &= \frac{1}{\mathcal{Q}}\left[-\frac{1}{\beta}\left(\frac{\partial \mathcal{Q}}{\partial \varepsilon}\right)_{z,\,T,\text{ all other } \varepsilon\text{'s}}\right] \\ &\equiv -\frac{1}{\beta}\left(\frac{\partial q}{\partial \varepsilon}\right)_{z,\,T,\text{ all other } \varepsilon\text{'s}} \\ &= \frac{1}{z^{-1}e^{\beta\varepsilon}+a}, \end{aligned} \tag{22}$$

which indeed fits with eqns. (20) and (21). Comparing the final result (22) with its counterpart (6.1.18a), we find that the *mean* value $\langle n \rangle$ and the *most probable* value n^* of the occupation number of a single-particle state are just identical. This suggests that the relative fluctuations in the actual values of this number are statistically negligible. However, in view of the fundamental importance of these fluctuations, we would like to study them in some detail here.

6.3. Statistics of the occupation numbers

Equation (6.2.22) gives the *mean occupation number* of a single-particle state with energy ε as an explicit function of the quantity $(\varepsilon - \mu)/kT$:

$$\langle n_\varepsilon \rangle = \frac{1}{e^{(\varepsilon-\mu)/kT}+a}. \tag{1}$$

The functional behavior of this number is shown in Fig. 6.2. In the Fermi–Dirac case ($a = +1$) the mean occupation number never exceeds unity; this is natural because the variable n_ε itself cannot have a value other than 0 or 1. Moreover, for $\varepsilon < \mu$ and $|\varepsilon - \mu| \gg kT$, the mean occupation number tends to its maximum possible value, namely 1. In the Bose–Einstein case ($a = -1$), we must have: $\mu <$ all ε, for otherwise some of the mean occupation numbers would become negative, which is unphysical. In fact, when μ becomes equal to the lowest value of ε, say ε_0, the occupancy of that particular level becomes infinitely high; this, in turn, leads to the famous phenomenon of *Bose–Einstein*

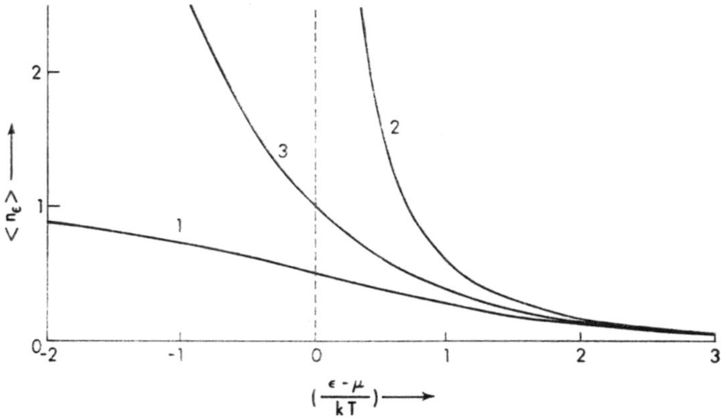

FIG. 6.2. The mean occupation number $\langle n_\varepsilon \rangle$ of a single-particle energy state ε in a system of noninteracting particles: curve 1 is for fermions, curve 2 for bosons and curve 3 for the Maxwell–Boltzmann particles.

condensation (see Sec. 7.1). For $\mu < \varepsilon_0$, the values of $(\varepsilon - \mu)$ are all positive and the behavior of all the $\langle n_\varepsilon \rangle$'s is nonsingular. Finally, in the Maxwell–Boltzmann case ($a = 0$), the mean occupation number takes the form

$$\langle n_\varepsilon \rangle_{\text{M.B.}} = \exp\{(\mu - \varepsilon)/kT\} \propto \exp(-\varepsilon/kT), \tag{2}$$

a familiar result of the classical statistical mechanics. The important thing to note here is that the distinction between the quantum statistics ($a = \mp 1$) and the classical statistics ($a = 0$) becomes imperceptible when the role of the parameter a in eqn. (1) becomes unimportant. This happens if, for all values of ε that are of practical interest,

$$\exp\{(\varepsilon - \mu)/kT\} \gg 1. \tag{3}$$

In that event, eqn. (1) essentially reduces to eqn. (2) and we may write, instead of (3),

$$\langle n_\varepsilon \rangle \ll 1. \tag{4}$$

The condition (4) is quite understandable, for it implies that the probability of any of the n_ε's being greater than unity is quite negligible, as a result of which the classical weight factors $g\{n_\varepsilon\}$, as given by eqn. (6.2.7), become essentially equal to 1. The distinction between the classical treatment (which is basically defective) and the quantum-mechanical treatment (which is basically sound) then becomes physically insignificant. Correspondingly, we find, in Fig. 6.2, that for large values of $(\varepsilon-\mu)/kT$ the quantum curves 1 and 2 simply merge into the classical curve 3. Since we already know that the higher the temperature of the system the better the validity of the classical treatment, condition (3) is essentially equivalent to the requirement that μ, the chemical potential of the system, must be negative and large in magnitude. This means that the fugacity $z[\equiv \exp(\mu/kT)]$ of the system must be much smaller than unity; see also eqn. (6.2.22). It is easy to see that this is further equivalent to the requirement:

$$\frac{N\lambda^3}{V} \ll 1; \tag{5}$$

see eqn. (4.4.6). It is satisfying to note that the foregoing condition is identical with the one established in Chapter 5; see eqn. (5.5.20).

We shall now consider the statistical fluctuations in the values of the variable n_ε. Going a step further from the calculation that led to eqn. (6.2.22), we have

$$\langle n_\varepsilon^2 \rangle = \frac{1}{\mathcal{Q}}\left[\left(-\frac{1}{\beta}\frac{\partial}{\partial\varepsilon}\right)^2 \mathcal{Q}\right]_{z,\,T,\,\text{all other }\varepsilon\text{'s}}; \tag{6}$$

it then follows that

$$\langle n_\varepsilon^2\rangle - \langle n_\varepsilon\rangle^2 = \left[\left(-\frac{1}{\beta}\frac{\partial}{\partial\varepsilon}\right)^2 \ln \mathcal{Q}\right]_{z,\,T,\,\text{all other }\varepsilon\text{'s}}$$

$$= \left[\left(-\frac{1}{\beta}\frac{\partial}{\partial\varepsilon}\right)\langle n_\varepsilon\rangle\right]_{z,\,T}. \tag{7}$$

For the relative mean square fluctuation, we obtain (irrespective of the statistics obeyed by the particles)

$$\frac{\langle n_\varepsilon^2\rangle - \langle n_\varepsilon\rangle^2}{\langle n_\varepsilon\rangle^2} = \left(\frac{1}{\beta}\frac{\partial}{\partial\varepsilon}\right)\left\{\frac{1}{\langle n_\varepsilon\rangle}\right\} = z^{-1}e^{\beta\varepsilon}. \tag{8}$$

Numerically, this quantity will depend upon the statistics of the particles because, for a given particle density (N/V) and temperature T, the value of z will be different for different statistics.

It seems more instructive to write eqn. (8) in the form

$$\frac{\langle n_\varepsilon^2\rangle - \langle n_\varepsilon\rangle^2}{\langle n_\varepsilon\rangle^2} = \frac{1}{\langle n_\varepsilon\rangle} - a. \tag{9}$$

In the classical case $(a = 0)$, the fluctuation is *normal*, so to say. In the Fermi–Dirac case, the fluctuation is given by $1/\langle n_\varepsilon\rangle - 1$, which is rather *infranormal*; in fact, the fluctuation tends to vanish as $\langle n_\varepsilon\rangle \to 1$, which is true for states with $\varepsilon < \mu$ and at temperatures sufficiently small in comparison with $(\mu-\varepsilon)/k$. In the Bose–Einstein case, the fluctuation is given by $1/\langle n_\varepsilon\rangle + 1$, which is rather *above normal.** Obviously, this result would apply to a gas of photons and, hence, to the oscillator states in the black-body radiation. In the latter context, Einstein derived this result as early as 1909 following Planck's approach and even

* The special case of the fluctuations in the *ground state occupation number*, n_0, of a Bose–Einstein system has been discussed by Wergeland (1969) and by Fujiwara, ter Haar and Wergeland (1970).

pointed out that the term 1 in the expression for the fluctuation may be attributed to the wave character of the radiation and the term $1/\langle n_\varepsilon \rangle$ to the particle character of the photons; for details, see Kittel (1958), ter Haar (1968). Closely related to the subject of fluctuations is the problem of "statistical correlations in photon beams", which have been observed experimentally (see Hanbury Brown and Twiss, 1956–8) and have been explained theoretically in terms of the quantum-statistical nature of these fluctuations (see Purcell, 1956; Kothari and Auluck, 1957). For further details, refer to Mandel, Sudarshan and Wolf (1964); Holliday and Sage (1964).

For a greater understanding of the statistics of the occupation numbers, let us evaluate the quantity $p_\varepsilon(n)$, which denotes the probability that there are exactly n particles in the state of energy ε. The answer is pretty simple in the case of fermions. Observing that

$$\langle n_\varepsilon \rangle_{\text{F.D.}} = \sum_{n=0}^{1} n p_\varepsilon(n) = p_\varepsilon(1), \tag{10}$$

we readily obtain

$$p_\varepsilon(0) = 1 - \langle n_\varepsilon \rangle \quad \text{and} \quad p_\varepsilon(1) = \langle n_\varepsilon \rangle. \tag{11a}$$

To determine $p_\varepsilon(n)$ in the Bose–Einstein case, let us look at the corresponding expression (6.2.15) for the grand partition function $\mathcal{Q}$. The probability $p_\varepsilon(n)$ is naturally proportional to the quantity $(ze^{-\beta\varepsilon})^n$. On normalization, it becomes

$$p_\varepsilon(n) = (ze^{-\beta\varepsilon})^n [1 - ze^{-\beta\varepsilon}].$$

Finally, making use of the corresponding expression for $\langle n_\varepsilon \rangle$, it takes the form

$$\begin{aligned} p_\varepsilon(n) &= \left(\frac{\langle n_\varepsilon \rangle}{\langle n_\varepsilon \rangle + 1}\right)^n \frac{1}{\langle n_\varepsilon \rangle + 1} \\ &= \frac{(\langle n_\varepsilon \rangle)^n}{(\langle n_\varepsilon \rangle + 1)^{n+1}}. \end{aligned} \tag{11b}$$

In the Maxwell–Boltzmann case, $p_\varepsilon(n)$ will be proportional to the quantity $(ze^{-\beta\varepsilon})^n/n!$ instead; see eqn. (6.2.8). On normalization, it becomes

$$p_\varepsilon(n) = \frac{(ze^{-\beta\varepsilon})^n/n!}{\exp(ze^{-\beta\varepsilon})}.$$

Making use of the corresponding expression for $\langle n_\varepsilon \rangle$, it takes the form

$$p_\varepsilon(n) = \frac{(\langle n_\varepsilon \rangle)^n}{n!} e^{-\langle n_\varepsilon \rangle}. \tag{11c}$$

The distribution (11c) is clearly a *Poisson distribution*, for which the mean square deviation of the variable is equal to the mean value itself; cf. eqn. (9), with $a = 0$. It also resembles the distribution of the total particle number N in a grand canonical ensemble consisting of ideal, classical systems; cf. Problem 4.4. We also note that the ratio $p_\varepsilon(n)/p_\varepsilon(n-1)$ in this case decreases inversely with n, which is the "normal" statistical behavior of uncorrelated events. On the other hand, the distribution in the Bose–Einstein case is *geometric*, with a constant common ratio $\langle n_\varepsilon\rangle/(\langle n_\varepsilon\rangle+1)$. This means that the probability of a state ε acquiring one more particle for itself is independent of the number of particles already occupying that state; thus, in comparison with the normal statistical behavior, bosons exhibit a larger tendency of "bunching" together, i.e. a *positive* statistical correlation. In contrast, fermions exhibit a *negative* statistical correlation.

6.4. Kinetic considerations

The thermodynamic pressure of an ideal gas is given by eqn. (6.1.23) or by eqn. (6.2.18). In view of the largeness of volume V, the single-particle energy states ε would be so close to one another that a summation over them may be replaced by an integration. One thereby gets

$$P = \frac{kT}{a}\int_0^\infty \ln\,[1+aze^{-\beta\varepsilon(p)}]\frac{4\pi p^2\,dp}{h^3}$$

$$= \frac{4\pi kT}{ah^3}\left[\frac{p^3}{3}\ln\{1+aze^{-\beta\varepsilon(p)}\}\Big|_0^\infty + \int_0^\infty \frac{p^3}{3}\frac{aze^{-\beta\varepsilon(p)}}{1+aze^{-\beta\varepsilon(p)}}\beta\frac{d\varepsilon}{dp}\,dp\right].$$

The integrated part vanishes at both the limits while the rest of the expression reduces to

$$P = \frac{4\pi}{3h^3}\int_0^\infty \frac{1}{z^{-1}e^{\beta\varepsilon(p)}+a}\left(p\frac{d\varepsilon}{dp}\right)p^2\,dp. \tag{1}$$

Now, the total number of particles in the gas is given by

$$N = \int n_p\frac{V\,d^3p}{h^3} = \int_0^\infty \frac{1}{z^{-1}e^{\beta\varepsilon(p)}+a}\,\frac{V4\pi p^2\,dp}{h^3}. \tag{2}$$

Comparing (1) and (2), we can write

$$P = \frac{1}{3}\frac{N}{V}\left\langle p\frac{d\varepsilon}{dp}\right\rangle = \frac{1}{3}n\langle pu\rangle, \tag{3}$$

where n is the particle density in the gas and u the velocity of an individual particle. If the relationship between the energy ε and the momentum p is of the type $\varepsilon \propto p^s$, then

$$P = \frac{s}{3}n\langle\varepsilon\rangle = \frac{s}{3}\frac{E}{V}. \tag{4}$$

The particular cases $s = 1$ and $s = 2$ are pretty easy to recognize. It must, of course, be noted that the results (3) and (4) hold independently of the statistics obeyed by the particles.

The structure of formula (3) strongly suggests that the pressure of the gas arises essentially from the physical motion of the particles; it should, therefore, be derivable from kinetic considerations alone. To do this, we consider the bombardment, by the particles of the gas, on a section of the walls of the container. Let us take, for example, an element of area dA on one of the walls normal to the z-axis (Fig. 6.3), and focus our attention on those particles in the gas whose velocity lies between $\boldsymbol{u}$ and $\boldsymbol{u}+d\boldsymbol{u}$; the number of such particles per unit volume may be denoted by $nf(\boldsymbol{u})\,d\boldsymbol{u}$, where

$$\int_{\text{all } \boldsymbol{u}} f(\boldsymbol{u})\, d\boldsymbol{u} = 1. \tag{5}$$

Now, the question is: how many of these particles will strike the area dA in time dt? The answer is: all those particles which happen to lie in a cylindrical region of base dA and height $\boldsymbol{u}\,dt$, as shown in Fig. 6.3. Since the volume of this region is $(d\boldsymbol{A}\cdot\boldsymbol{u})\,dt$, the relevant number of

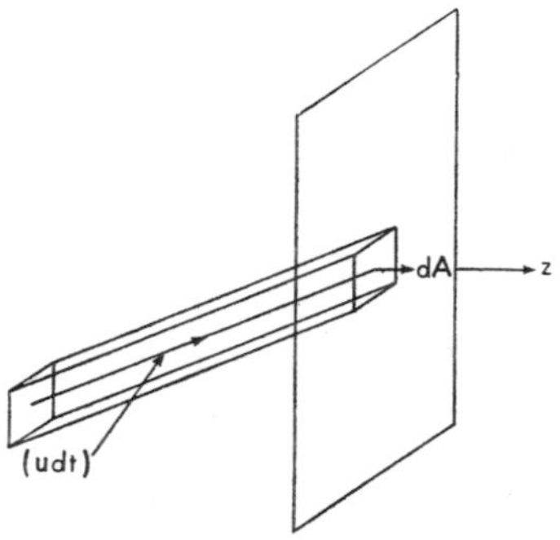

FIG. 6.3. The molecular bombardment on one of the walls of the container.

particles would be $\{(d\boldsymbol{A}\cdot\boldsymbol{u})\,dt\cdot nf(\boldsymbol{u})\,d\boldsymbol{u}\}$. On reflection from the wall, the normal component of the momentum of a particle would undergo a change from p_z to $-p_z$; as a result, the normal momentum imparted by these particles per unit time to a unit area of the wall would be $2p_z\{u_z nf(\boldsymbol{u})\,d\boldsymbol{u}\}$. Integrating this expression over all relevant values of $\boldsymbol{u}$, we obtain the total normal momentum imported per unit time to a unit area of the wall by all the particles of the gas, which is, by definition, the *kinetic* pressure of the gas:

$$P = 2n \int_{u_x=-\infty}^{\infty} \int_{u_y=-\infty}^{\infty} \int_{u_z=0}^{\infty} p_z u_z f(\boldsymbol{u})\, du_x\, du_y\, du_z. \tag{6*}$$

Since (i) $f(\boldsymbol{u})$ is a function of u alone and (ii) the product $(p_z u_z)$ is an even function of u_z, the foregoing result can be written as

$$P = n \int_{\text{all } \boldsymbol{u}} (p_z u_z)\, f(\boldsymbol{u})\, d\boldsymbol{u}. \tag{7}$$

Comparing (7) with (5), we obtain

$$P = n\langle p_z u_z\rangle = n\langle pu \cos^2\theta\rangle \tag{8}$$

$$= \tfrac{1}{3}n\langle pu\rangle, \tag{9}$$

which is identical with (3).

* Clearly, only those velocities are relevant for which $u_z > 0$.

In a similar manner, we can determine the rate of *effusion* of the gas particles through a hole (of unit area) in the wall. This is given by, cf. (6),

$$R = n \int_{u_x=-\infty}^{\infty} \int_{u_y=-\infty}^{\infty} \int_{u_z=0}^{\infty} u_z f(\mathbf{u})\, du_x\, du_y\, du_z$$

$$= n \int_{\varphi=0}^{2\pi} \int_{\theta=0}^{\pi/2} \int_{u=0}^{\infty} \{u \cos \theta f(\mathbf{u})\} (u^2 \sin \theta\, du\, d\theta\, d\varphi);$$

note that the condition $u_z > 0$ restricts the range of the angle θ between the values 0 and $\pi/2$. Carrying out integrations over θ and φ, we obtain

$$R = n\pi \int_0^{\infty} f(\mathbf{u}) u^3\, du. \tag{10}$$

In view of the fact that

$$\int_0^{\infty} f(\mathbf{u}) (4\pi u^2\, du) = 1, \tag{5a}$$

eqn. (10) can be written as

$$R = \tfrac{1}{4} n \langle u \rangle. \tag{11}$$

Again, this result holds independently of the statistics obeyed by the particles.

It is obvious that the velocity distribution among the effused particles is considerably different from the velocity distribution among the particles inside the container. This is due to the fact that firstly the velocity component u_z of the effused particles must be positive (which introduces an element of anisotropy into the distribution) and secondly the particles with larger values of u_z appear with an extra weightage, the weightage being directly proportional to the value of u_z; see the original expression for R. As a result of this, (i) the effused particles carry with them a net forward momentum, thus causing the container to experience a recoil force, and (ii) they carry away a relatively larger amount of energy per particle, thus leaving the gas in the container at not only a progressively decreasing pressure and density but also a progressively decreasing temperature; see Problem 6.14. For a quantitative study, we consider the case of a Maxwell–Boltzmann gas.

For the gas inside the container, we have

$$f(\mathbf{u})\, d\mathbf{u} = A e^{-\beta((1/2)mu^2)} (4\pi u^2\, du), \tag{12}$$

where A is a normalization constant to be fixed with the help of eqn. (5):

$$4\pi A \int_0^{\infty} e^{-\beta((1/2)mu^2)} u^2\, du = A\left(\frac{2\pi}{m\beta}\right)^{3/2} = 1,$$

so that

$$A = \left(\frac{m\beta}{2\pi}\right)^{3/2}. \tag{13}$$

We then have

$$f(\mathbf{u})\, d\mathbf{u} = \left(\frac{m}{2\pi kT}\right)^{3/2} e^{-(1/2)mu^2/kT}\, d^3u, \tag{14}$$

which is the famous Maxwell's law of distribution of velocities among the molecules of an ideal, classical gas. Clearly, the distribution (14) is isotropic in u_x, u_y and u_z; not only that, we can write it in the form of a product of three identical, statistically independent, distributions:

$$f(\mathbf{u})\,d\mathbf{u} = g(u_x)\,du_x \cdot g(u_y)\,du_y \cdot g(u_z)\,du_z\,,$$

where

$$g(u_x)\,du_x = \left(\frac{m}{2\pi kT}\right)^{1/2} e^{-(1/2)mu_x^2/kT}\,du_x\,, \tag{15}$$

with identical expressions for the y- and z-components. From these expressions, it follows that

$$\left\langle \tfrac{1}{2}mu_x^2 \right\rangle = \left\langle \tfrac{1}{2}mu_y^2 \right\rangle = \left\langle \tfrac{1}{2}mu_z^2 \right\rangle = \tfrac{1}{2}kT, \tag{16}$$

and hence

$$\left\langle \tfrac{1}{2}mu^2 \right\rangle = \tfrac{3}{2}kT, \tag{17}$$

which is consistent with the equipartition theorem. The *root-mean-square speed* of the molecules is, therefore, given by

$$u_{\text{r.m.s.}} = \sqrt{\langle u^2 \rangle} = \left(\frac{3kT}{m}\right)^{1/2}. \tag{18}$$

It would be of value to define the *mean speed* $\bar{u}$ as well:

$$\bar{u} = \frac{\int_0^\infty e^{-(1/2)mu^2/kT}u(4\pi u^2\,du)}{\int_0^\infty e^{-(1/2)mu^2/kT}(4\pi u^2\,du)} = \left(\frac{8kT}{\pi m}\right)^{1/2}. \tag{19}$$

In addition, we have the *most probable speed* $\tilde{u}$, which is given by the condition

$$\left(\frac{\partial F(u)}{\partial u}\right)_{u=\tilde{u}} = 0,$$

where

$$F(u)\,du = \left\{\left(\frac{m}{2\pi kT}\right)^{3/2} e^{-(1/2)mu^2/kT}4\pi u^2\right\}du. \tag{14a}$$

We readily obtain

$$\tilde{u} = \left(\frac{2kT}{m}\right)^{1/2}. \tag{20}$$

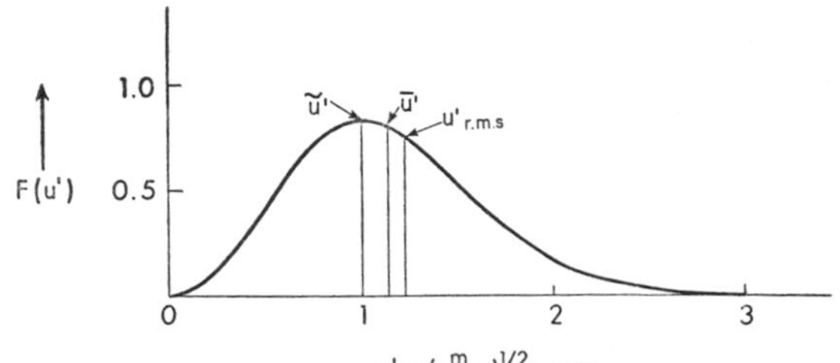

FIG. 6.4. The velocity distribution in a Maxwellian gas.

Figure 6.4 displays the Maxwell distribution function $F(u)$. The relative positions of $\tilde{u}$, $\bar{u}$ and $u_{\text{r.m.s.}}$, which are in the ratio of about 1 : 1.128 : 1.224, are also shown.

Among the effused particles, on the other hand, the distribution of velocities is given by

$$f^*(\boldsymbol{u})\, d\boldsymbol{u} = g(u_x)\, du_x \cdot g(u_y)\, du_y \cdot u_z g(u_z)\, du_z, \tag{21}$$

along with the condition: $u_z > 0$. The root-mean-square values of the three components of velocity are now given by

$$(u_x)^*_{\text{r.m.s.}} = (u_y)^*_{\text{r.m.s.}} = \left(\frac{kT}{m}\right)^{1/2}, \tag{22}$$

and

$$(u_z)^*_{\text{r.m.s.}} = \left(\frac{2kT}{m}\right)^{1/2}, \tag{23}$$

whence it follows that

$$\langle \tfrac{1}{2}mu^2 \rangle^* = 2kT, \tag{24}$$

instead of the customary $\frac{3}{2}kT$. Moreover, whereas both $\bar{u}_x$ and $\bar{u}_y$ are equal to zero, $\bar{u}_z$ turns out to be $(\pi kT/2m)^{1/2}$. Recalling that the rate of effusion of particles from an opening of area a is given by, see eqns. (11) and (19),

$$Ra = \tfrac{1}{4}na\langle u \rangle = na(kT/2\pi m)^{1/2}, \tag{25}$$

we obtain for the net rate of *momentum transfer* through the opening

$$Ra \cdot m\bar{u}_z = na\left(\frac{kT}{2\pi m}\right)^{1/2} \cdot m\left(\frac{\pi kT}{2m}\right)^{1/2} = \frac{1}{2}Pa, \tag{26}$$

where $P(= nkT)$ is the pressure of the gas inside the container. The physical interpretation of this result is straightforward.

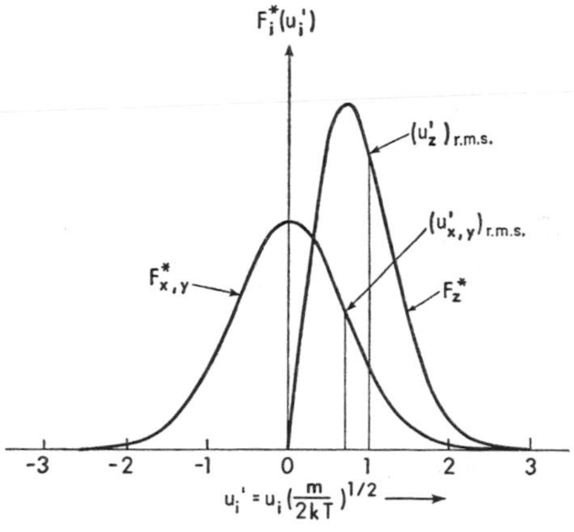

FIG. 6.5. The velocity distribution among the effused particles.

Figure 6.5 displays the velocity distributions for the components u_x, u_y and u_z among the effused particles. The special behavior of the component u_z underlines the difference between the statistics of these particles and of the particles inside the container.

6.5. A gaseous system in mass motion

We shall now study the statistical mechanics of an ideal, gaseous system in uniform mass motion. For this, we examine the system from the point of view of an observer in a reference frame K, with respect to which the *rest* frame K_0 of the system is moving with a uniform velocity $\boldsymbol{v} \neq 0$; see Fig. 6.6. It is understood that $\boldsymbol{v}$ is the velocity with which the

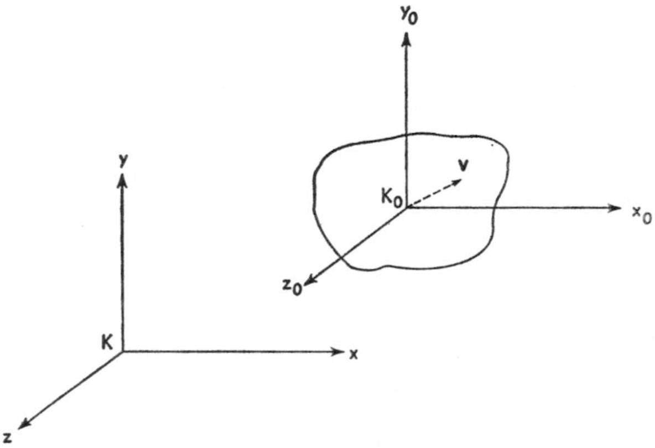

FIG. 6.6. A physical system in mass motion.

walls of the vessel enclosing the system are moving with respect to the frame K; accordingly, $\boldsymbol{v}$ must be equal to the "average drift velocity" of the particles constituting the system, as observed by an observer in K. Clearly, the statistical distribution of the particles over the various single-particle states of the system will no longer be isotropic; it will be governed by an additional constraint on the system, namely that of a fixed total momentum $\boldsymbol{P}$. The mean occupation number $\langle n(\boldsymbol{p})\rangle$ of the single-particle state $\varepsilon(\boldsymbol{p})$ is then given by the formula

$$\langle n(\boldsymbol{p})\rangle = \frac{1}{\exp(\alpha+\beta\varepsilon+\boldsymbol{\gamma}\cdot\boldsymbol{p})+a}; \tag{1}$$

cf. the isotropic formula (6.1.18a). The appearance of the undetermined multiplier $\boldsymbol{\gamma}$ in the foregoing expression is related to the conservation of the total momentum $\boldsymbol{P}$, in the same way as the appearance of α and β is related to the conservation of N and E. Proceeding as in Sec. 6.1, we obtain

$$\frac{1}{a}\sum_{\boldsymbol{p}} \ln\{1+ae^{-\alpha-\beta\varepsilon-\boldsymbol{\gamma}\cdot\boldsymbol{p}}\} = \frac{S}{k}-\alpha N-\beta E-\boldsymbol{\gamma}\cdot\boldsymbol{P}; \tag{2}$$

cf. (6.1.20). As suggested in Problem 6.21, the quantity on the left-hand side of this equation is to be identified with the pressure term $\beta\mathcal{P}V$, which leads to the following generalized statement of thermodynamics:

$$dE = \frac{1}{k\beta}\,dS-\mathcal{P}\,dV-\frac{\alpha}{\beta}\,dN-\frac{1}{\beta}\,\boldsymbol{\gamma}\cdot d\boldsymbol{P}. \tag{3*}$$

* In this section, the pressure of the system is denoted by the symbol $\mathcal{P}$, so as to distinguish it from the momentum $\boldsymbol{P}$.

This in turn leads to the correspondence:

$$\beta = \frac{1}{kT}, \quad \alpha = -\frac{\mu}{kT} \quad \text{and} \quad \boldsymbol{\gamma} = -\frac{\boldsymbol{v}}{kT}. \tag{4}$$

Substituting for β and $\boldsymbol{\gamma}$, the distribution (1) takes the form

$$\langle n(\boldsymbol{p})\rangle = \frac{1}{\exp\left\{\alpha + \dfrac{\varepsilon - \boldsymbol{v}\cdot\boldsymbol{p}}{kT}\right\} + a}, \tag{5}$$

the corresponding result in the rest frame K_0 being

$$\langle n_0(\boldsymbol{p}_0)\rangle = \frac{1}{\exp\left\{\alpha_0 + \dfrac{\varepsilon_0}{kT_0}\right\} + a}. \tag{6}$$

Now, the distribution function (5), or (6), is a measure of the relative abundance of particles that are found to be in the single-particle state $\varepsilon(\boldsymbol{p})$, or $\varepsilon_0(\boldsymbol{p}_0)$, averaged over a considerable span of time t, or t_0, as observed in the reference frame K, or K_0; by statistical hypothesis, this becomes a measure of the fraction of the particles that are found to be in the relevant single-particle state at any given time t, or t_0. Thus, by its very physical meaning, the distribution function (5), or (6), corresponds to an observation made *simultaneously* (in the reference frame K, or K_0) on all the constituent particles of the system. Hence, insofar as they refer only to single instants of time, the two distribution functions must be connected with one another through a straightforward Lorentz transformation.

In view of this, we require that the quantity $\alpha + (\varepsilon - \boldsymbol{v}\cdot\boldsymbol{p})/kT$ of eqn. (5) go over *quite generally* to the quantity $\alpha_0 + \varepsilon_0/kT_0$ of eqn. (6). However, for a single particle, it is well known that

$$\frac{\varepsilon - \boldsymbol{v}\cdot\boldsymbol{p}}{\sqrt{(1 - v^2/c^2)}} = \varepsilon_0; \tag{7}$$

therefore, we must have

$$\alpha = \alpha_0 \tag{8}$$

and

$$T = T_0(1 - v^2/c^2)^{1/2}. \tag{9}$$

That the absolute temperature of a thermodynamical system would be Lorentz-contracted was first established by Planck (1907) and by Einstein (1907).

Next, in view of the basic connection between the entropy of a system and the number of distinct complexions accessible to it, the entropy must be an invariant quantity:

$$S = S_0. \tag{10}$$

Combining (9) and (10), one obtains for the transformation of heat

$$\delta Q = \delta Q_0(1 - v^2/c^2)^{1/2}. \tag{11}$$

We now proceed to derive relevant expressions for other physical properties of the moving system by making use of the distribution function (5). We first of all calculate the average drift velocity $\langle \boldsymbol{u} \rangle$ of the particles, as viewed from the reference frame K. Obviously,

$$\langle \boldsymbol{u} \rangle = \frac{1}{N} \int \langle n(\boldsymbol{p}) \rangle \boldsymbol{u} \left(\frac{V\, d^3p}{h^3} \right)$$
$$= \frac{V}{Nh^3} \int \frac{\boldsymbol{u}}{\exp\left\{\alpha + \dfrac{\varepsilon - \boldsymbol{v}\cdot\boldsymbol{p}}{kT}\right\} + a}\, d^3p. \tag{12}$$

We make the transformation

$$\boldsymbol{p} = \boldsymbol{p}_0 + \frac{\boldsymbol{v}(\boldsymbol{v}\cdot\boldsymbol{p}_0)}{v^2}(\gamma - 1) + \gamma \boldsymbol{v} \frac{\varepsilon_0}{c^2} \qquad \{\gamma = (1 - v^2/c^2)^{-1/2}\}, \tag{13}$$

which implies

$$\varepsilon = \gamma(\varepsilon_0 + \boldsymbol{v}\cdot\boldsymbol{p}_0). \tag{14}$$

The relationships (13) and (14) will be recognized as the Lorentz transformation equations for the momentum and energy of a single particle between the reference frames K and K_0 (see, for example, Pathria, 1963). Under this transformation (see also eqns. (7)–(9)), our $\langle n(\boldsymbol{p}) \rangle$ just goes over to $\langle n_0(\boldsymbol{p}_0) \rangle$, which fits well with the requirements of covariance! The transformation of the element d^3p into the corresponding element d^3p_0, however, introduces the Jacobian

$$\frac{\partial(p_x, p_y, p_z)}{\partial(p_{x_0}, p_{y_0}, p_{z_0})} = \gamma\left(1 + \frac{\boldsymbol{v}\cdot\boldsymbol{p}_0}{\varepsilon_0}\right). \tag{15}$$

Moreover, while the total number of particles in the system is an invariant,

$$N = N_0, \tag{16}$$

the transformation of the volume of the system introduces the familiar factor $(1 - v^2/c^2)^{1/2}$:

$$V = V_0(1 - v^2/c^2)^{1/2}. \tag{17}$$

Finally, we have the transformation of the velocity $\boldsymbol{u}$, viz.

$$\boldsymbol{u} = \frac{\boldsymbol{u}_0 + \dfrac{\boldsymbol{v}(\boldsymbol{v}\cdot\boldsymbol{u}_0)}{v^2}(\gamma - 1) + \gamma \boldsymbol{v}}{\gamma\left(1 + \dfrac{\boldsymbol{v}\cdot\boldsymbol{u}_0}{c^2}\right)} \qquad \{\gamma = (1 - v^2/c^2)^{-1/2}\}. \tag{18}$$

Substituting these results into the right-hand side of (12), we obtain

$$\langle \boldsymbol{u} \rangle = \frac{V_0(1 - v^2/c^2)^{1/2}}{N_0 h^3} \int \frac{\left\{\boldsymbol{u}_0 + \dfrac{\boldsymbol{v}(\boldsymbol{v}\cdot\boldsymbol{u}_0)}{v^2}(\gamma - 1) + \gamma \boldsymbol{v}\right\}}{\exp\left\{\alpha_0 + \dfrac{\varepsilon_0}{kT_0}\right\} + a}\, d^3p_0. \tag{19}$$

In view of the isotropy of the denominator, the integrals of the first two terms identically vanish; the remaining integral is directly related to the number of particles per unit volume, as viewed from the reference frame K_0. The final result is

$$\langle \boldsymbol{u} \rangle = \boldsymbol{v}, \tag{20}$$

as it should have been. Incidentally, this demonstration vindicates our identification of the undetermined multiplier $\boldsymbol{\gamma}$ with the vector $-\boldsymbol{v}/kT$; see eqn. (4).

For the energy E of the moving system, we have

$$E = \int \langle n(\boldsymbol{p}) \rangle \, \varepsilon(\boldsymbol{p}) \left(\frac{V \, d^3 p}{h^3} \right). \tag{21}$$

Making the same transformation as before, we obtain

$$E = \int \langle n_0(\boldsymbol{p}_0) \rangle \, \gamma \left\{ \varepsilon_0 + 2(\boldsymbol{v} \cdot \boldsymbol{p}_0) + \frac{(\boldsymbol{v} \cdot \boldsymbol{p}_0)^2}{\varepsilon_0} \right\} \left(\frac{V_0 \, d^3 p_0}{h^3} \right). \tag{22}$$

Again, in view of the isotropy of the function $n_0(\boldsymbol{p}_0)$, the integral of the middle term vanishes identically. The remaining terms yield the following result:

$$E = \gamma N_0 \left\{ \langle \varepsilon_0 \rangle + \frac{\langle (\boldsymbol{v} \cdot \boldsymbol{p}_0)(\boldsymbol{v} \cdot \boldsymbol{u}_0) \rangle}{c^2} \right\}. \tag{23}$$

Comparing the second part of (23) with the expression (6.4.8) for the rest pressure $\mathcal{P}_0$ of the system, we finally obtain

$$E = \gamma \left(E_0 + \frac{v^2}{c^2} \mathcal{P}_0 V_0 \right). \tag{24}$$

In a similar manner, we obtain for the linear momentum $\boldsymbol{P}$ of the system

$$\boldsymbol{P} = \gamma \frac{\boldsymbol{v}}{c^2} (E_0 + \mathcal{P}_0 V_0). \tag{25}$$

We can also determine the pressure $\mathcal{P}$ of the system, as it would appear in the reference frame K. The calculation is slightly more involved than the ones for E and $\boldsymbol{P}$; the final result is, however, amazingly simple (see Pathria, 1957, 1967):

$$\mathcal{P} = \mathcal{P}_0. \tag{26}$$

Combining (24) and (26) with (17), we further obtain

$$(E + \mathcal{P} V) = \gamma (E_0 + \mathcal{P}_0 V_0). \tag{27}$$

From the relations (24), (25) and (27), we conclude that it is not the momentum $\boldsymbol{P}$ and the energy E of the system, but rather the momentum $\boldsymbol{P}$ and the enthalpy $(E + \mathcal{P} V)$, that transform like the components of a four-vector. This is due to the fact that the system under study is not a "closed" system, in that it consists of a thermodynamic fluid enclosed in a vessel under the influence of "external" pressure from the walls of the vessel. In fact, the very result that $\boldsymbol{P}$ and E do not transform like the components of a four-vector may, in general, be taken to indicate that the system under consideration is a "non-closed" one.

In the end, we would like to make certain comments on the transformation formulae (9) and (11) for temperature and heat, respectively. The fact that these quantities get Lorentz-contracted has been challenged by a number of authors, notably by Ott (1963) and by Arzeliès (1965). The crux of their contention is that, in view of the physical nature of these quantities (as customarily understood), they should *dilate* rather than *contract*. We, however, believe that the transformations obtained here are correct. To substantiate this belief, we present the following argument (see Pathria, 1966a):†

Assume that the given system absorbs a heat δQ_0, as observed in the rest frame K_0. Its inertial mass would thereby increase by an amount

$$\delta M_0 = \delta Q_0/c^2. \tag{28}$$

The corresponding increase δM, as observed in the laboratory frame K, would be given by

$$\delta M = \delta M_0\left(1-\frac{v^2}{c^2}\right)^{-1/2}. \tag{29}$$

The increase $(\delta Q)^*$ in the energy content of the system, as observed in the frame K, must then be

$$(\delta Q)^* = \delta M c^2 = \delta Q_0\left(1-\frac{v^2}{c^2}\right)^{-1/2}, \tag{30}$$

which is indeed a *dilated* amount. However, we must not fail to note that the quantity $(\delta Q)^*$ as such cannot be interpreted as the heat gained by the system because the process under consideration necessarily entails a change of momentum $\delta \boldsymbol{P}(= \boldsymbol{v}\,\delta M)$ in the system, in view of which a logical interpretation of the process in the frame K would be to associate an amount $\boldsymbol{v}\cdot\delta\boldsymbol{P}(= v^2\,\delta M)$ of the resulting energy change with the "ordered" motion of the system as a whole and only the remaining amount $(c^2-v^2)\,\delta M\{= (1-v^2/c^2)\,(\delta Q)^*\}$ with the "thermal" or "disordered" motion within the system. Thus, for an observer in the frame K, it is only the latter amount, which is precisely equal to $(1-v^2/c^2)^{1/2}\,\delta Q_0$, that will appear to have gone into increasing the irregular, thermal agitation of the system and which alone can be regarded as "heat". In this connection, it may also be noted that in the generalized form (3) of the laws of thermodynamics, the "ordered" part of the energy change, which is directly associated with the motion of the centre of mass of the system, is taken care of separately (by the last term); in fact, it appears most instructive to write it as

$$(dE-\boldsymbol{v}\cdot d\boldsymbol{P}) = T\,dS-\mathcal{P}\,dV+\mu\,dN. \tag{3a}$$

In this form, the statement becomes manifestly covariant under a Lorentz transformation; the left-hand side, as well as each term on the right-hand side, is precisely $(1-v^2/c^2)^{1/2}$ times its corresponding rest measure!

Another consideration to which we can appeal is the *equipartition theorem* of classical mechanics; see Sec. 2.6. As suggested in Problem 6.23, this theorem can be formulated

† Several papers have been written, in recent years, to determine an appropriate basis for relativistic thermodynamics. For a survey of the various viewpoints, the reader may refer to van Kampen (1969), Yuen (1970) and Stuart *et al.* (1970).

for a moving system as well; moreover, the reformulated theorem is covariant under a Lorentz transformation. The interesting thing to observe here is that the quantity $\langle \boldsymbol{p}\cdot(\dot{\boldsymbol{q}}-\boldsymbol{v})\rangle$ that appears in the statement of this theorem is precisely $(1-v^2/c^2)^{1/2}$ times its corresponding rest measure $\langle \boldsymbol{p}_0\cdot\dot{\boldsymbol{q}}_0\rangle$; this strongly suggests that the relative temperature T must also be $(1-v^2/c^2)^{1/2}$ times the rest temperature T_0. For details, see Pathria (1970).

6.6. Gaseous systems composed of molecules with internal motion

In most of our studies so far we have considered only the translational part of the molecular motion. Though this aspect of motion is invariably present in a gaseous system, other aspects, which are essentially concerned with the internal motion of the molecules, also exist. It is only natural that in the calculation of the physical properties of such a system, contributions arising from these motions are also taken into account. In doing so, we shall assume here that (i) the effects of the intermolecular interactions are negligible and (ii) the nondegeneracy criterion

$$\frac{nh^3}{(2\pi mkT)^{3/2}} \ll 1 \tag{5.5.20}$$

is fulfilled; effectively, this makes our system an *ideal, Boltzmannian* gas. Under these assumptions, which hold sufficiently well in a large number of practical applications, the partition function of the system is given by

$$Q_N(V, T) = \frac{1}{N!}[Q_1(V, T)]^N, \tag{1}$$

where

$$Q_1(V, T) = \left\{\frac{V}{h^3}(2\pi mkT)^{3/2}\right\} j(T); \tag{2}$$

cf. (3.5.19). The factor within the curly brackets is the familiar translational partition function of a molecule, while the factor $j(T)$ is supposed to be the partition function corresponding to the internal motions. The latter may be written as

$$j(T) = \sum_i g_i e^{-\varepsilon_i/kT}, \tag{3}$$

where ε_i is the molecular energy associated with an internal state of motion (which is characterized by the quantum numbers i), while g_i represents the degeneracy of that state.

The contributions made by the internal motions of the molecules to the various thermodynamic quantities of the system follow straightforwardly from the function $j(T)$. We obtain

$$A_{\text{int}} = -NkT \ln j, \tag{4}$$

$$\mu_{\text{int}} = -kT \ln j, \tag{5}$$

$$S_{\text{int}} = Nk\left(\ln j + T\frac{\partial}{\partial T}\ln j\right), \tag{6}$$

$$U_{\text{int}} = NkT^2\frac{\partial}{\partial T}\ln j \tag{7}$$

and

$$(C_V)_{\text{int}} = Nk \frac{\partial}{\partial T}\left\{T^2 \frac{\partial}{\partial T} \ln j\right\}. \tag{8}$$

Thus, the central problem in this study consists of deriving an explicit expression for the function $j(T)$ from a knowledge of the internal states of the molecules. For this purpose, we note that the internal state of a molecule is determined by (i) the electronic state, (ii) the state of the nuclei, (iii) the vibrational state and (iv) the rotational state. Rigorously speaking, these four modes of excitation mutually interact; in many cases, however, they can be treated independently of one another. We then write

$$j(T) = j_{\text{elec}}(T)\, j_{\text{nuc}}(T)\, j_{\text{vib}}(T)\, j_{\text{rot}}(T), \tag{3a}$$

with the result that the net contribution made by the internal motions to the various thermodynamic quantities of the system is given by a simple sum of the four respective contributions. There is one interaction, however, which plays a very special role in the case of homonuclear molecules, such as AA, and that is the one between the states of the nuclei and the rotational states. In such a case, we better write

$$j(T) = j_{\text{elec}}(T)\, j_{\text{nuc-rot}}(T)\, j_{\text{vib}}(T). \tag{3b}$$

We now examine this problem for various physical systems in the order of increasing complexity.

A. Monatomic molecules

At the very outset we should note that we cannot consider a monatomic gas except at temperatures such that the thermal energy kT is small in comparison with the ionization energy E_{ion}; for different atoms, this amounts to the condition: $T \ll E_{\text{ion}}/k \simeq 10^4$–$10^5$ °K. At these temperatures the number of ionized atoms in the gas would be quite insignificant. The same would be true for atoms in the excited states, for the simple reason that the separation of any of the excited states from the ground state of the atom is generally comparable to the ionization energy itself. Thus, we may regard all the atoms of the gas to be in their (electronic) ground state.

Now, there is a well-known class of atoms, namely He, Ne, A, . . ., which, in their ground state, possess neither orbital angular momentum nor spin ($L = S = 0$). Their (electronic) ground state is clearly a singlet: $g_e = 1$. The nucleus, however, possesses a degeneracy which arises from the possibility of different orientations of the nuclear spin.* If the value of this spin is S_n, the corresponding degeneracy factor $g_n = 2S_n+1$. Moreover, a monatomic molecule is incapable of having any vibrational or rotational states. The internal partition function (3a) of such a molecule is, therefore, given by

$$j(T) = (g)_{\text{gr.st.}} = g_e \cdot g_n = 2S_n+1. \tag{9}$$

* As is well known, the presence of the nuclear spin gives rise to the so-called *hyperfine structure* in the electronic states. However, the intervals of this structure are such that for practically all temperatures of interest they are small in comparison with kT; for concreteness, these intervals correspond to T-values of the order of 10^{-1}–10^0 °K. Accordingly, in the evaluation of the partition function, the hyperfine splitting of the electronic state may be disregarded while the multiplicity introduced by the nuclear spin may be taken care of through a degeneracy factor.

Equations (4)–(8) then tell us that the internal motions in this case contribute only towards properties such as the chemical potential and the entropy of the gas; they do not make contribution towards properties such as the internal energy and the specific heat.

If, on the other hand, the ground state does not possess orbital angular momentum but possesses spin ($L = 0$, $S \neq 0$—as, for example, in the case of alkali atoms), then the ground state will still have no fine structure; nevertheless, it will have a definite degeneracy: $g_e = 2S+1$. As a result, the internal partition function $j(T)$ will get multiplied by a factor of $(2S+1)$ and the properties such as the chemical potential and the entropy of the gas will get accordingly modified.

In other cases, the ground state of the atom may possess both orbital angular momentum and spin ($L \neq 0$, $S \neq 0$); the ground state would then possess a definite fine structure. The intervals of this structure are, in general, comparable with kT; hence, in the evaluation of the partition function, the energies of the various components of the fine structure must be taken into account. Since these components differ from one another in the value of the total angular momentum J, the relevant partition function may be written as

$$j_{\text{elec}}(T) = \sum_J (2J+1)e^{-\varepsilon_J/kT}. \tag{10}$$

The foregoing expression simplifies considerably in the following limiting cases:

(a) $kT \gg$ all ε_J; then

$$j_{\text{elec}}(T) \simeq \sum_J (2J+1) = (2L+1)(2S+1). \tag{10a}$$

(b) $kT \ll$ all ε_J; then

$$j_{\text{elec}}(T) \simeq (2J_0+1)e^{-\varepsilon_0/kT}, \tag{10b}$$

where J_0 is the total angular momentum, and ε_0 the energy, of the atom in the *lowest* state. In either case, the electronic motion makes no contribution towards the specific heat of the gas. Of course, at intermediate temperatures we do obtain a contribution towards this property. And, in view of the fact that both at high and low temperatures the specific heat tends to be equal to the translational value $\frac{3}{2}Nk$, it must be passing through a maximum at a temperature comparable to the separation of the fine structure levels.*

Needless to say, the multiplicity $(2S_n+1)$ introduced by the nuclear spin must be taken into account in each case.

B. Diatomic molecules

Now, just as we could not consider a monatomic gas except at temperatures for which kT is small compared with the energy of ionization, for similar reasons one may not consider a diatomic gas except at temperatures for which kT is small compared with the energy of dissociation; for different molecules this amounts once again to the condition: $T \ll E_{\text{diss}}/k \simeq 10^4$–$10^5$ °K. At these temperatures the number of dissociated molecules in the gas would be quite insignificant. At the same time, in most cases, there would be

* It may be worth while to note here that the values of $\Delta\varepsilon_J/k$ for the components of the normal triplet term of oxygen are 230°K and 320°K, while those for the normal quintuplet term of iron range from 600°K to 1400°K.

practically no molecules in the excited states as well, for the separation of any of these states from the ground state of the molecule is in general comparable to the dissociation energy itself.* Accordingly, in the evaluation of $j(T)$, we have to take into account only the lowest electronic state of the molecule.

The lowest electronic state, in most cases, is non-degenerate: $g_e = 1$. We then need not consider any further the question of the electronic state making a contribution towards the thermodynamic properties of the gas. However, certain molecules (though not very many) have, in their lowest electronic state, either (i) a nonzero orbital angular momentum ($\Lambda \neq 0$) or (ii) a nonzero spin ($S \neq 0$) or (iii) both Λ and S nonzero. In case (i), the electronic state acquires a two-fold degeneracy corresponding to the two possible orientations of the orbital angular momentum relative to the molecular axis;† as a result, we have: $g_e = 2$. In case (ii), the state acquires a degeneracy of $2S+1$ corresponding to the space quantization of the spin.‡ In both these cases the chemical potential and the entropy of the gas are modified by the multiplicity of the electronic state, while the energy and the specific heat remain unaffected. In case (iii), we encounter a fine structure which necessitates a rather detailed study because the intervals of this structure are generally of the same order of magnitude as kT. In particular, for a doublet fine-structure term, such as the one that arises in the molecule NO ($\Pi_{1/2,\,3/2}$ with a separation of 178°K, the components themselves being Λ-doublets), we have for the electronic partition function

$$j_{\text{elec}}(T) = g_0 + g_1 e^{-\Delta/kT}, \tag{11}$$

where g_0 and g_1 are the degeneracy factors of the two components while Δ is their separation energy. The contribution made by (11) towards the various thermodynamic properties of the gas can be readily calculated with the help of the formulae (4)–(8). In particular, we obtain for the contribution towards the specific heat

$$(C_V)_{\text{elec}} = Nk \frac{(\Delta/kT)^2}{\left(1+(g_0/g_1)e^{\Delta/kT}\right)\left(1+(g_1/g_0)e^{-\Delta/kT}\right)}. \tag{12}$$

We note that this contribution vanishes both for $T \ll \Delta/k$ and for $T \gg \Delta/k$ and has a maximum value for a certain temperature $\approx \Delta/k$; cf. the corresponding situation in the case of monatomic molecules.

Let us now consider the effect of the vibrational states of the molecules on the thermodynamic properties of the gas. To have an idea of the temperature range over which this effect would be significant, we note that the magnitude of the corresponding quantum of energy, namely $\hbar\omega$, for different diatomic gases is of order of 10^3 °K. Thus, we would obtain full contributions (consistent with the dictates of the equipartition theorem) at temperatures of the order of 10^4 °K or more, and practically no contribution at temperatures of the order of 10^2 °K or less. We assume, however, that the temperature is not high enough

* An odd case arises with oxygen. The separation between its normal term $^3\Sigma$ and the first excited term $^1\Delta$ is about 11,250°K, whereas the dissociation energy is about 55,000°K. The relevant factor $e^{-\varepsilon_1/kT}$, therefore, can be quite significant even when the factor $e^{-E_{\text{diss}}/kT}$ is not, say for $T \approx$ 2000–6000°K.

† Strictly speaking, the term in question splits up into two levels—the so-called Λ-doublet. The separation of the levels, however, is such that we can safely neglect it.

‡ The separation of the resulting levels is again negligible from the thermodynamic point of view; as an example, one may cite the very narrow triplet term of O_2.

to excite vibrational states of large energy; the oscillations of the nuclei are then small in amplitude and hence harmonic. The energy levels for a mode of frequency ω are then given by the well-known expression, viz. $(n+\frac{1}{2})\hbar\omega$.* The evaluation of the vibrational partition function $j_{\text{vib}}(T)$ is quite elementary; see Sec. 3.7. In view of the rapid convergence of the series involved, the summation may formally be extended to $n = \infty$. The corresponding contributions towards the various thermodynamic properties of the system are given by eqns. (3.7.16) to (3.7.21). In particular, we have

$$(C_V)_{\text{vib}} = Nk\left(\frac{\Theta_v}{T}\right)^2 \frac{e^{\Theta_v/T}}{(e^{\Theta_v/T}-1)^2}; \qquad \Theta_v = \frac{\hbar\omega}{k}. \tag{13}$$

We note that for $T \gg \Theta_v$ the vibrational specific heat is very nearly equal to the equipartition value Nk; otherwise, it is always less than Nk. In particular, for $T \ll \Theta_v$, the specific heat tends to zero (see Fig. 6.7); the vibrational degrees of freedom are then said to be "frozen".

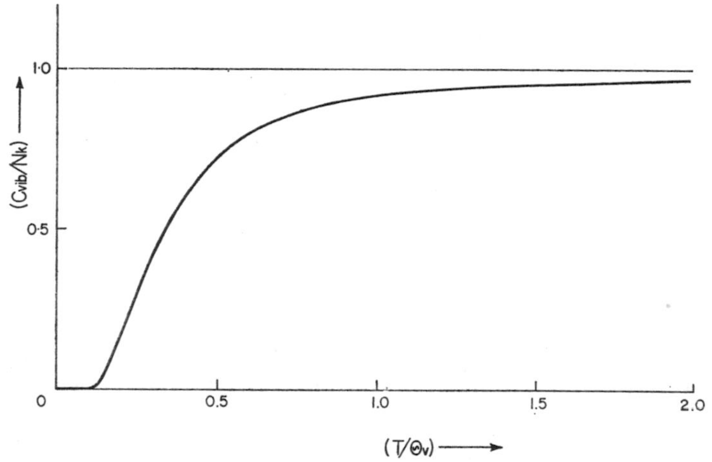

FIG. 6.7. The vibrational specific heat of a gas of diatomic molecules. At $T = \Theta_v$, the specific heat is already about 93 per cent of the equipartition value.

At sufficiently high temperatures, when vibrations with large n are also excited, the effects of anharmonicity and of interaction between the vibrational and the rotational modes of the molecule can become important.† However, since this happens only at large n, the appropriate corrections to the thermodynamic quantities can be determined even classically; see Problems 3.24 and 3.25. One finds that the first-order correction to C_{vib} is directly proportional to the temperature of the gas.

Finally, we consider the effect of (i) the states of the nuclei and (ii) the rotational states of the molecule; wherever necessary, we shall take into account the mutual interaction of

* It may be pointed out that the vibrational motion of a molecule is influenced by the centrifugal force arising from the molecular rotation. This leads to an interaction between the rotational and the vibrational modes. However, unless the temperature is too high, this interaction can be neglected and the two modes treated independently of one another.

† In principle, these two effects are of the same order of magnitude.

these modes. This interaction is of no relevance in the case of *heteronuclear* molecules, such as *AB*; it is, however, important in the case of *homonuclear* molecules, such as *AA*. We may, therefore, consider the former ones first.

The states of the nuclei in this case may be treated separately from the rotational states of the molecule. Proceeding in the some manner as for monatomic molecules, we conclude that the effect of the nuclear states is adequately taken care of through a degeneracy factor g_n. Denoting the spins of the two nuclei by S_A and S_B, this factor is given by

$$g_n = (2S_A+1)(2S_B+1). \tag{14}$$

As before, we obtain a finite contribution towards the chemical potential and the entropy of the gas but none towards the internal energy and the specific heat.

Now, the rotational levels of a *linear* "rigid" rotator, with two degrees of freedom (for the axis of rotation) and the principal moments of inertia $(I, I, 0)$, are given by

$$\varepsilon_{\text{rot}} = l(l+1)\hbar^2/2I; \tag{15}$$

here $I = \mu r_0^2$, where $\mu[= m_1m_2/(m_1+m_2)]$ is the *reduced mass* of the nuclei and r_0 the *equilibrium distance* between them. The rotational partition function of the molecule is then given by

$$\begin{aligned} j_{\text{rot}}(T) &= \sum_{l=0}^{\infty} (2l+1) \exp\left\{-l(l+1)\frac{\hbar^2}{2IkT}\right\} \\ &= \sum_{l=0}^{\infty} (2l+1) \exp\left\{-l(l+1)\frac{\Theta_r}{T}\right\}; \qquad \Theta_r = \frac{\hbar^2}{2Ik}. \end{aligned} \tag{16}$$

The values of Θ_r, for all gases except the ones involving the isotopes H and D, are much smaller than the room temperature. For example, the value of Θ_r for HCl is about 15°K, for N_2, O_2 and NO it lies between 2° and 3°K, while for Cl_2 it is about one-third of a degree. On the other hand, the values of Θ_r for H_2, D_2 and HD are, respectively, 85°K, 43°K and 64°K. These values give us an idea of the respective temperature ranges in which the effects arising from the discreteness of the rotational states are expected to be important.

For $T \gg \Theta_r$, the spectrum of the rotational states may be approximated by a continuum. The summation in (16) is then replaced by an integration:

$$j_{\text{rot}}(T) \simeq \int_0^{\infty} (2l+1) \exp\left\{-l(l+1)\frac{\Theta_r}{T}\right\} dl = \frac{T}{\Theta_r}. \tag{17}$$

The rotational specific heat is then given by

$$(C_V)_{\text{rot}} = Nk, \tag{18}$$

which is indeed consistent with the equipartition theorem.

A better evaluation of the sum (16) can be made with the help of the Euler–Maclaurin formula, viz.

$$\sum_{n=0}^{\infty} f(n) = \int_0^{\infty} f(x)\,dx + \tfrac{1}{2}f(0) - \tfrac{1}{12}f'(0) + \tfrac{1}{720}f'''(0) - \tfrac{1}{30240}f^{\text{v}}(0) + \ldots. \tag{19}$$

Putting

$$f(x) = (2x+1)\exp\{-x(x+1)\Theta_r/T\},$$

one obtains

$$j_{\rm rot}(T) = \frac{T}{\Theta_r}+\frac{1}{3}+\frac{1}{15}\frac{\Theta_r}{T}+\frac{4}{315}\left(\frac{\Theta_r}{T}\right)^2+\ldots, \tag{20}$$

which is the so-called *Mulholland's formula*; as expected, the main term of this formula is identical with the classical partition function (17). The corresponding result for the specific heat is

$$(C_V)_{\rm rot} = Nk\left\{1+\frac{1}{45}\left(\frac{\Theta_r}{T}\right)^2+\frac{16}{945}\left(\frac{\Theta_r}{T}\right)^3+\ldots\right\}, \tag{21}$$

which shows that at high temperatures the rotational specific heat *decreases* with temperatures and ultimately tends to the classical value Nk. Thus, at high (but finite) temperatures the rotational specific heat of a diatomic gas is greater than the classical value. On the other hand, it must go to zero as $T \to 0$. We, therefore, conclude that it must pass through at least one maximum. Detailed investigation based on numerical analysis shows that there is only one maximum which appears at a temperature of about 0.81 Θ_r and has a value of about 1.1 Nk; see Fig. 6.8.

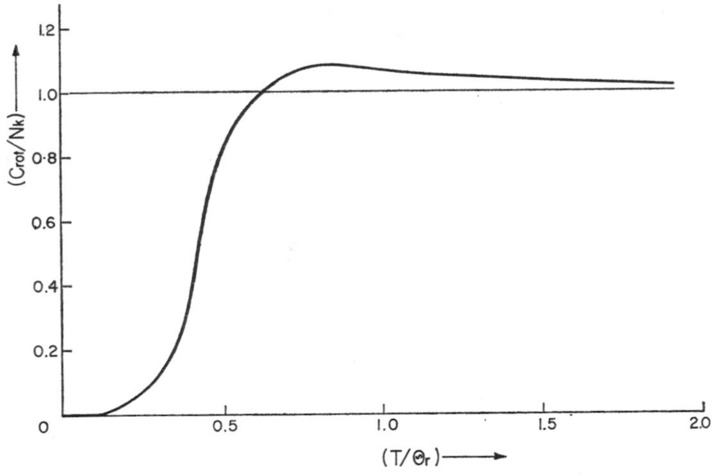

FIG. 6.8. The rotational specific heat of a gas of heteronuclear diatomic molecules.

In the opposite limiting case, namely for $T \ll \Theta_r$, one may retain only the first few terms of the sum (16); then

$$j_{\rm rot}(T) = 1+3e^{-2\Theta_r/T}+5e^{-6\Theta_r/T}+\ldots, \tag{22}$$

whence one obtains, in the *lowest* approximation,

$$(C_V)_{\rm rot} \simeq 12Nk\left(\frac{\Theta_r}{T}\right)^2 e^{-2\Theta_r/T}. \tag{23}$$

Thus, as $T \to 0$, the specific heat drops exponentially to zero; see again Fig. 6.8. We, therefore, conclude that at low enough temperatures the rotational degrees of freedom of the molecules are also "frozen".

At this stage it appears worth while to remark that since the internal motions of the molecules do not make any contribution towards the pressure of the gas (A_{int} being independent of V), the quantity $(C_P - C_V)$ is the same for a diatomic gas as for a monatomic one. Moreover, under the assumptions made in the very beginning of this section, the value of this quantity at all temperatures of interest would be equal to the classical value, namely Nk. Thus, at sufficiently low temperatures (when rotational as well as vibrational degrees of freedom of the molecules are "frozen"), we have by virtue of the translational motion alone

$$C_V = \tfrac{3}{2}Nk, \quad C_P = \tfrac{5}{2}Nk; \quad \gamma = \tfrac{5}{3}. \tag{24}$$

As temperature rises, the rotational degrees of freedom begin to "loosen", until we reach temperatures that are much larger than Θ_r but much smaller than Θ_v; the rotational degrees of freedom are then fully excited while the vibrational degrees of freedom are still "frozen". Accordingly, for $\Theta_r \ll T \ll \Theta_v$,

$$C_V = \tfrac{5}{2}Nk, \quad C_P = \tfrac{7}{2}Nk; \quad \gamma = \tfrac{7}{5}. \tag{25}$$

As temperature rises further, the vibrational degrees of freedom as well start loosening, until we reach temperatures that are much larger than Θ_v. Then, the vibrational degrees of freedom are also fully excited and we have

$$C_V = \tfrac{7}{2}Nk, \quad C_P = \tfrac{9}{2}Nk; \quad \gamma = \tfrac{9}{7}. \tag{26}$$

All these features are displayed in Fig. 6.9 where the experimental results for the specific heat at constant pressure have been plotted for the gases HD, HT and DT. We note

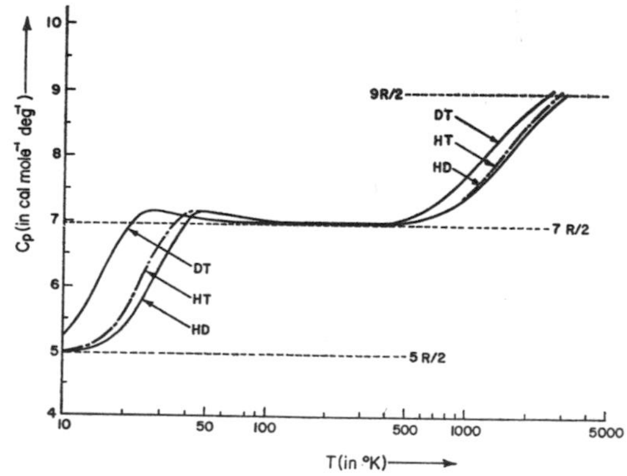

FIG. 6.9. The rotational-vibrational specific heat, C_P, of the diatomic gases HD, HT and DT.

that, in view of the considerable difference between the respective values of Θ_r and Θ_v, the situation depicted by (25) obtains over a considerably large range of temperatures. In passing, it may be pointed out that, for most diatomic gases, the situation at room temperatures corresponds to the one depicted by (25).

We shall now study the case of *homonuclear* molecules, such as AA. To start with, we consider the limiting case of high temperatures where classical approximation is admissible. The rotational motion of the molecule may then be visualized as a rotation of the molecular axis, i.e. the line joining the two nuclei, about an "axis of rotation" which is perpendicular to the molecular axis and passes through the center of mass of the molecule. Then, the two opposing positions of the molecular axis, viz. the ones that correspond to the azimuthal angles φ and $\varphi+\pi$, differ simply by an interchange of the two identical nuclei and, hence, correspond to only *one* distinct state of the molecule. Therefore, in the evaluation of the (classical) partition function, the range of the angle φ should be taken as $(0, \pi)$ instead of the customary one $(0, 2\pi)$. Moreover, since the energy of rotational motion does not depend upon the angle φ, the only effect of this on the partition function of the molecule is to reduce it by a factor of 2. We thus obtain, *in the classical approximation*,*

$$j_{\text{nuc-rot}}(T) = (2s_A+1)^2 \frac{T}{2\Theta_r}. \tag{27}$$

Obviously enough, the factor 2 here will not affect the specific heat of the gas; in the classical approximation, therefore, the specific heat of a gas of homonuclear molecules is the same as that of a corresponding gas of heteronuclear molecules.

In contrast, significant changes result at relatively lower temperatures when the states of rotational motion have to be treated as *discrete*. These changes arise from the coupling between the nuclear and the rotational states, which in turn arises from the symmetry character of the nuclear-rotational wave function. As discussed in Sec. 5.4, the total wave function of physical state must be either symmetric or antisymmetric (depending upon the statistics obeyed by the particles involved) with respect to an interchange of two identical particles. Now, the rotational wave function of a diatomic molecule is symmetric or antisymmetric according as the quantum number l is even or odd. The nuclear wave function, on the other hand, consists of a linear combination of the spin functions of the two nuclei and its symmetry character depends upon the manner in which the linear combination is formed. It is not difficult to see that, of the $(2s_A+1)^2$ different combinations that can in principle be constructed, exactly $(s_A+1)(2s_A+1)$ are symmetric with respect to an interchange of the nuclei and the remaining $s_A(2s_A+1)$ antisymmetric.† In constructing the total wave function as a product of the nuclear and the rotational wave functions, we then have to proceed as follows:

* It seems instructive to outline here the purely classical derivation of the rotational partition function. Specifying the rotation of the molecule by the angles (θ, φ) and the corresponding momenta by (p_θ, p_φ), the kinetic energy assumes the form:

$$\varepsilon_{\text{rot}} = \frac{1}{2I} p_\theta^2 + \frac{1}{2I \sin^2\theta} p_\varphi^2,$$

whence

$$j_{\text{rot}}(T) = \frac{1}{h^2} \int e^{-\varepsilon_{\text{rot}}/kT} (dp_\theta\, dp_\varphi\, d\theta\, d\varphi) = \frac{IkT}{\pi\hbar^2} \int_0^{\varphi_{\max}} d\varphi.$$

For heteronuclear molecules $\varphi_{\max} = 2\pi$, while for homonuclear ones $\varphi_{\max} = \pi$.

† See, for example, Schiff (1968), sec. 41.

(i) If the nuclei are fermions $\left(s_A = \frac{1}{2}, \frac{3}{2}, \ldots\right)$, as in the molecule H_2, the total wave function must be *antisymmetric*. To secure this, we may associate any one of the $s_A(2s_A+1)$ antisymmetric nuclear wave functions with any one of the even-l rotational wave functions or any one of the $(s_A+1)(2s_A+1)$ symmetric nuclear wave functions with any one of the odd-l rotational wave functions. Accordingly, the nuclear-rotational partition function of such a molecule would be

$$j_{\text{nuc-rot}}^{(\text{F.D.})}(T) = s_A(2s_A+1)r_{\text{even}} + (s_A+1)(2s_A+1)r_{\text{odd}}, \tag{28}$$

where

$$r_{\text{even}} = \sum_{l=0,2,\ldots}^{\infty} (2l+1)\exp\{-l(l+1)\Theta_r/T\} \tag{29}$$

and

$$r_{\text{odd}} = \sum_{l=1,3,\ldots}^{\infty} (2l+1)\exp\{-l(l+1)\Theta_r/T\}. \tag{30}$$

(ii) If the nuclei are bosons ($s_A = 0, 1, 2, \ldots$), as in the molecule D_2, the total wave function must be symmetric. To secure this, we may associate any one of the $(s_A+1)(2s_A+1)$ symmetric nuclear wave functions with any one of the even-l rotational wave functions or any one of the $s_A(2s_A+1)$ antisymmetric nuclear wave functions with any one of the odd-l rotational wave functions. We then have

$$j_{\text{nuc-rot}}^{(\text{B.E.})}(T) = (s_A+1)(2s_A+1)r_{\text{even}} + s_A(2s_A+1)r_{\text{odd}}. \tag{31}$$

At high temperatures, it is the large values of l that contribute most to the sums (29) and (30). The difference between the two sums then becomes negligibly small and we have

$$r_{\text{even}} \simeq r_{\text{odd}} \simeq \tfrac{1}{2} j_{\text{rot}}(T) = T/2\Theta_r; \tag{32}$$

cf. eqns. (16) and (17). Consequently, we have

$$j_{\text{nuc-rot}}^{(\text{B.E.})} \simeq j_{\text{nuc-rot}}^{(\text{F.D.})} = (2s_A+1)^2\, T/2\Theta_r, \tag{33}$$

which is identical with our previous result (27). Under these circumstances, the statistics governing the nuclei does not make significant difference to the thermodynamic behavior of the gas.

Things become very different if the temperature of the gas happens to be in a range comparable to the value of Θ_r. It seems most reasonable then to regard the gas as a mixture of two components, generally referred to as *ortho-* and *para-*, whose relative concentrations in equilibrium are determined by the relative magnitudes of the two parts of the partition function (28) or (31), as the case may be. Customarily, the name ortho- is given to that component which carries the larger statistical weight. Thus, in the case of fermions (as in H_2), the ortho- to para- ratio is given by

$$n^{(\text{F.D.})} = \frac{(s_A+1)r_{\text{odd}}}{s_A r_{\text{even}}}, \tag{34}$$

while in the case of bosons (as in D_2), the corresponding ratio is given by

$$n^{(\text{B.E.})} = \frac{(s_A+1)r_{\text{even}}}{s_A r_{\text{odd}}}. \tag{35}$$

As temperature rises, the factor r_{odd}/r_{even} tends to unity and the ratio n, in each case, approaches the temperature-independent value $(s_A+1)/s_A$. In the case of H_2, this limiting value is $3\left(\because s_A = \frac{1}{2}\right)$ while in the case of D_2, it is $2\,(\because s_A = 1)$. At sufficiently low temperatures, one may retain only the main terms of the sums (29) and (30), with the result

$$\frac{r_{odd}}{r_{even}} \simeq 3 \exp\left(-\frac{2\Theta_r}{T}\right) \qquad (T \ll \Theta_r), \tag{36}$$

which tends to zero as $T \to 0$. The ratio n, then, tends to zero in the case of fermions and to infinity in the case of bosons. Hence, as $T \to 0$, the hydrogen gas is wholly para-, while deuterium is wholly ortho-; of course, in each case, the molecules settle down in the rotational state $l = 0$.

At intermediate temperatures, one has to work with the equilibrium ratio (34), or (35), and with the composite partition function (28), or (31), in order to compute the thermodynamic properties of the gas. One finds, however, that the theoretical results so obtained do not *generally* agree with the ones obtained experimentally. The discrepancy was resolved by Dennison (1927) who pointed out that the samples of hydrogen, or deuterium, ordinarily subjected to experiment are not in thermal equilibrium as regards the relative magnitudes of the ortho- and para- components. These samples are ordinarily prepared and kept at room temperatures which are well above Θ_r, with the result that the ortho- to para- ratio in them is very nearly equal to the limiting value $(s_A+1)/s_A$. If now the temperature is lowered, one would expect the ratio to change in accordance with (34), or (35). However, it does not do so for the following reason. Since the transition of a molecule from one form of existence to another involves the flipping of the spin of one of its nuclei, the transition probability of the process is quite small. Actually, the periods involved are of the order of a year. Obviously, one cannot expect to attain the equilibrium ratio n during the short periods available. Consequently, even at lower temperatures, what one generally has is a *nonequilibrium* mixture of two independent substances, the relative concentration of which is fixed and preassigned. The partition functions (28) and (31) as such are, therefore, inapplicable. We rather have directly for the specific heat

$$C^{(\text{F.D.})} = \frac{s_A}{2s_A+1} C_{even} + \frac{s_A+1}{2s_A+1} C_{odd} \tag{37}$$

and

$$C^{(\text{B.E.})} = \frac{s_A+1}{2s_A+1} C_{even} + \frac{s_A}{2s_A+1} C_{odd}, \tag{38}$$

where

$$C_{even/odd} = Nk \frac{\partial}{\partial T} \{T^2 (\partial/\partial T) \ln r_{even/odd}\}. \tag{39}$$

We have, therefore, to compute C_{even} and C_{odd} separately and then derive the net value of the rotational specific heat with the help of the formula (37) or (38), as the case may be. Figure 6.10 shows the relevant results for hydrogen. Curves 1 and 2 correspond to the para-hydrogen (C_{even}) and the ortho-hydrogen (C_{odd}), respectively, while curve 3 represents the weighted mean, as given by eqn. (37). The experimental results are also shown in the figure; agreement between theory and experiment is clearly good.

Further evidence in favor of Dennison's explanation is obtained by performing experiments with ortho–para mixtures of different relative concentration. This can be done by speeding up the ortho–para conversion by passing hydrogen over activated charcoal. By doing this at various temperatures, and afterwards removing the catalyst, one can fix the ratio n at any desired value. The specific heat then follows a curve obtained by mixing

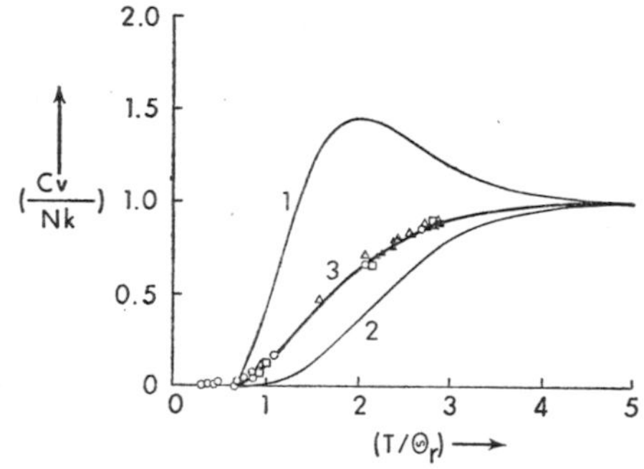

FIG. 6.10. The theoretical specific heat of a 1 : 3 mixture of para-hydrogen and ortho-hydrogen. The experimental points originate from various sources listed in Wannier (1966).

C_{even} and C_{odd} with appropriate weight factors. Further, if one measures the specific heat of the gas in such a way that the ratio n, at every temperature T, has the value given by the formula (34), it indeed follows the curve obtained from expression (28) for the partition function.

C. Polyatomic molecules

Once again, the translational degrees of freedom of the molecules contribute their usual share, viz. $\frac{3}{2}k$ per molecule, towards the specific heat of the gas. As regards the lowest electronic state, it is, in most cases, far below any of the excited states; nevertheless, it generally possesses a multiplicity (depending upon the orbital and spin angular momenta of the state) which can be taken care of by a degeneracy factor g_e. As regards the rotational states, they can be treated classically because the large values of the moments of inertia obtaining in polyatomic molecules make the quanta of rotational energy $(\hbar^2/2I_i)$ much smaller than the thermal energy (kT) at practically all temperatures of interest. Consequently, the interaction between the rotational states and the states of the nuclei can also be treated classically. As a result, the nuclear-rotational partition function is given by the product of the respective partition functions, divided by a symmetry number γ which denotes the number of physically indistinguishable configurations realized during one complete rotation of the molecule:*

$$j_{\text{nuc-rot}}(T) = \frac{g_{\text{nuc}} j^C_{\text{rot}}(T)}{\gamma}; \tag{40}$$

cf. eqn. (27). Here, $j^C_{\text{rot}}(T)$ is the rotational partition function of the molecule evaluated in the classical approximation (without paying regard to the presence of identical nuclei, if any);

* As examples, the symmetry number γ for H_2O (isosceles triangle) is 2, for NH_3 (regular triangular pyramid) it is 3, for CH_4 (tetrahedron) and C_6H_6 (regular hexagon) it is 12. For heteronuclear molecules, the symmetry number is unity.

it is given by

$$j^{C}_{\text{rot}}(T) = \pi^{1/2}\left(\frac{2I_1kT}{\hbar^2}\right)^{1/2}\left(\frac{2I_2kT}{\hbar^2}\right)^{1/2}\left(\frac{2I_3kT}{\hbar^2}\right)^{1/2}, \tag{41}$$

where I_1, I_2 and I_3 are the principal moments of inertia of the molecule; see Problems 6.31a and 6.31b.* The rotational specific heat is then given by

$$C_{\text{rot}} = Nk\frac{\partial}{\partial T}\left\{T^2\frac{\partial}{\partial T}\ln j^{C}_{\text{rot}}(T)\right\} = \frac{3}{2}Nk, \tag{42}$$

in agreement with the equipartition theorem.

As regards the vibrational states, we first note that, unlike a diatomic molecule, a polyatomic molecule has not one but several vibrational degrees of freedom. In particular, a non-collinear molecule consisting of n atoms has $3n-6$ vibrational degrees of freedom, six degrees of freedom out of the total number $3n$ having gone into the translational and the rotational motions. On the other hand, a collinear molecule consisting of n atoms would have $3n-5$ vibrational degrees of freedom, for the rotational motion in this case has only two, not three, degrees of freedom. The vibrational degrees of freedom correspond to a set of normal modes of vibration which in turn are characterized by a set of respective frequencies ω_i. It might happen that some of these frequencies have identical values; we then speak of *degenerate frequencies*.†

In the harmonic approximation, these normal modes may be treated independently of one another. The vibrational partition function of the molecule is then given by the product of the partition functions corresponding to individual normal modes,

$$j_{\text{vib}}(T) = \prod_i \frac{e^{-\Theta_i/2T}}{1-e^{-\Theta_i/T}}; \qquad \Theta_i = \frac{\hbar\omega_i}{k}, \tag{43}$$

and the vibrational specific heat by the sum of the contributions arising from the individual modes,

$$C_{\text{vib}} = Nk\sum_i\left\{\left(\frac{\Theta_i}{T}\right)^2\frac{e^{\Theta_i/T}}{(e^{\Theta_i/T}-1)^2}\right\}. \tag{44}$$

In general, the various Θ's are of the order of 10^3 °K; for instance, in the case of CO_2, which was cited in the last footnote, $\Theta_1 = \Theta_2 = 960$°K, $\Theta_3 = 1990$°K and $\Theta_4 = 3510$°K. For temperatures large in comparison with all the Θ_i, the specific heat would be given by the equipartition value, namely Nk for each of the normal modes. In practice, however, this

* In the case of a collinear molecule, such as N_2O and CO_2, there are only two degrees of freedom for rotation; consequently, $j^{C}_{\text{rot}}(T)$ is given by $(2I\,kT/\hbar^2)$, where I is the (common) value of the two moments of inertia of the molecule; cf. eqn. (17). Of course, we must also take into account the symmetry number γ. In the examples quoted here, the molecule N_2O, being spatially asymmetric (NNO), has symmetry number 1, while the molecule CO_2, being spatially symmetric (OCO), has symmetry number 2.

† For example, of the four frequencies characterizing the normal modes of vibration of the collinear molecule OCO, two that correspond to the (transverse) bending modes, namely $\underset{\downarrow}{\mathrm{O}}\,\overset{\uparrow}{\mathrm{C}}\,\underset{\downarrow}{\mathrm{O}}$, are equal, while the others that correspond to (longitudinal) oscillations along the molecular axis, namely $\leftarrow\mathrm{O}\ \mathrm{C}\rightarrow\ \leftarrow\mathrm{O}$ and $\leftarrow\mathrm{O}\ \mathrm{C}\ \mathrm{O}\rightarrow$, are different; see Problem 6.32.

limit can hardly be realized because the polyatomic molecules generally break up well before these high temperatures are reached. Secondly, the different frequencies ω_i of a polyatomic molecule are generally spread over a rather wide range of values. Consequently, as temperature rises, different modes of vibration get gradually "included" into the process; in between these "inclusions" the specific heat of the gas may stay constant over considerably large stretches of temperature.

Problems

6.1. Starting from eqn. (5.5.17), which determines the diagonal elements of the density matrix of an ideal gas obeying quantum statistics, obtain eqns. (6.2.16) for the grand partition function of the gas.

6.2. Show that the entropy of an ideal gas in thermal equilibrium is given by the formula

$$S = k \sum_{\varepsilon} [\langle n_\varepsilon + 1\rangle \ln \langle n_\varepsilon + 1\rangle - \langle n_\varepsilon\rangle \ln \langle n_\varepsilon\rangle]$$

in the case of *bosons* and by the formula

$$S = k \sum_{\varepsilon} [-\langle 1 - n_\varepsilon\rangle \ln \langle 1 - n_\varepsilon\rangle - \langle n_\varepsilon\rangle \ln \langle n_\varepsilon\rangle]$$

in the case of *fermions*. Verify that these results are consistent with the general formula

$$S = -k \sum_{\varepsilon} \left\{\sum_{n} p_\varepsilon(n) \ln p_\varepsilon(n)\right\},$$

where $p_\varepsilon(n)$ is the probability that there are exactly n particles in the state ε.

6.3. Derive, for all three statistics, the relevant expressions for the quantity $\langle n_\varepsilon^2\rangle - \langle n_\varepsilon\rangle^2$ of Sec. 6.3 from the respective expressions for the probabilities $p_\varepsilon(n)$. Further show that, quite generally,

$$\langle n_\varepsilon^2\rangle - \langle n_\varepsilon\rangle^2 = kT\left(\frac{\partial\langle n_\varepsilon\rangle}{\partial\mu}\right)_T;$$

compare with the corresponding result, namely (4.5.5), for a system embedded in a grand canonical ensemble.

6.4. The potential energy of a system of charged particles, characterized by particle charge e and number density $n(\boldsymbol{r})$, is given by

$$U = \frac{e^2}{2}\int \frac{n(\boldsymbol{r})\, n(\boldsymbol{r}')}{|\boldsymbol{r}-\boldsymbol{r}'|}\, d\boldsymbol{r}\, d\boldsymbol{r}' + e\int n(\boldsymbol{r})\, \varphi_{\text{ext}}(\boldsymbol{r})\, d\boldsymbol{r},$$

where $\varphi_{\text{ext}}(\boldsymbol{r})$ is the potential of the external field. Assume that the entropy of the system, apart from an additive constant, is given by the formula

$$S = -k\int n(\boldsymbol{r}) \ln n(\boldsymbol{r})\, d\boldsymbol{r};$$

cf. formula (3.3.20). Using these expressions, derive the equilibrium equations satisfied by the number density $n(\boldsymbol{r})$ and the total potential $\varphi(\boldsymbol{r})$ of the system, the latter being

$$\varphi_{\text{ext}}(\boldsymbol{r}) + e\int \frac{n(\boldsymbol{r}')}{|\boldsymbol{r}-\boldsymbol{r}'|}\, d\boldsymbol{r}'.$$

6.5. Show that the root-mean-square deviation in the value of the molecular energy ε, in a system obeying Maxwell–Boltzmann distribution, is $\sqrt{\frac{2}{3}}$ times the mean molecular energy $\bar{\varepsilon}$. Compare this result with that for an ideal gaseous system embedded in a canonical ensemble (Problem 3.18).

6.6. (a) Show that, for any law of distribution of molecular speeds, the inequality

$$\left\{\langle u\rangle \left\langle \frac{1}{u}\right\rangle\right\} \geq 1$$

must hold. Verify this result for the *Maxwellian distribution*.

(b) Estimate the fraction of the molecules, in a *Maxwellian distribution*, whose speed is less than the root-mean-square value.

6.7. Through a small window in a furnace, which contains a gas at a high temperature T, the spectral lines emitted by the gas molecules are observed. Because of molecular motions, each spectral line exhibits *Doppler broadening*. Prove that the variation of the relative intensity $I(\lambda)$ with the wavelength λ is given by the formula

$$I(\lambda) \propto \exp\left\{-\frac{mc^2(\lambda-\lambda_0)^2}{2\lambda_0^2 kT}\right\},$$

where m is the molecular mass, c the velocity of light and λ_0 the mean wavelength of the line.

6.8. An ideal classical gas composed of N particles of mass m is enclosed in an infinitely long cylinder placed in a uniform gravitational field (of acceleration g) and is in thermal equilibrium. Evaluate the partition function of the gas and derive expressions for the major thermodynamic quantities. Explain physically why the specific heat of this system is larger than that of a corresponding system in free space.

6.9. (a) Show that, *if the temperature is uniform*, the pressure of a classical gas in a uniform gravitational field decreases with height according to the *barometric formula*

$$P(z) = P(0)\exp\{-mgz/kT\},$$

where the various symbols have their usual meaning.*

(b) Derive the corresponding formula for an *adiabatic* atmosphere, i.e. the one in which (PV^{γ}), rather than (PV), stays constant. Also study the variation, with height, of the temperature and the density of the atmosphere.

6.10. A cylinder of radius R and height L rotates about its axis of symmetry with a constant angular velocity ω. Evaluate the partition function, and derive the density distribution, of an ideal classical gas enclosed in the cylinder, assuming that the gas is in thermal equilibrium at temperature T and that the effects of gravitation are negligible.

[Note that the Hamiltonian describing a physical system in a rotating frame of reference is given by $H' = H-\omega L$, where H is the Hamiltonian of the system in the laboratory frame and L its angular momentum.]

6.11. (a) Determine the momentum distribution of particles in a relativistic Boltzmannian gas: $\varepsilon = c(p^2+m_0^2c^2)^{1/2}$. Hence derive the limiting distributions for (i) a nonrelativistic gas and (ii) an extreme relativistic one.

(b) Verify that, quite generally, $\langle p_x u_x\rangle = \langle p_y u_y\rangle = \langle p_z u_z\rangle = kT$.

6.12. (a) Considering the loss of translational energy suffered by the molecules of a gas on reflection from a *receding* wall, derive, for a *quasi-static* adiabatic expansion of an ideal *nonrelativistic* gas, the well-known equation

$$PV^{\gamma} = \text{const.},$$

where $\gamma = (3a+2)/3a$, a being the ratio of the total energy to the translational energy of the gas.

(b) Show that, in the case of an ideal *extreme relativistic* gas, $\gamma = (3a+1)/3a$.

6.13. (a) Determine the number of impacts, made by gas molecules on a unit area of the wall in a unit time, for which the angle of incidence lies between θ and $\theta+d\theta$.

(b) Determine the number of impacts, made by gas molecules on a unit area of the wall in a unit time, for which the speed of the molecules lies between u and $u+du$.

(c) A molecule AB dissociates if it hits the surface of a solid catalyst with a translational energy greater than 10^{-19} J. Show that the rate of the dissociative reaction $AB \rightarrow A+B$ is more than doubled by raising the temperature of the gas from 300°K to 310°K.

6.14. Assuming that the effusion of gas molecules through an opening of area a in the wall of a vessel of volume V is so slow that the gas inside is always in a state of *quasi-static* equilibrium, determine the manner in

---

* This formula was first given by Boltzmann (1879). For a critical study of its derivation, see Walton (1969).

which the density, the temperature and the pressure of the gas would vary with time. Take the gas to be Boltzmannian.

[*Hint.* While the molecules inside the vessel have a mean energy $\frac{3}{2}kT$, the effused ones have a mean energy $2kT$, T being the *quasi-static* equilibrium temperature.]

6.15. A polyethylene balloon at an altitude of 30,000 m is filled with helium gas at a pressure of 10^{-2} atm and a temperature of 300°K. The balloon has a diameter of 10 m, and has numerous pinholes of diameter 10^{-5} m each. How many pinholes per square meter of the surface must there be if 1 per cent of the gas is to leak out in 1 hour?

6.16. Consider two Boltzmannian gases A and B, at pressures P_A and P_B and temperatures T_A and T_B respectively, contained in two regions of space which are in communication through a very narrow opening

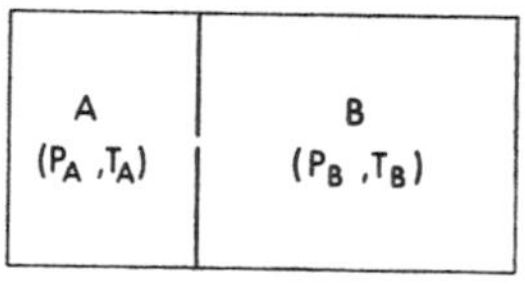

FIG. 6.11. The molecules of the gases A and B undergoing a two-way effusion.

in the partitioning wall; see Fig. 6.11. Show that the dynamic equilibrium set up by the mutual effusion of the two kinds of molecules satisfies the condition

$$P_A/P_B = (T_A/T_B)^{1/2},$$

rather than the condition $P_A = P_B$ (which would be satisfied if the equilibrium were set up under hydrodynamic flow).

[*Note.* The above considerations apply only when the collisions between the molecules are effectively inconsequential, i.e. when the gas is so rarefied that the mean free path of the molecules is comparable with the linear dimensions of the opening. A gas in such a state of dilution is said to be in the *Knudsen region.*]

6.17. A *small* sphere, with initial temperature T, is immersed in an ideal Boltzmannian gas at a temperature T_0. Assuming that the molecules incident on the sphere are first absorbed and then emitted with the temperature of the sphere, determine the variation of the temperature of the sphere with time.

[*Note.* The radius of the sphere may be assumed to be much smaller than the mean free path of the molecules.]

6.18. Study the statistical distribution of the relative velocity u' of a pair of molecules in a Boltzmannian gas and determine the average number of collisions a molecule has with all others in a unit time. Assume the molecules to be rigid spheres of diameter d.

[Note that the total number of intermolecular collisions in the gas in a unit time will be given by this number, multiplied by $N/2$.]

6.19. Show that the mean value of the *relative speed* of two molecules in a Maxwellian gas is $\sqrt{2}$ times the mean speed of a molecule with respect to the walls of the container.

[Note that a similar result for the root-mean-square speeds (instead of the mean speeds) holds under much more general conditions.]

6.20. What is the probability that two molecules picked up at random from a Maxwellian gas will have a total energy between E and $E+dE$? Verify that $\langle E \rangle = 3kT$.

6.21. On the basis of kinetic considerations, such as those developed in Sec. 6.4, identify the left-hand side of eqn. (6.5.2) with the pressure term $\beta(\mathcal{P}V)$.

6.22. Verify the mathematical result embodied in eqn. (6.5.15).

6.23. Show that the equipartition theorem in the case of a gaseous system in *mass motion* takes the form (Pathria, 1970)

$$\langle \boldsymbol{p}\cdot(\dot{\boldsymbol{q}}-\boldsymbol{v}) \rangle = 3kT.$$

Comparing this equation with its counterpart in the rest frame of the system, establish the covariance of the theorem under Lorentz transformation.

6.24. The energy difference between the lowest electronic state $^1S_0$ and the first excited state $^3S_1$ of the helium atom is 159,843 cm^{-1}. Evaluate the relative fraction of the excited atoms in helium gas at a temperature of 6000°K.

6.25. Derive an expression for the equilibrium constant $K(T)$ for the reaction $H_2+D_2 \rightleftarrows 2HD$ at temperatures high enough to allow classical approximation for the rotational motion. Show that $K(\infty) = 4$.
[*Note*. For the definition of the equilibrium constant, see Problem 3.14.]

6.26. With the help of the Euler–Maclaurin formula (6.6.19), derive *high-temperature* expansions for r_{even} and r_{odd}, as defined by eqns. (6.6.29) and (6.6.30), and obtain corresponding expansions for C_{even} and C_{odd}, as defined by eqn. (6.6.39). Compare the mathematical trend of these results with the nature of the corresponding curves in Fig. 6.10. Also study the *low-temperature* behavior of the two specific heats and again compare your results with the relevant parts of the aforementioned curves.

6.27. The potential energy between the atoms of a hydrogen molecule is given by the (semi-empirical) *Morse potential*

$$V(r) = V_0\{e^{-2(r-r_0)/a} - 2e^{-(r-r_0)/a}\},$$

where $V_0 = 7\times10^{-12}$ erg, $r_0 = 8\times10^{-9}$ cm and $a = 5\times10^{-9}$ cm. Calculate the rotational and vibrational quanta of energy and estimate the temperatures at which the rotational and vibrational modes of the molecules begin to contribute towards the specific heat of the hydrogen gas.

6.28. Show that the fractional change in the equilibrium value of the internuclear distance of a diatomic molecule, as a result of rotation, is given by

$$\frac{\Delta r_0}{r_0} \simeq \left(\frac{\hbar}{\mu r_0^2 \omega}\right)^2 J(J+1) = 4\left(\frac{\Theta_r}{\Theta_v}\right)^2 J(J+1);$$

here, ω is the angular frequency of the vibrational state in which the molecule happens to be. Estimate the numerical value of this fraction in a typical case.

6.29. The ground state of an oxygen atom is a triplet, with the following *fine structure*:

$$\varepsilon_{J=2} = \varepsilon_{J=1} - 158.5\ \text{cm}^{-1} = \varepsilon_{J=0} - 226.5\ \text{cm}^{-1}.$$

Calculate the relative fractions of the atoms occupying different J-levels in a sample of atomic oxygen at 300°K.

6.30. Calculate the contribution of the first excited electronic state, viz. $^1\Delta$ with $g_e = 2$, of the O_2 molecule towards the Helmholtz free energy and the specific heat of oxygen gas at a temperature of 5000°K; the separation of this state from the ground state, viz. $^3\Sigma$ with $g_e = 3$, is 7824 cm^{-1}. In what way would these results be affected if the parameters Θ_r and Θ_v of the O_2 molecule had different values in the two electronic states?

6.31. (a) The rotational kinetic energy of a rotator with three degrees of freedom can be written as

$$\varepsilon_{\text{rot}} = \frac{M_\xi^2}{2I_1} + \frac{M_\eta^2}{2I_2} + \frac{M_\zeta^2}{2I_3},$$

where (ξ, η, ζ) are the coordinates in a rotating frame of reference whose axes coincide with the principal axes of the rotator, while (M_ξ, M_η, M_ζ) are the corresponding angular momenta. Carrying out integrations in the phase space of the rotator, derive expression (6.6.41) for the partition function in the classical approximation.

(b) In terms of the Eulerian angles (θ, φ, ψ) and the corresponding momenta $(p_\theta, p_\varphi, p_\psi)$, the kinetic energy of the rotator is given by

$$\varepsilon_{\text{rot}} = \frac{1}{2I_1 \sin^2\theta}\{(p_\varphi - \cos\theta p_\psi)\cos\psi - \sin\theta \sin\psi p_\theta\}^2$$
$$+ \frac{1}{2I_2 \sin^2\theta}\{(p_\varphi - \cos\theta p_\psi)\sin\psi + \sin\theta\cos\psi p_\theta\}^2 + \frac{1}{2I_3}p_\psi^2.$$

Carrying out integrations in the phase space appropriate to these coordinates, derive the same expression for the partition function of the rotator.
[*Hint*. Carry out integrations in the order $p_\theta, p_\varphi, p_\psi$].

6.32. Determine the translational, rotational and vibrational contributions towards the molar entropy and the molar specific heat of carbon dioxide at n.t.p. Assume the ideal-gas formulae and use the following data: molecular weight $M = 44.01$; moment of inertia I of a CO_2 molecule $= 71.67 \times 10^{-40}$ g cm^2; wave numbers of the different modes of vibration: $\bar{\nu}_1 = \bar{\nu}_2 = 667.3$ cm^{-1}, $\bar{\nu}_3 = 1383.3$ cm^{-1} and $\bar{\nu}_4 = 2439.3$ cm^{-1}.

6.33. Determine the molar specific heat of ammonia at a temperature of 300°K. Assume the ideal-gas formulae and use the following data:

The principal moments of inertia: $I_1 = 4.44 \times 10^{-40}$ g cm^2 and $I_2 = I_3 = 2.816 \times 10^{-40}$ g cm^2; wave numbers of the different modes of vibration: $\bar{\nu}_1 = \bar{\nu}_2 = 3336$ cm^{-1}, $\bar{\nu}_3 = \bar{\nu}_4 = 950$ cm^{-1}, $\bar{\nu}_5 = 3414$ cm^{-1} and $\bar{\nu}_6 = 1627$ cm^{-1}.

CHAPTER 7

IDEAL BOSE SYSTEMS

IN CONTINUATION of Secs. 6.1–6.3, we shall now investigate in detail the physical behavior of a class of systems in which, while the intermolecular interactions are still negligible, the effects of quantum statistics (which arise from the indistinguishability of the particles) assume an increasingly important role. This automatically means that the temperature T and the particle density n of the system do not conform to the criterion

$$n\lambda^3 \equiv \frac{nh^3}{(2\pi mkT)^{3/2}} \ll 1, \tag{6.3.5}$$

where $\lambda \{\equiv h/(2\pi mkT)^{1/2}\}$ is the *mean thermal wavelength* of the particles. In fact, the degeneracy discriminant $(n\lambda^3)$ turns out to be a very appropriate parameter, in terms of which the various physical properties of the system can be adequately expressed. In the limit $(n\lambda^3) \to 0$, all physical properties go over smoothly to their classical counterparts. For small, but not negligible, values of $(n\lambda^3)$, the various quantities pertaining to the system can be expanded as power series in this parameter; from these expansions one obtains the first glimpse of the analytical manner in which a departure from the classical behavior sets in. When $(n\lambda^3)$ becomes of the order of unity, the behavior of the system becomes significantly different from the classical one and is characterized by typical quantum effects. A study of the system under these circumstances brings one face to face with a set of phenomena unknown in classical statistics!

It is evident that a system is better prepared to display quantum behavior when it is at a relatively low temperature and/or has a relatively high density of particles.* Moreover, the smaller the particle mass the larger the quantum effects.

Now, when $(n\lambda^3)$ is of the order of unity, then not only does the behavior of a physical system exhibit a significant departure from typical classical behavior but it is also influenced by the fact whether the particles constituting the system obey Bose–Einstein statistics or Fermi–Dirac statistics. Under these circumstances, the properties of the two kinds of systems

* It is indeed the ratio $n/T^{3/2}$, rather than the quantities n and T separately, that determines the degree of degeneracy prevailing in the system. For instance, white dwarf stars, even at temperatures of the order of 10^7 °K, constitute statistically degenerate systems; see Sec. 8.4.

are themselves very different. In the present chapter we propose to consider systems belonging to the first category while the succeeding chapter will deal with systems belonging to the second category.

7.1. Thermodynamic behavior of an ideal Bose gas

We obtained, in Secs. 6.1 and 6.2, the following formulae for an ideal Bose gas:

$$\frac{PV}{kT} \equiv \ln \mathcal{Q} = -\sum_{\varepsilon} \ln (1-ze^{-\beta\varepsilon}) \tag{1}$$

and

$$N \equiv \sum_{\varepsilon} \langle n_\varepsilon \rangle = \sum_{\varepsilon} \frac{1}{z^{-1}e^{\beta\varepsilon}-1}, \tag{2}$$

where $\beta = 1/kT$, while z is the fugacity of the gas; the latter is related to the chemical potential μ through the formula

$$z \equiv \exp (\mu/kT). \tag{3}$$

In view of the fact that, for large V, the spectrum of the single-particle states $\varepsilon(\boldsymbol{p})$ of the system is almost a continuous one, the summations on the right-hand sides of eqns. (1) and (2) may be replaced by integrations. In doing this, we make use of the asymptotic expression (2.4.7) for the density of states $a(\varepsilon)$ in the neighborhood of a given energy value ε, namely*

$$a(\varepsilon)\, d\varepsilon = \{(2\pi V/h^3)(2m)^{3/2}\varepsilon^{1/2}\}\, d\varepsilon. \tag{4}$$

We, however, note that by substituting the foregoing expression into our integrals we are inadvertently giving a weight *zero* to the energy level $\varepsilon = 0$. This is clearly wrong because in a quantum-mechanical treatment we must give a statistical weight *unity* to every non-degenerate single-particle state in the system. It is, therefore, advisable to take this particular state out of the main sequence and include the rest of them into the respective integrals. We thus obtain

$$\frac{P}{kT} = -\frac{2\pi}{h^3}(2m)^{3/2}\int_0^\infty \varepsilon^{1/2} \ln (1-ze^{-\beta\varepsilon})\, d\varepsilon - \frac{1}{V}\ln (1-z) \tag{5}$$

and

$$\frac{N}{V} = \frac{2\pi}{h^3}(2m)^{3/2}\int_0^\infty \frac{\varepsilon^{1/2}\, d\varepsilon}{z^{-1}e^{\beta\varepsilon}-1} + \frac{1}{V}\frac{z}{1-z}; \tag{6}$$

of course, the lower limit of the integrals can still be taken as 0, because the state $\varepsilon = 0$ is not going to contribute towards the integrals anyway.

Before proceeding further a word may be said about the relative importance of the last terms in eqns. (5) and (6). For $z \ll 1$, which corresponds to situations not far removed from

* The theory of this section is restricted to a system of *nonrelativistic* particles. For the general case, see Kothari and Singh (1941) and Landsberg and Dunning-Davis (1965).

the classical limit, each of these terms is of the order of $1/N$ and, therefore, negligible. However, as z increases and assumes values close to unity, the term $z/(1-z)V$ in eqn. (6), which is identically equal to N_0/V (N_0 being the number of particles in the ground state, viz. $\varepsilon = 0$), can well become a significant fraction of the quantity N/V; this accumulation of a macroscopic fraction of the given particles into a single state $\varepsilon = 0$ leads to the famous phenomenon of *Bose–Einstein condensation*. Nevertheless, since the quantity $z/(1-z)$ is identically equal to N_0 and hence z is identically equal to $N_0/(N_0+1)$, the term $\{-V^{-1}\ln(1-z)\}$ in eqn. (5) is identically equal to $\{V^{-1}\ln(N_0+1)\}$ and is at most $0(N^{-1}\ln N)$; this term is, therefore, negligible for *all* values of z and may be dropped altogether.

We now obtain from eqns. (5) and (6), on substituting $\beta\varepsilon = p^2/(2mkT) = x$,

$$\frac{P}{kT} = -\frac{2\pi(2mkT)^{3/2}}{h^3}\int_0^\infty x^{1/2}\ln(1-ze^{-x})\,dx = \frac{1}{\lambda^3}g_{5/2}(z) \tag{7}$$

and

$$\frac{N-N_0}{V} = \frac{2\pi(2mkT)^{3/2}}{h^3}\int_0^\infty \frac{x^{1/2}\,dx}{z^{-1}e^x-1} = \frac{1}{\lambda^3}g_{3/2}(z), \tag{8}$$

where

$$\lambda = h/(2\pi mkT)^{1/2} \tag{9}$$

and

$$g_n(z) = \frac{1}{\Gamma(n)}\int_0^\infty \frac{x^{n-1}\,dx}{z^{-1}e^x-1}; \tag{10}$$

it may be noted that to write eqn. (7) in terms of the function $g_{5/2}(z)$ we have first carried out an integration by parts. Equations (7) and (8) are our basic results; on elimination of z, they give us the *equation of state* of the ideal Bose gas.

The internal energy of this system is given by

$$U \equiv -\left(\frac{\partial}{\partial\beta}\ln\mathcal{Q}\right)_{z,V} = kT^2\left\{\frac{\partial}{\partial T}\left(\frac{PV}{kT}\right)\right\}_{z,V}$$

$$= kT^2Vg_{5/2}(z)\left\{\frac{d}{dT}\left(\frac{1}{\lambda^3}\right)\right\} = \frac{3}{2}kT\frac{V}{\lambda^3}g_{5/2}(z); \tag{11}$$

here, use has been made of eqn. (7) and of the fact that $\lambda \propto T^{-1/2}$. Thus, quite generally, our system satisfies the relationship

$$P = \tfrac{2}{3}(U/V). \tag{12}$$

For small values of z, we can make use of the expansion

$$g_n(z) = \sum_{l=1}^{\infty}\frac{z^l}{l^n} = z + \frac{z^2}{2^n} + \frac{z^3}{3^n} + \ldots; \tag{13}$$

see Appendix D. At the same time, we can neglect N_0 in comparison with N. An elimination of z between eqns. (7) and (8) can then be carried out by first inverting the series appearing in (8) to obtain an expansion for z in the powers of $(n\lambda^3)$ and then substituting this expansion into the series appearing in (7). The equation of state thereby takes the form of the *virial expansion*, namely

$$\frac{PV}{NkT} = \sum_{l=1}^{\infty} a_l \left(\frac{\lambda^3}{v}\right)^{l-1}, \tag{14}$$

where $v(\equiv 1/n)$ is the volume per particle; the coefficients a_l, which are referred to as the *virial coefficients* of the system, turn out to be

$$\left.\begin{aligned} a_1 &= 1, \\ a_2 &= -\frac{1}{4\sqrt{2}} = -0.17678, \\ a_3 &= -\left(\frac{2}{9\sqrt{3}} - \frac{1}{8}\right) = -0.00330, \\ a_4 &= -\left(\frac{3}{32} + \frac{5}{32\sqrt{2}} - \frac{1}{2\sqrt{6}}\right) = -0.00011, \end{aligned}\right\} \tag{15}$$

and so on. For the specific heat of the gas we obtain

$$\begin{aligned} \frac{C_V}{Nk} &\equiv \frac{1}{Nk}\left(\frac{\partial U}{\partial T}\right)_{N,V} = \frac{3}{2}\left\{\frac{\partial}{\partial T}\left(\frac{PV}{Nk}\right)\right\}_v \\ &= \frac{3}{2}\sum_{l=1}^{\infty} \frac{5-3l}{2} a_l \left(\frac{\lambda^3}{v}\right)^{l-1} \\ &= \frac{3}{2}\left[1 + 0.0884\left(\frac{\lambda^3}{v}\right) + 0.0066\left(\frac{\lambda^3}{v}\right)^2 + 0.0004\left(\frac{\lambda^3}{v}\right)^3 + \dots\right]. \end{aligned} \tag{16}$$

As $T \to \infty$ (and hence $\lambda \to 0$), both the pressure and the specific heat of the gas approach their classical values, namely nkT and $\frac{3}{2}Nk$, respectively. We also note that at finite, but large, temperatures the specific heat of the gas is larger than its limiting value; in other words, the (C_V, T)-curve has a negative slope at high temperatures. On the other hand, as $T \to 0$, the specific heat must tend to zero. Consequently, it must pass through a maximum. A detailed study shows that this maximum is in the nature of a cusp which appears at a characteristic temperature T_c; the derivative of the specific heat is found to be discontinuous at this temperature (see Fig. 7.4).

As the temperature of the system falls (and hence the value of the parameter λ^3/v grows), expansions of the type (14) and (16) do not remain useful. We then have to work with the formulae (7), (8) and (11) as such. The precise value of z is then obtained from eqn. (8), which may be rewritten as

$$N_e = V\frac{(2\pi mkT)^{3/2}}{h^3} g_{3/2}(z), \tag{17}$$

where N_e is the number of particles in the excited states $(\varepsilon \neq 0)$; of course, unless z gets extremely close to unity, N_e is practically identical with the total number of particles N.*

* It must be noted here that the largest value z can have *in principle* is unity, for otherwise some of the occupation numbers $\langle n_\varepsilon \rangle$ will have to be negative, which is not acceptable. Actually, we find that, as $T \to 0$, $z \to N/(N+1)$, which is very nearly the same as unity (but certainly on the right side of it).

It is obvious that, for $0 \leqslant z \leqslant 1$, the function $g_{3/2}(z)$ increases monotonically with z and is *bounded*, its largest value being

$$g_{3/2}(1) = 1+\frac{1}{2^{3/2}}+\frac{1}{3^{3/2}}+\ldots \equiv \zeta\left(\frac{3}{2}\right) \simeq 2.612; \tag{18}$$

see eqn. (D.6). Hence, for all z of interest,

$$g_{3/2}(z) \leqslant \zeta(\tfrac{3}{2}). \tag{19}$$

Consequently, for given values of V and T, the total number of particles in all the excited states taken together is also bounded, i.e.

$$N_e \leqslant V\frac{(2\pi mkT)^{3/2}}{h^3}\zeta\left(\frac{3}{2}\right). \tag{20}$$

Now, so long as the actual number of particles in the system is less than this limiting value, everything is well and good; practically all the particles in the system are distributed over the excited states and the precise value of z is determined by eqn. (17), with $N_e \simeq N$. On the other hand, if the actual number of particles exceeds this limiting value, then it is natural that the excited states will receive as many particles as they can hold, namely

$$N_e = V\frac{(2\pi mkT)^{3/2}}{h^3}\zeta\left(\frac{3}{2}\right), \tag{21a}$$

while the rest of them will be pushed *en masse* into the ground state $\varepsilon = 0$ (whose capacity, under the circumstances, is practically unlimited):

$$N_0 = N-\left\{V\frac{(2\pi mkT)^{3/2}}{h^3}\zeta\left(\frac{3}{2}\right)\right\}. \tag{21b}$$

The value of z is then determined by the formula

$$z = \frac{N_0}{N_0+1} \simeq 1-\frac{1}{N_0}, \tag{22}$$

which, for all practical purposes, would be equal to unity. This curious phenomenon of a macroscopically large number of particles accumulating in a single quantum state ($\varepsilon = 0$) is generally referred to as the phenomenon of *Bose–Einstein condensation.* In a certain sense, this phenomenon is akin to the familiar process of a vapor condensing into the liquid state, which takes place in the ordinary physical space. Conceptually, however, the two processes are very different. Firstly, the Bose–Einstein condensation is purely of a quantum origin (occurring even in the absence of intermolecular forces); secondly, it takes place at best in the momentum space and not in the coordinate space.*

* Of course, the repercussions of this phenomenon in the coordinate space are no less curious It. prepares the stage for the onset of *superfluidity*, a quantum manifestation about which we propose to say something in Sec. 7.5.

The condition for the appearance of Bose–Einstein condensation is

$$N > VT^{3/2}\frac{(2\pi mk)^{3/2}}{h^3}\zeta\left(\frac{3}{2}\right) \tag{23}$$

or, if we hold N and V constant and vary T,

$$T < T_c = \frac{h^2}{2\pi mk}\left\{\frac{N}{V\zeta(\frac{3}{2})}\right\}^{2/3}; \tag{24*}$$

the symbol T_c denotes a characteristic temperature that depends upon the particle mass m and the particle density N/V in the system. Accordingly, for $T < T_c$, the system may be looked upon as a mixture of two "phases":

(i) a *gaseous* phase, consisting of $N_e\{= N(T/T_c)^{3/2}\}$ particles distributed over the excited states ($\varepsilon \neq 0$), and
(ii) a *condensed* phase, consisting of $N_0\{= (N-N_e)\}$ particles accumulated in the ground state ($\varepsilon = 0$).

Figure 7.1 shows the manner in which the complementary fractions (N_e/N) and (N_0/N) vary with the temperature of the system. For $T > T_c$, we have the gaseous phase alone; the number of particles in the ground state, viz. $z/(1-z)$, is then $0(1)$ which is completely negligible in comparison with the total number N.

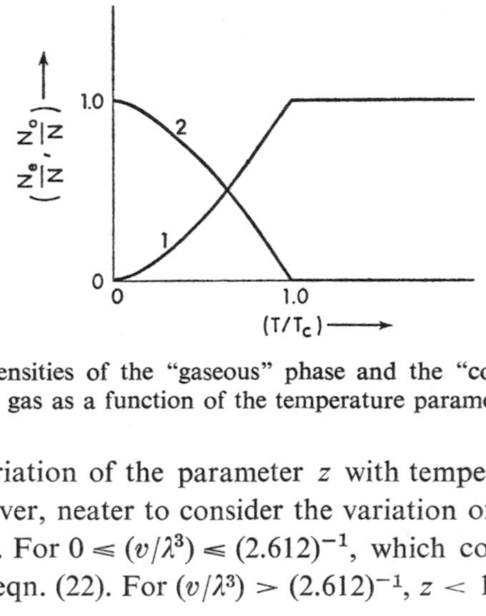

FIG. 7.1. The partial densities of the "gaseous" phase and the "condensed" phase in an ideal Bose gas as a function of the temperature parameter (T/T_c).

A knowledge of the variation of the parameter z with temperature T is also of considerable interest. It is, however, neater to consider the variation of z with (v/λ^3), the latter being proportional to $T^{3/2}$. For $0 \leqslant (v/\lambda^3) \leqslant (2.612)^{-1}$, which corresponds to $0 \leqslant T \leqslant T_c$, the parameter $z \simeq 1$; see eqn. (22). For $(v/\lambda^3) > (2.612)^{-1}$, $z < 1$ and is determined by the relationship

$$g_{3/2}(z) = (\lambda^3/v) < 2.612; \tag{25†}$$

* For a rigorous discussion of the onset of Bose–Einstein condensation, see Landsberg (1954b) where an attempt has also been made to coordinate much of the previously published work on this topic. A mention may be made of the work of Temperley (1949), in particular, who has studied the problem of Bose–Einstein condensation on the basis of the "partition theory of numbers"; see also Nanda (1953).

† An equivalent relationship is $g_{3/2}(z)/g_{3/2}(1) = (T_c/T)^{3/2} < 1$.

see eqn. (8). For $(v/\lambda^3) \gg 1$, we have: $g_{3/2}(z) \ll 1$ and, hence, $z \ll 1$. Under these circumstances, $g_{3/2}(z) \simeq z$; see eqn. (13). Therefore, in the asymptotic region, $z \simeq (v/\lambda^3)^{-1}$, which is in agreement with the classical case.* Figure 7.2 depicts graphically the variation of z with (v/λ^3).

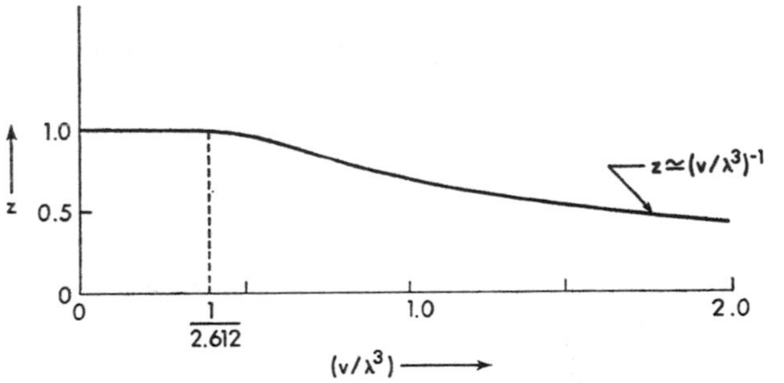

FIG. 7.2. The fugacity of an ideal Bose gas as a function of the quantity (v/λ^3).

Next, we examine the (P, T)-diagram of this system, i.e. the variation of pressure with temperature, *keeping v fixed*. Now, for $T < T_c$, the pressure is given by eqn. (7), with z replaced by unity, i.e.

$$P(T) = \frac{kT}{\lambda^3}\zeta\left(\frac{5}{2}\right), \tag{26}$$

which is proportional to $T^{5/2}$ and is *completely independent* of v. At the transition point the value of the pressure is

$$P(T_c) = \left(\frac{2\pi m}{h^2}\right)^{3/2}(kT_c)^{5/2}\zeta\left(\frac{5}{2}\right); \tag{27}$$

with the help of (24) this can be written as

$$P(T_c) = \frac{\zeta(\frac{5}{2})}{\zeta(\frac{3}{2})}\left(\frac{N}{V}kT_c\right) \simeq 0.5134\left(\frac{N}{V}kT_c\right). \tag{28}$$

Thus, the pressure (exerted by the particles) of an ideal Bose gas at the transition temperature T_c is about one-half of that (exerted by the particles) of an equivalent Boltzmannian gas.† For $T > T_c$, the pressure is given by

$$P(T) = \frac{N}{V}kT\frac{g_{5/2}(z)}{g_{3/2}(z)}, \tag{29}$$

where $z(T)$ is determined by the implicit relationship

$$g_{3/2}(z) = \frac{\lambda^3}{v} = \frac{N}{V}\frac{h^3}{(2\pi mkT)^{3/2}}; \tag{30}$$

* Equation (6.2.12) gives, for an ideal Boltzmannian gas, $\ln \mathcal{Q} = zV/\lambda^3$. Accordingly, $N \equiv z(\partial \ln \mathcal{Q}/\partial z) = z(V/\lambda^3)$, whence it follows that $z = (\lambda^3/v)$.

† Actually, for all $T \leq T_c$, we can write

$$P(T) = P(T_c)\cdot(T/T_c)^{5/2} \simeq 0.5134(N_e kT/V).$$

We can, therefore, say that whereas the particles in the condensed phase do not exert any pressure at all, the particles in the excited states are about half as effective as in the Boltzmannian case.

unless T is very large, the expression for $P(T)$ cannot be expressed in any simpler terms; however, for $T \gg T_c$, the virial expansion (14) can be made use of. Finally, as $T \to \infty$, the pressure approaches the classical value NkT/V. All these features are shown in Fig. 7.3. The transition line in the figure portrays eqn. (26). The actual (P, T)-curve follows this line from $T = 0$ up to $T = T_c$ and thereafter departs, tending asymptotically to the classical limit. It may be pointed out that the region to the right of the transition line belongs to the pure gaseous phase, the line itself belongs to the mixed phase (namely, gaseous plus condensed), while the region to the left of the line is inaccessible to the system.

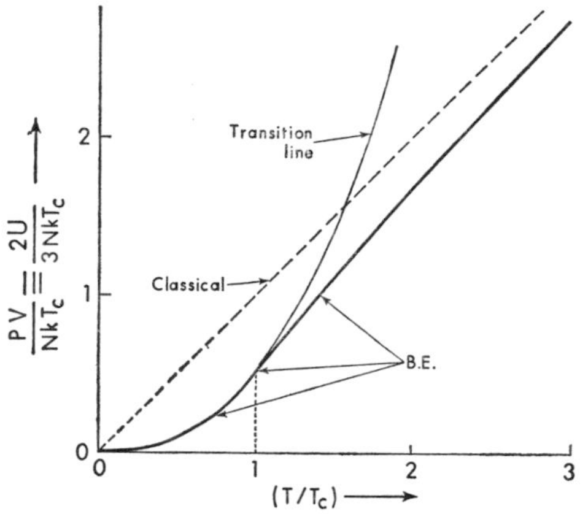

FIG. 7.3. The pressure and the internal energy of an ideal Bose gas as a function of the temperature parameter (T/T_c).

In view of the direct relationship between the internal energy of the gas and its pressure, see eqn. (12), Fig. 7.3 depicts equally well the variation of U with T (of course, with v fixed). Its slope should, therefore, be a measure of the specific heat $C_V(T)$ of the gas. We readily observe that the specific heat is vanishingly small at low temperatures, rises with T until it reaches a maximum value at $T = T_c$, and thereafter decreases, tending asymptotically to the constant classical value. Analytically, we obtain, for $T \leqslant T_c$,

$$\frac{C_V}{Nk} = \frac{3}{2}\frac{V}{N}\zeta\left(\frac{5}{2}\right)\frac{d}{dT}\left(\frac{T}{\lambda^3}\right) = \frac{15}{4}\zeta\left(\frac{5}{2}\right)\frac{v}{\lambda^3}, \tag{31}$$

which is proportional to $T^{3/2}$. At $T = T_c$, the value of this number is

$$\frac{C_V(T_c)}{Nk} = \frac{15}{4}\frac{\zeta(\frac{5}{2})}{\zeta(\frac{3}{2})} \simeq 1.925, \tag{32}$$

which is significantly higher than the classical value of 1.5. For $T > T_c$, we obtain an implicit formula. First of all,

$$\frac{C_V}{Nk} = \left[\frac{\partial}{\partial T}\left(\frac{3}{2}T\frac{g_{5/2}(z)}{g_{3/2}(z)}\right)\right]_v; \tag{33}$$

see eqns. (11) and (30). To carry out the differentiation, we need to know $(\partial z/\partial T)_v$. This obtains from eqn. (30), with the help of the formula (D.7). On the one hand,

$$\left[\frac{\partial}{\partial T} g_{3/2}(z)\right]_v = -\frac{3}{2T} g_{3/2}(z); \tag{34}$$

on the other,

$$\frac{\partial}{\partial z} g_{3/2}(z) = \frac{1}{z} g_{1/2}(z). \tag{35}$$

Combining the two, we obtain

$$\left(\frac{\partial z}{\partial T}\right)_v = -\frac{3z}{2T}\frac{g_{3/2}(z)}{g_{1/2}(z)}. \tag{36}$$

With the help of this result, eqn. (33) takes the form

$$\frac{C_V}{Nk} = \frac{15}{4}\frac{g_{5/2}(z)}{g_{3/2}(z)} - \frac{9}{4}\frac{g_{3/2}(z)}{g_{1/2}(z)}; \tag{37}$$

the value of z, as a function of T, is again to be determined from eqn. (30). In the limit $z \to 1$, the second term of (37) vanishes because of the divergence of $g_{1/2}(1)$, while the first term gives exactly the result as in (32). Thus, the value of the specific heat is continuous at the transition point. Its derivative is, however, discontinuous, the magnitude of the discontinuity being

$$\left(\frac{\partial C_V}{\partial T}\right)_{T=T_c-0} - \left(\frac{\partial C_V}{\partial T}\right)_{T=T_c+0} = \frac{27Nk}{16\pi T_c}\left\{\zeta\left(\frac{3}{2}\right)\right\}^2 = 3.665\frac{Nk}{T_c}; \tag{38}$$

see Problem 7.4. Beyond $T = T_c$, the specific heat decreases steadily towards the limiting value

$$\left(\frac{C_V}{Nk}\right)_{z\to 0} = \frac{15}{4} - \frac{9}{4} = \frac{3}{2}. \tag{39}$$

Figure 7.4 shows all these features of the (C_V, T)-relationship. It may be noted that it was the similarity of this curve with the experimental one for liquid He^4 (Fig. 7.5) that

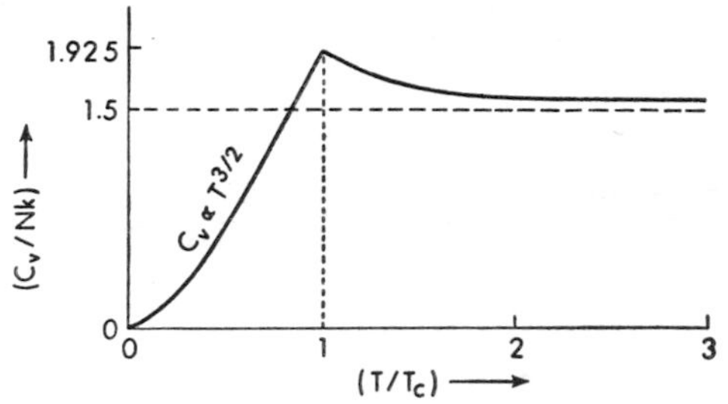

FIG. 7.4. The specific heat of an ideal Bose gas as a function of the temperature parameter (T/T_c).

prompted F. London to suggest, in 1938, that the curious phase transition that occurs in liquid He^4 at a temperature of about 2.19°K might be a manifestation of the *Bose–Einstein condensation* taking place in the liquid. Indeed, if we substitute, in (24), data for liquid He^4, namely $m = 6.65\times 10^{-24}$ g and $V = 27.6$ cm^3/mole, we obtain for T_c a value of about

3.13°K, which is not drastically different from the observed transition temperature of the liquid. Moreover, the interpretation of the phase transition in liquid He^4 as a Bose–Einstein condensation provides a theoretical basis for the *two-fluid model* of this liquid, which was empirically put forward by Tisza (1938) in order to explain the physical behavior of the liquid below the transition temperature. According to London, the N_0 particles that occupy a single, entropyless state ($\varepsilon = 0$) could be identified with the "superfluid fraction" of the liquid and the N_e particles that occupy the excited states ($\varepsilon \neq 0$) with the "normal fraction"! As was required in the empirical model of Tisza, the superfluid fraction makes its appearance at the transition temperature T_c, and builds up at the cost of the normal fraction until at $T = 0$ the whole of the fluid becomes superfluid; cf. Fig. 7.1. Of course, the temperature dependence of these fractions, and of other physical quantities pertaining to liquid He^4, is significantly different from what the simple-minded ideal Bose gas suggests.

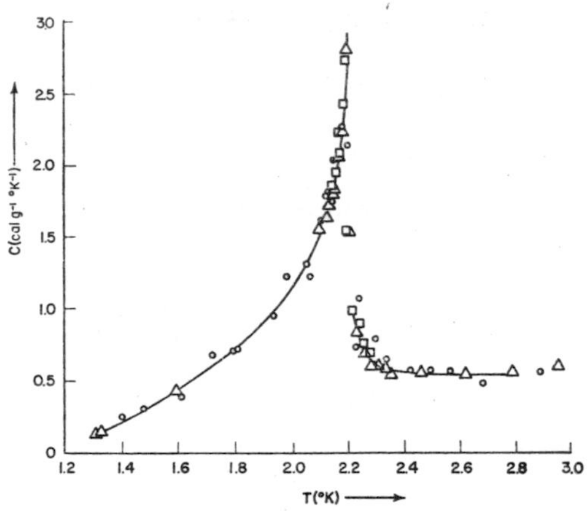

FIG. 7.5. The specific heat of liquid He^4 under its own vapor pressure (after Keesom and coworkers).

London had expected that the inclusion of the intermolecular interactions would improve the quantitative agreement. Although this expectation has been partially vindicated (see Chaps. 10 and 11), there have also been other advances in the field which provide alternative ways of looking at the helium problem; see Sec. 7.5. Nevertheless, some of the features provided by London's interpretation of this phenomenon are still of value.

Historically, the experimental measurements of the specific heat of liquid He^4, which led to the discovery of this so-called He I–He II transition, were first made by Keesom in 1927–8. Impressed by the shape of the (C_V, T)-curve, Keesom gave this transition the name λ-transition; as a result, the term transition temperature (or transition point) also came to be known as λ-temperature (or λ-point).

We shall now study the *isotherms* of the ideal Bose gas, i.e. the variation of the pressure of the gas with its volume, *keeping T fixed.* The Bose–Einstein condensation now sets in at a characteristic volume v_c, which is given by the formula

$$v_c = \lambda^3/\zeta(\tfrac{3}{2}); \tag{40}$$

see (23). We note that $v_c \propto T^{-3/2}$. For $v < v_c$, the pressure of the gas is independent of the volume and is given by the formula

$$P_0 = \frac{kT}{\lambda^3}\zeta\left(\frac{5}{2}\right), \tag{41}$$

which is proportional to $T^{5/2}$. The region of the mixed phase, in the (P, v)-diagram, is marked by a boundary line (or the *transition line*) given by the equation

$$P_0 v_c^{5/3} = \frac{h^2}{2\pi m}\frac{\zeta(\frac{5}{2})}{\{\zeta(\frac{3}{2})\}^{5/3}} = \text{const.}; \tag{42}$$

see Fig. 7.6. Clearly, the region to the left of the line belongs to the mixed phase, while the region to the right belongs to the gaseous phase.

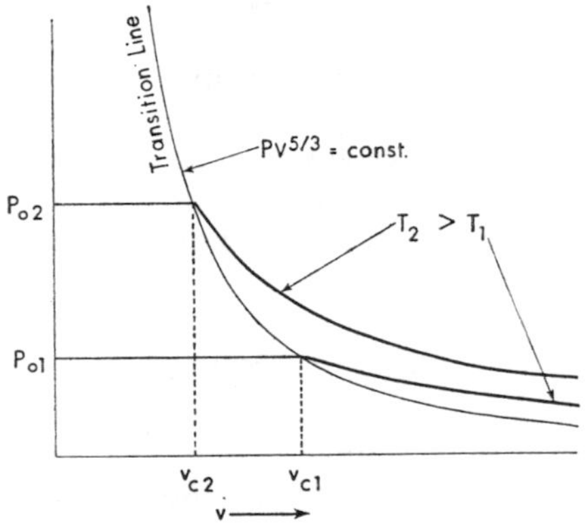

FIG. 7.6. The isotherms of an ideal Bose gas.

In the first region, as v decreases from v_c down to zero the number of particles in the excited states ($\varepsilon \neq 0$) also decreases from N down to zero. At all stages, a strict proportionality holds between the two:

$$\frac{N_e}{N} = \frac{v}{\lambda^3}\zeta\left(\frac{3}{2}\right) = \frac{v}{v_c}, \tag{43}$$

whence it follows that

$$N_e v_c = Nv = V. \tag{44}$$

It is plausible, therefore, to associate, in a *formal* manner, a volume v_c with a particle belonging to the excited states and a volume zero with a particle belonging to the ground state. Thus, we may look upon the condensation process as a "sinking down" of the particles from a region in which they have a volume v_c available to each one of them into a region in which they are "packed into a space of volume zero":

$$|\Delta v| = v_c. \tag{45}$$

At the same time, we obtain from eqns. (40) and (41)

$$\frac{dP_0}{dT} = \frac{5}{2}\frac{P_0}{T} = \frac{5}{2}\frac{k}{v_c}\frac{\zeta(\frac{5}{2})}{\zeta(\frac{3}{2})},$$

which, with the help of (45), becomes

$$\frac{dP_0}{dT} = \frac{\frac{5}{2}kT\zeta(\frac{5}{2})/\zeta(\frac{3}{2})}{T\,|\Delta v|}. \tag{46}$$

We thus obtain, *formally* though, the Clausius–Clapeyron equation for the Bose–Einstein transition. As it stands, eqn. (46) prompts us to interpret the quantity

$$\tfrac{5}{2}kT\zeta(\tfrac{5}{2})/\zeta(\tfrac{3}{2}) \tag{47}$$

as the *latent heat* (per particle) associated with the transition.

In the second region, viz. for $v > v_c$, we have:

$$P(v) \propto g_{5/2}(z), \tag{48}$$

where z, as usual, is determined by the relationship

$$g_{3/2}(z) = \lambda^3/v. \tag{49}$$

The pressure indeed decreases as v increases; however, for $v \gg v_c$, the variation becomes extremely simple, for then $z \ll 1$ and hence

$$P(v) \propto z \propto 1/v, \tag{50}$$

in agreement with the classical *Boyle's law*.

Finally, we shall study the *adiabatics* of the ideal Bose gas. For this, we ought to have an expression for the entropy of the system. Making use of the thermodynamic identity

$$U - TS + PV \equiv N\mu$$

and the expressions for U and P obtained above, we get

$$\frac{S}{Nk} \equiv \frac{U+PV}{NkT} - \frac{\mu}{kT} = \begin{cases} \dfrac{5}{2}\dfrac{g_{5/2}(z)}{g_{3/2}(z)} - \ln z & \text{for} \quad T \geqslant T_c, \\[2ex] \dfrac{5}{2}\dfrac{v}{\lambda^3}\zeta\left(\dfrac{5}{2}\right) & \text{for} \quad T \leqslant T_c. \end{cases} \tag{51}$$

Again, the value of $z(T)$, for $T \geqslant T_c$, is to be obtained from eqn. (49). Now, an adiabatic process implies the constancy of S and N. For $T > T_c$, this implies the constancy of z and in turn, by (49), the constancy of (v/λ^3). For $T \leqslant T_c$, it again implies, more directly though, the constancy of (v/λ^3). We thus obtain, quite generally, the following relationship between the volume and the temperature of the system when it undergoes an adiabatic process:

$$vT^{3/2} = \text{const.} \tag{52}$$

The corresponding relationship between the pressure and the temperature is

$$P/T^{5/2} = \text{const.}; \tag{53}$$

see eqns. (7) and (26). Eliminating T, we obtain

$$Pv^{5/3} = \text{const.} \tag{54}$$

as the desired equation for an adiabatic of the ideal Bose gas.

Incidentally, the foregoing results are exactly the same as for the classical ideal gas. There is, however, a significant difference between the two cases, i.e. while the exponent $\frac{5}{3}$ in the formula (54) is identically equal to the ratio of the specific heats C_P and C_V in the case of the ideal classical gas, it is not so in the case of the ideal Bose gas. In the latter case, the ratio γ is given by

$$\begin{aligned}\gamma = \frac{C_P}{C_V} &= 1+\frac{4}{9}\frac{C_V}{Nk}\frac{g_{1/2}(z)}{g_{3/2}(z)}\\ &= \frac{5}{3}\frac{g_{5/2}(z)\,g_{1/2}(z)}{\{g_{3/2}(z)\}^2};\end{aligned} \tag{55}$$

see Problem 7.3 and eqn. (37). It is only for $T \gg T_c$ that γ is equal to 5/3. At any finite temperature, $\gamma > 5/3$ and as $T \to T_c$, $\gamma \to \infty$. Equation (54), on the other hand, holds for *all* values of T.

Considering for a while the region of the mixed phase ($T \leqslant T_c$), we have for the entropy of the gas

$$S = N_e \cdot \frac{5}{2} k \frac{\zeta(\frac{5}{2})}{\zeta(\frac{3}{2})} \propto N_e; \tag{56}$$

see eqns. (21a) and (51). As expected, the N_0 particles (that constitute the "condensate") do not make a contribution towards the entropy of the system, while the N_e particles (that constitute the gaseous part) contribute an amount of $\frac{5}{2}k\zeta(\frac{5}{2})/\zeta(\frac{3}{2})$ per particle. We note that this result is consistent with our earlier finding that the phase transition taking place in the ideal Bose gas may be regarded as accompanied by a latent heat (per particle), as given by the expression (47).

7.2. Thermodynamics of the black-body radiation

One of the most important applications of Bose–Einstein statistics is to investigate the equilibrium properties of the *black-body radiation*. We consider a radiation cavity of volume V at temperature T. Historically, this system has been looked upon from two, practically identical but conceptually different, points of view:

(i) as an assembly of *harmonic oscillators* with quantized energies $(n_s+\frac{1}{2})\hbar\omega_s$, where $n_s = 0, 1, 2, \ldots$, and ω_s is the (angular) frequency of an oscillator, or
(ii) as a gas of identical and indistinguishable quanta—the so-called *photons*—the energy of a photon (corresponding to the (angular) frequency ω_s of the radiation mode) being $\hbar\omega_s$.

The first point of view is essentially the same as adopted by Planck (1900), except that we have also included here the zero-point energy of the oscillator; for the thermodynamics of the radiation, this energy is of no great consequence and may be dropped altogether.

The oscillators, being distinguishable from one another (by the very values of ω_s), would obey Maxwell–Boltzmann statistics; however, the expression for the single-oscillator partition function $Q_1(V, T)$ would be different from the classical expression because now the energies accessible to the oscillator are discrete, rather than continuous; cf. eqns. (3.7.2) and (3.7.14). The expectation value of the energy of a Planck oscillator of frequency ω_s is then given by eqn. (3.7.20), excluding the zero-point term $\frac{1}{2}\hbar\omega_s$:

$$\langle \varepsilon_s \rangle = \frac{\hbar\omega_s}{e^{\hbar\omega_s/kT}-1}. \tag{1}$$

Now, the number of normal modes of vibration per unit volume of the enclosure in the frequency range $(\omega, \omega+d\omega)$ is given by the *Rayleigh expression*

$$2\cdot 4\pi\left(\frac{1}{\lambda}\right)^2 d\left(\frac{1}{\lambda}\right) = \frac{\omega^2\,d\omega}{\pi^2c^3}, \tag{2}$$

where the factor 2 has been included to take into account the duplicity of the transverse modes;* the symbol c here denotes the velocity of light. By eqns. (1) and (2), the energy density associated with the frequency range $(\omega, \omega+d\omega)$ is given by

$$u(\omega)\,d\omega = \frac{\hbar}{\pi^2c^3}\,\frac{\omega^3\,d\omega}{e^{\hbar\omega/kT}-1}, \tag{3}$$

which is the famous *Planck's formula* for the distribution of energy over the black-body spectrum. Integrating (3) over all values of ω, we obtain an expression for the total energy density in the radiation cavity.

The second point of view originated with Bose (1924) and Einstein (1924, 1925). Bose investigated the problem of the "distribution of *photons* over the various energy levels" in the system; however, instead of worrying about the allocation of the various photons to the various energy levels (as one would have ordinarily done), he concentrated on the statistics of the energy levels themselves! He studied questions such as the "probability of an energy level ε_s $(=\hbar\omega_s)$ being occupied by n_s photons at a time", "the mean values of n_s and ε_s", etc. The statistics of the energy levels is indeed Boltzmannian; the mean values of n_s and ε_s, however, turn out to be

$$\langle n_s \rangle = \sum_{n_s=0}^{\infty} n_s\, e^{-n_s\hbar\omega_s/kT} \Bigg/ \sum_{n_s=0}^{\infty} e^{-n_s\hbar\omega_s/kT} = \frac{1}{e^{\hbar\omega_s/kT}-1} \tag{4}$$

and

$$\langle \varepsilon_s \rangle = \hbar\omega_s \langle n_s \rangle = \frac{\hbar\omega_s}{e^{\hbar\omega_s/kT}-1}, \tag{5}$$

which is identical with our earlier result (1). To obtain the number of the photon states with momenta lying between $\hbar\omega/c$ and $\hbar(\omega+d\omega)/c$, Bose made use of the connection be-

* As is well known, the longitudinal modes do not appear in the case of radiation.

tween this number and the "volume of the relevant region of the phase space", with the result

$$g(\omega)\,d\omega \simeq 2\cdot\frac{V}{h^3}\left\{4\pi\left(\frac{\hbar\omega}{c}\right)^2\left(\frac{\hbar\,d\omega}{c}\right)\right\} = \frac{V\omega^2\,d\omega}{\pi^2c^3}, \tag{6}^*$$

which is also identical with our earlier result (2). Thus, he finally obtained the distribution formula of Planck. It must be noted here that, *although emphasis lay elsewhere*, the mathematical steps that led Bose to the final result went literally parallel to the ones occurring in the oscillator approach!

Einstein, on the other hand, went into the deeper details of the distribution problem and pondered over the statistics of both the photons and the energy levels, *taken together*. He concluded (from Bose's treatment) that the basic fact to remember during the process of distributing the various photons over the various energy levels is that the photons are *indistinguishable*—a fact that had been implicitly taken care of in Bose's treatment. Einstein's derivation of the desired distribution was essentially the same as in Sec. 6.1, with one important difference, i.e. since the total number of photons in any given volume is indefinite, the constraint of a *fixed* value of N is no longer present. As a result, the Lagrange multiplier α does not enter into the discussion and to that extent the final formula for $\langle n_\varepsilon\rangle$ is simpler:†

$$\langle n_\varepsilon\rangle = \frac{1}{e^{\varepsilon/kT}-1}; \tag{7}$$

cf. eqn. (6.1.18a) or (6.2.22). The foregoing result is identical with (4), with $\varepsilon = \hbar\omega_s$. The subsequent steps in Einstein's treatment are the same as in Bose's.

Looking back at the two approaches, we note that there is a complete correspondence between them: "an oscillator in the eigenstate n_s, with energy $(n_s+\frac{1}{2})\hbar\omega_s$" in the first approach corresponds to "the occupation of the energy level $\hbar\omega_s$ by n_s photons" in the second approach, "the average energy $\langle\varepsilon_s\rangle$ of an oscillator" corresponds to "the mean occupation number $\langle n_s\rangle$ of the corresponding energy level", and so on.

Figure 7.7 shows a plot of the distribution function (3), which may as well be written in the dimensionless form:

$$u'(x)\,dx = \frac{x^3\,dx}{e^x-1}, \tag{8}$$

where

$$u'(x) = \frac{\pi^2\hbar^3c^3}{(kT)^4}\,u(x) \quad \text{and} \quad x = \frac{\hbar\omega}{kT}. \tag{9}$$

For long wavelengths ($x \ll 1$), the formula (8) reduces to the classical approximation of

* The factor 2 in this expression arises essentially from the same cause as in the Rayleigh expression (2). However, in the present context, it would be more appropriate to regard it as representing the two states of polarization of the photon spin.

† When compared with the standard Bose–Einstein result, say (7.1.2), the formula (7) suggests that we are dealing here with a case for which $z \equiv 1$. It is not difficult to see that this is solely due to the fact that the total number of particles in the present case is *indefinite*. For then, their equilibrium number $\bar{N}$ has to be determined by the condition that the free energy of the system is a minimum, i.e. $\{(\partial A/\partial N)_{N=\bar{N}}\}_{V,T} = 0$, which, by definition, implies that $\mu = 0$ and hence $z = 1$.

Rayleigh (1900) and Jeans (1905), namely*

$$u'(x) \simeq x^2, \tag{10}$$

while for short wavelengths ($x \gg 1$), it reduces to the rival formula of Wien (1896), namely

$$u'(x) \simeq x^3 e^{-x}. \tag{11}$$

The limiting forms (10) and (11) are also included in the figure. We note that the areas under the Planck curve and the Wien curve are $\pi^4/15$ ($\simeq$ 6.49) and 6, respectively. The Rayleigh–Jeans curve, however, suffers from a short-wavelength catastrophe!

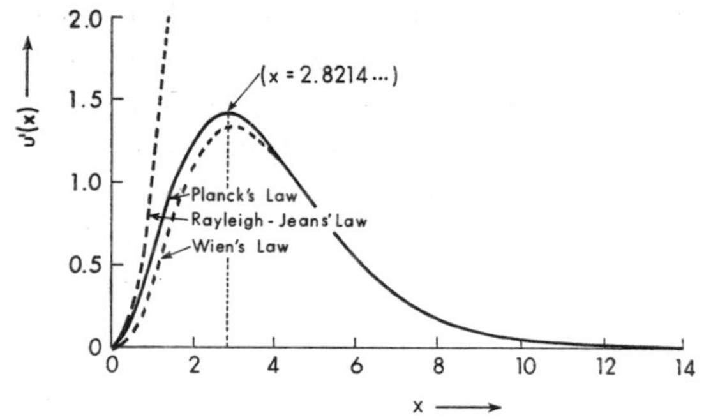

FIG. 7.7. The spectral distribution of energy in the black-body radiation. The solid curve represents the quantum-theoretical formula of Planck. The long-wavelength approximation of Rayleigh–Jeans and the short-wavelength approximation of Wien are also shown.

For the total energy density in the radiation cavity, we obtain from eqns. (8) and (9)

$$\frac{U}{V} = \int_0^\infty u(x)\,dx = \frac{(kT)^4}{\pi^2\hbar^3c^3}\int_0^\infty \frac{x^3\,dx}{e^x-1}$$

$$= \frac{\pi^2k^4}{15\hbar^3c^3}T^4. \tag{12}^\dagger$$

If there is a small opening in the walls of the cavity, the photons will "effuse" through it. The net rate of flow of the radiation, per unit area of the opening, will be given by, cf. eqn. (6.4.11),

$$\frac{1}{4}\frac{U}{V}c = \frac{\pi^2k^4}{60\hbar^3c^2}T^4 = \sigma T^4, \tag{13}$$

where

$$\sigma = \frac{\pi^2k^4}{60\hbar^3c^2} = 5.670\times10^{-8}\ \text{W m}^{-2}\ {}^\circ\text{K}^{-4}. \tag{14}$$

Equation (13) embodies the *Stefan–Boltzmann law* of black-body radiation, σ being the

* The Rayleigh–Jeans formula is *directly* obtained if we use for $\langle\varepsilon_s\rangle$ the equipartition value (kT) rather than the quantum-theoretical value (1).

† Here, use has been made of the fact that the value of the definite integral is $6\zeta(4) = \pi^4/15$; see Appendix D.

Stefan constant. This law was deduced from experimental observations by Stefan in 1879; five years later, Boltzmann derived it from thermodynamic considerations.

For a further study of thermodynamics, we may evaluate the grand partition function of the photon gas. We obtain, from eqn. (6.1.23) with $\alpha = 0$,

$$\ln \mathcal{Q}(V, T) \equiv \frac{PV}{kT} = -\sum_{\varepsilon} \ln (1 - e^{-\varepsilon/kT}). \tag{15}$$

Replacing summation by integration and making use of the extreme relativistic formula

$$a(\varepsilon)\, d\varepsilon = 2V \frac{4\pi p^2\, dp}{h^3} = \frac{8\pi V}{h^3 c^3} \varepsilon^2\, d\varepsilon, \tag{16}$$

we obtain, after an integration by parts,

$$\ln \mathcal{Q}(V, T) \equiv \frac{PV}{kT} = \frac{8\pi V}{3h^3c^3} \frac{1}{kT} \int_0^\infty \frac{\varepsilon^3\, d\varepsilon}{e^{\varepsilon/kT} - 1}.$$

By a change of variable, it becomes

$$PV = \frac{8\pi V}{3h^3c^3} (kT)^4 \int_0^\infty \frac{x^3\, dx}{e^x - 1}$$

$$= \frac{8\pi^5 V}{45h^3c^3} (kT)^4 = \frac{1}{3} U. \tag{17}$$

We thus obtain the well-known result of the radiation theory, viz. that the pressure of the radiation is equal to one-third of the energy density; see also eqn. (6.4.9). Next, since the chemical potential of the system is identically zero, the Helmholtz free energy is identically equal to $-PV$:

$$A = -PV = -\tfrac{1}{3} U, \tag{18}$$

whence it follows that

$$S \equiv \frac{U - A}{T} = \frac{4}{3} \frac{U}{T} \propto VT^3 \tag{19}$$

and

$$C_V = T \left(\frac{\partial S}{\partial T} \right)_V = 3S. \tag{20}$$

If the radiation undergoes an adiabatic change, the law governing the variation of T with V follows readily from eqn. (19):

$$VT^3 = \text{const.} \tag{21}$$

Combining (21) with the fact that $P \propto T^4$, we obtain an equation for the *adiabatics* of the system, namely

$$PV^{4/3} = \text{const.} \tag{22}$$

It must be noted, however, the ratio C_P/C_V of the photon gas is not 4/3; it is infinite!

Finally, we derive an expression for the equilibrium number $\bar{N}$ of the photons in the radiation cavity. We obtain, by making use of the relevant expressions,

$$\bar{N} = \frac{V}{\pi^2 c^3} \int_0^\infty \frac{\omega^2 \, d\omega}{e^{\hbar\omega/kT}-1}$$

$$= V \frac{2\zeta(3)\,(kT)^3}{\pi^2 \hbar^3 c^3} \propto VT^3. \tag{23}$$

Instructive though it may be, formula (23) cannot be taken at its face value, because in the present problem the magnitude of the *fluctuations* in the variable N, which is determined by the quantity $(\partial P/\partial V)^{-1}$, is infinitely large; see eqn. (4.5.10). This is also related to the fact that the isotherms of the photon gas are flat.

7.3. The field of sound waves

A problem, mathematically similar to the one discussed in Sec. 7.2, arises from the vibrational modes of a macroscopic body, specifically a solid. As in the case of the black-body radiation, the problem of the vibrational modes of a solid can be studied equally well by regarding the system as a collection of harmonic oscillators *or* by regarding it as an enclosed region containing a gas of sound quanta—the so-called *phonons*. To illustrate this point, we may consider the Hamiltonian of a *classical* solid composed of N atoms whose positions in space are specified by the coordinates $(x_1, x_2, \ldots, x_{3N})$. In the state of lowest energy, the values of these coordinates may be denoted by $(\bar{x}_1, \bar{x}_2, \ldots, \bar{x}_{3N})$. Denoting the displacements $(x_i - \bar{x}_i)$ of the atoms from their equilibrium positions by the variables ξ_i $(i = 1, 2, \ldots, 3N)$, the kinetic energy of the system in the configuration (x_i) is given by

$$K = \tfrac{1}{2} m \sum_{i=1}^{3N} \dot{x}_i^2 = \tfrac{1}{2} m \sum_{i=1}^{3N} \dot{\xi}_i^2, \tag{1}$$

and the potential energy by

$$\Phi \equiv \Phi(x_i) = \Phi(\bar{x}_i) + \sum_i \left(\frac{\partial \Phi}{\partial x_i}\right)_{(x_i)=(\bar{x}_i)} (x_i - \bar{x}_i) + \sum_{i,j} \frac{1}{2} \left(\frac{\partial^2 \Phi}{\partial x_i\, \partial x_j}\right)_{(x_i)=(\bar{x}_i)} (x_i - \bar{x}_i)(x_j - \bar{x}_j) + \ldots. \tag{2}$$

The main term in this expansion represents the (minimum) energy of the solid when all the N atoms are at rest at their mean positions $\bar{x}_i$; this energy may be denoted by the symbol Φ_0. The next set of terms in the expansion is identically equal to zero, because the function $\Phi(x_i)$ has its minimum value at $(x_i) = (\bar{x}_i)$ and hence all its first derivatives must vanish there. The second-order terms of the expansion represent the *harmonic component* of the atomic vibrations. If we assume that the overall amplitudes of the atomic vibrations are not very large we may retain only the harmonic terms of the expansion and neglect all the successive ones; we are then working in the so-called *harmonic approximation*. Thus, we may write

$$H = \Phi_0 + \left\{ \sum_i \tfrac{1}{2} m \dot{\xi}_i^2 + \sum_{i,j} \alpha_{ij} \xi_i \xi_j \right\}, \tag{3}$$

where

$$\alpha_{ij} = \frac{1}{2}\left(\frac{\partial^2 \Phi}{\partial x_i \partial x_j}\right)_{(x_i)=(\bar{x}_i)}. \tag{4}$$

We now introduce a linear transformation, from the coordinates ξ_i to the so-called *normal coordinates* q_i, and choose the transformation matrix in such a way that the new expression for the Hamiltonian does not contain any cross terms, i.e.

$$H = \Phi_0 + \sum_i \tfrac{1}{2} m(\dot{q}_i^2 + \omega_i^2 q_i^2), \tag{5}$$

where ω_i $(i = 1, 2, \ldots, 3N)$ are the *characteristic frequencies* of the so-called *normal modes* of the system and are determined essentially by the quantities α_{ij} or, in turn, by the nature of the potential energy function $\Phi(x_i)$. The expression (5) suggests that the energy of the solid, over and above the (minimum) value Φ_0, may be considered as arising from a set of $3N$ one-dimensional, *noninteracting*, harmonic oscillators, whose characteristic frequencies ω_i are determined by the nature of the interatomic interactions in the system.

Classically, each of the $3N$ normal modes of vibration corresponds to a wave of distortion of the lattice points, i.e. a sound wave. Quantum-mechanically, these modes give rise to quanta, called phonons, in very much the same way as the vibrational modes of the electromagnetic field give rise to photons. There is one important difference, however, i.e. while the number of normal modes in the cases of electromagnetic field is indefinite, the number of normal modes (or the number of phonon energy levels) in the case of a solid is fixed by the number of lattice sites in it.* This introduces certain differences in the thermodynamic behavior of the sound field in contrast to the thermodynamic behavior of the radiation field; however, at low temperatures, when the high-frequency modes of the solid are not very likely to be excited, these differences become rather insignificant and we obtain a striking similarity between the two sets of results.

The thermodynamics of the solid can now be studied along the lines of Sec. 3.7. First of all, we note that the quantum-mechanical eigenvalues of the Hamiltonian (5) would be

$$E\{n_i\} = \Phi_0 + \sum_i \left(n_i + \tfrac{1}{2}\right)\hbar\omega_i, \tag{6}$$

where the numbers n_i denote the "states of excitation" of the various oscillators (or, equally well, the occupation numbers of the various phonon levels). The internal energy of the system is then given by

$$U(T) = \left\{\Phi_0 + \sum_i \frac{1}{2}\hbar\omega_i\right\} + \sum_i \frac{\hbar\omega_i}{e^{\hbar\omega_i/kT} - 1}. \tag{7}$$

The expression within the curly brackets gives the energy of the solid at absolute zero. The term Φ_0 is necessarily negative, and larger in magnitude than the total zero-point energy, $\Sigma_i \frac{1}{2}\hbar\omega_i$, of the oscillators; together they determine the *binding energy* of the lattice. The last

* Of course, the number of phonons is again indefinite. As a result, the chemical potential of the phonon gas is likewise zero.

term in the formula represents the temperature-dependent part of the energy,* which determines the specific heat of the solid:

$$C_V(T) \equiv \left(\frac{\partial U}{\partial T}\right)_V = k \sum_i \frac{(\hbar\omega_i/kT)^2 \, e^{\hbar\omega_i/kT}}{(e^{\hbar\omega_i/kT}-1)^2}. \tag{8}$$

To proceed further, we must have a knowledge of the frequency spectrum of the solid. To acquire this knowledge from first principles is not an easy task. Accordingly, one either obtains this spectrum through experiment or else makes certain plausible assumptions about it. Einstein, who was the first to apply quantum concept to the theory of solids (1907), assumed, for simplicity, that the frequencies ω_i are all equal in value! Denoting this (common) value by ω_E, the specific heat of the solid is given by

$$C_V(T) = 3NkE(x), \tag{9}$$

where $E(x)$ is the so-called *Einstein function*:

$$E(x) = \frac{x^2 \, e^x}{(e^x-1)^2}, \tag{10}$$

with

$$x = \hbar\omega_E/kT = \Theta_E/T. \tag{11}$$

The dashed curve in Fig. 7.8 depicts the variation of the specific heat with temperature, as given by the Einstein formula (9). At sufficiently high temperatures, when $T \gg \Theta_E$ and hence $x \ll 1$, the Einstein result tends towards the classical one, viz. $C_V = 3Nk$.† At sufficiently

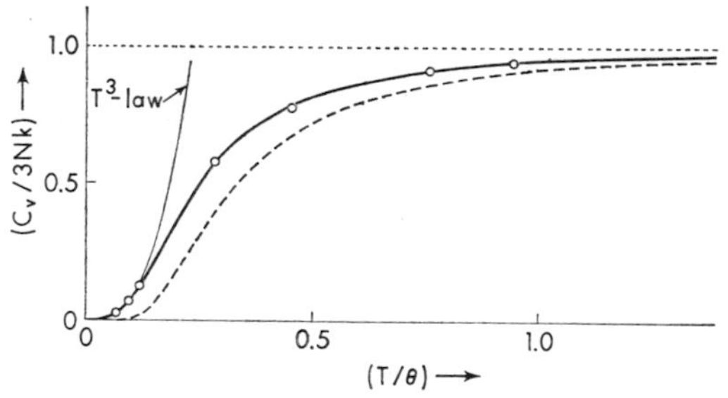

FIG. 7.8. The specific heat of a solid, according to the Einstein model: ------, and according to the Debye model: ———. The circles denote the experimental results for copper.

low temperatures, when $T \ll \Theta_E$ and hence $x \gg 1$, the specific heat falls at an exponentially fast rate and tends to zero as $T \to 0$. The theoretical rate of fall, however, turns out to be rather too fast in comparison with the observed rate. Nevertheless, Einstein's approach to the problem did at least provide a theoretical basis for understanding the observed departure

* The thermal energy of the solid may as well be written as $\Sigma_i \langle n_i \rangle \hbar\omega_i$, where $\langle n_i \rangle \{= (e^{\hbar\omega_i/kT}-1)^{-1}\}$ is the *mean occupation number* of the phonon energy level ε_i. Clearly, the phonons, like photons, obey Bose–Einstein statistics!

† Actually, when the temperature is high enough, so that *all* $(\hbar\omega_i/kT) \ll 1$, the general formula (8) itself reduces to the classical one. This corresponds to the situation when each of the $3N$ modes of vibration possesses a thermal energy equal to kT.

of the specific heat of solids from the classical law of Dulong and Petit, whereby $C_V = 3R \simeq 5.96$ calories per mole of the substance.

Debye (1912), on the other hand, allowed a *continuous spectrum* of frequencies, cut off at an upper limit ω_D such that the total number of normal modes of vibration is equal to $3N$, that is

$$\int_0^{\omega_D} g(\omega)\, d\omega = 3N, \tag{12}$$

where $g(\omega)\, d\omega$ denotes the number of normal modes of vibration whose frequency lies in the range $(\omega, \omega + d\omega)$. For $g(\omega)$, Debye adopted the Rayleigh expression (7.2.2), modified so as to suit the problem under study. Writing c_L for the velocity of propagation of the longitudinal modes and c_T for the velocity of propagation of the transverse modes (and noting that, for any frequency ω, the transverse mode is doubly degenerate), eqn. (12) becomes

$$\int_0^{\omega_D} V\left(\frac{\omega^2\, d\omega}{2\pi^2 c_L^3} + \frac{\omega^2\, d\omega}{\pi^2 c_T^3}\right) = 3N, \tag{13}$$

whence we obtain for the cut-off frequency

$$\omega_D^3 = 18\pi^2 \frac{N}{V}\left(\frac{1}{c_L^3} + \frac{2}{c_T^3}\right)^{-1}. \tag{14}$$

Accordingly, the Debye spectrum may be written as

$$g(\omega) = \begin{cases} \dfrac{9N}{\omega_D^3}\,\omega^2 & \text{for} \quad \omega \leqslant \omega_D, \\ 0 & \text{for} \quad \omega > \omega_D. \end{cases} \tag{15}$$

Before we proceed further to calculate the specific heat of solids on the basis of the Debye spectrum, two remarks appear in order. First, the Debye spectrum is only an idealization of the actual situation obtaining in a solid; it may be compared with a typical spectrum, such as the one shown in Fig. 7.9. While for low-frequency modes (the so-called *acoustical*

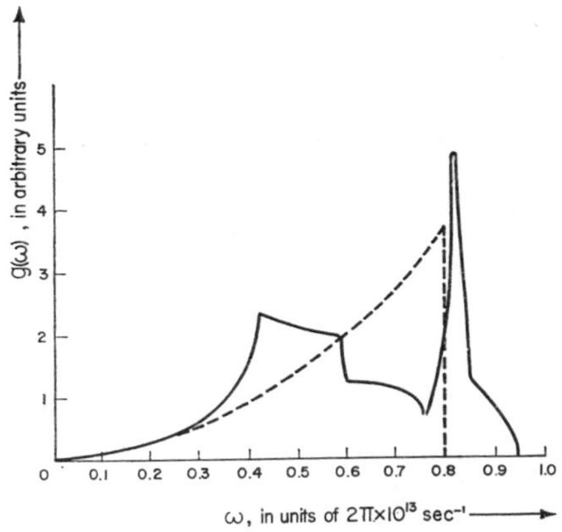

FIG. 7.9. The normal-mode frequency distribution $g(\omega)$ for aluminum. The solid curve is derived from x-ray scattering measurements (Walker, 1956) while the dashed curve represents the corresponding Debye approximation.

modes) the Debye approximation is reasonably valid, there are serious discrepancies in the case of high-frequency modes (the so-called *optical* modes). At any rate, for "averaged" quantities, such as the specific heat, the finer details of the spectrum are not very important. In fact, Debye approximation serves the purpose reasonably well; things indeed improve if we take account of the various peaks in the spectrum by including in our result a number of "suitably weighted" Einstein terms. Second, the longitudinal and the vibrational modes of the solid should have their own cut-off frequencies, $\omega_{D,L}$ and $\omega_{D,T}$ say, rather than having a common cut-off at ω_D, for the simple reason that, of the $3N$ normal modes of the lattice, N are longitudinal and $2N$ transverse. Accordingly, we should have, instead of (13),

$$\int_0^{\omega_{D,L}} V\frac{\omega^2\,d\omega}{2\pi^2 c_L^3} = N \quad \text{and} \quad \int_0^{\omega_{D,T}} V\frac{\omega^2\,d\omega}{\pi^2 c_T^3} = 2N. \tag{16}$$

We note that the two cut-offs $\omega_{D,L}$ and $\omega_{D,T}$ correspond to a *common wavelength* $\lambda_{\min}\{= (4\pi V/3N)^{1/3}\}$, which is comparable to the *mean interatomic distance* in the solid. This is quite reasonable because, for wavelengths shorter than $\lambda_{\min}$, it would be rather meaningless to speak of a wave of atomic displacements. The correction that arises from the replacement of the condition (13) by the conditions (16) forms the subject-matter of Problem 7.20.

In the Debye approximation, formula (8) for the specific heat of the solid becomes

$$C_V(T) = 3NkD(x_0), \tag{17}$$

where $D(x_0)$ is the so-called *Debye function*:

$$D(x_0) = \frac{3}{x_0^3}\int_0^{x_0} \frac{x^4 e^x\,dx}{(e^x-1)^2}, \tag{18}$$

with

$$x_0 = \frac{\hbar\omega_D}{kT} = \frac{\Theta_D}{T}, \tag{19}$$

Θ_D being the so-called *Debye temperature* of the solid. Integrating by parts, the expression for the Debye function becomes

$$D(x_0) = -\frac{3x_0}{e^{x_0}-1} + \frac{12}{x_0^3}\int_0^{x_0} \frac{x^3\,dx}{e^x-1}. \tag{20}$$

For $T \gg \Theta_D$, which means $x_0 \ll 1$, the function $D(x_0)$ may be expressed as a power series in x_0:

$$D(x_0) = 1 - \frac{x_0^2}{20} + \dots \,. \tag{21}$$

Thus, as $T \to \infty$, $C_V \to 3Nk$; moreover, according to this theory, the classical result should be applicable to within $\frac{1}{2}$ per cent so long as $T > 3\Theta_D$. For $T \ll \Theta_D$, which means $x_0 \gg 1$, the function $D(x_0)$ may be written as

$$D(x_0) = \frac{12}{x_0^3}\int_0^\infty \frac{x^3\,dx}{e^x-1} + 0(e^{-x_0}),$$

whence

$$D(x_0) \simeq \frac{4\pi^4}{5x_0^3} = \frac{4\pi^4}{5}\left(\frac{T}{\Theta_D}\right)^3. \tag{22}$$

Thus, at low temperatures the specific heat of the solid follows the *Debye T^3-law*:

$$C_V = Nk\frac{12\pi^4}{5}\left(\frac{T}{\Theta_D}\right)^3 = 464.4\left(\frac{T}{\Theta_D}\right)^3 \quad \text{cal/mole/°K}. \tag{23}$$

Thus, while in the limit $T \to \infty$ we recover the well-known classical behavior (C_V = const.), in the limit $T \to 0$ we obtain the typical phonon behavior ($C_V \propto T^3$). It is clear from eqn. (23) that a measurement of the low-temperature specific heat of a solid should enable us not only to check the validity of the T^3-law but also to obtain an empirical value of the Debye temperature Θ_D.* The value of Θ_D can also be obtained by computing the cut-off frequency ω_D from a knowledge of the parameters N/V, c_L and c_T; see formulae (14) and (19). The closeness of these estimates is another evidence in favor of Debye's theory. Once Θ_D is known, the whole of the temperature range can be covered theoretically by making use of the tabulated values of the function $D(x_0)$.† A typical case was shown in Fig. 7.8. We note that not only was the T^3-law obeyed at low temperatures, the agreement between theory and experiment was good throughout the range of observations.

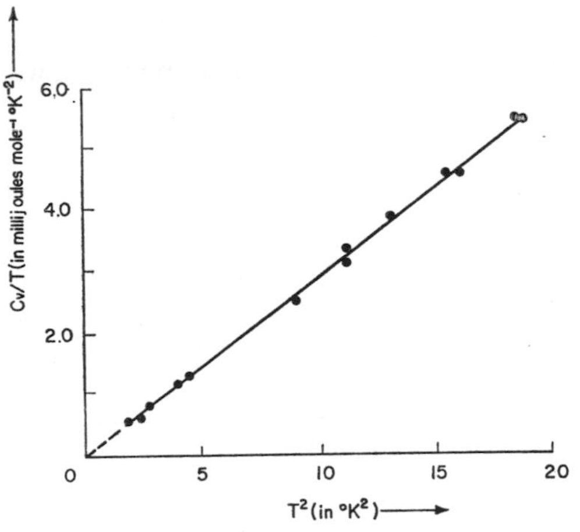

FIG. 7.10. A plot of (C_V/T) versus T^2 for KCl, showing the validity of the Debye T^3-law. The experimental points are due to Keesom and Pearlman (1953).

* One can show that, according to this theory, deviations from the T^3-law should not exceed 2 per cent so long as $T < \Theta_D/10$. However, in the case of *metals*, one can hardly expect to reach a true T^3-region because, well before that, the specific heat of the electron gas might become a dominant contribution (see Sec. 8.3); unless the two contributions are properly separated out, one is likely to obtain a rather suppressed value of Θ_D from these observations.

† See, for example, Fowler and Guggenheim (1960), p. 144.

As another illustration of agreement in the low-temperature region, we include here another plot, see Fig. 7.10, which is based on data obtained with the KCl crystal at temperatures below 5°K (Keesom and Pearlman, 1953). In this plot, the observed values of C_V/T are plotted against T^2. It is evident that the data fall quite well on a straight line (whose slope, in units of joule/mole/°K^2, is 15.37 ± 0.09). This corresponds to a value of 233 ± 3°K for Θ_D, which is in fair agreement with the values of 230–246°K obtaining from the various estimates of the relevant elastic constants.

In Table 7.1 are given the values of Θ_D for several crystals, as derived from the specific heat measurements and from the values of the elastic constants.

TABLE 7.1. THE VALUES OF THE DEBYE TEMPERATURE Θ_D FOR DIFFERENT CRYSTALS

Crystal	Pb	Ag	Zn	Cu	Al	C	NaCl	KCl	MgO
Θ_D from the specific heat measurements	88	215	308	345	398	~1850	308	233	~850
Θ_D from the elastic constants	73	214	305	332	402	—	320	240	~950

In general, if the specific heat measurements of a given system conform to a T^3-law one may conclude that the (thermal) excitations in the system are accounted for *solely* by the phonons. We expect something similar to happen in liquids as well, with two important differences though. First, since liquids cannot withstand sheer stress they cannot sustain transverse modes of vibration; a liquid composed of N atoms will, therefore, have only N (longitudinal) modes of vibration. Second, the normal modes of a liquid cannot be expected to be strictly harmonic; consequently, in addition to phonons, we might have other types of excitation in the liquid, such as *vortex flow*, *turbulence* (or even a modified kind of particle excitations, such as *rotons* in liquid He^4).

Now, helium is the only substance that remains liquid at temperatures low enough to exhibit the T^3-behavior. In the case of the lighter isotope, viz. liquid He^3, the results are strongly influenced by the Fermi–Dirac statistics; as a result, a specific heat proportional to the *first* power of T dominates the scene (see Sec. 8.1). In the case of the heavier isotope, viz. liquid He^4, the low-temperature situation is completely governed by phonons; accordingly, we expect its specific heat to be given by, cf. eqn. (23),

$$C_V = Nk\frac{4\pi^4}{5}\left(\frac{kT}{\hbar\omega_D}\right)^3, \tag{24}$$

where

$$\omega_D = \left(\frac{6\pi^2 N}{V}\right)^{1/3} c, \tag{25}$$

c being the velocity of sound in the liquid. The specific heat per gram of the liquid is then given by

$$c_V = \frac{2\pi^2 k^4}{15\varrho\hbar^3 c^3}T^3. \tag{26}$$

Substituting $\varrho = 0.1455$ g/cm³ and $c = 238$ m/sec, the foregoing result becomes

$$c_V = 0.0209T^3 \text{ J/g/°K}. \tag{27}$$

The experimental measurements of Wiebes *et al.* (1957), for $0 < T < 0.6$°K, conformed to the formula

$$c_V = (0.0204 \pm 0.0004)T^3 \text{ J/g/°K}. \tag{28}$$

The agreement between the theoretical result based on the theory of phonons and the experimental observations made on the liquid at low temperatures is clearly good.

7.4. Inertial density of the sound field

For a further understanding of the low-temperature behavior of liquid He^4, we should determine the "inertial mass" associated with a gas of sound quanta in thermal equilibrium. To do this, we may consider "a phonon gas in mass motion", because by determining the relationship between the momentum $\boldsymbol{P}$ of the gas and the velocity $\boldsymbol{v}$ of its mass motion we can readily evaluate the property in question. Now, since the total number of phonons in any given system is indefinite, the problem is free from the constraint of a fixed N; consequently, the undetermined multiplier α may be taken as identically equal to zero. However, we now have a new constraint on the system, namely that of a fixed total momentum $\boldsymbol{P}$, which is additional to the constraint of the fixed total energy E. Under these constraints, the mean occupation number of the phonon level $\varepsilon(\boldsymbol{p})$ would be

$$\langle n(\boldsymbol{p})\rangle = \frac{1}{\exp(\beta\varepsilon + \boldsymbol{\gamma}\cdot\boldsymbol{p}) - 1}; \tag{1}$$

cf. eqn. (6.5.1).

As usual, the parameter β is identically equal to $1/kT$. To determine the physical meaning of the parameter $\boldsymbol{\gamma}$, it seems natural to evaluate the drift velocity of the gas. Choosing z-axis to be in the direction of mass motion, the magnitude v of the drift velocity will be given by "the *mean* value of the component u_z of the individual phonon velocities":

$$v = \langle u \cos\theta\rangle. \tag{2}$$

Now, for phonons,

$$\varepsilon = pc \quad \text{and} \quad u \equiv \frac{d\varepsilon}{dp} = c, \tag{3}$$

where c is the *velocity of sound* in the medium. Moreover, by reasons of symmetry, we expect the undetermined vector $\boldsymbol{\gamma}$ to be either parallel or antiparallel to the direction of mass motion; hence, we may write

$$\boldsymbol{\gamma}\cdot\boldsymbol{p} = \gamma p_z = \gamma p \cos\theta. \tag{4}$$

In view of eqns. (1), (3) and (4), eqn. (2) becomes

$$v = \frac{\int_0^\infty \int_0^\pi [\exp\{\beta pc(1+(\gamma/\beta c)\cos\theta)\} - 1]^{-1}(c\cos\theta)(p^2\,dp\,2\pi\sin\theta\,d\theta)}{\int_0^\infty \int_0^\pi [\exp\{\beta pc(1+(\gamma/\beta c)\cos\theta)\} - 1]^{-1}(p^2\,dp\,2\pi\sin\theta\,d\theta)}. \tag{5}$$

Making the substitutions

$$\cos\theta = \eta, \quad p(1+(\gamma/\beta c)\eta) = p'$$

and cancelling away the integrations over p', we obtain

$$v = c\,\frac{\int_{-1}^{1}(1+(\gamma/\beta c)\eta)^{-3}\,\eta\,d\eta}{\int_{-1}^{1}(1+(\gamma/\beta c)\eta)^{-3}\,d\eta} = -\gamma/\beta.$$

It then follows that

$$\boldsymbol{\gamma} = -\beta \boldsymbol{v}. \tag{6}$$

Accordingly, the expression for the mean occupation number becomes

$$\langle n(\boldsymbol{p})\rangle = \frac{1}{\exp\{\beta(\varepsilon - \boldsymbol{v}\cdot\boldsymbol{p})\}-1}\,; \tag{7}$$

cf. eqn. (6.5.5). A comparison of (7) with the corresponding result in the rest frame of the gas, namely

$$\langle n_0(\boldsymbol{p}_0)\rangle = \frac{1}{\exp(\beta\varepsilon_0)-1}, \tag{8}$$

reveals that the formal change introduced by the imposition of mass motion on the system is nothing but a straightforward manifestation of *Galilean transformation* between the two frames of reference.

Alternatively, eqn. (7) may be written as

$$\langle n(\boldsymbol{p})\rangle = \frac{1}{\exp(\beta p'c)-1} = \frac{1}{\exp\{\beta pc(1-(v/c)\cos\theta)\}-1}. \tag{9}$$

As such, formula (9) lays down a severe restriction on the drift velocity $\boldsymbol{v}$, viz. it must not exceed c, the velocity of the phonons, for otherwise some of the occupation numbers will have to be negative! Actually, as our subsequent results will show, the formalism developed in this section breaks down as soon as $\boldsymbol{v}$ approaches c. The velocity c may, therefore, be looked upon as a *critical velocity* for the flow of the phonon gas:

$$(v_c)_{\text{ph}} = c. \tag{10)*}$$

The relevance of the foregoing result to the problem of superfluidity in liquid helium II will be brought out in the succeeding section.

We shall now calculate the total momentum $\boldsymbol{P}$ of the phonon gas:

$$\boldsymbol{P} = \sum_{\boldsymbol{p}}\langle n(\boldsymbol{p})\rangle \boldsymbol{p}. \tag{11}$$

* Since c is the velocity of *sound*, the condition $v \ll c$ ensures that we need not worry about any relativistic effects in this study. Thus, we may take $\beta = \beta_0$ and $V = V_0$ without any reservation.

Indeed, the vector $\boldsymbol{P}$ will be parallel to the vector $\boldsymbol{v}$, the latter being already in the direction of the z-axis. We have, therefore, to calculate only the z-component of the momentum:

$$
\begin{aligned}
P = P_z &= \sum_{p} \langle n(\boldsymbol{p})\rangle p_z \\
&= \int_0^\infty \int_0^\pi \frac{p \cos\theta}{\exp\{\beta pc(1-(v/c)\cos\theta)\}-1} \left(\frac{V\, p^2 dp\, 2\pi \sin\theta\, d\theta}{h^3}\right) \\
&= \frac{2\pi V}{h^3} \int_0^\infty \frac{p'^3\, dp'}{\exp(\beta p' c)-1} \int_0^\pi \{1-(v/c)\cos\theta\}^{-4} \cos\theta \sin\theta\, d\theta \\
&= V \frac{16\pi^5}{45 h^3 c^3 \beta^4} \cdot \frac{v/c^2}{(1-v^2/c^2)^3}.
\end{aligned} \tag{12}
$$

The total energy E of the gas is given by

$$
\begin{aligned}
E &= \sum_{p} \langle n(\boldsymbol{p})\rangle pc \\
&= \frac{2\pi V c}{h^3} \int_0^\infty \frac{p'^3\, dp'}{\exp(\beta p' c)-1} \int_0^\pi \{1-(v/c)\cos\theta\}^{-4} \sin\theta\, d\theta \\
&= V \frac{4\pi^5}{15 h^3 c^3 \beta^4} \cdot \frac{1+\frac{1}{3}(v/c)^2}{(1-v^2/c^2)^3}.
\end{aligned} \tag{13}
$$

It is, of course, natural to regard the ratio P/v as the "inertial mass" of the gas. The corresponding mass density ϱ is, therefore, given by

$$
\varrho = \frac{P}{vV} = \frac{16\pi^5 k^4 T^4}{45 h^3 c^5} \frac{1}{(1-v^2/c^2)^3}. \tag{14}
$$

For $(v/c) \ll 1$, which is very generally true, the mass density of the phonon gas is given by

$$
(\varrho_0)_{\text{ph}} = \frac{16\pi^5 k^4}{45 h^3 c^5} T^4 = \frac{4}{3c^2}(E_0/V). \tag{15}
$$

Substituting the value of c for liquid He^4 (at low temperatures), the phonon mass density, as a fraction of the actual density of the liquid, is given by

$$
(\varrho_0)_{\text{ph}}/\varrho_{\text{He}} = 1.22 \times 10^{-4}\, T^4; \tag{16}
$$

thus, for example, at $T = 0.3$°K the value of this fraction turns out to be about 9.9×10^{-7}. Now, at a low temperature like 0.3°K, the phonons are the only excitations in liquid He^4 that need to be considered; the calculated result must, therefore, correspond to the "ratio of the density ϱ_n of the *normal fluid* in the liquid to the total density ϱ of the liquid". It is practically impossible to make a *direct* determination of a fraction as small as that; however,

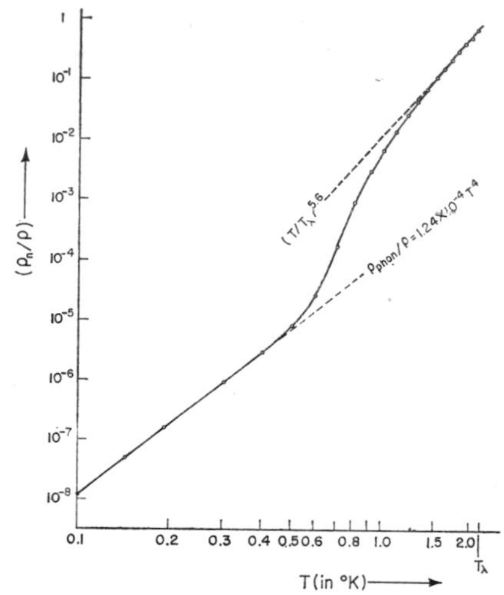

FIG. 7.11. The "normal" fraction (ϱ_n/ϱ), as obtained from the experimental data on (i) the velocity of second sound and (ii) the entropy of liquid He II (after de Klerk, Hudson and Pellam, 1953).

indirect evaluations, which make use of other experimentally studied properties of the liquid, provide a striking confirmation of the foregoing theoretical result; see Fig. 7.11.

The pressure $\mathcal{P}$ of the phonon gas can be obtained by calculating its grand partition function $\mathcal{Q}$:

$$\mathcal{P} \equiv \frac{kT}{V} \ln \mathcal{Q} = -\frac{kT}{V} \sum_{\boldsymbol{p}} \ln \left[1 - \exp\left\{-\beta(\varepsilon - \boldsymbol{v}\cdot\boldsymbol{p})\right\}\right]; \tag{17}$$

cf. eqn. (7.2.15). In terms of the variable p', we obtain

$$\begin{aligned}\mathcal{P} &= -\frac{2\pi kT}{h^3} \int_0^\infty \ln\left[1 - \exp\left(-\beta p' c\right)\right] p'^2\, dp' \int_0^\pi \{1 - (v/c)\cos\theta\}^{-3} \sin\theta\, d\theta \\ &= \frac{4\pi^5 k^4 T^4}{45 h^3 c^3} \frac{1}{(1 - v^2/c^2)^2}.\end{aligned} \tag{18}$$

Combining (13) and (18), we obtain

$$(E + \mathcal{P}V) = V \frac{16\pi^5 k^4 T^4}{45 h^3 c^3} \frac{1}{(1 - v^2/c^2)^3}. \tag{19}$$

Equation (12) may then be written as

$$\boldsymbol{P} = (E + \mathcal{P}V)\boldsymbol{v}/c^2. \tag{20}$$

The entropy of the gas can be calculated with the help of the formula

$$S = \frac{(E - \boldsymbol{v}\cdot\boldsymbol{P}) + \mathcal{P}V}{T} \qquad (\because\ \mu = 0), \tag{21}$$

whence it follows that

$$S = V\frac{16\pi^5k^4T^3}{45h^3c^3}\,\frac{1}{(1-v^2/c^2)^2}. \tag{22}$$

That completes our study of the phonon gas in mass motion.

In passing, we note that the foregoing results can be readily amended to become applicable to the radiation field of a moving black body, i.e. to a gas of photons in mass motion. Of course, one has to bear in mind that (i) the constant c would now stand for the velocity of *light*, and hence (ii) the condition $v \leqslant c$ would now correspond to the familiar condition met with in relativistic physics. This in turn requires that we take into account the Lorentz transformations of the volume V and the temperature T of the gas, namely

$$V = V_0(1-v^2/c^2)^{1/2} \quad \text{and} \quad T = T_0(1-v^2/c^2)^{1/2}; \tag{23}$$

see Sec. 6.5. Moreover, we should multiply our results by a factor of 2, for the well-known fact that the vibrational modes in the case of radiation are transverse and, hence, *doubly degenerate*. Accordingly, we obtain

$$E = V_0\frac{8\pi^5k^4T_0^4}{15h^3c^3}\,\frac{1+\frac{1}{3}v^2/c^2}{(1-v^2/c^2)^{1/2}} = E_0\,\frac{1+\frac{1}{3}v^2/c^2}{(1-v^2/c^2)^{1/2}}, \tag{13a}$$

$$\mathcal{P} = \frac{8\pi^5k^4T_0^4}{45h^3c^3} = \mathcal{P}_0, \tag{18a}$$

$$(E+\mathcal{P}V) = V_0\frac{32\pi^5k^4T_0^4}{45h^3c^3}\,\frac{1}{(1-v^2/c^2)^{1/2}} = \frac{\frac{4}{3}E_0}{(1-v^2/c^2)^{1/2}}, \tag{19a}$$

$$\boldsymbol{P} = \frac{\frac{4}{3}E_0(\boldsymbol{v}/c^2)}{(1-v^2/c^2)^{1/2}} \tag{20a}$$

and

$$S = V_0\frac{32\pi^5k^4T_0^3}{45h^3c^3} = S_0. \tag{22a}$$

These results were first obtained by Mosengeil (1907) on the basis of the electromagnetic theory of radiation. A few months later, Planck (1907) and Einstein (1907) derived more general results for the transformation of thermodynamic quantities pertaining to moving physical systems; the black-body radiation, with $\mathcal{P}_0V_0 = \frac{1}{3}E_0$, constituted a special case of these studies. In Sec. 6.5, we outlined a treatment based on statistical mechanics (Pathria, 1966, 1967) and obtained results in complete agreement with those of Planck and Einstein; in particular, the formulae obtained in this section bear one-to-one correspondence with the formulae (6.5.24), (6.5.26), (6.5.27), (6.5.25) and (6.5.10) of Sec. 6.5. We also note that the inertial mass density associated with the radiation field is given by

$$\varrho \equiv \frac{P}{vV} = \frac{\frac{4}{3}E_0/c^2}{V_0(1-v^2/c^2)} = \frac{\varrho_0}{(1-v^2/c^2)}, \tag{24}$$

where

$$\varrho_0 = \frac{32\pi^5k^4T_0^4}{45h^3c^5}. \tag{25}$$

7.5. Elementary excitations in liquid helium II

Landau (1941, 1947) developed a simple theoretical scheme which explains reasonably well the behavior of liquid helium II at temperatures not too close to the λ-point. According to this scheme, the liquid is treated as a weakly excited quantum-mechanical system, in which deviations from the ground state ($T = 0°K$) are described in terms of "a gas of elementary excitations" of the system against a *quiescent* background. The gas of excitations corresponds to the "normal fluid", while the quiescent background represents the "superfluid". At $T = 0°K$, there are no excitations at all ($\varrho_n = 0$) and the whole of the fluid constitutes the superfluid background ($\varrho_s = \varrho_{\mathrm{He}}$). At higher temperatures, we may write

$$\varrho_s(T) = \varrho_{\mathrm{He}}(T) - \varrho_n(T), \tag{1}$$

until at $T = T_\lambda$, $\varrho_n = \varrho_{\mathrm{He}}$ and $\varrho_s = 0$. Thereafter, the liquid behaves in all respects as a normal fluid. The latter phase of the liquid is commonly known as liquid helium I.

Guided by purely empirical considerations, Landau also proposed an energy–momentum relationship $\varepsilon(p)$ for the elementary excitations in liquid helium II. At low momenta, the relationship between ε and p was linear (which is characteristic of phonons), while at higher momenta it exhibited a non-monotonic character. The excitations were assumed to be bosons and, at low temperatures (when their number is not very large), mutually noninteracting; the macroscopic properties of the liquid could then be calculated by following a straightforward statistical-mechanical approach. It was found that Landau's theory could explain quite successfully the observed properties of liquid helium II over a temperature range of about 0–2°K; however, it still remained to be verified that the actual excitations in the liquid do, in fact, conform to the proposed energy spectrum.

Following a suggestion by Cohen and Feynman (1957), a number of experimental workers set out to investigate the spectrum of excitations in liquid helium II by scattering long-wavelength neutrons ($\lambda \gtrsim 4$ Å) from the liquid. At temperatures below 2°K, the most important scattering process is the one in which a neutron creates a *single* excitation in the liquid. By measuring the modified wavelength λ_f of the neutrons scattered at an angle ϕ, the energy ε and the momentum p of the excitation created in the scattering process could be determined on the basis of the relevant conservation laws:

$$\varepsilon = h^2(\lambda_i^{-2} - \lambda_f^{-2})/2m, \tag{2}$$

$$p^2 = h^2(\lambda_i^{-2} + \lambda_f^{-2} - 2\lambda_i^{-1}\lambda_f^{-1}\cos\varphi), \tag{3}$$

where λ_i is the initial wavelength of the neutrons and m the neutron mass. By varying φ, or λ_i, one could map the entire spectrum of the excitations.

The first exhaustive investigation along these lines was carried out by Yarnell *et al.* (1959); their results, shown in Fig. 7.12, possess a striking resemblance to the empirical spectrum proposed by Landau. The more important features of the spectrum, which was obtained at a temperature of 1.1°K, are the following:

(i) If we fit a linear, *phonon-like* spectrum ($\varepsilon = pc$) to points in the vicinity of $p/\hbar = 0.55$ Å^{-1}, we obtain for c a value of (239 ± 5) m/sec, which is in excellent agreement with the measured value of the sound velocity in the liquid, viz. about 238 m/sec.

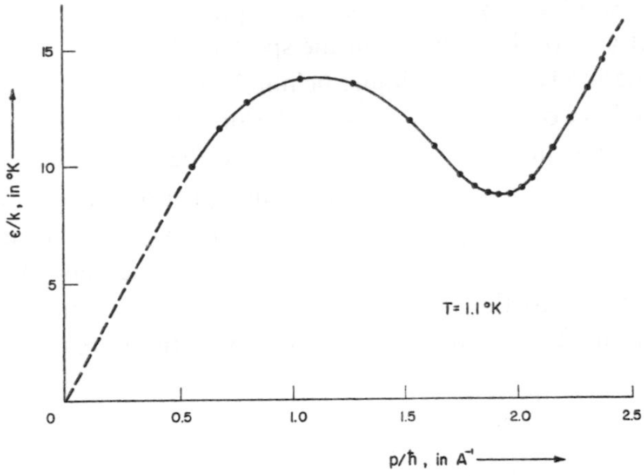

FIG. 7.12. The energy spectrum of the elementary excitations in liquid He II at 1.1°K (after Yarnell *et al.*, 1959); the dashed line emanating from the origin has a slope corresponding to the velocity of sound in the liquid, viz. (239 ± 5) m/sec.

(ii) The spectrum passes through a maximum value of $\varepsilon/k = (13.92 \pm 0.10)$°K at $p/\hbar = (1.11 \pm 0.02)$ Å^{-1}.

(iii) This in turn is followed by a minimum at $p/\hbar = (1.92 \pm 0.01)$ Å^{-1}, which may be represented by Landau's *roton* spectrum:

$$\varepsilon(p) = \Delta + \frac{(p-p_0)^2}{2\mu}, \tag{4}$$

with

$$\left.\begin{aligned} \Delta/k &= (8.65 \pm 0.04)^\circ\text{K} \\ p_0/\hbar &= (1.92 \pm 0.01)\ \text{Å}^{-1} \\ \mu &= (0.16 \pm 0.01) m_{\text{He}}. \end{aligned}\right\} \tag{5*}$$

and

(iv) Above $p/\hbar \simeq 2.18$ Å^{-1}, the spectrum rises linearly, again with a slope equal to c.

Data were also obtained at temperatures of 1.6°K and 1.8°K. The spectrum was found to be of the same general shape as at 1.1°K; only the value of Δ was slightly lower.

In a later investigation, Henshaw and Woods (1961) extended the range of observation at both ends of the spectrum; their results are shown in Fig. 7.13. On the lower side, they carried out measurements down to $p/\hbar = 0.26$ Å^{-1} and found that the experimental points indeed lie on a straight line (of slope 237 m/sec). On the upper side, they pushed their measurements up to $p/\hbar = 2.68$ Å^{-1} and found that, after passing through a minimum at 1.91 Å^{-1}, the curve rises with an increasing slope up to about 2.4 Å^{-1} at which point the

* The term "roton" for these excitations was coined by Landau who had originally thought that these excitations might, in some way, represent local disturbances of a *rotational* character in the liquid. However, subsequent theoretical work, especially that of Feynman (1953, 1954) and of Brueckner and Sawada (1957), does not support this contention. Nevertheless, the name "roton" has stayed on.

second derivative $\partial^2\varepsilon/\partial p^2$ changes sign; the subsequent trend of the curve suggests the possible existence of a second maximum in the spectrum!*

To evaluate the thermodynamics of liquid helium II, we first of all note that at sufficiently low temperatures we have only the low-lying excitations, namely the phonons. The thermodynamic behavior of the liquid is then governed by the formulae of Secs. 7.3 and 7.4. At temperatures higher than about 0.5°K, the second group of excitations, namely rotons (with momenta in the vicinity of p_0), also shows up. Between 0.5°K and about 1°K, the behavior of the liquid is governed by phonons and rotons together. Above 1°K, however, the phonon contributions to the various thermodynamic properties of the liquid become relatively less important; then, rotons are the only excitations that need to be considered.

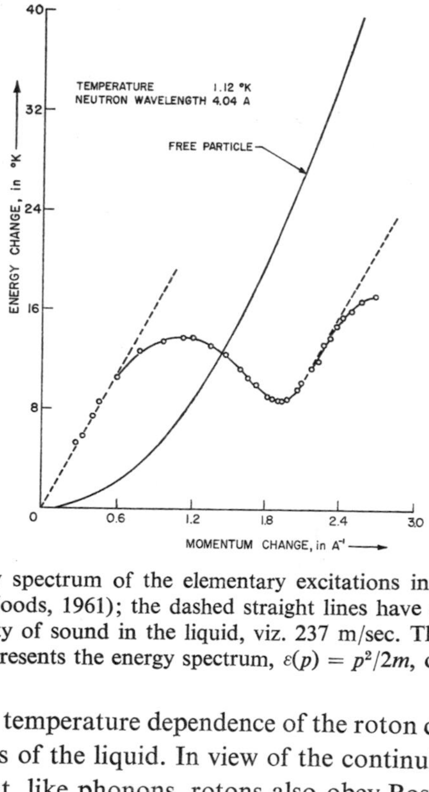

FIG. 7.13. The energy spectrum of the elementary excitations in liquid He II at 1.12°K (after Henshaw and Woods, 1961); the dashed straight lines have a common slope corresponding to the velocity of sound in the liquid, viz. 237 m/sec. The parabolic curve rising from the origin represents the energy spectrum, $\varepsilon(p) = p^2/2m$, of *free* helium atoms.

We shall now study the temperature dependence of the roton contributions to the various thermodynamic properties of the liquid. In view of the continuity of the energy spectrum, it is natural to expect that, like phonons, rotons also obey Bose–Einstein statistics. Moreover, their total number N in the system is quite indefinite; consequently, their chemical potential μ is identically equal to zero. We then have for the mean occupation numbers of the rotons

$$\langle n(\boldsymbol{p})\rangle = \frac{1}{\exp\{\beta\varepsilon(p)\}-1}, \tag{6}$$

where the energy $\varepsilon(p)$ is given by the formulae (4) and (5). Now, at all temperatures of inter-

* This seems to confirm the prediction of Pitaevskii (1959) that an end point in the spectrum might occur at a "critical" value p_c of the excitation momentum, where ε_c would be equal to 2Δ and $(\partial\varepsilon/\partial p)_c$ would be equal to zero.

est (viz. $T \lesssim 2°K$), the minimum value of the term $\exp\{\beta\varepsilon(p)\}$, namely $\exp(\Delta/kT)$, is considerably larger than unity. We may, therefore, write

$$\langle n(\boldsymbol{p})\rangle \simeq \exp\{-\beta\varepsilon(p)\}; \tag{7}$$

thus, the roton gas in liquid helium II is a fairly nondegenerate system and may be treated according to the Maxwell–Boltzmann statistics. The grand potential of the system is, therefore, given by

$$q(V, T) \equiv \frac{PV}{kT} = \sum_{p} \exp\{-\beta\varepsilon(p)\} = \bar{N}, \tag{8}$$

where $\bar{N}$ is the "equilibrium" value of the number of rotons in the system. The summation over $\boldsymbol{p}$ may be replaced by an integration, with the result

$$\frac{PV}{kT} = \bar{N} = \frac{V}{h^3}\int_0^\infty e^{-\left\{\Delta+\frac{(p-p_0)^2}{2\mu}\right\}/kT}(4\pi p^2\,dp). \tag{9}$$

Substituting $p = p_0+(2\mu kT)^{1/2}x$, we get

$$\frac{PV}{kT} = \bar{N} = \frac{4\pi p_0^2 V}{h^3} e^{-\Delta/kT}(2\mu kT)^{1/2}\int e^{-x^2}\left\{1+\frac{(2\mu kT)^{1/2}}{p_0}x\right\}^2 dx. \tag{10}$$

The "relevant" range of the variable x, the one that makes significant contribution towards the integral, is fairly symmetric about the value $x = 0$; consequently, the net effect of the *linear* term in the integrand is vanishingly small. The *quadratic* term too is unimportant because its coefficient $(2\mu kT)/p_0^2 \ll 1$. Thus, all we have to consider is the integral of $\exp(-x^2)$. Now, one can readily verify that the limits of this integral are such that, without seriously affecting the value of the integral, they may be taken as $-\infty$ and $+\infty$; the value of the integral is then simply $\pi^{1/2}$. We thus obtain

$$\frac{PV}{kT} = \bar{N} = \frac{4\pi p_0^2 V}{h^3}(2\pi\mu kT)^{1/2}e^{-\Delta/kT}. \tag{11*}$$

The free energy of the roton gas is given by ($\because\ \mu = 0$)

$$A(V, T) = -PV = -\bar{N}kT \propto T^{3/2}\,e^{-\Delta/kT}, \tag{12}$$

whence we obtain

$$S = -\left(\frac{\partial A}{\partial T}\right)_V = -A\left\{\frac{3}{2T}+\frac{\Delta}{kT^2}\right\} = \bar{N}k\left\{\frac{3}{2}+\frac{\Delta}{kT}\right\}, \tag{13}$$

$$U = A+TS = \bar{N}(\Delta+\tfrac{1}{2}kT) \tag{14†}$$

* Looking back at the integral (9), we observe that what we have done here simply amounts to having replaced p^2 in the integrand of (9) by its mean value p_0^2 and then carried out integration over the "complete" range of the variable $(p-p_0)$.

† This result is highly suggestive of the fact that for rotons there is only *one* degree of freedom, namely the *magnitude* of the roton momentum, which is thermally effective!

and

$$C_V = \left(\frac{\partial U}{\partial T}\right)_V = \bar{N}k\left\{\frac{3}{4}+\frac{\Delta}{kT}+\left(\frac{\Delta}{kT}\right)^2\right\}. \tag{15}$$

Clearly, as $T \to 0$, all these results tend to zero (essentially exponentially).

We shall now determine the inertia of the roton gas. Proceeding as in Sec. 7.4, we obtain for a gas of elementary excitations with energy spectrum $\varepsilon(p)$

$$\varrho_0 = \frac{M_0}{V} = \lim_{v\to 0}\frac{1}{v}\int n(\varepsilon-\boldsymbol{v}\cdot\boldsymbol{p})p\,\frac{d^3p}{h^3}, \tag{16}$$

where $n(\varepsilon-\boldsymbol{v}\cdot\boldsymbol{p})$ is the mean occupation number of the state $\varepsilon(\boldsymbol{p})$, as observed in a frame of reference K with respect to which the gas is in *mass motion* with a drift velocity $\boldsymbol{v}$;* the quantity $(d^3p)/h^3$ is the number of single-excitation states (*per unit volume* of the system) with momenta lying between $\boldsymbol{p}$ and $\boldsymbol{p}+d\boldsymbol{p}$. For small values of v, the function $n(\varepsilon-\boldsymbol{v}\cdot\boldsymbol{p})$ might be expanded as a Taylor series in $\boldsymbol{v}$ and only the terms $n(\varepsilon)-(\boldsymbol{v}\cdot\boldsymbol{p})\,\partial n(\varepsilon)/\partial\varepsilon$ retained. The integral over the first part denotes the momentum density of the system, as observed in the rest frame of the system, and is identically equal to zero. We are thus left with

$$\begin{aligned}\varrho_0 &= -\frac{1}{h^3}\int p^2\cos^2\theta\,\frac{\partial n(\varepsilon)}{\delta\varepsilon}\,(p^2\,dp\,2\pi\sin\theta\,d\theta)\\ &= -\frac{4\pi}{3h^3}\int_0^\infty \frac{\partial n(\varepsilon)}{\partial\varepsilon}\,p^4\,dp,\end{aligned} \tag{17}$$

which holds for *any* kind of energy spectrum and for *any* statistics. For phonons, in particular, we obtain

$$\begin{aligned}(\varrho_0)_{\text{ph}} &= -\frac{4\pi}{3h^3c}\int_0^\infty \frac{dn(p)}{dp}\,p^4\,dp\\ &= -\frac{4\pi}{3h^3c}\left[n(p)\cdot p^4\Big|_0^\infty - \int_0^\infty n(p)\cdot 4p^3\,dp\right]\\ &= \frac{4}{3c^2}\int_0^\infty n(p)\cdot pc\left(\frac{4\pi p^2\,dp}{h^3}\right) = \frac{4}{3c^2}\,(E_0)_{\text{ph}}/V,\end{aligned} \tag{18}$$

which is identical with our earlier result, namely (7.4.15).

* The drift velocity $\boldsymbol{v}$ must satisfy the condition: $(\boldsymbol{v}\cdot\boldsymbol{p}) \leqslant \varepsilon$, for otherwise some of the occupation numbers will have to be negative! This leads to the existence of a *critical velocity* v_c for the excitations, such that for values of v exceeding v_c the formalism developed here would simply break down. It is not difficult to see that this (critical) velocity is given by the relation: $v_c = (\varepsilon/p)_{\min}$; cf. eqn. (24).

For rotons, $n(\varepsilon) \simeq \exp(-\beta\varepsilon)$. Hence, $\partial n(\varepsilon)/\partial\varepsilon \simeq -\beta n(\varepsilon)$. Accordingly, eqn. (17) gives

$$(\varrho_0)_{\text{rot}} = \frac{4\pi\beta}{3h^3}\int n(\varepsilon)\, p^4\, dp$$

$$= \frac{\beta}{3}\langle p^2\rangle \frac{\bar{N}}{V} \simeq \frac{p_0^2}{3kT}\frac{\bar{N}}{V} \tag{19}$$

$$= \frac{4\pi p_0^4}{3h^3}\left(\frac{2\pi\mu}{kT}\right)^{1/2} e^{-\Delta/kT}; \tag{20}$$

see eqn. (11). At very low temperatures ($T < 0.3$°K), the roton contribution towards the inertia of the fluid is negligible in comparison with the phonon contribution. At relatively higher temperatures ($T \approx 0.6$°K), the two contributions become comparable. At temperatures above 1°K, the roton contribution is far more dominant than the phonon contribution; at such temperatures, the roton density alone accounts for the density ϱ_n of the normal fluid.

It would indeed be instructive to determine the *critical temperature* T_c at which the theoretical value of the density ϱ_n became equal to the actual density ϱ_{He} of the liquid; this would correspond to the disappearance of *superfluid component* from the liquid (and hence to the transition from liquid He II to liquid He I). In this manner, we find: $T_c \simeq 2.5$°K, as opposed to the experimental value of T_λ, which is $\simeq 2.19$°K. The comparison is not too bad, in view of the fact that in the present calculation we have assumed the roton gas to be a *noninteracting* system right up to the transition point; however, due to the presence of an exceedingly large number of excitations at higher temperatures, this assumption cannot remain plausible.

Equation (19) suggests that a roton excitation possesses an *effective mass*, which is given by $p_0^2/3kT$. Numerically, this is about 10–15 times the mass of a helium atom (and, hence, orders of magnitude larger than the parameter μ of the roton spectrum). However, the more important aspect of the roton effective mass is that it is *inversely* proportional to the temperature of the roton gas! Historically, this aspect was first discovered empirically by Landau (1947) on the basis of the experimental data on the velocity of second sound in liquid He II and its specific heat. Now, since the effective mass of an excitation is determined by the quantity $\langle p^2\rangle/3kT$, Landau concluded that the quantity $\langle p^2\rangle$ of the relevant excitations in the liquid must be temperature-independent. Thus, as the temperature of the liquid rises the mean value of p^2 (of the excitations) must stay constant; this value may be denoted by p_0^2. The mean value of ε, on the other hand, must rise with temperature. The only way to reconcile the two things was for Landau to invoke a *nonmonotonic* spectrum of excitations, as we have in eqn. (4).

Finally, we would like to touch upon the question of the *critical velocity* of superflow. For this we consider a mass M of excitation-free superfluid in mass motion; its kinetic energy E and momentum $\boldsymbol{P}$ are given by $\frac{1}{2}Mv^2$ and $M\boldsymbol{v}$, respectively. Any changes in these quantities are related as follows:

$$\delta E = (\boldsymbol{v}\cdot\delta\boldsymbol{P}). \tag{21}$$

Supposing that these changes have come about as a result of the creation of an excitation

$\varepsilon(\boldsymbol{p})$ in the fluid, we must have, by the principle of conservation,

$$\delta E = -\varepsilon \quad \text{and} \quad \delta \boldsymbol{P} = -\boldsymbol{p}. \tag{22}$$

Equations (21) and (22) lead to the result

$$\varepsilon = (\boldsymbol{v}\cdot\boldsymbol{p}) \leqslant vp. \tag{23}$$

Thus, it is impossible to create the excitation $\varepsilon(\boldsymbol{p})$ in the fluid unless the drift velocity v of the fluid is greater than, or at least equal to, the quantity (ε/p). Accordingly, if v is less than even the lowest value of ε/p, no excitation whatsoever can be created in the fluid, which will therefore maintain its superfluid character. We thus obtain a condition for the maintenance of superfluidity:

$$v < v_c = (\varepsilon/p)_{\min}, \tag{24}$$

which is known as the *Landau criterion* for superflow. The velocity v_c is called the *critical velocity* of superflow; it marks an "upper limit" to the flow velocities at which the fluid exhibits superfluid behavior. The observed magnitude of the critical velocity varies significantly with the geometry of the channel employed; as a rule, the narrower the channel the larger the critical velocity. It may be mentioned that the observed values of v_c range from about 0.1 cm/sec to about 70 cm/sec.

The theoretical estimates of v_c are clearly of interest. On the one hand, we find that if the excitations obey the ideal-gas relationship, viz. $\varepsilon = p^2/2m$, then the critical velocity turns out to be exactly zero! Any velocity v is then greater than the critical velocity; accordingly, no superflow is at all possible. This is an extremely significant result, for it brings out very clearly the fact that the interatomic interactions in the liquid, which give rise to an excitation spectrum different from the one characteristic of the ideal gas, play a very fundamental role in bringing about the phenomenon of superfluidity. Thus, while an ideal gas does undergo the phenomenon of Bose–Einstein condensation, it cannot support the phenomenon of superfluidity as such. On the other hand, we find that (i) for phonons, $v_c = c \simeq 2.4\times10^4$ cm/sec and (ii) for rotons, $v_c = \{(p_0^2+2\mu\Delta)^{1/2} - p_0\}/\mu \simeq \Delta/p_0 \simeq 6.3\times 10^3$ cm/sec, which are too high in comparison with the observed values of v_c. As will be seen in Sec. 10.8, there is another kind of collective excitations which can appear in liquid helium II, viz. the *quantized vortex rings*, with an energy–momentum relationship of the form: $\varepsilon \propto p^{1/2}$. The critical velocity for the creation of vortex rings turns out to be numerically consistent with the experimental findings; not only that, the variation of v_c with the geometry of the channel can also be understood in terms of the physical size of the rings created.

Problems

7.1. By considering the order-of-magnitude values of the occupation numbers $\langle n_\varepsilon \rangle$, show that it makes no difference to the individual parts of the right-hand side of eqn. (7.1.6) whether we combine a *finite* number of $(\varepsilon \neq 0)$-terms, of the series (7.1.2), with the $(\varepsilon = 0)$-part or include them into the integral over ε.

7.2. Deduce the virial expansion (7.1.14) from eqns. (7.1.7) and (7.1.8), and verify the quoted values of the virial coefficients.

7.3. Show that for the ideal Bose gas

$$\frac{C_P - C_V}{Nk} = \left(\frac{C_V}{\frac{3}{2}Nk}\right)^2 \frac{g_{1/2}(z)}{g_{3/2}(z)}.$$

Discuss *physically* the behavior of the foregoing quantity at temperatures below T_c.

7.4. Show that for the ideal Bose gas, studied in Sec. 7.1, the temperature derivative of the specific heat C_V is given by

$$\frac{1}{Nk}\left(\frac{\partial C_V}{\partial T}\right) = \begin{cases} \dfrac{1}{T}\left[\dfrac{45}{8}\dfrac{g_{5/2}(z)}{g_{3/2}(z)} - \dfrac{9}{4}\dfrac{g_{3/2}(z)}{g_{1/2}(z)} - \dfrac{27}{8}\dfrac{\{g_{3/2}(z)\}^2 g_{-1/2}(z)}{\{g_{1/2}(z)\}^3}\right] & \text{for} \quad T > T_c, \\ \dfrac{45}{8}\dfrac{v}{T\lambda^3}\zeta\left(\dfrac{5}{2}\right) & \text{for} \quad T < T_c. \end{cases}$$

With the help of these results and the formula (D.8), verify eqn. (7.1.38).

7.5. Investigate the statistical thermodynamics of an ideal Bose gas in a uniform gravitational field (of acceleration g). Show, in particular, that the phenomenon of Bose–Einstein condensation sets in at a temperature T_c given by

$$T_c \simeq T_c^0\left[1 + \frac{8}{9}\frac{1}{\zeta(\frac{3}{2})}\left(\frac{\pi m g L}{kT_c^0}\right)^{1/2}\right],$$

where L is the height of the container and $(mgL) \ll (kT_c^0)$. Also show that the condensation is accompanied by a discontinuity in the specific heat of the gas:

$$(\Delta C_V)_{T=T_c} \simeq -\frac{9}{8\pi}\zeta\left(\frac{3}{2}\right) Nk\left(\frac{\pi m g L}{kT_c^0}\right)^{1/2};$$

see Eisenschitz (1958).

7.6. Consider an ideal Bose gas composed of molecules with internal degrees of freedom. Assuming that, besides the ground state $\varepsilon_0 = 0$, it is only the first excited state ε_1 of the *internal* spectrum that needs to be taken into account, determine the condensation temperature of the gas as a function of ε_1. Show in particular that, for $(\varepsilon_1/kT_c^0) \gg 1$,

$$T_c/T_c^0 \simeq 1 - \frac{\frac{2}{3}}{\zeta(\frac{3}{2})} e^{-\varepsilon_1/kT_c^0}$$

and, for $(\varepsilon_1/kT_c^0) \ll 1$,

$$T_c/T_c^0 \simeq \left(\frac{1}{2}\right)^{2/3}\left[1 + \frac{2^{4/3}}{3\zeta(\frac{3}{2})}\left(\frac{\pi\varepsilon_1}{kT_c^0}\right)^{1/2}\right].$$

[*Hint*. To obtain the last result, one makes use of an expansion for the function $g_{3/2}(\alpha)$, for small α, which is given in Appendix D.]

7.7. Consider an ideal Bose gas in the grand canonical ensemble and study the extent of the fluctuations in the total number of particles N and the total energy E. Discuss, in particular, the situation when the gas becomes highly degenerate.

7.8. (a) Evaluate the grand partition function of a two-dimensional ideal Bose gas and derive an expression for the (equilibrium) number of particles *per unit area* of the system as a function of the parameters z and T. Show that this system does not exhibit the phenomenon of Bose–Einstein condensation.

(b) To elucidate this result, carry out a similar study of an n-dimensional Bose gas, whose single-particle energy spectrum is given by $\varepsilon \propto p^s$, where s is some positive number. Discuss the onset of the phenomenon of Bose–Einstein condensation in this system, especially its dependence on the numbers n and s. Study the essential features of the thermodynamic behavior of this system and show that, quite generally,

$$P = \frac{s}{n}\frac{U}{V} \quad \text{and} \quad C_V(T \to \infty) = \frac{n}{s}Nk.$$

Next, if the system admits of Bose–Einstein condensation, then

$$(C_V)_{T=T_c-} = \frac{n}{s}\left(\frac{n}{s}+1\right)\frac{\zeta(n/s+1)}{\zeta(n/s)}Nk,$$

$$(C_V)_{T=T_c-} - (C_V)_{T=T_c+} = \left(\frac{n}{s}\right)^2 \frac{\zeta(n/s)}{\zeta(n/s-1)}Nk$$

and

$$(C_P/C_V)_{T_c+} = \frac{n+s}{n}\zeta\left(\frac{n}{s}+1\right)\zeta\left(\frac{n}{s}-1\right)\Big/\zeta^2\left(\frac{n}{s}\right).$$

Further

$$\left(\frac{\partial C_V}{\partial T}\right)_{T_c-} = \left(\frac{n}{s}\right)^2\left(\frac{n}{s}+1\right)\frac{\zeta(n/s+1)}{\zeta(n/s)}\frac{Nk}{T_c}$$

and

$$\left(\frac{\partial C_V}{\partial T}\right)_{T_c-} - \left(\frac{\partial C_V}{\partial T}\right)_{T_c+} = \left(\frac{n}{s}\right)^2\frac{Nk}{T_c}\left[\frac{\zeta(n/s)}{\zeta(n/s-1)} + \frac{n}{s}\frac{\zeta^2(n/s)\,\zeta(n/s-2)}{\zeta^3(n/s-1)}\right].$$

(c) Study, with the help of these formulae, the case of an extreme relativistic Bose gas ($\varepsilon = pc$) and compare your results with those for the black-body radiation.

7.9. Evaluate the (canonical) partition function of the photon gas, compare it with the grand partition function of the gas, as given by eqn. (7.2.15), and interpret the result physically.

7.10. Show that the mean energy per photon in a black-body radiation cavity is very nearly equal to $2.7kT$.

7.11. Considering the dependence of the frequencies ω of the vibrational modes of the radiation field on the volume V of the system, establish the relation (7.2.17) between the pressure P and the energy density (U/V).

7.12. The sun may be regarded as a black-body at a temperature of 5800°K. Its diameter is about 1.4×10^9 m while its distance from the earth is about 1.5×10^{11} m.

(i) Calculate the total radiant intensity (in W/m^2) of sunlight at the surface of the earth.

(ii) What pressure would it exert on a perfectly absorbing surface placed normal to the rays (of the sun)?

(iii) If a flat surface on a satellite, which faces the sun, were an ideal absorber and radiator, what equilibrium temperature would it ultimately attain?

7.13. Write down the (canonical) partition function of a Debye solid and derive expressions for the various thermodynamic properties of the system.

7.14. Figure 7.14 is a plot of the function $C_V(T)$ against T for a solid, the limiting value $C_V(\infty)$ being the classical result given by Dulong and Petit. Show that the shaded area in the figure, namely

$$\int_0^\infty \{C_V(\infty) - C_V(T)\}\,dT,$$

is exactly equal to the zero-point energy of the solid. Interpret the result physically.

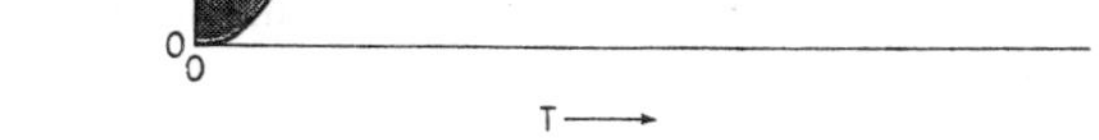

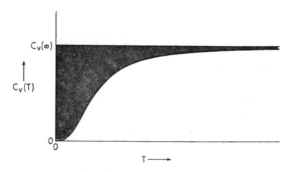

FIG. 7.14.

7.15. Show that the total *zero-point energy* of a Debye solid is equal to $\frac{9}{8}Nk\Theta_D$.
[Note that this implies, for each vibrational mode of the solid, a mean zero-point energy of $\frac{3}{8}k\Theta_D$, i.e. $\hbar\bar{\omega} = \frac{3}{4}k\Theta_D$.]

7.16. Show that, for $T \ll \Theta_D$, the quantity $(C_P - C_V)$ of a Debye solid varies as T^7 and hence the ratio $(C_P/C_V) \simeq 1$.

7.17. Determine the temperature T, in terms of the Debye temperature Θ_D, at which one-half of the oscillators in a Debye solid are expected to be in the excited states.

7.18. Determine the value of the parameter Θ_D for liquid He^4 from the empirical result (7.3.28).

7.19. (a) Compare the "mean thermal wavelength" λ_T of neutrons at a typical room temperature with the "minimum wavelength" $\lambda_{\min}$ of phonons in a typical crystal.

(b) Show that the frequency ω_D for a sodium chloride crystal is of the same order of magnitude as the frequency of an electromagnetic wave in the infrared.

7.20. Proceeding under the conditions (7.3.16) rather than the condition (7.3.13), prove that

$$C_V(T) = Nk\{D(x_{0,L}) + 2D(x_{0,T})\},$$

where $x_{0,L} = (\hbar\omega_{D,L}/kT)$ and $x_{0,T} = (\hbar\omega_{D,T}/kT)$. Compare this result with the one given by eqn. (7.3.17) and estimate the nature and the magnitude of the error involved in the latter.

7.21. A mechanical system of n identical masses (of mass m) connected in a straight line by identical springs (of stiffness K) has the natural vibrational frequencies given by

$$\omega_r = 2\sqrt{\left(\frac{K}{m}\right)} \sin\left(\frac{r}{n}\cdot\frac{\pi}{2}\right); \qquad r = 1, 2, \ldots, (n-1).$$

Correspondingly, a *linear* molecule composed of n identical atoms may be expected to have a vibrational spectrum given by

$$\nu_r = \nu_m \sin\left(\frac{r}{n}\cdot\frac{\pi}{2}\right); \qquad r = 1, 2, \ldots, (n-1),$$

where ν_m is a characteristic vibrational frequency of the molecule. Show that this leads to a vibrational specific heat, per molecule, which varies as T^1 at low temperatures and tends to the limiting value $(n-1)k$ at high temperatures.

7.22. Assuming the dispersion relation $\omega = Ak^s$ (where ω is the angular frequency and k the wave number) for wave motions existing in a solid, show that the respective contribution towards the specific heat of the solid *at low temperatures* is proportional to $T^{3/s}$.

[Note that while the case $s = 1$ corresponds to the elastic waves in the lattice, the case $s = 2$ applies to the spin waves propagating in a ferromagnetic system.]

7.23. Assuming the excitations to be phonons ($\omega = Ak$), show that their contribution towards the specific heat of an n-dimensional Debye system is proportional to T^n.

[Note that the elements selenium and tellurium form crystals in which atomic chains are arranged in parallel. The systems, in a certain sense, behave as one-dimensional; accordingly, the T^1-law holds (over a certain temperature range). At very low temperatures, however, it is replaced by a T^3-law. For a similar reason, graphite obeys the T^2-law over a certain range of temperatures.]

7.24. The (minimum) potential energy of a solid, when all its atoms are at rest at their equilibrium positions, may be denoted by the symbol $\Phi_0(V)$, where V is the volume of the solid. Similarly, the normal frequencies of vibration, ω_i ($i = 1, 2, \ldots, 3N-6$), may be denoted by the symbols $\omega_i(V)$. Show that the pressure of the solid is given by

$$P = -\frac{\partial\Phi_0}{\partial V} + \gamma\frac{U}{V},$$

where U is the internal energy of the solid, arising from the vibrations of the atoms, while γ is the *Grüneisen constant* of the solid:

$$\gamma = -\frac{\partial \ln \omega}{\partial \ln V} \approx \frac{1}{3}.$$

Assuming that, for $V \simeq V_0$,

$$\Phi_0(V) = \frac{(V-V_0)^2}{2\varkappa_0 V_0},$$

where $\varkappa_0$ and V_0 are constants and $\varkappa_0 C_V T \ll V_0$, show that the coefficient of thermal expansion (at constant pressure $P \simeq 0$) is given by

$$\alpha \equiv \frac{1}{V}\left(\frac{\partial V}{\partial T}\right)_{N,P} = \frac{\gamma\varkappa_0 C_V}{V_0}.$$

Also show that

$$C_P - C_V = \frac{\gamma^2 \varkappa_0 C_V^2 T}{V_0}.$$

7.25. Apply the general formula for the kinetic pressure of a gas, namely

$$P = \tfrac{1}{3} n \langle pu \rangle$$

to a gas of rotons and verify that the result agrees with the general Boltzmannian relationship: $P = nkT$.

7.26. Show that the free energy A and the inertial density ϱ of a roton gas in *mass motion* are given by

$$A(v) = A(0) \frac{\sinh x}{x}$$

and

$$\varrho(v) = \varrho(0) \frac{3(x \cosh x - \sinh x)}{x^3},$$

where $x = (vp_0/kT)$.

7.27. Integrating (7.5.17) by parts, show that the effective mass of an excitation, whose energy–momentum relationship is denoted by $\varepsilon(p)$, is given by

$$m_{\text{eff}} = \left\langle \frac{1}{3p^2} \left\{ \frac{d}{dp} \left(p^4 \frac{dp}{d\varepsilon} \right) \right\} \right\rangle.$$

Check the validity of this result by considering the examples of (i) an ideal-gas particle, (ii) a phonon and (iii) a roton.

CHAPTER 8

IDEAL FERMI SYSTEMS

8.1. Thermodynamic behavior of an ideal Fermi gas

According to Secs. 6.1 and 6.2, we obtain for an ideal Fermi gas

$$\frac{PV}{kT} \equiv \ln \mathcal{Q} = \sum_{\varepsilon} \ln (1+ze^{-\beta\varepsilon}) \tag{1}$$

and

$$N \equiv \sum_{\varepsilon} \langle n_\varepsilon \rangle = \sum_{\varepsilon} \frac{1}{z^{-1}e^{\beta\varepsilon}+1}, \tag{2}$$

where $\beta = 1/kT$ and $z = \exp(\mu/kT)$. Unlike what happens in the Bose case, the parameter z in the Fermi case can take on *unrestricted* values: $0 \leqslant z < \infty$. Moreover, in view of the Pauli exclusion principle, the question of a large number of particles getting together into a single energy state does not arise in this case; hence, we do not encounter a phenomenon like Bose–Einstein condensation. Nevertheless, at sufficiently low temperatures, the Fermi gas displays its own brand of quantal behavior, a detailed study of which is of great physical interest.

If we replace summations over ε by corresponding integrations, eqns. (1) and (2) become

$$\frac{P}{kT} = \frac{g}{\lambda^3} f_{5/2}(z) \tag{3}$$

and

$$\frac{N}{V} = \frac{g}{\lambda^3} f_{3/2}(z), \tag{4}$$

where g is a weight factor that arises from the "internal structure" of the particles (such as *spin*), λ is the mean thermal wavelength of the particles

$$\lambda = h/(2\pi mkT)^{1/2}, \tag{5}$$

while

$$f_n(z) = \frac{1}{\Gamma(n)} \int_0^\infty \frac{x^{n-1}\,dx}{z^{-1}e^x+1}. \tag{6}$$

Eliminating z between eqns. (3) and (4), we obtain the *equation of state* of the ideal Fermi gas.

The internal energy U of the ideal Fermi gas is given by

$$U \equiv -\left(\frac{\partial}{\partial \beta} \ln \mathcal{Q}\right)_{z, V} = kT^2 \left(\frac{\partial}{\partial T} \ln \mathcal{Q}\right)_{z, V}$$

$$= \frac{3}{2} kT \frac{gV}{\lambda^3} f_{5/2}(z) = \frac{3}{2} NkT \frac{f_{5/2}(z)}{f_{3/2}(z)}; \tag{7}$$

thus, quite generally, this system satisfies the relationship

$$p = \tfrac{2}{3}(U/V). \tag{8}$$

The specific heat C_V of the gas can be obtained by differentiating (7) with respect to T, keeping N and V constant, and making use of the relationship

$$\left(\frac{\partial z}{\partial T}\right)_v = -\frac{3}{2} \frac{z}{T} \frac{f_{3/2}(z)}{f_{1/2}(z)}, \tag{9}$$

which follows from eqn. (4) and the reduction formula (E.5). The final result is

$$\frac{C_V}{Nk} = \frac{15}{4} \frac{f_{5/2}(z)}{f_{3/2}(z)} - \frac{9}{4} \frac{f_{3/2}(z)}{f_{1/2}(z)}. \tag{10}$$

For the Helmholtz free energy of the gas, we get

$$A \equiv N\mu - PV = NkT\left\{\ln z - \frac{f_{5/2}(z)}{f_{3/2}(z)}\right\}, \tag{11}$$

and for the entropy

$$S \equiv \frac{U-A}{T} = Nk\left\{\frac{5}{2} \frac{f_{5/2}(z)}{f_{3/2}(z)} - \ln z\right\}. \tag{12}$$

In order to determine the various properties of the Fermi gas in terms of the particle density $n(= N/V)$ and the temperature T, we must have a knowledge of the functional dependence of the parameter z on n and T; this knowledge is formally contained in the implicit relationship (4). For detailed studies, one is sometimes obliged to make use of the numerical tables of the functions $f_n(z)$, which are indeed available (see Appendix E); for physical understanding, however, the various limiting forms of these functions can serve the necessary purpose.

Now, if the density of the gas is very low and its temperature very high, then the situation might correspond to

$$f_{3/2}(z) = \frac{nh^3}{g(2\pi mkT)^{3/2}} \ll 1; \tag{13}$$

we then speak of the gas as being highly *nondegenerate*. In view of the expansion (E.4), namely

$$f_n(z) = z - \frac{z^2}{2^n} + \frac{z^3}{3^n} - \dots, \tag{14}$$

the condition (13) implies that z itself is much smaller than unity and, in consequence, $f_n(z) \simeq z$. The equations for the various thermodynamic properties of the gas then become

$$P = NkT/V; \quad U = \tfrac{3}{2}NkT; \quad C_V = \tfrac{3}{2}Nk, \tag{15}$$

$$A = NkT\left\{\ln \frac{(n\lambda^3)}{g} - 1\right\} \tag{16}$$

and

$$S = Nk\left\{\frac{5}{2} - \ln \frac{(n\lambda^3)}{g}\right\}; \tag{17}$$

these results are the standard ones holding for an *ideal classical* gas.

If the parameter z is small in comparison with unity but not extremely small, then we should make a fuller use of the series (14) in order to eliminate z between eqns. (3) and (4). The procedure is just the same as followed in the corresponding Bose case, i.e. we first invert the series appearing in (4) to obtain an expansion for z in powers of $(n\lambda^3/g)$ and then substitute this expansion into the series appearing in (3). The equation of state then takes the form of the *virial expansion*:

$$\frac{PV}{NkT} = \sum_{l=1}^{\infty} (-1)^{l-1} a_l \left(\frac{\lambda^3}{gv}\right)^{l-1}, \tag{18}$$

where $v = 1/n$, while the coefficients a_l are the same as quoted in (7.1.15). For the specific heat, in particular, we obtain

$$\begin{aligned} C_V &= \frac{3}{2} Nk \sum_{l=1}^{\infty} (-1)^{l-1} \frac{5-3l}{2} a_l \left(\frac{\lambda^3}{gv}\right)^{l-1} \\ &= \frac{3}{2} Nk \left[1 - 0.0884\left(\frac{\lambda^3}{gv}\right) + 0.0066\left(\frac{\lambda^3}{gv}\right)^2 - 0.0004\left(\frac{\lambda^3}{gv}\right)^3 + \ldots\right]. \end{aligned} \tag{19}$$

Thus, at finite temperatures, the value of the specific heat is smaller than its limiting value, which is $\frac{3}{2}Nk$. As will be seen in the sequel, the specific heat of the ideal Fermi gas decreases *monotonically* as the temperature of the gas falls; see Fig. 8.2 and compare it with the corresponding Fig. 7.4 for the ideal Bose gas.

If the density and the temperature of the gas are such that the parameter $(n\lambda^3/g)$ is of the order of unity, the foregoing expansions cannot be of much use. In that case, one may have to make recourse to numerical calculation. However, if $(n\lambda^3/g) \gg 1$, the functions involved can again be expressed as *asymptotic* expansions in powers of $(\ln z)^{-1}$; we then speak of the gas as being *degenerate*. As $(n\lambda^3/g) \to \infty$, the functions assume a closed form, with the result that the expressions for the various thermodynamic quantities pertaining to the system become highly simplified; we then speak of the gas as being *completely degenerate*. For simplicity, we shall first discuss the main features of the system in a state of complete degeneracy.

In the limit $T \to 0$, which corresponds to the limit $(n\lambda^3/g) \to \infty$, the mean occupation numbers of the single-particle states $\varepsilon(\boldsymbol{p})$ become

$$\langle n_\varepsilon \rangle \equiv \frac{1}{e^{\{\varepsilon(\boldsymbol{p})-\mu\}/kT}+1} = \begin{cases} 1 & \text{for} \quad \varepsilon(\boldsymbol{p}) < \mu_0, \\ 0 & \text{for} \quad \varepsilon(\boldsymbol{p}) > \mu_0 \end{cases} \tag{20}$$

where μ_0 is the chemical potential of the system at $T = 0$. The function $\langle n_\varepsilon \rangle$ is clearly a *step function*, which stays constant at the (highest) value 1 right from $\varepsilon = 0$ to $\varepsilon = \mu_0$ and then suddenly drops down to the (lowest) value 0; see Fig. 8.1. Thus, at $T = 0$, all single-particle states up to $\varepsilon = \mu_0$ are "completely" filled, with one particle per state (in accordance with the Pauli principle), while all single-particle states with $\varepsilon > \mu_0$ are empty. The limiting energy μ_0 is generally referred to as the *Fermi-energy* of the system and is denoted by the symbol ε_F; the corresponding value of the single-particle momentum is referred to as the *Fermi*

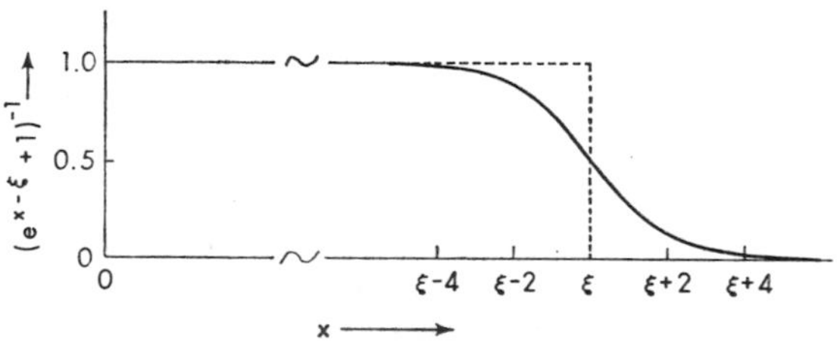

FIG. 8.1. Fermi distribution at low temperatures, with $x = \varepsilon/kT$ and $\xi = \mu/kT$. The rectangle denotes the limiting distribution as $T \to 0$; in that case, the Fermi function is equal to unity for $\varepsilon < \mu_0$ and zero for $\varepsilon > \mu_0$.

momentum and is denoted by the symbol p_F. The defining equation for these parameters is

$$\int_0^{\varepsilon_F} a(\varepsilon)\, d\varepsilon = N, \tag{21}$$

where $a(\varepsilon)$ denotes the *density of states* of the system and is given by the general expression

$$a(\varepsilon) = \frac{gV}{h^3} 4\pi p^2 \frac{dp}{d\varepsilon}. \tag{22}$$

We readily obtain

$$N = \frac{4\pi g V}{3h^3} p_F^3, \tag{23}$$

whence it follows that

$$p_F = \left(\frac{3N}{4\pi g V}\right)^{1/3} h; \tag{24}$$

accordingly, in the *nonrelativistic* case,

$$\varepsilon_F = \left(\frac{3N}{4\pi g V}\right)^{2/3} \frac{h^2}{2m} = \left(\frac{6\pi^2 n}{g}\right)^{2/3} \frac{\hbar^2}{2m}. \tag{25}$$

The ground state, or the zero-point, energy of the system is then given by

$$E_0 = \frac{4\pi g V}{h^3} \int_0^{p_F} \left(\frac{p^2}{2m}\right) p^2\, dp$$

$$= \frac{2\pi g V}{5mh^3} p_F^5, \tag{26}$$

whence we obtain for the energy per particle

$$\frac{E_0}{N} = \frac{3p_F^2}{10m} = \frac{3}{5}\varepsilon_F. \tag{27}$$

The ground-state pressure of the system is in turn given by

$$P_0 = \tfrac{2}{3}(E_0/V) = \tfrac{2}{5}n\varepsilon_F. \tag{28}$$

Substituting for ε_F, the foregoing expression takes the form

$$P_0 = \left(\frac{6\pi^2}{g}\right)^{2/3} \frac{\hbar^2}{5m} n^{5/3} \propto n^{5/3}. \tag{29}$$

The zero-point motion seen here is clearly a quantum effect arising because of the Pauli principle, according to which, even at $T = 0°$K, the particles constituting the system cannot settle down into a single energy state (as we had in the Bose case) and are therefore spread over a requisite number of lowest available energy states. As a result, the Fermi system, even at absolute zero, is quite live!

For a discussion of properties such as the specific heat and the entropy of the system, it is essential to extend our study to *finite* temperatures. If we decide to restrict ourselves to low temperatures, the deviations from the ground-state results will not be too large; accordingly, an analysis based on the asymptotic expansions of the functions $f_n(z)$ would be quite appropriate. However, before we study the problem analytically it seems useful to carry out a physical assessment of the situation with the help of the expression

$$\langle n_\varepsilon \rangle = \frac{1}{e^{(\varepsilon-\mu)/kT}+1} \tag{30}$$

for the mean occupation numbers of the single-particle states of the system. The situation corresponding to $T = 0$ is summarized in eqn. (20) and is shown as a step function in Fig. 8.1. Deviations from this (singular) situation, when T is finite (though still much smaller than the *characteristic temperature* μ_0/k), will be significant *only* for those values of ε for which the magnitude of the quantity $(\varepsilon - \mu)/kT$ is of the order of unity (for otherwise the value of the exponential term in (30) will not be much different from its ground-state value, viz. $e^{\pm\infty}$); see the solid curve in Fig. 8.1. We, therefore, conclude that a thermal reshuffling of the particles takes place *only* in a small energy range which is located around the energy value $\varepsilon = \mu_0$ and has a width $0(kT)$. The fraction of the particles that are thermally excited is, therefore, $0(kT/\varepsilon_F)$—the major part of the system remaining uninfluenced by the rise in temperature.* This is the most characteristic feature of a degenerate Fermi system and is essentially responsible for the qualitative as well as the quantitative differences between the physical behavior of this system and that of a corresponding classical system.

* We, therefore, speak of the totality of the energy levels filled at $T = 0$ as "the Fermi sea" and the small fraction of the particles that are excited near the top, when $T > 0$, as a "mist above the sea". Physically speaking, the origin of this behavior lies in the Pauli exclusion principle, according to which a fermion of energy ε cannot absorb a quantum of thermal excitation ε_T if the energy level $\varepsilon + \varepsilon_T$ is already filled. Since $\varepsilon_T = 0(kT)$, only those fermions which occupy energy states near the top level ε_F, up to a depth $0(kT)$, can be thermally excited to the available energy levels.

To conclude the argument, we observe that since the thermal energy per "excited" particle must be $0(kT)$, the thermal energy of the whole system must be $0(Nk^2T^2/\varepsilon_F)$; accordingly, the specific heat of the system must be $0(Nk \cdot kT/\varepsilon_F)$. Thus, the low-temperature specific heat of a Fermi system differs from the classical value $\frac{3}{2}Nk$ by a factor which not only reduces it considerably in magnitude but also makes it temperature-dependent (varying as T^1). It will be seen repeatedly that the first-power dependence of C_V on T is a typical feature of Fermi systems at low temperatures.

For an analytical study of Fermi gas at finite, but low, temperatures, we observe that the value of z, which was infinitely large at absolute zero, is now finite, though still large in comparison with unity. The functions $f_n(z)$ can, therefore, be expressed as asymptotic expansions in powers of the quantity $(\ln z)^{-1}$; see Sommerfeld's lemma (E.13). For the values of n we are immediately interested in, namely $n = \frac{5}{2}, \frac{3}{2}$ and $\frac{1}{2}$, we obtain *to the first approximation*

$$f_{5/2}(z) = \frac{8}{15\pi^{1/2}} (\ln z)^{5/2} \left[1+\frac{5\pi^2}{8}(\ln z)^{-2}+ \dots\right], \tag{31}$$

$$f_{3/2}(z) = \frac{4}{3\pi^{1/2}} (\ln z)^{3/2} \left[1+\frac{\pi^2}{8}(\ln z)^{-2}+ \dots\right] \tag{32}$$

and

$$f_{1/2}(z) = \frac{2}{\pi^{1/2}} (\ln z)^{1/2} \left[1-\frac{\pi^2}{24}(\ln z)^{-2}+ \dots\right]; \tag{33}$$

as expected, these expansions are mutually related through the reduction formula (E.5), namely

$$f_{n-1}(z) = \frac{\partial}{\partial \ln z} f_n(z). \tag{34}$$

Substituting (32) into (4), we obtain

$$\frac{N}{V} = \frac{4\pi g}{3}\left(\frac{2m}{h^2}\right)^{3/2} (kT \ln z)^{3/2} \left[1+\frac{\pi^2}{8}(\ln z)^{-2}+ \dots\right]. \tag{35}$$

In the zeroth approximation, this gives

$$(kT \ln z) \equiv \mu \simeq \left(\frac{3N}{4\pi g V}\right)^{2/3} \frac{h^2}{2m} = \varepsilon_F, \tag{36}$$

which is identical with the ground-state result (25). In the next approximation, we obtain

$$(kT \ln z) \equiv \mu \simeq \varepsilon_F\left[1-\frac{\pi^2}{12}\left(\frac{kT}{\varepsilon_F}\right)^2\right]. \tag{37}$$

Substituting (31) and (32) into (7), we obtain

$$\frac{U}{N} = \frac{3}{5}(kT \ln z)\left[1+\frac{\pi^2}{2}(\ln z)^{-2}+ \dots\right];$$

with the help of (37), this becomes

$$\frac{U}{N} = \frac{3}{5}\varepsilon_F\left[1+\frac{5\pi^2}{12}\left(\frac{kT}{\varepsilon_F}\right)^2+\dots\right]. \tag{38}$$

The pressure of the gas is then given by

$$P = \frac{2}{3}\frac{U}{V} = \frac{2}{5}n\varepsilon_F\left[1+\frac{5\pi^2}{12}\left(\frac{kT}{\varepsilon_F}\right)^2+\dots\right]. \tag{39}$$

As expected, the main terms of eqns. (38) and (39) are identical with the ground-state results (27) and (28). From the temperature-dependent part of (38), we readily obtain for the low-temperature specific heat of the gas:

$$\frac{C_V}{Nk} = \frac{\pi^2}{2}\frac{kT}{\varepsilon_F}+\dots\ . \tag{40}$$

Thus, for $T \ll T_F$, where $T_F (= \varepsilon_F/k)$ is the *Fermi temperature* of the system, the specific heat varies as the first power of temperature; moreover, in magnitude, it is considerably smaller than the classical value $\frac{3}{2}Nk$. The overall variation of C_V with T is shown in Fig. 8.2.

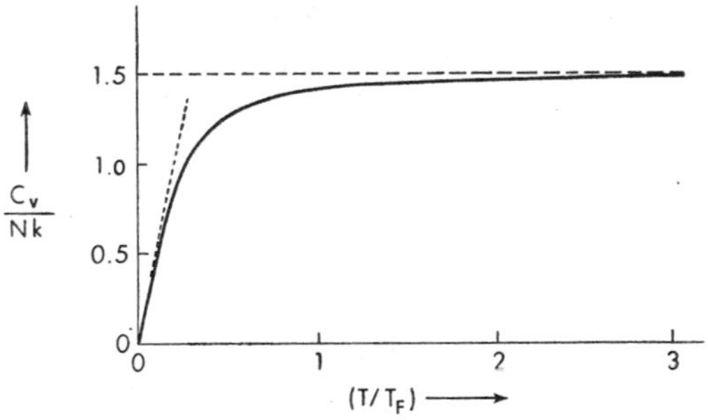

FIG. 8.2. The specific heat of an ideal Fermi gas; the dotted line depicts the *linear* trend at low temperatures.

The Helmholtz free energy of the system follows directly from eqns. (37) and (39):

$$\begin{aligned}\frac{A}{N} &= \mu - \frac{PV}{N}\\ &= \frac{3}{5}\varepsilon_F\left[1-\frac{5\pi^2}{12}\left(\frac{kT}{\varepsilon_F}\right)^2+\dots\right],\end{aligned} \tag{41}$$

whence we obtain for the low-temperature entropy of the system:

$$\frac{S}{Nk} = \frac{\pi^2}{2}\frac{kT}{\varepsilon_F}+\dots\ . \tag{42}$$

Thus, as $T \to 0$, $S \to 0$, in accordance with the third law of thermodynamics.

In passing, it may be noted that eqns. (40) and (42) for the specific heat and the entropy could as well be obtained from the general expressions (10) and (12) by making a direct use of the Sommerfeld series (31–33), with the result

$$\frac{C_V}{Nk} \simeq \frac{S}{Nk} \simeq \frac{\pi^2}{2}(\ln z)^{-1} \simeq \frac{\pi^2}{2}\left(\frac{kT}{\varepsilon_F}\right); \tag{43}$$

for a rigorous derivation of this result, see Weinstock (1969).

8.2. Magnetic behavior of an ideal Fermi gas

We now turn our attention to studying the equilibrium state of a gas of noninteracting fermions in the presence of an external magnetic field $\boldsymbol{B}$. The main problem here consists in determining the net magnetic moment $\boldsymbol{M}$ acquired by the gas (as a function of the temperature T) and then calculating the susceptibility $\chi(T)$. The answer naturally depends upon the value of the intrinsic magnetic moment μ^* of the particles and their multiplicity factor $(2J+1)$; see, for instance, the treatment given in Sec. 3.8, especially eqn. (3.8.26). According to the Boltzmannian treatment, one obtains a (positive) susceptibility $\chi(T)$ which at high temperatures obeys the *Curie law*: $\chi \propto T^{-1}$; at low temperatures, one obtains a state of magnetic saturation. However, if we treat the problem on the basis of Fermi statistics we obtain significantly different results, especially at low temperatures. In particular, since the Fermi gas is pretty live even at absolute zero, no magnetic saturation ever results; we rather obtain a limiting susceptibility χ_0 which is independent of temperature but strongly dependent upon the density of the gas. Studies along these lines were first made by Pauli, in 1927, when he suggested that the conduction electrons in alkali metals should be regarded as a "highly degenerate Fermi gas"; these studies in turn enabled him to uncover the physics behind the *feeble* and *temperature-independent* character of the paramagnetism of metals. Accordingly, this phenomenon is referred to as the *Pauli paramagnetism*—in contrast to the classical Langevin paramagnetism.

In quantum statistics, however, we encounter another magnetic effect which is totally absent in classical statistics. This is diamagnetic in character and arises from the quantization of the orbits of the *charged* particles in the presence of an external magnetic field (or, one may say, from the quantization of the kinetic energy of the charged particles associated with their motion perpendicular to the direction of the field). The existence of this effect was first established by Landau (1930); so, we refer to this phenomenon as the *Landau diamagnetism*. This leads to an additional susceptibility $\chi(T)$ which, though negative in sign, behaves somewhat similarly to the paramagnetic susceptibility, in that it obeys the Curie law at high temperatures and tends to a temperature-independent but density-dependent limiting value as $T \to 0$. Moreover, if the field is fairly strong, then at finite, but low, temperatures there appear, in addition, oscillatory terms which make the susceptibility of the system periodic in $1/B$. This effect was first discussed by Peierls in 1933, though it had been observed earlier by de Haas and van Alphen, in 1930, for the electron gas in bismuth. This effect is now commonly known as the *de Haas–van Alphen effect*.

In general, the magnetic behavior of the Fermi gas is determined jointly by the intrinsic magnetic moment of the particles and the quantization of their orbits. If the spin–orbit interaction is negligible, the resultant behavior is given by a simple addition of the two effects.

A. Pauli paramagnetism

The energy of a particle, in the presence of an external magnetic field $\boldsymbol{B}$, is given by

$$\varepsilon = \frac{p^2}{2m} - \boldsymbol{\mu}^* \cdot \boldsymbol{B}, \tag{1}$$

where $\boldsymbol{\mu}^*$ is the intrinsic magnetic moment of the particle and m its mass. For simplicity, we assume that the particle spin is $\frac{1}{2}$; the vector $\boldsymbol{\mu}^*$ must then be either parallel to the vector $\boldsymbol{B}$ or antiparallel. We thus have two groups of particles in the gas:

(i) those having $\boldsymbol{\mu}^*$ parallel to $\boldsymbol{B}$, with $\varepsilon = p^2/2m - \mu^* B$, and
(ii) those having $\boldsymbol{\mu}^*$ antiparallel to $\boldsymbol{B}$, with $\varepsilon = p^2/2m + \mu^* B$.

At absolute zero, all energy levels up to the Fermi level ε_F will be filled up, while all levels beyond ε_F will be empty. Accordingly, the kinetic energy of the particles in the first group will range between 0 and $(\varepsilon_F + \mu^* B)$, while the kinetic energy of the particles in the second group will range between 0 and $(\varepsilon_F - \mu^* B)$. The respective numbers of the occupied energy levels (and hence of the particles in the two groups) will, therefore, be

$$N^+ = \frac{4\pi V}{3h^3}\{2m(\varepsilon_F + \mu^* B)\}^{3/2} \tag{2}$$

and

$$N^- = \frac{4\pi V}{3h^3}\{2m(\varepsilon_F - \mu^* B)\}^{3/2}. \tag{3}$$

The net magnetic moment acquired by the gas is then given by

$$M = \mu^*(N^+ - N^-) = \frac{4\pi\mu^* V(2m)^{3/2}}{3h^3}\{(\varepsilon_F + \mu^* B)^{3/2} - (\varepsilon_F - \mu^* B)^{3/2}\}. \tag{4}$$

We thus obtain for the low-field susceptibility (per unit volume) of the gas

$$\chi_0 = \lim_{B \to 0}\left(\frac{M}{VB}\right) = \frac{4\pi\mu^{*2}(2m)^{3/2}\varepsilon_F^{1/2}}{h^3}. \tag{5}$$

Making use of the formula (8.1.25), with $g = 2$, the foregoing result can be written as

$$\chi_0 = \tfrac{3}{2} n\mu^{*2}/\varepsilon_F. \tag{6}$$

The corresponding high-temperature result is given by eqn. (3.8.31), with $J = \frac{1}{2}$:

$$\chi_\infty = n\mu^{*2}/kT. \tag{7}$$

To obtain an expression for χ that holds for all values of T, we may proceed as follows. Denoting the number of particles with momentum $\boldsymbol{p}$ and magnetic moment parallel (or antiparallel) to the field by the symbol n_p^+ (or n_p^-), the total energy of the gas can be written as

$$E_n = \sum_p \left[\left(\frac{p^2}{2m} - \mu^* B\right) n_p^+ + \left(\frac{p^2}{2m} + \mu^* B\right) n_p^-\right]$$

$$= \sum_p (n_p^+ + n_p^-) \frac{p^2}{2m} - \mu^* B(N^+ - N^-), \tag{8}$$

where N^+ and N^- denote the total number of particles in the two groups, respectively. The partition function $Q(N)$ of the system is then given by

$$Q(N) = \sum_{\{n_p^+\}, \{n_p^-\}}' \exp(-\beta E_n), \tag{9}$$

where the primed summation is subject to the conditions

$$n_p^+, \; n_p^- = 0 \text{ or } 1, \tag{10}$$

and

$$\sum_p n_p^+ + \sum_p n_p^- = N^+ + N^- = N. \tag{11}$$

To evaluate the sum in (9), we first fix an arbitrary value of the number N^+ (which automatically fixes the value of the number N^- as well) and sum over *all* n_p^+ and n_p^- that conform to the fixed values of the numbers N^+ and N^- as well as to the condition (10). Next, we sum over all possible values of N^+, viz. from $N^+ = 0$ to $N^+ = N$. We thus have

$$Q(N) = \sum_{N^+=0}^{N} \left[e^{\beta\mu^* B(2N^+ - N)} \left\{ \sum_{\{n_p^+\}}'' \exp\left(-\beta \sum_p \frac{p^2}{2m} n_p^+\right) \sum_{\{n_p^-\}}''' \exp\left(-\beta \sum_p \frac{p^2}{2m} n_p^-\right) \right\} \right]; \tag{12}$$

here, Σ'' is subject to the restriction: $\Sigma_p n_p^+ = N^+$, while Σ''' is subject to the restriction: $\Sigma_p n_p^- = N^- = N - N^+$.

Now, let $Q_0(\mathcal{N})$ denote the partition function of an ideal Fermi gas of $\mathcal{N}$ "spinless" particles of mass m; then, obviously,

$$Q_0(\mathcal{N}) = \sum_{\{n_p\}}' \exp\left(-\beta \sum_p \frac{p^2}{2m} n_p\right) \equiv \exp\{-\beta A_0(\mathcal{N})\}, \tag{13}$$

where $A_0(\mathcal{N})$ is the free energy of this fictitious system. Equation (12) can then be written as

$$Q(N) = e^{-\beta\mu^* BN} \sum_{N^+=0}^{N} [e^{2\beta\mu^* BN^+} Q_0(N^+) Q_0(N - N^+)], \tag{14}$$

whence it follows that

$$\frac{1}{N} \ln Q(N) = -\beta\mu^* B + \frac{1}{N} \ln \sum_{N^+=0}^{N} [\exp\{2\beta\mu^* BN^+ - \beta A_0(N^+) - \beta A_0(N - N^+)\}]. \tag{15}$$

As has been done on several occasions, the logarithm of the sum Σ_{N^+} may be replaced by the logarithm of the largest term in the sum; the error committed in doing so would be $O(\ln N)$, which is negligible in comparison with the term being retained. Now, the value $\overline{N^+}$ of N^+, which corresponds to the largest term in the sum, can be ascertained by setting the differential coefficient of the general term, with respect to N^+, equal to zero; this gives

$$2\mu^* B - \left[\frac{\partial A_0(N^+)}{\partial N^+}\right]_{N^+=\overline{N^+}} - \left[\frac{\partial A_0(N-N^+)}{\partial N^+}\right]_{N^+=\overline{N^+}} = 0,$$

that is

$$\mu_0(\overline{N^+}) - \mu_0(N-\overline{N^+}) = 2\mu^* B, \tag{16}$$

where $\mu_0(\mathcal{N})$ is the chemical potential of the fictitious system of $\mathcal{N}$ "spinless" fermions.

The physical interpretation of eqn. (16) is straightforward. If we regard the given system as a thermodynamic mixture of two components—one composed of particles with $\boldsymbol{\mu}^*$ parallel to $\boldsymbol{B}$ and the other composed of particles with $\boldsymbol{\mu}^*$ antiparallel—in which an exchange of particles from one component to the other is continually taking place, then the equilibrium state of the composite system is governed by the condition that the chemical potentials of the two components in the system be equal.† However, when we compare these components with the corresponding fictitious systems (composed of "spinless" particles), we must remember that whereas the single-particle energy levels of the first component are all lower than those of the fictitious system(s) by an amount $\mu^* B$, those of the second component are all higher by the same amount. The difference between the chemical potentials of the fictitious systems must, therefore, be

$$\begin{aligned}\mu_0(\overline{N^+}) - \mu_0(N-\overline{N^+}) &= (\mu^+ + \mu^* B) - (\mu^- - \mu^* B)\\ &= (\mu^+ - \mu^-) + 2\mu^* B\\ &= 2\mu^* B,\end{aligned}$$

which is identical with (16).

The foregoing equation contains the general solution sought for. To obtain an explicit expression for χ, we introduce a dimensionless parameter r, defined by

$$M = \mu^*(\overline{N^+} - \overline{N^-}) = \mu^*(2\overline{N^+} - N) = \mu^* N r \qquad (0 \leqslant r \leqslant 1); \tag{17}$$

eqn. (16) then becomes

$$\mu_0\left(\frac{1+r}{2}N\right) - \mu_0\left(\frac{1-r}{2}N\right) = 2\mu^* B. \tag{18}$$

If $B = 0$, $r = 0$ (which, as expected, corresponds to a *completely random* orientation of the elementary moments). For small B, r would also be small; so, we may carry out a Taylor expansion of the left-hand side of (18) about $r = 0$. Retaining only the first term of the expansion, we obtain

$$r \simeq \frac{2\mu^* B}{\left.\dfrac{\partial \mu_0(xN)}{\partial x}\right|_{x=1/2}}. \tag{19}$$

† Physically, this means the following. Since the equilibrium state of a given system corresponds to a *minimum* value of its free energy, the process of transferring a particle from one of its components to any other should not make any difference whatsoever to the value of the free energy. This in turn implies that a new particle added to the system would raise its free energy by the *same* amount, irrespective of the component to which it actually goes. We must, therefore, have: $\mu^+ = \mu^-$.

The low-field susceptibility (per unit volume) of the system is then given by

$$\chi = \frac{M}{VB} = \frac{\mu^* N r}{VB} = \frac{2n\mu^{*2}}{\left.\dfrac{\partial \mu_0(xN)}{\partial x}\right|_{x=1/2}}, \tag{20}$$

which is the desired result.

As $T \to 0$, the chemical potential of the fictitious system can be obtained from eqn. (8.1.36), with $g = 1$:

$$\mu_0(xN) = \left(\frac{3xN}{4\pi V}\right)^{2/3} \frac{h^2}{2m},$$

whence it follows that

$$\left.\frac{\partial \mu_0(xN)}{\partial x}\right|_{x=1/2} = \frac{2^{4/3}}{3}\left(\frac{3N}{4\pi V}\right)^{2/3} \frac{h^2}{2m}. \tag{21}$$

On the other hand, the Fermi energy of the actual system is given by eqn. (8.1.25), with $g = 2$:

$$\varepsilon_F = \left(\frac{3N}{8\pi V}\right)^{2/3} \frac{h^2}{2m}. \tag{22}$$

Making use of eqns. (21) and (22), we obtain from eqn. (20)

$$\chi_0 = \frac{2\pi\mu^{*2}}{\frac{4}{3}\varepsilon_F} = \frac{3}{2} n\mu^{*2}/\varepsilon_F, \tag{23}$$

which is in complete agreement with our earlier result (6). For finite, but low, temperatures, one has to use eqn. (8.1.37) instead of (8.1.36). The final result turns out to be

$$\chi \simeq \chi_0\left[1 - \frac{\pi^2}{12}\left(\frac{kT}{\varepsilon_F}\right)^2\right]. \tag{24}$$

On the other hand, as $T \to \infty$, the chemical potential of the fictitious system follows directly from eqn. (8.1.4), with $g = 1$ and $f_{3/2}(z) \simeq z$:

$$\mu_0(xN) = kT \ln (xN\lambda^3/V),$$

whence it follows that

$$\left.\frac{\partial \mu_0(xN)}{\partial x}\right|_{x=1/2} = 2kT. \tag{25}$$

Equation (20) then gives

$$\chi_\infty = n\mu^{*2}/kT, \tag{26}$$

which is in complete agreement with our earlier result (7). For large, but finite, temperatures, one has to take $f_{3/2}(z) \simeq z - (z^2/2^{3/2})$. The final result then turns out to be

$$\chi \simeq \chi_\infty\left(1 - \frac{n\lambda^3}{2^{5/2}}\right); \tag{27}$$

the correction term here is proportional to $(T_F/T)^{3/2}$ and tends to zero as $T \to \infty$.

B. Landau diamagnetism and de Haas–van Alphen effect

We shall now study the magnetism arising from the quantization of the orbital motion of (charged) particles in the presence of an external magnetic field. In a uniform field of intensity $\boldsymbol{B}$, directed along the z-axis, a charged particle would follow a helical path whose axis lies along the z-axis and whose projection on the (x, y)-plane is a circle. Motion along the z-direction proceeds with a constant linear velocity u_z, while that in the (x, y)-plane proceeds with a constant angular velocity eB/mc; the latter arises from the Lorentz force, $e(\boldsymbol{u}\times\boldsymbol{B})/c$, experienced by the particle. Quantum-mechanically, the energy associated with the circular motion in the (x, y)-plane is *quantized*, in the units of $e\hbar B/mc$. The energy associated with the linear motion along the z-axis is also quantized; but, in view of the smallness of the energy intervals, this may be taken as a continuous variable. We thus have for the total energy of the particle*

$$\varepsilon = \frac{e\hbar B}{mc}\left(j+\frac{1}{2}\right)+\frac{p_z^2}{2m} \qquad (j = 0, 1, 2, \ldots). \tag{28}$$

Now, these quantized energy levels are necessarily degenerate because they result from a "coalescing together" of an almost continuous set of the zero-field levels. A little reflection shows that all those levels for which the value of the quantity $(p_x^2+p_y^2)/2m$ lay between $e\hbar Bj/mc$ and $e\hbar B(j+1)/mc$ now "coalesce together" into a single level characterized by the quantum number j. The number of these levels is given by

$$\frac{L_x L_y}{h^2}\iint dp_x\, dp_y = \frac{L_x L_y}{h^2}\pi\left[2m\frac{e\hbar B}{mc}\{(j+1)-j\}\right]$$
$$= L_x L_y \frac{eB}{hc}, \tag{29}$$

which is independent of j. The *multiplicity factor* (29) is a quantum-mechanical measure of the freedom available to the particle for the centre of its orbit to be "located" in the total area $L_x L_y$ of the physical space. Figure 8.3 depicts the manner in which the zero-field

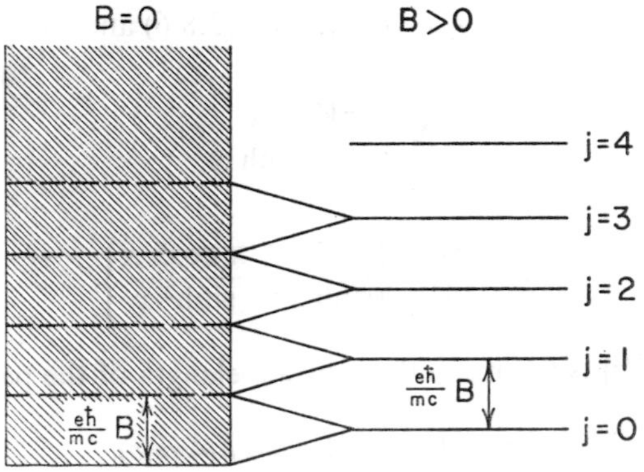

FIG. 8.3. The single-particle energy levels, for a two-dimensional motion, (i) in the absence of an external magnetic field ($B = 0$) and (ii) in the presence of an external magnetic field ($B > 0$).

* See, for instance, Gol'dman *et al.* (1960); Problem 6.3.

energy levels of the particle group themselves into a spectrum of oscillator-like levels on the application of an external magnetic field.

The grand partition function of the gas is given by the formula

$$\ln \mathcal{Q} = \sum_{\varepsilon} \ln (1+z\, e^{-\beta\varepsilon}), \tag{30}$$

where the summation has to be carried over all single-particle states of the system. Substituting (28) for ε, making use of the multiplicity factor (29) and replacing the summation over p_z by an integration, we get

$$\ln \mathcal{Q} = \int_{-\infty}^{\infty} \frac{L_z\, dp_z}{h} \left[\sum_{j=0}^{\infty} \left(L_x L_y \frac{eB}{hc} \right) \ln \{1+z\, e^{-\beta e\hbar B[j+(1/2)]/mc}\, e^{-\beta p_z^2/2m}\} \right]. \tag{31}$$

At high temperatures, $z \ll 1$; so, the system is effectively *Boltzmannian*. The grand partition function then reduces to

$$\begin{aligned} \ln \mathcal{Q} &= \frac{zVeB}{h^2c} \int_{-\infty}^{\infty} e^{-\beta p_z^2/2m}\, dp_z \sum_{j=0}^{\infty} e^{-\beta e\hbar B[j+(1/2)]/mc} \\ &= \frac{zVeB}{h^2c} \left(\frac{2\pi m}{\beta} \right)^{1/2} \left\{ 2 \sinh\left(\frac{\beta e\hbar B}{2mc} \right) \right\}^{-1}. \end{aligned} \tag{32}$$

The equilibrium number of particles $\bar{N}$ and the magnetic moment M of the gas are then given by

$$\bar{N} = \left(z \frac{\partial}{\partial z} \ln \mathcal{Q} \right)_{B, V, T} \tag{33}$$

and

$$M = \left\langle -\frac{\partial H}{\partial B} \right\rangle = \frac{1}{\beta} \left(\frac{\partial}{\partial B} \ln \mathcal{Q} \right)_{z, V, T}, \tag{34}$$

where H is the Hamiltonian of the system; cf. eqn. (3.8.6) and the accompanying footnote. We thus obtain

$$\bar{N} = \frac{zV}{\lambda^3} \frac{x}{\sinh x} \tag{35}$$

and

$$M = \frac{zV}{\lambda^3} \mu \left\{ \frac{1}{\sinh x} - \frac{x \cosh x}{\sinh^2 x} \right\}, \tag{36}$$

where $\lambda \{= h/(2\pi mkT)^{1/2}\}$ is the mean thermal wavelength of the particles, while

$$x = \beta\mu B \qquad (\mu = eh/4\pi mc). \tag{37}$$

Clearly, if e and m are taken to be the electronic charge and the electronic mass, then μ is the familiar Bohr magneton μ_B. Combining (35) and (36), we get

$$M = -\bar{N}\mu L(x), \tag{38}$$

where $L(x)$ is the *Langevin function*, viz.

$$L(x) = \coth x - \frac{1}{x}. \tag{39}$$

This result is very much similar to the one obtained in the Langevin theory of paramagnetism; see Sec. 3.8. The presence of the negative sign, however, means that the effect obtained in the present study is *diamagnetic* in character. We also note that this effect is a direct consequence of quantization; it vanishes if we let $h \to 0$. This is in complete accord with the *Bohr–van Leeuwen theorem*, according to which the phenomenon of diamagnetism does not arise in classical physics; see Problem 3.38.

If the field intensity B and the temperature T are such that $\mu B \ll kT$, then the foregoing results become

$$\bar{N} \simeq \frac{zV}{\lambda^3} \tag{40}$$

and

$$M \simeq -\bar{N}\mu^2 B/3kT. \tag{41}$$

Equation (40) is in agreement with the zero-field formula $z \simeq n\lambda^3$, while eqn. (41) leads to the diamagnetic counterpart of the Curie law:

$$\chi_\infty = \frac{M}{VB} = -\bar{n}\mu^2/3kT; \tag{42}$$

cf. eqn. (3.8.15). It must be noted here that the diamagnetic character of this phenomenon is independent of the sign of the charge on the particle. For an electron gas, in particular, the net susceptibility at high temperatures is directly given by the sum of the expression (7), with μ^* replaced by μ_B, and the expression (42):

$$\chi_\infty = \frac{n(\mu_B^2 - \frac{1}{3}\mu_B'^2)}{kT}, \tag{43}$$

where $\mu_B' = eh/4\pi m'c$, m' being the *effective mass* of the electron in the given system.

We shall now study the problem in the other extreme, viz. when $kT \ll \varepsilon_F$ and hence $z \gg 1$; once again, we assume the magnetic field to be weak, so that $\mu B \ll kT$. In view of the latter, the summation in (31) may be handled with the help of the Euler summation formula,

$$\sum_{j=0}^{\infty} f(j+\tfrac{1}{2}) \simeq \int_0^\infty f(x)\,dx + \tfrac{1}{24}f'(0), \tag{44}$$

with the result

$$\ln \mathcal{Q} = \frac{VeB}{h^2 c}\left[\int_0^\infty dx \int_{-\infty}^{\infty} dp_z \ln\{1 + ze^{-\beta(2\mu Bx + p_z^2/2m)}\} - \frac{1}{12}\beta\mu B \int_{-\infty}^{\infty} \frac{dp_z}{z^{-1}e^{\beta p_z^2/2m}+1}\right]. \tag{45}$$

In the first integral, we replace the variable x by a new variable

$$\varepsilon = 2\mu Bx + p_z^2/2m \tag{46}$$

and carry out integration over p_z, *keeping* ε *fixed*; this gives

$$\int_{-\infty}^{\infty} dp_z[\ldots] = 2\int_{0}^{\infty} dp_z[\ldots] = 2\int_{\varepsilon/(2\mu B)}^{0} \frac{-2\mu Bm\,dx}{\sqrt{[2m(\varepsilon-2\mu Bx)]}}[\ldots]$$
$$= 2(2m\varepsilon)^{1/2}[\ldots]. \tag{47}$$

Now, replacing dx by $d\varepsilon/2\mu B$, substituting $\mu = eh/4\pi mc$ and making use of (47), the first part of (45) becomes

$$\frac{2\pi V(2m)^{3/2}}{h^2}\int_{0}^{\infty}(\varepsilon^{1/2}\,d\varepsilon)\ln\{1+ze^{-\beta\varepsilon}\}, \tag{48}$$

which is identically the zero-field expression for $\ln \mathcal{Q}$; see eqn. (8.1.1) and the expression (7.1.4) for the density of states $a(\varepsilon)$. Of course, the value of the parameter z in (48) must correspond to the presence of the magnetic field ($B > 0$). In the second part of (45), we may write $\beta p_z^2/2m = y$, whence it becomes

$$-\frac{\pi V(2m)^{3/2}}{6h^3}(\mu B)^2\beta^{1/2}\int_{0}^{\infty}\frac{y^{-1/2}\,dy}{z^{-1}e^y+1}. \tag{49}$$

This represents the lowest-order correction arising from the quantization of the particle orbits under the conditions laid down above. The low-field susceptibility (per unit volume) of the gas is then given by

$$\chi = \frac{M}{VB} = \frac{1}{\beta VB}\left(\frac{\partial}{\partial B}\ln \mathcal{Q}\right)_{z,V,T}$$
$$= -\frac{\pi(2m)^{3/2}\mu^2}{3h^3\beta^{1/2}}\int_{0}^{\infty}\frac{y^{-1/2}\,dy}{z^{-1}e^y+1}. \tag{50}$$

For $kT \ll \varepsilon_F$, the integral here is very nearly equal to $2(\ln z)^{1/2}$; see eqn. (E.11), with $n = \frac{1}{2}$. Hence, we obtain

$$\chi_0 = -\frac{2\pi(2m)^{3/2}\mu^2\varepsilon_F^{1/2}}{3h^3} = -\frac{1}{2}n\mu^2/\varepsilon_F; \tag{51}$$

here, use has also been made of the fact that $(\beta^{-1}\ln z) \simeq \varepsilon_F$. This is the desired result for the special case considered here. It may be noted that, as before, the effect is diamagnetic in character, irrespective of the sign of the charge on the particle. Moreover, in magnitude, it is precisely one-third of the corresponding paramagnetic result (8.2.6), provided that we take the μ^* of that expression to be equal to the μ of this one.

Finally, we consider the case when the temperature of the system is still quite small in comparison with the Fermi temperature but the magnetic field is fairly strong, so that $kT \approx \mu B \ll \varepsilon_F$. Under these circumstances, a series of oscillatory terms appear in the grand partition function of the system and consequently its susceptibility exhibits a periodic character, which is the so-called *de Haas–van Alphen effect*. The origin of the oscillatory terms can be seen as follows. The grand partition function of the system is given by the usual formula

$$\ln \mathcal{Q} \equiv \frac{PV}{kT} = \int_0^\infty \ln\{1+z\,e^{-\beta\varepsilon}\}\,a(\varepsilon)\,d\varepsilon, \tag{52}$$

where $a(\varepsilon)$ denotes the density of states of the system around the energy value ε. Integrating twice by parts, we get

$$\ln \mathcal{Q} = -\beta \int_0^\infty \Omega(\varepsilon) \left\{ \frac{\partial}{\partial \varepsilon} \left(\frac{1}{z^{-1}\,e^{\beta\varepsilon}+1} \right) \right\} d\varepsilon, \tag{53}$$

where $\Omega(\varepsilon)$ is the double integral of $a(\varepsilon)$ over ε. Using (3.4.7), as adapted for a single particle, we can write

$$\Omega(\varepsilon) = \frac{1}{2\pi i} \int_{\beta'-i\infty}^{\beta'+i\infty} \frac{e^{\beta\varepsilon}}{\beta^2}\, Q_1(\beta)\, d\beta, \tag{54}$$

where $Q_1(\beta)$ is the single-particle partition function in the *Boltzmann statistics* while $\beta'(>0)$ is its convergence abscissa. The function $Q_1(\beta)$ in the present case can be obtained from eqns. (35) and (37), because in the *Boltzmann statistics*

$$\ln \mathcal{Q} \equiv \frac{PV}{kT} = \langle \bar{N} \rangle = zQ_1; \tag{55}$$

see Sec. 44. Hence

$$\Omega(\varepsilon) = V\mu B \frac{(2\pi m)^{3/2}}{h^3} \frac{1}{2\pi i} \int_{\beta'-i\infty}^{\beta'+i\infty} \frac{e^{\beta\varepsilon}\, d\beta}{\beta^{5/2}\sinh(\beta\mu B)}. \tag{56}$$

The oscillatory terms arise from that part of the integration which goes around the poles of the integrand, viz. the points $\beta = il\pi/\mu B$ $(l = \pm 1, \pm 2, \ldots)$. These terms can be evaluated from the relevant residues, with the result

$$\Omega_{\text{osc}}(\varepsilon) = -V \frac{2(2\pi m)^{3/2}}{h^3} \left(\frac{\mu B}{\pi} \right)^{5/2} \sum_{l=1}^{\infty} \frac{(-1)^l}{l^{5/2}} \cos\left(\frac{l\pi\varepsilon}{\mu B} - \frac{\pi}{4} \right); \tag{57}$$

note that the smooth part of the function $\Omega(\varepsilon)$, which is irrelevant for the present discussions, is being omitted here. Substituting (57) into (53) and making use of the fact that, for $kT \ll \varepsilon_F$,

$$\int_0^\infty \cos\left(\frac{l\pi\varepsilon}{\mu B} - \frac{\pi}{4} \right) \frac{\partial}{\partial \varepsilon} \left(\frac{1}{z^{-1}\,e^{\beta\varepsilon}+1} \right) d\varepsilon \simeq -\frac{\pi^2 l/\beta\mu B}{\sinh\{\pi^2 l/\beta\mu B\}} \cos\left(\frac{l\pi\varepsilon_F}{\mu B} - \frac{\pi}{4} \right), \tag{58}$$

we obtain

$$(\ln \mathcal{Q})_{\text{osc}} \simeq -V \frac{2(2\pi m)^{3/2}}{h^3} \frac{(\mu B)^{3/2}}{\pi^{1/2}} \sum_{l=1}^{\infty} \frac{(-1)^l}{l^{3/2}} \frac{\cos\{(l\pi\varepsilon_F/\mu B)-\pi/4\}}{\sinh\{\pi^2 l/\beta\mu B\}}. \tag{59}$$

Accordingly, the oscillatory part of the susceptibility (per unit volume) of the gas is given by

$$\chi_{\text{osc}} = \frac{1}{\beta VB}\frac{\partial}{\partial B}(\ln \mathcal{Q})_{\text{osc}}. \tag{60}$$

We observe that, because of the presence of the hyperbolic function in the denominator, the oscillatory terms are negligibly small if $\mu B \ll kT$; we then obtain only smooth expressions for the various properties of the system, such as in the Landau diamagnetism. In order that the oscillatory terms be significant, we must have $\mu B \approx \pi^2 kT$, which requires field intensities of the order of $(10^5 T)$ oersted; thus, the observation of the de Haas–van Alphen effect is quite difficult except at sufficiently low temperatures.

Going back to (59), we note that the most significant variation of this expression, with the variation of B, comes from the cosine term in the numerator. Disregarding the variation coming from other factors and retaining only the main term of the series, we may write

$$\chi_{\text{osc}} \simeq \frac{2\pi^2(2m)^{3/2}}{h^3}\frac{\mu^{1/2}\varepsilon_F}{\beta B^{3/2}}\frac{\sin\{(\pi\varepsilon_F/\mu B)-\pi/4\}}{\sinh\{\pi^2/\beta\mu B\}}. \tag{61}$$

Thus, the quantity $\{B^{3/2}\sinh(\pi^2/\beta\mu B)\}\chi$ is a periodic function of $1/B$, with period $2\mu/\varepsilon_F$. An experimental determination of the period of these oscillations, therefore, enables one to determine the Fermi energy of the given system. Substituting from (8.1.25), the foregoing expression takes the form

$$\chi_{\text{osc}} \simeq \pi\frac{3n\mu^2}{2\varepsilon_F}\frac{(kT)\varepsilon_F^{1/2}}{(\mu B)^{3/2}}\frac{\sin\{(\pi\varepsilon_F/\mu B)-\pi/4\}}{\sinh\{\pi^2/\beta\mu B\}}. \tag{62}$$

The expressions (61) and (62) may be compared with the corresponding nonoscillatory part (51), and also with the corresponding paramagnetic susceptibility (6).

8.3. The electron gas in metals

One physical system where the application of Fermi–Dirac statistics helped remove a number of inconsistencies and discrepancies is that of the conduction electrons in metals. Historically, the electron theory of metals was developed by Drude (1900) and Lorentz (1904–5), who applied the statistical mechanics of Maxwell and Boltzmann to the electron gas and derived theoretical results for the various properties of metals. The Drude–Lorentz model did provide a reasonable theoretical basis for a partial understanding of the physical behavior of metals; however, it encountered a number of serious discrepancies of a qualitative as well as a quantitative nature. For instance, the observed specific heat of metals appeared to be almost completely accountable by the lattice vibrations alone and practically no contribution seemed to be coming from the electron gas. The theory, however, demanded that, on the basis of the equipartition theorem, every electron in the gas must possess a mean thermal energy equal to $\frac{3}{2}kT$ and hence make a contribution of magnitude $\frac{3}{2}k$ to the specific heat of the metal. Similarly, one expected the electron gas to exhibit the phenomenon of paramagnetism, as arising from the intrinsic magnetic moment μ_B of the electrons. According to the classical theory, the paramagnetic susceptibility would be given by (8.2.7), with μ^* replaced by μ_B. Instead, one found that the susceptibility of a normal non-ferromagnetic metal was not only independent of temperature but had a magnitude which, at room temperatures, was hardly 1 per cent of the expected value.

The Drude–Lorentz theory was also applied to study transport properties of metals, such as the thermal conductivity K and the electrical conductivity σ. While the results for the individual conductivities were not very encouraging, their ratio did conform to the empirical law of Wiedemann and Franz (1853), as formulated by Lorenz (1872), namely that the quantity $K/\sigma T$ was a (universal) constant. The theoretical value of this quantity, which is generally known as the *Lorenz number*, turned out to be $3(k/e)^2 \simeq 2.48\times 10^{-13}$ e.s.u./deg^2; the corresponding experimental values for most alkali and alkaline–earth metals were, however, found to be scattered around a mean value of 2.72×10^{-13} e.s.u./deg^2. A still more uncomfortable feature of the classical theory was the uncertainty involved in assigning an appropriate numerical value to the mean free path of the electrons in a given metal and in ascribing to it an appropriate temperature dependence. For these reasons, the problem of the transport properties of metals remained in a rather unsatisfactory state until the correct lead was provided by Sommerfeld (1928).

The most significant change introduced by Sommerfeld was the replacement of the Maxwell–Boltzmann statistics by the Fermi–Dirac statistics for describing the electron gas in a metal. And, with a single stroke of genius, he was able to set most of the things right. To see how it worked, we must first of all estimate the Fermi energy ε_F of the electron gas in a typical metal, say sodium. Referring to eqn. (8.1.25), with $g = 2$, we obtain

$$\varepsilon_F = \left(\frac{3N}{8\pi V}\right)^{2/3} \frac{h^2}{2m'}, \tag{1}$$

where m' is the *effective mass* of an electron in the gas.† The electron density N/V, in the case of a cubic lattice, may be written as

$$\frac{N}{V} = \frac{n_e n_a}{a^3}, \tag{2}$$

where n_e is the number of conduction electrons per atom, n_a the number of atoms per unit cell and a the lattice constant (or the cell length).‡ For sodium, $n_e = 1$, $n_a = 2$ and $a = 4.29$ Å. Substituting these numbers into (2) and thereafter into (1), and also writing $m' = 0.98\, m_e$, we obtain

$$(\varepsilon_F)_{Na} = 5.03\times 10^{-12} \text{ erg} = 3.14 \text{ eV}. \tag{3}$$

Accordingly, we obtain for the Fermi temperature of the gas

$$\begin{aligned}(T_F)_{Na} &= (1.16\times 10^4)\times \varepsilon_F \quad (\text{in eV}) \\ &= 3.64\times 10^4\ {}^\circ\text{K},\end{aligned} \tag{4}$$

† To justify the assumption that "the conduction electrons in a metal may be treated as 'free' electrons", it is necessary to use for the electronic mass an effective value $m' \neq m$. This is an indirect way of accounting for the fact that the electrons in a metal are not really free; the ratio m'/m accordingly depends upon the structural details of the metal and, therefore, varies from metal to metal. In sodium, $m'/m \simeq 0.98$.

‡ Another way of expressing the electron density is to write: $N/V = f\varrho/M$, where f is the valency of the metal, ϱ its mass density and M the mass of an atom (ϱ/M, thus, being the number density of the atoms).

which is considerably larger than a typical room temperature T ($\approx 3\times10^2$ °K). The ratio T/T_F being of the order of 1 per cent, the conduction electrons in sodium constitute a *highly degenerate* Fermi system. The statement, in fact, applies to *all* metals because their Fermi temperatures are invariably of the order of 10^4–10^5 °K.

Now, the very fact that the electron gas in metals is a highly degenerate Fermi system is sufficient to explain away some of the basic difficulties of the Drude–Lorentz theory. For instance, the specific heat of this gas would no longer be given by the classical formula $C_V = \frac{3}{2}Nk$, but rather by the formula (8.1.43), viz.

$$C_V = \frac{\pi^2}{2} Nk(kT/\varepsilon_F); \tag{5}$$

obviously, the new result is much smaller in value because, at ordinary temperatures, the ratio $(kT/\varepsilon_F) \equiv (T/T_F) = 0(10^{-2})$. It is then hardly surprising that, at ordinary temperatures, the specific heat of metals is almost completely determined by the vibrational modes of the lattice and very little contribution is made by the conduction electrons. Of course, as temperature decreases the specific heat due to the lattice vibrations also decreases. Finally, this too becomes considerably smaller than the classical value; see Sec. 7.3, especially Fig. 7.8. A stage comes when the two contributions, both non-classical, become comparable in value. Ultimately, at extremely low temperatures, the specific heat due to the lattice vibrations, being proportional to T^3, becomes *even smaller* than the electronic specific heat, which is proportional to T^1. In general, we may write, for the low-temperature specific heat of a metal,

$$C_V = \gamma T + \delta T^3, \tag{6}$$

where the coefficient γ is given by eqn. (5), or better by the general formula stated in Problem 8.11, while the coefficient δ is given by eqn. (7.3.23). An experimental determination of the

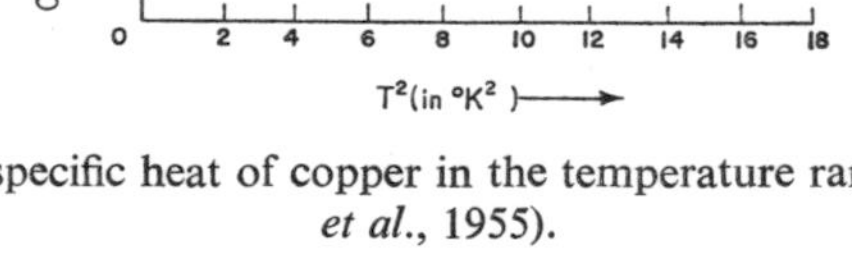

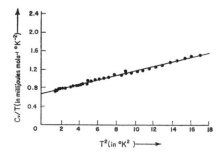

FIG. 8.4. The observed specific heat of copper in the temperature range 1 – 4°K (after Corak *et al.*, 1955).

specific heat of metals at low temperatures is, therefore, expected not only to verify the theoretical result based on quantum statistics but also to evaluate some of the parameters of the problem. Such determinations have been made, among others, by Corak *et al.* (1955) who have worked with the metals copper, silver and gold in the temperature range 1–5°K. Their results for copper are shown in Fig. 8.4. The very fact that the (C_V/T) vs. T^2 plot is well approximated by a straight line amply vindicates the theoretical formula (6). Further, the slope of this line directly gives the value of the coefficient δ, from which one can obtain the Debye temperature, Θ_D, of the metal. One thus obtains, for copper, $\Theta_D = (343.8 \pm 0.5)°K$ which compares favorably with Leighton's theoretical estimate of 345°K (which was based on the elastic constants of the metal). The intercept on the (C_V/T)-axis yields the value of the coefficient γ, viz. (0.688 ± 0.002) millijoule mole^{-1} deg^{-2}; this agrees favorably with Jones' estimate of 0.69 millijoule mole^{-1} deg^{-2} (which was based on a density-of-states calculation).

The general pattern of the magnetic behavior of the electron gas in non-ferromagnetic metals can be understood likewise. In view of the high degree of degeneracy prevailing in the gas, the magnetic susceptibility χ is given by the Pauli result (8.2.6) plus the Landau result (8.2.51), and not by the classical result (8.2.7). In complete agreement with observation, the new result is (i) independent of temperature and (ii) considerably smaller in value than the classical one. Finally, at sufficiently low temperatures and under sufficiently strong fields, the magnetic susceptibility of metals does exhibit a periodicity in $1/B$, viz. the *de Haas–van Alphen effect*; for details, see Shoenberg (1957).

As regards the transport properties, viz. K and σ, the new theory again led to the *Wiedemann–Franz law*; the Lorenz number, however, became $(\pi^2/3)(k/e)^2$, instead of the classical $3(k/e)^2$. The resulting theoretical value, viz. 2.71×10^{-13} e.s.u./deg^2, thus turned out to be much closer to the experimental mean value quoted earlier. Of course, the position regarding the individual conductivities of a metal and the mean free path of the electrons did not improve until Bloch (1928) developed a theory that took into account the interactions among the electron gas and the ion system in the metal. The theory of metals has steadily become more and more sophisticated; the important point, however, is that the development has all along been governed by the new statistics!

Before leaving this topic, we would like to include here a brief discussion of the fascinating phenomena of *thermionic* and *photoelectric* emission (of electrons) from metals. In view of the fact that electronic emission does not take place spontaneously, we conclude that the electrons inside a metal find themselves caught in some sort of a "potential well" created by the metallic ions. The detailed features of the potential energy of an electron in this well must depend upon the structure of the given metal. For a study of the electronic emission, however, we need not worry about the detailed features of the potential energy function and may instead assume that the potential energy of an electron stays constant (at a negative value: $-W$, say) throughout the interior of the metal and changes discontinuously to zero at the surface. Thus, while inside the metal, the electrons move about freely and independently of one another; however, as soon as any one of them approaches the surface of the metal and tries to escape, it encounters a potential barrier of height W. Accordingly, only those electrons whose kinetic energy (associated with the motion *perpendicular* to the surface) is greater than W can expect to escape through the surface barrier. At ordinary temperatures, especially in the absence of any external stimulus, such electrons are too few in any given metal, with the result that there is practically no spontaneous emission of electrons from the metals. At high temperatures, and more so if there is an external stimulus present, the population of such electrons in a given metal could become large enough to yield a sizeable emission. We then speak of phenomena such as the *thermionic effect* and the *photoelectric effect*.

Strictly speaking, these phenomena are not equilibrium phenomena because the electrons are flowing out *steadily* through the surface of the metal. However, if the number of electrons lost in a given interval of time is much too small in comparison with their total population in the metal, the magnitude of the emission current may be calculated on the assumption that the gas inside the metal continues to be in a state of *quasi-static* thermal equilibrium. The mathematical procedure for this calculation would be very much the same as the one followed in Sec. 6.4 (for determining the *rate of effusion R* of the particles of a gas through an opening in the walls of the container). There is one difference, however: whereas in the case of effusion any particle that reached the opening with $u_z > 0$ could escape unquestioned, in the present case we must have $u_z > (2W/m)^{1/2}$, so that the particle concerned could cross over the potential barrier at the surface of the metal. Moreover, even if this condition is satisfied, there is no guarantee that the particle *will* really escape, because the possibility of an inward reflection cannot be ruled out. In the following discussion, we propose to disregard this possibility; however, if one is looking for a detailed numerical comparison of theory with experiment, the results derived here must be multiplied by a factor $(1-r)$, where r is the *reflection coefficient* of the surface.

A. *Thermionic emission (the Richardson effect)*

The number of electrons leaving per unit area of the metal surface per unit time is given by

$$R = \int_{p_z=(2mW)^{1/2}}^{\infty} \int_{p_x=-\infty}^{\infty} \int_{p_y=-\infty}^{\infty} \left\{ \frac{2dp_x\, dp_y\, dp_z}{h^3} \frac{1}{e^{(\varepsilon-\mu)/kT}+1} \right\} u_z; \tag{7}$$

cf. the corresponding expression in Sec. 6.4. Integration over the variables p_x and p_y may be carried out by changing over to the corresponding polar coordinates (p', φ), with the result

$$\begin{aligned} R &= \frac{2}{h^3} \int_{p_z=(2mW)^{1/2}}^{\infty} \frac{p_z}{m}\, dp_z \int_{p'=0}^{\infty} \frac{2\pi p'\, dp'}{\exp\{[(p'^2/2m)+(p_z^2/2m)-\mu]/kT\}+1} \\ &= \frac{4\pi kT}{h^3} \int_{p_z=(2mW)^{1/2}}^{\infty} p_z\, dp_z \ln\left[1+\exp\{(\mu-p_z^2/2m)/kT\}\right] \\ &= \frac{4\pi mkT}{h^3} \int_{\varepsilon_z=W}^{\infty} d\varepsilon_z \ln\left[1+e^{(\mu-\varepsilon_z)/kT}\right]. \end{aligned} \tag{8}$$

As will be clear from the numbers given below, the exponential term inside the logarithm, at all temperatures of interest, is much smaller than unity; see footnote on p. 239. We may, therefore, write: $\ln(1+x) \simeq x$, with the result

$$\begin{aligned} R &= \frac{4\pi mkT}{h^3} \int_{\varepsilon_z=W}^{\infty} d\varepsilon_z e^{(\mu-\varepsilon_z)/kT} \\ &= \frac{4\pi mk^2T^2}{h^3} e^{(\mu-W)/kT}. \end{aligned} \tag{9}$$

The thermionic current density is then given by

$$J = eR = \frac{4\pi mek^2}{h^3} T^2 e^{\frac{\mu - W}{kT}}. \tag{10}$$

It is only now that the difference between the classical statistics and the Fermi statistics *really* shows up. In the case of classical statistics, the fugacity of the gas is given by (see eqn. (8.1.4), with $f_{3/2}(z) \simeq z$)

$$z \equiv e^{\mu/kT} = \frac{n\lambda^3}{g} = \frac{nh^3}{2(2\pi mkT)^{3/2}}; \tag{11}$$

accordingly,

$$J_{\text{class}} = ne\left(\frac{k}{2\pi m}\right)^{1/2} T^{1/2} e^{-\varphi/kT} \qquad (\varphi = W), \tag{12}$$

which corresponds to the very first formula of Richardson (1902). In the case of Fermi statistics, the chemical potential of the (highly degenerate) gas is practically independent of temperature and is given by the Fermi energy of the gas ($\mu \simeq \mu_0 \equiv \varepsilon_F$); accordingly,

$$J_{\text{F.D.}} = \frac{4\pi mek^2}{h^3} T^2 e^{-\varphi/kT} \qquad (\varphi = W - \varepsilon_F). \tag{13}$$

The quantity φ is generally referred to as the *work function* of the metal; according to (13), it represents the height of the surface barrier *over and above the Fermi level*; see Fig. 8.5.

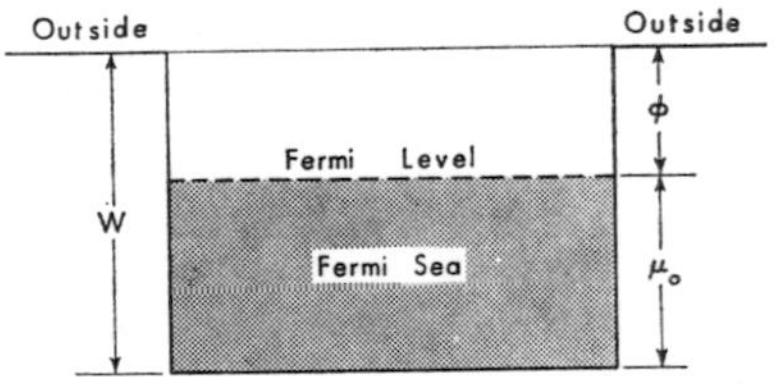

FIG. 8.5. The work function φ of a metal represents the height of the surface barrier *over and above the Fermi level.*

At this stage, it appears worth while to point out that the *rate of emission R* of the electrons from the surface of a metal can also be determined by computing instead the *rate of return* R' of the electrons to the metal from the space charge that builds outside the metal. If there are no electrodes set up to collect the thermionic current, then a dynamic equilibrium will ultimately set in between the interior of the metal and the space charge; the numbers R and R' will then be equal. Now, R' will be given correctly by the classical formula, because the electron density in the space charge will be low enough for the classical statistics to be valid; thus, we will have

$$R' = \frac{1}{4} n\langle u\rangle = \frac{1}{4}\frac{P}{kT}\left(\frac{8kT}{\pi m}\right)^{1/2}$$
$$= \frac{P}{(2\pi mkT)^{1/2}}; \tag{14}$$

see Sec. 6.4, especially eqns. (6.4.11) and (6.4.19). Here, P is the pressure in the space charge and is given by the formula (see eqn. (8.1.3), with $f_{5/2}(z) \simeq z$)

$$P = \frac{2kT(2\pi mkT)^{3/2}}{h^3} e^{\mu'/kT}, \tag{15}$$

where μ' is the chemical potential of the space charge. Combining (14) and (15), and taking into account the fact that the zero of energy in the space charge is different from the zero of energy inside the metal by an amount W (so that, in equilibrium, $\mu' = \mu - W$), we obtain

$$R' = \frac{4\pi m(kT)^2}{h^3} e^{(\mu-W)/kT}, \tag{16}$$

which is in complete agreement with formula (9). The mode of derivation presented in this paragraph is originally due to von Laue (1919).

The theoretical results embodied in eqns. (12) and (13) differ in certain important respects. The most striking difference seems to be in regard to the temperature dependence of the thermionic current density J. Actually, the major part of this dependence comes through the factor exp $(-\varphi/kT)$—so much so that whether we plot ln $(J/T^{1/2})$ against $(1/T)$ or ln (J/T^2) against $(1/T)$ we obtain, in each case, a fairly good straight-line fit. This means that, from the point of view of temperature dependence of J, a choice between formulae (12) and (13) is rather hard to make. However, the numerical value of the slope of the experimental line should give us *directly* the value of W if formula (12) applies or the value of $(W - \varepsilon_F)$ if formula (13) applies! Now, the value of W can be determined independently, for instance, by studying the "refractive index of the given metal for the de Broglie waves associated with an electron beam impinging upon the metal". For a beam of electrons whose initial kinetic energy is E, we have

$$\lambda_{\text{out}} = \frac{h}{\sqrt{(2mE)}} \quad \text{and} \quad \lambda_{\text{in}} = \frac{h}{\sqrt{[2m(E+W)]}}, \tag{17}$$

so that the refractive index of the metal is given by

$$n = \frac{\lambda_{\text{out}}}{\lambda_{\text{in}}} = \left(\frac{E+W}{E}\right)^{1/2}. \tag{18}$$

By studying electron diffraction for various values of E, one can derive the relevant value of W. In this manner, Davisson and Germer (1927) derived the value of W for a number of metals. For instance, they obtained for tungsten: $W \simeq 13.5$ eV. The experimental results on thermionic emission from tungsten are shown in Fig. 8.6. The value of φ resulting from the slope of the experimental line is about 4.5 eV. The large difference between the two values clearly shows that the classical formula (12) does not apply. That the quantum-statistical formula (13) does apply is shown by the fact that the Fermi energy of tungsten is very nearly 9 eV; so, the value 4.5 eV for the work function of tungsten is correctly given by the difference between the depth W of the potential well and the Fermi energy ε_F. To quote another example, the experimental value of the work function for nickel is found to be about 5.0 eV, while the theoretical estimate for its Fermi energy turns out to be about 11.8

eV. Accordingly, the depth of the potential well in the case of nickel should be about 16.8 eV. The experimental value of W, obtained by studying electron diffraction in nickel, is (17 ± 1) eV.*

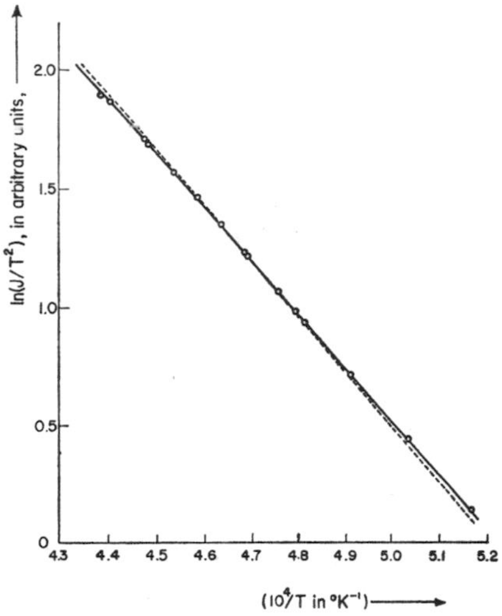

FIG. 8.6. Thermionic current from tungsten, as a function of the temperature of the metal. The continuous line corresponds to $r = \frac{1}{2}$ while the broken line corresponds to $r = 0$, r being the reflection coefficient of the surface.

The second point of difference between the formulae (12) and (13) relates to the actual numerical value of the saturation current obtained. In this respect, the classical formula turns out to be a complete failure while the quantum-statistical formula fares reasonably well. The value of the numerical factor in the latter formula is

$$\frac{4\pi mek^2}{h^3} = 120.4 \text{ amp cm}^{-2} \deg^{-2}; \tag{19}$$

of course, this has yet to be multiplied by the *transmission coefficient* $(1-r)$. The corresponding experimental number, for most metals with clean surfaces, turns out to be in the range 60–120 amp cm^{-2} deg^{-2}.

Finally, we shall study the influence of a moderately strong electric field on the thermionic emission from a metal—the so-called *Schottky effect*. Denoting the strength of the electric field by the symbol F (and assuming the field to be uniform and directed perpendicular to the metal surface), the difference Δ between the potential energy of an electron at a distance x from the surface and of one inside the metal is given by

$$\Delta(x) = W - eFx - \frac{e^2}{4x} \qquad (x > 0), \tag{20}$$

* In the light of the numbers quoted here, one can readily see that the quantity $e^{(\mu-\varepsilon_z)/kT}$ in eqn. (8), being *at most* equal to $e^{(\mu_0-W)/kT} \equiv e^{-\varphi/kT}$, is, at all temperatures of interest, much smaller than unity. This fact was made use of in simplifying the integral appearing there.

where the first term arises due to the potential well of the metal, the second due to the (attractive) field present and the third due to the attraction between the departing electron and the "image" induced in the metal; see Fig. 8.7. The *largest* value of the function $\Delta(x)$ occurs at $x = (e/4F)^{1/2}$, where

$$\Delta_{\text{max}}(x) = W \quad e^{3/2}F^{1/2}; \tag{21}$$

thus, the field has effectively reduced the height of the potential barrier by an amount $e^{3/2}F^{1/2}$. A corresponding reduction would take place in the work function as well.

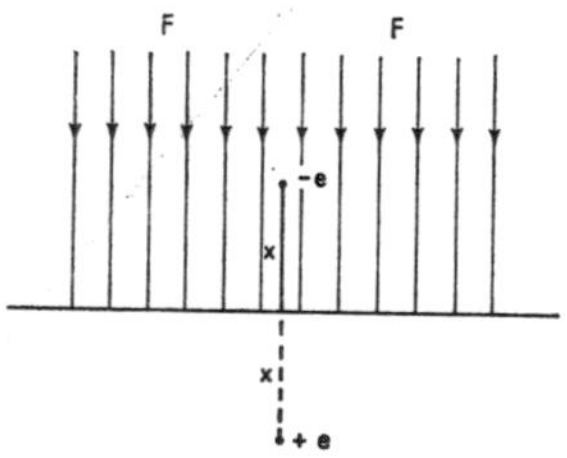

FIG. 8.7. A schematic diagram to illustrate the Schottky effect.

Accordingly, the thermionic current density in the presence of the field would be higher than the one in the absence of the field:

$$J_F = J_0 \exp(e^{3/2}F^{1/2}/kT). \tag{22}$$

A plot of ln (J_F/J_0) against $(F^{1/2}/T)$ should, therefore, be a straight line, with slope $e^{3/2}/k$. Working along these lines, de Bruyne (1928) obtained for the electronic charge a value of 4.84×10^{10} e.s.u., which is remarkably close to the actual value of e.

The theory of the Schottky effect, as outlined here, holds for field strengths up to about 10^6 volts/cm. For fields stronger than that, one obtains the so-called *cold emission*, which means that the electric field is now strong enough to make the potential barrier practically ineffective; for details, see Fowler and Nordheim (1928).

B. Photoelectric emission (the Hallwachs effect)

The physical situation in the case of photoelectric emission is different from that in the case of thermionic emission, in that there exists now an *external* agency, namely the photon in the incoming beam of light, that helps an electron inside the metal in overcoming the potential barrier at the surface. The condition to be satisfied by the momentum component p_z of an electron in order that it could escape from the metal now reads:

$$(p_z^2/2m) + h\nu > W, \tag{23}^*$$

where ν is the frequency of the incoming light (assumed monochromatic). Proceeding in the

* In writing this condition, we have tacitly assumed that the components p_x and p_y of the electron remain unchanged on the absorption of a photon.

same manner as in the case of thermionic emission, we obtain, instead of (8),

$$R = \frac{4\pi mkT}{h^3} \int_{\varepsilon_z = W - h\nu}^{\infty} d\varepsilon_z \ln [1 + e^{(\mu - \varepsilon_z)/kT}]. \tag{24}$$

We cannot, in general, approximate this integral in the way we did there; the integrand here stays as it is. It is advisable, however, to change over to a new variable x, defined by

$$x = (\varepsilon_z - W + h\nu)/kT, \tag{25}$$

whereby eqn. (24) becomes

$$R = \frac{4\pi m(kT)^2}{h^3} \int_0^\infty dx \ln \left[1 + \exp\left\{ \frac{h(\nu - \nu_0)}{kT} - x \right\} \right], \tag{26}$$

where

$$h\nu_0 = W - \mu \simeq W - \varepsilon_F = \varphi. \tag{27}$$

The quantity φ will be recognized as the (thermionic) work function of the metal; the characteristic frequency $\nu_0 (= \varphi/h)$ is generally referred to as the *threshold frequency* for photoelectric emission from the metal concerned. The current density of the photoelectric emission is, therefore, given by

$$J = \frac{4\pi mek^2}{h^2} T^2 \int_0^\infty dx \ln (1 + e^{\delta - x}), \tag{28}$$

where

$$\delta = h(\nu - \nu_0)/kT. \tag{29}$$

Integrating by parts, we find that

$$\int_0^\infty dx \ln (1 + e^{\delta - x}) = \int_0^\infty \frac{x\, dx}{e^{x - \delta} + 1} \equiv f_2(e^\delta); \tag{30}$$

see eqn. (E.3). Accordingly,

$$J = \frac{4\pi mek^2}{h^3} T^2 f_2(e^\delta). \tag{31}$$

For $h(\nu - \nu_0) \gg kT$, $e^\delta \gg 1$ and the function $f_2(e^\delta) \simeq \delta^2/2$; see the Sommerfeld lemma (E.13). Equation (31) then becomes

$$J \simeq \frac{2\pi me}{h} (\nu - \nu_0)^2, \tag{32}$$

which is completely independent of temperature; thus, when the energy of the light quantum is much greater than the work function of the metal, the temperature of the electron gas is a "dead" parameter of the problem. At the other extreme, when $\nu < \nu_0$ and $h|\nu - \nu_0|$

$\gg kT$, then $e^\delta \ll 1$ and the function $f_2(e^\delta) \simeq e^\delta$. Equation (31) then becomes

$$J \simeq \frac{4\pi mek^2}{h^3} T^2 e^{(h\nu-\varphi)/kT}, \tag{33}$$

which is just the thermionic current density (13), enhanced by the photon factor exp $(h\nu/kT)$; in other words, the situation now is very much the same as in the case of thermionic emission, but with a diminished work function $\varphi'(= \varphi - h\nu)$. At the threshold frequency $(\nu = \nu_0)$, $\delta = 0$ and the function $f_2(e^\delta) = f_2(1) = \frac{1}{2}\zeta(2) = \pi^2/12$. Equation (31) then gives

$$J_0 = \frac{\pi^3 mek^2}{3h^3} T^2. \tag{34}$$

Figure 8.8 shows a plot of the experimental results for photoelectric emission from palladium (φ = 4.97 eV). The agreement with theory is excellent. It will be noted that the plot includes some observations with $\nu < \nu_0$. The fact that we obtain a *finite* photo-current even for frequencies less than the so-called threshold frequency is fully consistent with the theory developed here. The reason for this lies in the fact that, at any *finite* temperature T, there exist in the gas a reasonable fraction of the electrons whose energy exceeds the Fermi energy ε_F by amounts $O(kT)$. Therefore, if the light quantum $h\nu$ gets absorbed by one of these electrons the condition (23) for photo-emission could be satisfied even if $h\nu < (W - \varepsilon_F) = h\nu_0$. Of course, the energy difference $h(\nu_0 - \nu)$ must not be much more than a few times kT, for otherwise the availability of the right kind of electrons will be extremely low. We, therefore, do expect to obtain a finite photo-current for radiation frequencies less than the threshold frequency, provided that $h(\nu_0 - \nu) \approx kT$.

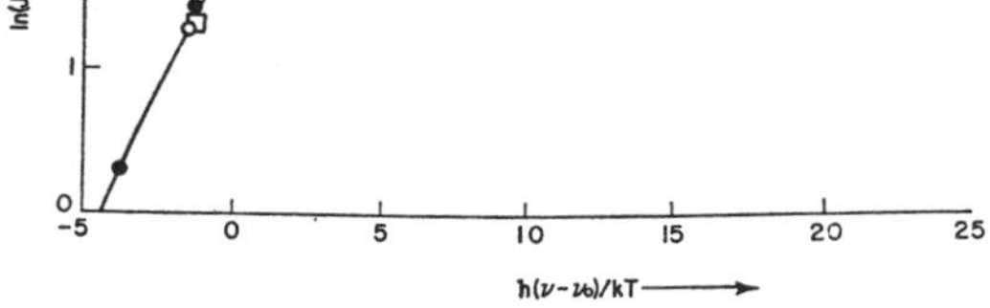

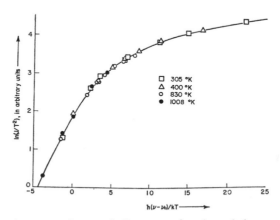

FIG. 8.8. Photoelectric current from palladium, as a function of the quantity $h(\nu - \nu_0)/kT$. The plot includes data taken at several temperatures T for several frequencies ν.

The kind of plot shown in Fig. 8.8, viz. ln (J/T^2) vs. δ, is generally referred to as the "Fowler plot". Adjusting the observed photoelectric data to this plot, one can obtain the characteristic frequency ν_0 and hence the work function φ of the given metal. We have previously seen that the work function of a metal can be derived from thermionic data as well. It is gratifying that there is complete agreement between the two sets of values obtained for the work function of the various metals.

8.4. Statistical equilibrium of white dwarf stars

Historically, the first application of Fermi statistics appeared in the field of astrophysics (Fowler, 1926). It related to the study of thermodynamic equilibrium of *white dwarf stars*—the small-sized stars which are abnormally faint for their (white) color. The general pattern of color–brightness relationship among stars shows that, by and large, a star with red color is expected to be a "dull" star, while a star with white color is expected to be a "brilliant" star. However, the red giants on one hand and the white dwarfs on the other constitute glaring exceptions to this general rule. The reason for this lies in the fact that (i) the red giant stars are relatively young stars whose hydrogen content is more or less intact, with the result that the thermonuclear reactions converting hydrogen into helium are going on at a rather high speed, thus making these stars a lot more bright than one would expect them to be on the basis of their color, while (ii) the white dwarf stars are relatively old stars whose hydrogen content is more or less used up, with the result that the thermonuclear reactions in them are proceeding at a rather low pace, thus making these stars a lot less bright than one would expect them to be on the basis of their color. The material content of the white dwarf stars, in the present state of their career, is mostly helium. And whatever little brightness they presently have derives from the gravitational energy released in the process of a slow contraction of these stars—a mechanism first proposed by Kelvin, in 1861, as a "possible" source of energy for *all* stars!

A typical, though somewhat idealized, model of a white dwarf star consists of a mass $M(\approx 10^{33}$ g) of helium, packed into a ball of high density $\varrho(\approx 10^7$ g cm^{-3}), at a central temperature $T(\approx 10^7$ °K). Now, a temperature of the order of 10^7 °K corresponds to a mean thermal energy (per particle) of the order of 10^3 eV, which is much too large in comparison with the energy required for ionizing a helium atom. Thus, practically the whole of the helium in the star exists in a state of complete ionization. The microscopic constituents of the star may, therefore, be taken as N electrons (each of mass m) and $N/2$ helium nuclei (each of mass $\simeq 4m_p$). The mass of the star is then given by

$$M \simeq N(m+2m_p) \simeq 2Nm_p \tag{1}$$

and, hence, the electron density by

$$n = \frac{N}{V} \simeq \frac{M/2m_p}{M/\varrho} = \frac{\varrho}{2m_p}. \tag{2}$$

A typical value of the electron density in white dwarfs would, therefore, be $0(10^{30})$ electrons per cm³. We then obtain for the Fermi momentum of the electron gas (see eqn. (8.1.24), with $g = 2$)

$$p_F = \left(\frac{3n}{8\pi}\right)^{1/3} h = 0(10^{-17}) \text{ g cm sec}^{-1}, \tag{3}$$

which is rather comparable with the characteristic momentum mc of an electron. The Fermi energy ε_F of the electron gas will, accordingly, be comparable with the rest energy mc^2 of an electron, i.e. $\varepsilon_F = 0(10^6)$ eV, and hence the Fermi temperature $T_F = 0(10^{10})$ °K. In view of these order-of-magnitude estimates, (i) the dynamics of the electrons in this problem is *relativistic* and (ii) the electron gas, though at a temperature large in comparison with terrestrial standards, is, statistically speaking, in a state of (*almost*) *complete degeneracy*: $(T/T_F) \approx 10^{-3}$. The second point was fully appreciated, and duly taken into account, by Fowler himself; the first one was taken care of later, by Anderson (1929) and by Stoner (1929–30). The problem, in full generality, was attacked in a series of papers by Chandrasekhar (1931–5) to whom the final picture of the theory of white dwarf stars is chiefly due; for details, see Chandrasekhar (1939), where a complete bibliography of the subject is also given.

Now, the helium nuclei do not contribute as significantly to the dynamics of the problem as do the electrons; in the first approximation, therefore, we may neglect the presence of the nuclei in the system. For a similar reason, we may neglect the effect of the radiation. We may thus consider the electron gas alone. Further, for simplicity, we may assume that the electron gas is *uniformly* distributed over the body of the star; we may then ignore the spatial variation of the various parameters of the problem—a variation which is physically essential for the very stability of the star! The contention here is that, in spite of neglecting the spatial variation of the parameters involved, we expect the results obtained to be correct, *at least* in a qualitative sense.

Let us then study the *ground-state* properties of a Fermi gas composed of N *relativistic* electrons ($g = 2$). First of all, we have

$$N = \frac{8\pi V}{h^3}\int_0^{p_F} p^2\,dp = \frac{8\pi V}{3h^3}p_F^3, \tag{4}$$

whence it follows that

$$p_F = \left(\frac{3n}{8\pi}\right)^{1/3} h. \tag{5}$$

The energy–momentum relationship for a relativistic particle is

$$\varepsilon = mc^2[\{1+(p/mc)^2\}^{1/2}-1], \tag{6}$$

the speed of the particle being

$$u \equiv \frac{d\varepsilon}{dp} = c\frac{(p/mc)}{\{1+(p/mc)^2\}^{1/2}}. \tag{7}$$

The total kinetic energy of the gas is then given by

$$E_0 = N\langle\varepsilon\rangle_0 = \frac{8\pi V}{h^3}\int_0^{p_F} mc^2[\{1+(p/mc)^2\}^{1/2}-1]p^2\,dp, \tag{8}$$

and its pressure by

$$P_0 = \frac{1}{3}\frac{N}{V}\langle pu\rangle_0 = \frac{8\pi}{3h^3}\int_0^{p_F} mc^2 \frac{(p/mc)^2}{\{1+(p/mc)^2\}^{1/2}} p^2\, dp. \tag{9}$$

We may introduce a dimensionless variable θ, defined by

$$p = mc \sinh \theta, \tag{10}$$

which makes

$$\varepsilon = mc^2(\cosh \theta - 1) \quad \text{and} \quad u = c \tanh \theta. \tag{11}$$

Equations (4), (8) and (9) then become

$$N = \frac{8\pi V m^3 c^3}{3h^3} \sinh^3 \theta_F = \frac{8\pi V m^3 c^3}{3h^3} x^3, \tag{12}$$

$$E_0 = \frac{8\pi V m^4 c^5}{h^3} \int_0^{\theta_F} (\cosh \theta - 1) \sinh^2 \theta \cosh \theta \, d\theta,$$

$$= \frac{\pi V m^4 c^5}{3h^3} B(x) \tag{13}$$

and

$$P_0 = \frac{8\pi m^4 c^5}{3h^3} \int_0^{\theta_F} \sinh^4 \theta \, d\theta$$

$$= \frac{\pi m^4 c^5}{3h^3} A(x), \tag{14}$$

where

$$A(x) = x(x^2+1)^{1/2}(2x^2-3)+3 \sinh^{-1} x \tag{15}$$

and

$$B(x) = 8x^3\{(x^2+1)^{1/2}-1\} - A(x), \tag{16}$$

with

$$x = \sinh \theta_F = p_F/mc = (3n/8\pi)^{1/3}(h/mc). \tag{17}$$

The functions $A(x)$ and $B(x)$ can be computed for any desired value of x; they have also been tabulated by Chandrasekhar (1939), p. 361. However, asymptotic results (for $x \ll 1$ and $x \gg 1$) are sometimes useful. These are given by (see Kothari and Singh, 1942)

$$\left.\begin{aligned} A(x) &= \tfrac{8}{5}x^5 - \tfrac{4}{7}x^7 + \tfrac{1}{3}x^9 - \tfrac{5}{22}x^{11} + \ldots && \text{for} \quad x \ll 1 \\ &= 2x^4 - 2x^2 + 3(\ln 2x - \tfrac{7}{12}) + (\tfrac{5}{4}x^{-2}) + \ldots && \text{for} \quad x \gg 1 \end{aligned}\right\} \tag{18}$$

and

$$\left.\begin{aligned} B(x) &= \tfrac{12}{5}x^5 - \tfrac{3}{7}x^7 + \tfrac{1}{6}x^9 - \tfrac{15}{176}x^{11} + \ldots && \text{for} \quad x \ll 1 \\ &= 6x^4 - 8x^3 + 6x^2 - 3(\ln 2x - \tfrac{1}{4}) - (\tfrac{3}{4}x^{-2}) + \ldots && \text{for} \quad x \gg 1. \end{aligned}\right\} \tag{19}$$

Of course, in the extreme case when $x \to 0$, the ratio A/B, which is identically equal to the ratio of the pressure P_0 to the energy density E_0/V, tends to the nonrelativistic value $\frac{2}{3}$, while in the opposite case when $x \to \infty$, it tends to the extreme relativistic value $\frac{1}{3}$; cf. Problem 8.8, part (i).

We shall now consider, in a crude manner, the equilibrium configuration of this model. In the absence of gravitation, it will be necessary to have "external walls" in order to keep the electron gas at a given density n. The gas will exert a pressure $P_0(n)$ on the walls and any compression or expansion (of the gas) will involve an expenditure of work. Assuming the configuration to be spherical, an adiabatic change in V will cause a change in the energy of the gas, as given by

$$dE_0 = -P_0(n)\,dV = -P_0(R)\cdot 4\pi R^2\,dR. \tag{20}$$

In the presence of gravitation, no external walls are needed, but the change in the *kinetic* energy of the gas, as a result of a change in the size of the sphere, will still be given by the formula (20); of course, the expression for P_0, as a function of the "mean" density n, must now take into account the nonuniformity of the system—a fact which will be disregarded in the present simple-minded treatment. However, eqn. (20) alone no longer gives us the net change in the energy of the system; if it were so, the system would expand indefinitely until both n and $P_0(n) \to 0$. Actually, there arises now a change in the *potential* energy as well; this is given by

$$dE_g = \left(\frac{dE_g}{dR}\right) dR = \alpha \frac{GM^2}{R^2}\,dR, \tag{21}$$

where M is the total mass of the gas, G the constant of gravitation, while α is a number (of the order of unity) whose value depends upon the nature of the spatial variation of n inside the sphere. If the configuration is in equilibrium, then the net change in the value of its energy, for an infinitesimal change in its size, should be identically zero; thus

$$P_0(R) = \frac{\alpha}{4\pi}\frac{GM^2}{R^4}. \tag{22}$$

For $P_0(R)$ we may substitute from eqn. (14), where the parameter x is now given by

$$x = \left(\frac{3n}{8\pi}\right)^{1/3}\frac{h}{mc} = \left(\frac{9N}{32\pi^2}\right)^{1/3}\frac{h/mc}{R}$$

or, in view of (1), by

$$x = \left(\frac{9M}{64\pi^2 m_p}\right)^{1/3}\frac{h/mc}{R} = \left(\frac{9\pi M}{8m_p}\right)^{1/3}\frac{\hbar/mc}{R}. \tag{23}$$

Equation (22) then takes the form

$$\begin{aligned} A\left(\left\{\frac{9\pi M}{8m_p}\right\}^{1/3}\frac{\hbar/mc}{R}\right) &= \frac{3\alpha h^3}{4\pi^2 m^4 c^5}\frac{GM^2}{R^4} \\ &= 6\pi\alpha\left(\frac{\hbar/mc}{R}\right)^3\frac{GM^2/R}{mc^2}; \end{aligned} \tag{24}$$

the function $A(x)$ is given by eqns. (15) and (18).

Equation (24) establishes a one-to-one correspondence between the masses and the radii of white dwarfs; it is, therefore, generally known as the *mass–radius relationship* for these stars. It is rather interesting to look at the combinations of parameters that appear in this relationship; we have here (i) the mass of the star in terms of the proton mass, (ii) the radius of the star in terms of the Compton wavelength of the electron, and (iii) the gravitational energy of the star in terms of the rest energy of the electron. The relationship, therefore, exhibits a remarkable combination of quantum-mechanics, relativity and gravitation. In view of the implicit character of the relationship, we cannot express the radius of the star as an explicit function of its mass, except in two extreme cases. To see this, we note that since $M \approx 10^{33}$ g, $m_p \approx 10^{-24}$ g and $\hbar/mc \approx 10^{-11}$ cm, the argument of the function $A(x)$ in eqn. (24) will be of the order of unity when $R \approx 10^8$ cm. Hence, we can define the two extreme cases as follows:

(i) $R \gg 10^8$ cm, which makes $x \ll 1$ and hence $A(x) \simeq \frac{8}{5}x^5$, with the result

$$R \simeq \frac{3(9\pi)^{2/3}}{40\alpha} \frac{\hbar^2 M^{-1/3}}{Gmm_p^{5/3}} \propto M^{-1/3}. \tag{25}$$

(ii) $R \ll 10^8$ cm, which makes $x \gg 1$ and hence $A(x) \simeq 2x^4 - 2x^2$, with the result

$$R \simeq \frac{(9\pi)^{1/3}}{2} \frac{\hbar}{mc} \left(\frac{M}{m_p}\right)^{1/3} \left\{1 - \left(\frac{M}{M_0}\right)^{2/3}\right\}^{1/2}, \tag{26}$$

where

$$M_0 = \frac{9}{64} \left(\frac{3\pi}{\alpha^3}\right)^{1/2} \frac{(\hbar c/G)^{3/2}}{m_p^2}. \tag{27}$$

We thus find that the greater the mass of a white dwarf, the smaller its size! Not only that, there exists a limiting mass M_0, given by eqn. (27), that corresponds to a vanishing size of the star. Obviously, for $M > M_0$, our mass–radius relationship does not possess any real solution. We, therefore, conclude that all white dwarf stars must have a mass less than M_0—a conclusion fully upheld by observation.

The correct limiting mass of a white dwarf star is generally referred to as the *Chandrasekhar limit*. The physical reason for the existence of this limit is that for a mass exceeding this limit the ground-state pressure P_0 of the system (that arises essentially from the Pauli exclusion principle) would be insufficient to support the gas against its "tendency towards a gravitational collapse". The numerical value of the limiting mass, as given by eqn. (27), turns out to be $\approx 10^{33}$ g. Detailed investigations by Chandrasekhar lead to the result:

$$M_0 = \frac{5.75}{\mu_e^2} \odot, \tag{28}$$

where $\odot$ denotes the mass of the sun, which is very nearly 2×10^{33} g, while μ_e is a number that represents the degree of ionization in the gas. By definition, $\mu_e = M/Nm_H$; cf. eqn. (1). Thus, in most cases, $\mu_e \simeq 2$. Accordingly, we obtain for the Chandrasekhar limit: $M_0 \simeq 1.44 \odot$.

Figure 8.9 shows a plot of the theoretical relationship between the masses and the radii of the white dwarf stars. It is based on Chandrasekhar's rigorous work. Nonetheless, the behavior in the two extreme regions, viz. for $R \gg l$ and $R \ll l$, is simulated quite well by the formulae (25) and (26) of the treatment given here.

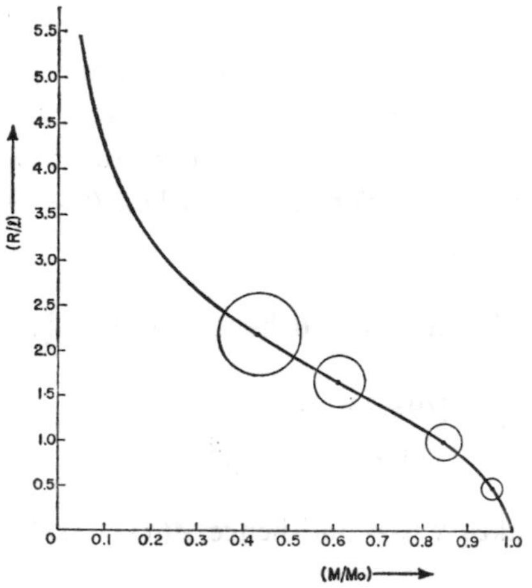

FIG. 8.9. The mass–radius relationship for white dwarfs (after Chandrasekhar, 1939). The masses are expressed in terms of the limiting mass M_0 and the radii in terms of a characteristic length l, which is given by $7.71\mu_e^{-1}\times 10^8$ cm $\simeq 3.86\times 10^8$ cm.

8.5. Statistical model of the atom

Another application of the Fermi statistics was made by Thomas (1927) and Fermi (1928) for calculating the electron distribution and electric field in the extra-nuclear space of a heavy atom. Their approach was based on the observation that the electrons in this system could be regarded as a completely degenerate Fermi gas of *nonuniform* density $n(\boldsymbol{r})$. By considering the equilibrium state of the configuration, one arrives at a differential equation whose solution gives directly the electric potential $\varphi(\boldsymbol{r})$ and the electron density $n(\boldsymbol{r})$ at the point $\boldsymbol{r}$. By the very nature of the model, which is generally referred to as the *Thomas–Fermi model* of the atom, the resulting function $n(\boldsymbol{r})$ is a smoothly varying function of $\boldsymbol{r}$, devoid of the "peaks" that would remind one of the electron orbits of the Bohr theory. Nevertheless, the model has proved quite useful in deriving composite properties such as the binding energy of an atom. And, after suitable modifications, it has been successfully applied to molecules, solids and nuclei as well. Here, we propose to outline only the simplest treatment of the model, as applied to an atomic system; for further details and for other applications, see Gombás (1949, 1952) and March (1957), where references to other contributions to the subject are also given.

According to the statistics of a completely degenerate Fermi gas, we have exactly two electrons (with opposite spins) in each elementary cell, of the phase space, with $p \leq p_F$; the Fermi momentum p_F of the electron gas is determined by the electron density n, according to the formula

$$p_F = (3\pi^2 n)^{1/3}\hbar. \tag{1}$$

In the given system, the electron density varies from point to point; so would the value of p_F. We have, therefore, to speak of the limiting momentum p_F as a function of $\boldsymbol{r}$, which is clearly a "quasi-classical" description of the situation. Such a description is justifiable if the de Broglie wavelength of the electrons in a given region of space is much smaller than the distance over which the functions $p_F(\boldsymbol{r})$, $n(\boldsymbol{r})$ and $\varphi(\boldsymbol{r})$ undergo a significant variation; later on, it will be seen that this requirement is satisfied reasonably well for the heavier atoms.

Now, the total energy ε of an electron at the top of the Fermi sea *at the point* $\boldsymbol{r}$ is given by

$$\varepsilon(\boldsymbol{r}) = \frac{1}{2m}p_F^2(\boldsymbol{r}) - e\varphi(\boldsymbol{r}), \tag{2}$$

where e stands for the magnitude of the electronic charge. When the system is in the *stationary* state, the value of $\varepsilon(\boldsymbol{r})$ should be the same throughout, so that electrons anywhere in the system do not have a tendency to "flow away" towards other parts of the system. Now, at the boundary of the system, p_F must be zero; also, by a suitable choice of the zero of energy, we can also have there: $\varphi = 0$. Thus, the value of ε at the boundary of the system turns out to be zero; so must, then, be the value of ε throughout the system. We thus have, for all $\boldsymbol{r}$,

$$\frac{1}{2m}p_F^2(\boldsymbol{r}) - e\varphi(\boldsymbol{r}) = 0. \tag{3}$$

Substituting from (1) and making use of the Poisson equation

$$\nabla^2\varphi(\boldsymbol{r}) = -4\pi\varrho(\boldsymbol{r}) = 4\pi en(\boldsymbol{r}), \tag{4}$$

we obtain

$$\nabla^2\varphi(\boldsymbol{r}) = \frac{4e(2me)^{3/2}}{3\pi\hbar^3}\{\varphi(\boldsymbol{r})\}^{3/2}. \tag{5}$$

Assuming spherical symmetry (which implies $J = 0$), eqn. (5) takes the form

$$\frac{1}{r^2}\frac{d}{dr}\left\{r^2\frac{d}{dr}\varphi(r)\right\} = \frac{4e(2me)^{3/2}}{3\pi\hbar^3}\{\varphi(r)\}^{3/2}, \tag{6}$$

which is known as the *Thomas–Fermi equation* of the system. Introducing dimensionless variables x and Φ, as defined by

$$x = 2\left(\frac{4}{3\pi}\right)^{2/3} Z^{1/3}\frac{me^2}{\hbar^2}r = \frac{Z^{1/3}}{0.88534a_B}r \tag{7}$$

and

$$\Phi(x) = \frac{\varphi(r)}{Ze/r}, \tag{8}$$

where Z is the atomic number of the system and a_B the first Bohr radius of the hydrogen atom, eqn. (6) reduces to

$$\frac{d^2\Phi}{dx^2} = \frac{\Phi^{3/2}}{x^{1/2}}. \tag{9}$$

Equation (9) is the *dimensionless Thomas–Fermi equation* of the system. The boundary conditions on the solution of this equation can be obtained as follows. As we approach the nucleus of the system ($r \to 0$), the potential $\varphi(r)$ approaches the unscreened value Ze/r; accordingly, we must have: $\Phi(x \to 0) = 1$. On the other hand, as we approach the boundary of the system ($r \to r_0$), $\varphi(r)$ in the case of a neutral atom must tend to zero; accordingly, we must have: $\Phi(x \to x_0) = 0$. In principle, these two conditions are sufficient to determine the function $\Phi(x)$ completely. However, it would be helpful if one knew the initial slope of the function as well, which in turn would depend upon the precise location of the boundary. Choosing the boundary to be at infinity ($r_0 = \infty$), the appropriate initial slope of the function $\Phi(x)$ turns out to be -1.5886; in fact, the nature of the solution near the origin is

$$\Phi(x) = 1 - 1.5886x + \tfrac{4}{3}x^{3/2} + \ldots . \tag{10}$$

For $x > 10$, the approximate solution has been determined by Sommerfeld (1932):

$$\Phi(x) \simeq \left\{1 + \left(\frac{x^3}{144}\right)^{\lambda}\right\}^{-1/\lambda}, \tag{11}$$

where

$$\lambda = \frac{\sqrt{(73)} - 7}{6} \simeq 0.257. \tag{12}$$

As $x \to \infty$, the solution tends to the simple form: $\Phi(x) = 144/x^3$. The complete numerical solution, which is a monotonically decreasing function of x, has been tabulated by Bush and Caldwell (1931). As a check on the numerical results, we note that the solution must satisfy the integral condition

$$\int_0^\infty \Phi^{3/2} x^{1/2}\, dx = 1, \tag{13}$$

which expresses the fact that the integral of the electron density $n(r)$ over the whole of the space available to the system must be equal to the total number Z of the electrons present.

From the function $\Phi(x)$, one readily obtains the electric potential $\varphi(r)$ and the electron density $n(r)$:

$$\varphi(r) = \frac{Ze}{r}\,\Phi\left(\frac{rZ^{1/3}}{0.88534a_B}\right) \propto Z^{4/3}, \tag{14}$$

and

$$n(r) = \frac{(2me)^{3/2}}{3\pi^2\hbar^3}\{\varphi(r)\}^{3/2} \propto Z^2. \tag{15}$$

Figure 8.10 shows a Thomas–Fermi plot of the electron distribution function $D(r)$ $\{= n(r) \times 4\pi r^2\}$ for an atom of mercury; the actual "peaked" distribution, which conveys unmistakably the preference of the electrons to be in the vicinity of the semi-classical orbits, is also shown in the figure.

In order to calculate the *binding energy* of the atom, we must determine the total energy of the electron cloud. Now, the mean kinetic energy of an electron at the point $\boldsymbol{r}$ would be $\frac{3}{5}\varepsilon_F(r)$; by eqn. (3), this is equal to $\frac{3}{5}e\varphi(r)$. The total kinetic energy of the electron cloud is,

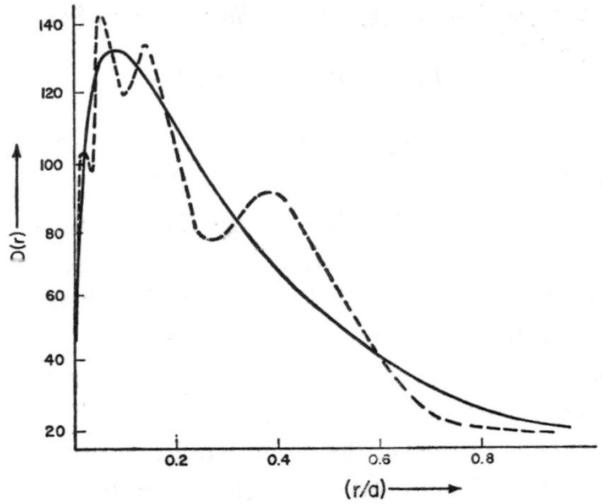

FIG. 8.10. The electron distribution function $D(r)$ for an atom of mercury. The distance r is expressed in terms of the atomic unit of length $a\,(=\hbar^2/me^2)$.

therefore, given by

$$\tfrac{3}{5}e\int_0^\infty \varphi(r)\,n(r)\cdot 4\pi r^2\,dr. \tag{16}$$

For the determination of the total potential energy of the cloud, we note that a part of the potential $\varphi(r)$ at the point $\boldsymbol{r}$ is due to the nucleus of the atom while the rest of it is due to the electron cloud itself; the former is clearly (Ze/r), so the latter must be $\{\varphi(r)-Ze/r\}$. The total potential energy of the cloud is, therefore, given by

$$-e\int_0^\infty \left[\frac{Ze}{r}+\frac{1}{2}\left\{\varphi(r)-\frac{Ze}{r}\right\}\right] n(r)\cdot 4\pi r^2\,dr. \tag{17}$$

We thus obtain for the *total energy* of the cloud

$$E_0=\int_0^\infty \left\{\frac{1}{10}\,e\varphi(r)-\frac{1}{2}\,\frac{Ze^2}{r}\right\} n(r)\cdot 4\pi r^2\,dr; \tag{18}$$

of course, the electron density $n(r)$, in terms of the potential function $\varphi(r)$, is given by eqn. (15).

Now, Milne (1927) has shown that the integrals

$$\int_0^\infty \{\varphi(r)\}^{5/2}\,r^2\,dr \quad \text{and} \quad \int_0^\infty \{\varphi(r)\}^{3/2}\,r\,dr, \tag{19}$$

that appear in the expression for E_0, can be expressed directly in terms of the *initial* slope of the function $\varphi(r)$, i.e. in terms of the number -1.5886 of eqn. (10). After a little calculus, one finds that

$$E_0=\frac{1.5886}{0.88534}\left(\frac{e^2}{2a_B}\right)Z^{7/3}\left\{\frac{1}{7}-1\right\}, \tag{20}$$

whence one obtains for the Thomas–Fermi *binding energy* of the atom:

$$E_B = -E_0 = 1.538 Z^{7/3}\chi, \tag{21}$$

where χ ($= e^2/2a_B \simeq 13.6$ eV) is the actual binding energy of the hydrogen atom. It is clear that the statistical result (21) cannot give us anything more than just the first term of an "asymptotic expansion" of the binding energy E_B in powers of the parameter $Z^{1/3}$. For practical values of Z, other terms of the expansion are also important; however, they cannot be obtained from the simple-minded treatment given here. More interested readers may refer to the review article by March (1957).

In the end we observe that since the total energy of the electron cloud is proportional to $Z^{7/3}$ the *mean* energy per electron would be proportional to $Z^{4/3}$; accordingly, the mean de Broglie wavelength of the electrons in the cloud would be proportional to $Z^{-2/3}$. At the same time, the overall linear dimensions of the cloud are proportional to $Z^{-1/3}$; see eqn. (15), whereby the effective volume of the cloud is proportional to Z^{-1}. The same result follows from eqn. (7), whereby the distances over which the values of the various physical functions of the problem undergo significant variations are proportional to $Z^{-1/3}$. We thus find that the quasi-classical description adopted in the Thomas–Fermi model is more appropriate for heavier atoms ($Z^{-2/3} \ll Z^{-1/3}$). Otherwise, too, the statistical nature of the approach demands that we have in the system as large a number of particles as possible.

Problems

8.1. Let the Fermi distribution at low temperatures be represented by a *broken line*, as shown in Fig. 8.11, the line being tangential to the actual curve at $\varepsilon = \varepsilon_F$. Show that this approximate representation gives a "correct" result for the low-temperature specific heat of the Fermi gas, except that the numerical factor concerned turns out to be $\frac{4}{3}$ instead of $\pi^2/3$. Discuss, in a qualitative manner, the origin of the numerical discrepancy.

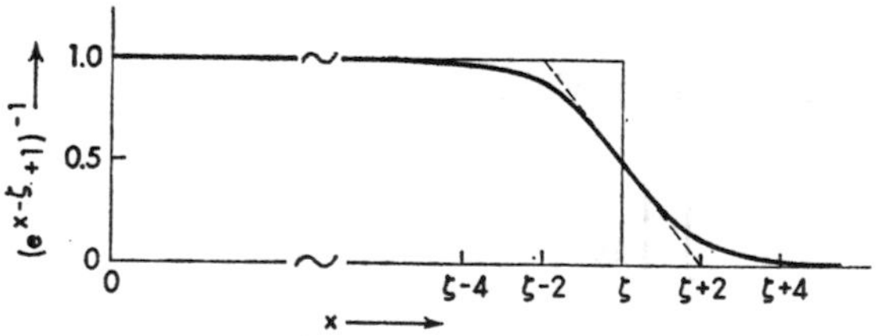

FIG. 8.11. An *approximate* representation of the Fermi distribution at low temperatures: here, $x = \varepsilon/kT$ and $\zeta = \mu/kT$.

8.2. Evaluate the quantity $\{\langle u\rangle\langle 1/u\rangle\}$ for a degenerate Fermi gas, both in the nonrelativistic and the extreme relativistic cases, and check the validity of the inequality stated in Problem 6.6.

8.3. For a Fermi–Dirac gas, we may define a temperature T_0 at which the chemical potential of the gas is zero ($z = 1$). Express T_0 in terms of the Fermi temperature T_F of the gas.

8.4. Derive an expression for the *low-temperature* isothermal compressibility of an ideal Fermi gas, and estimate its numerical value for the gas of conduction electrons in a sodium crystal.

8.5. Determine the velocity of sound in a degenerate Fermi gas, and compare your result with the Fermi velocity u_F $\{\equiv (\partial\varepsilon/\partial p)_{p=p_F}\}$.

8.6. Obtain numerical estimates for the Fermi energy (in eV) and the Fermi temperature (in °K) for the following systems:

(i) conduction electrons in silver, lead and aluminum,
(ii) nucleons in a heavy nucleus, such as $_{80}Hg^{200}$, and
(iii) He^3 atoms in liquid helium three (atomic volume: 63 Å^3 per atom).

8.7. Making use of another term of the Sommerfeld lemma (E.13), show that *in the second approximation* the chemical potential of a degenerate Fermi gas is given by

$$\mu \simeq \varepsilon_F\left[1-\frac{\pi^2}{12}\left(\frac{kT}{\varepsilon_F}\right)^2-\frac{\pi^4}{80}\left(\frac{kT}{\varepsilon_F}\right)^4\right] \tag{8.1.37a}$$

and the mean energy per particle by

$$\frac{U}{N} \simeq \frac{3}{5}\varepsilon_F\left[1+\frac{5\pi^2}{12}\left(\frac{kT}{\varepsilon_F}\right)^2-\frac{\pi^4}{16}\left(\frac{kT}{\varepsilon_F}\right)^4\right]. \tag{8.1.38a}$$

Hence, determine the T^3-correction to the customary T^1-result for the specific heat of an electron gas. Compare the magnitude of the T^3-term, in a typical metal such as copper, with the low-temperature specific heat that arises from the Debye modes of the lattice.

8.8. Consider an ideal Fermi gas, with energy spectrum $\varepsilon \propto p^s$, contained in a box of "volume" V in space of n dimensions. Show that, for this system,

(i) $PV = \frac{s}{n}U$;

(ii) $\frac{C_V}{Nk} = \frac{n}{s}\left(\frac{n}{s}+1\right)\frac{f_{(n/s)+1}(z)}{f_{n/s}(z)}-\left(\frac{n}{s}\right)^2\frac{f_{n/s}(z)}{f_{(n/s)-1}(z)}$;

(iii) $\frac{C_P-C_V}{Nk} = \left(\frac{sC_V}{nNk}\right)^2\frac{f_{(n/s)-1}(z)}{f_{n/s}(z)}$;

(iv) the equation of an *adiabatic* is: $PV^{1+(s/n)}$ = const., and

(v) the index $(1+(s/n))$ in the foregoing equation agrees with the ratio (C_P/C_V) of the gas only when $T \gg T_F$. On the other hand, when $T \ll T_F$, the ratio $(C_P/C_V) \simeq 1+(\pi^2/3)(kT/\varepsilon_F)^2$, irrespective of the values of the numbers s and n.

8.9. Study the results (ii) and (iii) of the previous problem both in the high-temperature limit ($T \gg T_F$) and in the low-temperature limit ($T \ll T_F$), and compare your results with the ones pertaining to a nonrelativistic gas and an extreme relativistic gas in *three* dimensions.

8.10. Prove that, in *two* dimensions, the specific heat $C_V(N, T)$ of an ideal nonrelativistic gas of fermions is identical with the specific heat of a corresponding gas of bosons, for *all* values of N and T. Further show that in the extreme relativistic case the same result holds in *one* dimension.

8.11. Prove that, *quite generally*, the low-temperature values of the chemical potential, the specific heat and the entropy of an ideal Fermi gas are given by

$$\mu \simeq \varepsilon_F\left[1-\frac{\pi^2}{6}\left(\frac{\partial \ln a(\varepsilon)}{\partial \ln \varepsilon}\right)_{\varepsilon=\varepsilon_F}\left(\frac{kT}{\varepsilon_F}\right)^2\right]$$

and

$$C_V \simeq S \simeq \frac{\pi^2}{3}k^2T\,a(\varepsilon_F),$$

where $a(\varepsilon)$ is the *density of (the single-particle) states* in the system. Examine these results for a gas with energy spectrum $\varepsilon \propto p^s$, confined to a space of n dimensions, and discuss the special cases: s = 1 and 2, with n = 2 and 3.

8.12. Investigate the Pauli paramagnetism of an ideal gas of fermions with intrinsic magnetic moment μ^* and spin $J\hbar$ ($J = \frac{1}{2}, \frac{3}{2}, \ldots$), and derive expressions for the low-temperature and high-temperature susceptibilities of the gas.

8.13. Verify eqns. (8.2.24) and (8.2.27) for the magnetic susceptibility of an ideal Fermi gas.

8.14. Complete the mathematical steps leading to eqn. (8.2.57) and verify the asymptotic formula (8.2.58).

8.15. The observed value of γ, see eqn. (8.3.6), for sodium is 4.3×10^{-4} cal mole^{-1} °K^{-2}. Evaluate thereby the Fermi energy ε_F and the number density n of the conduction electrons in the sodium metal. Compare the latter result with the number density of the atoms (given that, for sodium, $\varrho = 0.954$ g cm^{-3} and $M = 23$).

8.16. (i) Show that the average energy of the thermionic electrons, associated with the z-component of their motion, is equal to kT. Indeed, by the z-component of the motion we mean "the component *normal* to the surface of the emitter".

(ii) Also show that the average energy of the thermionic electrons, associated with the other two components of the motion, is simply $\frac{1}{2}kT$.

8.17. Calculate the fraction of the conduction electrons in tungsten ($\varepsilon_F = 9.0$ eV) at 3000°K, whose kinetic energy $\varepsilon\,(=\frac{1}{2}mu^2)$ is greater than W ($= 13.5$ eV). Also calculate the fraction of the electrons whose kinetic energy associated with the z-component of their motion, namely $(\frac{1}{2}mu_z^2)$, is greater than 13.5 eV.

8.18. Prove that the function $f_2(e^\delta)$ of eqn. (8.3.30) satisfies the identity

$$f_2(e^\delta)+f_2(e^{-\delta}) = \frac{\delta^2}{2}+2f_2(1) = \frac{\delta^2}{2}+\frac{\pi^2}{6}.$$

Hence, for $e^\delta \gg 1$,

$$f_2(e^\delta) = \frac{\delta^2}{2}+\frac{\pi^2}{6}-\left(e^{-\delta}-\frac{e^{-2\delta}}{2^2}+\frac{e^{-3\delta}}{3^2}-\dots\right).$$

Compare this (exact) result with the *asymptotic* one following from the Sommerfeld lemma (E.13), and explain the difference between the two results in terms of the footnote to Appendix E.

8.19. Verify the expressions (8.4.13) and (8.4.14) for the ground-state energy E_0 and the ground-state pressure P_0 of a relativistic Fermi gas, and show that these expressions satisfy the relationships

$$P_0 = -\left(\frac{\partial E_0}{\partial V}\right)_N \quad \text{and} \quad E_0+P_0V = N\mu_0 \equiv N\varepsilon_F.$$

8.20. (a) Show that the low-temperature specific heat of the relativistic Fermi gas, studied in Sec. 8.4, is given by

$$\frac{C_V}{Nk} = \pi^2\frac{(x^2+1)^{1/2}}{x^2}\frac{kT}{mc^2} \qquad \left(x = \frac{p_F}{mc}\right).$$

Check that this formula gives correct results for the nonrelativistic case as well as for the extreme relativistic case.

(b) Investigate the T^3-correction to the foregoing result and derive once again the special case considered in Problem 8.7.

8.21. Express the integrals (8.5.19) in terms of the initial slope of the function $\varphi(r)$, and verify eqn. (8.5.20).

8.22. The total energy E of the electron cloud in an atom can be written as

$$E = K+V_{ne}+V_{ee},$$

where K is the kinetic energy of the electrons, V_{ne} the interaction energy between the electrons and the nucleus and V_{ee} the mutual interaction energy of the electrons. Show that, according to the Thomas–Fermi model of a neutral atom,

$$K = -E, \quad V_{ne} = +\tfrac{7}{3}E \quad \text{and} \quad V_{ee} = -\tfrac{1}{3}E,$$

so that $V = (V_{ne}+V_{ee}) = +2E$. Note that these results are consistent with the *virial theorem* (see Problem 2.12, with $n = -1$).

CHAPTER 9

STATISTICAL MECHANICS OF INTERACTING SYSTEMS: THE METHOD OF CLUSTER EXPANSIONS

ALL the physical systems considered in the previous chapters were composed of, or could be regarded as composed of, *noninteracting* entities. Consequently, the results obtained, though of considerable intrinsic importance, suffer from severe limitations when applied to systems that actually exist in nature. For a real contact between the theory and experiment, one must take into account the interparticle interactions prevailing in the given system. This can be done with the help of the formalism developed in Chapters 3–5 which, in principle, can be applied to an unlimited variety of physical systems and problems; in practice, however, one encounters in most cases serious mathematical difficulties of analysis. These difficulties are less stringent in the case of systems such as (dilute) real gases, for which a corresponding noninteracting system can serve as an approximation. The mathematical expressions for the various physical quantities pertaining to such a system can be written in the form of *series expansions*, whose main terms denote the corresponding ideal-system results while the subsequent terms denote the corrections arising from the interparticle interactions in the system. A systematic method of carrying out these expansions, in the case of real gases obeying classical statistics, was developed by Mayer and his collaborators (1937 onward) and is known as the method of *cluster expansions*. Its generalization, which applies as well to gases obeying quantum statistics, was initiated by Kahn and Uhlenbeck (1938) and has been perfected by Lee and Yang (1959 a, b; 1960).

9.1. Cluster expansion for a classical gas

We start with a relatively simple physical system, namely a single-component, classical, monatomic gas whose potential energy is given by a sum of the two-particle interactions u_{ij}. The Hamiltonian of the system is then given by

$$H = \sum_i \left(\frac{1}{2m} p_i^2\right) + \sum_{i<j} u_{ij} \qquad (i, j = 1, 2, \ldots, N); \tag{1}$$

the summation in the second part goes over all the $N(N-1)/2$ pairs of particles in the system. In general, the potential u_{ij} is a function of the relative position vector $\mathbf{r}_{ij}(=\mathbf{r}_j-\mathbf{r}_i)$; however, if the two-body force is a central one, then the function u_{ij} depends only on the distance r_{ij} between the particles.

With the above Hamiltonian, the partition function of the system would be given by

$$Q_N(V,T)=\frac{1}{N!\,h^{3N}}\int\exp\left\{-\beta\sum_i\left(\frac{1}{2m}p_i^2\right)-\beta\sum_{i<j}u_{ij}\right\}d^{3N}p\,d^{3N}r. \tag{2}$$

The integration over the momenta of the particles can be carried out straightforwardly, with the result

$$Q_N(V,T)=\frac{1}{N!\,\lambda^{3N}}\int\exp\left\{-\beta\sum_{i<j}u_{ij}\right\}d^{3N}r=\frac{1}{N!\,\lambda^{3N}}Z_N(V,T), \tag{3}$$

where $\lambda\,\{=h/(2\pi mkT)^{1/2}\}$ is the *mean thermal wavelength* of the particles, while the function $Z_N(V,T)$ stands for the integral over the space coordinates $\mathbf{r}_1,\mathbf{r}_2,\ldots,\mathbf{r}_N$:

$$Z_N(V,T)=\int\exp\left[-\beta\sum_{i<j}u_{ij}\right]d^{3N}r=\int\prod_{i<j}(e^{-\beta u_{ij}})\,d^{3N}r. \tag{4}$$

The function $Z_N(V,T)$ is generally referred to as the *configuration integral* of the system For a gas of noninteracting particles, the integrand in (4) would be identically equal to unity; we would then have

$$Z_N^{(0)}(V,T)=V^N\quad\text{and}\quad Q_N^{(0)}(V,T)=\frac{V^N}{N!\,\lambda^{3N}}, \tag{5}$$

in agreement with our earlier result (3.5.9).

To treat the non-ideal case we introduce, after Mayer, the two-particle function f_{ij}, defined by the relationship

$$f_{ij}=e^{-\beta u_{ij}}-1. \tag{6}$$

In the absence of interactions, the function f_{ij} is identically equal to zero; in the presence of interactions, it is non-zero, but at sufficiently high temperatures it is quite small in comparison with unity. We, therefore, expect that the functions f_{ij} would be quite appropriate for carrying out a high-temperature expansion of the integrand in (4).

A typical plot of the functions u_{ij} and f_{ij} is shown in Fig. 9.1; we note that (i) the function f_{ij} is everywhere bounded and (ii) it becomes negligibly small as the interparticle distance becomes large in comparison with the "effective" range of the potential.

Now, to evaluate the configuration integral (4), we expand the integrand in ascending powers of the functions f_{ij}:

$$\begin{aligned}Z_N(V,T)&=\int\prod_{i<j}(1+f_{ij})\,d^3r_1\ldots d^3r_N\\&=\int[1+\Sigma f_{ij}+\Sigma f_{ij}f_{kl}+\ldots]\,d^3r_1\ldots d^3r_N.\end{aligned} \tag{7}$$

A convenient way of enumerating the various terms in (7) is to associate each term with a corresponding N-particle graph. For example, if N were equal to 8, the terms

$$t_A=\int f_{34}f_{68}\,d^3r_1\ldots d^3r_8\quad\text{and}\quad t_B=\int f_{12}f_{14}f_{67}\,d^3r_1\ldots d^3r_8, \tag{8}$$

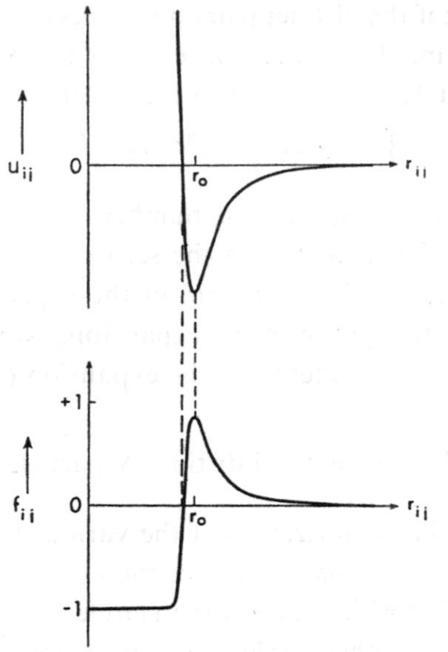

FIG. 9.1. A typical plot of the two-body potential function u_{ij} and the corresponding Mayer function f_{ij}.

in the expansion of the configuration integral Z_8 of the 8-particle system, could be associated with the 8-particle graphs

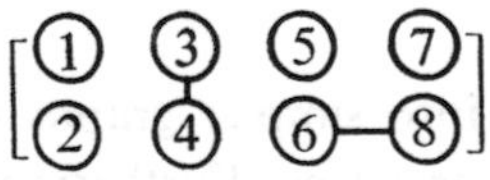　　and　　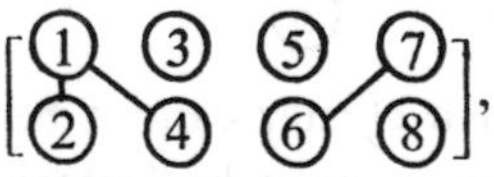,　　(9)

respectively. A closer look at the terms t_A and t_B (and at the corresponding graphs) suggests that we better regard these terms as suitably factorized (and the graphs correspondingly decomposed), that is,

$$t_A = \int d^3r_1 \int d^3r_2 \int d^3r_5 \int d^3r_7 \int f_{34}\, d^3r_3\, d^3r_4 \int f_{68}\, d^3r_6\, d^3r_8$$
$$\equiv [\textcircled{1}]\cdot[\textcircled{2}]\cdot[\textcircled{5}]\cdot[\textcircled{7}]\cdot[\textcircled{3}\!-\!\!-\!\textcircled{4}]\cdot[\textcircled{6}\!-\!\!-\!\textcircled{8}] \tag{10}$$

and similarly

$$t_B = \int d^3r_3 \int d^3r_5 \int d^3r_8 \int f_{12} f_{14}\, d^3r_1\, d^3r_2\, d^3r_4 \int f_{67}\, d^3r_6\, d^3r_7$$
$$\equiv [\textcircled{3}]\cdot[\textcircled{5}]\cdot[\textcircled{8}]\cdot[\textcircled{6}\!-\!\!-\!\textcircled{7}]\cdot[$$ 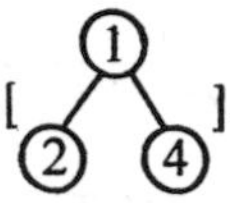 $$]. \tag{11}$$

We then say that the term t_A in the expansion of the integral Z_8 represents a "configuration" in which there are four "clusters" of one particle each and two "clusters" of two particles each, while the term t_B represents a "configuration" in which there are three "clusters" of one particle each, one "cluster" of two particles and one "cluster" of three particles.

In view of this, we may introduce the notion of an *N-particle graph* which, by definition, is a "collection of N distinct circles, numbered 1, 2, . . ., N, with a number of lines linking some (or all) of the circles"; if the distinct pairs (of circles), which are linked through these lines, are denoted by the symbols α, β, ..., λ (each of these symbols denoting a distinct *pair* of indices out of the set 1, 2, ..., N), then the graph represents the term

$$\int (f_\alpha f_\beta \ldots f_\lambda)\, d^3r_1 \ldots d^3r_N \tag{12}$$

of the expansion (7). A graph having the same number of linked pairs as the foregoing one but with the set (α', β', ..., λ') *distinct* from the set (α, β, ..., λ) has to be counted as a distinct graph, for it represents a different term in the expansion; of course, these terms will belong to one and the same group in the expansion. Now, in view of the one-to-one correspondence between the various terms in the expansion (7) and the various N-particle graphs, we must have

$$Z_N(V, T) = \text{sum of all distinct } N\text{-particle graphs.} \tag{13}$$

Further, in view of the possible factorization of the various terms (or the possible decomposition of the various graphs), we may introduce the notion of an *l-cluster* which, by definition, is an "l-particle graph in which each of the l circles, numbered 1, 2, ..., l, is directly or indirectly linked with every other circle". As an example, we write here a 5-particle graph which is also a 5-cluster:

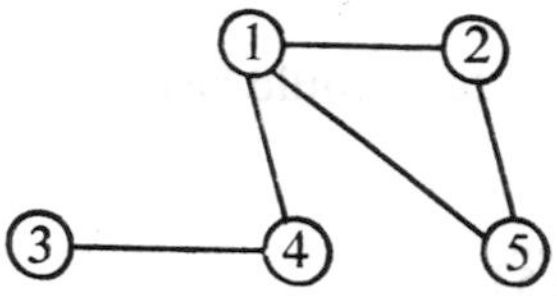

$$\equiv \int f_{12} f_{14} f_{15} f_{25} f_{34}\, d^3r_1 \ldots d^3r_5. \tag{14}$$

It is obvious that a cluster cannot be decomposed into simpler graphs inasmuch as the corresponding term cannot be factorized into simpler terms. Furthermore, a group of l particles (except when $l = 1$ or 2) can lead to a variety of l-clusters, some of which may be equal in value; for instance, a group of three particles leads to four different 3-clusters, namely

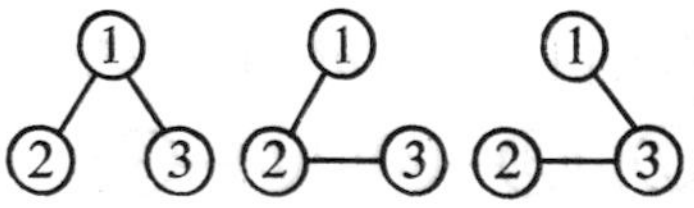

and

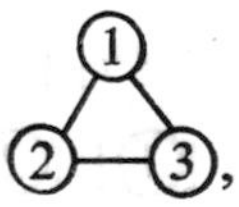

, (15)

of which the first three are equal in value. In view of the variety of ways in which an l-cluster can appear, we may introduce the notion of a *cluster integral* b_l, defined by

$$b_l(V, T) = \frac{1}{l!\, \lambda^{3(l-1)} V} \times (\text{the sum of all possible } l\text{-clusters}). \tag{16}$$

So defined, the cluster integral $b_l(V, T)$ is dimensionless and, in the limit $V \to \infty$, approaches a finite value $\bar{b}_l(T)$ which is independent of the size and the shape of the container (unless the latter is unduly queer). The first property is fairly obvious. The second one follows from the fact that if we hold one of the l particles fixed, at the point $\mathbf{r}_1$ say, and carry out integration over the coordinates of the remaining $(l-1)$ particles, then, because of the fact that the functions f_{ij} extend over a small finite range of distances, this integration would extend only over a *limited* region of the space available—a region whose linear dimensions are of the order of the range of the functions f_{ij}.* The result of this integration will be practically independent of the volume of the container.† Finally, we integrate over the coordinates $\boldsymbol{r}_1$ of the particle that was held fixed and obtain a straight factor of V; this cancels out the V that appears in the denominator of the defining formula (16). Thus, the dependence of the cluster integral $b_l(V, T)$ on the volume V of the container is no more than a mere "surface effect"—a dependence that disappears as $V \to \infty$, and we end up with a volume-independent number $\bar{b}_l(T)$.

Some of the simpler cluster integrals are

$$b_1 = \frac{1}{V}[\text{①}] = \frac{1}{V}\int d^3r_1 \equiv 1, \tag{17}$$

$$\begin{aligned} b_2 &= \frac{1}{2\lambda^3 V}[\text{①—②}] \equiv \frac{1}{2\lambda^3 V}\iint f_{12}\, d^3r_1\, d^3r_2 \\ &= \frac{1}{2\lambda^3}\int f_{12}\, d^3r_{12} = \frac{2\pi}{\lambda^3}\int_0^\infty f(r) r^2\, dr \\ &= \frac{2\pi}{\lambda^3}\int_0^\infty (e^{-u(r)/kT} - 1) r^2\, dr, \end{aligned} \tag{18}$$

$$\begin{aligned} b_3 &= \frac{1}{6\lambda^6 V} \times [\text{sum of the clusters (15)}] \\ &= \frac{1}{6\lambda^6 V}\int \underbrace{(f_{12}f_{13} + f_{12}f_{23} + f_{13}f_{23} + f_{12}f_{13}f_{23})}\, d^3r_1\, d^3r_2\, d^3r_3 \\ &= \frac{1}{6\lambda^6 V}\left[3V\iint f_{12}f_{13}\, d^3r_{12}\, d^3r_{13} + V\iint f_{12}f_{13}f_{23}\, d^3r_{12}\, d^3r_{13}\right] \\ &= 2b_2^2 + \frac{1}{6\lambda^6}\iint f_{12}f_{13}f_{23}\, d^3r_{12}\, d^3r_{13}, \end{aligned} \tag{19}$$

and so on.

We now proceed to the evaluation of the expression (13). Now, an N-particle graph will consist of a "number of clusters" of which, say, m_1 are 1-clusters, m_2 are 2-clusters, m_3 are 3-clusters, and so on; clearly, the set of numbers $\{m_l\}$ must satisfy the condition

$$\sum_{l=1}^{N} l m_l = N, \qquad m_l = 0, 1, 2, \ldots, N. \tag{20}$$

However, a given set of numbers $\{m_l\}$ does not specify a unique, single graph; rather it represents a "collection of graphs", the sum total of which may be denoted by the symbol $S\{m_l\}$. We may then write

$$Z_N(V, T) = \sum_{\{m_l\}}{}' S\{m_l\}, \tag{21}$$

* Hence the name "cluster".

† Of course, some dependence on the geometry of the container *will* arise if the fixed particle happens to be close to the walls of the container. This is, however, unimportant if $V \to \infty$.

where the primed summation Σ' goes over all the sets $\{m_l\}$ that conform to the restrictive condition (20). Equation (21) represents a systematic regrouping of the graphs, as opposed to the simple-minded grouping that first appeared in (7).

Our next task consists in evaluating the sum $S\{m_l\}$. To do this, we observe that the "collection of graphs" under the set $\{m_l\}$ arises essentially from the following two causes:

(i) there are, in general, many different ways of *assigning* the N particles of the system to the $\Sigma_l m_l$ clusters, and

(ii) for any given assignment, there are, in general, many different ways of *forming* the various clusters, for the reason that, even with a given group of particles, an l-cluster (if $l > 2$) can be formed in a number of different ways; see, for example, the four different ways of forming a 3-cluster with a given group of three particles, as listed in (15).

For the cause (i), we obtain a straightforward factor of

$$\frac{N!}{(1!)^{m_1}(2!)^{m_2}\ldots} = \frac{N!}{\prod_l (l!)^{m_l}}. \tag{22}$$

Now, if the cause (ii) were not there, i.e. if all the l-clusters were unique in their formation, the sum $S\{m_l\}$ would have been given by the product of the combinatorial factor (22) with "the value of any one graph in the set-up, viz.

$$\prod_l \text{(the value of an } l\text{-cluster)}^{m_l}\text{''}, \tag{23}$$

further corrected for the fact that any two arrangements which differ merely in the exchange of *all* the particles in one cluster with *all* the particles in another cluster of the same size, *must not* be counted as distinct, the corresponding correction factor being

$$\prod_l (1/m_l!). \tag{24}$$

A little reflection now shows that the cause (ii) is completely and correctly taken care of if we replace the product of the expressions (23) and (24) by the expression*

$$\prod_l [\text{(the } sum \text{ of the values of all possible } l\text{-clusters)}^{m_l}/m_l!], \tag{25}$$

which, with the help of eqn. (16), can be written as

$$\prod_l [(b_l l!\, \lambda^{3(l-1)} V)^{m_l}/m_l!]. \tag{26}$$

The sum $S\{m_l\}$ is now given by the product of the factor (22) and the expression (26). Substituting this result into eqn. (21), we obtain for the configuration integral

$$Z_N(V, T) = N!\, \lambda^{3N} \sum_{\{m_l\}}{}' \left[\prod_l \left\{\left(b_l \frac{V}{\lambda^3}\right)^{m_l} \frac{1}{m_l!}\right\}\right]. \tag{27}$$

* To appreciate the logic of this replacement, one should consider the expression [ ] in (25) as a multinomial expansion and interpret the various terms of the expansion in terms of the variety of the l-clusters.

Here, use has been made of the fact that

$$\prod_l (\lambda^{3l})^{m_l} = \lambda^{3\Sigma_l lm_l} = \lambda^{3N}; \tag{28}$$

see the restrictive condition (20). The partition function of the system follows from eqns. (3) and (27), with the result

$$Q_N(V, T) = \sum_{\{m_l\}}' \left[\prod_{l=1}^{N} \left\{ \left(b_l \frac{V}{\lambda^3} \right)^{m_l} \frac{1}{m_l!} \right\} \right]. \tag{29}$$

The evaluation of the primed sum in (29) is made complicated by the restrictive condition (20) which must be obeyed by every set $\{m_l\}$. We, therefore, move over to the grand partition function of the system:

$$\mathcal{Q}(z, V, T) = \sum_{N=0}^{\infty} z^N Q_N(V, T). \tag{30}$$

Writing

$$z^N = z^{\Sigma_l lm_l} = \prod_l (z^l)^{m_l}, \tag{31}$$

substituting for $Q_N(V, T)$ from (29), and noting that a *restricted* summation over the sets $\{m_l\}$, which are subject to the condition $\Sigma_l lm_l = N$, followed by a summation over all values of N (from $N = 0$ to $N = \infty$) is equivalent to an *unrestricted* summation over *all possible* sets $\{m_l\}$, we obtain

$$\begin{aligned} \mathcal{Q}(z, V, T) &= \sum_{m_1, m_2 \ldots = 0}^{\infty} \left[\prod_{l=1}^{\infty} \left\{ \left(b_l z^l \frac{V}{\lambda^3} \right)^{m_l} \frac{1}{m_l!} \right\} \right] \\ &= \prod_{l=1}^{\infty} \left[\sum_{m_l=0}^{\infty} \left\{ \left(b_l z^l \frac{V}{\lambda^3} \right)^{m_l} \frac{1}{m_l!} \right\} \right] \\ &= \prod_{l=1}^{\infty} \left[\exp \left(b_l z^l \frac{V}{\lambda^3} \right) \right], \end{aligned} \tag{32}$$

and, hence,

$$\frac{1}{V} \ln \mathcal{Q} = \frac{1}{\lambda^3} \sum_{l=1}^{\infty} b_l z^l. \tag{33}$$

In the limit $V \to \infty$, we obtain

$$\frac{P}{kT} \equiv \operatorname*{Lim}_{V \to \infty} \left(\frac{1}{V} \ln \mathcal{Q} \right) = \frac{1}{\lambda^3} \sum_{l=1}^{\infty} \bar{b}_l z^l, \tag{34}$$

and

$$\frac{N}{V} \equiv \operatorname*{Lim}_{V \to \infty} \left(\frac{z}{V} \frac{\partial \ln \mathcal{Q}}{\partial z} \right) = \frac{1}{\lambda^3} \sum_{l=1}^{\infty} l \bar{b}_l z^l, \tag{35}$$

which are the famous *cluster expansions* of the Mayer–Ursell formalism. Eliminating the fugacity parameter z, between eqns. (34) and (35), we obtain the *equation of state* of the system.

9.2. Virial expansion of the equation of state

The approach developed in the preceding section leads to rigorous results so long as we apply it to the gaseous phase alone. If we attempt to include in our study the phenomena of condensation, the critical point and the liquid phase, we encounter serious difficulties which are related to (i) the limiting procedure involved in eqns. (9.1.34) and (9.1.35), (ii) the convergence of the summations over l and (iii) the volume dependence of the cluster integrals. We, therefore, restrict our study to the gaseous phase alone. Then, the equation of state may be written in the form

$$\frac{Pv}{kT} = \sum_{l=1}^{\infty} a_l(T)\left(\frac{\lambda^3}{v}\right)^{l-1}, \tag{1}$$

where $v\,(= V/N)$ denotes the volume per particle in the system. The expansion (1), which is supposed to have been obtained by eliminating z between eqns. (9.1.34) and (9.1.35), is known as the *virial expansion* of the system while the numbers $a_l(T)$ are referred to as the *virial coefficients.** To determine the relationship between the coefficients a_l and the cluster integrals $\bar{b}_l$, we substitute from eqns. (9.1.34) and (9.1.35) into eqn. (1), with the result

$$\sum_{l=1}^{\infty} \bar{b}_l z^l \Bigg/ \sum_{l=1}^{\infty} l\bar{b}_l z^l = \sum_{l=1}^{\infty}\left[a_l(T)\left\{\sum_{n=1}^{\infty} n\bar{b}_n z^n\right\}^{l-1}\right]. \tag{2}$$

After cross-multiplication, we equate the coefficients of like powers of z on the two sides of the resulting equation; on rearrangement, we finally obtain

$$a_1 = \bar{b}_1 \equiv 1, \tag{3}$$

$$a_2 = -\bar{b}_2 = -\frac{2\pi}{\lambda^3}\int_0^{\infty} (e^{-u(r)/kT} - 1)\, r^2\, dr, \tag{4}$$

$$a_3 = 4\bar{b}_2^2 - 2\bar{b}_3 = -\frac{1}{3\lambda^6}\int_0^{\infty}\int_0^{\infty} f_{12}f_{13}f_{23}\, d^3r_{12}\, d^3r_{13}, \tag{5}$$

$$a_4 = -20\bar{b}_2^3 + 18\bar{b}_2\bar{b}_3 - 3\bar{b}_4 = \ldots, \tag{6}$$

and so on; here, use has also been made of the formulae (9.1.17–19). We note that the coefficient a_l is completely determined by the quantities $b_1, b_2, \ldots, b_l$, i.e. by the sequence of configuration integrals $Z_1, Z_2, \ldots, Z_l$; see also eqns. (9.4.5–8).

From eqn. (5) we observe that the third virial coefficient of the gas is determined solely by the 3-cluster 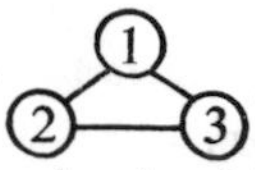

. This suggests that the higher-order virial coefficients ($l > 3$) may also be determined solely by a special "subgroup" of the various l-clusters. This is indeed true, and the relevant result is:†

$$a_l = -\frac{l-1}{l}\beta_{l-1}, \qquad (l \geq 2) \tag{7}$$

* For various manipulations with the virial equation of state, see Kilpatrick and Ford (1969).

† For proof, see Hill (1956), Secs. 24 and 25; also our own Sec. 9.4.

where β_{l-1} is the so-called *irreducible cluster integral*, defined by the formula

$$\beta_{l-1} = \frac{1}{(l-1)!\,\lambda^{3(l-1)}V} \times (\text{the sum of all irreducible } l\text{-clusters}); \tag{8}$$

by an *irreducible l*-cluster we mean an "l-particle graph which is multiply-connected (in the sense that there are at least two entirely independent, nonintersecting paths linking each pair of particles in the graph)". For instance, of the four possible 3-clusters, see (9.1.15), only the last one is irreducible. Indeed, if we write eqn. (5) in terms of this particular cluster and make use of the definition (8) for β_2, we obtain for the third virial coefficient

$$a_3 = -\tfrac{2}{3}\beta_2, \tag{9}$$

in agreement with the general result (7).*

The quantities β_{l-1}, like the quantities b_l, are dimensionless and, in the limit $V \to \infty$, approach finite values which are independent of the size and the shape of the container (unless the latter is unduly queer). Moreover, the two sets of quantities are mutually related. As will be shown in Sec. 9.4,

$$b_l = \frac{1}{l^2} \sum_{\{m_k\}}{}' \left[\prod_{k=1}^{l-1} \frac{(l\beta_k)^{m_k}}{m_k!} \right], \tag{10}$$

where the primed summation goes over all the sets $\{m_k\}$ that conform to the condition

$$\sum_{k=1}^{l-1} k m_k = l-1; \qquad m_k = 0, 1, 2, \ldots. \tag{11}$$

The formula (10) was first obtained by Mrs. Mayer in 1937. Its inverse, however, was established much later (Mayer, 1942; Kilpatrick, 1953):

$$\beta_{l-1} = \sum_{\{m_i\}}{}' (-1)^{\Sigma_i m_i - 1} \frac{(l-2+\Sigma_i m_i)!}{(l-1)!} \prod_i \frac{(ib_i)^{m_i}}{m_i!}, \tag{12}$$

where the primed summation goes over all the sets $\{m_i\}$ that conform to the condition

$$\sum_{i=2}^{l} (i-1)\, m_i = l-1; \qquad m_i = 0, 1, 2, \ldots. \tag{13}$$

It is not difficult to see that the highest value of the index i in the set $\{m_i\}$ would be l (the corresponding set having all its m's equal to 0, except m_l which would be equal to 1); accordingly, the highest order to which the quantities b_i would appear in the expression for β_{l-1} is that of b_l. We thus see once again that the virial coefficient a_l is completely determined by the quantities $b_1, b_2, \ldots, b_l$.

* It may be mentioned here that a 2-cluster is also regarded as an *irreducible* cluster. Accordingly, we have: $\beta_1 = 2b_2$; see eqns. (9.1.16) and (9.2.8). Equation (4) then gives: $a_2 = -b_2 = -\frac{1}{2}\beta_1$, which again agrees with the general result (7).

9.3. Evaluation of the virial coefficients

If a given physical system does not show great departures from the ideal-gas behavior, the equation of state of the system is given adequately by the first few virial coefficients. Now, since $a_1 \equiv 1$, the lowest-order virial coefficient that we need to consider here is a_2, which is given by the formula (9.2.4):

$$a_2 = -\bar{b}_2 = \frac{2\pi}{\lambda^3} \int_0^\infty (1 - e^{-u(r)/kT})\, r^2\, dr, \tag{1}$$

$u(r)$ being the potential energy of interparticle interaction. A typical plot of the function $u(r)$ was shown in Fig. 9.1; a typical *semi-empirical* formula is given by (Lennard-Jones, 1924)

$$u(r) = 4\varepsilon\left[\left(\frac{\sigma}{r}\right)^{12} - \left(\frac{\sigma}{r}\right)^{6}\right]. \tag{2}$$

It will be noted that the most significant features of the actual interparticle potential are well simulated by the Lennard-Jones formula. For instance, the function $u(r)$ given by this formula exhibits a "minimum", of value $-\varepsilon$, at a distance $r_0 (= 2^{1/6}\sigma)$, and rises to an infinitely large (positive) value for $r < \sigma$ and to a vanishingly small (negative) value for $r \gg \sigma$. The portion to the left of the "minimum" is dominated by the *repulsive* interaction that comes into play when two particles come too close to one another, while the portion to the right of the "minimum" is dominated by the *attractive* interaction that operates between particles when they are separated by a respectable distance. For most practical purposes, the precise form of the repulsive part of the potential is not very important; it may as well be replaced by the crude approximation

$$u(r) = +\infty \qquad (\text{for } r < r_0), \tag{3}$$

which amounts to attributing an *impenetrable* core, of diameter r_0, to each particle. The precise form of the attractive part of the potential is, however, generally significant; in view of the fact that there exists good theoretical basis for the sixth-power attractive potential (see Problem 3.32), this part may be simply expressed as

$$u(r) = -u_0(r_0/r)^6 \qquad (\text{for } r \geq r_0). \tag{4}$$

The potential given by the expressions (3) and (4) may, therefore, be used if one is only interested in a qualitative assessment of the situation and not in a quantitative comparison between the theory and the experiment.

Substituting (3) and (4) into (1) we obtain for the second virial coefficient

$$a_2 = \frac{2\pi}{\lambda^3}\left[\int_0^{r_0} r^2\, dr + \int_{r_0}^{\infty} \left[1 - \exp\left\{\frac{u_0}{kT}\left(\frac{r_0}{r}\right)^6\right\}\right] r^2\, dr\right]. \tag{5}$$

The first integral is quite straightforward; the second one is considerably simplified if we assume that

$$(u_0/kT) \ll 1, \tag{6}$$

which makes the integrand very nearly equal to $-(u_0/kT)(r_0/r)^6$. Equation (5) then gives

$$a_2 \simeq \frac{2\pi r_0^3}{3\lambda^3}\left(1-\frac{u_0}{kT}\right). \tag{7}$$

Substituting this expression for a_2 into the expansion (9.2.1), we obtain as a *first-order* improvement on the ideal-gas law

$$P \simeq \frac{kT}{v}\left\{1+\frac{2\pi r_0^3}{3v}\left(1-\frac{u_0}{kT}\right)\right\} \tag{8}$$

$$= \frac{kT}{v}\left\{1+\frac{B_2(T)}{v}\right\}, \text{ say.} \tag{9}$$

The coefficient B_2, which is also sometimes referred to as the second virial coefficient of the system, is given by

$$B_2 \equiv a_2\lambda^3 \simeq \frac{2\pi r_0^3}{3}\left(1-\frac{u_0}{kT}\right). \tag{10}$$

In our derivation it was explicitly assumed that (i) the potential function $u(r)$ is given by the simplified expressions (3) and (4), and (ii) $(u_0/kT) \ll 1$. We cannot, therefore, expect formula (10) to be a faithful representation of the second virial coefficient of a real gas. Nevertheless, it does correspond, *almost exactly*, to the *van der Waals approximation* to the equation of state of a real gas. This can be seen by rewriting eqn. (8) in the form

$$\left(P+\frac{2\pi r_0^3 u_0}{3v^2}\right) \simeq \frac{kT}{v}\left(1+\frac{2\pi r_0^3}{3v}\right) \simeq \frac{kT}{v}\left(1-\frac{2\pi r_0^3}{3v}\right)^{-1},$$

which readily leads to the van der Waals equation of state

$$\left(P+\frac{a}{v^2}\right)(v-b) \simeq kT, \tag{11}$$

where

$$a = \frac{2\pi r_0^3 u_0}{3} \quad \text{and} \quad b = \frac{2\pi r_0^3}{3} \equiv 4v_0. \tag{12}$$

We note that the parameter b in the van der Waals equation of state is four times the actual *molecular volume* v_0, the latter being the "volume of a sphere of diameter r_0"; cf. Problem 1.4. We also note that in this derivation we have assumed that $b \ll v$, which means that the gas is sufficiently dilute for the mean interparticle distance to be much larger than the effective range of the interparticle interaction. Finally, we observe that, according to this simple-minded calculation, the van der Waals constants a and b are temperature-independent, which in reality is not true.

A realistic study of the second virial coefficient requires the use of a realistic potential, such as the one given by Lennard-Jones, for evaluating the integral appearing in (1). This

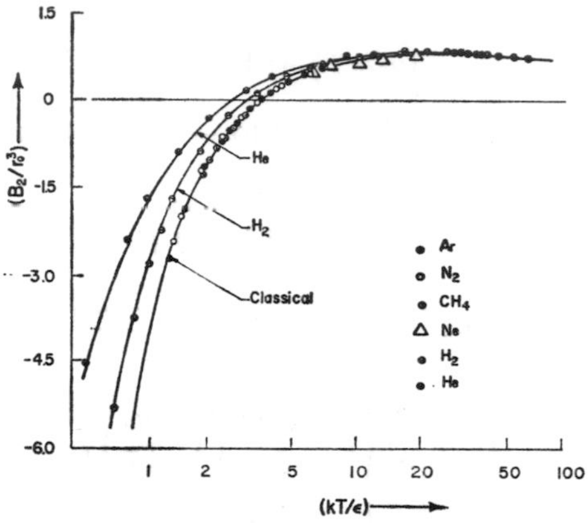

FIG. 9.2. A dimensionless plot showing the temperature dependence of the second virial coefficient of several gases (after Hirschfelder *et al.*, 1954).

has indeed been done and the results obtained are shown in Fig. 9.2, where the reduced coefficient B_2' ($= B_2/r_0^3$) has been plotted against the reduced temperature T' ($= kT/\varepsilon$):

$$B_2'(T') = 2\pi \int_0^\infty (1 - e^{-u'(r')/T'})\, r'^2\, dr', \tag{13}$$

with

$$u'(r') = \left\{\left(\frac{1}{r'}\right)^{12} - 2\left(\frac{1}{r'}\right)^{6}\right\}, \tag{14}$$

r' being equal to (r/r_0). Expressed in this form, the quantity B_2' is a *universal* function of the quantity T'. Included in the plot are the experimental results for several gases. We note that for most gases the agreement is reasonably good; this is especially satisfying in view of the fact that in each case we had only two adjustable parameters, r_0 and ε, against a much larger number of experimental points available. In the first place, this agreement vindicates the adequacy of the Lennard-Jones potential for providing an analytical description of a typical interparticle potential. In the second place, it enables one to derive empirical values of the respective parameters of the potential; for instance, one obtains for argon: $r_0 = 3.82$ Å and $\varepsilon/k = 120$°K.* We cannot fail to observe that the lighter gases, hydrogen and helium, constitute glaring exceptions to the general rule of agreement between the theory and the experiment. The reason for this discrepancy lies in the fact that in the case of these gases the quantum-mechanical effects assume a lot of importance—more so at low temperatures. To substantiate this point, we have included in Fig. 9.2 theoretical curves for H_2 and He after taking into account the quantum-mechanical effects; as a result, we find once again a fairly good agreement between the theory and the experiment.

We shall now make a few remarks on the qualitative nature of the function $B_2(T)$. When the temperature is sufficiently low, the energy of the system is mainly potential. In search of a state of *minimum* energy, the system is then most likely to be in a configuration which keeps the interparticle distances r_{ij} in the neighborhood of the equilibrium value r_0. The most

* Values for various other gases have been summarized in Hill (1960), p. 484.

operative part of the potential then is the "attractive part", which naturally reduces the pressure of the gas and thereby makes the coefficient B_2 negative. At high temperatures the particles possess a lot of kinetic energy, are scarcely affected by the "attractive part" of the potential and, by and large, feel the presence of the "repulsive part" alone. This increases the pressure of the gas and in turn makes B_2 positive. As the temperature rises, the coefficient B_2, according to the simple-minded formula (10), should approach a limiting value of $2\pi r_0^3/3$ which arises solely from the *impenetrable* core of diameter r_0. In an actual case, however, the repulsive core, though hard, is not literally impenetrable; in fact, due to the large kinetic energy (at high temperatures), the particles can come comparatively closer to one another, which implies a core of a comparatively smaller diameter. This results in a relative lowering of the value of B_2, though it continues to remain positive in sign; accordingly, our plots show a rather broad maximum in the neighborhood of $T' = 20$.

As regards the higher-order virial coefficients ($l > 2$), we shall confine our discussion to a gas of hard spheres alone. We then have

$$\left.\begin{aligned} u(r) &= +\infty \quad \text{for} \quad r < r_0 \\ &= \ \ 0 \quad \text{for} \quad r > r_0 \end{aligned}\right\} \tag{15}$$

and, hence,

$$\left.\begin{aligned} f(r) &= -1 \quad \text{for} \quad r < r_0 \\ &= \ \ 0 \quad \text{for} \quad r > r_0. \end{aligned}\right\} \tag{16}$$

The second virial coefficient of the gas is given exactly by the expression

$$a_2 = \frac{2\pi r_0^3}{3\lambda^3} = 4\frac{v_0}{\lambda^3}. \tag{17}$$

The third virial coefficient can be determined with the help of eqn. (9.2.5), viz.

$$a_3 = -\frac{1}{3\lambda^6}\int_0^\infty\int_0^\infty f_{12}f_{13}f_{23}\, d^3r_{12}\, d^3r_{13}. \tag{18}$$

To evaluate this integral, we first fix the positions of the particles 1 and 2 (such that $r_{12} < r_0$) and let the particle 3 take all possible positions so that we can effect an integration over the

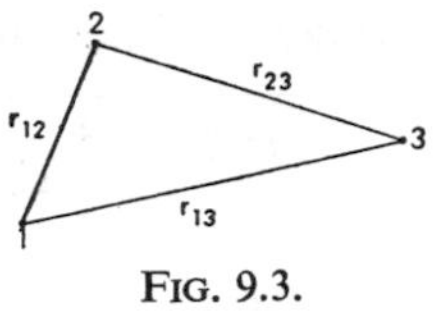

FIG. 9.3.

variable r_{13}; see Fig. 9.3. Since our integrand is equal to -1 if each of the distances r_{13} and r_{23} (like the distance r_{12}) is less than r_0 and is equal to 0 otherwise, we can write

$$a_3 = \frac{1}{3\lambda^6}\int_{r_{12}=0}^{r_0}\left\{\int' d^3r_{13}\right\} d^3r_{12}, \tag{19}$$

where the primed integration arises from the particle 3 taking all possible positions of interest. In view of the conditions $r_{13} < r_0$ and $r_{23} < r_0$, this integral is precisely equal to the "volume common to the spheres S_1 and S_2, each of radius r_0, centered at the *fixed* points 1 and 2"; see Fig. 9.4. This in turn can be computed by calculating the "volume swept by

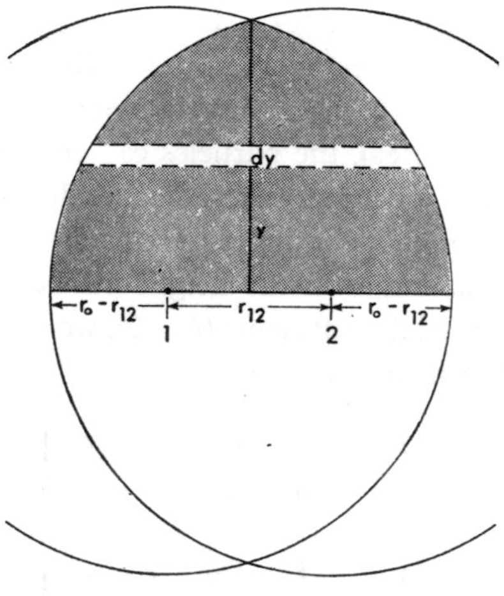

FIG. 9.4.

the shaded area in the figure on going through a complete revolution about the line of centres". One obtains:

$$\int' d^3r_{13} = \int_0^{\sqrt{[r_0^2-(r_{12}/2)^2]}} \{2(r_0^2-y^2)^{1/2}-r_{12}\}2\pi y\, dy.$$

While the quantity within the curly brackets stands for the length of the strip, which is marked out in the figure, the element of area $2\pi y\, dy$ arises from the revolution given; the limits of integration for the variable y can be checked quite easily. The evaluation of this integral is straightforward; we obtain

$$\int' d^3r_{13} = \frac{4\pi}{3}\left\{r_0^3-\frac{3r_0^2r_{12}}{4}+\frac{r_{12}^3}{16}\right\}.$$

Substituting this result in (19) and carrying out integration over r_{12}, we finally obtain

$$a_3 = \frac{5\pi^2 r_0^6}{18\lambda^6} = \frac{5}{8}a_2^2. \tag{20}$$

The fourth virial coefficient of the hard-sphere gas has also been evaluated exactly. It is given by (Boltzmann, 1899; Majumdar, 1929)*

$$\begin{aligned} a_4 &= \left\{\frac{1283}{8960}+\frac{3}{2}\cdot\frac{73\sqrt{(2)}+1377(\tan^{-1}\sqrt{(2)}-\pi/4)}{1120\pi}\right\}a_2^3 \\ &= 0.28695a_2^3. \end{aligned} \tag{21}$$

* See also Katsura (1959).

The fifth and the sixth virial coefficients of this system have been computed numerically, using the Monte Carlo method for the evaluation of the integrals involved. The results so obtained are (Ree and Hoover, 1964)

$$a_5 = (0.1103 \pm 0.003)a_2^4 \tag{22}$$

and

$$a_6 = (0.0386 \pm 0.004)a_2^5. \tag{23}$$

Ree and Hoover's estimate of the seventh virial coefficient is $0.0127a_2^6$, but the error in the evaluation of this coefficient has not been specified.

9.4. General remarks on cluster expansions

Shortly after the pioneering work of Mayer, Kahn and Uhlenbeck (1938) initiated the development of a similar treatment for quantum-mechanical systems. Of course, their treatment applies to the limiting case of classical systems as well. However, it faces certain inherent difficulties of analysis. Some of these difficulties have now been removed by the formal methods developed by Lee and Yang (1959 a, b; 1960). We propose to discuss these developments in the remaining sections of this chapter. However, before doing that we would like to make a few general observations on the problem of cluster expansions. These observations, due primarily to Ono (1951) and Kilpatrick (1953), are of considerable interest insofar as they hold for a very large class of physical systems. For instance, the system may be quantum-mechanical or classical, it may be a multi-component one or single-component, its molecules may be polyatomic or monatomic, its potential energy may or may not be given by a sum of the two-particle interactions, etc. All we have to assume is that (i) the system is gaseous in state and (ii) the partition functions $Q_N(V, T)$, *for some low values of* N, can somehow be obtained. We can then calculate the "cluster integrals" b_l, and the virial coefficients a_l, of the system in the following straightforward manner.

Quite generally, the *grand partition function* of the system can be written as

$$\mathcal{Q}(z, V, T) \equiv \sum_{N=0}^{\infty} Q_N(V, T)z^N = \sum_{N=0}^{\infty} \frac{Z_N(V, T)}{N!}\left(\frac{z}{\lambda^3}\right)^N, \tag{1}$$

where we have introduced the "configuration integral" $Z_N(V, T)$, defined in analogy with eqn. (9.1.3) of the classical treatment:

$$Z_N(V, T) \equiv N!\,\lambda^{3N} Q_N(V, T). \tag{2}$$

Dimensionally, the quantity Z_N is like (a volume)^N; moreover, the quantity Z_0 (like Q_0) is supposed to be identically equal to 1, while the quantity Z_1 ($\equiv \lambda^3 Q_1$) is identically equal to V. We then have, in the limit $V \to \infty$,

$$\frac{P}{kT} \equiv \frac{1}{V}\ln \mathcal{Q} = \frac{1}{V}\ln\left\{1 + \frac{Z_1}{1!}\left(\frac{z}{\lambda^3}\right)^1 + \frac{Z_2}{2!}\left(\frac{z}{\lambda^3}\right)^2 + \dots\right\} \tag{3}$$

$$= \frac{1}{\lambda^3}\sum_{l=1}^{\infty} \mathfrak{b}_l z^l, \quad \text{say.} \tag{4}$$

Again, the last expression has been written in analogy with the classical cluster expansion (9.1.34); the coefficients $\mathfrak{b}_l$ may, therefore, be looked upon as the *cluster integrals* of the given system. Expanding (3) as a power series in z and equating the respective coefficients with the $\mathfrak{b}_l$'s of (4), we obtain

$$\mathfrak{b}_1 = \frac{1}{V} Z_1 \equiv 1, \tag{5}$$

$$\mathfrak{b}_2 = \frac{1}{2!\,\lambda^3 V}(Z_2 - Z_1^2), \tag{6}$$

$$\mathfrak{b}_3 = \frac{1}{3!\,\lambda^6 V}(Z_3 - 3Z_2Z_1 + 2Z_1^3), \tag{7}$$

$$\mathfrak{b}_4 = \frac{1}{4!\,\lambda^9 V}(Z_4 - 4Z_3Z_1 - 3Z_2^2 + 12Z_2Z_1^2 - 6Z_1^4), \tag{8}$$

and so on. We note that, for all $l > 1$, the sum of the numerical coefficients appearing within the parentheses is identically equal to zero. Consequently, in the case of an ideal classical gas, for which $Z_i \equiv V^i$, see eqn. (9.1.4), all cluster integrals with $l > 1$ identically vanish. This in turn implies the vanishing of all the virial coefficients of the gas (except, of course, a_1 which is identically equal to unity).

Comparing eqns. (6)–(8) with eqn. (9.1.16), which defines the classical cluster integrals b_l, we find that in the general case the expressions involving the products of the various Z_i's and appearing within the parentheses play the same role here as "the sum of all possible l-clusters" in the classical case. We, therefore, expect that the $\mathfrak{b}_l$'s here would also be (i) dimensionless and (ii) independent of the size and the shape of the container (unless the latter is unduly queer). This in turn requires that, in the limit $V \to \infty$, the various combinations of the Z_i's appearing within the parentheses must invariably be proportional to the *first* power of V. This observation leads to the very interesting result, first noticed by Rushbrooke, namely

$$\mathfrak{b}_l = \frac{1}{l!\,\lambda^{3(l-1)}} \times (\text{the coefficient of } V^1 \text{ in the volume expansion of } Z_l). \tag{9}$$

At this stage, it seems worth while to point out that the expressions appearing within the parentheses of eqns. (6)–(8) are well known in mathematical statistics as the *semi-invariants* of Thiele. The general formula for these expressions is

$$\begin{aligned}(\ldots)_l &\equiv \mathfrak{b}_l\{l!\,\lambda^{3(l-1)}V\} \\ &= l! \sum_{\{m_i\}}{}' (-1)^{\sum_i m_i - 1}\left[\left(\sum_i m_i - 1\right)!\,\prod_i \left\{\frac{(Z_i/i!)^{m_i}}{m_i!}\right\}\right],\end{aligned} \tag{10}$$

where the primed summation goes over all the sets $\{m_i\}$ that conform to the condition

$$\sum_{i=1}^{l} i m_i = l; \qquad m_i = 0, 1, 2, \ldots. \tag{11}$$

The general expression for the inverse relationships can be written down by referring to eqn. (9.1.29) of the classical treatment; we thus have

$$Z_M \equiv M!\,\lambda^{3M} Q_M = M!\,\lambda^{3M} \sum_{\{m_l\}}' \prod_{l=1}^{M} \left\{ \frac{(V\bar{b}_l/\lambda^3)^{m_l}}{m_l!} \right\}, \tag{12}$$

where the primed summation goes over all the sets $\{m_l\}$ that conform to the condition

$$\sum_{l=1}^{M} l m_l = M; \qquad m_l = 0, 1, 2, \ldots. \tag{13}$$

The calculation of the virial coefficients a_l now consists of a straightforward mathematical step which involves a use of the formulae (5)–(8) in conjunction with the formulae (9.2.3–6). It appears, however, of interest to demonstrate here the manner in which the general relationship (9.2.7) between the virial coefficients a_l and the "irreducible cluster integrals" β_{l-1} arises mathematically. An added advantage of this demonstration is that we will acquire yet another interpretation of the β's.

Now, in view of the basic formulae

$$\frac{P}{kT} \equiv \lim_{V\to\infty} \left(\frac{1}{V} \ln \mathcal{Q} \right) = \frac{1}{\lambda^3} \sum_{l=1}^{\infty} \bar{b}_l z^l \tag{14}$$

and

$$\frac{1}{v} \equiv \lim_{V\to\infty} \left(\frac{z}{V} \frac{\partial \ln \mathcal{Q}}{\partial z} \right) = \frac{1}{\lambda^3} \sum_{l=1}^{\infty} l\bar{b}_l z^l, \tag{15}$$

we can write

$$\frac{P(z)}{kT} = \int_0^z \frac{1}{v(z)} \frac{dz}{z}. \tag{16}$$

Equation (15) tells us that, in the limit $z \to 0$, the integrand $1/\{zv(z)\}$ of (16) tends to the value $1/\lambda^3$; consequently, the quantity $P(z)/kT$ would tend to the value z/λ^3, which is in complete agreement with the expansion (14) for this quantity. This also means that, in the limit $z \to 0$, the quantity P/kT approaches $1/v$.

Let us introduce a new variable x, defined as

$$x = n\lambda^3 = \lambda^3/v. \tag{17}$$

In terms of this variable, eqn. (15) becomes

$$x(z) = \sum_{l=1}^{\infty} l\bar{b}_l z^l, \tag{18}$$

the inverse of which may be written as (see Mayer and Harrison, 1938; also Kahn, 1938)

$$z(x) = x \exp\{-\varphi(x)\}. \tag{19}$$

Now, in view of the fact that, for $z \ll 1$, the variables z and x are practically the same, the function $\varphi(x)$ must tend to zero as $x \to 0$; it may, therefore, be expressed as a power series in x:

$$\varphi(x) = \sum_{k=1}^{\infty} \beta_k x^k. \tag{20}$$

It may be mentioned beforehand that the coefficients β_k of the foregoing expansion are ultimately going to be identified as the "irreducible cluster integrals" β_{l-1}. Substituting from eqns. (17), (19) and (20) into (16), we get

$$\frac{P(x)}{kT} = \int_0^x \frac{x}{\lambda^3}\left\{\frac{1}{x} - \varphi'(x)\right\} dx = \frac{1}{\lambda^3}\left[x - \int_0^x \left\{\sum_{k=1}^{\infty} k\beta_k x^k\right\} dx\right]$$

$$= \frac{x}{\lambda^3}\left[1 - \sum_{k=1}^{\infty}\left(\frac{k}{k+1}\beta_k x^k\right)\right]. \tag{21}$$

Combining (17) and (21), we obtain

$$\frac{Pv}{kT} = 1 - \sum_{k=1}^{\infty}\left(\frac{k}{k+1}\beta_k x^k\right). \tag{22}$$

Comparing this result with the virial expansion (9.2.1), we arrive at the desired relationship:

$$a_l = -\frac{l-1}{l}\beta_{l-1} \qquad (l > 1). \tag{23}$$

It seems that, in view of the generality of our approach, the β's occurring here may be regarded as a *generalization* of the irreducible cluster integrals of Mayer.

Finally, we would like to derive a relationship between the β's and the $\bar{b}$'s. For this, we make use of an important theorem due to Lagrange which, for the present purpose, states that "the solution $x(z)$ of the equation

$$z(x) = x/f(x) \tag{24}$$

is given by

$$x(z) = \sum_{j=1}^{\infty}\frac{z^j}{j!}\left[\frac{d^{j-1}}{d\xi^{j-1}}\{f(\xi)\}^j\right]_{\xi=0}\text{"}; \tag{25}$$

it is obvious that the expression within the square brackets is $(j-1)!$ times "the coefficient of ξ^{j-1} in the Taylor expansion of the function $\{f(\xi)\}^j$ around the point $\xi = 0$". Applying this theorem to the function

$$f(x) = \exp\{\varphi(x)\} = \exp\left\{\sum_{k=1}^{\infty}\beta_k x^k\right\} = \prod_{k=1}^{\infty}\exp(\beta_k x^k), \tag{26}$$

we obtain

$$x(z) = \sum_{j=1}^{\infty}\frac{z^j}{j!}(j-1)!\times\left\{\text{the coefficient of } \xi^{j-1} \text{ in the Taylor expansion of } \prod_{k=1}^{\infty}\exp(j\beta_k\xi^k) \text{ around } \xi = 0\right\}.$$

Comparing this with eqn. (18), we obtain

$$\bar{b}_j = \frac{1}{j^2}\times\left\{\text{the coefficient of } \xi^{j-1} \text{ in } \prod_{k=1}^{\infty}\left[\sum_{m_k \geq 0}\frac{(j\beta_k)^{m_k}}{m_k!}\xi^{km_k}\right]\right\}$$

$$= \frac{1}{j^2}\sum_{\{m_k\}}{}' \prod_{k=1}^{j-1}\frac{(j\beta_k)^{m_k}}{m_k!}, \tag{27}$$

where the primed summation goes over all the sets $\{m_k\}$ that conform to the condition

$$\sum_{k=1}^{j-1} km_k = j-1; \qquad m_k = 0, 1, 2, \ldots. \tag{28}$$

We note that the relationship (27) is identical with our earlier version, viz. (9.2.10). The inverse relationship, which expresses an irreducible cluster integral in terms of the reducible ones, has also been quoted earlier; see eqn. (9.2.12).

9.5. Exact treatment of the second virial coefficient

We now present a formulation, due originally to Beth and Uhlenbeck (1936, 1937), that enables us to carry out an exact calculation of the second virial coefficient of a quantum-mechanical system from a knowledge of the two-body interaction potential $u(r)$.* We have, in view of (9.4.6),

$$\mathfrak{b}_2 = -a_2 = \frac{1}{2\lambda^3 V}(Z_2 - Z_1^2). \tag{1}$$

For the corresponding *noninteracting* system, one would have

$$\mathfrak{b}_2^{(0)} = -a_2^{(0)} = \frac{1}{2\lambda^3 V}(Z_2^{(0)} - Z_1^{(0)2}); \tag{2}$$

the superscript (0) on the various symbols implies that they pertain to the noninteracting system. Taking the difference of (1) and (2), and remembering that $Z_1 \equiv V \equiv Z_1^{(0)}$, we obtain

$$\mathfrak{b}_2 - \mathfrak{b}_2^{(0)} = \frac{1}{2\lambda^3 V}(Z_2 - Z_2^{(0)}), \tag{3}$$

which, by virtue of the relation (9.4.2), becomes

$$\mathfrak{b}_2 - \mathfrak{b}_2^{(0)} = \frac{\lambda^3}{V}(Q_2 - Q_2^{(0)}) \equiv \frac{\lambda^3}{V}\,\mathrm{Tr}\,(e^{-\beta \hat{H}_2} - e^{-\beta \hat{H}_2^{(0)}}). \tag{4}$$

For evaluating the trace we should know the eigenvalues of the two-body Hamiltonian, which in turn requires the solving of the Schrödinger equation†

$$\hat{H}_2 \Psi_\alpha(\boldsymbol{r}_1, \boldsymbol{r}_2) = E_\alpha \Psi_\alpha(\boldsymbol{r}_1, \boldsymbol{r}_2), \tag{5}$$

where

$$\hat{H}_2 = -\frac{\hbar^2}{2m}(\nabla_1^2 + \nabla_2^2) + u(r_{12}). \tag{6}$$

Transforming to the *centre-of-mass* coordinates $\boldsymbol{R}\left\{= \frac{1}{2}(\boldsymbol{r}_1 + \boldsymbol{r}_2)\right\}$ and the *relative* coordinates $\boldsymbol{r}\,\{= (\boldsymbol{r}_2 - \boldsymbol{r}_1)\}$, we have

$$\Psi_\alpha(\boldsymbol{R}, \boldsymbol{r}) = \psi_j(\boldsymbol{R})\,\psi_n(\boldsymbol{r}) = \left\{\frac{1}{V^{1/2}}\,e^{i(\boldsymbol{P}_j \cdot \boldsymbol{R})/\hbar}\right\}\psi_n(\boldsymbol{r}), \tag{7}$$

* For a discussion of the third virial coefficient, see Pais and Uhlenbeck (1959).
† For simplicity, we assume the particles to be "spinless". For the influence of the spin, see Problems 9.13–9.16.

with

$$E_\alpha = \frac{P_j^2}{2(2m)} + \varepsilon_n. \tag{8}$$

Here, $\boldsymbol{P}$ denotes the total momentum of the two particles and $2m$ their total mass, while ε denotes the energy associated with the relative motion of the particles; the symbol α refers to the set of quantum numbers j and n that determine the actual values of the variables $\boldsymbol{P}$ and ε. The wave equation for the relative motion will be

$$\left\{-\frac{\hbar^2}{2(\frac{1}{2}m)}\nabla_r^2 + u(r)\right\}\psi_n(\boldsymbol{r}) = \varepsilon_n \psi_n(\boldsymbol{r}), \tag{9}$$

$\frac{1}{2}m$ being the reduced mass of the particles; the normalization condition for the relative wave function will be

$$\int |\psi_n(\boldsymbol{r})|^2\, d^3r = 1. \tag{10}$$

Equation (4) thus becomes

$$\begin{aligned} b_2 - b_2^{(0)} &= \frac{\lambda^3}{V}\sum_\alpha \left(e^{-\beta E_\alpha} - e^{-\beta E_\alpha^{(0)}}\right) \\ &= \frac{\lambda^3}{V}\sum_j e^{-\beta P_j^2/4m} \sum_n \left\{e^{-\beta\varepsilon_n} - e^{-\beta\varepsilon_n^{(0)}}\right\}. \end{aligned} \tag{11}$$

For the first sum we obtain

$$\sum_j e^{-\beta P_j^2/4m} \simeq \frac{4\pi V}{h^3}\int_0^\infty e^{-(\beta P^2/4m)} P^2\, dP = \frac{8^{1/2} V}{\lambda^3}, \tag{12}$$

whence eqn. (11) becomes

$$b_2 - b_2^{(0)} = 8^{1/2}\sum_n \left\{e^{-\beta\varepsilon_n} - e^{-\beta\varepsilon_n^{(0)}}\right\}. \tag{13}$$

The next step consists in examining the energy spectra, ε_n and $\varepsilon_n^{(0)}$, of the two systems. In the case of noninteracting system, all we have is a "continuum"

$$\varepsilon_n^{(0)} = \frac{p^2}{2(\frac{1}{2}m)} = \frac{\hbar^2 k^2}{m} \qquad (k = p/\hbar), \tag{14}$$

with the standard density of states $g^{(0)}(k)$. In the case of interacting system, we may have a set of *discrete eigenvalues* ε_B (that correspond to the two-body "bound" states), along with a "continuum"

$$\varepsilon_n = \frac{\hbar^2 k^2}{m} \qquad (k = p/\hbar), \tag{15}$$

with a characteristic density of states $g(k)$. Consequently, eqn. (13) can be written as

$$b_2 - b_2^{(0)} = 8^{1/2}\sum_B e^{-\beta\varepsilon_B} + 8^{1/2}\int_0^\infty e^{-\beta\hbar^2k^2/m}\left\{g(k) - g^{(0)}(k)\right\} dk, \tag{16}$$

where the summation in the first part goes over all possible bound states of the two-body system with interactions.

The next thing to consider is the density of states $g(k)$. For this, we note that, since the two-body potential has been assumed to be central, the wave function $\psi_n(\boldsymbol{r})$ for the relative motion can be written as a product of a radial function $\chi(r)$ and a spherical harmonic $Y(\theta, \varphi)$:

$$\psi_{klm}(\boldsymbol{r}) = A_{klm} \frac{\chi_{kl}(r)}{r} Y_{l,m}(\theta, \varphi). \tag{17}$$

Moreover, the requirement of symmetry, namely $\psi(-\boldsymbol{r}) = \psi(\boldsymbol{r})$ for bosons and $\psi(-\boldsymbol{r}) = -\psi(\boldsymbol{r})$ for fermions, lays down the restriction that the quantum number l must be even for bosons and odd for fermions. The (external) boundary condition on the wave function may be written as

$$\chi_{kl}(R_0) = 0, \tag{18}$$

where R_0 is a fairly large value (of the variable r) that ultimately tends to infinity. The asymptotic form of the function $\chi_{kl}(r)$ is well known:

$$\chi_{kl}(r) \propto \sin\left\{kr - \frac{l\pi}{2} + \eta_l(k)\right\}; \tag{19}$$

accordingly, we must have

$$kR_0 - \frac{l\pi}{2} + \eta_l(k) = n\pi, \qquad n = 0, 1, 2, \ldots. \tag{20}$$

The symbol $\eta_l(k)$ here stands for the *scattering phase shift*, due to the two-body potential $u(r)$, for the lth partial wave of wave number k.

Equation (20) determines the complete spectrum of the *partial waves*. To obtain an expression for the respective density of state $g_l(k)$, we observe that the wave number difference Δk between the consecutive states n and $(n+1)$ is given by the formula

$$\left\{R_0 + \frac{d\eta_l(k)}{dk}\right\} \Delta k = \pi, \tag{21}$$

whence it follows that

$$g_l(k) = \frac{2l+1}{\Delta k} = \frac{2l+1}{\pi}\left\{R_0 + \frac{\partial \eta_l(k)}{\partial k}\right\}; \tag{22}$$

the factor $(2l+1)$ has been included here to account for the fact that each eigenvalue k pertaining to an lth partial wave is $(2l+1)$-fold degenerate (because the magnetic quantum number m can take any of the values l, $(l-1)$, $\ldots$, $-l$, without affecting the eigenvalues). The total density of state $g(k)$, of all *relevant* partial waves of wave numbers around the value k, is then given by

$$g(k) = \sum_l{}' g_l(k) = \frac{1}{\pi} \sum_l{}' (2l+1)\left\{R_0 + \frac{\partial \eta_l(k)}{\partial k}\right\}. \tag{23}$$

Note that the primed summation Σ' goes over $l = 0, 2, 4, \ldots$ in the case of bosons and over $l = 1, 3, 5, \ldots$ in the case of fermions.

For the corresponding noninteracting case, we have (since all $\eta_l(k) = 0$)

$$g^{(0)}(k) = \frac{R_0}{\pi} \sum_l{}' (2l+1). \tag{24}$$

Combining (23) and (24), we obtain the desired formula:

$$g(k) - g^{(0)}(k) = \frac{1}{\pi} \sum_l{}' (2l+1) \frac{\partial \eta_l(k)}{\partial k}. \tag{25}$$

Substituting (25) into (16), we obtain the final result

$$b_2 - b_2^{(0)} = 8^{1/2} \sum_B e^{-\beta \varepsilon_B} + \frac{8^{1/2}}{\pi} \sum_l{}' (2l+1) \int_0^\infty e^{-\beta \hbar^2 k^2/m} \frac{\partial \eta_l(k)}{\partial k} dk, \tag{26}$$

which, in principle, is calculable for any given potential $u(r)$ through the respective phase shifts $\eta_l(k)$.

Equation (26) can be used for determining the quantity $b_2 - b_2^{(0)}$. In order to determine b_2 by itself, we must know the value of $b_2^{(0)}$. This has already been obtained in Chapter 7 for bosons and in Chapter 8 for fermions; see eqns. (7.1.14) and (8.1.18). Thus

$$b_2^{(0)} = -a_2^{(0)} = \pm \frac{1}{2^{5/2}}, \tag{27}$$

where the upper sign stands for bosons and the lower sign for fermions. It may be noted that the foregoing result can be obtained directly from the relationship

$$b_2^{(0)} = \frac{1}{2\lambda^3 V} (Z_2^{(0)} - Z_1^{(0)2}) \equiv \frac{\lambda^3}{V} \left(Q_2^{(0)} - \frac{1}{2} Q_1^{(0)2} \right)$$

by substituting for $Q_2^{(0)}$ the exact expression (5.5.25):

$$b_2^{(0)} = \frac{\lambda^3}{V} \left[\left\{ \frac{1}{2} \left(\frac{V}{\lambda^3} \right)^2 \pm \frac{1}{2^{5/2}} \left(\frac{V}{\lambda^3} \right)^1 \right\} - \frac{1}{2} \left(\frac{V}{\lambda^3} \right)^2 \right] = \pm \frac{1}{2^{5/2}}. \tag{28*}$$

It is of interest to note that this result can also be obtained by using the classical formula (9.1.18) and substituting for the two-body potential $u(r)$ the "statistical potential" (5.5.28):

$$b_2^{(0)} = \frac{2\pi}{\lambda^3} \int_0^\infty (e^{-u_s(r)/kT} - 1)\, r^2\, dr$$

$$= \pm \frac{2\pi}{\lambda^3} \int_0^\infty e^{-2\pi r^2/\lambda^2} r^2\, dr = \pm \frac{1}{2^{5/2}}. \tag{29}$$

* This calculation incidentally verifies the general formula (9.4.9) for the case $l = 2$. By that formula, the "cluster integral" b_2 of a given system would be equal to $1/(2\lambda^3)$ times the coefficient of V^1 in the volume expansion of the "configuration integral" Z_2 of the system. In the case under study, this coefficient is equal to $\pm \lambda^3/2^{3/2}$; hence the result.

As an illustration, we shall calculate here the second virial coefficient of a gas of hard spheres. The two-body potential in this case can be written as

$$\left.\begin{aligned} u(r) &= +\infty \quad \text{for} \quad r < r_0 \\ &= 0 \quad \text{for} \quad r > r_0. \end{aligned}\right\} \tag{30}$$

The scattering phase shifts $\eta_l(k)$ can be readily determined by making use of the (internal) boundary condition, namely that the radial function $\chi(r)$ for the relative motion must vanish at $r = r_0$. We thus obtain (see, for example, Schiff, 1968)

$$\eta_l(k) = \tan^{-1}\frac{j_l(kr_0)}{n_l(kr_0)}, \tag{31}$$

where the functions $j_l(x)$ and $n_l(x)$ are, respectively, the "spherical Bessel functions" and the "spherical Neumann functions":

$$j_0(x) = \frac{\sin x}{x}, \quad j_1(x) = \frac{\sin x - x\cos x}{x^2}, \quad j_2(x) = \frac{(3-x^2)\sin x - 3x\cos x}{x^3}, \ldots$$

and

$$n_0(x) = -\frac{\cos x}{x}, \quad n_1(x) = -\frac{\cos x + x\sin x}{x^2}, \quad n_2(x) = -\frac{(3-x^2)\cos x + 3x\sin x}{x^3}, \ldots.$$

Accordingly, we have

$$\eta_0(k) = \tan^{-1}\{-\tan(kr_0)\} = -kr_0, \tag{32}$$

$$\begin{aligned} \eta_1(k) &= \tan^{-1}\left\{-\frac{\tan(kr_0) - kr_0}{1 + kr_0\tan(kr_0)}\right\} = -\{kr_0 - \tan^{-1}(kr_0)\} \\ &= -\frac{(kr_0)^3}{3} + \frac{(kr_0)^5}{5} - \ldots, \end{aligned} \tag{33}$$

$$\begin{aligned} \eta_2(k) &= \tan^{-1}\left\{-\frac{\tan(kr_0) - 3(kr_0)/[3-(kr_0)^2]}{1 + 3(kr_0)\tan(kr_0)/[3-(kr_0)^2]}\right\} = -\left\{kr_0 - \tan^{-1}\frac{3(kr_0)}{3-(kr_0)^2}\right\} \\ &= -\frac{(kr_0)^5}{45} + \ldots, \end{aligned} \tag{34}$$

and so on. We now have to substitute these results into the formula (26). However, before doing that we would like to point out that, in the case of hard-sphere interaction, (i) we cannot have any bound states whatsoever and (ii) since, for all values of l, $\eta_l^{(0)} = 0$, the integral appearing in formula (26) can be simplified by a prior integration by parts. Thus, we have

$$b_2 - b_2^{(0)} = \frac{8^{1/2}\lambda^2}{\pi^2}\sum_l{}' (2l+1)\int_0^\infty e^{-\beta\hbar^2k^2/m}\eta_l(k)k\,dk. \tag{35}$$

Substituting for $l = 0$ and 2 in the case of bosons and for $l = 1$ in the case of fermions, we obtain (to the fifth power of r_0/λ)

$$b_2 - b_2^{(0)} = -2\left(\frac{r_0}{\lambda}\right)^1 - \frac{10\pi^2}{3}\left(\frac{r_0}{\lambda}\right)^5 - \dots \qquad \text{(Bose)} \tag{36}$$

$$= -6\pi\left(\frac{r_0}{\lambda}\right)^3 + 18\pi^2\left(\frac{r_0}{\lambda}\right)^5 - \dots \qquad \text{(Fermi).} \tag{37}$$

9.6. Cluster expansion for a quantum-mechanical system

When it comes to calculating b_l for $l > 2$ we do not obtain any formula comparable in simplicity to the formula (9.5.26) for b_2. This is due to the fact that we do not have any treatment of the l-body problem (for $l > 2$) which is as neat as the phase-shift analysis of the two-body problem. Nevertheless, a formal theory for the calculation of the higher-order "cluster integrals" has been developed by Kahn and Uhlenbeck (1938); an elaboration by Lee and Yang (1959 a, b; 1960) has made this theory good enough to treat a quantum-mechanical system almost as comprehensively as Mayer's theory could treat a classical gas. The basic approach in this theory is to evolve a scheme of expansion for the grand partition function of the given system in essentially the same way as one carries out the cluster expansion of the grand partition function of a classical gas. However, because of the interplay of the quantum-statistical effects and the effects arising from the interparticle interactions, the mathematical structure of this theory is considerably involved.

We consider here a quantum-mechanical system of N identical particles enclosed in a box of volume V. The Hamiltonian of the system is assumed to be of the form

$$\hat{H}_N = -\frac{\hbar^2}{2m}\sum_{i=1}^{N} \nabla_i^2 + \sum_{i<j} u(r_{ij}). \tag{1}$$

Now, the partition function of the system is given by

$$Q_N(V, T) \equiv \mathrm{Tr}\,(e^{-\beta\hat{H}_N}) = \sum_\alpha e^{-\beta E_\alpha}$$
$$= \sum_\alpha \int_V \{\Psi_\alpha^*(1, \dots, N)\, e^{-\beta\hat{H}_N}\Psi_\alpha(1, \dots, N)\}\, d^{3N}r, \tag{2}$$

where the functions Ψ_α are supposed to form a complete set of (properly symmetrized) orthonormal wave functions of the system, while the numbers $1, \dots, N$ denote the position coordinates $\boldsymbol{r}_1, \dots, \boldsymbol{r}_N$, respectively. We may as well introduce the *probability density operator* $\hat{W}_N$ of the system through the matrix elements

$$\langle 1', \dots, N' | \hat{W}_N | 1, \dots, N\rangle \equiv N!\,\lambda^{3N} \sum_\alpha \{\Psi_\alpha(1', \dots, N')\, e^{-\beta\hat{H}_N}\Psi_\alpha^*(1, \dots, N)\}$$
$$= N!\,\lambda^{3N} \sum_\alpha \{\Psi_\alpha(1', \dots, N')\, \Psi_\alpha^*(1, \dots, N)\}\, e^{-\beta E_\alpha}. \tag{3}$$

We denote the diagonal elements of the operator $\hat{W}_N$ by the symbols $W_N(1, \dots, N)$; thus

$$W_N(1, \dots, N) = N!\,\lambda^{3N} \sum_\alpha \{\Psi_\alpha(1, \dots, N)\, \Psi_\alpha^*(1, \dots, N)\}\, e^{-\beta E_\alpha}, \tag{4}$$

whence eqn. (2) takes the form

$$Q_N(V,T) = \frac{1}{N!\,\lambda^{3N}} \int_V W_N(1, \ldots, N)\, d^{3N}r = \frac{1}{N!\,\lambda^{3N}} \operatorname{Tr}(\hat{W}_N). \tag{5}$$

A comparison of eqn. (5) with eqns. (9.1.3) and (9.4.2) shows that the "trace of the probability density operator $\hat{W}_N$" is the true analogue of the "configuration integral" Z_N and that the quantity $W_N(1, \ldots, N)\, d^{3N}r$ is a measure of the probability that the "configuration" of the given system is found to be within the interval

$$\{(\boldsymbol{r}_1, \ldots, \boldsymbol{r}_N), (\boldsymbol{r}_1 + d\boldsymbol{r}_1, \ldots, \boldsymbol{r}_N + d\boldsymbol{r}_N)\}.$$

Before we proceed further, let us acquaint ourselves with some of the important properties of the matrix elements (3):

(i)
$$\begin{aligned}\langle 1' | \hat{W}_1 | 1 \rangle &= \lambda^3 \sum_p \left\{ \frac{1}{\sqrt{V}} e^{i(\boldsymbol{p}\cdot\boldsymbol{r}_1')/\hbar} \frac{1}{\sqrt{V}} e^{-i(\boldsymbol{p}\cdot\boldsymbol{r}_1)/\hbar} \right\} e^{-\beta p^2/2m} \\ &= \frac{\lambda^3}{V} \iiint_{-\infty}^{+\infty} \frac{V\, d^3p}{h^3} e^{\{i\boldsymbol{p}\cdot(\boldsymbol{r}_1' - \boldsymbol{r}_1)/\hbar - \beta p^2/2m\}} \\ &= e^{-\pi(\boldsymbol{r}_1' - \boldsymbol{r}_1)^2/\lambda^2};\end{aligned} \tag{6}$$

cf. eqn. (5.3.16) for density matrix of a single particle. The foregoing result is a manifestation of the quantum-mechanical, *not quantum-statistical*, correlation between the positions $\boldsymbol{r}$ and $\boldsymbol{r}'$ of a given particle (or, for that matter, any particle in a given system). Obviously, this correlation extends up to distances of the order of λ which is, therefore, a measure of the linear dimensions of the wave packet representing the particle. As $T \to \infty$, and hence $\lambda \to 0$, the matrix element (6) tends to zero for all finite values of $|\boldsymbol{r}_1' - \boldsymbol{r}_1|$.

(ii)
$$\langle 1 | \hat{W}_1 | 1 \rangle = 1; \tag{7}$$

consequently, by eqn. (5), we have

$$Q_1(V,T) = \frac{1}{\lambda^3} \int_V 1\, d^3r = \frac{V}{\lambda^3}. \tag{8}$$

(iii) Whatever the symmetry character of the wave functions Ψ, the diagonal elements $W_N(1, \ldots, N)$ of the probability density operator are *symmetric* in respect of a permutation among the arguments $(1, \ldots, N)$.

(iv) The elements $W_N(1, \ldots, N)$ are *invariant* under a unitary transformation of the set $\{\Psi_\alpha\}$.

(v) Suppose that the values of the coordinates $\boldsymbol{r}_1, \ldots, \boldsymbol{r}_N$ are such that they can be divided into two groups, A and B, with the property that *any* two coordinates, say $\boldsymbol{r}_i$ and $\boldsymbol{r}_j$, of which one belongs to the group A and the other to the group B, satisfy the conditions that

(a) the separation r_{ij} is much larger than the mean thermal wavelength λ of the particles, and

(b) it is also much larger than the effective range $\tilde{r}$ of the two-body potential, then

$$W_N(\boldsymbol{r}_1, \ldots, \boldsymbol{r}_N) \simeq W_A(\boldsymbol{r}_A)\, W_B(\boldsymbol{r}_B), \tag{9}$$

where the symbols $\boldsymbol{r}_A$ and $\boldsymbol{r}_B$ denote *collectively* the coordinates in the group A and the group B, respectively. It is not easy to furnish here a rigorous mathematical proof of this property, though physically it is quite understandable. One can see this by noting that, in view of the conditions (a) and (b), there does not exist any spatial correlation between the particles of the group A on one hand and the particles of the group B on the other (either by virtue of the statistics or by virtue of the interparticle interactions). The two groups, therefore, behave towards each other like two *independent* entities. It is then natural that, to a very good approximation, the probability density W_N of the composite configuration should be given by the product of the probability densities W_A and W_B.

Let us now proceed with the formulation. First of all, to fix our ideas about the approach to be followed, we may consider the simple case with $N = 2$. In that case, as $r_{12} \to \infty$, we expect, in view of the property (v), that

$$W_2(1, 2) \to W_1(1)\, W_1(2) \equiv 1. \tag{10}$$

In general, however, $W_2(1, 2)$ will not be equal to $W_1(1)W_1(2)$. Now, if we denote the difference between $W_2(1, 2)$ and $W_1(1)W_1(2)$ by the symbol $U_2(1, 2)$, then, as $r_{12} \to \infty$,

$$U_2(1, 2) \to 0. \tag{11}$$

It is not difficult to see that the quantity $U_2(1, 2)$ is the quantum-mechanical analogue of the Mayer function f_{ij}. With this in mind, we introduce a sequence of *cluster functions* $\hat{U}_l$, defined by the hierarchy*

$$\langle 1' | \hat{W}_1 | 1 \rangle = \langle 1' | \hat{U}_1 | 1 \rangle, \tag{12}$$

$$\langle 1', 2' | \hat{W}_2 | 1, 2 \rangle = \langle 1' | \hat{U}_1 | 1 \rangle \langle 2' | \hat{U}_1 | 2 \rangle + \langle 1', 2' | \hat{U}_2 | 1, 2 \rangle, \tag{13}$$

$$\begin{aligned}\langle 1', 2', 3' | \hat{W}_3 | 1, 2, 3 \rangle = {} & \langle 1' | \hat{U}_1 | 1 \rangle \langle 2' | \hat{U}_1 | 2 \rangle \langle 3' | \hat{U}_1 | 3 \rangle \\ & + \langle 1' | \hat{U}_1 | 1 \rangle \langle 2', 3' | \hat{U}_2 | 2, 3 \rangle \\ & + \langle 2' | \hat{U}_1 | 2 \rangle \langle 1', 3' | \hat{U}_2 | 1, 3 \rangle \\ & + \langle 3' | \hat{U}_1 | 3 \rangle \langle 1', 2' | \hat{U}_2 | 1, 2 \rangle \\ & + \langle 1', 2', 3' | \hat{U}_3 | 1, 2, 3 \rangle,\end{aligned} \tag{14}$$

and so on. A particular function $\hat{U}_l$ is thus defined with the help of the first l equations of the hierarchy. The last equation in the hierarchy will be (writing the diagonal elements alone)

$$W_N(1, \ldots, N) = \sum_{\{m_l\}}' \Big\{ \sum_P [\underbrace{U_1(\;) \ldots U_1(\;)}_{m_1 \text{ factors}}] [\underbrace{U_2(\;) \ldots U_2(\;)}_{m_2 \text{ factors}}] \ldots \Big\}, \tag{15}$$

where the primed summation goes over all the sets $\{m_l\}$ that conform to the condition

$$\sum_{l=1}^{N} l m_l = N; \qquad m_l = 0, 1, 2, \ldots. \tag{16}$$

* The functions U_l were first introduced by Ursell, in 1927, in order to simplify the classical configuration integral. Their introduction into the quantum-mechanical formalism is due to Kahn and Uhlenbeck (1938).

Moreover, in selecting the arguments of the various U's appearing in eqn. (15), out of the numbers 1, ..., N, one must remember that a permutation of the arguments within the same bracket is *not* regarded as leading to anything distinctly different from what one had before the permutation; the symbol Σ_P then denotes a summation over all *distinct* ways of selecting the arguments under the set $\{m_l\}$.

Relations inverse to the foregoing ones are

$$\langle 1' | \hat{U}_1 | 1 \rangle = \langle 1' | \hat{W}_1 | 1 \rangle, \tag{17}$$

$$\langle 1', 2' | \hat{U}_2 | 1, 2 \rangle = \langle 1', 2' | \hat{W}_2 | 1, 2 \rangle - \langle 1' | \hat{W}_1 | 1 \rangle \langle 2' | \hat{W}_1 | 2 \rangle, \tag{18}$$

$$\begin{aligned}\langle 1', 2', 3' | \hat{U}_3 | 1, 2, 3 \rangle = & \langle 1', 2', 3' | \hat{W}_3 | 1, 2, 3 \rangle \\ & - \langle 1' | \hat{W}_1 | 1 \rangle \langle 2', 3' | \hat{W}_2 | 2, 3 \rangle \\ & - \langle 2' | \hat{W}_1 | 2 \rangle \langle 1', 3' | \hat{W}_2 | 1, 3 \rangle \\ & - \langle 3' | \hat{W}_1 | 3 \rangle \langle 1', 2' | \hat{W}_2 | 1, 2 \rangle \\ & + 2 \langle 1' | \hat{W}_1 | 1 \rangle \langle 2' | \hat{W}_1 | 2 \rangle \langle 3' | \hat{W}_1 | 3 \rangle, \end{aligned} \tag{19}$$

and so on. We note that (i) the coefficient of a general term on the right-hand side of these equations is

$$(-1)^{\Sigma_l m_l - 1} (\Sigma_l m_l - 1)!, \tag{20}$$

where $\Sigma_l m_l$ is the number of W's in the term, and (ii) the sum of the coefficients of all the terms on the right-hand side of eqns. (18), (19), is identically equal to zero. Moreover, the diagonal elements $U_l(1, \ldots, l)$, just like the diagonal elements of the operators $\hat{W}_N$, are symmetric in respect of the permutations among the arguments $(1, \ldots, l)$ and are determined by the sequence of the diagonal elements $W_1, W_2, \ldots, W_l$. Finally, in view of the property (v) of the W's, as embodied in formula (9), the U's possess the following important property:

$$U_l(1, \ldots, l) \simeq 0 \quad \text{if} \quad r_{ij} \gg \lambda, \tilde{r}; \tag{21}$$

here, r_{ij} is the separation between *any* two of the coordinates $(1, \ldots, l)$.*

We now define the "cluster integral" b_l by the formula

$$b_l(V, T) = \frac{1}{l!\, \lambda^{3(l-1)} V} \int U_l(1, \ldots, l)\, d^{3l}r; \tag{22}$$

cf. eqn. (9.1.16). Clearly, the quantity $b_l(V, T)$ is dimensionless and, by virtue of the property (21) of the elements $U_l(1, \ldots, l)$, is independent of V (so long as V is large). Accordingly, in the limit $V \to \infty$, the quantity $b_l(V, T)$ tends to a finite volume-independent value, which may be denoted by $\bar{b}_l(T)$. We then obtain for the partition function of the system, see eqns. (5) and (15),

$$Q_N(V, T) = \frac{1}{N!\, \lambda^{3N}} \int d^{3N}r \left\{ \sum_{\{m_l\}}' \sum_P [U_1 \ldots U_1][U_2 \ldots U_2] \ldots \right\} \tag{23}$$

$$= \frac{1}{N!\, \lambda^{3N}} \sum_{\{m_l\}}' \frac{N!}{(1!)^{m_1} (2!)^{m_2} \ldots m_1!\, m_2! \ldots} \int d^{3N}r \{[U_1 \ldots U_1][U_2 \ldots U_2] \ldots\}. \tag{24}$$

* This can be seen by examining the break-up of the structure on the right-hand side of any equation in the hierarchy (18, 19, ...) when one or more of the l coordinates in the "cluster" get sufficiently separated from the rest of the coordinates.

In writing the last result we have made use of the fact that, since a permutation among the arguments of the functions U_l does not affect the value of the integral concerned, the summation over P may be replaced by any term of the summation, multiplied by the number of *distinct* permutations allowed by the set $\{m_l\}$; cf. the corresponding product of the numbers (9.1.22) and (9.1.24). Making use of the definition (22), eqn. (24) can now be written as

$$Q_N(V,T) = \frac{1}{\lambda^{3N}} \sum_{\{m_l\}}' \left[\prod_{l=1}^{N} \{(b_l \lambda^{3(l-1)} V)^{m_l}/m_l!\} \right]$$

$$= \sum_{\{m_l\}}' \left[\prod_{l=1}^{N} \left\{ \left(b_l \frac{V}{\lambda^3} \right)^{m_l} \frac{1}{m_l!} \right\} \right]; \tag{25}$$

again, use has been made of the fact that

$$\prod_l (\lambda^{3l})^{m_l} = \lambda^{3\Sigma_l l m_l} = \lambda^{3N}. \tag{26}$$

Equation (25) is formally identical with eqn. (9.1.29) of Mayer's theory. Naturally, then, the subsequent development of the formalism, leading to the equation of state of the system, will also be formally identical with that of Mayer's theory. Thus, we should finally obtain the familiar cluster expansions:

$$\frac{P}{kT} = \frac{1}{\lambda^3} \sum_{l=1}^{\infty} \mathfrak{b}_l z^l \quad \text{and} \quad \frac{1}{v} = \frac{1}{\lambda^3} \sum_{l=1}^{\infty} l \mathfrak{b}_l z^l. \tag{27}$$

There are, however, important physical differences. We may recall that the calculation of the cluster integrals $\mathfrak{b}_l$ in the classical case involves the evaluation of a number of finite, though difficult, integrals. The corresponding calculation in the quantum-mechanical case necessitates a knowledge of the functions U_l and hence of *all* the functions W_n for $n \leqslant l$; this in turn requires solutions of the n-body Schrödinger equations for *all* $n \leqslant l$. The case $l = 2$ can be handled neatly, as was done in Sec. 9.5. For $l > 2$, the mathematical procedure is rather cumbersome. Nevertheless, Lee and Yang have evolved a scheme that enables us to calculate the higher $\mathfrak{b}_l$'s in *successive approximations.* We propose to outline the essential features of this scheme in the next section. In passing, however, we may note another important difference between the quantum-mechanical case and the classical case. In the latter case, if the interparticle interactions are absent, then all the $\mathfrak{b}_l$'s, with $l \geqslant 2$, identically vanish. This is not true in the quantum-mechanical case; we rather have in this case (see Secs. 7.1 and 8.1)

$$\mathfrak{b}_l^{(0)} = (\pm 1)^{l-1} l^{-5/2}, \tag{28}$$

of which eqn. (9.5.27) was only a special case. At this point it seems worth while to mention that the nonzero values of the $\mathfrak{b}_l^{(0)}$'s arise solely from the *statistical correlations* among the particles, i.e. from the *symmetry properties* of the many-body wave functions; see Sec. 9.8.

9.7. The binary collision method of Lee and Yang*

The essence of the method proposed by Lee and Yang is that the problem of evaluating the functions U_l of a given physical system may be solved by "separating out" the effects of statistics from the effects of interparticle interactions, i.e. we may first take care of the statis-

* For a critical discussion, and generalization, of this method, see Mohling (1964–5).

tical aspect of the problem and thereafter tackle the dynamical aspect of it. Thus, the whole feat is accomplished in two distinct steps. First, the U-functions pertaining to the given physical system are expressed in terms of U-functions pertaining to a corresponding *quantum-mechanical system obeying Boltzmann statistics*, i.e. a (fictitious) system described by *unsymmetrized wave functions*. This expansion is supposed to take care of the statistics of the actual system, i.e. of the symmetry properties of the wave functions describing the system. Next, the U-functions of the (fictitious) Boltzmannian system are expanded, loosely speaking, in powers of a *binary kernel* B (which is obtainable from a solution of the two-body problem with the given interaction). A commendable feature of the method is that it can be applied even if the given interaction contains a singular, repulsive core, i.e. even if the potential energy for certain configurations of the system becomes infinitely large.

We begin by introducing certain definitions for the (fictitious) *Boltzmannian* system that corresponds to the actual physical system under study. By implication, the Hamiltonian of the Boltzmannian system is precisely the same as that for the actual system, namely

$$\hat{H}_N = -\frac{\hbar^2}{2m}\sum_{i=1}^{N}\nabla_i^2 + \sum_{i<j} u(r_{ij}). \tag{9.6.1}$$

However, in view of the supposed *distinguishability* of the particles, this system will be described by *unsymmetrized* wave functions. Consequently, some of the definitions in the formalism will have to be modified; for instance, the partition function will now be given by

$$Q_N(V,T) = \frac{1}{N!}\sum_i e^{-\beta E_i} = \frac{1}{N!}\sum_i \int_V \{\Psi_i^*(1,\ldots,N)\, e^{-\beta\hat{H}_N}\Psi_i(1,\ldots,N)\}\, d^{3N}r. \tag{9.6.2a}$$

We recall that the introduction of the *Gibbs correction factor* $(1/N!)$ puts the foregoing expression for $Q_N(V, T)$ at par with the expression pertaining to the actual system; in other words, this factor cancels out the major error arising from an undue multiplicity of the microstates of the system which in turn is caused by the supposed distinguishability of the particles (or, what is the same thing, by our failure to symmetrize the wave functions). For the same reason, the probability density operator $\hat{W}_N$ of this system is defined as

$$\langle 1',\ldots,N'|\hat{W}_N|1,\ldots,N\rangle = \lambda^{3N}\sum_i \{\Psi_i(1',\ldots,N')\,\Psi_i^*(1,\ldots,N)\}e^{-\beta E_i}, \tag{9.6.3a}$$

whence

$$W_N(1,\ldots,N) = \lambda^{3N}\sum_i \{\Psi_i(1,\ldots,N)\,\Psi_i^*(1,\ldots,N)\}e^{-\beta E_i}. \tag{9.6.4a}$$

Nevertheless, the relationship between the partition function of the system and the trace of the probability density operator remains the same:

$$Q_N(V,T) = \frac{1}{N!\,\lambda^{3N}}\int W_N(1,\ldots,N)\, d^{3N}r = \frac{1}{N!\,\lambda^{3N}}\,\mathrm{Tr}\,(\hat{W}_N). \tag{9.6.5}$$

We now introduce the U-functions for this system through the same hierarchy of equations as we had for the actual system, viz. eqns. (9.6.12)–(9.6.16). Naturally, then, the inverse relationships, as embodied in eqns. (9.6.17)–(9.6.20), will also hold as such. There will not be any modifications in the subsequent analysis, so the final results will again be

$$\frac{P}{kT} = \frac{1}{\lambda^3}\sum_{l=1}^{\infty} \mathfrak{b}_l z^l \quad \text{and} \quad \frac{1}{v} = \frac{1}{\lambda^3}\sum_{l=1}^{\infty} l\mathfrak{b}_l z^l, \tag{9.6.27}$$

where

$$\mathfrak{b}_l = \frac{1}{l!\,\lambda^{3(l-1)}} \lim_{V\to\infty} \left\{\frac{1}{V}\operatorname{Tr}(\hat{U}_l)\right\}. \tag{9.6.22a}$$

From now onward we shall adopt the following notation: the superscript S will be placed on every physical quantity pertaining to the given system if the given system obeys *symmetric* (B.E.) statistics, the superscript A if the given system obeys *antisymmetric* (F.D.) statistics; the quantities pertaining to the Boltzmannian analogue will go without such a superscript. Thus, the quantities appearing in Sec. 9.6 are now supposed to be carrying a definite superscript—S or A, as the case may be. The first step of the Lee–Yang method then consists in expressing the functions U_l^S, or U_l^A, of Sec. 9.6 in terms of the functions U_l of the present section. This is made possible by the fact that the functions W_n^S, and W_n^A, are related to the functions W_n in a rather straightforward manner (see Feynman, 1953). If we compare eqns. (9.6.3) and (9.6.3a), which define $W_N^{S/A}$ and W_N respectively, and remember that the functions Ψ_α in the former case are duly symmetrized while the functions Ψ_i in the latter case are not, we arrive at the conclusion

$$\langle 1', \ldots, N' | \hat{W}_N^S | 1, \ldots, N\rangle = \sum_{P'} P' \langle 1', \ldots, N' | \hat{W}_N | 1, \ldots, N\rangle \tag{1}$$

and

$$\langle 1', \ldots, N' | \hat{W}_N^A | 1, \ldots, N\rangle = \sum_{P'} (-1)^{[P']} P' \langle 1', \ldots, N' | \hat{W}_N | 1, \ldots, N\rangle, \tag{2}$$

where P' denotes any of the $N!$ operators that permute the coordinates $r_1', \ldots, r_N'$, while $[P']$ denotes the order of the permutation P'. One may recall here that a given permutation is said to be even (or odd) provided that the original order of the entities permuted is regained by an even (or odd) number of pairwise exchanges; thus, if $P'(1', 2', 3') = (2', 1', 3')$ or $(1', 3', 2')$ or $(3', 2', 1')$ then P' is an odd permutation, and if $P'(1', 2', 3') = (2', 3', 1')$ or $(1', 2', 3')$ or $(3', 1', 2')$ then P' is an even permutation; in the first three cases the relevant coefficient in the expansion (2) will be -1, while in the last three cases it will be $+1$. As examples, we note the following:

$$\langle 1' | \hat{W}_1^{S/A} | 1\rangle = \langle 1' | \hat{W}_1 | 1\rangle \left\{= e^{-\pi(r_1'-r_1)^2/\lambda^2}\right\}, \tag{3}$$

$$\langle 1', 2' | \hat{W}_2^{S/A} | 1, 2\rangle = \langle 1', 2' | \hat{W}_2 | 1, 2\rangle \pm \langle 2', 1' | \hat{W}_2 | 1, 2\rangle, \tag{4}$$

$$\begin{aligned}\langle 1', 2', 3' | \hat{W}_3^{S/A} | 1, 2, 3\rangle = {} & \langle 1', 2', 3' | \hat{W}_3 | 1, 2, 3\rangle \pm \langle 2', 1', 3' | \hat{W}_3 | 1, 2, 3\rangle \\ & + \langle 2', 3', 1' | \hat{W}_3 | 1, 2, 3\rangle \pm \langle 1', 3', 2' | \hat{W}_3 | 1, 2, 3\rangle \\ & + \langle 3', 1', 2' | \hat{W}_3 | 1, 2, 3\rangle \pm \langle 3', 2', 1' | \hat{W}_3 | 1, 2, 3\rangle,\end{aligned} \tag{5}$$

and so on. Combining these results with the hierarchies of eqns. (9.6.12)–(9.6.14) and of eqns. (9.6.17)–(9.6.19), which hold for *all* statistics, we obtain

$$\langle 1' | \hat{U}_1^{S/A} | 1 \rangle = \langle 1' | \hat{U}_1 | 1 \rangle \{= e^{-\pi(r_1'-r_1)^2/\lambda^2}\}, \tag{6}$$

$$\langle 1', 2' | \hat{U}_2^{S/A} | 1, 2 \rangle = \langle 1', 2' | \hat{U}_2 | 1, 2 \rangle \pm \langle 2', 1' | \hat{U}_2 | 1, 2 \rangle \pm \langle 2' | \hat{U}_1 | 1 \rangle \langle 1' | \hat{U}_1 | 2 \rangle, \tag{7}$$

$$\begin{aligned}\langle 1', 2', 3' | \hat{U}_3^{S/A} | 1, 2, 3 \rangle = &\sum_{P'} (\pm 1)^{[P']} P' \langle 1', 2', 3' | \hat{U}_3 | 1, 2, 3 \rangle \\ &+ \langle 2' | \hat{U}_1 | 1 \rangle \{ \langle 3', 1' | \hat{U}_2 | 2, 3 \rangle \pm \langle 1', 3' | \hat{U}_2 | 2, 3 \rangle \} \\ &+ \langle 3' | \hat{U}_1 | 1 \rangle \{ \langle 1', 2' | \hat{U}_2 | 2, 3 \rangle \pm \langle 2', 1' | \hat{U}_2 | 2, 3 \rangle \} \\ &+ \text{two similar sets of terms} \\ &+ \langle 2' | \hat{U}_1 | 1 \rangle \langle 3' | \hat{U}_1 | 2 \rangle \langle 1' | \hat{U}_1 | 3 \rangle \\ &+ \langle 3' | \hat{U}_1 | 1 \rangle \langle 1' | \hat{U}_1 | 2 \rangle \langle 2' | \hat{U}_1 | 3 \rangle,\end{aligned} \tag{8}$$

and so on. The general formula for expressing the functions U_l^S and U_l^A in terms of the functions $U_n (n \leqslant l)$ can now be stated in the form of a (rather lengthy) rule:

(i) First of all, we divide the l integers $(1, 2, \ldots, l)$ into groups such that there are m_α groups containing α integers each:

$$\sum_{\alpha=1}^{l} \alpha m_\alpha = l; \qquad m_\alpha = 0, 1, 2, \ldots . \tag{9}$$

Such a grouping may be written as

$$\underbrace{\{(a)(b)\ldots\}}_{m_1 \text{ groups}} \qquad \underbrace{\{(cd)(ef)\ldots\}}_{m_2 \text{ groups}} \qquad \underbrace{\{(ghi)\ldots\}}_{m_3 \text{ groups}} \ldots, \tag{10}$$

where $a, b, c, \ldots$ are the various integers. We take care that (i) within each round bracket the integers are arranged in the ascending order, and (ii) within each curly bracket the round brackets are so arranged that their first integers are in the ascending order.

(ii) We then form the sum

$$\begin{aligned}\sum_{P'}{}' (\pm 1)^{[P']} [\{ \langle A' | \hat{U}_1 | a \rangle \langle B' | \hat{U}_1 | b \rangle \ldots \} \\ \times \{ \langle C'D' | \hat{U}_2 | cd \rangle \langle E'F' | \hat{U}_2 | ef \rangle \ldots \} \\ \times \{ \langle G'H'I' | \hat{U}_3 | ghi \rangle \ldots \} \times \ldots],\end{aligned} \tag{11}$$

where the sequence $[A', B', C', \ldots]$ is a permutation P' of the coordinates $[1', 2', \ldots, l']$, while $[P']$ denotes the order of this permutation. The primed summation in (11) goes over only those permutations which satisfy the condition that on putting $r_i' = r_i$ (for all i) the summand does not "break up" into factors that depend upon *mutually exclusive* coordinates!

(iii) Finally, we sum up expressions (11) over all possible groupings of the l integers $(1, 2, \ldots, l)$. The resulting sum is equal to $\langle 1', \ldots, l' | \hat{U}_l^{S/A} | 1, \ldots, l \rangle$.

We might hasten to add that the above rule is much simpler to apply than it seems at first sight. Thus, when $l = 2$, the grouping $\{(12)\}$, for which $m_2 = 1$ and every other $m_\alpha = 0$, leads to the first two terms of eqn. (7), while the grouping {(1) (2)}, for which $m_1 = 2$ and every other $m_\alpha = 0$, leads to the last term there; we note that in the latter case the term $\langle 1' | \hat{U}_1 | 1 \rangle \langle 2' | \hat{U}_1 | 2 \rangle$ could not be included because on putting $r_1' = r_1$ and $r_2' = r_2$ it would "break up" into factors, one of which depends exclusively on the coordinate r_1 and the other exclusively on the coordinate r_2. Similarly, when $l = 3$, the grouping {(123)} leads to the first set of (six) terms in eqn. (8), the grouping {(1)} {(23)} leads to the next set of (four) terms (again the terms $\langle 1' | \hat{U}_1 | 1 \rangle \{\langle 2', 3' | \hat{U}_2 | 2, 3 \rangle \pm \langle 3', 2' | \hat{U}_2 | 2, 3 \rangle\}$ had to be left out), the groupings {(2)} {(13)} and {(3)} {(12)} lead to four terms each (which have not been written down explicitly), while the grouping {(1) (2) (3)} leads to the last pair of terms in the equation; once again, we note that of the six terms possible with the grouping {(1) (2) (3)} only two have been included, the rest having been disqualified for the reason given above.

The second step of the Lee–Yang method consists in formulating a procedure whereby the functions U_l pertaining to the ***Boltzmannian*** system are "expanded" in powers of a binary kernel B which, in turn, is calculable from a solution of the two-body problem with the given interaction. This procedure is purely dynamical in nature, free from any statistical complications. The functions W_N and U_l are indeed operators, and we have

$$W_N(\beta) = \lambda^{3N} \exp(-\beta \hat{H}_N); \tag{12}$$

see eqn. (9.6.3a). It may be noted that we are now explicitly indicating the temperature dependence of the functions W_N. Writing $\hat{H}_N = \hat{T}_N + \hat{\Omega}_N$, where $\hat{T}_N$ and $\hat{\Omega}_N$ are, respectively, the operators for the kinetic and the potential energies of the system, we obtain (for a noninteracting system)

$$\begin{aligned} W_N^{(0)}(\beta) &= \lambda^{3N} \exp(-\beta \hat{T}_N) = \prod_{i=1}^{N} \left\{ \lambda^3 \exp\left(\frac{\beta \hbar^2}{2m} \nabla_i^2\right) \right\} \\ &= \prod_{i=1}^{N} w(\beta; i), \end{aligned} \tag{13}$$

where

$$w(\beta; i) \equiv \lambda^3 \exp\left(\frac{\beta \hbar^2}{2m} \nabla_i^2\right). \tag{14}$$

The explicit form of the operator $w(\beta; 1)$, in the coordinate representation, is given by the matrix elements $\langle 1' | \hat{W}_1 | 1 \rangle$, which are also equal to the matrix elements $\langle 1' | \hat{U}_1 | 1 \rangle$; see eqns. (3) and (6). Thus, our $W_N^{(0)}(\beta)$ is a product of N operators, each of which operates on the coordinates of a single particle in the system. If interactions are present, we may expand $W_N(\beta)$ into an "exponential series" in powers of $\hat{\Omega}_N$:

$$\begin{aligned} W_N(\beta) = W_N^{(0)}(\beta) &+ \int_0^\beta d\beta' \{ W_N^{(0)}(\beta - \beta')(-\hat{\Omega}_N) W_N^{(0)}(\beta') \} \\ &+ \int_0^\beta d\beta' \int_0^{\beta'} d\beta'' \{ W_N^{(0)}(\beta - \beta')(-\hat{\Omega}_N) W_N^{(0)}(\beta' - \beta'')(-\hat{\Omega}_N) W_N^{(0)}(\beta'') \} \\ &+ \ldots . \end{aligned} \tag{15}$$

We note that $W_N(\beta) \to W_N^{(0)}(\beta)$ not only as $\hat{\Omega}_N \to 0$ but also as $\beta \to 0$.

Now, as it stands, the series (15) becomes meaningless if, for any configuration of the system, $\hat{\Omega}_N = \infty$. For the time being, therefore, we may regard $\hat{\Omega}_N$ as finite everywhere. However, after a suitable rearrangement of terms in the series, the cases involving $\hat{\Omega}_N = \infty$ can also be brought within the scope of this scheme; this will enable us to tackle successfully a system of particles interacting through a potential that may include a singular, repulsive core! With this in mind, we rewrite (15) in terms of a series of diagrams which are more or less self-explanatory:

$W_1(\beta) \equiv$ 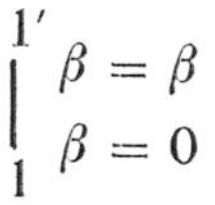

$\equiv w(\beta;\, 1),$

(16)

$W_2(\beta) \equiv$ 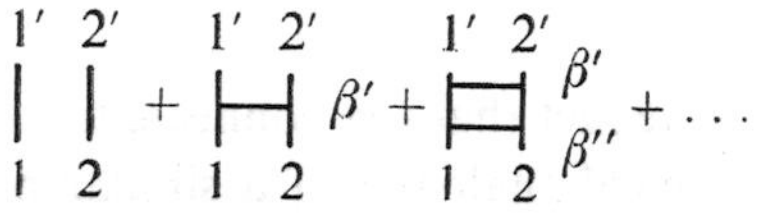

$$= w(\beta;1)\, w(\beta;2)$$
$$+\int_0^{\beta} d\beta' \{w(\beta-\beta';1)\, w(\beta-\beta';2)\, (-\hat{\Omega}_{12})\, w(\beta';1)\, w(\beta';2)\}$$
$$+\int_0^{\beta} d\beta' \int_0^{\beta'} d\beta'' \{w(\beta-\beta';1)\, w(\beta-\beta';2)\, (-\hat{\Omega}_{12})\, w(\beta'-\beta'';1)$$
$$w(\beta'-\beta'';2)\, (-\hat{\Omega}_{12})\, w(\beta'';1)\, w(\beta'';2)\}$$
$$+\ \ldots, \qquad (17)$$

$W_3(\beta) \equiv$ 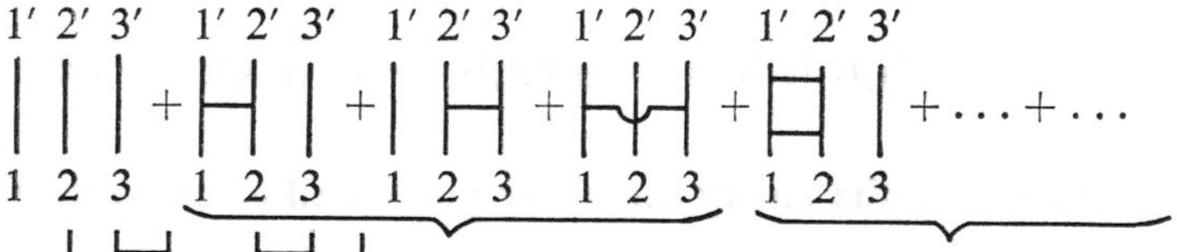

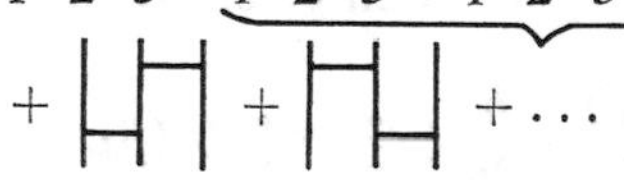

$$= w(\beta;1)\, w(\beta;2)\, w(\beta;3)$$
$$+\left[\int_0^{\beta} d\beta' \{w(\beta-\beta';1)\, w(\beta-\beta';2)\, (-\hat{\Omega}_{12})\, w(\beta';1)\, w(\beta';2)\}\right] w(\beta;3)$$

+ two similar terms

+ terms of higher orders, (18)

and so on. In terms of these diagrams, eqn. (15) takes the form:

$W_N(\beta) =$ the sum of all different diagrams with parameter β and with N particles. (19)

As can be seen from the examples given here, some of these diagrams may contain "unconnected parts". The "unconnected parts" of a diagram represent "commuting operators" into which the operator corresponding to the whole diagram may be factorized. If we now define a *connected* diagram as one "in which all the parts are mutually connected through vertical lines and horizontal links" and then compare the diagrammatic representations (16–18) with the corresponding Ursell expansions (9.6.12–14), we readily conclude that

$U_l(\beta) =$ the sum of all different *connected* diagrams with parameter β and with l particles. (20)

Thus

$$U_2(\beta) = \text{[diagram]} \beta' + \text{[diagram]} \beta' + \ldots, \qquad (21)$$

$$U_3(\beta) = \text{[diagram]} + \text{[diagram]} + \ldots$$

$$+ \text{[diagram]} + \text{[diagram]} + \ldots, \qquad (22)$$

and so on. We must not lose sight of the fact that the connectedness of the *l*-particle diagrams here is the exact quantum-mechanical analogue of a similar property possessed by the integrands of the *l*-body *cluster integrals* of the Mayer theory.

Now, by eqns. (17) and (21),

$$U_2(\beta) = W_2(\beta) - W_2^{(0)}(\beta)$$

$$= \lambda^6 \left[\exp\left(-\beta \hat{H}_2\right) - \exp\left(\frac{\beta \hbar^2}{2m} \nabla_1^2\right) \exp\left(\frac{\beta \hbar^2}{2m} \nabla_2^2\right) \right]. \qquad (23)$$

We introduce the *binary kernel* $B(\beta; 1, 2)$, defined by the formula

$$B(\beta; 1, 2) \equiv -\hat{\Omega}_{12} W_2(\beta) = -\lambda^6 \hat{\Omega}_{12} \exp\left(-\beta \hat{H}_2\right). \qquad (24)$$

It can be verified that the binary kernel is equally well given by the formula

$$B(\beta; 1, 2) = \frac{\partial U_2(\beta)}{\partial \beta} - \frac{\hbar^2}{2m} \left(\nabla_1^2 + \nabla_2^2\right) U_2(\beta). \qquad (25)$$

We observe that by making use of the solutions of the two-body problem (in a box of volume V) we can readily compute the function $\exp\left(-\beta \hat{H}_2\right)$; eqns. (23) and (25) then enable us to compute the function $U_2(\beta)$ and the kernel $B(\beta; 1, 2)$, *without* directly encountering the operator $\hat{\Omega}_{12}$ which appears so explicitly in the defining formula (24). We, therefore, expect that the matrix B can be evaluated explicitly even if, for certain configurations of the two-body system, the matrix Ω_{12} becomes infinite.

The diagrammatic representation of the binary kernel can be inferred directly from the diagrammatic representation of $W_2(\beta)$:

$$B(\beta; 1, 2) \equiv -\hat{\Omega}_{12} W_2(\beta) = \text{[diagram]} + \text{[diagram]} + \text{[diagram]} + \ldots$$

$$= \text{[diagram]} , \text{ say.} \qquad (26)$$

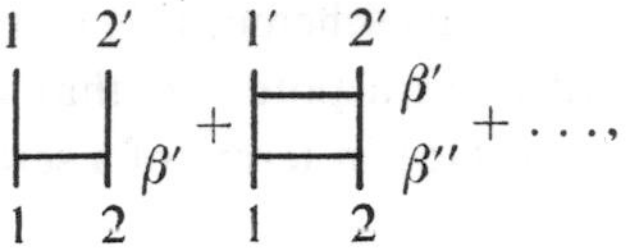

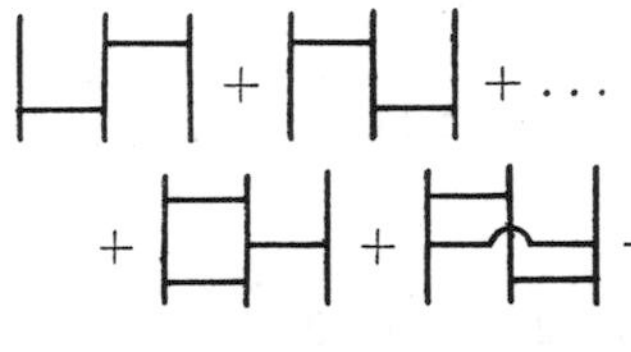

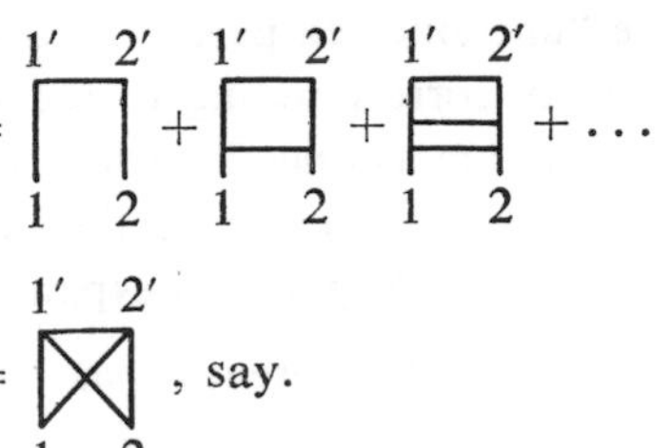

The top horizontal link in each diagram here represents the common factor $(-\hat{\Omega}_{12})$, while the cross in the last diagram denotes the "totality of diagrams with *any* number of horizontal links at *any* height(s) between the values 0 and β of the temperature parameter". In a similar manner, we can write

$$B(\beta'-\beta''; 1, 2) = \qquad (27)$$

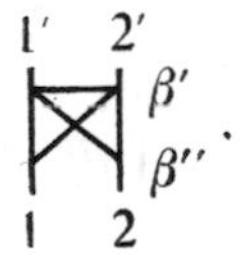

which denotes the "totality of diagrams with *one* horizontal link at the value β' of the temperature parameter and *any* number of horizontal links at *any* height(s) between the values β'' and β' of this parameter".

We can now regroup the diagrams appearing in (20) in such a manner that in the various terms of the rearranged sum only B's appear, i.e. *no horizontal links are left isolated.* So far as the function $U_2^{(\beta)}$ is concerned, all the diagrams appearing in (21) group together into a single term,

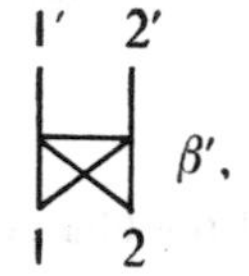

(28)

in which the position of the top horizontal link can be anywhere between 0 and β. Analytically, this means that

$$U_2(\beta) = \int_0^\beta d\beta' \{w(\beta-\beta'; 1)\, w(\beta-\beta'; 2)\, B(\beta'; 1, 2)\}, \tag{29}$$

which is consistent with eqn. (25). In a similar manner, the diagrams appearing in (22) can be grouped into terms such as

and

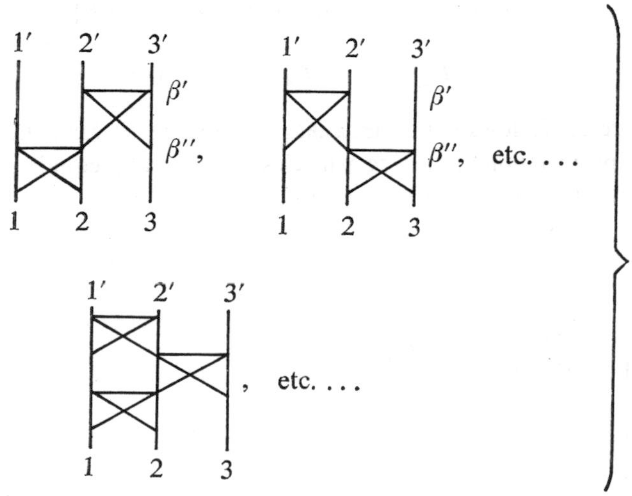

(30)

This leads to the following result for the function $U_3(\beta)$:

$$U_3(\beta) = \int_0^\beta d\beta' \int_0^{\beta'} d\beta'' \{w(\beta-\beta''; 1)\, w(\beta-\beta'; 2)\, w(\beta-\beta'; 3)$$
$$\times B(\beta'-\beta''; 2, 3)\, B(\beta''; 1, 2) \times w(\beta''; 3)\}$$
$$+\int_0^\beta d\beta' \int_0^{\beta'} d\beta'' \{w(\beta-\beta'; 1)\, w(\beta-\beta'; 2)\, w(\beta-\beta''; 3)$$
$$\times B(\beta'-\beta''; 1, 2)\, B(\beta''; 2, 3) \times w(\beta''; 1)\}$$
$$+\text{four other terms of order } B^2$$
$$+\text{terms of higher orders in } B. \tag{31}$$

Similarly, we can obtain the so-called *binary expansions* for the functions $U_4(\beta)$, $U_5(\beta)$, ... Since B vanishes for free particles, these expansions may be looked upon as *perturbation series* for the functions $U_l(\beta)$. The convergence properties of these expansions are not well understood. It is, however, hoped that, for interactions which do not lead to any three-body bound states, these expansions do converge.

9.8. Applications of the binary collision method*

A. A gas of noninteracting particles

In the Boltzmannian case, our system reduces to an *ideal, classical* gas. The operator $\hat{W}_N$ then factorizes into $\hat{W}_1$'s; see eqn. (9.7.13). Accordingly, for all N,

$$\langle 1', 2', \ldots, N' | \hat{W}_N | 1, 2, \ldots, N \rangle = \langle 1' | \hat{W}_1 | 1 \rangle \langle 2' | \hat{W}_1 | 2 \rangle \ldots \langle N' | \hat{W}_1 | N \rangle. \tag{1}$$

Substituting this result into the Ursell expansions (9.6.17–19), we readily obtain

$$\langle 1' | \hat{U}_1 | 1 \rangle = \langle 1' | \hat{W}_1 | 1 \rangle \{= e^{-\pi(r_1'-r_1)^2/\lambda^2}\}, \tag{2}$$

while

$$\hat{U}_2 = \hat{U}_3 = \hat{U}_4 = \ldots = 0. \tag{3}$$

Indeed, the last result follows from the binary expansions as well, since the binary kernel $B(\beta; 1, 2)$ for a pair of noninteracting particles is identically equal to zero. The "cluster integrals" $\mathfrak{b}_l$ of the Boltzmannian system are then given by

$$\mathfrak{b}_l = \frac{1}{l!\,\lambda^{3(l-1)}V} \operatorname{Tr}(\hat{U}_l) = 1 \quad \text{for} \quad l = 1$$
$$= 0 \quad \text{otherwise}. \tag{4}$$

In the Bose–Einstein or Fermi–Dirac case, the functions $\hat{U}_l^{S/A}$ (for $l > 1$) no longer

* For application to *multicomponent* systems, see Pathria and Kawatra (1963); also Mohling and Grandy (1965).

vanish. For instance, eqns. (9.7.7) and (9.7.8) now give

$$\langle 1', 2' | \hat{U}_2^{S/A} | 1, 2 \rangle = \pm \langle 2' | \hat{U}_1 | 1 \rangle \langle 1' | \hat{U}_1 | 2 \rangle, \tag{5}$$

$$\begin{aligned} \langle 1', 2', 3' | \hat{U}_3^{S/A} | 1, 2, 3 \rangle &= \langle 2' | \hat{U}_1 | 1 \rangle \langle 3' | \hat{U}_1 | 2 \rangle \langle 1' | \hat{U}_1 | 3 \rangle \\ &\quad + \langle 3' | \hat{U}_1 | 1 \rangle \langle 1' | \hat{U}_1 | 2 \rangle \langle 2' | \hat{U}_1 | 3 \rangle, \end{aligned} \tag{6}$$

and so on. It is clear that the nonvanishing character of these operators is due to the *exchange effect* that arises from the symmetry properties of the respective wave functions, irrespective of the presence or absence of the interparticle interactions. For general l, we have (see the rule stated on p. 285)

$$\langle 1', \ldots, l' | \hat{U}_l^{S/A} | 1, \ldots, l \rangle = \sum_{P'}{}' (\pm 1)^{[P']} \{ \langle A' | \hat{U}_1 | 1 \rangle \langle B' | \hat{U}_1 | 2 \rangle \ldots \}. \tag{7}$$

Note that, since $\hat{U}_2 = \hat{U}_3 = \ldots = 0$, we are concerned with only one grouping of the integers $(1, \ldots, l)$, namely

$$\{(1)(2)\ldots(l)\}, \tag{9.7.10a}$$

which corresponds to the distribution set $\{m_\alpha\}$, with

$$m_1 = l, m_2 = m_3 = \ldots = 0. \tag{9.7.9a}$$

Further, in view of the fact that we can include only those permutations P' (of the coordinates $1', \ldots, l'$) which satisfy the condition that on putting $\boldsymbol{r}_i' = \boldsymbol{r}_i$ (for all i) the summand does not "break up" into factors that depend upon *mutually exclusive* coordinates, we shall have exactly $(l-1)!$ terms in the sum (7);* moreover, these terms will carry a common sign, $(\pm 1)^{l-1}$, and will have a common value for their trace. Therefore,

$$\begin{aligned} \mathrm{Tr}\,(\hat{U}_l^{S/A}) &= (l-1)! \text{ times the trace of any term in the sum} \\ &= (l-1)!\,(\pm 1)^{l-1} \int_V \{ \langle l | \hat{U}_1 | 1 \rangle \langle 1 | \hat{U}_1 | 2 \rangle \ldots \langle l-1 | \hat{U}_1 | l \rangle \}\, d^{3l}r \\ &= (l-1)!\,(\pm 1)^{l-1}\, \mathrm{Tr}\,(\hat{U}_1)^l. \end{aligned} \tag{8}$$

Accordingly,

$$\mathfrak{b}_l^{S/A} = \frac{(\pm 1)^{l-1}}{l \lambda^{3(l-1)} V} \mathrm{Tr}\,(\hat{U}_1)^l. \tag{9}$$

To evaluate the trace, we may go over to the momentum representation, for in tha representation the function U_1 is diagonal; actually, from the steps that led to eqn. (9.6.6) we obtain

$$\langle \boldsymbol{p}' | \hat{U}_1 | \boldsymbol{p} \rangle \equiv \langle \boldsymbol{p}' | \hat{W}_1 | \boldsymbol{p} \rangle = \lambda^3 \delta_{\boldsymbol{p}', \boldsymbol{p}}\, e^{-\beta p^2/2m}, \tag{10}$$

* In view of the condition stated, the coordinate A' cannot be $1'$; it can, of course, be *any* one of the remaining $(l-1)$ coordinates, say $2'$. Then, B' cannot be either $2'$ or $1'$; it can, however, be *any* one of the remaining $(l-2)$ coordinates. And so on. We thus have in all $(l-1)!$ different ways of choosing the coordinates $(A', B', \ldots)$ which are consistent with our restrictive condition; cf. eqn. (6) for $l = 3$.

where $\delta_{p',p}$ is the three-dimensional Kronecker delta. The desired trace then turns out to be

$$\operatorname{Tr}(\hat{U}_1)^l = \lambda^{3l} \sum_p e^{-l\beta p^2/2m} \to \lambda^{3l} \frac{4\pi V}{h^3} \int_0^\infty e^{-l\beta p^2/2m} p^2 \, dp$$

$$= \lambda^{3(l-1)} V / l^{3/2}. \tag{11}$$

Substituting (11) into (9), we obtain

$$b_l^{S/A} = (\pm 1)^{l-1} / l^{5/2}, \tag{12}$$

which is identical with the familiar ideal-gas result (9.6.28).

B. A gas of hard spheres

We now consider a gas of bosons or fermions interacting through a *repulsive* interaction characterized by a hard core of diameter r_0; see eqn. (9.5.30).† To obtain the binary kernel for this interaction, we first evaluate the matrix of the Boltzmannian function $W_2(\beta)$:

$$\langle 1', 2' | \hat{W}_2 | 1, 2 \rangle = \lambda^6 \sum_\alpha \{\Psi_\alpha(1', 2') \Psi_\alpha^*(1, 2)\} e^{-\beta E_\alpha}; \tag{13}$$

see eqn. (9.6.3a). Since the hard-sphere interaction does not admit bound states, we have to consider states of positive energy alone. In terms of the center-of-mass coordinates $\boldsymbol{R}$ and the relative coordinates $\boldsymbol{r}$, the eigenfunctions Ψ_α and the eigenvalues E_α of the two-body wave equation can be expressed as in eqns. (9.5.7) and (9.5.8), respectively. In view of the separation of variables, the right-hand side of eqn. (13) splits into two factors — the $\boldsymbol{R}$-factor, which is equal to

$$\lambda^3 \sum_P \left\{ \frac{1}{\sqrt{V}} e^{i(\boldsymbol{P}\cdot\boldsymbol{R}')/\hbar} \frac{1}{\sqrt{V}} e^{-i(\boldsymbol{P}\cdot\boldsymbol{R})/\hbar} \right\} e^{-\beta P^2/4m} \simeq 2^{3/2} e^{-2\pi(\boldsymbol{R}'-\boldsymbol{R})^2/\lambda^2}, \tag{14}$$

and the $\boldsymbol{r}$-factor, which is given by

$$\lambda^3 \sum_{l,m} \int_0^\infty dk \, \{\psi_{klm}(\boldsymbol{r}') \psi_{klm}^*(\boldsymbol{r})\} e^{-\beta\hbar^2k^2/m}; \tag{15}$$

for notation, see eqns. (9.5.15, 17). The orthonormal wave function $\psi_{klm}(\boldsymbol{r})$ is given by

$$\psi_{klm}(\boldsymbol{r}) = \left(\frac{2}{\pi}\right)^{1/2} Y_{l,m}(\theta, \varphi) \frac{\chi_{kl}(r)}{r}, \tag{16}$$

so that

$$\int \psi_{k'l'm'}^*(\boldsymbol{r}) \psi_{klm}(\boldsymbol{r}) \, d^3r = \delta_{l',l} \, \delta_{m',m} \, \delta(k'-k), \tag{17}$$

where $\delta(k'-k)$ is the Dirac delta function while the first two δ's are the Kronecker deltas.

† The problem of a one-dimensional quantum system of *hard lines*, obeying Boltzmann statistics, can be solved exactly. For details, see Henderson (1964).

Making use of the relation

$$\sum_{m=-l}^{l} Y_{l,m}(\theta', \varphi')\, Y^*_{l,m}(\theta, \varphi) = \frac{2l+1}{4\pi} P_l(\cos \Theta), \tag{18}$$

where

$$\cos \Theta = (\boldsymbol{r} \cdot \boldsymbol{r}')/(rr'), \tag{19}$$

the expression (15) becomes

$$\frac{\lambda^3}{2\pi^2 rr'} \sum_l \left[(2l+1)\, P_l(\cos \Theta) \int_0^\infty dk\, \{\chi_{kl}(r')\, \chi^*_{kl}(r)\}\, e^{-\beta\hbar^2k^2/m}\right]. \tag{20}$$

Multiplying (14) and (20), we obtain

$$\langle 1', 2' |\hat{W}_2| 1, 2\rangle = \frac{2^{1/2}\lambda^3}{\pi^2 rr'} \exp\{-\pi(\boldsymbol{r}_1' + \boldsymbol{r}_2' - \boldsymbol{r}_1 - \boldsymbol{r}_2)^2/2\lambda^2\}$$

$$\times \sum_l \left[(2l+1)\, P_l(\cos \Theta) \int_0^\infty dk\, \{\chi_{kl}(r')\, \chi^*_{kl}(r)\}\, e^{-\beta\hbar^2k^2/m}\right]. \tag{21}$$

To obtain U_2, we have to subtract from (21) the corresponding expression for a pair of *free* particles; see eqn. (9.7.23).. In other words, we have to replace the factor $\{\chi_{kl}(r') \times \chi^*_{kl}(r)\}$ by the factor

$$\{\chi_{kl}(r')\, \chi^*_{kl}(r)\} - \{\chi_{kl}(r')\, \chi^*_{kl}(r)\}^{(0)}. \tag{22}$$

Now, in view of the exponential factor appearing in (21), the important values of k in the process of integration are of the order of λ^{-1}. Therefore, if $r_0 \ll \lambda$, the important values of k would be much smaller than r_0^{-1}. The phase shifts $\eta_l(k)$ will then be

$$\eta_l(k) = \tan^{-1} \frac{j_l(kr_0)}{n_l(kr_0)} = 0(kr_0)^{2l+1}; \tag{23}$$

see eqns. (9.5.32–34). Consequently, if we are prepared to neglect contributions of an order higher than $(r_0/\lambda)^2$, we need not consider any states other than the S-states ($l = 0$). Expression (22) may then be replaced by, see also eqns. (9.5.19) and (9.5.32),

$$\{\sin (kr' - kr_0) \sin (kr - kr_0) - \sin (kr') \sin (kr)\}$$
$$= \tfrac{1}{2}\{\cos (kr' + kr) - \cos (kr' + kr - 2kr_0)\}$$

if both $r', r > r_0$; otherwise, by

$$\{0 - \sin (kr') \sin (kr)\}$$
$$= \tfrac{1}{2}\{\cos (kr' + kr) - \cos (kr' - kr)\}.$$

Substituting the foregoing expression(s) for the factor $\{\chi(r')\, \chi^*(r)\}$ in eqn. (21) and taking no other values of l except $l = 0$, we obtain

$$\langle 1', 2' | \hat{U}_2 | 1, 2 \rangle = \frac{2^{1/2}\lambda^3}{\pi^2 rr'} \exp\{-\pi(\boldsymbol{r}_1'+\boldsymbol{r}_2'-\boldsymbol{r}_1-\boldsymbol{r}_2)^2/2\lambda^2\} \int_0^\infty dk \{\ldots\} e^{-\beta\hbar^2k^2/m}$$

$$= \frac{\lambda^2}{2\pi rr'} \exp\{-\pi(\boldsymbol{r}_1'+\boldsymbol{r}_2'-\boldsymbol{r}_1-\boldsymbol{r}_2)^2/2\lambda^2\}$$

$$\times \begin{cases} \exp\{-\pi(r'+r)^2/2\lambda^2\} - \exp\{-\pi(r'+r-2r_0)^2/2\lambda^2\} & \text{if both } r', r > r_0 \\ \exp\{-\pi(r'+r)^2/2\lambda^2\} - \exp\{-\pi(r'-r)^2/2\lambda^2\} & \text{otherwise.} \end{cases} \quad (24)$$

It may be noted here that $r' = |\boldsymbol{r}_2' - \boldsymbol{r}_1'|$ and $r = |\boldsymbol{r}_2 - \boldsymbol{r}_1|$.

As a check, we may evaluate the coefficient b_2^S for a Bose gas of hard spheres; we have

$$b_2^S = \frac{1}{2\lambda^3 V} \iint \langle 1, 2 | \hat{U}_2^S | 1, 2 \rangle \, d^3r_1 \, d^3r_2. \quad (25)$$

Making use of the formula (9.7.7), we obtain

$$b_2^S = \frac{1}{2\lambda^3 V} \iint [\langle 2 | \hat{U}_1 | 1 \rangle \langle 1 | \hat{U}_1 | 2 \rangle + \langle 1, 2 | \hat{U}_2 | 1, 2 \rangle + \langle 2, 1 | \hat{U}_2 | 1, 2 \rangle] \, d^3r_1 \, d^3r_2. \quad (26)$$

The first term will yield the ideal-gas result, viz. $2^{-5/2}$; see eqns. (5) and (12). The second and third terms, on the other hand, will yield results arising from the interparticle interactions; making use of the expression (24) and changing over to the coordinates $\boldsymbol{R}$ and $\boldsymbol{r}$, we obtain

$$\{b_2 - b_2^{(0)}\}^S = \frac{1}{2\pi\lambda} \left[\int_{r_0}^\infty \frac{1}{r^2} (e^{-2\pi r^2/\lambda^2} - e^{-2\pi(r-r_0)^2/\lambda^2}) \, 4\pi r^2 \, dr + \int_0^{r_0} \frac{1}{r^2} (e^{-2\pi r^2/\lambda^2} - 1) \, 4\pi r^2 \, dr \right]$$

$$= -2\left(\frac{r_0}{\lambda}\right), \quad (27)$$

it will be noted that this result is identical with the first-order term of the Beth–Uhlenbeck result (9.5.36).

To study the influence of interparticle interactions on the higher-order coefficients, we first observe that in the general expression

$$U_l^S = \underset{l \text{ factors}}{\Sigma'\{U_1 \ldots U_1\}} + \underset{(l-2) \text{ factors}}{\Sigma'\{U_1 \ldots U_1\}}\{U_2\}$$

$$+ \underset{(l-4) \text{ factors}}{\Sigma'\{U_1 \ldots U_1\}}\{U_2 U_2\} + \underset{(l-3) \text{ factors}}{\Sigma'\{U_1 \ldots U_1\}}\{U_3\} + \ldots \quad (28)$$

the first sum will yield the ideal-gas result, the second sum will yield the first-order correction, the third and the fourth sums will yield the second-order correction, and so on; see the binary expansions (9.7.29) and (9.7.31) for U_2 and U_3. Consequently, if we are interested in the first-order correction alone, we have only to consider the sum $\Sigma'\{U_1 \ldots U_1\}\{U_2\}$, which in fact can be handled without going into the binary expansion at all. For calculating the contribution of this sum towards the trace of the operator $\hat{U}_l^S$ and hence towards the coefficient b_l^S, we note that the trace of the various terms in this sum will be of the type

$$\mathrm{Tr}\,\{(\hat{U}_1)^{n_1}\,\hat{U}_2(\hat{U}_1)^{n_2}\}, \tag{29}$$

where n_1 and n_2 are integers such that $n_1 \geqslant 0$, $n_2 \geqslant 0$ and $n_1+n_2 = l-2$; cf. eqn. (8), in which we were concerned with $\mathrm{Tr}\,(\hat{U}_1)^l$ instead. The corresponding multiplicity factor, arising due to the "allowed" choices of the coordinates $(1', \ldots, l')$ will now be $(l-2)!$. The desired contribution towards $\mathrm{Tr}\,(\hat{U}_l^S)$ is, therefore, given by

$$(l-2)!\sum_{i<j}\left[\sum_{n_1,n_2}{}' \mathrm{Tr}\,\{\hat{U}_1^{n_1}[\langle I', J'|\hat{U}_2|i, j\rangle+\langle J', I'|\hat{U}_2|i, j\rangle]\hat{U}_1^{n_2}\}\right], \tag{30}$$

where the first sum goes over all possible choices of the numbers i and j out of the set $(1, \ldots, l)$. Since the number of these choices is $l(l-1)/2$ and they all contribute equally in the end, our result is finally given by

$$(\tfrac{1}{2})l!\left[\sum_{n_1,n_2}{}' \mathrm{Tr}\,\{\hat{U}_1^{n_1}[\langle I', J'|\hat{U}_2|1, 2\rangle+\langle J', I'|\hat{U}_2|1, 2\rangle]\hat{U}_1^{n_2}\}\right]. \tag{31}$$

To evaluate this expression, we better go over to the momentum representation. In this representation,

$$\langle \boldsymbol{k}_1'|\hat{U}_1|\boldsymbol{k}_1\rangle = \lambda^3\delta_{\boldsymbol{k}_1', \boldsymbol{k}_1}e^{-\beta\hbar^2k_1^2/2m} \tag{32}$$

and

$$\langle \boldsymbol{k}_1', \boldsymbol{k}_2'|\hat{U}_2|\boldsymbol{k}_1, \boldsymbol{k}_2\rangle = \frac{\lambda^6 r_0}{2\pi^2(k^2-k'^2)}\delta_{\boldsymbol{k}_1'+\boldsymbol{k}_2', \boldsymbol{k}_1+\boldsymbol{k}_2}\{e^{-\beta E}-e^{-\beta E'}\}+0(r_0^2); \tag{33}$$

see eqn. (10) and Problem 9.10. We note that in eqn. (33)

$$k = \tfrac{1}{2}|\boldsymbol{k}_1-\boldsymbol{k}_2|; \qquad k' = \tfrac{1}{2}|\boldsymbol{k}_1'-\boldsymbol{k}_2'| \tag{34}$$

and

$$E = \frac{\hbar^2}{2m}(k_1^2+k_2^2); \qquad E' = \frac{\hbar^2}{2m}(k_1'^2+k_2'^2). \tag{35}$$

In view of the structure of the expression (31) and the presence of the δ-functions in eqns. (32) and (33), we must consider the combination

$$\langle \boldsymbol{k}_1, \boldsymbol{k}_2|\hat{U}_2|\boldsymbol{k}_1, \boldsymbol{k}_2\rangle+\langle \boldsymbol{k}_2, \boldsymbol{k}_1|\hat{U}_2|\boldsymbol{k}_1, \boldsymbol{k}_2\rangle. \tag{36}$$

Making use of eqns. (33)–(35), we obtain for this combination the value

$$\begin{aligned}
&\frac{\lambda^6 r_0}{\pi^2}\lim_{\boldsymbol{k}_{1,2}'\to\boldsymbol{k}_{1,2}}\delta_{\boldsymbol{k}_1'+\boldsymbol{k}_2', \boldsymbol{k}_1+\boldsymbol{k}_2}\left\{\frac{e^{-\beta E}-e^{-\beta E'}}{k^2-k'^2}\right\}+0(r_0^2)\\
&= \frac{\lambda^8 r_0 e^{-\beta E}}{4\pi^3}\lim_{\boldsymbol{k}_{1,2}'\to\boldsymbol{k}_{1,2}}\delta_{\boldsymbol{k}_1'+\boldsymbol{k}_2', \boldsymbol{k}_1+\boldsymbol{k}_2}\left\{\frac{(k_1'^2+k_2'^2)-(k_1^2+k_2^2)}{k^2-k'^2}\right\}+0(r_0^2)\\
&= -\frac{\lambda^8 r_0}{2\pi^3}\exp\left\{-\beta\frac{\hbar^2}{2m}(k_1^2+k_2^2)\right\}+0(r_0^2).
\end{aligned} \tag{37}$$

To obtain the corresponding contribution towards the coefficient $\mathfrak{b}_l^S$, we have to evaluate the quantity

$$\lim_{V\to\infty} \frac{1}{l!\,\lambda^{3(l-1)}V} \int \langle \boldsymbol{r}_1, \boldsymbol{r}_2, \ldots, \boldsymbol{r}_l | \hat{U}_l^S | \boldsymbol{r}_1, \boldsymbol{r}_2, \ldots, \boldsymbol{r}_l \rangle \, d^3r_1\, d^3r_2 \ldots d^3r_l$$

$$= \frac{1}{l!\,\lambda^{3(l-1)}} \int \langle 0, \boldsymbol{r}_2, \ldots, \boldsymbol{r}_l | \hat{U}_l^S | 0, \boldsymbol{r}_2, \ldots, \boldsymbol{r}_l \rangle \, d^3r_2 \ldots d^3r_l. \tag{38}$$

Now, when we go over to the (continuum) momentum representation, the Kronecker δ-functions δ_{k_i', k_i} become Dirac δ-functions $\delta^3(\boldsymbol{k}_i' - \boldsymbol{k}_i)$ and the expression (38) takes the form

$$\frac{1}{l!\,\lambda^{3(l-1)}(2\pi)^3} \int \langle \boldsymbol{k}_1, \ldots, \boldsymbol{k}_l | u_l^S | \boldsymbol{k}_1, \ldots, \boldsymbol{k}_l \rangle \, d^3k_1 \ldots d^3k_l, \tag{39}$$

where the function u_l^S denotes the function $\hat{U}_l^S$, *apart from the conservation factor* $\delta^3(\Sigma \boldsymbol{k}_i' - \Sigma \boldsymbol{k}_i)$, while the factor $(2\pi)^3$ in the denominator arises from the difference between the numbers of variables of integration in the expressions (38) and (39), respectively; see Lee and Yang (1959a). The first-order contribution towards the coefficient $\mathfrak{b}_l^S$ is then given by*

$$\frac{1}{l!\,\lambda^{3(l-1)}8\pi^3} \left(\frac{1}{2}(l!)\right) \left(\frac{-\lambda^8 r_0}{2\pi^3}\right) \sum_{n_1, n_2}' \lambda^{3(n_1+n_2)} \iint \exp\left(-n_1\beta \frac{\hbar^2 k_1^2}{2m}\right) \exp\left\{-\frac{\beta\hbar^2}{2m}(k_1^2+k_2^2)\right\}$$

$$\times \exp\left(-n_2\beta \frac{\hbar^2 k_2^2}{2m}\right) d^3k_1\, d^3k_2$$

$$= -\frac{\lambda^5 r_0}{32\pi^6} \iint \sum_{n_1, n_2}' \exp\left\{-(n_1+1)\beta \frac{p_1^2}{2m}\right\} \exp\left\{-(n_2+1)\beta \frac{p_2^2}{2m}\right\} \frac{d^3p_1\, d^3p_2}{\hbar^6}. \tag{40}$$

The double summation over the indices n_1 and n_2 is rather difficult to carry out because it is restricted by the condition: $n_1 + n_2 = l - 2$. Let us, therefore, consider the complete fugacity series $\Sigma_l \mathfrak{b}_l^S z^l$ rather than the single coefficient $\mathfrak{b}_l^S$. The first-order contribution towards this series will be obtained by multiplying the expression (40) by the factor $z^l (= z^{n_1+1}\, z^{n_2+1})$ and summing over *all* values of l; this clearly amounts to making the summations over n_1 and n_2 *unrestricted*. The integrand in (40) thus becomes

$$\frac{1}{\hbar^6} \frac{1}{z^{-1}\exp(\beta p_1^2/2m) - 1} \frac{1}{z^{-1}\exp(\beta p_2^2/2m) - 1},$$

whence we obtain for the integral concerned

$$\left[\int_0^\infty \frac{1}{\hbar^3} \frac{4\pi p^2\, dp}{z^{-1}\exp(\beta p^2/2m) - 1}\right]^2 = \left\{\frac{8\pi^3}{\lambda^3} g_{3/2}(z)\right\}^2;$$

see Sec. 7.1, especially eqn. (7.1.8). Substituting into (40), we obtain for the first-order

* It should not be difficult to check that the presence of the various δ-functions in the integrand of (39) ultimately reduces the integral to one over d^3k_1 and d^3k_2 only.

contribution towards the fugacity series

$$-\frac{2r_0}{\lambda}\{g_{3/2}(z)\}^2. \tag{41}$$

Thus, the fugacity series for a hard-sphere Bose gas is given by

$$\sum_l b_l^S z^l \equiv \frac{P\lambda^3}{kT} = g_{5/2}(z) - \frac{2r_0}{\lambda}\{g_{3/2}(z)\}^2 + 0\left(\frac{r_0}{\lambda}\right)^2, \tag{42}$$

the main term of the series being the result, (7.1.7), for an *ideal* Bose gas. Once again we note that, to the first-order in (r_0/λ), the expansion (42) is consistent with the corresponding term of the Beth–Uhlenbeck result (9.5.36); see also our earlier result (27).

Problems

9.1. For imperfect-gas calculations one sometimes employs the *Sutherland potential*:

$$\begin{aligned} u(r) &= \infty && \text{for} \quad r < r_0 \\ &= -\varepsilon\left(\frac{\sigma}{r}\right)^6 && \text{for} \quad r > r_0. \end{aligned}$$

Using this potential, determine the second virial coefficient of a classical gas. Also determine the first-order corrections to the gas law and to the various thermodynamic properties of the gas.

9.2. According to Lennard-Jones, the physical behavior of most real gases can be well understood if the intermolecular potential is assumed to be of the form

$$u(r) = \frac{A}{r^m} - \frac{B}{r^n},$$

where n is very nearly equal to 6 while m ranges between 11 and 13. Determine the second virial coefficient of a Lennard-Jones gas and compare your result with that for a van der Waals gas; see eqn. (9.3.10).

9.3. (a) Show that, for a gas obeying van der Waals equation of state (9.3.11),

$$C_P - C_V = Nk\left\{1 - \frac{2a}{kTv^3}(v-b)^2\right\}^{-1}.$$

(b) Also show that, for a van der Waals gas with *constant* specific heat C_V, the equation for an adiabatic process is

$$(v-b)T^{C_V/Nk} = \text{const.}$$

[Compare this result with the ideal-gas one, viz. $vT^{3/2} = \text{const.}$]

(c) Further show that the change in temperature resulting from an expansion of the gas (into vacuum) from volume V_1 to volume V_2 is given by

$$T_2 - T_1 = \frac{N^2 a}{C_V}\left(\frac{1}{V_2} - \frac{1}{V_1}\right).$$

9.4. The coefficient of volume expansion α and the isothermal bulk modulus B of a gas are given by the empirical expressions

$$\alpha = \frac{1}{T}\left(1 + \frac{3a'}{vT^2}\right) \quad \text{and} \quad B = P\left(1 + \frac{a'}{vT^2}\right)^{-1},$$

where a' is a constant parameter. Show that the foregoing expressions are mutually compatible. Also derive the equation of state of this gas.

9.5. Show that the first-order Joule–Thomson coefficient of a gas is given by the formula

$$\left(\frac{\partial T}{\partial P}\right)_H = \frac{1}{C_P}\left(T\frac{\partial(a_2\lambda^3)}{\partial T} - a_2\lambda^3\right),$$

where $a_2(T)$ is the second virial coefficient of the gas; see eqn. (9.2.4). Derive an explicit expression for this coefficient in the case of a gas with interparticle interaction

$$u(r) = \begin{cases} +\infty & \text{for} \quad 0 < r < r_0, \\ -u_0 & \text{for} \quad r_0 < r < r_1, \\ 0 & \text{for} \quad r_1 < r < \infty, \end{cases}$$

and discuss the temperature dependence of the coefficient.

9.6. Assume that the molecules of the nitrogen gas interact through the potential of the previous problem. Making use of the experimental data given below, determine the "best" empirical values for the parameters r_0, r_1 and u_0/k:

T (in °K)	100	200	300	400	500
$a_2\lambda^3$ (in °K per atm)	-1.80	-4.26×10^{-1}	-5.49×10^{-2}	$+1.12\times10^{-1}$	$+2.05\times10^{-1}$.

9.7. Determine the lowest-order corrections to the *ideal*-gas values of the Helmholtz free energy, the Gibbs free energy, the entropy, the internal energy, the enthalpy and the (constant-volume and constant-pressure) specific heats of a *real* gas.

Discuss the temperature dependence of these corrections in the case of a gas whose molecules interact through the potential of Problem 9.5.

9.8. The molecules of a solid attract one another with a force $F(r) = \alpha(l/r)^5$. Two semi-infinite solids composed of n molecules per unit volume are separated by a distance d, i.e. the solids fill the whole of the space where $x \leqslant 0$ or $x \geqslant d$. Calculate the force of attraction, per unit area of the surface, between the two solids.

9.9. Choosing the wave functions

$$u_p(\boldsymbol{r}) = \frac{1}{\sqrt{V}} e^{i(\boldsymbol{p}\cdot\boldsymbol{r})/\hbar}$$

to describe the motion of a free particle, write down the *symmetrized* wave functions for a pair of noninteracting bosons/fermions and derive, from first principles, the formula

$$\langle 1', 2' \mid \hat{U}_2^{S/A} \mid 1, 2\rangle = \pm\langle 2' \mid \hat{W}_1 \mid 1\rangle\langle 1' \mid \hat{W}_1 \mid 2\rangle.$$

9.10. Starting from the coordinate representation (9.8.24) for the operator $\hat{U}_2$, derive the momentum representation (9.8.33).

9.11. Show that, in the momentum representation, the binary kernel is given by the expression

$$\langle \boldsymbol{k}_1', \boldsymbol{k}_2' \mid \hat{B} \mid \boldsymbol{k}_1, \boldsymbol{k}_2\rangle = \frac{\partial}{\partial\beta}\langle \boldsymbol{k}_1', \boldsymbol{k}_2' \mid \hat{U}_2 \mid \boldsymbol{k}_1, \boldsymbol{k}_2\rangle + E'\langle \boldsymbol{k}_1', \boldsymbol{k}_2' \mid \hat{U}_2 \mid \boldsymbol{k}_1, \boldsymbol{k}_2\rangle,$$

where E' is defined by eqn. (9.8.35). Now, making use of the matrix (9.8.33) for the operator $\hat{U}_2$, show that

$$\langle \boldsymbol{k}_1', \boldsymbol{k}_2' \mid \hat{B} \mid \boldsymbol{k}_1, \boldsymbol{k}_2\rangle = -\frac{\lambda^6 r_0\hbar^2}{2\pi^2 m}\,\delta_{\boldsymbol{k}_1'+\boldsymbol{k}_2', \boldsymbol{k}_1+\boldsymbol{k}_2}\, e^{-\beta E} + O(r_0^2).$$

Clearly, the first-order term in this expression may be written as

$$-\lambda^6 \Omega_{12}' \exp\left\{\beta\frac{\hbar^2}{2m}(\nabla_1^2+\nabla_2^2)\right\},$$

where

$$\Omega_{12}' = \frac{r_0\hbar^2}{2\pi^2 m} 8\pi^3\,\delta(\boldsymbol{r}_1-\boldsymbol{r}_2) = \frac{4\pi r_0\hbar^2}{m}\,\delta(\boldsymbol{r}_1-\boldsymbol{r}_2);$$

compare this result with the defining equation (9.7.24), and note that in this form our result is closely related to the pseudopotential approach formulated in the next chapter; see eqn. (10.2.5).

9.12. Complete the mathematical steps leading to eqn. (9.8.37).

9.13. (i) Show that for a Bose gas of hard spheres, with spin J, we have for the fugacity series

$$\sum_l b_l^S z^l \equiv \frac{P\lambda^3}{kT} = (2J+1)\, g_{5/2}(z) - 2(J+1)(2J+1)\frac{r_0}{\lambda}\{g_{3/2}(z)\}^2 + 0\left(\frac{r_0}{\lambda}\right)^2.$$

Note that for $J = 0$ this series reduces to eqn. (9.8.42).

(ii) In a similar manner, show that for a Fermi gas of hard spheres, with spin J, the fugacity series will be

$$\sum_l b_l^A z^l \equiv \frac{P\lambda^3}{kT} = (2J+1)\, f_{5/2}(z) - 2J(2J+1)\frac{r_0}{\lambda}\{f_{3/2}(z)\}^2 + 0\left(\frac{r_0}{\lambda}\right)^2.$$

Note that for $J = 0$ the first-order term disappears; cf. the corresponding eqn. (9.5.37) of Beth and Uhlenbeck.

9.14. Show that, quite generally,

$$b_2^S(J) = (J+1)(2J+1)\, b_2^S(0) + J(2J+1)\, b_2^A(0)$$

and

$$b_2^A(J) = J(2J+1)\, b_2^S(0) + (J+1)(2J+1)\, b_2^A(0).$$

Verify that the values of $b_2^S(J)$ and $b_2^A(J)$, as obtained from the fugacity series of the previous problem, are consistent with these formulae.

9.15. Defining the virial coefficients $B(T)$, $C(T)$, ... by the *Kamerlingh Onnes expansion*

$$P = \frac{NkT}{V}\left\{1 + \frac{B(T)}{V} + \frac{C(T)}{V^2} + \ldots\right\},$$

it follows that

$$B(T) = -N\lambda^3 b_2/b_1^2, \qquad C(T) = N^2\lambda^6(-2b_1 b_3 + 4b_2^2)/b_1^4, \ldots .$$

Substituting for the coefficients b_l, as obtained from the series expansions of Problem 9.13, show that for a gas of hard-sphere bosons/fermions, with spin J,

$$B(T) = -N\frac{\lambda^3}{2J+1}\left[\pm\frac{1}{2^{5/2}} - (2J+1\pm 1)\frac{r_0}{\lambda} + 0\left(\frac{r_0}{\lambda}\right)^2\right]$$

and

$$C(T) = N^2\frac{\lambda^6}{(2J+1)^2}\left[\left(\frac{1}{8} - \frac{2}{3^{5/2}}\right) + 0\left(\frac{r_0}{\lambda}\right)^2\right].$$

9.16. Show that the coefficient b_2 for a quantum-mechanical *Boltzmannian* gas composed of "spinless" particles satisfies the following results:

$$b_2 = \lim_{J\to\infty}\left\{\frac{1}{(2J+1)^2}\, b_2^S(J)\right\} = \lim_{J\to\infty}\left\{\frac{1}{(2J+1)^2}\, b_2^A(J)\right\}$$

$$= \frac{1}{2}\{b_2^S(0) + b_2^A(0)\}.$$

Obtain the value of b_2, to the fifth order in (r_0/λ), by using the Beth–Uhlenbeck results (9.5.36, 37), and compare your answer with the classical value of b_2, namely $-(2\pi/3)(r_0/\lambda)^3$.

9.17. Study the analytic behavior of the series (9.8.42) near $z = 1$ and show that a Bose gas of hard spheres exhibits the phenomenon of Bose–Einstein condensation when the fugacity of the gas assumes the critical value

$$z_c = 1 + 4\zeta\left(\frac{3}{2}\right)\frac{r_0}{\lambda} + 0\left(\frac{r_0}{\lambda}\right)^2.$$

Further show that the pressure of the gas at the critical point is given by (Lee and Yang, 1958; 1960b)

$$\frac{P_c}{kT} = \frac{1}{\lambda^3}\left[\zeta\left(\frac{5}{2}\right) + 2\left\{\zeta\left(\frac{3}{2}\right)\right\}^2\frac{r_0}{\lambda} + 0\left(\frac{r_0}{\lambda}\right)^2\right].$$

CHAPTER 10

STATISTICAL MECHANICS OF INTERACTING SYSTEMS: THE METHOD OF PSEUDOPOTENTIALS

IN THIS chapter we propose to investigate the *low-temperature* behavior of imperfect gases by a method based on the concept of "pseudopotentials". This concept was first introduced by Fermi, in 1936, but the development for the purpose of the present application is essentially due to Huang and Yang (1957). Consequently, the treatment in the first four sections of this chapter runs very much parallel to the one given by Huang (1963).

By an imperfect gas we mean here an *extremely dilute* system of particles interacting through a potential which has a *finite* range and is such that there do not exist any two-body bound states. For such a system we have three basic parameters of the problem, each of which has the dimensions of a [length]:

(i) The *scattering length* a, which is an overall measure of the strength and range of the two-body potential. We recall that, for *low-energy* scattering of a particle by a center of force, the total scattering cross-section is given by $4\pi a^2$, irrespective of the detailed shape of the potential. The scattering length is, therefore, a "gross" representative of the interparticle interaction. For hard-sphere interaction, the scattering length is just the same as the hard-sphere diameter (which may be denoted by r_0).

(ii) The *mean thermal wavelength* λ, which is a measure of the average spatial extent of the wave packets that represent the particles of the system. This parameter represents a pure *quantum effect*: $\lambda = h/(2\pi mkT)^{1/2}$. At sufficiently low temperatures, the value of λ can be quite large—in fact so large that the role of the interparticle interactions might become a matter of secondary importance. Clearly, this will happen only when λ becomes much larger than a. In that case, the ideal-gas system ($a = 0$) will constitute our *unperturbed system* while the interparticle interactions will act as a *perturbation*. We will, therefore, work under the assumption: $(a/\lambda) \ll 1$.

(iii) The *mean interparticle separation* l, which is of the order of $v^{1/3}$ [$v(\equiv V/N)$ being the volume per particle in the system]. In an extremely dilute system, it is natural to assume that $v^{1/3} \gg a$ [that is, $na^3 \ll 1$, $n(\equiv N/V)$ being the particle density in the system]. Of course, the mean thermal wavelength λ and the mean interparticle separation l will be assumed to be of comparable magnitudes; consequently, the symmetry properties of the wave functions will be of considerable importance. Moreover, unless a statement is made to the contrary, the parameters a, λ and l will all be assumed to be much smaller than the linear dimensions of the container.

The problem now consists in formulating a scheme for calculating the thermodynamic functions of the given system to first few orders in the small quantities (a/λ) and $(a/v^{1/3})$. In the method of pseudopotentials, this is accomplished by replacing the actual Hamiltonian of the system (which contains the real potential) by an "effective" Hamiltonian (which contains a "pseudopotential" instead), such that the ground state and the low-lying energy levels of the system are given equally well by the new Hamiltonian. The mathematical simplicity arises from the fact that the pseudopotential does not concern itself with the details of the real potential—it only depends upon certain "gross" parameters of the problem (such as the phase shifts, the scattering length, etc.).

10.1. The two-body pseudopotential

To begin with, we consider a pair of particles interacting through a hard-sphere interaction of diameter r_0. The actual two-body Hamiltonian is given by

$$\hat{H}(\boldsymbol{r}_1, \boldsymbol{r}_2) = -\frac{\hbar^2}{2m}(\nabla_1^2+\nabla_2^2)+u(r), \tag{1}$$

where

$$\left.\begin{aligned} u(r) &= +\infty \quad \text{for} \quad r \leqslant r_0 \\ &= \quad 0 \quad\ \text{for} \quad r > r_0. \end{aligned}\right\} \tag{2}$$

In terms of the *center-of-mass* coordinates $\boldsymbol{R} = \frac{1}{2}(\boldsymbol{r}_1+\boldsymbol{r}_2)$ and the *relative* coordinates $\boldsymbol{r} = (\boldsymbol{r}_2-\boldsymbol{r}_1)$, the two-body wave function may be written as

$$\Psi(\boldsymbol{R}, \boldsymbol{r}) = \frac{1}{\sqrt{V}} e^{i(\boldsymbol{P}\cdot\boldsymbol{R})/\hbar}\psi(\boldsymbol{r}), \tag{3}$$

the eigenvalue equation for the relative motion being

$$\left\{-\frac{\hbar^2}{m}\nabla_r^2+u(r)\right\}\psi(\boldsymbol{r}) = \varepsilon\psi(\boldsymbol{r}); \tag{4}$$

here, m stands for "twice the reduced mass of the particles". Writing $\varepsilon = \hbar^2k^2/m$, k being the wave number for relative motion, we obtain for the specific case of hard-spheres

$$(\nabla_r^2+k^2)\,\psi(\boldsymbol{r}) = 0 \quad \text{for} \quad r > r_0, \tag{5}$$

$$\psi(\boldsymbol{r}) = 0 \quad \text{for} \quad r \leqslant r_0. \tag{6}$$

Thus, our hard-sphere potential is no more than a *boundary condition* for the relative-motion wave function! Indeed, there will be another boundary condition for the wave function as r becomes very large, but that has nothing to do with the potential problem. Equations (5) and (6) constitute an eigenvalue problem for the quantity k.

Let us first of all consider *spherically symmetric* solutions (which refer to the case $l = 0$ and are usually termed as the S-wave solutions), especially the ones corresponding to very low energies, i.e. for $k \to 0$. We then have, from eqns. (5) and (6),

$$\frac{1}{r^2}\frac{d}{dr}\left(r^2\frac{d\psi}{dr}\right) = 0 \quad \text{for} \quad r > r_0 \tag{5a}$$

$$\psi(r) = 0 \quad \text{for} \quad r \leqslant r_0. \tag{6}$$

The general solution of eqn. (5a) is well known: $\psi(r) = A + B/r$. However, for continuity with (6), the ratio B/A must be equal to $-r_0$; accordingly, we should have

$$\psi(r) = \text{const.}\left(1 - \frac{r_0}{r}\right) \quad \text{for} \quad r > r_0, \tag{5b}$$

$$= 0 \qquad\qquad \text{for} \quad r \leqslant r_0. \tag{6}$$

We now define an "extended" wave function $\psi_{ex}(r)$, such that

$$(\nabla_r^2 + k^2)\,\psi_{ex}(r) = 0 \quad \text{for all } r \text{ (except } r = 0), \tag{7}$$

with the specific condition that

$$\psi_{ex}(r_0) = 0. \tag{8}$$

For $k \to 0$, we have the obvious result:

$$\psi_{ex}(r) = \chi\left(1 - \frac{r_0}{r}\right) \quad \text{for all } r \text{ (except } r = 0), \tag{9}$$

where

$$\chi = \frac{\partial}{\partial r}(r\psi_{ex}). \tag{10}$$

It must be noted here that eqn. (10) is a direct consequence of eqn. (9); it does not fix the value of the constant χ. Our aim now is to generalize the eigenvalue equation (7) in such a way that the specific condition (8) is automatically taken care of! This in turn requires that we replace eqn. (7) by one that holds for all values of r, *including* $r = 0$. To do this, we note that, for $k \to 0$, the left-hand side of this equation has the value

$$\nabla_r^2\left\{\chi\left(1 - \frac{r_0}{r}\right)\right\} = -\chi r_0\,\nabla_r^2\left\{\frac{1}{r}\right\} = 4\pi\chi r_0\,\delta(\boldsymbol{r}); \tag{11}$$

see eqn. (74) of Appendix B. This indeed vanishes for all values of r, except when $r = 0$ (in which case it diverges). This is hardly surprising because the wave function $\psi_{ex}(r)$ itself diverges at $r = 0$. Naturally then, if we insert expression (11) into the right-hand side of eqn. (7), the resulting equation will hold for $r = 0$ as well—of course, in the limit $k \to 0$. The new equation would be

$$(\nabla_r^2 + k^2)\,\psi_{ex}(r) = 4\pi r_0\,\delta(\boldsymbol{r})\frac{\partial}{\partial r}(r\psi_{ex}) \quad \text{for all } r. \tag{12}$$

It will be noted that (i) the constant χ has been replaced by the formal expression (10) and (ii) the boundary condition (8) is no longer required to be explicitly stated.

The eigenvalue equation (12) admits of a rather simple interpretation. What has happened here is that the requirement of the boundary condition (8), which arose from the real two-body potential, has been eliminated and in its place a *pseudopotential operator*,

$$4\pi r_0 \frac{\hbar^2}{m}\,\delta(\mathbf{r})\frac{\partial}{\partial r}r, \tag{13}$$

has been introduced into the eigenvalue equation. The pseudopotential, we note, is singularly concentrated at the origin and has a value such that the vanishing of the wave function $\psi_{ex}(r)$ at $r = r_0$ is automatically guaranteed! Moreover, since the pseudopotential vanishes for all $r \neq 0$, eqn. (12) is identical with the original eqn. (5) for all $r > r_0$. As a result, the extended wave function $\psi_{ex}(r)$ is identical with the actual wave function $\psi(r)$ for all $r \geqslant r_0$ (and the same would be true for the eigenvalues of k); see Fig. 10.1. At the same time, we should note that the mathematical solution $\psi_{ex}(r)$, for $r < r_0$, is of no physical relevance to us; in this sense, the method of pseudopotentials works in the same spirit as the method of *electrical images*. The important point here is that the low-lying energy levels of the two-body problem, with hard-sphere interaction, as well as the relevant wave functions in the region of physical interest are given correctly by the new eigenvalue equation (12), *with no boundary conditions (except the one for large r)*.

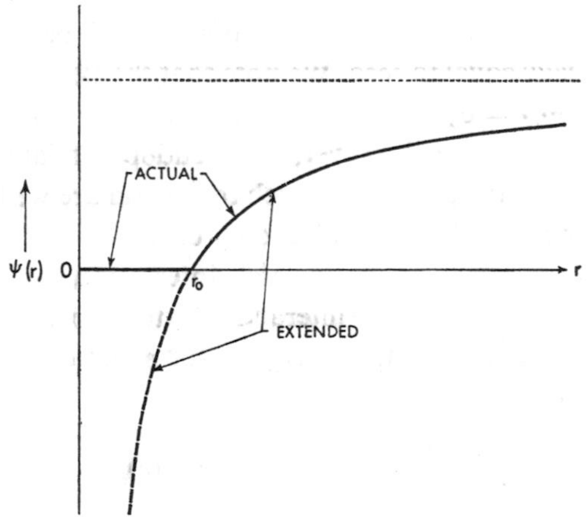

FIG. 10.1. The actual and the extended wave functions for the relative motion of two particles interacting through a hard-sphere interaction.

It is, however, important to note that eqn. (12) is only an approximation to truth, for its solutions coincide with the S-wave solutions of the actual two-body problem only in the limit $k \to 0$. For arbitrary values of k, the exact S-wave pseudopotential turns out to be

$$-4\pi \frac{\tan \eta_0(k)}{k}\frac{\hbar^2}{m}\,\delta(\mathbf{r})\frac{\partial}{\partial r}r, \tag{14}$$

where $\eta_0(k)$ is the relevant *phase shift* corresponding to the actual potential. In the case of hard sphere interaction, $\eta_0(k) = -kr_0$; see (9.5.32). The exact S-wave pseudopotential in this case may, therefore, be written as

$$4\pi \frac{\tan(kr_0)}{k} \frac{\hbar^2}{m} \delta(\boldsymbol{r}) \frac{\partial}{\partial r} r, \tag{15}$$

which agrees with (13) as $k \to 0$. We note that, by using (13) instead of (15), we neglect terms of the order of $(kr_0)^3$; the approximate pseudopotential should, therefore, be sufficient for calculating the various properties of the system correct to the *second* order in r_0. We also note that if we wish to include the P-wave solutions ($l = 1$), the D-wave solutions ($l = 2$), etc., a series of pseudopotential terms will have to be added to the right-hand side of the eigenvalue equation, the respective terms being proportional to r_0^3, r_0^5, etc. Thus, to the *second* order in r_0, it suffices to consider the S-wave solutions alone! Moreover, at sufficiently low temperatures, the assumption of k being small is quite reasonable, for it is physically equivalent to the condition: $(kr_0) \ll 1$ or $(r_0/\lambda) \ll 1$, which we have already assumed.

We shall now make an important observation on the operator character of the pseudopotential, namely that "the differential operator $(\partial/\partial r)r$ in the expressions (13)–(15) may be replaced by unity, provided that the wave function $\psi^{(0)}(\boldsymbol{r})$ on which the pseudopotential is supposed to operate is *well behaved* at $r = 0$". For, then,

$$\left[\frac{\partial}{\partial r}\{r\psi^{(0)}(\boldsymbol{r})\}\right]_{r=0} = \psi^{(0)}(\boldsymbol{r}=0) + \left[r\frac{\partial}{\partial r}\psi^{(0)}(\boldsymbol{r})\right]_{r=0} = \psi^{(0)}(\boldsymbol{r}=0); \tag{16}$$

the evaluation in (16) has been done only at the point $r = 0$ because everywhere else the pseudopotential is identically equal to zero. We note that the extended wave function, $\psi_{ex}(r)$, itself is not well behaved at $r = 0$; for this wave function, the differential operator $(\partial/\partial r)r$ will have to stay as such. In practice, however, the pseudopotential will be operating on the *unperturbed wave functions* of the problem, which in general are well behaved at $r = 0$; the differential operator $(\partial/\partial r)r$ may then be replaced by unity.

Finally, we consider a system with an interparticle interaction more general than the hard-sphere one. Once again we assume that the interaction extends over a finite range and is such that no two-body bound states can exist. The eigenvalue equation for the relative motion of two particles would then be

$$\frac{\hbar^2}{m}(\nabla_r^2 + k^2)\,\psi(r) = u(r)\,\psi(r), \tag{17}$$

with a suitable boundary condition on the solutions for large r; of course, the potential $u(r)$ might imply an *inner* boundary condition as well if, for example, a hard core is present. Again, for low energies ($k \to 0$), S-wave solutions alone will be important; we may, therefore, consider only those states that are spherically symmetric. Our equation then gives, for large values of r,

$$\begin{aligned}\psi_\infty(r) &= \text{const.}\,\frac{1}{r}\sin\{kr + \eta_0(k)\}\\ &= \text{const.}\,\frac{1}{r}\{\sin kr + \tan\eta_0(k)\cos kr\},\end{aligned} \tag{18}$$

where $\eta_0(k)$ is the *S-wave phase shift* for the potential of the problem. For $k \to 0$, our asymptotic solution takes the form

$$\psi_\infty(r) = \text{const.} \left\{1 + \frac{\tan \eta_0(k)}{kr}\right\}. \tag{19}$$

Now, for small k, we have the well-known expansion

$$\cot \eta_0(k) = -\frac{1}{ka} + \frac{1}{2} kr^* + \ldots, \tag{20}$$

a being the "scattering length" and r^* the "effective range" of the potential; for hard-sphere interaction $\{\eta_0(k) = -kr_0\}$, $a = r_0$ while $r^* = \frac{2}{3} r_0$. In general, the value of a is positive or negative according as the potential in question is predominantly repulsive or predominantly attractive.

In the "shape-independent" approximation, which holds for low values of k, we may take

$$\cot \eta_0(k) \simeq -\frac{1}{ka}; \tag{21}$$

eqn. (19) then takes the form

$$\psi_\infty(r) \simeq \text{const.} \{1 - a/r\}, \tag{22}$$

which may be compared with the expression (5b) for the hard-sphere case. Thus, for large r, the wave function of the problem is precisely the same as that with a hard-sphere interaction of diameter a. Now, since the most important values of r will be of the order of $v^{1/3} \simeq \lambda$, they will be large in comparison with the range of interaction; the conditions of the approximation are, therefore, fully met. Accordingly, we may take over the hard-sphere pseudopotential as such, except that the parameter r_0 is now replaced by the scattering length of the given potential. Thus, our two-body equation for the "extended" wave function would be

$$(\nabla_r^2 + k^2)\, \psi_{ex}(\boldsymbol{r}) = 4\pi a\, \delta(\boldsymbol{r}) \frac{\partial}{\partial r}(r\psi_{ex}), \tag{23}$$

the corresponding pseudopotential being

$$4\pi a \frac{\hbar^2}{m} \delta(\boldsymbol{r}) \frac{\partial}{\partial r} r. \tag{24}$$

Needless to say, the foregoing results hold irrespective of the *sign* of the scattering length.

10.2. The *N*-body pseudopotential and its eigenvalues

We now consider a generalization from the two-body problem to the *N*-body problem, with similar approximating conditions. In the case of hard-sphere interaction (with diameter r_0), the eigenvalue equation for the *N*-body problem would be

$$-\frac{\hbar^2}{2m}(\nabla_1^2 + \ldots + \nabla_N^2)\, \Psi(\boldsymbol{r}_1, \ldots, \boldsymbol{r}_N) = E\Psi(\boldsymbol{r}_1, \ldots, \boldsymbol{r}_N) \quad \text{if all } r_{ij} > r_0; \tag{1}$$

$$\Psi(\boldsymbol{r}_1, \ldots, \boldsymbol{r}_N) = 0 \quad \text{otherwise.} \tag{2}$$

Once again, the interaction is equivalent to an "inner" boundary condition on the wave function of the system, viz. that the wave function must vanish whenever the value of any of the $N(N-1)/2$ interparticle separations r_{ij} is less than or equal to the hard-sphere diameter r_0. In the ($3N$-dimensional) configuration space, this condition implies the vanishing of Ψ on and within a "tree-like" hypersurface whose "trunks", which are $N(N-1)/2$ in number, denote the regions where one of the interparticle separations is less than or equal to r_0; see Fig. 10.2. It is clear that the intersections of these "trunks" will represent configurations for which more interparticle separations than one are *simultaneously* less than or equal to r_0.

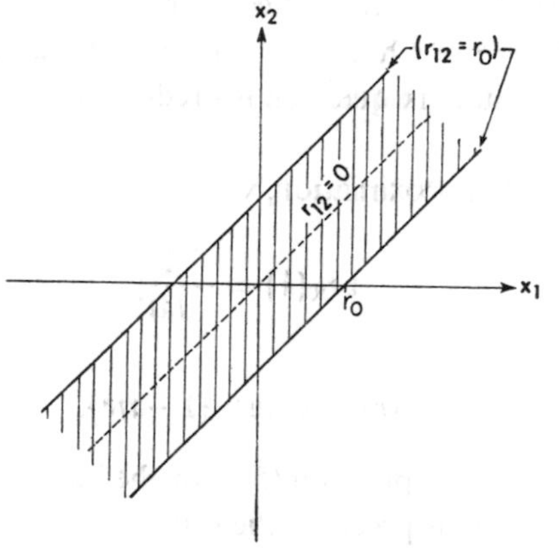

FIG. 10.2. The "tree-like" hypersurface in the configuration space of the particles 1 and 2. For configurations belonging to the shaded region ($r_{12} \leqslant r_0$), $\Psi = 0$.

In the spirit of our method, the low-energy wave functions $\Psi(\boldsymbol{r}_1, \ldots, \boldsymbol{r}_N)$, *throughout* the space outside the "tree", and the corresponding energy eigenvalues E will be given correctly if we replace the "trunks of the trees", $r_{ij} \leqslant r_0$, by a series of "multipoles" located at the "axes of the trunks", i.e. at the lines $r_{ij} = 0$. This amounts to introducing, into the eigenvalue equation, a series of two-body pseudopotentials, so that the extended wave functions $\Psi_{\text{ex}}(\boldsymbol{r}_1, \ldots, \boldsymbol{r}_N)$ are determined by the new Hamiltonian

$$\hat{H}' = -\frac{\hbar^2}{2m}\sum_i \nabla_i^2 + \frac{4\pi r_0 \hbar^2}{m}\sum_{i<j}\left\{\delta(\boldsymbol{r}_i - \boldsymbol{r}_j)\left(\frac{\partial}{\partial r_{ij}} r_{ij}\right)\right\}, \tag{3}$$

with no "inner" boundary conditions whatsoever: cf. the expression (10.1.13). Of course, this replacement will be somewhat in error in the vicinity of an intersection of two or more "trunks", i.e. when more than two particles get mutually scattered. In fact, in these regions we will have to place additional, higher-order pseudopotentials, which would take care of the fact that under these conditions the *binary collision approximation* does not hold. However, we can readily see that even the three-body pseudopotential, which has to be placed at the intersection of two trunks, say $r_{ij} \leqslant r_0$ and $r_{ik} \leqslant r_0$, will be $0(r_0^4)$, for it will appear along with a product of two delta functions, $\delta(\boldsymbol{r}_i - \boldsymbol{r}_j)$ and $\delta(\boldsymbol{r}_i - \boldsymbol{r}_k)$, and hence will require an additional factor of dimensions [length]^3 which, under the circumstances, can only be r_0^3. Similarly, the four-body pseudopotentials will be $0(r_0^7)$, and so on. Therefore, to the order of approximation we are aiming at, a sum of two-body pseudopotentials alone should suffice.

Finally, if the interaction potential is more general than the hard-sphere one (but has a finite range and does not permit two-body bound states), then the previous considerations can be taken over without much modification. All one has to do is replace the hard-sphere diameter r_0 by the scattering length a of the actual interaction, with the result

$$\hat{H}' = -\frac{\hbar^2}{2m}\sum_i \nabla_i^2 + \frac{4\pi a\hbar^2}{m}\sum_{i<j}\left\{\delta(\boldsymbol{r}_i-\boldsymbol{r}_j)\left(\frac{\partial}{\partial r_{ij}} r_{ij}\right)\right\}. \tag{4}$$

We now proceed to determine the eigenvalues of the Hamiltonian (4) in the *lowest* order of the perturbation. For this, we have to evaluae the diagonal matrix elements of the N-body pseudopotential, using free-particle wave functions to represent the "unperturbed states" of the system. Thus, the operators $(\partial/\partial r_{ij})\, r_{ij}$ will be operating on a set of functions which are *well behaved* for all values of r_{ij}; accordingly, these operators may be replaced by the *unit operators*. The pseudopotential then takes the form

$$\Omega(\boldsymbol{r}_1, \ldots, \boldsymbol{r}_N) = \sum_{i<j} \omega_{ij}; \qquad \omega_{ij} = \frac{4\pi a\hbar^2}{m}\,\delta(\boldsymbol{r}_i-\boldsymbol{r}_j). \tag{5}$$

Let the "unperturbed states" of the system be represented by the free-particle wave functions Φ_n, which are labeled by the set of (occupation) numbers $\{n_{p,s}\}$; here, $n_{p,s}$ denotes the number of particles with momentum $\boldsymbol{p}$ and spin quantum number s. Clearly, then,

$$\Phi_n(\boldsymbol{r}_1, \ldots, \boldsymbol{r}_N) = (N!)^{-1/2}\sum_P (\pm 1)^{[P]} P[u_{\alpha_1}(1) \ldots u_{\alpha_N}(N)] \tag{6a}$$

$$= (N!)^{-1/2}\sum_P (\pm 1)^{[P]} [u_{\alpha_1}(P1) \ldots u_{\alpha_N}(PN)] \tag{6b}$$

$$= (N!)^{-1/2}\sum_P (\pm 1)^{[P]} [u_{P\alpha_1}(1) \ldots u_{P\alpha_N}(N)]; \tag{6c}$$

cf. eqns. (5.5.7–10). Our notation here is the same as in Sec. 5.5, except that the α's here denote the spin as well as the momentum of the single-particle states; needless to say, the α's in (6) have to be consistent with the distribution set $\{n_{p,s}\}$ of the state Φ_n. The desired matrix element is then given by

$$\langle \Phi_n | \Omega | \Phi_n \rangle = \tfrac{1}{2}N(N-1)\int d^{3N}r(\Phi_n^*\omega_{12}\Phi_n); \tag{7}$$

in writing (7), use has been made of the fact that the various terms in the pseudopotential (5) make equal contributions towards the desired matrix element. Now, substituting (6c) into (7), we obtain

$$\langle \Phi_n | \Omega | \Phi_n \rangle = \frac{N(N-1)}{2N!}\sum_P\sum_Q (\pm 1)^{[P]+[Q]}\int d^{3N}r[u^*_{P\alpha_1}(1) \ldots u^*_{P\alpha_N}(N)]\,\omega_{12} \times [u_{Q\alpha_1}(1) \ldots u_{Q\alpha_N}(N)]$$

$$= \frac{1}{2(N-2)!}\sum_P\sum_Q (\pm 1)^{[P]+[Q]}[\langle P\alpha_1, P\alpha_2 | \omega_{12} | Q\alpha_1, Q\alpha_2\rangle \{\delta_{P\alpha_3, Q\alpha_3} \ldots \delta_{P\alpha_N, Q\alpha_N}\}], \tag{8}$$

where

$$\langle \alpha, \beta | \omega_{12} | \gamma, \delta\rangle = \int d^3r_1\, d^3r_2 \{u^*_\alpha(1)\, u^*_\beta(2)\, \omega_{12} u_\gamma(1)\, u_\delta(2)\}; \tag{9}$$

it will be noted that the $(N-2)$ Kronecker deltas have arisen from integrations over the coordinates $\boldsymbol{r}_3, \ldots, \boldsymbol{r}_N$. In view of these deltas, the permutation Q must be identical with the permutation P, except that the states $(Q\alpha_1, Q\alpha_2)$ may be either $(P\alpha_1, P\alpha_2)$ or $(P\alpha_2, P\alpha_1)$; in the former case, the order $[Q]$ of the permutation Q would be the same as the order $[P]$ of the permutation P, while in the latter case it would be opposite. Thus, we are left with

$$\langle \Phi_n | \Omega | \Phi_n \rangle = \frac{1}{2(N-2)!} \sum_P [\langle P\alpha_1, P\alpha_2 | \omega_{12} | P\alpha_1, P\alpha_2 \rangle \pm \langle P\alpha_1, P\alpha_2 | \omega_{12} | P\alpha_2, P\alpha_1 \rangle]. \quad (10)$$

Now, as we go through the various permutations of the states $\alpha_1, \ldots, \alpha_N$, the number of permutations for which the indices $(P\alpha_1, P\alpha_2)$ assume the specific values (l, m) will be $f_{lm}(N-2)!$, where the factor f_{lm} depends upon the occupation numbers n_l and n_m of the single-particle states l and m: if $l \neq m$, then $f_{lm} = n_l n_m$; if $l = m$, then $f_{lm} = \frac{1}{2} n_l(n_l - 1)$. Thus, we may write

$$\langle \Phi_n | \Omega | \Phi_n \rangle = \frac{1}{2} \sum_{l, m} f_{lm} [\langle l, m | \omega_{12} | l, m \rangle \pm \langle l, m | \omega_{12} | m, l \rangle]. \quad (11)$$

We now consider the following specific cases:

(i) *A gas of spinless bosons.* The single-particle states in this case are characterized by momenta alone; for instance,

$$u_p(\boldsymbol{r}) = \frac{1}{\sqrt{V}} \exp\{i(\boldsymbol{p}\cdot\boldsymbol{r})/\hbar\}. \quad (12)$$

Then, by virtue of eqns. (5), (9) and (12),

$$\begin{aligned}\langle \boldsymbol{p}_1, \boldsymbol{p}_2 | \omega_{12} | \boldsymbol{p}_1', \boldsymbol{p}_2' \rangle &= \frac{4\pi a \hbar^2}{mV^2} \int d^3r_1\, d^3r_2\, [e^{-i\{\boldsymbol{p}_1\cdot\boldsymbol{r}_1 + \boldsymbol{p}_2\cdot\boldsymbol{r}_2\}/\hbar}\, \delta(\boldsymbol{r}_1 - \boldsymbol{r}_2)\, e^{i\{\boldsymbol{p}_1'\cdot\boldsymbol{r}_1 + \boldsymbol{p}_2'\cdot\boldsymbol{r}_2\}/\hbar}] \\ &= \frac{4\pi a \hbar^2}{mV^2} \int d^3r_1\, [e^{-i\{(\boldsymbol{p}_1+\boldsymbol{p}_2) - (\boldsymbol{p}_1'+\boldsymbol{p}_2')\}\cdot\boldsymbol{r}_1/\hbar}] \\ &= \frac{4\pi a \hbar^2}{mV} \delta_{\boldsymbol{p}_1+\boldsymbol{p}_2,\, \boldsymbol{p}_1'+\boldsymbol{p}_2'}. \end{aligned} \quad (13)$$

Applying this result to eqn. (11), with the *upper* sign, and substituting for the factor f_{lm}, we obtain

$$\langle \Phi_n | \Omega | \Phi_n \rangle = \frac{4\pi a \hbar^2}{mV} \left[\sum_{p \neq k} n_p n_k + \frac{1}{2} \sum_p n_p(n_p - 1) \right]. \quad (14)$$

Further, in view of the fact that

$$\sum_{p \neq k} n_p n_k = \sum_p n_p \sum_k n_k - \sum_p (n_p)^2 = N^2 - \sum_p (n_p)^2,$$

eqn. (14) becomes

$$\langle \Phi_n | \Omega | \Phi_n \rangle = \frac{4\pi a \hbar^2}{mV} \left[N^2 - \frac{1}{2} N - \frac{1}{2} \sum_p (n_p)^2 \right]. \quad (15)$$

(ii) *A gas of spin-half fermions.* In this case, for any given value $\boldsymbol{p}$ of the particle momentum, we have two possibilities for its spin, namely that the spin is either up (which may be denoted by the symbol +) or down (which may be denoted by the symbol −). A little reflection then shows that

$$\langle l, m\,|\omega_{12}|\, l, m\rangle = \frac{4\pi a\hbar^2}{mV}, \tag{16}$$

while

$$\langle l, m\,|\omega_{12}|\, m, l\rangle = \frac{4\pi a\hbar^2}{mV}\,\delta_{s_l, s_m}. \tag{17}$$

Consequently, we obtain from eqn. (11), with the *lower* sign,

$$\langle \Phi_n\,|\Omega|\,\Phi_n\rangle = \frac{2\pi a\hbar^2}{mV}\sum_{l\neq m} n_l n_m[1-\delta_{s_l, s_m}] \tag{18a}$$

$$= \frac{2\pi a\hbar^2}{mV}\sum_{p\neq k}[n_p^+ n_k^- + n_p^- n_k^+] = \frac{4\pi a\hbar^2}{mV}N^+N^-, \tag{18b}$$

where $N^+(N^-)$ denotes the total number of particles with spin up (down):

$$N^+ = \sum_p n_p^+, \quad N^- = \sum_p n_p^-. \tag{19}$$

Finally, we obtain for the *first-order* energy eigenvalues E_n of the system

$$E_n \equiv \langle \Phi_n\,|\hat{H}'|\,\Phi_n\rangle = \sum_p n_p\frac{p^2}{2m} + \frac{4\pi a\hbar^2}{mV}\left[N^2 - \frac{1}{2}N - \frac{1}{2}\sum_p (n_p)^2\right] \tag{20}$$

in the case of *spinless bosons*, and

$$\sum_p (n_p^+ + n_p^-)\frac{p^2}{2m} + \frac{4\pi a\hbar^2}{mV}N^+N^- \tag{21}$$

in the case of *spin-half fermions.*

10.3. Low-temperature behavior of an imperfect Fermi gas

To the first order in a, the properties of an imperfect Fermi gas can be studied on the basis of the expression (10.2.21) for the energy eigenvalues E_n of the system. We must, however, remember that the expression (10.2.21) has been derived under the assumptions: $|a|/\lambda \ll 1$ and $|a|/v^{1/3} \ll 1$. Whereas the first assumption limits the applicability of the theory to *low temperatures*, the second assumption limits it to *low densities*. For a Fermi gas, however, these assumptions are closely related to one another. While the condition $\lambda \gg |a|$ implies that the important values of the single-particle wave numbers k, which will be $O(k_F)$, must be less than $|a^{-1}|$, i.e.

$$k \simeq k_F \equiv (3\pi^2 n)^{1/3} \ll |a^{-1}| \quad \text{or} \quad n\,|a^3| \ll 1, \tag{1}$$

the condition $|a| \ll v^{1/3}$ also implies that

$$|a| \ll n^{-1/3} \quad \text{or} \quad n\,|a^3| \ll 1. \tag{2}$$

Another point worth noting is that if $a > 0$, i.e. if the two-body interaction is predominantly repulsive, then the perturbation term in (10.2.21) favors configurations for which the product N^+N^- is small; in other words, it favors situations in which more and more particles are aligned in the same direction. One, therefore, expects that at sufficiently low temperatures a gas of fermions possessing an intrinsic magnetic moment μ^* and having $a > 0$ might display the phenomenon of *ferromagnetism*! The deeper reason for such a behavior lies in the fact that fermions with parallel spins prefer to remain physically apart,† which in turn helps to keep the (positive) potential energy of the system low; see the expression (10.2.5) for the pseudopotential Ω. Similarly, if $a < 0$, i.e. if the two-body interaction is predominantly attractive, then the perturbation term in (10.2.21) will favor configurations for which the product N^+N^- is large; in other words, it will favor situations in which $N^+ \simeq N^- \simeq N/2$. The physical reason for this kind of behavior is again understandable, namely that fermions with opposite spins prefer to remain physically close to one another,† which again helps to keep the (negative) potential energy of the system low. This in turn suggests the possibility of *antiferromagnetism*.

If, for the time being, we disregard the intrinsic magnetic moment of the particles, then the ground state of the gas will correspond to

$$\overline{N^+} = \overline{N^-} = \tfrac{1}{2}N$$

and, hence,

$$E_0 = E_0\,(\text{ideal}) + \frac{4\pi a\hbar^2}{mV}\left(\frac{N^2}{4}\right). \tag{3}$$

Substituting (8.1.27) for the ideal-gas result, we obtain for the ground state energy per particle

$$\frac{E_0}{N} = \frac{3}{5}\varepsilon_F\left\{1+\frac{5\pi a\hbar^2}{3m}\frac{n}{\varepsilon_F}\right\}. \tag{4}$$

Further, since $n = k_F^3/(3\pi^2)$ and $\varepsilon_F = \hbar^2k_F^2/(2m)$, the foregoing result takes the form

$$\frac{E_0}{N} = \frac{3}{5}\varepsilon_F\left\{1+\frac{10}{9\pi}(k_Fa)\right\}. \tag{5}$$

The velocity of sound waves in the gas, which may be denoted by the symbol c_0, is given by

$$c_0^2 = \frac{\partial P_0}{\partial(mn)}, \tag{6‡}$$

where P_0 denotes the ground state pressure of the gas:

$$P_0 \equiv -(\partial E_0/\partial V)_N = +n^2\frac{\partial(E_0/N)}{\partial n}. \tag{7}$$

† Note the factor $(1-\delta_{s_l, s_m})$ in eqn. (10.2.18a).

‡ Note that the ratio γ of the adiabatic compressibility to the isothermal compressibility of a Fermi gas tends to the value unity as $T \to 0$°K.

Substituting from (5) and remembering that $\varepsilon_F \propto n^{2/3}$ and $k_F \propto n^{1/3}$, we obtain for P_0

$$P_0 = \tfrac{2}{5} n\varepsilon_F \{1+(5/3\pi)(k_F a)\}; \tag{8}$$

note that for a *non-ideal* gas the pressure is no longer given by the two-thirds of the energy density! Finally, substituting (8) into (6), we obtain

$$c_0^2 = \frac{2\varepsilon_F}{3m}\{1+(2/\pi)(k_F a)\}. \tag{9}$$

We shall now study the magnetic behavior of a gas of fermions that possess an intrinsic magnetic moment $\boldsymbol{\mu}^*$ and interact through a two-body potential characterized by a *positive* scattering length a. In the presence of an external magnetic field $\boldsymbol{B}$, the energy levels of the system become

$$E_n = \sum_p (n_p^+ + n_p^-)\frac{p^2}{2m} + \frac{4\pi a\hbar^2}{mV} N^+N^- - \mu^*(N^+ - N^-)B, \tag{10}$$

where the last term denotes the potential energy that arises from the alignment of the elementary magnets along or against the direction of the applied field. The next step now consists in evaluating the partition function of the system which in turn would enable us to determine the degree of magnetization developed in the system. To do this, we follow the same procedure as adopted in Sec. 8.2; of course, we now have an interaction term as well in the expression for E_n. The minimization of the free energy of the system now leads to the condition

$$\left[\frac{\partial}{\partial N^+}\left\{\mu^* B(2N^+ - N) - \frac{4\pi a\hbar^2}{mV} N^+(N-N^+) - A_0(N^+) - A_0(N-N^+)\right\}\right]_{N^+=\overline{N^+}} = 0,$$

that is,

$$2\mu^* B - \frac{4\pi a\hbar^2}{mV}(N - 2\overline{N^+}) - \mu_0(\overline{N^+}) + \mu_0(N - \overline{N^+}) = 0, \tag{11}$$

where $A_0(\mathcal{N})$ and $\mu_0(\mathcal{N})$ are, respectively, the Helmholtz free energy and the chemical potential of a "fictitious" system of $\mathcal{N}$ "*spinless*" *noninteracting* fermions, of mass m, confined to the volume V of the given system. Equation (11) contains the complete solution of our problem.

To write down the final result in a more explicit form, we introduce a dimensionless variable r, defined by the formula

$$M \equiv \mu^*(\overline{N^+} - \overline{N^-}) = \mu^*(2\overline{N^+} - N) = \mu^* N r; \tag{12}$$

consequently, the susceptibility (per unit volume) of the gas will be given by

$$\chi = \frac{M}{VB} = \frac{\mu^* N r}{VB} = \mu^* n \frac{r}{B}. \tag{13}$$

Equation (11) may be written in terms of the variable r:

$$\mu_0\left(\frac{1+r}{2}N\right)-\mu_0\left(\frac{1-r}{2}N\right) = 2\mu^* B+\frac{4\pi a\hbar^2}{m}nr. \tag{14}$$

To study the possibility of *spontaneous magnetization*, we set $B = 0$ and look for any *nonzero* solutions of the resulting equation. At absolute zero, the resulting equation takes the form, see (8.1.25) with $g = 1$,

$$(6\pi^2 n)^{2/3}\frac{\hbar^2}{2m}\left\{\left(\frac{1+r}{2}\right)^{2/3}-\left(\frac{1-r}{2}\right)^{2/3}\right\} = \frac{4\pi a\hbar^2}{m}nr,$$

that is,

$$\{(1+r)^{2/3}-(1-r)^{2/3}\} = \zeta r, \tag{15}$$

where

$$\zeta = \frac{8\pi an}{(3\pi^2 n)^{2/3}} = \frac{8}{3\pi}(k_F a). \tag{16}$$

We note that eqn. (15) is insensitive to the sign of the variable r; consequently, for any given solution of this equation there exists another solution of equal and opposite value. This duplicity arises from the fact that, in the absence of an external field, no absolute meaning can be assigned to the directions "up" and "down". It is, therefore, superfluous to discuss all values of the variable r; it will suffice to consider positive values alone, viz. $0 \leqslant r \leqslant 1$.

Now, the solutions of eqn. (15) can be obtained graphically by determining the point(s) of intersection of the curve

$$y = (1+r)^{2/3}-(1-r)^{2/3} \tag{17}$$

with the straight line

$$y = \zeta r; \tag{18}$$

see Fig. 10.3. The curve (17) starts from the origin ($r = 0$) with an initial slope of $\frac{4}{3}$, rises with an ever-increasing slope and finally terminates at the point $r = 1$ (with value $4^{1/3} \simeq 1.5874$ and with infinite slope). The straight line (18), on the other hand, has a constant slope of value ζ. It thus follows that, for $\zeta \leqslant \frac{4}{3}$, there is no point of intersection other than $r = 0$; consequently, spontaneous magnetization is not possible unless $\zeta > \frac{4}{3}$, i.e. $(k_F a) > \pi/2$.

At this stage it is important to recall that the theory we are developing here is valid only if $(k_F a) \ll 1$; thus, within the framework of this theory, the possibility of spontaneous magnetization in an imperfect Fermi gas cannot be established with confidence. It is, however, instructive to see how spatial repulsion between fermions promotes an alignment of their spins. With this in view, we might carry on the use of our formalism to situations where $(k_F a) = 0(1)$; of course, the results following from such a study will have to be taken with caution. At any rate, if we allow the parameter $(k_F a)$ to exceed $\pi/2$ (and at the same time regard the first-order formalism to be applicable) the phenomenon of ferromagnetism does indeed set in. Referring to Fig. 10.3, we find that, for $\frac{4}{3} < \zeta < 4^{1/3}$, we do obtain a nonzero solution, $r(\zeta)$, of eqn. (15); this corresponds to a spontaneous magnetization of the system. For $\zeta = 4^{1/3}$, we have: $r = 1$, which corresponds to a state of magnetic saturation. For $\zeta > 4^{1/3}$, we do not obtain a proper solution in the mathematical sense, though physi-

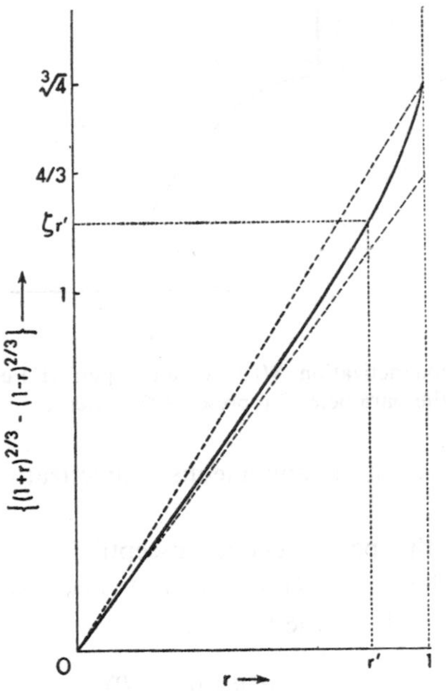

FIG. 10.3.

cally the situation continues to be one of magnetic saturation; under these conditions, the value $r = 1$ does ensure the minimization of the free energy $A(r)$ of the system, but this minimization is related to the value of the free energy being "lowest possible" and not to the "vanishing of the derivative $A'(r)$".

At finite but low temperatures ($T \ll T_F$), eqn. (15) is replaced by the following:

$$\{(1+r)^{2/3} - (1-r)^{2/3}\} - \frac{\pi^2}{12}\left(\frac{kT}{\varepsilon_F}\right)^2 \left\{\frac{1}{(1+r)^{2/3}} - \frac{1}{(1-r)^{2/3}}\right\} = \zeta r. \tag{19}$$

We observe that if $r(T = 0)$ is zero, then $r(T > 0)$ is also zero; this happens when $\zeta \leqslant \frac{4}{3}$. On the other hand, if $r(T = 0) > 0$, then (in view of the fact that the perturbation term in (19) is positive and the main term changes faster with r than does the term on the right-hand side) we would have: $r(T > 0) < r(T = 0)$. Thus, spontaneous magnetization decreases with rising temperature. There indeed exists a *critical temperature* $T_c(\zeta)$, above which spontaneous magnetization does not occur. If the critical temperature is itself much lower than the Fermi temperature T_F of the system, then we can determine it by equating the initial slope of the function $f(r)$ appearing on the left-hand side of eqn. (19) with the parameter ζ:

$$\frac{4}{3} + \frac{\pi^2}{9}\left(\frac{kT_c}{\varepsilon_F}\right)^2 = \zeta,$$

whence it follows that

$$\frac{kT_c}{\varepsilon_F} = \frac{3}{\pi}\left(\zeta - \frac{4}{3}\right)^{1/2} = \left\{\frac{24}{\pi^3}\left(k_F a - \frac{\pi}{2}\right)\right\}^{1/2}. \tag{20}$$

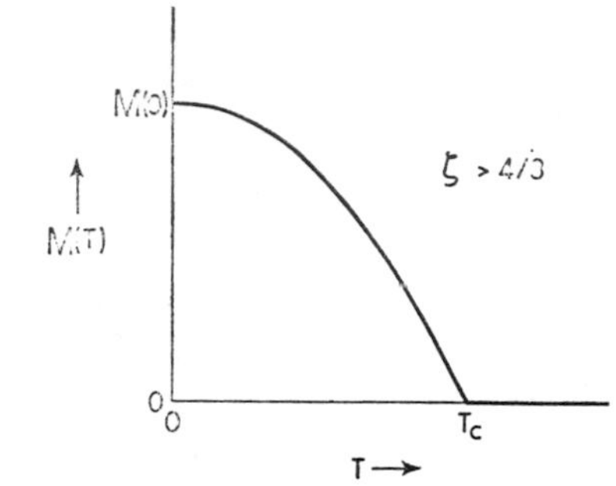

FIG. 10.4. Spontaneous magnetization $M(T)$ of an imperfect Fermi gas as a function of temperature; the parameter ζ represents the interparticle interactions.

Figure 10.4 depicts the variation of spontaneous magnetization with temperature T of the system.

Finally, we shall study the paramagnetic susceptibility of the imperfect Fermi gas. For this, we have to consider eqn. (14) with $B \neq 0$. The low-field susceptibility (per unit volume) of the system follows from the formula

$$\chi = \mu^* n \lim_{B \to 0} (r/B). \tag{21}$$

To evaluate (21), we carry out a Taylor expansion of the left-hand side of eqn. (14) about the value $r = 0$ and retain only the first term of the expansion, with the result

$$r \simeq \frac{2\mu^* B}{\left.\frac{\partial \mu_0(xN)}{\partial x}\right|_{x=\frac{1}{2}} - \frac{4\pi a \hbar^2}{m} n}, \tag{22}$$

whence we obtain

$$\chi = \frac{2n\mu^{*2}}{\left.\frac{\partial \mu_0(xN)}{\partial x}\right|_{x=\frac{1}{2}} - \frac{4\pi a \hbar^2}{m} n}; \tag{23}$$

cf. the corresponding ideal-gas results (8.2.19) and (8.2.20). We thus find that the paramagnetic susceptibility of the imperfect Fermi gas (with $a > 0$) is, at all temperatures, larger than the paramagnetic susceptibility of the corresponding ideal gas. In the limit $T \to 0$, the derivative appearing in the denominator of the expressions (22) and (23) takes the value $\frac{4}{3}\varepsilon_F$; hence,

$$\chi_{(T \to 0)} = \frac{\frac{3}{2}(n\mu^{*2}/\varepsilon_F)}{1-(2/\pi)(k_F a)}, \tag{24}$$

which may be compared with the Pauli result (8.2.23). In the limit $T \to \infty$, the same derivative takes the value $2kT$; hence

$$\chi_{(T \to \infty)} = \frac{n\mu^{*2}/kT}{1-(a\lambda^2/v)} \simeq n\mu^{*2}/kT = C/T, \tag{25}$$

where C is the *Curie constant* of the system. Here, λ is the mean thermal wavelength of the particles and v the volume per particle; obviously, the parameter $(a\lambda^2/v)$ would be negligibly small at high temperatures.

Figure 10.5 displays the dimensionless quantity $(\chi T/C)$, as a function of the dimensionless parameter (kT/ε_F), for an imperfect Fermi gas as well as for an ideal one. At high temperatures, both curves tend towards the common limiting value unity. At low temperatures, however, they differ considerably; the curve for the imperfect gas leaves the origin with a higher slope (indicating the fact that spatial repulsion between fermions enhances the probability of their spin alignment), attains values larger than unity and passes through a maximum in the vicinity of $T = T_F$.

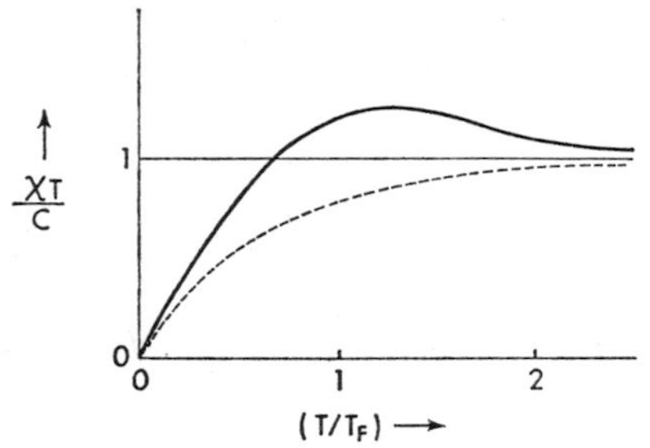

FIG. 10.5. Paramagnetic susceptibility of a Fermi gas (i) in the presence of a repulsive interaction (———) and (ii) in the absence of interactions (----------).

10.4. Low-temperature behavior of an imperfect Bose gas

The low-lying energy levels of a gaseous system composed of spinless bosons interacting through a two-body potential, characterized by the scattering length a, are given by eqn. (10.2.20):

$$E_n = \sum_p n_p \frac{p^2}{2m} + \frac{4\pi a\hbar^2}{mV}\left[N^2 - \frac{1}{2}N - \frac{1}{2}\sum_p (n_p)^2\right]; \tag{1}$$

unless a statement is made to the contrary, the scattering length a throughout this study will be assumed to be positive. Now, the validity of the foregoing expression requires that

$$\frac{a}{\lambda} \approx (ka) \ll 1 \quad \text{and} \quad \frac{a}{v^{1/3}} \ll 1. \tag{2}$$

As before, these conditions limit the validity of the formalism to systems at low temperatures and low densities.

We first of all examine the *ground state* of the given system, which corresponds to the distribution set

$$n_p = \begin{cases} 0 & \text{for } \boldsymbol{p} \neq 0, \\ N & \text{for } \boldsymbol{p} = 0. \end{cases} \tag{3}$$

Equation (1) then gives for the ground state energy of the system

$$E_0 = \frac{4\pi a\hbar^2}{mV}\frac{1}{2}N(N-1) \simeq N\frac{2\pi a\hbar^2 n}{m}. \tag{4}$$

We note that, unlike the Fermi gas, the ground state energy of the Bose gas arises *solely* from the interparticle interactions. Moreover, it appears from expression (4) as if each particle found itself floating over a "platform" of uniform potential φ, whose height is directly proportional to the particle density n in the system and to the scattering length a of the two-body interaction; obviously enough, the direct dependence of φ on a is consistent with the *shape-independent* approximation underlying the present treatment. It seems of interest to mention here that the result embodied in eqn. (4), in the special case of a hard-sphere Bose gas, had been obtained by Lenz as early as 1929.

Next, we consider the energy level E_1 of the system, which stands just above the ground level E_0. This will correspond to the transfer of *one* of the N particles of the system from the state $\varepsilon = 0$ to the state $\varepsilon = \varepsilon_1$, leaving all others in the state $\varepsilon = 0$. We then obtain from the general expression (1)

$$E_1 - E_0 = \varepsilon_1 + \frac{4\pi a\hbar^2}{mV}\left(-\frac{1}{2}\right)\left[\left(\sum_p n_p^2\right)_1 - \left(\sum_p n_p^2\right)_0\right]$$

$$= \varepsilon_1 - \frac{2\pi a\hbar^2}{mV}[\{(N-1)^2+1\} - N^2] \simeq \varepsilon_1 + \frac{4\pi a\hbar^2 n}{m}. \tag{5}$$

Now, $\varepsilon_1 \approx (\hbar^2/mV^{2/3}) \simeq 0$; therefore, the first excited state of the system seems to be separated from the ground state by an "energy gap" of magnitude $4\pi a\hbar^2 n/m$. In reality, if the theory is developed more rigorously, we do not obtain a finite energy gap for this system; we obtain instead a *diminution* of level density $g(E)$, not an *absence* of levels, just above the ground state E_0, which in turn changes the energy–momentum relationship of the low-lying excitations from $\varepsilon = p^2/2m$ to $\varepsilon = pc$, where c is the velocity of sound in the system (see Sec. 11.3). In this context, it seems worth while to derive here an expression for the sound velocity c in the limit $T \to 0$. To do this, we proceed in the same manner as we did in the case of an imperfect Fermi gas; see Sec. 10.3. Thus, we have

$$c_0^2 = \frac{\partial P_0}{\partial(mn)}, \quad \text{where} \quad P_0 = n^2\frac{\partial(E_0/N)}{\partial n}. \tag{6}$$

Substituting from (4), we obtain

$$P_0 = \frac{2\pi a\hbar^2 n^2}{m}, \quad \text{and} \quad c_0^2 = \frac{4\pi a\hbar^2 n}{m^2}. \tag{7}$$

Inserting numbers relevant to *liquid* He^4, viz. $a \simeq 2.2$ Å, $v = n^{-1} \simeq 45$ Å^3 per particle and $m \simeq 6.65\times10^{-24}$ g, we obtain $c_0 \simeq 125$ m/s. A comparison with the actual sound velocity in this liquid, which is about 240 m/s, should not be disheartening, for the theory developed here was never intended to be applicable to a liquid—it was only meant for a gaseous system, and that too a very dilute one!

We shall now discuss the low-lying excited states of the imperfect Bose gas, including the onset of *Bose–Einstein condensation*; see also Huang (1959, 1960). Remembering that, of all the n_p's, none except n_0 is expected to be macroscopically large, the sum Σn_p^2 in (1) might be replaced by n_0^2. At the same time, the term $\frac{1}{2}N$ might be dropped in comparison with the term N^2. Thus, we may write*

$$E_n \simeq \sum_p n_p \frac{p^2}{2m} + \frac{4\pi a\hbar^2}{mV}\left(N^2 - \frac{1}{2}n_0^2\right). \tag{8}$$

The partition function of the system will then be

$$\begin{aligned} Q(N) &\equiv \sum_n \exp(-\beta E_n) = \sum_{\{n_p\}} \exp\left[-\beta\left\{\sum_p n_p \frac{p^2}{2m} + \frac{4\pi a\hbar^2}{mV}\left(N^2 - \frac{1}{2}n_0^2\right)\right\}\right] \\ &= \exp\left(-N\frac{2a\lambda^2}{v}\right)\sum_{n_0=0}^{N}\left[\exp\left(n_0^2\frac{a\lambda^2}{V}\right)\sum_{\{n_p\}}{}' \exp\left(-\beta\sum_{p\neq 0} n_p \frac{p^2}{2m}\right)\right], \end{aligned} \tag{9}$$

where $\lambda\{=\hbar(2\pi\beta/m)^{1/2} = h/(2\pi mkT)^{1/2}\}$ is the mean thermal wavelength of the particles. For convenience, we have separated out the state $\boldsymbol{p} = 0$ from the states $\boldsymbol{p} \neq 0$, so that we first carry out the primed summation over all $n_p(\boldsymbol{p} \neq 0)$ with a *fixed* value $(N-n_0)$ of the partial sum $\Sigma_{p\neq 0} n_p$ and then carry out a summation over all possible values of n_0, viz. $n_0 = 0$ to $n_0 = N$. We also find it convenient to introduce a new variable ξ,

$$\xi \equiv \frac{n_0}{N} = 0, \frac{1}{N}, \frac{2}{N}, \ldots, 1, \tag{10}$$

which denotes the fraction of the particles lying in the state $\boldsymbol{p} = 0$. In terms of ξ, the expression for the partition function can be written as

$$Q(N) = \exp\left(-N\frac{2a\lambda^2}{v}\right)\sum_{\xi=0}^{\xi=1}\left[\exp\left(N\xi^2\frac{a\lambda^2}{v}\right)Q_0\{N(1-\xi)\}\right], \tag{11}$$

where $Q_0\{\mathcal{N}\}$ stands for the partition function of a "fictitious" system of $\mathcal{N}$ *noninteracting* bosons, of mass m, confined to the volume V of the actual system *(from which the state $p = 0$ has been "artificially removed" and hence all the $\mathcal{N}$ particles of the system are distributed over the states $p \neq 0$)*. We thus obtain

$$\begin{aligned} \frac{1}{N}\ln Q(N) &= -\frac{2a\lambda^2}{v} + \frac{1}{N}\ln\sum_{\xi=0}^{\xi=1}\left[\exp\left(N\frac{a\lambda^2}{v}\xi^2\right)Q_0\{N(1-\xi)\}\right] \\ &= -\frac{2a\lambda^2}{v} + \frac{1}{N}\ln\sum_{\xi=0}^{\xi=1}\left\{\exp\left[N\frac{a\lambda^2}{v}\xi^2 - \beta A_0\{N(1-\xi)\}\right]\right\}, \end{aligned} \tag{12}$$

where $A_0\{\mathcal{N}\}$ stands for the free energy of the "fictitious" system. Again, as has been done on several other occasions, the logarithm of the sum Σ_ξ in (12) may be replaced by the logarithm of the *largest* term in the sum, for the error committed in doing so will be statistically negligible. The value $\bar{\xi}$, of the variable ξ, which corresponds to the largest term in the sum can be obtained by setting the ξ-derivative of the general term in the sum equal to zero, i.e.

$$\left[N\frac{2a\lambda^2}{v}\xi - \beta\frac{\partial}{\partial\xi}A_0\{N(1-\xi)\}\right]_{\xi=\bar{\xi}} = 0; \tag{13}$$

* It should be noted here that in replacing the total sum $\Sigma_p n_p^2$ by the single term n_0^2 we are neglecting the partial sum $\Sigma_{p\neq 0} n_p^2$ in comparison with the number $(2N^2 - n_0^2)$ that has been retained in eqn. (8). Justification for this lies in the fact that, by the theory of fluctuations, the neglected part will be $O(N)$ and not $O(N^2)$.

consequently, the free energy of the actual system will be given by

$$A(N) \equiv -\frac{1}{\beta}\ln Q(N) \simeq \frac{2a\lambda^2}{\beta v}N - \frac{1}{\beta}\left[N\frac{a\lambda^2}{v}\bar{\xi}^2 - \beta A_0\{N(1-\bar{\xi})\}\right]$$

$$= A_0\{N(1-\bar{\xi})\} + N\frac{a\lambda^2}{\beta v}(2-\bar{\xi}^2). \tag{14}$$

In the limit $T \to 0$, $\bar{\xi} \to 1$; we, therefore, obtain for the ground state value of the free energy

$$N\frac{a\lambda^2}{\beta v} = N\frac{2\pi a\hbar^2 n}{m}, \tag{15}$$

which agrees, as it should, with the value (4) of the ground-state energy of the system.

Now, the free energy $A_0\{N(1-\xi)\}$ of the "fictitious" system can be obtained from the formulae derived in Sec. 7.1; we have

$$A_0\{N(1-\xi)\} = (G-PV)_0$$

$$= N(1-\xi)kT\ln z - V\frac{kT}{\lambda^3}g_{5/2}(z), \tag{16}$$

where z is the fugacity of the "fictitious" system (not of the real one) and is determined by the equation

$$N(1-\xi) = \frac{V}{\lambda^3}g_{3/2}(z) \qquad (0 \leqslant z \leqslant 1). \tag{17}$$

Equation (13), which determines $\bar{\xi}$, then takes the form

$$N\frac{2a\lambda^2}{v}\xi - \beta\left[-NkT\ln z + \left\{N(1-\xi)kT - V\frac{kT}{\lambda^3}g_{3/2}(z)\right\}\frac{\partial \ln z}{\partial \xi}\right]_{\substack{\xi=\bar{\xi}\\ z=\bar{z}}} = 0,$$

which, in view of the relationship (17), becomes

$$\frac{2a\lambda^2}{v}\bar{\xi} + \ln\bar{z} = 0; \tag{18}$$

of course, the quantities $\bar{z}$ and $\bar{\xi}$ are also related by the equation

$$(1-\bar{\xi}) = \frac{v}{\lambda^3}g_{3/2}(\bar{z}). \tag{19}$$

Equations (18) and (19) determine the values of the parameters $\bar{\xi}(T)$ and $\bar{z}(T)$, which in turn determine the free energy of the system:

$$A(N) = N(1-\bar{\xi})kT\ln\bar{z} - V\frac{kT}{\lambda^3}g_{5/2}(\bar{z}) + N\frac{a\lambda^2}{\beta v}(2-\bar{\xi}^2). \tag{20}$$

Equation (17) defines a family of (ξ, z)-curves for varying values of the quantity λ^3/v; see Fig. 10.6. The point $\xi = 1$, $z = 0$ is common to all the curves. The curve for $\lambda^3/v = g_{3/2}(1) \equiv \zeta(\frac{3}{2}) \simeq 2.612$ terminates at the point $\xi = 0$, $z = 1$. If $\lambda^3/v < g_{3/2}(1)$, then $\xi \to 0$

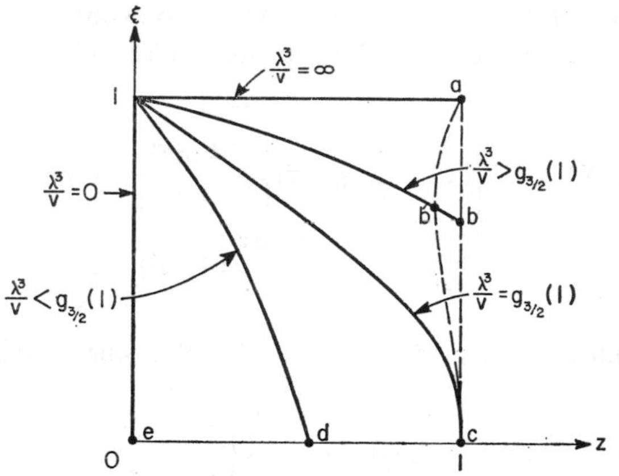

FIG. 10.6. The family of (ξ, z)-curves for a gas of bosons (after Huang). In the presence of interactions, the equilibrium values $\bar{\xi}$, of the variable ξ, lie on the curve *ab′cde* rather than on *abcde*.

for $z < 1$; the precise value of the latter is determined by the equation $g_{3/2}(z) = \lambda^3/v$. Similarly, if $\lambda^3/v > g_{3/2}(1)$, then $z \to 1$ for $\xi < 1$; the precise value of the latter is now determined by the equation $(1-\xi) = g_{3/2}(1)\, v/\lambda^3$.

Now, for a given value of the parameter λ^3/v, the equilibrium values $\bar{\xi}$ and $\bar{z}$, of the variables ξ and z, are those that satisfy eqn. (18) as well. For $\lambda^3/v > g_{3/2}(1)$, i.e. for $T < T_c$ (where T_c denotes the temperature at which Bose–Einstein condensation sets in, see eqn. (7.1.24)),

$$\bar{z} = \exp\left(-\frac{2a\lambda^2}{v}\,\bar{\xi}\right) \simeq 1-\frac{2a\lambda^2}{v}\,\bar{\xi}, \tag{21}$$

and

$$\bar{\xi} = 1-\frac{v}{\lambda^3}\,g_{3/2}(\bar{z}) \simeq 1-\frac{v}{\lambda^3}\,g_{3/2}(1)+0\left(\frac{a}{\lambda}\right). \tag{22}$$

Combining (21) and (22), we obtain*

$$\bar{z} \simeq 1-\frac{2a\lambda^2}{v}\left\{1-\frac{v}{\lambda^3}\,g_{3/2}(1)\right\}. \tag{23}$$

The foregoing situation is depicted by the point b' in Fig. 10.6; the corresponding situation in the case of an ideal gas is depicted by the point b. Clearly, when λ^3/v becomes infinitely high, i.e. when $T \to 0$, then $\bar{\xi} \to 1$ and $\bar{z} \to 1-2a\lambda^2/v$. On the other hand, when $\lambda^3/v = g_{3/2}(1)$, i.e. when $T = T_c$, then $\bar{\xi} \to 0$ and $\bar{z} \to 1$ exactly. The situation $\bar{\xi} > 0$ corresponds to the *condensed phase* of the system.

For $\lambda^3/v < g_{3/2}(1)$, i.e. $T > T_c$, $\bar{z}$ would be significantly less than unity. The condition (18), then, cannot be satisfied; instead, one has to accept that value of ξ which maximizes the general term in the sum (11) *numerically* rather than mathematically. This value turns out to be identically zero, so the relevant value of the parameter $\bar{z}$ is now determined by the equation $g_{3/2}(\bar{z}) = \lambda^3/v$. This situation corresponds to the *gaseous phase* of the system.

* Using the binary collision method, this result was also derived by Lee and Yang (1960c).

We are now in a position to write down an explicit expression for the free energy of the given system. This can be done by making proper substitutions into (20), whereby it follows that

$$\frac{1}{N}A(N) = kT\left\{\ln \bar{z} - \frac{v}{\lambda^3}g_{5/2}(\bar{z})\right\} + \frac{4\pi a\hbar^2}{mv} \quad \text{for} \quad \bar{\xi} = 0 \tag{24}$$

$$= -kT\frac{v}{\lambda^3}\zeta\left(\frac{5}{2}\right) + \frac{2\pi a\hbar^2}{mv}(2-\bar{\xi}^2) \quad \text{for} \quad \bar{\xi} > 0. \tag{25}$$

In writing the last result we have made use of the fact that since, for $\bar{\xi} > 0$, $\bar{z}$ is very close to unity, therefore

$$g_{5/2}(\bar{z}) \simeq g_{5/2}(1) + g_{3/2}(1)\ln \bar{z} \simeq \zeta\left(\frac{5}{2}\right) + \frac{\lambda^3}{v}(1-\bar{\xi})\ln \bar{z}$$

and, hence,

$$\left\{(1-\bar{\xi})\ln \bar{z} - \frac{v}{\lambda^3}g_{5/2}(\bar{z})\right\} \simeq -\frac{v}{\lambda^3}\zeta\left(\frac{5}{2}\right).$$

Substituting for $\bar{\xi}$ from eqn. (22), namely

$$\bar{\xi} = 1 - \frac{v}{v_c} \qquad \left\{v_c = \frac{\lambda^3}{g_{3/2}(1)}\right\}, \tag{26}$$

the expression for the free energy may be written as*

$$\frac{1}{N}A(N) = \frac{1}{N}A_{\text{id}}(N) + \frac{4\pi a\hbar^2}{mv} \quad \text{for} \quad \bar{\xi} = 0 \tag{27}$$

$$= \frac{1}{N}A_{\text{id}}(N) + \frac{2\pi a\hbar^2}{mv}\left(1 + \frac{2v}{v_c} - \frac{v^2}{v_c^2}\right) \quad \text{for} \quad \bar{\xi} > 0; \tag{28}$$

here, $A_{\text{id}}(N)$ denotes the ideal-gas value of the free energy $A(N)$. The pressure P of the imperfect Bose gas now follows straightforwardly:

$$P \equiv -\left(\frac{\partial A}{\partial V}\right)_{N,T} = -\left(\frac{\partial(A/N)}{\partial v}\right)_T = P_{\text{id}} + \frac{4\pi a\hbar^2}{mv^2} \quad \text{for} \quad \bar{\xi} = 0 \tag{29}$$

$$= P_{\text{id}} + \frac{2\pi a\hbar^2}{mv^2}\left(1 + \frac{v^2}{v_c^2}\right) \quad \text{for} \quad \bar{\xi} > 0. \tag{30}$$

We note that, unlike the ideal Bose gas, the pressure of the imperfect Bose gas continues to increase as v decreases below v_c; see Fig. 10.7. Moreover, in the limit $T \to 0$, since $P_{\text{id}} \to 0$ and $v_c \to \infty$, the pressure of the imperfect Bose gas tends to the limiting value

$$P_{T\to 0} = \frac{2\pi a\hbar^2}{mv^2}, \tag{31}$$

which agrees, as it should, with the relevant derivative of the limiting value of the free energy of the system; see expression (15).

* These results were first obtained by Lee and Yang (1958, 1960c).

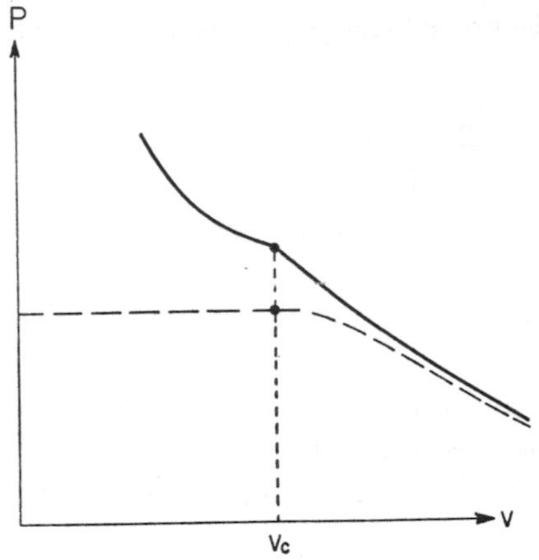

FIG. 10.7. The isotherms of a Bose gas (i) in the presence of a repulsive interaction (———) and (ii) in the absence of interactions (----------); after Huang.

Finally, we write down an explicit expression for the chemical potential μ of the system. For this, we simply combine the relevant expressions for the free energy A and the pressure P, with the result*

$$\mu \equiv \frac{A}{N} + Pv = \mu_{\text{id}} + \frac{8\pi a\hbar^2}{mv} \qquad \text{for} \quad \xi = 0 \tag{32}$$

$$= \mu_{\text{id}} + \frac{4\pi a\hbar^2}{mv}\left(1 + \frac{v}{v_c}\right) \qquad \text{for} \quad \xi > 0. \tag{33}$$

Thus, at the transition point (where $v = v_c$ and $\mu_{\text{id}} = 0$), the chemical potential of the gas is given by

$$\mu_c = \frac{8\pi a\hbar^2}{mv_c} = \frac{8\pi a\hbar^2}{m\lambda^3} g_{3/2}(1). \tag{34}$$

With the help of (34), we obtain for the corresponding value of the fugacity of the system

$$z_c = \exp(\mu_c/kT) \simeq 1 + \frac{4a}{\lambda} g_{3/2}(1); \tag{35}$$

cf. Problem 9.17. It should be noted here that the fugacity z of the actual system is not the same as the fugacity $\bar{z}$ of the "fictitious" system; the latter is always less than 1, as can be seen from eqn. (23). In passing, we note that, as $T \to 0$, the chemical potential tends to the limiting value

$$\mu_{T\to 0} = \frac{4\pi a\hbar^2}{mv}, \tag{36}$$

which again agrees with the relevant derivative of the limiting free energy (15).

* Using the binary collision method, this result was obtained by Lee and Yang (1960c).

10.5. The ground state wave function of a Bose fluid

In the preceding section we studied the influence of the interparticle interactions on the low-temperature behavior of a Bose gas. It is equally important to study the influence of these interactions on the wave functions of the system, for the properties of the wave functions of a system of bosons are of direct relevance to the phenomenon of superfluidity. To appreciate the significance of the role played by the interparticle interactions, one has only to note the "absurdity" of the density distribution in a system of noninteracting bosons at 0°K!

As is well known, the ground state of a system of noninteracting bosons is characterized by the fact that all the particles of the system are found to be in the *same* single-particle state. Denoting this state by the single-particle wave function $g^{(0)}(\boldsymbol{r})$, the total wave function of the system is given by

$$\Psi^{(0)}(\boldsymbol{r}_1, \ldots, \boldsymbol{r}_N) = \prod_{i=1}^{N} g^{(0)}(\boldsymbol{r}_i), \tag{1}$$

which indeed is a symmetric function of the arguments $\boldsymbol{r}_i$.* It is obvious that the density distribution in the system will be determined by the spatial variation of the quantity $|g^{(0)}(\boldsymbol{r})|^2$. Now, if we impose the condition that the wave function of the system vanish at the boundary of the container (which is assumed to be a cubical box of side L), then the density distribution will be determined by the function

$$\frac{8}{L^3} \sin^2\left(\frac{\pi x}{L}\right) \sin^2\left(\frac{\pi y}{L}\right) \sin^2\left(\frac{\pi z}{L}\right), \tag{2}$$

which possesses the "absurd" feature of an unduly large degree of concentration of the particles towards the center of the box! Physically, a situation of this kind does not make any sense. If, on the other hand, we impose *periodic* boundary conditions on the wave function, then the density distribution at 0°K is determined by

$$\left|\frac{1}{L^{3/2}} \exp\{i(\boldsymbol{k}\cdot\boldsymbol{r})\}\right|^2 = \frac{1}{L^3}, \tag{3}$$

which instead leads to a *uniform* distribution of particles throughout the system. The question now arises: why should a physical system be so sensitive to the boundary conditions imposed? The answer to this question rests with the interparticle interactions in the system, for a repulsive interaction, howsoever weak it may be, is sure to "even out" any nonuniformity of density distribution in the system, except possibly very near the walls where the boundary conditions might still leave some of their characteristic influence.

We, therefore, need a model which takes a proper account of the interparticle interactions in the system. One such model consists of particles interacting through a potential which, in the "shape-independent" approximation, may be replaced by a pseudopotential such as

$$\omega_{ij} = \frac{4\pi a\hbar^2}{m} \delta(\boldsymbol{r}_i - \boldsymbol{r}_j); \tag{4}$$

* It will be noted that, at absolute zero, a *Bose* system and a corresponding *quantum-mechanical Boltzmannian* system have identical properties!

see eqn. (10.2.5). We recall that, in the shape-independent approximation, the potential (4), which is *singular* in form, represents equally well a potential which may be *long- ranged* but weak enough (so that it has the same scattering length a). We may, therefore, adopt the Hartree approach, based on the self-consistent field approximation, in which one assumes that the range of the interparticle potential is several times the mean interparticle spacing (so that the rapidly fluctuating forces on a particle are much less important than the smoothly varying average force). The single-particle picture is again applicable—of course, with the difference that the single-particle wave functions are no longer the same as the free particle wave functions but rather the ones determined by the average force experienced by the particle(s). At any rate, the ground state wave function of the system is again written as

$$\Psi(\boldsymbol{r}_1, \ldots, \boldsymbol{r}_N) = \prod_{i=1}^{N} g(\boldsymbol{r}_i), \tag{5}$$

where $g(\boldsymbol{r})$ denotes the ground state wave function of a single particle while all the particles in the system are found to be in the same single-particle state. The quantity $|g(\boldsymbol{r})|^2$ will now determine the density distribution in the system.

We prefer to work with a function $\psi(\boldsymbol{r})$, which differs from the function $g(\boldsymbol{r})$ only through a constant factor, such that the quantity $|\psi(\boldsymbol{r})|^2$ gives precisely the particle density $n(\boldsymbol{r})$ in the system; the normalization condition then looks like

$$\int |\psi(\boldsymbol{r})|^2 \, d^3r = N = n_0 V, \tag{6}$$

where n_0 denotes the *mean particle density* in the system. The expectation value of the ground state energy of the system is now given by

$$E_0 = -\frac{\hbar^2}{2m}\int \psi^*(\boldsymbol{r}) \nabla^2\psi(\boldsymbol{r}) \, d^3r + \frac{1}{2}\iint |\psi(\boldsymbol{r}_1)|^2 u(\boldsymbol{r}_1-\boldsymbol{r}_2) |\psi(\boldsymbol{r}_2)|^2 \, d^3r_1 \, d^3r_2, \tag{7}$$

where $u(\boldsymbol{r}_1-\boldsymbol{r}_2)$ stands for the interparticle potential operating in the system. Replacing $u(\boldsymbol{r}_1-\boldsymbol{r}_2)$ by the pseudopotential (4), we obtain

$$E_0 = -\frac{\hbar^2}{2m}\int \psi^*(\boldsymbol{r}) \nabla^2\psi(\boldsymbol{r}) \, d^3r + \frac{2\pi a\hbar^2}{m}\int |\psi(\boldsymbol{r})|^4 \, d^3r. \tag{8}$$

If the particle density were uniform throughout the system, then the function $\psi(\boldsymbol{r})$ would be everywhere equal to the constant value $n_0^{1/2}$; see eqn. (6). The ground state energy would then be wholly potential:

$$E_0 = \frac{2\pi a\hbar^2}{m} n_0^2 V = N \frac{2\pi a\hbar^2 n_0}{m}, \tag{9}$$

which is identical with our earlier result (10.4.4). The energy associated with a nonuniformity of density (that might exist in the system) is, therefore, given by the difference between the expressions (8) and (9):

$$\begin{aligned} \mathcal{E} &= -\frac{\hbar^2}{2m}\int \psi^*(\boldsymbol{r}) \nabla^2\psi(\boldsymbol{r}) \, d^3r + \frac{2\pi a\hbar^2}{m}\int \{|\psi(\boldsymbol{r})|^4 - n_0^2\} \, d^3r \\ &= +\frac{\hbar^2}{2m}\int |\nabla\psi(\boldsymbol{r})|^2 \, d^3r + \frac{2\pi a\hbar^2}{m}\int \{|\psi(\boldsymbol{r})|^2 - n_0\}^2 \, d^3r, \end{aligned} \tag{10}$$

which, for $a > 0$, is positive definite. It may be noted that, in writing the last expression, we have made use of the fact that, by assumption, the wave function vanishes at the boundary of the container.

The central problem now consists in writing down a differential equation which determines the self-consistent wave function $\psi(\boldsymbol{r})$. It is obvious that this equation should be in the nature of a Schrödinger equation, in which the potential term is itself ψ-dependent. A little reflection shows that the desired equation is (see Ginzburg and Pitaevskii, 1958; Gross, 1958, 1960)

$$\left(-\frac{\hbar^2}{2m}\nabla^2+\frac{4\pi a\hbar^2}{m}|\psi|^2\right)\psi = \varepsilon\psi, \tag{11}$$

where ε is a *characteristic energy* associated with the ground state of the system. For a "uniform" system, the kinetic energy term vanishes and, since $\psi = n_0^{1/2}$, one obtains for ε

$$\varepsilon = \frac{4\pi a\hbar^2}{m}n_0, \tag{12}$$

which identifies ε with the *chemical potential* of the system; see eqn. (10.4.36). A real system, on the other hand, is essentially uniform throughout the space of the container (with particle density n^*, say) but there can exist regions (close to the walls or to certain singular locations within the system) where, by virtue of the mathematical conditions imposed on the wave function, the particle density might be different from the general value $n^*: 0 \leqslant n(r) < n^*$. It is clear that the value n^* would be slightly greater than the mean value n_0, for the deficiency of particles in the regions of nonuniformity will have to be compensated for by a slight excess of particles spread over the entire container; it is also clear that

$$\frac{n^*}{n_0} = 1+0\left(\frac{l}{L}\right), \tag{13}$$

where l is a measure of the spatial extent of the regions of nonuniformity while L is a measure of the physical dimensions of the container. The value of the characteristic energy ε is now obtained by noting that, in the overwhelming region of uniformity, the wave function ψ must tend to the contant value $n^{*1/2}$; eqn. (11) then gives

$$\varepsilon = \frac{4\pi a\hbar^2}{m}n^* \simeq \frac{4\pi a\hbar^2}{m}n_0. \tag{14}$$

To study the nature of the nonuniformities allowed by eqn. (11), we first consider a highly specialized situation, namely a system confined to the half-space $x \geqslant 0$, i.e. a system bounded on the left by the plane $x = 0$ and on the right by the plane $x = X \to \infty$, the remaining four walls being $y = \pm\infty$ and $z = \pm\infty$. The problem is clearly *one-dimensional*, because the particle density n and the wave function ψ will not depend upon the coordinates y and z. Now, substituting (14) into (11), one obtains

$$\frac{d^2\psi}{dx^2} - 8\pi a\psi^3+8\pi an^*\psi = 0. \tag{15}$$

Expressing x in terms of a "healing length" l,

$$x = l\xi \; \{l = (8\pi a n^*)^{-1/2} \simeq (8\pi a n_0)^{-1/2}\}, \tag{16}$$

eqn. (15) takes the form

$$\frac{d^2\psi}{d\xi^2} - \frac{1}{n^*}\psi^3 + \psi = 0, \tag{17}$$

whence one obtains the desired solution (Ginzburg and Pitaevskii, 1958; Wu, 1961)

$$\psi(\xi) = n^{*1/2} \tanh(\xi/\sqrt{2}), \tag{18}$$

which possesses the necessary features, namely

$$\psi(\xi)\Big|_{\xi=0} = 0 \quad \text{and} \quad \psi(\xi)\Big|_{\xi\to\infty} \to n^{*1/2}. \tag{19}$$

We note that the limiting density n^* is essentially reached as soon as ξ approaches a value of 5 or so; thus, the region of nonuniformity does not extend beyond about five healing lengths from the bounding wall. This result is highly satisfying because in a typical case, such as liquid helium (to which we may venture to apply the theory), the value of the healing length (with $a \simeq 2.2\times10^{-8}$ cm and $n_0 \simeq 2.2\times10^{22}$ particles/cm³) is very nearly 1 Å. Thus, the region of nonuniformity in this particular case extends up to about 5 Å from the wall; beyond that, the density of the fluid is essentially constant. This completely resolves the "absurdity" met with in the case of a noninteracting system.†

It seems worth while to mention here that the distribution of particles in a box, whose physical dimensions are much larger than the healing length ($L \gg l$), can be readily understood on the basis of the self-consistent wave function (18). Figure 10.8 shows the variation

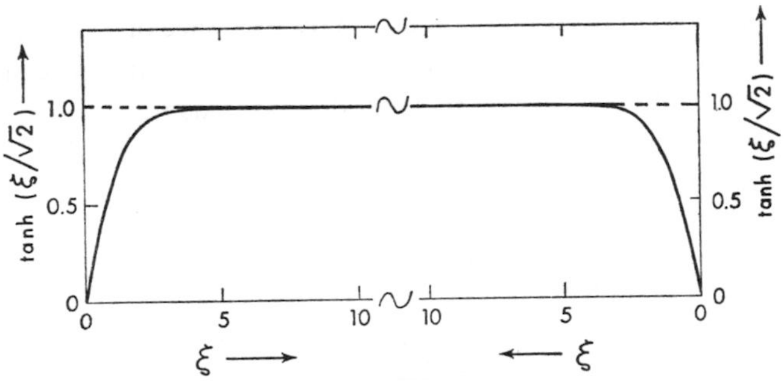

FIG. 10.8. The space variation of the "self-consistent" wave function $\psi(x)$ for a system of bosons interacting through a repulsive interaction ($a > 0$): $\xi = x(8\pi a n^*)^{1/2}$.

of the wave function, across the box, between two opposing walls. Over most of the space, the system is uniform; the regions of nonuniformity are confined to the immediate neighborhood of the walls. It is clear that if the distance between the walls were comparable to the healing length, the system would behave very differently; the solution in that case can be

† It should be noted here that if the scattering length a were zero, then our basic *nonlinear* equation (11) would reduce to a *linear* one: $\nabla^2\psi = -(2m\varepsilon/\hbar^2)\psi$, which would lead to the expression (2), with $\varepsilon = 3\pi^2\hbar^2/2mL^2$. The region of nonuniformity would then extend *throughout* the container!

employed to understand the limitations on the superfluid character of thin helium films (see Ginzburg and Pitaevskii, 1958).

The energy associated with the nonuniformity described by the wave function (18) can be calculated with the help of the formula (10). One obtains, per unit area of the wall,

$$\frac{\mathcal{E}}{A} = \frac{\hbar^2}{2m}\int_0^X \left(\frac{d\psi}{dx}\right)^2 dx + \frac{2\pi a\hbar^2}{m}\int_0^X (\psi^2 - n_0)^2\, dx, \tag{20}$$

where $\psi = n^{*1/2}\tanh(\xi/\sqrt{2})$ while the densities n^* and n_0 are connected by the normalization condition

$$\int_0^X \psi^2\, dx = \frac{N}{A} = n_0 X. \tag{21}$$

Carrying out the integrations and going over to the limit $X \to \infty$, we obtain

$$\frac{\mathcal{E}}{A} = \frac{\hbar^2}{2m}\left(\frac{\sqrt{2}n^*}{3l}\right) + \frac{2\pi a\hbar^2}{m}\left(\frac{2\sqrt{2}}{3}n^{*2}l\right) = \frac{4\sqrt{\pi}}{3}\,\frac{\hbar^2 n^{*3/2}a^{1/2}}{m}, \tag{22}$$

the energy being exactly half kinetic and half potential.† Clearly, the effect obtained here is a pure *quantum effect*, though it comes through the presence of the interparticle interactions. Again, substituting numbers pertaining to liquid helium, we obtain: $\mathcal{E}/A \simeq 0.19$ erg cm^{-2}. One feels tempted to compare this result with the low-temperature value of the surface tension of this liquid, which is about 0.37 erg cm^{-2}!

To calculate the surface tension, however, one has to study the nonuniformity of particle distribution in the region of the *free surface* of a Bose liquid rather than in the region close to a rigid wall. In this connection, we note that the existence of a free surface for a liquid necessarily requires that the particles in the surface region experience a predominantly *attractive* force towards the interior of the liquid; consequently, the effective scattering length of interaction for particles in the "surface layer" must be negative: $a = -|a|$. Then, applying our basic equation to the surface layer alone, we obtain, instead of eqn. (15),

$$\frac{d^2\psi}{dx^2} + 8\pi|a|\psi^3 - 8\pi|a|n_s\psi = 0, \tag{23}$$

where n_s is the value of $n(x)$ at the point of inflexion of the function $\psi(x)$; see Fig. 10.9. Introducing the dimensionless variable ξ, defined by

$$x = l\xi \; \{l = (8\pi|a|n_s)^{-1/2}\}, \tag{24}$$

eqn. (23) takes the form

$$\frac{d^2\psi}{d\xi^2} + \frac{1}{n_s}\psi^3 - \psi = 0, \tag{25}$$

† A more physical way of looking at this "equipartition" of energy is to note, from the expression (20) itself, that the quantity $\mathcal{E}/A$ must be of the form $Al^{-1}+Bl$. The *characteristic length* l may then be determined by minimizing the energy with respect to the variable l, which immediately gives $l = (A/B)^{1/2}$, whence the energy per unit area becomes $(AB)^{1/2}+(AB)^{1/2} = 2(AB)^{1/2}$. One can readily check that in the expression (22) the values of A and B are such that l is indeed equal to $(8\pi an^*)^{-1/2}$.

whence one obtains the desired solution (Brouwer and Pathria, 1967)

$$\psi(\xi) = \sqrt{(2n_s)} \operatorname{sech} \xi, \tag{26}$$

which possesses the necessary features, namely

$$\psi(\xi)\Big|_{\xi\to\infty} \to 0 \quad \text{and} \quad \psi(\xi)\Big|_{\xi=0} = \sqrt{(2n_s)} = \sqrt{n_l}, \tag{27}$$

where the quantity n_l may be identified with the particle density in the main body of the liquid. Moreover, since the derivative $d\psi/d\xi$ vanishes at $\xi = 0$, our solution joins smoothly with the "constant" wave function inside the liquid; see again Fig. 10.9.

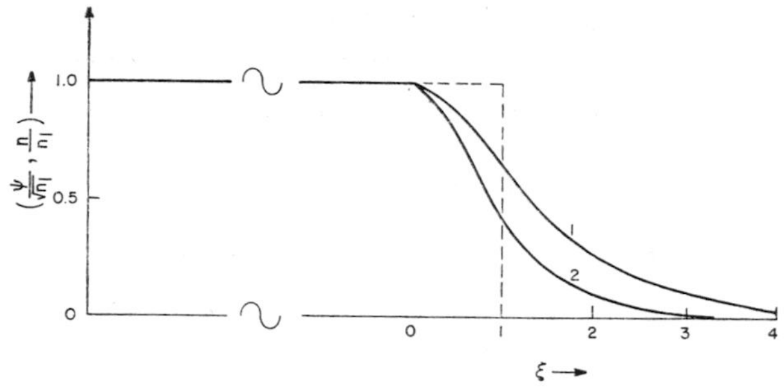

FIG. 10.9. Variation of the superfluid wave function (curve 1) and superfluid density (curve 2) at the free surface of a Bose liquid at absolute zero (after Brouwer and Pathria, 1967). The dashed line indicates the spatial extent of the liquid if the density were uniform throughout.

To evaluate the energy associated with this nonuniformity, we make use of the general formula (10), keeping in mind the fact that if the particles constituting the surface layer were distributed uniformly (with the limiting density n_l) they would precisely fill a slab of thickness l, because

$$\int_{x=0}^{x=\infty} (2n_s \operatorname{sech}^2 \xi)\, dx = n_l l. \tag{28}$$

The *structural energy* per unit area of the surface is, therefore, given by the difference between the energy per unit area of the surface layer as it is and the energy per unit area of the corresponding uniform slab. One obtains

$$\mathcal{E}/A = \frac{2\sqrt{\pi}}{3} \frac{\hbar^2 n_l^{3/2} |a|^{1/2}}{m} \simeq 0.09 \text{ erg cm}^{-2}, \tag{29}$$

which may be referred to as the "intrinsic" surface tension of liquid helium, as given by the imperfect Bose gas model.

Now, one knows that the existence of an "intrinsic" surface tension must lead to a set of quantized vibrational modes, or the so-called *ripplons*, on the free surface of the liquid, which in turn would give rise to a *zero-point energy* per unit area of the surface. This has also been evaluated by Brouwer and Pathria (1967), who obtained for it a value of 0.19 erg cm^{-2}. Combining this with (29), we obtain a net result of 0.28 erg cm^{-2} for the zero-point surface tension of liquid helium; considering the limitations of the model employed, this is not too bad an estimate in comparison with the extrapolated experimental estimate of 0.37 erg cm^{-2}.

10.6. States with quantized circulation

We now proceed to examine the possibility of an "organized motion" in the ground state of a Bose fluid. In this context, the most important concept is the one embodied in the *circulation theorem* of Feynman (1955), which establishes the physical basis for the existence of a "quantized vortex motion" in the fluid. In the case of liquid helium II, this motion has successfully resolved some of the vital questions which had been baffling the superfluid physicists for a long time.

The ground state wave function of a superfluid, composed of N bosons, may be denoted by a symmetric function $\Psi(\boldsymbol{r}_1, \ldots, \boldsymbol{r}_N)$; if the superfluid does not possess any kind of organized motion, then Ψ will be a pure real number. Now, suppose that the same fluid possessed a *uniform* mass motion with velocity $\boldsymbol{v}_s$; the appropriate wave function would then be

$$\Psi' = \Psi\, e^{i(\boldsymbol{P}_s \cdot \boldsymbol{R})/\hbar} = \Psi\, e^{im(\boldsymbol{v}_s \cdot \Sigma_i \boldsymbol{r}_i)/\hbar}, \tag{1}$$

where $\boldsymbol{P}_s$ denotes the total momentum of the fluid and $\boldsymbol{R}$ its center of mass:

$$\boldsymbol{P}_s = Nm\boldsymbol{v}_s; \quad \boldsymbol{R} = N^{-1} \sum_i \boldsymbol{r}_i. \tag{2}$$

Equation (1) gives the desired wave function *exactly* if the drift velocity $\boldsymbol{v}_s$ is uniform throughout the fluid. If $\boldsymbol{v}_s$ is nonuniform, then the wave function (1) would still be good locally—in the sense that the phase change $\Delta\varphi$ resulting from a "set of local displacements" of the atoms (over distances small enough for velocity variations to be appreciable) would be practically the same as the one following from the expression (1). Thus, for a given set of displacements $\Delta\boldsymbol{r}_i$ of the atoms constituting the fluid, the change in the phase of the wave function is very closely given by

$$\Delta\phi = \frac{m}{\hbar} \sum_i (\boldsymbol{v}_{si} \cdot \Delta\boldsymbol{r}_i), \tag{3}$$

where $\boldsymbol{v}_s$ is now a function of the variable $\boldsymbol{r}$.

The foregoing result may be applied for calculating the *net* phase change resulting from a displacement of atoms constituting a ring, from their original positions in the ring to the neighboring ones, so that on displacement we obtain a configuration which is physically identical with the initial configuration. In view of the symmetry character of the boson wave functions, the net phase change resulting from such a displacement must be an integral multiple of 2π (so that the wave function after the displacement is identical with the wave function before the displacement):

$$\frac{m}{\hbar} \sum_i{}' (\boldsymbol{v}_{si} \cdot \Delta\boldsymbol{r}_i) = 2\pi n, \qquad n = 0, \pm 1, \pm 2, \ldots; \tag{4}$$

the summation Σ' here goes over all the atoms constituting the ring. We note that for the foregoing result to be valid it is only the individual $\Delta\boldsymbol{r}_i$'s that have to be small, not the whole perimeter of the ring! Now, for a ring of a *macroscopic* size, one may regard the fluid as a continuum; eqn. (4) then becomes

$$\oint \boldsymbol{v}_s \cdot d\boldsymbol{l} = n\frac{h}{m}, \qquad n = 0, \pm 1, \pm 2, \ldots . \tag{5}$$

The quantity on the left-hand side of this equation is, by definition, the *circulation* (of the flow) associated with the circuit of integration and is clearly quantized, the "quantum of circulation" being h/m. Equation (5) is the mathematical statement of *Feynman's circulation theorem*.* It bears a striking resemblance to the quantum condition of Bohr, namely

$$\oint p\, dq = nh; \tag{6}$$

however, the region of application of the Feynman theorem is macroscopic rather than microscopic!

By Stokes's theorem, eqn. (5) may be written as

$$\int_S (\text{curl}\, \boldsymbol{v}_s)\cdot d\boldsymbol{S} = n\frac{h}{m}, \qquad n = 0, \pm 1, \pm 2, \ldots, \tag{7}$$

where $\boldsymbol{S}$ denotes the area enclosed by the circuit of integration. If this area is "simply-connected" and the velocity $\boldsymbol{v}_s$ is continuous throughout the area, then the domain of integration can be shrunk in a *continuous* manner without limit. The integral on the left-hand side is then expected to decrease *continuously* and finally tend to zero. The right-hand side, however, *cannot* vary continuously. We, therefore, conclude that in such a case the quantum number n must be zero, i.e. our integral must be *identically* vanishing. Accordingly, "in a simply-connected region, in which the velocity field is throughout continuous, the condition

$$\text{curl}\, \boldsymbol{v}_s = 0 \tag{8}$$

holds everywhere". This is precisely the condition postulated by Landau, in 1941, and has been a cornerstone of the theoretical understanding of the hydrodynamic behavior of superfluid helium.†

Clearly, the Landau condition is only a special case of the Feynman theorem. It is quite likely that in a "multiply-connected" domain, which cannot be shrunk *continuously* to zero (without encountering singularities in the velocity field), the Landau condition may not hold everywhere. A typical example of such a domain is provided by a *vortex flow*, which is a planar flow with cylindrical symmetry, such that

$$v_\varrho = 0, \quad v_\varphi = \frac{K}{2\pi\varrho}, \quad v_z = 0, \tag{9}$$

* That the vortices in a superfluid may be quantized, the quantum of circulation being h/m, was first suggested by Onsager (1949) in a footnote to a paper dealing with the classical vortex theory and the theory of turbulence!

† Drawing upon the well-known analogy between the phenomena of superfluidity and superconductivity, and the resulting correspondence between the mechanical momentum $m\boldsymbol{v}_s$ of a superfluid particle and the electromagnetic momentum $2e\boldsymbol{A}/c$ of a Cooper pair of electrons, we note that the relevant counterpart of the Landau condition (8) would be

$$\text{curl}\, \boldsymbol{A} \equiv \boldsymbol{B} = 0, \tag{8a}$$

which is precisely the *Meissner effect* in superconductors; furthermore, the relevant counterpart of the Feynman theorem (7) would be

$$\int_S \boldsymbol{B}\cdot d\boldsymbol{S} = n\frac{hc}{2e}, \tag{7a}$$

which leads to the "quantization of the magnetic flux", the quantum of flux being $hc/2e$.

where ϱ is the distance measured perpendicular to the axis of symmetry while K is the circulation of the flow:

$$\oint \boldsymbol{v} \cdot d\boldsymbol{l} = \oint v_\varphi(\varrho \, d\varphi) = K; \tag{10}$$

it is clear that the circuit of integration in (10) does enclose the axis of the vortex. Another version of the foregoing result is

$$\int_S (\text{curl } \boldsymbol{v}) \cdot d\boldsymbol{S} = \int_S \left\{ \frac{1}{\varrho} \frac{d}{d\varrho} (\varrho v_\varphi) \right\} (2\pi\varrho \, d\varrho) = K. \tag{11}$$

Now, at all $\varrho \neq 0$, curl $\boldsymbol{v} = 0$, but at $\varrho = 0$, where v_φ is singular, curl $\boldsymbol{v}$ appears to be rather indeterminate; it is not difficult to see that, at $\varrho = 0$, curl $\boldsymbol{v}$ actually diverges (in such a way that the integral (11) turns out to be finite). In this context, it seems worth while to point out that if we carry out the integration in (10) along a circuit which *does not* enclose the axis of the vortex, or the integration in (11) over a region that does not include the point $\varrho = 0$, the answer would be identically zero.

At this stage, we note that the energy associated with a unit length of a *classical* vortex is given by

$$\frac{\mathscr{E}}{L} = \int_a^b \frac{1}{2} \{2\pi\varrho \, d\varrho(mn_0)\} \left(\frac{K}{2\pi\varrho}\right)^2 = \frac{mn_0 K^2}{4\pi} \ln (b/a); \tag{12}$$

here, mn_0 is the mass density of the fluid (which is assumed to be uniform), b (the upper limit of integration) is related to the size of the container while a (the lower limit of integration) depends upon the structure of the vortex. In our study, a would be comparable to the interatomic separation.

In view of these results, we expect that, in a Bose fluid, vortex motions of the type suggested by the circulation theorem might indeed exist. To see this happen, we go back to the basic equation (10.5.11), which determines the self-consistent wave function $\psi(\boldsymbol{r})$ of the fluid. Since we are now looking for a planar flow with cylindrical symmetry, see eqn. (9), it seems natural to write

$$\psi(\boldsymbol{r}) = n^{*1/2} e^{is\varphi} f_s(\varrho), \tag{13}$$

so that

$$n(\boldsymbol{r}) \equiv |\psi(\boldsymbol{r})|^2 = n^* f_s^2(\varrho). \tag{14}$$

The velocity field associated with this wave function will be

$$\boldsymbol{v}(\boldsymbol{r}) = \frac{\hbar}{2im(\psi^*\psi)} (\psi^* \nabla\psi - \psi \nabla\psi^*)$$

$$= \frac{\hbar}{m} \nabla(s\varphi) = \left(0, s\frac{\hbar}{m\varrho}, 0\right). \tag{15†}$$

† It is of interest to see that the angular momentum per particle is given by

$$\frac{1}{\psi}\left(\frac{\hbar}{i}\frac{\partial}{\partial\varphi}\psi\right) = s\hbar = (mv_\varphi\varrho);$$

this may again be compared with the quantum condition of Bohr!

Comparing (15) with (9), we conclude that the circulation K in the present case is sh/m; by the circulation theorem, then, s must be an integer:

$$s = 0, \pm 1, \pm 2, \ldots. \tag{16}$$

Clearly, the value 0 is of no interest to us. Furthermore, the negative values differ from the positive ones only in the "sense of rotation" of the fluid. It is, therefore, sufficient to consider the positive values alone, viz.

$$s = 1, 2, 3, \ldots. \tag{17}$$

Now, if we substitute (13) into (10.5.11), we obtain

$$\left[\frac{1}{\varrho}\frac{d}{d\varrho}\left\{\varrho\frac{d}{d\varrho}f_s(\varrho)\right\} - \frac{s^2}{\varrho^2}f_s(\varrho)\right] - 8\pi a n^* f_s^3(\varrho) + \frac{2m\varepsilon}{\hbar^2}f_s(\varrho) = 0. \tag{18}$$

As $\varrho \to \infty$, $f_s(\varrho)$ must tend to unity; see eqn. (14) in which n^* stands for the limiting density in regions far away from the axis of the vortex. Equation (18) then demands that the characteristic energy ε be given by

$$\varepsilon = \frac{4\pi a\hbar^2}{m}n^*. \tag{19}$$

Substituting this into (18) and expressing ϱ in terms of a characteristic length l,

$$\varrho = l\varrho' \ \{l = (8\pi a n^*)^{-1/2} \simeq (8\pi a n_0)^{-1/2}\}, \tag{20}$$

we obtain

$$\frac{d^2f_s}{d\varrho'^2} + \frac{1}{\varrho'}\frac{df_s}{d\varrho'} + \left(1 - \frac{s^2}{\varrho'^2}\right)f_s - f_s^3 = 0. \tag{21}$$

Towards the axis of the vortex, where $\varrho \to 0$, the very high velocity of the fluid (and the very large centrifugal force accompanying it) must make the particles fly apart, thus causing an enormous decrease in the density of the fluid. Consequently, the function f_s must tend to zero as $\varrho \to 0$. This makes the last term in eqn. (21) negligible, whereby it reduces to the familiar Bessel's equation; accordingly, for small ϱ,

$$f_s(\varrho') \propto J_s(\varrho') \propto \varrho^s, \tag{22}$$

J_s being the *ordinary Bessel function* of order s. For $\varrho' \gg 1, f_s \simeq 1$; then, the first two terms of eqn. (21) become unimportant, with the result

$$f_s(\varrho') \simeq 1 - \frac{s^2}{2\varrho'^2}. \tag{23}$$

The actual solution is obtained by integrating the equation numerically; the results so obtained are shown in Fig. 10.10, where solutions for $s = 1$, 2 and 3 have been displayed.

We thus find that the model of the imperfect Bose gas does allow the presence of quantized vortices in the system. Not only that, we do not have to invoke here any special assumptions regarding the nature of the "core" of the vortex (as one has to do in the classical theory); the treatment leads naturally to a continual diminution of the particle density as the axial

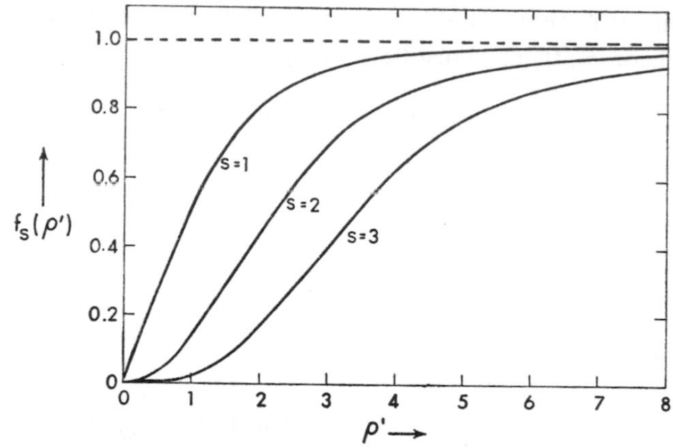

FIG. 10.10. Solutions of eqn. (21) for various values of s (after Kawatra and Pathria, 1966).

line is approached, so there does not exist any specific distribution of vorticity around this line. The distance scale, which governs the spatial variation of the particle density, is provided by the parameter l of eqn. (20).

Pitaevskii (1961), who was among the first to demonstrate the possibility of obtaining solutions whose natural interpretation lay in the quantized vortex motion (see also Gross, 1961; Weller, 1963), also evaluated the energy per unit length of the vortex. This is done by substituting (13), with the known values of the functions $f_s(\varrho)$, into the formula (10.5.10). The desired integrations have to be carried out numerically. For the three values of s, viz. $s = 1$, 2 and 3, Pitaevskii's results for the energy per unit length are, respectively,

$$\frac{\hbar^2 n_0 \pi}{m}\{1 \ln (1.46R/l), \quad 4 \ln (0.59R/l) \quad \text{and} \quad 9 \ln (0.38R/l)\}; \tag{24}$$

here, R denotes the outer radius of the domain concerned. These results may be compared with the "semi-classical" estimates,

$$\frac{\hbar^2 n_0 \pi}{m}\{1 \ln (R/a), \quad 4 \ln (R/a) \quad \text{and} \quad 9 \ln (R/a)\}, \tag{25}$$

which obtain directly from the formula (12), with K replaced by sh/m and b by R. It is obvious that vortices with $s > 1$ would be relatively unstable because *energetically* it would be cheaper for a system to have s vortices of unit circulation rather than a single vortex of circulation s.

The existence of *quantized vortex lines* in liquid helium II has been convincingly demonstrated by the ingenious experiments of Vinen (1958–61), where the circulation K around a fine wire immersed in the liquid was measured by means of the influence it exerts on the transverse vibrations of the wire. Vinen found that while vortices with unit circulation were exceptionally stable those with higher circulation too made appearance. Repeating Vinen's experiment with thicker wires, Whitmore and Zimmermann (1965) were able to observe stable vortices with circulation up to three quanta! For a survey of this and other aspects of superfluid behavior, see Vinen (1968) and Betts (1969).

10.7. "Rotation" of the superfluid

In view of the Landau condition, curl $\boldsymbol{v}_s = 0$, which was supposed to be the cornerstone of a theoretical understanding of the hydrodynamic behavior of a superfluid, it seemed natural to conclude that superfluid helium was incapable of acquiring a *rotational flow*, such as the one acquired by any normal fluid contained in a rotating vessel. In the normal case, a uniform rotation of a cylindrical vessel, with angular velocity $\boldsymbol{\omega}$ about the axis of symmetry of the vessel (which may be taken in the direction of the z-axis), results in the whole mass of the fluid rotating with the same angular velocity $\boldsymbol{\omega}$, the resulting velocity field $\boldsymbol{v}$ being

$$\boldsymbol{v} = \boldsymbol{\omega} \times \boldsymbol{r} = (0, \omega\varrho, 0), \tag{1}$$

with

$$\text{curl}\, \boldsymbol{v} = 2\boldsymbol{\omega} \neq 0. \tag{2}$$

The shape of the free surface of the rotating fluid is determined by the joint action of (i) the gravitational pull on the fluid and (ii) the centrifugal force resulting from rotation. In equilibrium, the surface is parabolic in shape, with curvature determined by the ratio ω^2/g:

$$z_1(\varrho) = z(0) + \frac{\omega^2}{2g}\varrho^2. \tag{3}$$

The question now arises: how would liquid helium II (which is a mixture of two fluids —a normal one comprising a fraction $x(T)$ of the whole and a superfluid one comprising the rest of it) behave on rotation? One feels tempted to argue that, since the superfluid seems incapable of rotation on account of the Landau condition, only the normal fraction $x(T)$ of the fluid will participate in the process of rotation. Consequently, one expects to have a situation intermediate between the full rotation on the one hand and the absence of rotation on the other—and, interestingly enough, the resulting picture will be a function of temperature. In particular, the shape of the free surface, though still parabolic, will have a lesser curvature:

$$z_2(\varrho) = z(0) + x(T)\frac{\omega^2}{2g}\varrho^2. \tag{4}$$

The first to examine this question experimentally was Osborne who, in 1950, studied the free surface of liquid helium II rotating at an angular velocity of 10 rad/sec. Surprisingly enough, the shape of the free surface was found to be in conformity with eqn. (3) rather than eqn. (4). In other words, the superfluid fraction as well was participating in the rotation of the liquid! In 1963, Osborne repeated his experiment with lower values of ω, viz. 3.1 rad/sec and 1.5 rad/sec, but the final result turned out to be the same as before. It may be mentioned here that, apart from Osborne's experiments, there exist a host of other experiments, very much different in approach, which have demonstrated the same fact, namely the superfluid component of liquid helium II "rotates" in practically the same manner as the normal component does.

Our problem then consists in reconciling two mutually contradicting elements. On one hand, we have the *observed rotation* of the superfluid which is very much like the one depicted by eqns. (1) and (2); on the other, we have the Landau condition, curl $\boldsymbol{v}_s = 0$, which advocates an *irrotational flow*! The desired reconciliation comes about through the "generation of quantized vortex lines in the superfluid", which gives the fluid an *overall* velocity field $\boldsymbol{v}$ identical with the one acquired by a normal fluid; at the same time, the quantity curl $\boldsymbol{v}_s$ vanishes practically throughout the fluid (but diverges at the axes of the vortices, so that its *average* value in the fluid is precisely 2ω). What is then the arrangement of the quantized vortex lines in the superfluid so that the fluid simulates so well the rotational pattern of a normal fluid? To determine an appropriate arrangement, we resort to an argument due to Lane (1962).

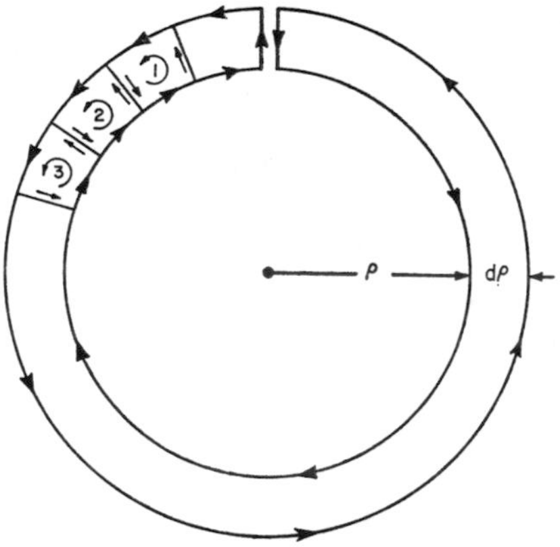

FIG. 10.11. A model for the "rotation" of the superfluid.

Referring to Fig. 10.11, consider the elemental area enclosed between the circles ϱ and $\varrho+d\varrho$. Let $n(\varrho)$ be the areal density of the vortex lines in the fluid, each line carrying a circulation of value h/m. Then, the net circulation in the contour shown will be equal to h/m times the number of lines enclosed, i.e.

$$\frac{h}{m}\{n(\varrho)\cdot 2\pi\varrho\, d\varrho\}. \tag{5}$$

On the other hand, the net circulation in the contour is also given by

$$(v+dv)\cdot 2\pi(\varrho+d\varrho)-v\cdot 2\pi\varrho \simeq 2\pi d(v\varrho), \tag{6}$$

where $v(\varrho)$ denotes the *overall* velocity field. Since vortex lines are the only source of circulation here, expressions (5) and (6) must be equal; it then follows that

$$n(\varrho) = \frac{m}{h\varrho}\frac{d}{d\varrho}(v\varrho). \tag{7}$$

Clearly, the velocity field in the absence of a vortex distribution, $n(\varrho) = 0$, will be

$$v = \text{const.}/\varrho, \tag{8}$$

which is indeed *irrotational* and satisfies the Landau condition *everywhere* (except at the axis $\varrho = 0$ where a source of circulation might still exist). On the other hand, the normal-fluid pattern, $v(\varrho) = \omega\varrho$, is possible if and only if we have a *uniform array* of vortex lines, with the areal density

$$n(\varrho) = \frac{2\omega m}{h} \tag{9}$$

and the consequent mean spacing between the lines

$$b \approx \frac{1}{n^{1/2}} \approx \left(\frac{h}{m\omega}\right)^{1/2} \propto \frac{1}{\omega^{1/2}}. \tag{10}$$

For $\omega \approx 1$ rad/sec, b turns out to be a fraction of a millimeter, which is much larger than the lateral extent of a vortex, the latter being only a few Angstrom units. Thus, *by and large*, curl $\boldsymbol{v}_s = 0$; it is only over (infinitesimally) narrow regions of space that curl $\boldsymbol{v}_s$ is (infinitely) large. Now, by definition, the *average* value of curl $\boldsymbol{v}_s$ is equal to the mean circulation per unit area; it must, therefore, be given by h/m times the number of vortex lines per unit area. By (9), this is precisely equal to 2ω.

We have thus obtained a workable model for the "rotation" of the superfluid in terms of quantized vortex lines. A word may now be said about the shape of the free surface, as follows from this model. Clearly, since the *overall* velocity field of the superfluid is the same as that of a normal fluid, viz. $v(\varrho) = \omega\varrho$, the free surface of the liquid will have a parabolic shape with regular curvature, as given by eqn. (3); this is precisely what the experiments found. However, if the vortex lines are strictly localized (which, in practice, may not be the case) then one may expect to observe a "depression" in the surface at the location of each line. It has been estimated that, if we neglect the smoothening influence of the surface tension of the liquid, this depression may be of the order of 10 microns at a distance of about 1 micron from the center of the vortex; however, the surface tension would probably "flatten" the depressions so much that they may not be much deeper than about 10 Å anywhere.

In the end, we would like to check our model on the basis of *energy considerations*. In other words, we would like to see how this model fares, in comparison with solid-like rotation, as regards the energy required to set up the flow. To investigate this point, we start with the assumption that the macroscopic velocity field $v(\varrho)$ is an arbitrary but well-behaved function of the radial distance ϱ. Then, the energy (per unit height) of the rotating fluid, as seen by a stationary observer, will be

$$E = \int_0^R \tfrac{1}{2}\{2\pi\varrho\, d\varrho(mn_0)\}\, v^2(\varrho), \tag{11}$$

while its angular momentum will be

$$J = \int_0^R \{2\pi\varrho\, d\varrho(mn_0)\}\, v(\varrho)\cdot\varrho; \tag{12}$$

here, mn_0 is the mass density of the fluid (which is assumed to be uniform) while R is the radius of the cylindrical vessel. Now, for any physical system, the state of thermodynamic equilibrium with a closed vessel in uniform rotation (of angular velocity ω_0) is obtained by minimizing the quantity $E' = (E-\omega_0 J)$; see, for example, Landau and Lifshitz (1958). Accordingly, to obtain the *equilibrium velocity field* in the present problem, we must minimize the quantity

$$E' = \pi(mn_0)\int_0^R \{\varrho v^2(\varrho)-2\omega_0\varrho^2 v(\varrho)\}\, d\varrho. \tag{13}$$

Taking a functional derivative of E' with respect to $v(\varrho)$ and requiring that the derivative vanish for ***arbitrary*** values of R, we obtain for the equilibrium velocity field

$$2\varrho v_{\text{eq}}(\varrho)-2\omega_0\varrho^2 = 0, \quad \text{i.e.} \quad v_{\text{eq}}(\varrho) = \omega_0\varrho, \tag{14}$$

which corresponds precisely to the solid-like rotation of the fluid. It is then natural that any normal fluid contained in a uniformly rotating vessel should acquire a solid-like rotation. Substituting (14) into (13), we obtain for the equilibrium value of E'

$$E'_{\text{eq}} = -\frac{\pi}{4}(mn_0)\omega_0^2 R^4. \tag{15}$$

Now, our vortex-line model for the "rotation" of the superfluid does possess an overall velocity field identical with the expression (14); its detailed pattern, however, is very very different. It is, therefore, necessary that a calculation be carried out for the quantity E'_{eq} for this model and a comparison made with the classical result (15). This has been done by Hall (1960), who considered an array of N vortex lines, each of circulation h/m, distributed *uniformly* over a circle of area πr^2 ($= Nh/2m\omega$, so that the average value of curl $\boldsymbol{v}_s$ be 2ω); the whole array of lines rotated with a uniform angular velocity ω. The space outside this circle was supposed to be vortex free, so the velocity pattern in this space was determined by the totality of the vortex lines within:

$$v(\varrho) = \frac{N\hbar}{m\varrho} \qquad (r < \varrho < R). \tag{16}$$

The parameters N and ω were the natural variables of the problem. Hall then wrote down general expressions for the energy E and the angular momentum J of the total fluid, each of the expressions consisting of the following three parts:

(i) the contribution arising from the *overall* solid-like rotation of the fluid in the region $0 < \varrho < r$,
(ii) the "extra" contribution arising from the presence of the *specific* localized fields of vortices in the same region, and
(iii) the contribution arising from the *irrotational* flow in the region $r < \varrho < R$.

Minimizing $E' = (E-\omega_0 J)$, where ω_0 is the angular velocity of the uniformly rotating vessel, with respect to the variables ω and N, Hall found the remarkable results

$$\omega_{\text{eq}} = \omega_0 \tag{17}$$

and

$$N_{\text{eq}} = \frac{2\pi R^2\omega_0}{h/m} - \left[\frac{4\pi R^2\omega_0}{h/m}\left\{\ln(b/l)-\frac{1}{2}\right\}\right]^{1/2} + \ldots, \tag{18}$$

where b is the mean spacing of the vortices, see eqn. (10), while l is the characteristic length that determines the spatial extent of a vortex; see eqn. (10.6.20). We thus find that (i) the superfluid, including the vortex-line pattern, "rotates" with the *same* angular velocity as the containing vessel, and (ii) the vortices fill practically the *whole* of the vessel, except a "monolayer" ring in the immediate neighborhood of the bounding wall.*

Finally, substituting the equilibrium values of ω and N into the original expression for E', one obtains for its equilibrium value

$$E'_{eq} = -\frac{\pi}{4}(mn_0)\omega_0^2 R^4\left[1-\frac{4\hbar}{m\omega_0 R^2}\left\{\ln(b/l)-\frac{1}{2}\right\}+\dots\right] \tag{19}$$

$$= -\frac{\pi}{4}(mn_0)\,\omega_0^2 R^4\left[1-0\left(\frac{1}{N_{eq}}\right)\right]. \tag{20}$$

Compared with the corresponding result for solid-like rotation, see eqn. (15), the value of E'_{eq} in the present case is larger by a term $0(1/N_{eq})$, which in all practical cases would be too small; in any case, this is a price, though not a high one, which has to be paid for electing a pattern of flow different from the one that claims to be the "natural one".

10.8. Quantized vortex rings and the breakdown of superfluidity

Feynman (1955) was the first to suggest that the formation of vortices in liquid helium II might provide the mechanism responsible for the breakdown of superfluidity in the liquid. He considered the flow of liquid helium II from an orifice of diameter D and, by tentative arguments, found that the velocity v_0, for which the flow energy would be just large enough to create quantized vortices in the liquid, is given by

$$v_0 = \frac{\hbar}{mD}\ln(D/l). \tag{1}$$

Thus, for an orifice of diameter 10^{-5} cm, v_0 would be of the order of 1 m/sec.† It seems natural to identify v_0 with v_c, the *critical velocity* of superflow through the given capillary, despite the fact that the theoretical estimate for v_0 is an order of magnitude higher than the corresponding experimental estimates of v_c; the latter, for instance, are 13 cm/sec, 8 cm/sec and 4 cm/sec for capillary diameters 1.2×10^{-5} cm, 7.9×10^{-5} cm and 3.9×10^{-4} cm, respectively. Nevertheless, the present estimate is far more acceptable than the prohibitively large estimates obtained earlier on the basis of a possible creation of phonons or rotons in the liquid; see Sec. 7.5. Moreover, one obtains here a definite dependence of the critical velocity of superflow *on* the width of the capillary employed which, on the whole, agrees reasonably well with the various experimental findings. In the following, we propose to develop Feynman's idea along the lines of the knowledge gained in Secs. 10.6 and 10.7.

So far we have been dealing with the so-called *linear vortices*, whose velocity field possesses cylindrical symmetry. More generally, however, a vortex line need not be straight—it may

* Physically, it means that the "image vortices" in the wall practically annul the outermost ring of the actual vortices in the fluid—an effect well known in classical hydrodynamics.

† We have taken here: $l \simeq 1$ Å, so that $\ln(D/l) \simeq 7$.

be curved and, if it does not terminate on the walls of the container or on the free surface of the liquid, may close on itself. We then speak of a *vortex ring*, which is very much like a smoke ring. Needless to say, the fundamental quantization condition (10.6.5) is as valid for a vortex ring as for a vortex line. However, the dynamical properties of a vortex ring are quite different from those of a vortex line; see, for example, Fig. 10.12 which shows *schematically* a vortex ring in cross-section, the radius r of the ring being much larger than

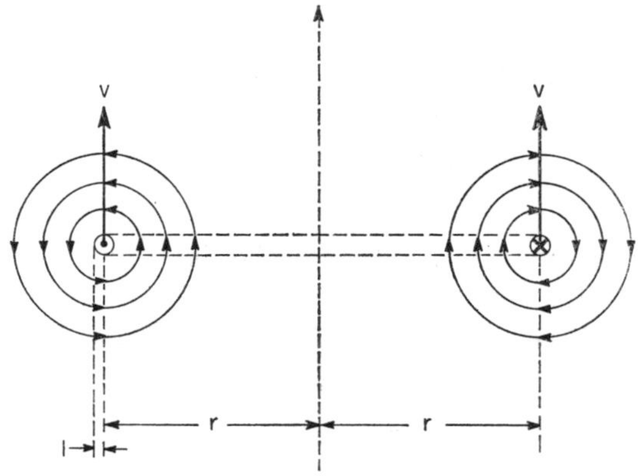

FIG. 10.12. Schematic illustration of a quantized vortex ring in cross-section.

the core dimension l. The flow velocity $\boldsymbol{v}_s$ at any point in the field is given by a superposition of the flow velocities due to the various elements of the ring. It is not difficult to see that the velocity field of the ring, including the ring itself, moves in a direction perpendicular to the plane of the ring, with a velocity*

$$v \approx \hbar/2mr; \tag{2}$$

see eqn. (10.6.15), with $s = 1$ and $\varrho \approx 2r$. An estimate for the energy associated with the flow may be obtained from the expression (10.6.12), with $L = 2\pi r$, $K = h/m$ and $b \approx r$; thus

$$\varepsilon \approx 2\pi^2\hbar^2 n_0 m^{-1} r \ln (r/l). \tag{3}$$

Clearly, the dependence of ε on r arises mainly from the factor r and only slightly from the factor $\ln (r/l)$. Therefore, with good approximation, $v \propto \varepsilon^{-1}$, i.e. a ring with larger energy moves slower! The physical reason behind this "startling" result is that in the first place a larger ring moves slower because, on the whole, there are larger distances between the various circulation-carrying elements of the ring, but since it carries along with it a much larger amount of the fluid ($M \propto r^3$), the total energy associated with the ring is larger (essentially proportional to Mv^2, i.e. $\propto r$). The product $v\varepsilon$, apart from the slowly varying factor $\ln (r/l)$, is a constant, which is equal to $\pi^2\hbar^3 n_0/m^2$.

It is gratifying to note that vortex rings such as the one discussed here have indeed been observed and circulation around them is found to be as close to the quantum unit h/m

* This would be an exact result if we had a pair of oppositely directed *linear* vortices, with the same cross-section as shown in Fig. 10.12. In the case of a ring, the velocity would be somewhat larger.

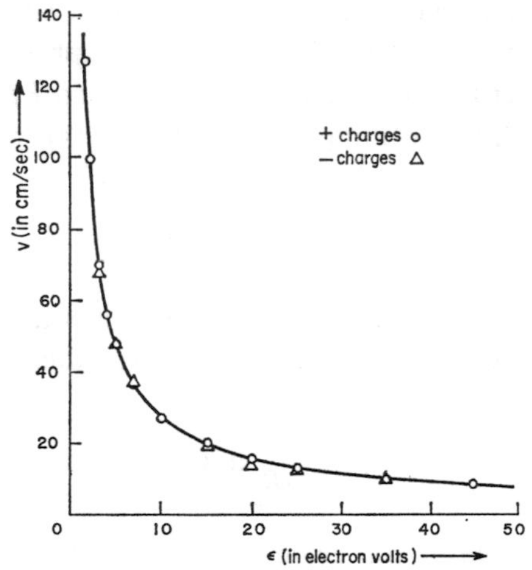

FIG. 10.13. The velocity–energy relationship of the vortex rings set up in liquid helium II (after Rayfield and Reif, 1964). The points indicate the experimental data, while the curve represents the theoretical relationship based on the "quantum of circulation" h/m.

as one could expect under the conditions of the experiment. Figure 10.13 shows the experimental results of Rayfield and Reif (1964) for the velocity–energy relationship of free-moving, charge-carrying vortex rings created in liquid helium II by suitably accelerated helium ions. Vortex rings carrying positive as well as negative charge were observed; dynamically, however, they behaved *alike*, as indeed expected because the velocity and energy of a vortex ring are determined by the properties of a large amount of fluid carried along with the ring rather than by the small charge coupled to it. Fitting experimental results with the notion of the vortex rings, Rayfield and Reif concluded that their rings carried a circulation of value $(1.00 \pm 0.03) \times 10^{-3}$ cm^2 sec^{-1}, which is pretty close to the Onsager–Feynman unit h/m ($= 0.997 \times 10^{-3}$ cm^2 sec^{-1}); moreover, they seemed to have a core radius of about 1.2 Å, which is again comparable with the characteristic parameter l of the fluid.

We shall now show that the dynamics of the quantized vortex rings is such that their creation in liquid helium II does provide a mechanism for the *breakdown of superfluidity*. To see this, it is simplest to consider the case of a superfluid flowing through a capillary of radius R. As the velocity of flow increases and approaches the critical value v_c, quantized vortex rings begin to form and energy dissipation sets in, which in turn brings about the rupture of the superflow. By geometrical symmetry, the rings will be so formed that their central plane will be perpendicular to the axis of the capillary and they will be moving in the direction of the main flow. Now, by the Landau criterion (7.5.24), the critical velocity of flow is directly determined by the energy spectrum of the excitations created:

$$v_c = (\varepsilon/p)_{\min}. \tag{4}$$

We, therefore, require an expression for the momentum p of the vortex ring. In analogy with a classical vortex ring, we may take

$$p = 2\pi^2 \hbar n_0 r^2, \tag{5}$$

which seems satisfactory because (i) it conforms to the general result: $v = (\partial\varepsilon/\partial p)$, though to a first approximation only, and (ii) it leads to the approximate dispersion relation: $\varepsilon \propto p^{1/2}$, which has been verified separately by Rayfield and Reif by subjecting these rings to a transverse electric field. Substituting (3) and (5) into (4), we obtain

$$v_c \simeq \left\{\frac{\hbar}{mr}\ln(r/l)\right\}_{\min}. \tag{6}$$

Now, since the r-dependence of the quantity (ε/p) arises mainly from the factor $(1/r)$, the minimum in (6) would obtain when r has its largest value, namely R, the radius of the capillary. We thus obtain

$$v_c \simeq \frac{\hbar}{mR}\ln(R/l), \tag{7}$$

which is very much the same as the original estimate of Feynman—of course, with D replaced by R. Naturally, then, the numerical values of v_c obtained from the new expression (7) continue to be larger than the corresponding experimental values; however, the theory is now much better founded.

Fetter (1963) was the first to account for the fact that as the radius r of the ring approaches the radius R of the capillary, the influence of the "image vortex" becomes extremely important. The energy of the flow falls below the asymptotic value given by the expression (3) by a factor of 10 or so. This, in turn, reduces the critical velocity by a similar factor; the actual result obtained by Fetter was

$$v_c \simeq \frac{11}{24}\frac{\hbar}{mR} = 0.46\frac{\hbar}{mR}. \tag{8}$$

Kawatra and Pathria (1966) extended Fetter's calculation to take into account the boundary effects arising explicitly from the walls of the capillary as well as the ones arising implicitly from the "image vortex"; moreover, in the computation of ε, they employed the actual wave functions, obtained by solving the *Gross–Pitaevskii equation* (10.6.21), rather than the analytical approximation employed by Fetter. They obtained

$$v_c \simeq 0.59\frac{\hbar}{mR}, \tag{9}$$

which is about 30 per cent higher than Fetter's value; for comments regarding the "most favorable" location for the formation of the vortex ring in the capillary, see the original reference.

Problems

10.1. Establish the pseudopotential (10.1.14), of which the pseudopotential (10.1.13) is only a limiting case.

10.2. Solve the eigenvalue equation

$$(\nabla_r^2 + k_n^2)\,\psi_n(r) = 0$$

in the region bounded by two concentric spheres of radii a and R $(R > a)$, with the boundary conditions

$$\psi(a) = \psi(R) = 0.$$

Expand the eigenvalues k_n^2 in powers of a, retaining terms up to the order a^2.

Next, calculate the eigenvalues k_n^2 by the method of pseudopotentials, up to the order a^2, and show that these results agree with the ones obtained earlier.

10.3. For the imperfect Bose gas discussed in Sec. 10.4, calculate the specific heat C_V near absolute zero, and show that, as $T \to 0$, the specific heat vanishes in a manner characteristic of a system with an "energy gap" Δ.

10.4. Show that, to the *first* order in the scattering length a, the discontinuity in the specific heat C_V of an imperfect Bose gas at the transition point T_c is given by

$$(C_V)_{T=T_c-} - (C_V)_{T=T_c+} = Nk\frac{9a}{2\lambda_c}\zeta\left(\frac{3}{2}\right),$$

while the discontinuity in the bulk modulus K is given by

$$(K)_{T=T_c-} - (K)_{T=T_c+} = -\frac{4\pi a\hbar^2}{mv_c^2}.$$

10.5. Carry out the sequence of calculations leading to (i) eqn. (10.5.22) and to (ii) eqn. (10.5.29).

10.6. Assume, after Atkins (1953, 1957) and Kuper (1956), that the *surface modes of vibration* of a liquid have a frequency-dependent phase velocity $u(\nu)$, given by

$$u(\nu) = (2\pi\sigma\nu/\varrho)^{1/3},$$

where σ is the surface tension of the liquid and ϱ its mass density. Show that this gives for the "ripplons", which are the quanta of these vibrations, the energy–momentum relationship

$$\varepsilon = (\sigma/\varrho\hbar)^{1/2}p^{3/2}.$$

Proceeding as in the Debye theory of solids, derive a low-temperature formula for the energy (per unit area) of the surface arising from the ripplon field.

10.7. (a) Applying the considerations of equilibrium, derive eqn. (10.7.3) for the shape of the free surface of a rotating fluid.

(b) Derive, along the same lines, the shape of the free surface of a superfluid in the presence of a linear vortex which terminates at the free surface. Verify the statement, made in the text, that in the case of a vortex carrying a circulation of one quantum the depression in the surface would be of the order of 10 microns at a distance of about 1 micron from the center of the vortex.

10.8. Using Fetter's analytical approximation,

$$f_1(\varrho') = \frac{\varrho'}{\sqrt{(1+\varrho'^2)}},$$

for the solution of eqn. (10.6.21) with $s = 1$, calculate the energy (per unit length) associated with a quantized vortex line of unit circulation. Compare your result with the one quoted in (10.6.24).

10.9. Study the nature of the velocity field arising from a pair of parallel vortex lines, with $s_1 = +1$ and $s_2 = -1$, separated by a distance $2d$. Derive and discuss the general equation of the stream lines. Again, using Fetter's analytical approximation for the functions $f(\varrho_1')$ and $f(\varrho_2')$, calculate the energy (per unit length) of the system and show that its limiting value, as $d \to 0$, is $\frac{11}{12}(\hbar^2\pi n_0/m)$. Making use of this result, derive expression (10.8.8) for the critical velocity of superflow.

CHAPTER 11

STATISTICAL MECHANICS OF INTERACTING SYSTEMS: THE METHOD OF QUANTIZED FIELDS

IN THIS chapter we present yet another method of dealing with a system composed of mutually interacting particles. This method is based on the concept of a *quantized field* which is characterized by the field operators $\psi(\boldsymbol{r})$, and their hermitian conjugates $\psi^{\dagger}(\boldsymbol{r})$, that satisfy a set of well-defined commutation rules. In terms of these operators, one defines a number operator $\hat{N}$ and a Hamiltonian operator $\hat{H}$ which provide an adequate representation for a system composed of any finite number of particles and possessing any finite amount of energy. In view of the formal similarity with the Schrödinger formulation, the formulation in terms of a quantized field is generally referred to as the *second quantization* of the system. For convenience of calculation, the field operators $\psi(\boldsymbol{r})$ and $\psi^{\dagger}(\boldsymbol{r})$ may be expressed as superpositions of a set of single-particle wave functions $\{u_{\alpha}(\boldsymbol{r})\}$, with coefficients a_{α} and $a_{\alpha}^{\dagger}$; the latter turn out to be the *annihilation* and *creation* operators which again satisfy a set of well-defined commutation rules. The operators $\hat{N}$ and $\hat{H}$ find a convenient expression in terms of the operators a_{α} and $a_{\alpha}^{\dagger}$, and the final formulation is well suited for a treatment based on the operator algebra; as a result, many calculations, which would otherwise be tedious, can be carried out in a more or less straightforward manner. We propose to demonstrate this by calculating the low-temperature properties of imperfect gases to the *second* order in a, where a is the scattering length of the interparticle potential.

11.1. The formalism of second quantization

To represent a system of particles by a *quantized field*, we invoke the field operators $\psi(\boldsymbol{r})$ and $\psi^{\dagger}(\boldsymbol{r})$, which are defined for all values of the position coordinate $\boldsymbol{r}$ and which operate upon a *Hilbert space*; a vector in this space corresponds to a particular state of the quantized field. The values of the quantities ψ and $\psi^{\dagger}$, at all possible points $\boldsymbol{r}$, represent the *degrees of freedom* of the field; since $\boldsymbol{r}$ is a continuous variable, the number of these degrees of freedom is innumerably infinite. Now, if the given system is composed of bosons, the field operators $\psi(\boldsymbol{r})$ and $\psi^{\dagger}(\boldsymbol{r})$ are supposed to satisfy the commutation rules

$$[\psi(\boldsymbol{r}), \psi^{\dagger}(\boldsymbol{r}')] = \delta(\boldsymbol{r}-\boldsymbol{r}') \tag{1a}$$

$$[\psi(\boldsymbol{r}), \psi(\boldsymbol{r}')] = [\psi^{\dagger}(\boldsymbol{r}), \psi^{\dagger}(\boldsymbol{r}')] = 0, \tag{1b}$$

where the symbol $[A, B]$ stands for the "commutator $(AB-BA)$ of the given operators A and B". If, on the other hand, the given system is composed of fermions, the field operators are supposed to satisfy the rules

$$\{\psi(\mathbf{r}), \psi^\dagger(\mathbf{r}')\} = \delta(\mathbf{r}-\mathbf{r}'), \tag{2a}$$

$$\{\psi(\mathbf{r}), \psi(\mathbf{r}')\} = \{\psi^\dagger(\mathbf{r}), \psi^\dagger(\mathbf{r}')\} = 0, \tag{2b}$$

where the symbol $\{A, B\}$ stands for the "anticommutator $(AB+BA)$ of the given operators A and B". In the case of fermions, the operators $\psi(\mathbf{r})$ and $\psi^\dagger(\mathbf{r})$ possess certain explicit properties which follow directly from the rules (2*b*), namely

$$\psi(\mathbf{r})\,\psi(\mathbf{r}') = -\psi(\mathbf{r}')\,\psi(\mathbf{r}), \quad \therefore \quad \psi(\mathbf{r})\,\psi(\mathbf{r}) = 0 \quad \text{for all } \mathbf{r}; \tag{2c}$$

similarly,

$$\psi^\dagger(\mathbf{r})\,\psi^\dagger(\mathbf{r}') = -\psi^\dagger(\mathbf{r}')\,\psi^\dagger(\mathbf{r}), \quad \therefore \; \psi^\dagger(\mathbf{r})\,\psi^\dagger(\mathbf{r}) = 0 \quad \text{for all } \mathbf{r}. \tag{2d}$$

This does not hold for the field operators pertaining to the bosons. In the sequel we shall see that the mathematical difference between the commutation rules (1) for the boson field operators and rules (2) for the fermion field operators is intimately related to the fundamental difference in the symmetry properties of the respective wave functions in the Schrödinger formulation. Of course, in their own place, both sets of rules, (1) and (2), are essentially axiomatic.

We now introduce two hermitian operators, the *particle-number operator* $\hat{N}$ and the *Hamiltonian operator* $\hat{H}$, through definitions that hold for bosons as well as fermions:

$$\hat{N} \equiv \int d^3r\,\psi^\dagger(\mathbf{r})\,\psi(\mathbf{r}) \tag{3}$$

and

$$\hat{H} \equiv -\frac{\hbar^2}{2m}\int d^3r\,\psi^\dagger(\mathbf{r})\,\nabla^2\psi(\mathbf{r}) + \frac{1}{2}\iint d^3r_1\,d^3r_2\,\psi^\dagger(\mathbf{r}_1)\,\psi^\dagger(\mathbf{r}_2)\,u(\mathbf{r}_1, \mathbf{r}_2)\,\psi(\mathbf{r}_2)\,\psi(\mathbf{r}_1), \tag{4}$$

where $u(\mathbf{r}_1, \mathbf{r}_2)$ denotes the two-body interaction potential. It is quite natural to interpret the product $\psi^\dagger(\mathbf{r})\,\psi(\mathbf{r})$ as the *number density operator* of the field. The similarity of the foregoing definitions with the mathematical expressions for the expectation values of the corresponding physical quantities in the Schrödinger formulation is quite obvious. However, the similarity is only "formal" because there we are concerned with the wave functions of the given system (which themselves are *c-numbers*) while here we are concerned with the *operators* of the corresponding matter field. We can easily verify that, irrespective of the commutation rules obeyed by the operators $\psi(\mathbf{r})$ and $\psi^\dagger(\mathbf{r})$, the operators $\hat{N}$ and $\hat{H}$ commute:

$$[\hat{N}, \hat{H}] = 0; \tag{5}$$

accordingly, the operators $\hat{N}$ and $\hat{H}$ can be simultaneously diagonalized.

We now choose a *complete orthonormal basis* of the Hilbert space, such that any vector $|\Phi_n\rangle$ of the basis is a simultaneous eigenstate of the operators $\hat{N}$ and $\hat{H}$. Accordingly, we may denote any particular member of the basis by the vector symbol $|\Psi_{NE}\rangle$, with the properties

$$\hat{N}\,|\Psi_{NE}\rangle = N\,|\Psi_{NE}\rangle, \quad \hat{H}\,|\Psi_{NE}\rangle = E\,|\Psi_{NE}\rangle \tag{6}$$

and

$$\langle\Psi_{NE}\,|\,\Psi_{NE}\rangle = 1. \tag{7}$$

The vector $|\Psi_{00}\rangle$, which represents the *vacuum state* of the field and is generally denoted by the symbol $|0\rangle$, is assumed to be unique; it possesses the properties

$$\hat{N}|0\rangle = \hat{H}|0\rangle = 0 \quad \text{and} \quad \langle 0|0\rangle = 1. \tag{8}$$

Next we observe that, whether we employ the boson commutation rules (1) or the fermion rules (2), the operator $\hat{N}$ and the operators $\psi(\boldsymbol{r})$ and $\psi^\dagger(\boldsymbol{r})$ satisfy the commutation properties

$$[\psi(\boldsymbol{r}), \hat{N}] = \psi(\boldsymbol{r}) \quad \text{and} \quad [\psi^\dagger(\boldsymbol{r}), \hat{N}] = -\psi^\dagger(\boldsymbol{r}), \tag{9}$$

whence it follows that

$$\hat{N}\psi(\boldsymbol{r})\,|\Psi_{NE}\rangle = \big(\psi(\boldsymbol{r})\hat{N} - \psi(\boldsymbol{r})\big)\,|\Psi_{NE}\rangle = (N-1)\,\psi(\boldsymbol{r})\,|\Psi_{NE}\rangle \tag{10}$$

and

$$\hat{N}\psi^\dagger(\boldsymbol{r})\,|\Psi_{NE}\rangle = \big(\psi^\dagger(\boldsymbol{r})\hat{N} + \psi^\dagger(\boldsymbol{r})\big)\,|\Psi_{NE}\rangle = (N+1)\,\psi^\dagger(\boldsymbol{r})\,|\Psi_{NE}\rangle. \tag{11}$$

Clearly, the state $\psi(\boldsymbol{r})\,|\Psi_{NE}\rangle$ is also an eigenstate of the operator $\hat{N}$, with eigenvalue $(N-1)$; thus, the application of the operator $\psi(\boldsymbol{r})$ on to the state $|\Psi_{NE}\rangle$ of the field *annihilates* one particle from the field. Similarly, the state $\psi^\dagger(\boldsymbol{r})\,|\Psi_{NE}\rangle$ is also an eigenstate of the operator $\hat{N}$, with eigenvalue $(N+1)$ instead; thus, the application of the operator $\psi^\dagger(\boldsymbol{r})$ on to the state $|\Psi_{NE}\rangle$ of the field *creates* a particle in the field. In each case, the process (of annihilation or creation) is tied down to the point $\boldsymbol{r}$ of the field; however, the energy associated with the process, which also means the change in the energy of the field, remains indeterminate (see eqns. (18) and (19)). By a repeated application of the operator $\psi^\dagger$ on to the vacuum state $|0\rangle$, we find that the eigenvalues of the operator $\hat{N}$ are 0, 1, 2, ... On the other hand, the application of the operator ψ on to the vacuum state $|0\rangle$ gives identically zero because, for obvious reasons, we cannot admit negative eigenvalues for the operator $\hat{N}$. Of course, if we apply the operator ψ on to the state $|\Psi_{NE}\rangle$ repeatedly N times, we end up with the vacuum state; we then have, by virtue of the orthonormality of the basis chosen,

$$\langle \Phi_n|\psi(\boldsymbol{r}_1)\,\psi(\boldsymbol{r}_2)\,\ldots\,\psi(\boldsymbol{r}_N)|\,\Psi_{NE}\rangle = 0 \tag{12}$$

unless the state $|\Phi_n\rangle$ is itself the vacuum state, in which case we will obtain a nonzero result instead. In terms of the latter result, we may define a function of the N coordinates $\boldsymbol{r}_1, \boldsymbol{r}_2, \ldots, \boldsymbol{r}_N$, namely

$$\Psi_{NE}(\boldsymbol{r}_1, \ldots, \boldsymbol{r}_N) = (N!)^{-1/2}\,\langle 0|\,\psi(\boldsymbol{r}_1)\,\ldots\,\psi(\boldsymbol{r}_N)\,|\Psi_{NE}\rangle. \tag{13}$$

Obviously, the function $\Psi_{NE}(\boldsymbol{r}_1, \boldsymbol{r}_2, \ldots, \boldsymbol{r}_N)$ has something to do with "an assemblage of N particles located at the points $\boldsymbol{r}_1, \ldots, \boldsymbol{r}_N$ of the field" because their annihilation from these very points of the field has led us to the vacuum state of the field. To obtain a precise meaning of this function, we first of all note that in the case of bosons (or fermions) this function is symmetric (or antisymmetric) with respect to an interchange between any two of the N coordinates; see eqns. (1b) and (2b, c), respectively. Secondly, its norm is equal to unity, which can be seen as follows:

By the very definition of $\Psi_{NE}(\mathbf{r}_1, \ldots, \mathbf{r}_N)$,

$$\begin{aligned}\int d^{3N}r\, \Psi^*_{NE}(\mathbf{r}_1, \ldots, \mathbf{r}_N)\, \Psi_{NE}(\mathbf{r}_1, \ldots, \mathbf{r}_N) &= (N!)^{-1} \int d^{3N}r \langle \Psi_{NE}|\, \psi^\dagger(\mathbf{r}_N) \ldots \psi^\dagger(\mathbf{r}_1)\,|0\rangle\langle 0|\, \psi(\mathbf{r}_1) \ldots \psi(\mathbf{r}_N)\,|\,\Psi_{NE}\rangle \\ &= (N!)^{-1} \int d^{3N}r \sum_n \langle \Psi_{NE}\,|\psi^\dagger(\mathbf{r}_N) \ldots \psi^\dagger(\mathbf{r}_1)|\Phi_n\rangle\langle \Phi_n|\,\psi(\mathbf{r}_1) \ldots \psi(\mathbf{r}_N)|\,\Psi_{NE}\rangle \\ &= (N!)^{-1} \int d^{3N}r \langle \Psi_{NE}|\,\psi^\dagger(\mathbf{r}_N) \ldots \psi^\dagger(\mathbf{r}_2)\,\psi^\dagger(\mathbf{r}_1)\,\psi(\mathbf{r}_1)\,\psi(\mathbf{r}_2) \ldots \psi(\mathbf{r}_N)|\,\Psi_{NE}\rangle;\end{aligned}$$

here, use has been made of eqn. (12) which holds for all eigenstates of the field except the vacuum state, and of the fact that the summation of $|\Phi_n\rangle\langle\Phi_n|$ over the complete orthonormal set of the basis chosen is equivalent to a unit operator. We now carry out integration over $\mathbf{r}_1$, the relevant factor being

$$\int d^3r_1 \psi^\dagger(\mathbf{r}_1)\,\psi(\mathbf{r}_1) \equiv \hat{N}.$$

Next, we carry out integration over $\mathbf{r}_2$, the corresponding factor being

$$\int d^3r_2 \psi^\dagger(\mathbf{r}_2)\,\hat{N}\psi(\mathbf{r}_2) = \int d^3r_2 \psi^\dagger(\mathbf{r}_2)\,\psi(\mathbf{r}_2)\,(\hat{N}-1) = \hat{N}(\hat{N}-1);$$

see eqn. (10). By iteration, we obtain

$$\begin{aligned}\int d^{3N}r\, \Psi^*_{NE}(\mathbf{r}_1, \ldots, \mathbf{r}_N)\, \Psi_{NE}(\mathbf{r}_1, \ldots, \mathbf{r}_N) &= (N!)^{-1} \langle \Psi_{NE}|\, \hat{N}(\hat{N}-1)\,(\hat{N}-2) \ldots \text{up to } N \text{ factors } |\,\Psi_{NE}\rangle \\ &= (N!)^{-1} N!\, \langle \Psi_{NE}|\,\Psi_{NE}\rangle = 1. \end{aligned} \tag{14}$$

Finally, we can show that, for bosons as well as fermions, the function $\Psi_{NE}(\mathbf{r}_1, \ldots, \mathbf{r}_N)$ satisfies the differential equation, see Problem 11.1,

$$\left(-\frac{\hbar^2}{2m}\sum_{i=1}^{N} \nabla_i^2 + \sum_{i<j} u_{ij}\right) \Psi_{NE}(\mathbf{r}_1, \ldots, \mathbf{r}_N) = E\Psi_{NE}(\mathbf{r}_1, \ldots, \mathbf{r}_N), \tag{15}$$

which is simply the *Schrödinger equation* of an N-particle system. The function $\Psi_{NE}(\mathbf{r}_1, \ldots, \mathbf{r}_N)$ is, therefore, to be identified with the Schrödinger wave function of the system, with energy eigenvalue E; accordingly, the quantity $\Psi^*_{NE}\Psi_{NE}$ denotes the probability density for the particles of the system to be in the vicinity of the coordinates $(\mathbf{r}_1, \ldots, \mathbf{r}_N)$ when the system itself happens to be in an eigenstate of energy E. This establishes a correspondence between the quantized field formulation on one hand and the Schrödinger formulation on the other. In passing, we write down the quantized-field expression for the function $\Psi^*_{NE}(\mathbf{r}_1, \ldots, \mathbf{r}_N)$, which is the complex conjugate of the wave function $\Psi_{NE}(\mathbf{r}_1, \ldots, \mathbf{r}_N)$, viz.

$$\Psi^*_{NE}(\mathbf{r}_1, \ldots, \mathbf{r}_N) = (N!)^{-1/2} \langle \Psi_{NE}|\,\psi^\dagger(\mathbf{r}_N) \ldots \psi^\dagger(\mathbf{r}_1)|0\rangle; \tag{16}$$

cf. eqn. (13).

We now introduce a complete orthonormal set of single-particle wave functions $u_\alpha(\mathbf{r})$, where the suffix α provides an identifying label for the various single-particle states; it could, for instance, be the energy eigenvalue of the state (or the momentum $\mathbf{p}$, along with the spin component σ pertaining to the state). In view of the orthonormality of the wave functions,

$$\int d^3r u^*_\alpha(\mathbf{r})\, u_\beta(\mathbf{r}) = \delta_{\alpha\beta}. \tag{17}$$

The field operators $\psi(\boldsymbol{r})$ and $\psi^{\dagger}(\boldsymbol{r})$ may now be expanded in terms of the functions $u_{\alpha}(\boldsymbol{r})$:

$$\psi(\boldsymbol{r}) = \sum_{\alpha} a_{\alpha} u_{\alpha}(\boldsymbol{r}) \tag{18}$$

and

$$\psi^{\dagger}(\boldsymbol{r}) = \sum_{\alpha} a_{\alpha}^{\dagger} u_{\alpha}^{*}(\boldsymbol{r}). \tag{19}$$

Relations inverse to (18) and (19) will be

$$a_{\alpha} = \int d^3r \psi(\boldsymbol{r})\, u_{\alpha}^{*}(\boldsymbol{r}) \tag{20}$$

and

$$a_{\alpha}^{\dagger} = \int d^3r \psi^{\dagger}(\boldsymbol{r})\, u_{\alpha}(\boldsymbol{r}). \tag{21}$$

The expansion coefficients a_{α} and $a_{\alpha}^{\dagger}$, like the field variables $\psi(\boldsymbol{r})$ and $\psi^{\dagger}(\boldsymbol{r})$, are operators, which operate upon the elements of the relevant Hilbert space, while the functions $u_{\alpha}(\boldsymbol{r})$ are ordinary, non-quantized functions. Indeed, the operators a_{α} and $a_{\alpha}^{\dagger}$ now take over the role of the *degrees of freedom* of the field.

Substituting (18) and (19) into the set of rules (1) or (2), and making use of the closure property of the u's, viz.

$$\sum_{\alpha} u_{\alpha}(\boldsymbol{r})\, u_{\alpha}^{*}(\boldsymbol{r}') = \delta(\boldsymbol{r}-\boldsymbol{r}'), \tag{22}$$

we obtain for the operators a_{α} and $a_{\alpha}^{\dagger}$:

$$[a_{\alpha}, a_{\beta}^{\dagger}] = \delta_{\alpha\beta} \tag{23a}$$

$$[a_{\alpha}, a_{\beta}] = [a_{\alpha}^{\dagger}, a_{\beta}^{\dagger}] = 0 \tag{23b}$$

in the case of *bosons*, and

$$\{a_{\alpha}, a_{\beta}^{\dagger}\} = \delta_{\alpha\beta} \tag{24a}$$

$$\{a_{\alpha}, a_{\beta}\} = \{a_{\alpha}^{\dagger}, a_{\beta}^{\dagger}\} = 0 \tag{24b}$$

in the case of *fermions*. In the latter case, the operators a_{α} and $a_{\alpha}^{\dagger}$ possess certain explicit properties which follow directly from the rules (24b):

$$a_{\alpha}a_{\beta} = -a_{\beta}a_{\alpha}, \quad \therefore\ a_{\alpha}a_{\alpha} = 0 \quad \text{for all } \alpha; \tag{24c}$$

similarly

$$a_{\alpha}^{\dagger}a_{\beta}^{\dagger} = -a_{\beta}^{\dagger}a_{\alpha}^{\dagger}, \quad \therefore\ a_{\alpha}^{\dagger}a_{\alpha}^{\dagger} = 0 \quad \text{for all } \alpha. \tag{24d}$$

No such properties are displayed by the operators pertaining to bosons. We will shortly see that this vital difference between the commutation rules for the boson operators and the fermion operators is closely linked with the fact that while fermions have to conform to the restrictions imposed by the Pauli exclusion principle, there are no such restrictions for bosons.‡

‡ It is now understandable why, in the commutation rules for the fermion operators, we have the anti-commutators { } and not the commutators [ ]. This is related to the following circumstance: because of the Pauli exclusion principle, which permits only one quantum in any given state, the fermion fields *do not* have a macroscopic classical limit (in which the commutators yield place to the Poisson brackets)!

We now proceed to express the operators $\hat{N}$ and $\hat{H}$ in terms of the operators a_α and $a_\alpha^\dagger$. Substituting (18) and (19) into (3), we obtain

$$\hat{N} \equiv \int d^3r \sum_{\alpha,\beta} a_\alpha^\dagger a_\beta u_\alpha^*(\boldsymbol{r})\, u_\beta(\boldsymbol{r}) = \sum_{\alpha,\beta} a_\alpha^\dagger a_\beta \delta_{\alpha\beta}$$
$$= \sum_\alpha a_\alpha^\dagger a_\alpha . \tag{25}$$

It seems natural to speak of the operator $a_\alpha^\dagger a_\alpha$ as the *particle-number operator* pertaining to the single-particle state α. We denote this operator by the symbol $\hat{N}_\alpha$:

$$\hat{N}_\alpha = a_\alpha^\dagger a_\alpha . \tag{26}$$

It is easy to verify that, for bosons as well as fermions, the operators $\hat{N}_\alpha$ commute with one another; hence, they can be simultaneously diagonalized. Accordingly, we may choose the complete orthonormal basis of the Hilbert space in such a way that any vector belonging to the basis is a simultaneous eigenstate of the operators $\hat{N}_\alpha$.‡ Let a particular member of the basis be denoted by the vector $|n_0, n_1, \ldots, n_\alpha, \ldots\rangle$, or by the shorter symbol $|\Phi_n\rangle$, with the properties

$$\hat{N}_\alpha |\Phi_n\rangle = n_\alpha |\Phi_n\rangle \tag{27}$$

and

$$\langle\Phi_n|\Phi_n\rangle = 1; \tag{28}$$

the number n_α, being the eigenvalue of the operator $\hat{N}_\alpha$ in the state $|\Phi_n\rangle$ of the field, denotes the number of particles in the single-particle state α of the given system. One particular member of the basis, for which $n_\alpha = 0$ for all α, will represent the *vacuum state* of the field; denoting the vacuum state by the symbol $|\Phi_0\rangle$, we have

$$\hat{N}_\alpha |\Phi_0\rangle = 0 \quad \text{for all } \alpha, \quad \text{and} \quad \langle\Phi_0|\Phi_0\rangle = 1. \tag{29}$$

Next we observe that, whether we employ the boson commutation rules (23) or fermion rules (24), the operator $\hat{N}_\alpha$ and the operators a_α and $a_\alpha^\dagger$ satisfy the commutation properties

$$[a_\alpha, \hat{N}_\alpha] = a_\alpha \quad \text{and} \quad [a_\alpha^\dagger, \hat{N}_\alpha] = -a_\alpha^\dagger , \tag{30}$$

whence it follows that

$$\hat{N}_\alpha a_\alpha |\Phi_n\rangle = (a_\alpha \hat{N}_\alpha - a_\alpha)|\Phi_n\rangle = (n_\alpha - 1)\, a_\alpha |\Phi_n\rangle \tag{31}$$

and

$$\hat{N}_\alpha a_\alpha^\dagger |\Phi_n\rangle = (a_\alpha^\dagger \hat{N}_\alpha + a_\alpha^\dagger)|\Phi_n\rangle = (n_\alpha + 1)\, a_\alpha^\dagger |\Phi_n\rangle. \tag{32}$$

Clearly, the state $a_\alpha|\Phi_n\rangle$ is also an eigenstate of the operator $\hat{N}_\alpha$, with eigenvalue $(n_\alpha - 1)$; thus, the application of the operator a_α on to the state $|\Phi_n\rangle$ of the field *annihilates* one particle from the field. Similarly, the state $a_\alpha^\dagger|\Phi_n\rangle$ is also an eigenstate of the operator $\hat{N}_\alpha$, with eigenvalue $(n_\alpha + 1)$; thus, the application of the operator $a_\alpha^\dagger$ on to the state $|\Phi_n\rangle$ *creates* a particle in the field. The operators a_α and $a_\alpha^\dagger$ are, therefore, referred to as the

‡ This representation of the field is generally referred to as the *particle-number representation*.

annihilation and *creation operators*. Of course, in each case the process (of annihilation or creation) is tied down to the single-particle state α; the precise location of the happening in the coordinate space remains indeterminate; see eqns. (20) and (21). Now, since the application of the operator a_α or $a_\alpha^\dagger$ on to the state $|\Phi_n\rangle$ of the field does not affect the eigenvalues of the particle-number operators other than $\hat{N}_\alpha$, we may write

$$a_\alpha\,|n_0, n_1, \ldots, n_\alpha, \ldots\rangle = A(n_\alpha)\,|n_0, n_1, \ldots, n_\alpha-1, \ldots\rangle \tag{33}$$

and

$$a_\alpha^\dagger\,|n_0, n_1, \ldots, n_\alpha, \ldots\rangle = B(n_\alpha)\,|n_0, n_1, \ldots, n_\alpha+1, \ldots\rangle, \tag{34}$$

where the factors $A(n_\alpha)$ and $B(n_\alpha)$ can be determined with the help of the commutation rules governing the operators a_α and $a_\alpha^\dagger$. For bosons,

$$A(n_\alpha) = \sqrt{n_\alpha}; \qquad B(n_\alpha) = \sqrt{(n_\alpha+1)}; \tag{35}$$

consequently, if we regard the state $|\Phi_n\rangle$ to have arisen from the vacuum state $|\Phi_0\rangle$ by a repeated application of creation operators, we can write

$$|\Phi_n\rangle = \frac{1}{\sqrt{(n_0!\,n_1!\,\ldots\,n_\alpha!\,\ldots)}}\,(a_0^\dagger)^{n_0}\,(a_1^\dagger)^{n_1}\,\ldots\,(a_\alpha^\dagger)^{n_\alpha}\,\ldots\,|\Phi_0\rangle. \tag{36}$$

In the case of fermions, the operators $a^\dagger$ anticommute, with the result: $a_\alpha^\dagger a_\beta^\dagger = -a_\beta^\dagger a_\alpha^\dagger$; accordingly, there remains an uncertainty of a phase factor ± 1 *unless* we specify the order in which the $a^\dagger$'s operate on the vacuum state. To be definite, we may agree that, as indicated in (36), the $a^\dagger$'s are arranged in the order of increasing subscripts and the phase factor is then $+1$. Secondly, since the product $a_\alpha^\dagger a_\alpha^\dagger$ now identically vanishes, none of the n_α's in (36) can exceed unity; the eigenvalues of the fermion operators $\hat{N}_\alpha$ are, therefore, restricted to the values 0 and 1, which is precisely the requirement of the Pauli exclusion principle.‡ Accordingly, the factor $[\Pi_\alpha(n_\alpha!)]^{-1/2}$ in (36) would be identically equal to unity. In passing, we note that in the case of fermions the operation (33) has meaning only if $n_\alpha = 1$ and the operation (34) has meaning only if $n_\alpha = 0$.

In the end, we note that the substitution of the expansions (18) and (19) into the expression (4) gives for the Hamiltonian operator of the field

$$\hat{H} = -\frac{\hbar^2}{2m}\sum_{\alpha,\beta}\langle\alpha\,|\nabla^2|\,\beta\rangle\,a_\alpha^\dagger a_\beta + \frac{1}{2}\sum_{\alpha,\beta,\gamma,\lambda}\langle\alpha\beta\,|u|\,\gamma\lambda\rangle\,a_\alpha^\dagger a_\beta^\dagger a_\gamma a_\lambda, \tag{37}$$

where

$$\langle\alpha\,|\nabla^2|\,\beta\rangle = \int d^3r\; u_\alpha^*(\boldsymbol{r})\,\nabla^2 u_\beta(\boldsymbol{r}) \tag{38}$$

and

$$\langle\alpha\beta\,|u|\,\gamma\lambda\rangle = \iint d^3r_1\,d^3r_2\; u_\alpha^*(\boldsymbol{r}_1)\,u_\beta^*(\boldsymbol{r}_2)\,u_{12}u_\gamma(\boldsymbol{r}_2)\,u_\lambda(\boldsymbol{r}_1). \tag{39}$$

Now, if the single-particle wave functions are chosen to be

$$u_\alpha(\boldsymbol{r}) = \frac{1}{\sqrt{V}}\,e^{i\boldsymbol{p}_\alpha\cdot\boldsymbol{r}/\hbar}, \tag{40}$$

‡ This can also be seen by noting that the fermion operators $\hat{N}_\alpha$ satisfy the identity

$$\hat{N}_\alpha^2 = a_\alpha^\dagger a_\alpha a_\alpha^\dagger a_\alpha = a_\alpha^\dagger(1-a_\alpha^\dagger a_\alpha)\,a_\alpha = a_\alpha^\dagger a_\alpha = \hat{N}_\alpha \qquad (\because\ a_\alpha^\dagger a_\alpha^\dagger a_\alpha a_\alpha \equiv 0).$$

The same would be true for the eigenvalues n_α. Hence, $n_\alpha^2 = n_\alpha$ which means that $n_\alpha = 0$ or 1.

where $\boldsymbol{p}_\alpha$ denotes the momentum of the particle (assumed to be "spinless"), then the matrix elements (38) and (39) become

$$\langle\alpha\,|\nabla^2|\,\beta\rangle = \frac{1}{V}\int d^3r\, e^{-i\boldsymbol{p}_\alpha\cdot\boldsymbol{r}/\hbar}\left(-\frac{p_\beta^2}{\hbar^2}\right)e^{i\boldsymbol{p}_\beta\cdot\boldsymbol{r}/\hbar} = -\frac{p_\beta^2}{\hbar^2}\,\delta_{\alpha\beta} \tag{41}$$

and

$$\langle\alpha\beta\,|u|\,\gamma\lambda\rangle = \frac{1}{V^2}\iint d^3r_1\, d^3r_2\, e^{-i(\boldsymbol{p}_\alpha-\boldsymbol{p}_\lambda)\cdot\boldsymbol{r}_1/\hbar}u(\boldsymbol{r}_2-\boldsymbol{r}_1)\, e^{-i(\boldsymbol{p}_\beta-\boldsymbol{p}_\gamma)\cdot\boldsymbol{r}_2/\hbar}. \tag{42}$$

In view of the fact that total momentum is conserved in the collision, i.e.

$$\boldsymbol{p}_\alpha+\boldsymbol{p}_\beta = \boldsymbol{p}_\gamma+\boldsymbol{p}_\lambda, \tag{43}$$

the matrix element (42) takes the form

$$\begin{aligned}\langle\alpha\beta\,|u|\,\gamma\lambda\rangle &= \frac{1}{V^2}\iint d^3r_1\, d^3r_2\, e^{-i(\boldsymbol{p}_\beta-\boldsymbol{p}_\gamma)\cdot(\boldsymbol{r}_2-\boldsymbol{r}_1)/\hbar}u(\boldsymbol{r}_2-\boldsymbol{r}_1)\\ &= \frac{1}{V}\int d^3r e^{i\boldsymbol{p}\cdot\boldsymbol{r}/\hbar}u(\boldsymbol{r}),\end{aligned} \tag{44}$$

where $\boldsymbol{p}$ denotes the *momentum transfer* in the collision:

$$\boldsymbol{p} = (\boldsymbol{p}_\gamma-\boldsymbol{p}_\beta) = -(\boldsymbol{p}_\lambda-\boldsymbol{p}_\alpha). \tag{45}$$

Substituting (41) and (44) into (37), we finally obtain

$$\hat{H} = \sum_{\boldsymbol{p}}\frac{p^2}{2m}a_{\boldsymbol{p}}^\dagger a_{\boldsymbol{p}} + \frac{1}{2}\sum{}' u_{\boldsymbol{p}_1,\boldsymbol{p}_2}^{\boldsymbol{p}_1',\boldsymbol{p}_2'}a_{\boldsymbol{p}_1'}^\dagger a_{\boldsymbol{p}_2'}^\dagger a_{\boldsymbol{p}_2}a_{\boldsymbol{p}_1}, \tag{46}$$

where $u_{\boldsymbol{p}_1,\boldsymbol{p}_2}^{\boldsymbol{p}_1',\boldsymbol{p}_2'}$ denotes the matrix element (44), with

$$\boldsymbol{p} = (\boldsymbol{p}_2-\boldsymbol{p}_2') = -(\boldsymbol{p}_1-\boldsymbol{p}_1'); \tag{47}$$

it will be noted that the primed summation in the second term goes only over those values of the momenta $\boldsymbol{p}_1, \boldsymbol{p}_2, \boldsymbol{p}_1'$ and $\boldsymbol{p}_2'$ that conserve the total momentum of the particles: $\boldsymbol{p}_1'+\boldsymbol{p}_2' = \boldsymbol{p}_1+\boldsymbol{p}_2$. It is obvious that the main term of the expression (46) represents the *kinetic energy operator* of the field ($a_{\boldsymbol{p}}^\dagger a_{\boldsymbol{p}}$ being the particle-number operator pertaining to the single-particle state $\boldsymbol{p}$), while the second term represents the *potential energy operator*.

In the case of spin-half fermions, the single-particle states have to be characterized not only by the value $\boldsymbol{p}$ of the particle momentum but also by the value σ of the z-component of its spin; accordingly, the creation and annihilation operators would carry double indices. The operator $\hat{H}$ then takes the form

$$\hat{H} = \sum_{\boldsymbol{p},\sigma}\frac{p^2}{2m}a_{\boldsymbol{p}\sigma}^\dagger a_{\boldsymbol{p}\sigma} + \frac{1}{2}\sum{}' u_{\boldsymbol{p}_1\sigma_1,\boldsymbol{p}_2\sigma_2}^{\boldsymbol{p}_1'\sigma_1',\boldsymbol{p}_2'\sigma_2'}a_{\boldsymbol{p}_1'\sigma_1'}^\dagger a_{\boldsymbol{p}_2'\sigma_2'}^\dagger a_{\boldsymbol{p}_2\sigma_2}a_{\boldsymbol{p}_1\sigma_1}; \tag{48}$$

again, the summation in the second term goes only over those states (of the two particles) that conform to the conditions of both momentum conservation and spin conservation.

In the following sections we propose to apply the formalism of second quantization to investigate the low-temperature properties of gaseous systems composed of mutually interacting particles. We shall study these systems under those very approximating conditions as were assumed in Chapter 10, namely $a/\lambda \ll 1$ and $na^3 \ll 1$, where a is the *scattering length* of the two-body interaction, λ the *mean thermal wavelength* of the particles and n the *particle density* in the system. However, in the present study we propose to carry out calculations to the *second* order in a. Now, the effective scattering cross-section for the collision of two particles of mass m is primarily determined by the "scattering amplitude" a, where

$$a = \frac{m}{4\pi\hbar^2}\int u(\mathbf{r})\, e^{i\mathbf{p}\cdot\mathbf{r}/\hbar}\, d^3r; \tag{49}$$

here, $\mathbf{p}$ denotes the momentum transfer in the collision. For *low-energy* scattering (which implies "slow" collisions)

$$a \simeq \frac{mu_0}{4\pi\hbar^2}; \qquad u_0 = \int u(\mathbf{r})\, d^3r. \tag{50*}$$

11.2. Low-lying states of an imperfect Bose gas

The Hamiltonian of the quantized field for spinless bosons is given by the expression (11.1.46), where the matrix element $u_{p_1, p_2}^{p_1', p_2'}$ is a function of the momentum $\mathbf{p}$ transferred during the collision and is given by the formula (11.1.44). At low temperatures the particle momenta are small, so we may insert for the matrix elements $u(\mathbf{p})$ their value at $\mathbf{p} = 0$, namely u_0/V; see eqns. (11.1.44) and (11.1.50). We then have

$$\hat{H} = \sum_p \frac{p^2}{2m} a_p^\dagger a_p + \frac{u_0}{2V}\sum{}' a_{p_1'}^\dagger a_{p_2'}^\dagger a_{p_2} a_{p_1}, \tag{1}$$

where the primed summation goes only over those momenta that conform to the principle of momentum conservation. Let N be the number of particles constituting the system; then

$$\sum_p a_p^\dagger a_p = \hat{N}. \tag{2}$$

Now, we know that in the ground state of an *ideal* Bose gas all the particles are found to be in the state $\mathbf{p} = 0$, i.e. n_0 is equal to N while n_p, for all $\mathbf{p} \neq 0$, is equal to zero. We, therefore, expect that in the low-lying states, as well as in the ground state, of a slightly imperfect gas the occupation numbers n_p, for $\mathbf{p} \neq 0$, would still be small in comparison with the number n_0, while the latter would still be close to the total number N:

$$a_0^\dagger a_0 = n_0 = O(N). \tag{3}$$

However,

$$a_0 a_0^\dagger - a_0^\dagger a_0 = 1 \ll N; \tag{4}$$

therefore,

$$a_0 a_0^\dagger = n_0 + 1 \simeq a_0^\dagger a_0. \tag{5}$$

* This result is consistent with the *pseudopotential approach*, where $u(\mathbf{r})$ is replaced by $(4\pi a\hbar^2/m)\,\delta(\mathbf{r})$, whereby the integral u_0 becomes $4\pi a\hbar^2/m$. The parameter a is, therefore, the same as the scattering length of Chapter 10.

Thus, we may disregard the noncommutability of the operators a_0, $a_0^\dagger$ and treat them as c-numbers (each equal to $n_0^{1/2}$); cf. eqns. (11.1.33)–(11.1.35).

To apply the perturbation theory, we expand the quadruple sum of eqn. (1) in terms of the small quantities a_p and $a_p^\dagger$ ($p \neq 0$). The lowest-order term of the expansion is

$$a_0^\dagger a_0^\dagger a_0 a_0 = a_0^\dagger (a_0 a_0^\dagger - 1)\, a_0 = a_0^\dagger a_0 a_0^\dagger a_0 - a_0^\dagger a_0 = n_0^2 - n_0; \tag{6}$$

accordingly, the lowest-order result for the ground state energy of the system is

$$E_0 = \frac{u_0}{2V}(n_0^2 - n_0) \simeq \frac{2\pi a\hbar^2}{mV} N^2, \tag{7}$$

which is in agreement with our earlier result (10.4.4). To go to the next order of approximation, we have to do two things: (i) to express the term $a_0^\dagger a_0^\dagger a_0 a_0$ more accurately than has been done in (7), and (ii) to include a further set of terms from the quadruple sum of eqn. (1). The first task is accomplished by making use of the relationship

$$n_0 = N - \sum_{p \neq 0} a_p^\dagger a_p,$$

whence it follows that

$$a_0^\dagger a_0^\dagger a_0 a_0 = n_0^2 - n_0 \simeq N^2 - 2N \sum_{p \neq 0} a_p^\dagger a_p. \tag{8}$$

For the second, we note that the next-order terms to be included are

$$\sum_{p \neq 0} (a_p^\dagger a_{-p}^\dagger a_0 a_0 + a_0^\dagger a_0^\dagger a_p a_{-p} + a_0^\dagger a_p^\dagger a_0 a_p + a_0^\dagger a_p^\dagger a_p a_0 + a_p^\dagger a_0^\dagger a_p a_0 + a_p^\dagger a_0^\dagger a_0 a_p)$$
$$\simeq N \sum_{p \neq 0} (a_p^\dagger a_{-p}^\dagger + a_p a_{-p} + 4 a_p^\dagger a_p). \tag{9}$$

Combining (8) and (9) with the kinetic energy part of the Hamiltonian, we obtain in the desired approximation

$$\hat{H} = \frac{N^2 u_0}{2V} + \frac{N u_0}{2V} \sum_{p \neq 0} (a_p^\dagger a_{-p}^\dagger + a_p a_{-p} + 2 a_p^\dagger a_p) + \sum_p \frac{p^2}{2m} a_p^\dagger a_p. \tag{10}$$

Our next task consists in expressing the quantity u_0, which was defined in (11.1.50), in terms of the scattering length a. The approximate result

$$u_0 \simeq 4\pi a\hbar^2/m \tag{11}$$

is accurate enough to evaluate the second term of the expression (10); for the evaluation of the first term, however, we need a better approximation than the one given in (11). For this, we note that "if the probability of a particular quantum transition in a given system under the influence of a constant perturbation $\hat{V}$ is, in the *first* approximation, determined by the matrix element V_0^0, then in the *second* approximation we have instead

$$V_0^0 + \sum_{n \neq 0} \frac{V_n^0 V_0^n}{E_0 - E_n},$$

the summation being performed over the various states of the unperturbed system".

In the present case, we are dealing with a collision process in the two-particle system (with reduced mass $\frac{1}{2}m$) and the role of V_0^0 is played by the quantity

$$u_{00}^{00} = \frac{1}{V}\int u(\boldsymbol{r})\, d^3\boldsymbol{r} = \frac{u_0}{V};$$

see eqn. (11.1.44) for the matrix element $u_{p_1, p_2}^{p_1', p_2'}$. Making use of the other matrix elements, we find that for going over from the first to the second approximation we have to replace u_0/V by

$$\frac{u_0}{V} + \frac{1}{V^2}\sum_{p\neq 0}\frac{\left|\int d^3r e^{ip\cdot r/\hbar}u(r)\right|^2}{-p^2/m} \simeq \frac{u_0}{V} - \frac{u_0^2 m}{V^2}\sum_{p\neq 0}\frac{1}{p^2}.$$

Equating the foregoing result with the expression $4\pi a\hbar^2/mV$, we obtain, instead of (11),

$$u_0 \simeq \frac{4\pi a\hbar^2}{m}\left(1+\frac{4\pi a\hbar^2}{V}\sum_{p\neq 0}\frac{1}{p^2}\right). \tag{12}$$

Substituting (12) into (10), we get

$$\hat{H} = \frac{2\pi a\hbar^2}{m}\frac{N^2}{V}\left(1+\frac{4\pi a\hbar^2}{V}\sum_{p\neq 0}\frac{1}{p^2}\right) + \frac{2\pi a\hbar^2}{m}\frac{N}{V}\sum_{p\neq 0}\left(a_p^\dagger a_{-p}^\dagger + a_p a_{-p} + 2a_p^\dagger a_p\right) + \sum_{p\neq 0}\frac{p^2}{2m}a_p^\dagger a_p. \tag{13}$$

To evaluate the energy levels of the system it is n ecessary to diagonalize the Hamiltonia, (13). This is accomplished by introducing a *linear transformation* of the operators a_p and $a_p^\dagger$n first employed by Bogoliubov (1947):

$$b_p = \frac{a_p + \alpha_p a_{-p}^\dagger}{\sqrt{(1-\alpha_p^2)}}, \quad b_p^\dagger = \frac{a_p^\dagger + \alpha_p a_{-p}}{\sqrt{(1-\alpha_p^2)}}, \tag{14}$$

where

$$\alpha_p = \frac{mV}{4\pi a\hbar^2 N}\left\{\frac{4\pi a\hbar^2 N}{mV} + \frac{p^2}{2m} - \varepsilon(\boldsymbol{p})\right\}, \tag{15}$$

with

$$\varepsilon(\boldsymbol{p}) = \left\{\frac{4\pi a\hbar^2 N}{mV}\frac{p^2}{m} + \left(\frac{p^2}{2m}\right)^2\right\}^{1/2}; \tag{16}$$

clearly, each $\alpha_p < 1$. Relations inverse to (14) are

$$a_p = \frac{b_p - \alpha_p b_{-p}^\dagger}{\sqrt{(1-\alpha_p^2)}}, \quad a_p^\dagger = \frac{b_p^\dagger - \alpha_p b_{-p}}{\sqrt{(1-\alpha_p^2)}}. \tag{17}$$

It is easy to check that the new operators b_p and $b_p^\dagger$ satisfy the same commutation rules as did the old operators a_p and $a_p^\dagger$, namely

$$[b_p, b_{p'}^\dagger] = \delta_{pp'} \tag{18a}$$

$$[b_p, b_{p'}] = [b_p^\dagger, b_{p'}^\dagger] = 0. \tag{18b}$$

Substituting (17) into (13), we obtain the Hamiltonian in the diagonalized form:

$$\hat{H} = E_0 + \sum_{p\neq 0} \varepsilon(\boldsymbol{p})\, b_p^\dagger b_p\,, \tag{19}$$

where

$$E_0 = \frac{2\pi a\hbar^2 N^2}{mV} + \frac{1}{2}\sum_{p\neq 0}\left\{\varepsilon(\boldsymbol{p}) - \frac{p^2}{2m} - \frac{4\pi a\hbar^2 N}{mV} + \left(\frac{4\pi a\hbar^2 N}{mV}\right)^2 \frac{m}{p^2}\right\}. \tag{20}$$

In view of the commutation rules (18) and the expression (19) for the Hamiltonian operator $\hat{H}$, it seems natural to infer that the operators b_p and $b_p^\dagger$ are the annihilation and creation operators of certain "quasi-particles"—which represent the *elementary excitations* of the system—with the energy–momentum relationship given by (16); moreover, it is clear that these quasi-particles obey *Bose–Einstein statistics.* The quantity $b_p^\dagger b_p$ then becomes the particle-number operator for the quasi-particles (or elementary excitations) of momentum $\boldsymbol{p}$, whereby the second part of the Hamiltonian (19) becomes the energy operator corresponding to the elementary excitations present in the system. The first part of the Hamiltonian, which is explicitly stated in eqn. (20), is therefore the ground state energy of the system. Replacing the summation over $\boldsymbol{p}$ by an integration and introducing a dimensionless variable x, defined by

$$x = p\left(\frac{V}{8\pi a\hbar^2 N}\right)^{1/2},$$

we obtain for the ground state energy of the system

$$E_0 = \frac{2\pi a\hbar^2 N^2}{mV}\left[1 + \left(\frac{128\, Na^3}{\pi V}\right)^{1/2} \int_0^\infty dx \left[x^2\left(\frac{1}{2x^2} + x\sqrt{(x^2+2)} - 1 - x^2\right)\right]\right]. \tag{21}$$

The value of the integral turns out to be $(128)^{1/2}/15$, with the result

$$\frac{E_0}{N} = \frac{2\pi a\hbar^2 n}{m}\left[1 + \frac{128}{15\pi^{1/2}}(na^3)^{1/2}\right], \tag{22}$$

where n denotes the particle density in the system. Equation (22) represents the first two terms of the expansion of the quantity E_0/N in terms of the low-density parameter $(na^3)^{1/2}$.*

The foregoing result was first obtained by Lee and Yang (1957) by using the binary collision method of Chapter 9; the details of this calculation, however, appeared much later (see Lee and Yang, 1960a; also Problem 11.3). Using the pseudopotential method, this result was rederived by Lee, Huang and Yang (1957).

* The evaluation of higher-order terms of this expansion necessitates the consideration of three-body collisions as well; hence, in general, they cannot be expressed in terms of the scattering length alone. The exceptional case of a hard-sphere gas has been pursued by Wu (1959), who obtained (using the pseudopotential method)

$$E_0/N = \frac{2\pi a\hbar^2 n}{m}\left[1 + \frac{128}{15\pi^{1/2}}(na^3)^{1/2} + 8\left(\frac{4\pi}{3} - \sqrt{3}\right)(na^3)\ln(12\pi na^3) + 0(na^3)^1\right],$$

which shows that the expansion does not proceed in *simple powers* of $(na^3)^{1/2}$.

The ground state pressure of the system can be obtained as follows:

$$P_0 = -\left(\frac{\partial E_0}{\partial V}\right)_N = +n^2\frac{\partial(E_0/N)}{\partial n}$$
$$= \frac{2\pi a\hbar^2 n^2}{m}\left[1+\frac{64}{5\pi^{1/2}}(na^3)^{1/2}\right], \tag{23}$$

whence one obtains for the velocity of sound in the system

$$c_0^2 = \frac{\partial P_0}{\partial(mn)} = \frac{4\pi a\hbar^2 n}{m^2}\left[1+\frac{16}{\pi^{1/2}}(na^3)^{1/2}\right]. \tag{24}$$

Equations (23) and (24) are an *improved* version of the lowest-order results obtained earlier; see eqns. (10.4.7).

The ground state of the system is characterized by a complete absence of excitations; accordingly, the eigenvalue of the number operator $b_p^\dagger b_p$ of the "quasi-particles" must be zero for all $\boldsymbol{p} \neq 0$. As for the real particles, there must be some that possess nonzero energies even at absolute zero, for only then can the system have a finite energy in the ground state. The momentum distribution of the real particles can be determined by evaluating the ground state expectation values of the number operators $a_p^\dagger a_p$. Now, in the ground state of the system,

$$a_p|\Psi_0\rangle = \frac{1}{\sqrt{(1-\alpha_p^2)}}(b_p-\alpha_p b_{-p}^\dagger)|\Psi_0\rangle = \frac{-\alpha_p}{\sqrt{(1-\alpha_p^2)}}b_{-p}^\dagger|\Psi_0\rangle \tag{25}$$

because $b_p|\Psi_0\rangle \equiv 0$. Constructing the hermitian conjugate of (25) and remembering that α_p is real, we have

$$\langle\Psi_0|a_p^\dagger = \frac{-\alpha_p}{\sqrt{(1-\alpha_p^2)}}\langle\Psi_0|b_{-p}. \tag{26}$$

The scalar product of the expressions (25) and (26) gives

$$\langle\Psi_0|a_p^\dagger a_p|\Psi_0\rangle = \frac{\alpha_p^2}{1-\alpha_p^2}\langle\Psi_0|b_{-p}b_{-p}^\dagger|\Psi_0\rangle = \frac{\alpha_p^2}{1-\alpha_p^2}; \tag{27}$$

here, use has been made of the facts that $b_p b_p^\dagger - b_p^\dagger b_p \equiv 1$ and in the ground state, for all $\boldsymbol{p} \neq 0$, $b_p^\dagger b_p = 0$ (and hence $b_p b_p^\dagger = 1$). Thus, we obtain (for $\boldsymbol{p} \neq 0$)

$$\bar{n}_p = \frac{\alpha_p^2}{1-\alpha_p^2} = \frac{1+x^2}{2x\sqrt{(x^2+2)}}-\frac{1}{2}, \tag{28}$$

where $x = p(8\pi a\hbar^2 n)^{-1/2}$. The total number of "excited" particles in the ground state of the system is, therefore, given by

$$\sum_{p\neq 0}\bar{n}_p = \sum_{p\neq 0}\frac{\alpha_p^2}{1-\alpha_p^2} = \sum_{x>0}\frac{1}{2}\left(\frac{1+x^2}{x\sqrt{(x^2+2)}}-1\right)$$
$$\simeq N\left\{\frac{32}{\pi}(na^3)\right\}^{1/2}\int_0^\infty dx\left[x\left(\frac{1+x^2}{\sqrt{(x^2+2)}}-x\right)\right]. \tag{29}$$

The value of the integral turns out to be $(2)^{1/2}/3$, with the result

$$\sum_{p\neq 0} \bar{n}_p \simeq N \frac{8}{3\pi^{1/2}} (na^3)^{1/2}. \tag{30}$$

Accordingly,

$$\bar{n}_0 = N - \sum_{p\neq 0} \bar{n}_p \simeq N\left[1 - \frac{8}{3\pi^{1/2}} (na^3)^{1/2}\right]. \tag{31}$$

The foregoing result was first obtained by Lee, Huang and Yang (1957), using the pseudopotential method. It may be noted that the significance of the *real-particle* occupation numbers n_p for the study of an interacting Bose system had been emphasized earlier by Penrose and Onsager (1956).

11.3. Energy spectrum of a Bose liquid

In this section we propose to study the most essential features of the energy spectrum of a Bose liquid and to discuss the relevance of this study to the problem of liquid He^4. In this context we recall that the low-lying states of a *dilute gaseous* system composed of *weakly interacting* bosons are characterized by the presence of the so-called *elementary excitations* (or "quasi-particles"), which are themselves bosons and whose energy spectrum is given by

$$\varepsilon(p) = \{p^2c^2 + (p^2/2m)^2\}^{1/2}, \tag{1}$$

where

$$c = (4\pi a\hbar^2 n)^{1/2}/m; \tag{2}$$

see eqns. (11.2.16) and (11.2.18).* For low values of p, the spectrum is essentially linear: $\varepsilon \simeq pc$. The initial slope of the (ε, p)-curve is, therefore, given by the parameter c, which is identical with the limiting value of the velocity of sound in the system; compare (2) with (11.2.24). It is then natural that the low-momentum excitations be identified with *phonons*—the quanta of the sound field. For large values of p, the spectrum approaches the classical limit: $\varepsilon \simeq p^2/2m$. It is important to note that, all along, the energy–momentum relationship is strictly *monotonic* and does not display any "dip" of the kind propounded by Landau (for liquid He^4) and observed by Yarnell *et al.* and by Henshaw and Woods; see Sec. 7.5. Thus, the spectrum provided by the theory of the preceding sections simulates the Landau spectrum only to the extent of phonons; it does not account for rotons. This is hardly surprising because the theory in question was intended for a dilute Bose gas ($na^3 \ll 1$) and not for liquid He^4 ($na^3 \simeq 0.2$).

Subsequently, Brueckner and Sawada (1957) developed a theory of the interacting Bose gas which could be applicable at higher densities as well. They obtained, for the energy spectrum of the elementary excitations,†

$$\varepsilon(p) = \left\{\left(\frac{p^2}{2m}\right)^2 + \frac{1}{2}\Lambda\left(\frac{p\hbar}{ma}\right)^2 \frac{\sin(pa/\hbar)}{(pa/\hbar)}\right\}^{1/2}, \tag{3}$$

* The spectrum (1) was first obtained by Bogoliubov (1947) by the method outlined in Secs. 11.1 and 11.2. Using the pseudopotential method, this spectrum was rederived by Lee, Huang and Yang (1957).

† See also Liu, Liu and Wong (1964).

where parameter Λ is given by the implicit relationship

$$4\pi^2 na^3 = \Lambda \int_0^\infty \frac{x \sin^2 x}{x^3+\Lambda \sin x}\, dx. \tag{4}$$

Figure 11.1, taken from the original reference, depicts the variation of the parameter Λ with

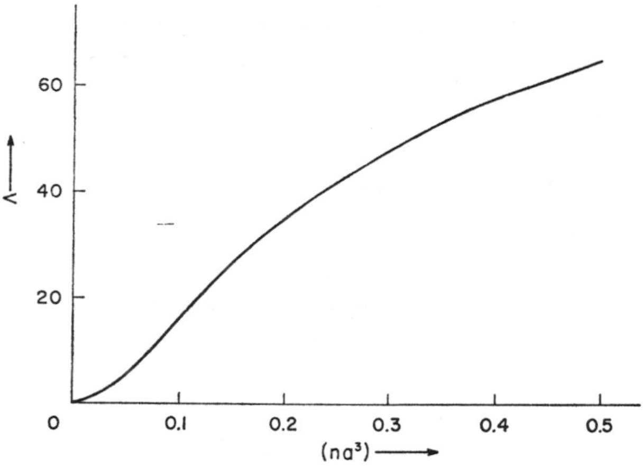

FIG. 11.1. The parameter Λ of an interacting Bose gas, as a function of the quantity na^3 (after Brueckner and Sawada, 1957).

the quantity na^3. For $na^3 \ll 1$, Λ is also much less than 1; relationship (4) then reduces to

$$4\pi^2 na^3 \simeq \Lambda \int_0^\infty \frac{\sin^2 x}{x^2}\, dx = \Lambda \frac{\pi}{2},$$

whence it follows that

$$\Lambda \simeq 8\pi na^3 \tag{5}$$

and, hence,

$$\varepsilon(p) \simeq \left\{\left(\frac{p^2}{2m}\right)^2 + \frac{4\pi na\hbar^2}{m^2} p^2 \frac{\sin (pa/\hbar)}{(pa/\hbar)}\right\}^{1/2}. \tag{6}$$

For low momenta, $\sin (pa/\hbar) \simeq (pa/\hbar)$ and the spectrum finally reduces to the Bogoliubov spectrum (1). Thus, the phonon part is once again accounted for, though for general values of na^3 the sound velocity is given by

$$c = \left(\frac{\varepsilon}{p}\right)_{p\to 0} = \left(\frac{\Lambda}{2}\right)^{1/2} \frac{\hbar}{ma}; \tag{7}$$

for $na^3 \ll 1$, $\Lambda \simeq 8\pi na^3$ and the expression (7) reduces to the Bogoliubov value (2).

For larger values of p, the spectrum of Brueckner and Sawada does lead to a "dip", which simulates quite well the roton part of the Landau spectrum; see Fig. 11.2. In this connection we note that, due to the presence of the sinusoidal term, the energy spectrum (3)

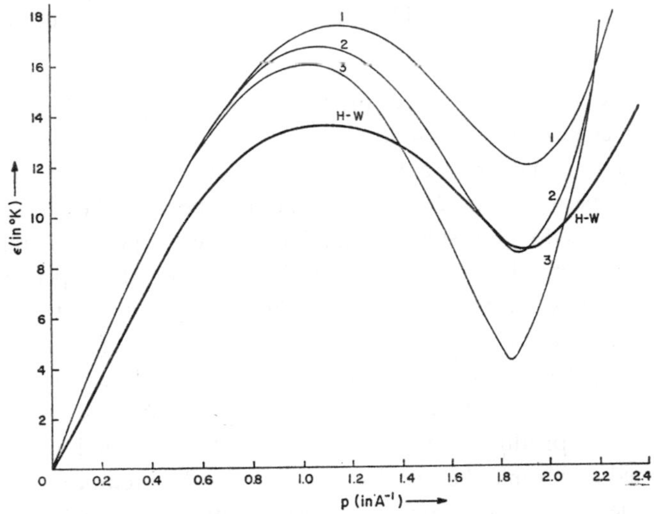

FIG. 11.2. The energy spectrum of the elementary excitations in an interacting Bose gas (after Brueckner and Sawada, 1957). The curves marked 1, 2 and 3 are the theoretical curves with $a = 2.0$ Å, 2.1 Å and 2.2 Å, respectively; in each case, n was equal to the liquid-helium value 2.2×10^{22} cm^{-3}. The experimental curve of Henshaw and Woods (1961), for the elementary excitations in liquid helium II, is also included.

passes through a minimum at $p = p_0$, where p_0 is determined by the equation

$$2(p_0 a/\hbar)^3 + \Lambda\,\{\sin(p_0 a/\hbar) + (p_0 a/\hbar)\cos(p_0 a/\hbar)\} = 0, \tag{8}$$

while the energy at the minimum is given by

$$\Delta = \frac{\hbar^2}{2ma^2}\left\{\left(\frac{p_0 a}{\hbar}\right)^4 + 2\Lambda\left(\frac{p_0 a}{\hbar}\right)\sin\left(\frac{p_0 a}{\hbar}\right)\right\}^{1/2}. \tag{9}$$

In fact, the energy spectrum in the vicinity of the minimum can be written in the Landau form, namely

$$\varepsilon(p) \simeq \Delta + \frac{(p-p_0)^2}{2\mu}; \tag{10}$$

one finds that the parameter μ, which determines the curvature of the (ε, p)-curve at the point $p = p_0$, is given by the expression

$$\mu = \left(\frac{d^2\varepsilon}{dp^2}\right)^{-1}_{p=p_0} = 2m\,\frac{\{(p_0 a/\hbar)^4 + 2\Lambda(p_0 a/\hbar)\sin(p_0 a/\hbar)\}^{1/2}}{6(p_0 a/\hbar)^2 + \Lambda\{2\cos(p_0 a/\hbar) - (p_0 a/\hbar)\sin(p_0 a/\hbar)\}}. \tag{11}$$

Thus, all the parameters of the "roton dip", namely p_0, Δ and μ, are determined in terms of the interaction parameter a and the particle density n. Pathria and Singh (1960) have shown that

(i) Equation (8) does not possess a solution for p_0 if Λ is less than a critical value Λ_1 which turns out to be about 18.93, corresponding to a value of about 0.11 for the quantity na^3,* and

(ii) Equation (9) yields a positive value for Δ only if Λ is less than another critical value Λ_2 which turns out to be about 42.10, corresponding to a value of about 0.25 for na^3.

Thus, the Brueckner–Sawada spectrum does possess the desired *nonmonotonic* character, but only within the above-mentioned limits. It is heartening that the case of liquid He^4 ($\Lambda \simeq$ 30–40) does fall within these limits.

The thermodynamic behavior of a Bose gas, for which the energy spectrum is of the *Brueckner–Sawada* type, or of the *Bogoliubov* type, has been studied in detail by Pathria and Singh (1960) and by Singh (1968); for the latter type of spectrum, reference may also be made to Lee and Yang (1958). A similar study, based on the experimentally observed spectrum, has been carried out by Bendt *et al.* (1959). The flow properties of the gas, including the phenomenon of critical velocity, and the propagation of first and second sounds in the gas have been investigated by Lee and Yang (1959c), while Wu (1961) has discussed the question of the appearance of "vorticity" in the flow of the gas.

In the end, we shall reproduce, in a summary form, the main findings of Feynman who, in 1953–4, developed an atomic theory of a *Bose liquid* at low temperatures—in particular, liquid He^4. In a series of three fundamental papers starting from first principles, Feynman established the following important results.†

(i) In spite of the presence of interatomic forces, a Bose liquid does undergo a transition analogous to the momentum-space condensation occurring in the ideal Bose gas; in other words, the original suggestion of London (1938) regarding liquid He^4, see Sec. 7.1, is essentially correct.

(ii) At sufficiently low temperatures, the only excited states possible in the liquid are the ones related to compressional waves, viz. *phonons*. Long-range motions, which leave the density of the liquid unaltered (and consequently imply nothing more than a simple "stirring" of the liquid), do not constitute excited states because they differ from the ground state only in the "permutation" of certain atoms. Motions on an atomic scale are indeed possible, but they require a *minimum* energy Δ for their excitation; clearly, these excitations would show up only at comparatively higher temperatures ($T \approx \Delta/k$), and might turn out to be Landau's *rotons*.

(iii) The wave function of the liquid, in the presence of *an* excitation, should be approximately of the form

$$\Psi = \sum_i f(\boldsymbol{r}_i)\Phi, \qquad (12)$$

* In the light of this observation, it would be wrong to suggest (as has sometimes been done) that the limiting form (6) of the energy spectrum might exhibit a nonmonotonic character; in fact, it cannot, because in the limiting case of $na^3 \ll 1$ (and hence $\Lambda \ll 1$) the coefficient of the sinusoidal term would be too small to make the spectrum nonmonotonic. In fact, even if we agree to apply the spectrum (6) *as such* to higher values of na^3, such as 0.2 for liquid He^4, the value of Λ, as obtained from (5), would still be of the order of 5 or so, which is far below the limit set by Pathria and Singh. The spectrum (6), therefore, continues to be monotonic under all realistic situations.

† The reader interested in pursuing Feynman's line of argument, which at times becomes highly qualitative, should refer to the original papers.

where Φ denotes the ground state wave function of the system while the summation of $f(\boldsymbol{r}_i)$ goes over all the N coordinates $\boldsymbol{r}_1, \ldots, \boldsymbol{r}_N$; the wave function Ψ is, clearly, symmetric in its arguments. The exact character of the function $f(\boldsymbol{r})$ can be determined by making use of a variational principle, which requires the energy of the state Ψ (and hence the energy associated with the excitation in question) to be a minimum.

The optimum choice of $f(r)$ turns out to be, see Problem 11.5,

$$f(\boldsymbol{r}) = \exp i(\boldsymbol{k} \cdot \boldsymbol{r}), \tag{13}$$

with the (minimized) energy value

$$\varepsilon(\boldsymbol{k}) = \frac{\hbar^2 k^2}{2mS(k)}, \tag{14}$$

where $S(k)$, the so-called *structure factor* of the liquid, is the Fourier transform of the *pair distribution function* $p(\boldsymbol{r})$:

$$S(k) = \int p(\boldsymbol{r}) \exp i(\boldsymbol{k} \cdot \boldsymbol{r})\, d^3r; \tag{15}$$

it will be recalled that the function $p(\boldsymbol{r}_1 - \boldsymbol{r}_2)$ denotes the probability density for finding an atom in the neighborhood of the point $\boldsymbol{r}_2$ if another one is known to be at the point $\boldsymbol{r}_1$. The optimum wave function is, therefore, given by

$$\Psi = \sum_i e^{i(\boldsymbol{k} \cdot \boldsymbol{r}_i)} \Phi. \tag{16}$$

Now, the momentum associated with this state is equal to $\hbar \boldsymbol{k}$, because

$$\boldsymbol{P}\Psi = \left(\frac{\hbar}{i} \sum_i \nabla_i\right) \Psi = \hbar \boldsymbol{k} \Psi, \tag{17}$$

$\boldsymbol{P}\Phi$ being identically equal to zero. Naturally, this would be interpreted as the momentum $\boldsymbol{p}$ associated with the excitation. One thus obtains, from first principles, the energy–momentum relationship for the elementary excitations in a Bose liquid.

On physical grounds, one can show that, for small k, the structure factor $S(k)$ rises *linearly* as $\hbar k/2mc$, reaches a maximum near $k = 2\pi/r_0$ (corresponding to a maximum in the pair distribution function at the nearest-neighbor spacing r_0, which for liquid He^4 is about 3.6 Å) and thereafter decreases to approach, with minor oscillations (corresponding to the subsidiary maxima in the pair distribution function at the spacings of the next nearest neighbors), the limiting value of unity for large k; the limiting value of unity arises due to the presence of a delta function in the expression for $p(\boldsymbol{r})$ (because, as $\boldsymbol{r}_2 \to \boldsymbol{r}_1$, one is sure to find an atom there).* Accordingly, the energy $\varepsilon(k)$ of an elementary excitation in liquid He^4 would start linearly as $\hbar kc$, would show a "dip" at $k_0 \simeq 2$ Å^{-1}† and would rise again to approach the eventual limit of $\hbar^2 k^2/2m$. These features are well displayed in Fig. 11.3.

* For a microscopic study of the structure factor $S(k)$, see Huang and Klein (1964); also Jackson and Feenberg (1962).

† It is natural that at some value of $k < k_0$, the (ε, k)-curve must pass through a *maximum*; clearly, this will happen at that value of k for which $dS/dk = 2S/k$.

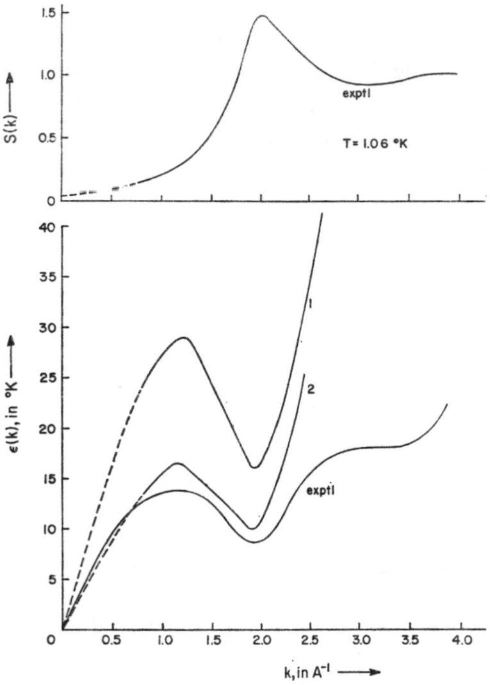

FIG. 11.3. The energy spectrum of the elementary excitations in liquid He[4]. The upper portion shows the structure factor of the liquid, as derived by Henshaw (1960) from the experimental data on neutron diffraction. Curve 1 in the lower portion shows the energy–momentum relationship based on the Feynman formula (11.3.14) while curve 2 is based on an improved formula due to Feynman and Cohen (1956). For comparison, we have included the experimental results of Woods (1966) obtained direct from neutron scattering.

11.4. Low-lying states of an imperfect Fermi gas

The Hamiltonian of the quantized field for spin-half fermions ($\sigma = +\frac{1}{2}$ or $-\frac{1}{2}$) is given by the expression (11.1.48), viz.

$$\hat{H} = \sum_{p,\sigma} \frac{p^2}{2m} a^{\dagger}_{p\sigma} a_{p\sigma} + \frac{1}{2} \sum{}' u^{p_1'\sigma_1',\, p_2'\sigma_2'}_{p_1\sigma_1,\, p_2\sigma_2} a^{\dagger}_{p_1'\sigma_1'} a^{\dagger}_{p_2'\sigma_2'} a_{p_2\sigma_2} a_{p_1\sigma_1}, \tag{1}$$

where the matrix elements u are related to the scattering length a of the two-body interaction; the summation in the second part of this expression goes only over those states (of the two particles) which conform to the principles of momentum and spin conservation. As in the Bose case, the matrix elements u in the second sum may be replaced by their values at $\boldsymbol{p} = 0$, i.e.

$$u^{p_1'\sigma_1',\, p_2'\sigma_2'}_{p_1\sigma_1,\, p_2\sigma_2} \simeq u^{0\sigma_1',\, 0\sigma_2'}_{0\sigma_1,\, 0\sigma_2}. \tag{2}$$

At the same time we note that, in view of the anticommuting character of the fermion operators $a_{p_1\sigma_1}$ and $a_{p_2\sigma_2}$, all those terms (of the second sum) which contain identical indices σ_1 and σ_2 vanish. Similarly, all those terms which contain identical indices σ_1' and σ_2' also vanish.* Thus, for a given set of values of the particle momenta, the only possible choices of the spin components are

$$\begin{array}{llll}
\text{(i)} \ \sigma_1 = +\tfrac{1}{2}, & \sigma_2 = -\tfrac{1}{2}; & \sigma_1' = +\tfrac{1}{2}, & \sigma_2' = -\tfrac{1}{2} \\
\text{(ii)} \ \sigma_1 = +\tfrac{1}{2}, & \sigma_2 = -\tfrac{1}{2}; & \sigma_1' = -\tfrac{1}{2}, & \sigma_2' = +\tfrac{1}{2} \\
\text{(iii)} \ \sigma_1 = -\tfrac{1}{2}, & \sigma_2 = +\tfrac{1}{2}; & \sigma_1' = -\tfrac{1}{2}, & \sigma_2' = +\tfrac{1}{2} \\
\text{(iv)} \ \sigma_1 = -\tfrac{1}{2}, & \sigma_2 = +\tfrac{1}{2}; & \sigma_1' = +\tfrac{1}{2}, & \sigma_2' = -\tfrac{1}{2}.
\end{array}$$

It is not difficult to see that the contribution arising from choice (i) will be identically equal to the one arising from choice (iii), while the contribution arising from choice (ii) will be identically equal to the one arising from choice (iv). We may, therefore, write

$$\hat{H} = \sum_{p,\sigma} \frac{p^2}{2m} a_{p\sigma}^\dagger a_{p\sigma} + \frac{u_0}{V} \Sigma' a_{p_1'+}^\dagger a_{p_2'-}^\dagger a_{p_2-} a_{p_1+}, \tag{3}$$

where

$$\frac{u_0}{V} = \left(u_{0+;\,0-}^{0+;\,0-} - u_{0+;\,0-}^{0-;\,0+}\right), \tag{4}$$

while the indices + and − denote the spin states $\sigma = +\frac{1}{2}$ and $\sigma = -\frac{1}{2}$, respectively; the summation in the second part of (3) goes over all such momenta as conform to the conservation law

$$\boldsymbol{p}_1' + \boldsymbol{p}_2' = \boldsymbol{p}_1 + \boldsymbol{p}_2. \tag{5}$$

To evaluate the eigenvalues of the Hamiltonian (3) we shall employ the technique of the *perturbation theory.*

First of all, we note that the main term in $\hat{H}$ is already diagonal, and its eigenvalues are

$$E^{(0)} = \sum_{p,\sigma} \frac{p^2}{2m} n_{p\sigma}, \tag{6}$$

where $n_{p\sigma}$ is the occupation number of the single-particle state $(\boldsymbol{p}, \sigma)$; its mean value, in equilibrium, is given by the Fermi distribution function

$$\bar{n}_{p\sigma} = \frac{1}{z_0^{-1} \exp(p^2/2mkT) + 1}. \tag{7}$$

The sum in (6) may be replaced by an integral, with the result (see Sec. 8.1)

$$E^{(0)} = V \frac{3kT}{\lambda^3} f_{5/2}(z_0), \tag{8}$$

where

$$\lambda = h/(2\pi mkT)^{1/2} \tag{9}$$

and

$$f_n(z_0) = \frac{1}{\Gamma(n)} \int_0^\infty \frac{x^{n-1}\,dx}{z_0^{-1} e^x + 1} = \sum_{l=1}^{\infty} (-1)^{l-1} \frac{z_0^l}{l^n}; \tag{10}$$

the ideal-gas fugacity z_0 is determined by the total number of particles in the system:

$$N = \sum_{p,\sigma} n_{p\sigma} = V \frac{2}{\lambda^3} f_{3/2}(z_0). \tag{11}$$

* Physically, this means that in the limiting case of *slow collisions* only particles with opposite spins interact with one another.

The first-order correction to the energy of the system is given by the diagonal matrix elements of the interaction term, namely the elements for which $\boldsymbol{p}_1' = \boldsymbol{p}_1$ and $\boldsymbol{p}_2' = \boldsymbol{p}_2$; thus

$$E^{(1)} = \frac{u_0}{V} \sum_{\boldsymbol{p}_1, \boldsymbol{p}_2} n_{\boldsymbol{p}_1^+} n_{\boldsymbol{p}_2^-} = \frac{u_0}{V} N^+ N^-, \tag{12}$$

where N^+ (or N^-) denotes the total number of particles with spin up (or down). Substituting the equilibrium values $\overline{N^+} = \overline{N^-} = \frac{1}{2}N$, we obtain for the first-order correction

$$E^{(1)} = \frac{u_0}{4V} N^2 = V u_0 \frac{1}{\lambda^6} \{f_{3/2}(z_0)\}^2. \tag{13}$$

Substituting $u_0 \simeq 4\pi a\hbar^2/m$, see eqn. (11.1.50), we obtain to the *first order* in a

$$E_1^{(1)} = \frac{\pi a\hbar^2}{m} \frac{N}{V} N = V \frac{2kT}{\lambda^3} \left(\frac{a}{\lambda}\right) \{f_{3/2}(z_0)\}^2, \tag{14}$$

which may be compared with our earlier results (10.3.3, 4).

The second-order correction to the energy of the system obtains from the formula

$$E_n^{(2)} = \sum_{m \neq n} \frac{|V_{nm}|^2}{E_n - E_m}, \tag{15}$$

where the indices n and m represent the unperturbed states of the system. A simple calculation yields

$$E^{(2)} = 2 \frac{u_0^2}{V^2} \sum_{\boldsymbol{p}_1, \boldsymbol{p}_2, \boldsymbol{p}_1'}{}' \frac{n_{\boldsymbol{p}_1^+} n_{\boldsymbol{p}_2^-} (1 - n_{\boldsymbol{p}_1'^+})(1 - n_{\boldsymbol{p}_2'^-})}{(p_1^2 + p_2^2 - p_1'^2 - p_2'^2)/2m}, \tag{16}$$

where the summation goes over *all* values of the momenta $\boldsymbol{p}_1, \boldsymbol{p}_2$ and $\boldsymbol{p}_1'$ (the value of $\boldsymbol{p}_2'$ being fixed by the requirement of momentum conservation); it is obvious that we must not include in the sum Σ' any terms for which $p_1^2 + p_2^2 = p_1'^2 + p_2'^2$. It will be noted that the numerator of the summand in (16) is closely related to the fact that the squared matrix element for the transition $(\boldsymbol{p}_1, \boldsymbol{p}_2) \rightarrow (\boldsymbol{p}_1', \boldsymbol{p}_2')$ is directly proportional to the probability that "the states $\boldsymbol{p}_1$ and $\boldsymbol{p}_2$ are occupied and at the same time the states $\boldsymbol{p}_1'$ and $\boldsymbol{p}_2'$ are unoccupied". Now, the expression (16) does not in itself exhaust the second-order terms in the energy. A contribution of the same order of magnitude also arises from the expression (12) if for u_0 we employ an expression more accurate than the one employed before. The desired expression can be obtained in the same manner as in the Bose case; check the steps leading to eqn. (11.2.12). In the present case, we obtain

$$\frac{4\pi a\hbar^2}{mV} \simeq \frac{u_0}{V} + 2 \frac{u_0^2}{V^2} \sum_{\boldsymbol{p}_1'} \frac{1}{(p_1^2 + p_2^2 - p_1'^2 - p_2'^2)/2m},$$

whence it follows that

$$u_0 \simeq \frac{4\pi a\hbar^2}{m} \left[1 - \frac{8\pi a\hbar^2}{mV} \sum_{\boldsymbol{p}_1'} \frac{1}{(p_1^2 + p_2^2 - p_1'^2 - p_2'^2)/2m}\right]. \tag{17}$$

Substituting (17) into (12), we obtain, apart from the first-order term quoted in (14), a second-order term, namely

$$E_2^{(1)} = -2 \left(\frac{4\pi a\hbar^2}{mV}\right)^2 \sum_{\boldsymbol{p}_1, \boldsymbol{p}_2, \boldsymbol{p}_1'}{}' \frac{n_{\boldsymbol{p}_1^+} n_{\boldsymbol{p}_2^-}}{(p_1^2 + p_2^2 - p_1'^2 - p_2'^2)/2m}. \tag{18}$$

For the second-order term in (16), the approximation $u_0 \simeq 4\pi a\hbar^2/m$ would suffice, with the result

$$E_2^{(2)} = 2\left(\frac{4\pi a\hbar^2}{mV}\right)^2 \sum_{p_1, p_2, p_1'}{}' \frac{n_{p_1^+} n_{p_2^-} (1-n_{p_1'^+})(1-n_{p_2'^-})}{(p_1^2+p_2^2-p_1'^2-p_2'^2)/2m}. \tag{19}$$

Combining (18) and (19), we obtain*

$$E_2 = E_2^{(1)}+E_2^{(2)} = -2\left(\frac{4\pi a\hbar^2}{mV}\right)^2 \sum_{p_1, p_2, p_1'}{}' \frac{n_{p_1^+} n_{p_2^-} (n_{p_1'^+}+n_{p_2'^-})}{(p_1^2+p_2^2-p_1'^2-p_2'^2)/2m}. \tag{20}$$

To evaluate (20), it is preferable to write it as a *symmetrical* summation over the four momenta p_1, p_2, p_1' and p_2' by introducing a Kronecker delta to take care of the momentum conservation:

$$E_2 = -2\left(\frac{4\pi a\hbar^2}{mV}\right)^2 \sum_{p_1, p_2, p_1', p_2'} \frac{n_{p_1^+} n_{p_2^-} (n_{p_1'^+}+n_{p_2'^-})\,\delta_{p_1+p_2,\, p_1'+p_2'}}{(p_1^2+p_2^2-p_1'^2-p_2'^2)/2m}. \tag{21}$$

It is obvious that the two parts of the sum (21), the one arising from the factor $n_{p_1'^+}$ and the other from the factor $n_{p_2'^-}$, would give identically equal results. We may, therefore, write

$$E_2 = -4\left(\frac{4\pi a\hbar^2}{mV}\right)^2 \sum_{p_1, p_2, p_1', p_2'} \frac{n_{p_1^+} n_{p_2^-} n_{p_1'^+}\delta_{p_1+p_2,\, p_1'+p_2'}}{(p_1^2+p_2^2-p_1'^2-p_2'^2)/2m}. \tag{22}$$

The sum in (22) can be evaluated in the same manner as has been done by Huang, Yang and Luttinger (1957) for the Bose gas, with the result†

$$E_2 = V\frac{8kT}{\lambda^3}\left(\frac{a^2}{\lambda^2}\right) F(z_0), \tag{23}$$

where

$$F(z_0) = -\sum_{r,s,t=1}^{\infty} \frac{(-z_0)^{r+s+t}}{\sqrt{(rst)}\,(r+s)\,(r+t)}. \tag{24}$$

Combining (8), (14) and (23), we obtain to the *second order* in a

$$E = V\frac{kT}{\lambda^3}\left[3f_{5/2}(z_0)+\frac{2a}{\lambda}\{f_{3/2}(z_0)\}^2+\frac{8a^2}{\lambda^2}F(z_0)\right], \tag{25}$$

where z_0 is determined by the relation (11).

It is now straightforward to obtain the *ground state energy* of the gas ($z_0 \to \infty$); we have only to know the asymptotic behavior of the functions involved. For the f's we obtain from the Sommerfeld lemma

$$f_n(z_0) \simeq (\ln z_0)^n/\Gamma(n+1), \tag{26}$$

* We have omitted here the term involving a product of four n's because of the following reason; in view of the fact that the numerator of this term is symmetric and the denominator antisymmetric with respect to the exchange operation $(p_1, p_2) \leftrightarrow (p_1', p_2')$, the sum of this term over the variables p_1, p_2, p_1' (and p_2') identically vanishes.

† For a direct evaluation of the sum (22), in the limit $T \to 0°$K, see Abrikosov and Khalatnikov (1957). See also Problem 11.7.

so that

$$f_{5/2}(z_0) \simeq \frac{8}{15\pi^{1/2}} (\ln z_0)^{5/2}; \qquad f_{3/2}(z_0) \simeq \frac{4}{3\pi^{1/2}} (\ln z_0)^{3/2}. \tag{27}$$

Equation (11) then gives

$$n = \frac{N}{V} \simeq \frac{8}{3\pi^{1/2}\lambda^3} (\ln z_0)^{3/2}, \tag{28}$$

whence it follows that

$$\ln z_0 \simeq \lambda^2 \left(\frac{3\pi^{1/2} n}{8} \right)^{2/3}. \tag{29}$$

The asymptotic behavior of $F(z_0)$ is given by

$$F(z_0) \simeq \frac{16(11-2\ln 2)}{105\pi^{3/2}} (\ln z_0)^{7/2}; \tag{30}$$

see Problem 11.7. Substituting (27) and (30) into (25) and making use of the relation (29), we obtain

$$\frac{E_0}{N} = \frac{3}{10} \frac{\hbar^2}{m} (3\pi^2 n)^{2/3} + \frac{\pi a \hbar^2}{m} n \left\{ 1 + \frac{6}{35} (11 - 2\ln 2) \left(\frac{3}{\pi} \right)^{1/3} n^{1/3} a \right\}. \tag{31}$$

The ground state pressure of the gas is then given by

$$\begin{aligned} P_0 &\equiv n^2 \frac{\partial (E_0/N)}{\partial n} \\ &= \frac{1}{5} \frac{\hbar^2}{m} (3\pi^2 n)^{2/3} n + \frac{\pi a \hbar^2}{m} n^2 \left\{ 1 + \frac{8}{35} (11 - 2\ln 2) \left(\frac{3}{\pi} \right)^{1/3} n^{1/3} a \right\}. \end{aligned} \tag{32}$$

We may also calculate the velocity of sound, which directly involves the compressibility of the system,

$$\begin{aligned} c_0^2 &\equiv \frac{\partial P_0}{\partial (mn)} \\ &= \frac{1}{3} \frac{\hbar^2}{m^2} (3\pi^2 n)^{2/3} + \frac{2\pi a \hbar^2}{m^2} n \left\{ 1 + \frac{4}{15} (11 - 2\ln 2) \left(\frac{3}{\pi} \right)^{1/3} n^{1/3} a \right\}. \end{aligned} \tag{33}$$

The leading terms of the foregoing expressions represent the ground state results for an ideal Fermi gas. The first-order corrections to the ideal-gas results were obtained in Sec. 10.3 as well; see, for example, eqns. (10.3.5, 8 and 9).

The result embodied in eqn. (31) was first obtained by Huang and Yang (1957) by the method of pseudopotentials; Martin and De Dominicis (1957) were the first to attempt an estimate of the third-order correction.* Lee and Yang (1957) obtained (31) on the basis of the binary collision method; for the details of their calculation, see Lee and Yang (1959b, 1960a). The same result was derived later by Galitskii (1958) who employed the method of Green's functions.

* The third-order correction has also been discussed by Mohling (1961).

11.5. Energy spectrum of a Fermi liquid; Landau's phenomenological theory*

In Sec. 11.3 we discussed the main features of the energy spectrum of a Bose liquid; such a spectrum is generally referred to as a *Bose type* spectrum. A liquid consisting of spin-half particles, such as liquid He^3, is expected to have a different kind of spectrum which, by contrast, may be called a *Fermi type* spectrum. We must, however, emphasize that a liquid consisting of Fermi particles may not necessarily possess a spectrum of the Fermi type; the type of spectrum actually possessed by such a liquid depends crucially on the nature of the interparticle interactions operating in the liquid. For instance, if the interparticle interactions are such that they tend to associate particles into "pairs", then in a particular limit we may expect to obtain a "molecular" liquid whose constituents possess "integral" spin and hence a "Bose type" spectrum. This indeed happens with the electron gas in superconductors which, at low temperatures, displays a remarkable preference for the formation of the so-called "Cooper pairs" of electrons. The formation of these pairs is essentially due to the electron–phonon–electron interaction operating in the material and lends to the electron gas the same properties of "superfluidity" as are witnessed in the case of liquid He^4 (which is a Bose liquid in its own right). For details of the mechanism of pair formation, reference may be made to the original articles of Cooper (1956) and Bardeen, Cooper and Schrieffer (1957); a simpler version of the theory is given in Cooper (1960). Again, on the basis of interatomic interactions, one expects that at extremely low temperatures liquid He^3 might also become a superfluid, though such a possibility has so far eluded experimental confirmation.† In the present section we propose to discuss the main features of a spectrum which is characteristically of the Fermi type. For a general theory of quantum liquids, see Pines and Nozières (1966).

According to Landau (1956), whose work provides the basic framework for our discussion, the Fermi type spectrum of a quantum liquid is constructed analogously to the spectrum of an ideal Fermi gas. As is well known, the ground state of the latter system corresponds to a "complete filling up of the single-particle states with $p \leqslant p_F$ and a complete absence of particles in the states with $p > p_F$"; the excitation of the system corresponds to a transition of one or more particles from the occupied states to the unoccupied states. The limiting momentum p_F is directly related to the particle density in the system and for spin-half particles is given by

$$p_F = \hbar(3\pi^2 N/V)^{1/3}. \tag{1}$$

In a liquid, we cannot speak of quantum states for *individual* particles. However, as a basis for constructing the desired spectrum, we assume that as the interparticle interactions are gradually "switched on" and a transition made from the gaseous to the liquid state, the classification of the levels remains unchanged. Of course, in this classification, the role of the gas particles passes on to the "elementary excitations" of the liquid (also referred to as the "quasi-particles"), *whose number coincides with the number of particles in the liquid and which also obey Fermi statistics.* Each "quasi-particle" possesses a definite momentum,

* For a *microscopic* theory of a Fermi liquid, see Nozières (1964); also Tuttle and Mohling (1966).

† See, in this connection, Emery and Sessler (1960).

so we can speak of a *distribution function* $n(\boldsymbol{p})$ which satisfies the obvious condition

$$\int n(\boldsymbol{p})\, d\tau = N/V, \tag{2}$$

where $d\tau = 2\, d^3p/h^3$. We then expect that the specification of the function $n(\boldsymbol{p})$ uniquely determines the total energy E of the liquid. Of course, E will not be given by a simple sum of the energies $\varepsilon(\boldsymbol{p})$ of the quasi-particles; it will rather be a *functional* of the distribution function $n(\boldsymbol{p})$. In other words, the energy E will not reduce to the simple integral $\int \varepsilon(\boldsymbol{p})\, n(\boldsymbol{p})\, V\, d\tau$, though in the first approximation a *variation* in its value may be written as

$$\delta E = V \int \varepsilon(\boldsymbol{p})\, \delta n(\boldsymbol{p})\, d\tau, \tag{3}$$

where $\delta n(\boldsymbol{p})$ is an *assumed variation* in the distribution function of the "quasi-particles". The reason why E does not reduce to an integral of the quantity $\varepsilon(\boldsymbol{p})\, n(\boldsymbol{p})$ is related to the fact that the quantity $\varepsilon(\boldsymbol{p})$ is itself a functional of the distribution function. If the initial distribution function is a step function (which corresponds to the ground state of the system), then the variation in $\varepsilon(\boldsymbol{p})$ due to a *small* deviation of the distribution function from the step function (which implies only *low-lying* excited states of the system) would be given by a *linear* functional relationship:

$$\delta\varepsilon(\boldsymbol{p}) = \int f(\boldsymbol{p}, \boldsymbol{p}')\, \delta n(\boldsymbol{p}')\, d\tau'. \tag{4}$$

Thus, the quantities $\varepsilon(\boldsymbol{p})$ and $f(\boldsymbol{p}, \boldsymbol{p}')$ are the first and second *functional derivatives* of the total energy E with respect to the distribution function $n(\boldsymbol{p})$. Inserting spin dependence, if any, we may as well write

$$\delta E = \sum_{\boldsymbol{p},\sigma} \varepsilon(\boldsymbol{p}, \sigma)\, \delta n(\boldsymbol{p}, \sigma) + \frac{1}{2V} \sum_{\boldsymbol{p},\sigma;\, \boldsymbol{p}',\sigma'} f(\boldsymbol{p}, \sigma; \boldsymbol{p}', \sigma')\, \delta n(\boldsymbol{p}, \sigma)\, \delta n(\boldsymbol{p}', \sigma'), \tag{5}$$

where δn's are *small* variations in the distribution function $n(\boldsymbol{p})$ from the step function that characterizes the ground state of the system; it is obvious that these variations will be significant only in the vicinity of the limiting momentum p_F, which continues to be given by eqn. (1). It is implied that the quantity $\varepsilon(\boldsymbol{p}, \sigma)$ in (5) corresponds to the distribution function $n(\boldsymbol{p}, \sigma)$ being infinitely close to the step function (of the ground state). One may also note that the function $f(\boldsymbol{p}, \sigma; \boldsymbol{p}', \sigma')$, being a second functional derivative of E, must be symmetric in its arguments; often, it is of the form $a + b\hat{s}_1 \cdot \hat{s}_2$, where the coefficients a and b depend only upon the angle between the momentum vectors $\boldsymbol{p}$ and $\boldsymbol{p}'$.* The function f plays a central role in the theory of the Fermi liquid; in the limit of an ideal gas, $f \to 0$.

To discover the formal dependence of the distribution function $n(\boldsymbol{p})$ on the quantity $\varepsilon(\boldsymbol{p})$, we note that, in view of the one-to-one correspondence between the energy levels of the liquid and of the ideal gas, the number of complexions (and hence the entropy) of the liquid system must be given by the same expression as for the ideal gas; see eqn. (6.1.15) with all $g_i = 1$ and $a = +1$:

$$\frac{S}{k} = -\sum_{\boldsymbol{p}} \{n \ln n + (1-n) \ln (1-n)\} \simeq -\int \{n \ln n + (1-n) \ln (1-n)\}\, d\tau. \tag{6}$$

* Of course, if the functions involved are spin-dependent, then the factor 2 in the element $d\tau$ (as well as in the element $d\tau'$) must be replaced by a summation over the spin variable(s).

Maximizing this expression, under the constraints $\delta E = 0$ and $\delta N = 0$, we obtain for the *equilibrium distribution function*

$$\bar{n} = \frac{1}{\exp\{(\varepsilon-\mu)/kT\}+1}. \tag{7}$$

It must be noted here that, despite its formal similarity with the familiar expression for the Fermi–Dirac distribution function, formula (7) is different insofar as the quantity ε appearing here is itself a function of $\bar{n}$; consequently, this formula gives only an *implicit*, and probably a very complicated, definition of the function $\bar{n}$.

A word may be said about the quantity ε appearing in eqn. (5). Since this ε corresponds to the *limiting case* of n being a step function, this is expected to be a completely defined function of p. Equation (7) then reduces to the usual Fermi–Dirac distribution function, which is indeed an *explicit* function of ε. It is not difficult to see that this reduction remains valid so long as expression (5) is valid, i.e. so long as the variations δn are small, which in turn means that so long as $T \ll T_F$. As was mentioned before, the variation δn will be significant only in the vicinity of the Fermi momentum p_F; accordingly, we will not have much to do with the function $\varepsilon(\boldsymbol{p})$ except when $p \simeq p_F$. We may, therefore, write

$$\varepsilon(p \simeq p_F) = \varepsilon_F + \left(\frac{\partial\varepsilon}{\partial p}\right)_{p=p_F}(p-p_F)+ \;\ldots\; \simeq \varepsilon_F + u_F(p-p_F), \tag{8}$$

where u_F denotes the "velocity" of the quasi-particles at the Fermi surface. In the case of an ideal gas ($\varepsilon = p^2/2m$), $u_F = p_F/m$. By analogy, we define a parameter m^* as

$$m^* \equiv \frac{p_F}{u_F} = \frac{p_F}{(\partial\varepsilon/\partial p)_{p=p_F}} \tag{9}$$

and call it the *effective mass* of the quasi-particles with the limiting momentum p_F (or with $p \simeq p_F$). Another way of looking at the parameter m^* is due to Brueckner and Gammel (1958), who write

$$\varepsilon(p \simeq p_F) = \frac{p^2}{2m} + V(p) = \frac{p^2}{2m^*} + \text{const.}; \tag{10}$$

the philosophy behind this expression is that "for quasi-particles with $p \simeq p_F$, the modification, $V(p)$, brought into the quantity $\varepsilon(\boldsymbol{p})$ by the interparticle interactions operating in the liquid may be replaced by a constant term, while the kinetic energy part, $p^2/2m$, may be replaced by a similar expression in which the mass m of the actual particle is replaced by the effective mass m^* of the quasi-particle"; in other words, we adopt an *effective field* point of view. Differentiating (10) with respect to p and putting $p = p_F$, we obtain

$$\frac{1}{m^*} = \frac{1}{m} + \frac{1}{p_F}\left(\frac{dV(p)}{dp}\right)_{p=p_F}. \tag{11}$$

The quantity m^*, in particular, determines the low-temperature specific heat of a Fermi liquid. It is easily seen that, for $T \ll T_F$, the ratio of the specific heat of a Fermi liquid to that of an ideal Fermi gas is precisely equal to the ratio m^*/m:

$$\frac{(C_V)\text{ real}}{(C_V)\text{ ideal}} = \frac{m^*}{m}. \tag{12}$$

This follows from the facts that (i) the expression (6) for the entropy S, in terms of the distribution function n, is the same for the liquid as for the gas, (ii) the same is true for the relation (7) between $\bar{n}$ and ε, and (iii) for the evaluation of the integral (6) *at low temperatures* only momenta close to p_F are important. Consequently, the result proved in Problem 8.11, namely

$$C_V \simeq S \simeq \frac{\pi^2}{3} k^2 T a(\varepsilon_F), \tag{13}$$

continues to hold, with the sole difference that in the expression for the density of states $a(\varepsilon_F)$, in the vicinity of the Fermi surface, the particle mass m gets replaced by the effective mass m^* of the relevant quasi-particles.

We now proceed to establish a relationship between the parameters m and m^* in terms of the characteristic function f. In doing this, we neglect the spin-dependence of f, if any; the necessary modification can be introduced without much difficulty. The guiding principle in this derivation is that, in the absence of external forces, the momentum density of the liquid must be equal to the density of mass transfer. The former is given by $\int \boldsymbol{p} n\, d\tau$, while the latter is given by $m\int (\partial\varepsilon/\partial\boldsymbol{p}) n\, d\tau$; here, $(\partial\varepsilon/\partial\boldsymbol{p})$ is the "velocity" of the quasi-particle with momentum $\boldsymbol{p}$ and energy ε.† Thus

$$\int \boldsymbol{p} n\, d\tau = m \int \frac{\partial\varepsilon}{\partial\boldsymbol{p}} n\, d\tau. \tag{14}$$

Varying the distribution function by δn and making use of the formula (4), we obtain

$$\begin{aligned}\int \boldsymbol{p}\, \delta n\, d\tau &= m \int \frac{\partial\varepsilon}{\partial\boldsymbol{p}} \delta n\, d\tau + m \iint \left\{ \frac{\partial f(\boldsymbol{p}, \boldsymbol{p}')}{\partial\boldsymbol{p}} \delta n'\, d\tau' \right\} n\, d\tau \\ &= m \int \frac{\partial\varepsilon}{\partial\boldsymbol{p}} \delta n\, d\tau - m \iint f(\boldsymbol{p}, \boldsymbol{p}') \frac{\partial n'}{\partial\boldsymbol{p}'} \delta n\, d\tau\, d\tau';\end{aligned} \tag{15}$$

in obtaining the last expression, we have interchanged the variables $\boldsymbol{p}$ and $\boldsymbol{p}'$ and also carried out an integration *by parts*. In view of the arbitrariness of the variation δn, eqn. (15) requires that

$$\frac{\boldsymbol{p}}{m} = \frac{\partial\varepsilon}{\partial\boldsymbol{p}} - \int f(\boldsymbol{p}, \boldsymbol{p}') \frac{\partial n'}{\partial\boldsymbol{p}'} d\tau'. \tag{16}$$

We apply this result to quasi-particles with momenta close to p_F; at the same time, we replace the distribution function n' by a "step" function, whereby

$$\frac{\partial n'}{\partial\boldsymbol{p}'} = -\frac{\boldsymbol{p}'}{p'} \delta(p' - p_F).$$

† Since the total number of quasi-particles in the liquid is the same as the total number of real particles, it is clear that to obtain the net transport of mass by the quasi-particles one has to multiply their number by the mass m of the *real* particle.

This enables us to carry out integration over the magnitude p' of the momentum, so that we have

$$\int f(\boldsymbol{p}, \boldsymbol{p}') \frac{\partial n'}{\partial p'} \frac{2p'^2 \, dp' \, d\omega'}{h^3} = -\frac{2p_F}{h^3} \int f(\theta) \boldsymbol{p}'_F \, d\omega' , \tag{17}$$

where $d\omega'$ denotes the element of the solid angle; note that we have contracted the arguments of the function f because it depends only upon the angle between the two momenta. Inserting (17) into (16), with $\boldsymbol{p} = \boldsymbol{p}_F$, making a scalar product with $\boldsymbol{p}_F$ and dividing by p_F^2, we obtain the desired result

$$\frac{1}{m} = \frac{1}{m^*} + \frac{p_F}{2h^3} \cdot 4 \int f(\theta) \cos\theta \, d\omega' . \tag{18}$$

If the function f depends upon the spins s_1 and s_2 of the particles involved, the factor 4 before the integral will have to be replaced by a summation over the spin variables.

We now derive a formula for the velocity of sound at absolute zero. From first principles, we have†

$$c_0^2 = \frac{\partial P_0}{\partial(mN/V)} = -\frac{V^2}{mN} \left(\frac{\partial P_0}{\partial V}\right)_N .$$

In the present context, it is preferable to have an expression in terms of the chemical potential of the liquid. This can be obtained by making use of the formula $N \, d\mu_0 = V \, dP_0$, whence it follows that‡

$$\left(\frac{\partial \mu_0}{\partial N}\right)_V \equiv -\frac{V}{N} \left(\frac{\partial \mu_0}{\partial V}\right)_N = -\frac{V^2}{N^2} \left(\frac{\partial P_0}{\partial V}\right)_N$$

and hence

$$c_0^2 = \frac{N}{m} \left(\frac{\partial \mu_0}{\partial N}\right)_V . \tag{19}$$

Now, $\mu_0 = \varepsilon(p_F) = \varepsilon_F$; therefore, the change $\delta\mu_0$ arising from a change δN in the total number of particles in the system is given by

$$\delta\mu_0 = \frac{\partial \varepsilon_F}{\partial p_F} \delta p_F + \int f(\boldsymbol{p}_F, \boldsymbol{p}') \, \delta n' \, d\tau' . \tag{20}$$

† At $T = 0$, $S = 0$; so there is no need to distinguish between the isothermal and adiabatic compressibilities of the liquid.

‡ Since μ_0 is an *intensive* quantity and therefore depends upon N and V only through the ratio N/V, we can write: $\mu_0 = \mu_0(N/V)$. Consequently,

$$\left(\frac{\partial \mu_0}{\partial N}\right)_V = \mu_0' \left(\frac{\partial(N/V)}{\partial N}\right)_V = \mu_0' \frac{1}{V}$$

and

$$\left(\frac{\partial \mu_0}{\partial V}\right)_N = \mu_0' \left(\frac{\partial(N/V)}{\partial V}\right)_N = -\mu_0' \frac{N}{V^2} .$$

Hence

$$\left(\frac{\partial \mu_0}{\partial N}\right)_V = -\frac{V}{N} \left(\frac{\partial \mu_0}{\partial V}\right)_N .$$

The first part in (20) arises due to the fact that a change in the total number of particles in the system inevitably alters the value of the limiting momentum p_F; see eqn. (1), whence it follows that (for constant V)

$$\delta p_F/p_F = \tfrac{1}{3}\,\delta N/N$$

and hence

$$\frac{\partial \varepsilon_F}{\partial p_F}\,\delta p_F = \frac{p_F^2}{3m^*}\,\frac{\delta N}{N}. \tag{21}$$

The variation $\delta n'$ appearing in the integral of (20) is significant only for $p' \simeq p_F$; we may, therefore, write

$$\int f(\boldsymbol{p}_F, \boldsymbol{p}')\,\delta n'\,d\tau' \simeq \int f(\theta)\,d\omega' \int \delta n'\,\frac{d\tau'}{4\pi} = \frac{\delta N}{4\pi V}\int f(\theta)\,d\omega'. \tag{22}$$

Substituting (21) and (22) into (20), we obtain

$$\left(\frac{\partial \mu_0}{\partial N}\right)_V = \frac{p_F^2}{3m^* N} + \frac{1}{4\pi V}\int f(\theta)\,d\omega'. \tag{23}$$

Making use of eqns. (18) and (1), we finally obtain

$$c_0^2 = \frac{N}{m}\left(\frac{\partial \mu_0}{\partial N}\right)_V = \frac{p_F^2}{3m^2} + \frac{p_F^3}{6mh^3}\cdot 4\int f(\theta)\,(1-\cos\theta)\,d\omega'. \tag{24}$$

Again, if the function f depends upon the spins of the particles, then the factor 4 before the integral will have to be replaced by a summation over the spin variables.

It seems instructive to apply this theory to the imperfect Fermi gas previously studied in Sec. 11.4. To calculate $f(\boldsymbol{p}, \sigma; \boldsymbol{p}', \sigma')$, we have to differentiate *twice* the sum of the expression (11.4.12), with $u_0 = 4\pi a\hbar^2/m$, and the expression (11.4.22) with respect to the distribution function $n(\boldsymbol{p}, \sigma)$ and then substitute $p = p' = p_F$. Performing the desired calculations, then changing summations into integrations and carrying out integrations by simple means, we find that the function f is spin-dependent—the spin-dependent term being in the nature of an *exchange term*, proportional to $\hat{\boldsymbol{s}}\cdot\hat{\boldsymbol{s}}_2$. The complete result is (see Abrikosov and Khalatnikov, 1957):

$$f(\boldsymbol{p}, \sigma; \boldsymbol{p}', \sigma') = A(\theta) + B(\theta)\hat{\boldsymbol{s}}_1\cdot\hat{\boldsymbol{s}}_2, \tag{25}$$

where

$$A(\theta) = \frac{2\pi a\hbar^2}{m}\left[1 + 2a\left(\frac{3N}{\pi V}\right)^{1/3}\left\{2 + \frac{\cos\theta}{2\sin(\theta/2)}\ln\frac{1+\sin(\theta/2)}{1-\sin(\theta/2)}\right\}\right]$$

and

$$B(\theta) = -\frac{8\pi a\hbar^2}{m}\left[1 + 2a\left(\frac{3N}{\pi V}\right)^{1/3}\left\{1 - \frac{1}{2}\sin\left(\frac{\theta}{2}\right)\ln\frac{1+\sin(\theta/2)}{1-\sin(\theta/2)}\right\}\right],$$

a being the scattering length of the two-body potential and θ the angle between the momentum vectors $\boldsymbol{p}_F$ and $\boldsymbol{p}_F'$. Substituting (25) into formulae (18) and (24), in which the factor 4 is supposed to be replaced by a summation over the spin variables, we find that while the spin-dependent term $B(\theta)\,\hat{s}_1 \cdot \hat{s}_2$ does not make any contribution towards the final results, the spin-independent term $A(\theta)$ leads to‡

$$\frac{1}{m^*} = \frac{1}{m} - \frac{8}{15m}(7\ln 2 - 1)\left(\frac{3N}{\pi V}\right)^{2/3} a^2, \tag{26}$$

which agrees with the results quoted in Problems 11.8 and 11.9, and

$$c_0^2 = \frac{p_F^2}{3m^2} + \frac{2\pi a\hbar^2}{m^2}\frac{N}{V}\left[1 + \frac{4}{15}a\left(\frac{3N}{\pi V}\right)^{1/3}(11 - 2\ln 2)\right], \tag{27}$$

which is identical with the result (11.4.33) obtained in the previous section. Proceeding backward, one can obtain from (27) the relevant expressions for the ground state pressure P_0 and the ground state energy E_0, namely (11.4.32) and (11.4.31), as well as the ground state chemical potential μ_0, as quoted in Problem 11.10.

Problems

11.1. (a) Prove that, for bosons as well as fermions,

$$[\psi(\boldsymbol{r}_j), \hat{H}] = \left(-\frac{\hbar^2}{2m}\nabla_j^2 + \int d^3r\, \psi^\dagger(\boldsymbol{r})\, u(\boldsymbol{r}, \boldsymbol{r}_j)\, \psi(\boldsymbol{r})\right)\psi(\boldsymbol{r}_j),$$

where $\hat{H}$ is the Hamiltonian operator defined by eqn. (11.1.4).

(b) Making use of the foregoing result, show that the equation

$$\frac{1}{\sqrt{N!}}\langle 0 \,|\, \psi(\boldsymbol{r}_1) \ldots \psi(\boldsymbol{r}_N)\hat{H} \,|\, \Psi_{NE}\rangle = E\frac{1}{\sqrt{N!}}\langle 0 \,|\psi(\boldsymbol{r}_1) \ldots \psi(\boldsymbol{r}_N)|\, \Psi_{NE}\rangle$$
$$= E\Psi_{NE}(\boldsymbol{r}_1, \ldots, \boldsymbol{r}_N)$$

is equivalent to the Schrödinger equation (11.1.15).

11.2. (a) Complete the mathematical steps leading to eqns. (11.2.21) and (11.2.22).
(b) Complete the mathematical steps leading to eqns. (11.2.29) and (11.2.30).

11.3. The ground state pressure of an interacting Bose gas turns out to be (see Lee and Yang, 1960a)

$$P_0 = \frac{\mu_0^2 m}{8\pi a\hbar^2}\left[1 - \frac{64}{15\pi}\frac{\mu_0^{1/2}m^{1/2}a}{\hbar} + \ldots\right],$$

where μ_0 is the ground state chemical potential of the gas. It then follows that

$$n \equiv \left(\frac{dP_0}{d\mu_0}\right) = \frac{\mu_0 m}{4\pi a\hbar^2}\left[1 - \frac{16}{3\pi}\frac{\mu_0^{1/2}m^{1/2}a}{\hbar} + \ldots\right]$$

and

$$\frac{E_0}{V} \equiv (n\mu_0 - P_0) = \frac{\mu_0^2 m}{8\pi a\hbar^2}\left[1 - \frac{32}{5\pi}\frac{\mu_0^{1/2}m^{1/2}a}{\hbar} + \ldots\right].$$

Eliminating μ_0, derive eqns. (11.2.22) and (11.2.23).

11.4. Show that the mean occupation number $\bar{n}_p$ of the *real* particles and the mean occupation number $\bar{N}_p$ of the *quasi*-particles, in an interacting Bose gas, are connected by the relationship

$$\bar{n}_p = \frac{\bar{N}_p + \alpha_p^2(\bar{N}_p + 1)}{1 - \alpha_p^2} \qquad (\boldsymbol{p} \neq 0),$$

where α_p is given by eqns. (11.2.15) and (11.2.16).

Note that eqn. (11.2.28) corresponds to the special case: $\bar{N}_p \equiv 0$.

‡ In a dense system, such as liquid He^3, the ratio m^*/m must be significantly larger than unity. The experimental work of Roberts and Sydoriak (1955), on the specific heat of liquid He^3, and the theoretical work of Brueckner and Gammel (1958), on the thermodynamics of a dense Fermi gas, suggest that the ratio $(m^*/m)_{He^3} \simeq 1.85$.

11.5. The excitation energy of liquid He^4, carrying a *single* excitation above the ground state, is given by the *minimum* value of the quantity

$$\varepsilon = \int \Psi^* \left\{ -\frac{\hbar^2}{2m} \sum_i \nabla_i^2 + V - E_0 \right\} \Psi \, d^{3N}r \Big/ \int \Psi^* \Psi \, d^{3N}r,$$

where E_0 denotes the ground state energy of the liquid while Ψ, according to Feynman, is given by (11.3.12). Show that the process of minimization leads to the result (11.3.14) for the energy of excitation.

[Hint: First of all, express ε in the form

$$\varepsilon = \frac{\hbar^2}{2m} \int |\nabla f(\boldsymbol{r})|^2 \, d^3r \Big/ \int f^*(\boldsymbol{r}_1) f(\boldsymbol{r}_2) \, p(\boldsymbol{r}_1 - \boldsymbol{r}_2) \, d^3r_1 \, d^3r_2.$$

Then show that ε is *minimum* if $f(\boldsymbol{r})$ is of the form (11.3.13).]

11.6. Show that for a *sufficiently large* momentum (in fact, such that the slope $d\varepsilon/dk$ is greater than the initial slope $\hbar c$), a state of *double excitation* in liquid He^4 is energetically more favorable than a state of *single excitation*, i.e. there exist momenta $\boldsymbol{k}_1$ and $\boldsymbol{k}_2$ such that, while $\boldsymbol{k}_1 + \boldsymbol{k}_2 = \boldsymbol{k}$, $\varepsilon(k_1) + \varepsilon(k_2) < \varepsilon(k)$.

[The importance of this result is brought home by the fact that the slope of the experimentally obtained (ε, p)-curve does not anywhere exceed the initial slope. Actually, the rise in its slope after the roton "dip" continues steadily until it attains a value precisely equal to the initial value; after that, the slope again declines.]

11.7. Establish the *asymptotic* formula (11.4.30) for the function $F(z_0)$.

[Hint: Write the coefficient of the expansion (11.4.24) in the form

$$\frac{1}{\sqrt{(rst)}\,(r+s)\,(r+t)} = \left(\frac{2}{\sqrt{\pi}}\right)^3 \int_0^\infty e^{-X^2 r - Y^2 s - Z^2 t - \xi(r+s) - \eta(r+t)} \, dX \, dY \, dZ \, d\xi \, d\eta.$$

Insert this expression into the defining expansion (11.4.24) and carry out summations over r, s and t, with the result

$$F(z_0) = \frac{8}{\pi^{3/2}} \int_0^\infty \frac{1}{z_0^{-1} e^{X^2+\xi+\eta}+1} \, \frac{1}{z_0^{-1} e^{Y^2+\xi}+1} \, \frac{1}{z_0^{-1} e^{Z^2+\eta}+1} \, dX \, dY \, dZ \, d\xi \, d\eta.$$

In the limit $z_0 \to \infty$, the integrand $\simeq 1$ in the region R defined by

$$X^2+\xi+\eta \leqslant \ln z_0, \qquad Y^2+\xi \leqslant \ln z_0 \qquad \text{and} \qquad Z^2+\eta \leqslant \ln z_0;$$

outside this region, the integrand $\simeq 0$. Hence, the dominant term of the *asymptotic* expansion would be

$$\frac{8}{\pi^{3/2}} \int_R 1 \cdot dX \, dY \, dZ \, d\xi \, d\eta,$$

which, in turn, reduces to the double integral

$$\frac{8}{\pi^{3/2}} \iint (\ln z_0 - \xi - \eta)^{1/2} (\ln z_0 - \xi)^{1/2} (\ln z_0 - \eta)^{1/2} \, d\xi \, d\eta;$$

the limits of integration here must be such that not only $\xi \leqslant (\ln z_0)$ and $\eta \leqslant (\ln z_0)$ but also $(\xi + \eta) \leqslant (\ln z_0)$. The rest of the calculation is straightforward.]

11.8. The grand partition function of a gaseous system composed of spin-half fermions has been evaluated by Lee and Yang (1957), with the result†

$$\ln \mathcal{Q} \equiv \frac{PV}{kT} = \frac{V}{\lambda^3} \left[2f_{5/2}(z) - \frac{2a}{\lambda} \{f_{3/2}(z)\}^2 + \frac{4a^2}{\lambda^2} f_{1/2}(z) \{f_{3/2}(z)\}^2 - \frac{8a^2}{\lambda^2} F(z) + \ldots \right],$$

where z is the fugacity of the *actual* system (not of the corresponding noninteracting system, which was denoted by the symbol z_0 in the text); the functions $f_n(z)$ and $F(z)$ are defined in a manner similar to eqns.

† For the details of this calculation, see Lee and Yang (1959b) where the case of bosons, as well as of fermions, with spin J has been treated on the basis of the *binary collision method*. The second-order result for the case of *spinless* bosons was first obtained by Huang, Yang and Luttinger (1957), using the *method of pseudopotentials*.

(11.4.10) and (11.4.24). From here, one can derive expressions for the quantities $E(z, V, T)$ and $N(z, V, T)$:

$$E(z, V, T) \equiv kT^2 \frac{\partial}{\partial T}(\ln \mathcal{Q}) \quad \text{and} \quad N(z, V, T) \equiv \frac{\partial(\ln \mathcal{Q})}{\partial(\ln z)}\left\{= \frac{2V}{\lambda^3} f_{3/2}(z_0)\right\}.$$

(a) Eliminating z between these two results, derive eqn. (11.4.25) for E.

(b) Obtain the zero-point value of the chemical potential μ, correct to the *second* order in (a/λ), and verify, with the help of eqns. (11.4.31, 32), that

$$(E+PV)_{T=0} = N(\mu)_{T=0}.$$

[Hint: At $T = 0°$K, $\mu = (\partial E/\partial N)_V$.]

(c) Show that the low-temperature specific heat and the low-temperature entropy of the gas are given by (see Pathria and Kawatra, 1962)

$$\frac{C_V}{Nk} \simeq \frac{S}{Nk} \simeq \frac{\pi^2}{2}\left(\frac{kT}{\varepsilon_F}\right)\left[1+\frac{8}{15\pi^2}(7 \ln 2-1)(k_F a)^2+ \ldots\right],$$

where $k_F = (3\pi^2 n)^{1/3}$. Clearly, the factor within the long brackets is to be identified with the ratio m^*/m; see eqn. (11.5.12).

[Hint: To determine C_V to the *first* power in T, we must know E to the *second* power in T. For this, we require higher order terms of the asymptotic expansions of the functions $f_n(z)$ and $F(z)$; these are

$$f_{5/2}(z) = \frac{8}{15\sqrt{\pi}}(\ln z)^{5/2}+\frac{\pi^{3/2}}{3}(\ln z)^{1/2}+0(1),$$

$$f_{3/2}(z) = \frac{4}{3\sqrt{\pi}}(\ln z)^{3/2}+\frac{\pi^{3/2}}{6}(\ln z)^{-1/2}+0(\ln z)^{-5/2},$$

$$f_{1/2}(z) = \frac{2}{\sqrt{\pi}}(\ln z)^{1/2}-\frac{\pi^{3/2}}{12}(\ln z)^{-3/2}+0(\ln z)^{-7/2}$$

and

$$F(z) = \frac{16(11-2 \ln 2)}{105\pi^{3/2}}(\ln z)^{7/2}-\frac{2(2 \ln 2-1)}{3}\pi^{1/2}(\ln z)^{3/2}+0(\ln z)^{5/4}.]$$

The first three results follow from the Sommerfeld lemma (E.13); for the last one, see Yang (1962).

11.9. The energy spectrum $\varepsilon(\boldsymbol{p})$ of mutually interacting, spin-half fermions is given by (Galitskii, 1958; Mohling, 1961)

$$\frac{\varepsilon(\boldsymbol{p})}{p_F^2/2m} \simeq x^2+\frac{4}{3\pi}(k_F a)+\frac{4}{15\pi^2}(k_F a)^2\left[11+2x^4 \ln \frac{x^2}{|x^2-1|}-10\left(x-\frac{1}{x}\right)\ln\left|\frac{x+1}{x-1}\right|\right.$$
$$\left.-\frac{(2-x^2)^{5/2}}{x}\ln\left|\frac{1+x\sqrt{(2-x^2)}}{1-x\sqrt{(2-x^2)}}\right|\right],$$

where $x = p/p_F \leqslant \sqrt{2}$ and $k = p/\hbar$. Show that, for k close to k_F, the spectrum reduces to

$$\frac{\varepsilon(\boldsymbol{p})}{p_F^2/2m} \simeq x^2+\frac{4}{3\pi}(k_F a)+\frac{4}{15\pi^2}(k_F a)^2\left[(11-2 \ln 2)-4(7 \ln 2-1)\left(\frac{k}{k_F}-1\right)\right].$$

Using (11.5.10, 11), verify that

$$\frac{m^*}{m} \simeq 1+\frac{8}{15\pi^2}(7 \ln 2-1)(k_F a)^2.$$

11.10. In the ground state of a Fermi system, the chemical potential is identical with the Fermi energy: $(\mu)_{T=0} = \varepsilon(p_F)$. Making use of the energy spectrum $\varepsilon(\boldsymbol{p})$ of the previous problem, we obtain

$$(\mu)_{T=0} \simeq \frac{p_F^2}{2m}\left[1+\frac{4}{3\pi}(k_F a)+\frac{4}{15\pi^2}(11-2 \ln 2)(k_F a)^2\right].$$

Integrating this result, rederive eqn. (11.4.31) for the ground state energy of the system.

CHAPTER 12

THEORY OF PHASE TRANSITIONS

VARIOUS physical phenomena to which the formalism of statistical mechanics has been applied may, in a general sense, be divided into two categories. In the first category, the microscopic constituents of the given system are regarded as practically noninteracting; as a result, the thermodynamic functions of the macroscopic system follow straightforwardly from a knowledge of the energy levels of the individual microscopic constituents. Notable examples of phenomena belonging to this category are the specific heats of gases (Secs. 1.4 and 6.6), the specific heats of solids (Sec. 7.3), chemical reactions and equilibrium constants (Problem 3.14), the condensation of an ideal Bose gas (Sec. 7.1), the spectral distribution of the black-body radiation (Sec. 7.2), the elementary electron theory of metals (Sec. 8.3), the phenomenon of paramagnetism (Secs. 3.8 and 8.2), etc. In the case of solids, the interatomic interaction does, in fact, play an important physical role; however, since the actual positions of the atoms, over a substantial range of temperatures, do not depart significantly from their mean values, we can transform our problem to the so-called *normal* coordinates and treat the solid as an "assembly of practically noninteracting harmonic oscillators". We note that the most important feature of the phenomena falling in this category is that, with the *sole* exception of Bose–Einstein condensation, the thermodynamic functions of the systems involved are smooth and continuous!

Phenomena belonging to the second category, however, present a very different situation. In most cases, one encounters analytic discontinuities or singularities in the thermodynamic functions of the given system, which in turn correspond to the occurrence of various kinds of *phase transitions*. Notable examples of phenomena belonging to this category are the condensation of gases, the melting of solids, phenomena associated with the liquid state (such as the coexistence of phases, especially in the neighborhood of the critical point), the behavior of mixtures and solutions (including the onset of phase separation), phenomena of ferromagnetism and antiferromagnetism, the famous family of lambda-transitions (such as the order–disorder transitions in alloys, the transition from liquid He I to liquid He II or the transition from a normal to a superconducting material), etc. The characteristic feature of the interparticle interactions in these systems is that they cannot be "removed" by means of a transformation of the coordinates of the problem; accordingly, the energy levels of the total system cannot, in any simple manner, be related to the energy levels of the microscopic constituents. One rather finds that under favorable circumstances a large number of microscopic constituents of the system exhibit a tendency of interacting with one another in a *strong, cooperative* fashion. This cooperative behavior assumes macroscopic significance at a particular temperature T_c, known as the *transition temperature* or the *critical temperature* of the system, and gives rise to the kind of phenomena listed above.

Mathematical problems associated with the study of cooperative phenomena are extremely formidable.† To facilitate calculations, one is forced to introduce models in which the interparticle interactions are considerably simplified, rctaining at the same time the cooperative characteristics essential to the problem. One then hopes that a statistical study of the simplified models, which still involves serious difficulties of analysis, might simulate the basic features of the phenomena exhibited by actual physical systems. Notable examples of these simplified models are based on

(i) the use of *hard-sphere* molecules in elucidating the fundamental properties of (a) gas–liquid transition or (b) Bose–Einstein condensation in an imperfect Bose gas, and
(ii) the neglect of all but *nearest-neighbor* interactions in studying phase transitions in lattice structures.

12.1. General remarks on the problem of condensation

We consider an N-particle system, obeying classical or quantum statistics, with the *proviso* that the total potential energy of the system is given by the sum of *two-particle* terms, such as $u(r_{ij})$ where $i < j$. The function $u(r)$ is supposed to satisfy the conditions:

$$\begin{aligned} u(r) &= +\infty \quad &&\text{for} \quad r \leqslant \sigma \\ 0 > u(r) &> -\varepsilon \quad &&\text{for} \quad \sigma < r < r^* \\ u(r) &= 0 \quad &&\text{for} \quad r \geqslant r^*; \end{aligned} \tag{1}$$

see Fig. 12.1. Thus, each particle may be looked upon as a hard sphere of diameter σ, surrounded by an attractive potential of range r^* and of (maximum) depth ε.‡ From a practi-

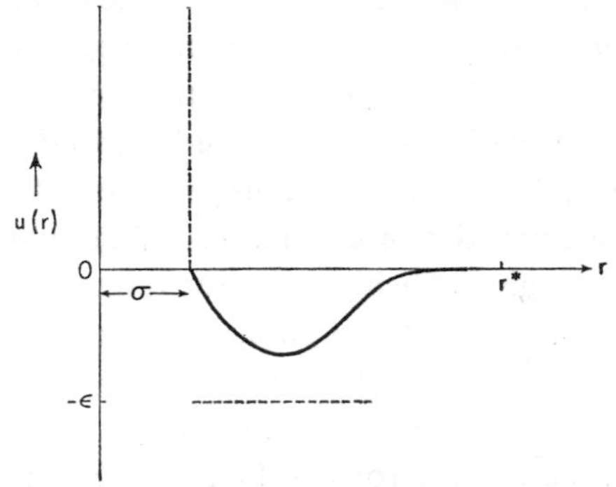

FIG. 12.1. A sketch of the interparticle potential (12.1.1).

† In this connection, one should note that the mathematical schemes developed in Chapters 9–11 give useful results only if the interactions among the microscopic constituents of the given system are sufficiently weak—in fact too weak to bring about *cooperative transitions*.

‡ If we let $\varepsilon \to 0$, we are left with a *bare* hard sphere (of diameter σ).

cal point of view, the conditions (1) cannot be regarded as a "serious restriction" on the two-body potential because the interparticle potentials ordinarily met with are not materially different from the one satisfying these conditions. We, therefore, expect that conclusions resulting from the use of this potential will not be very far from the realities of the actual physical phenomena.

Suppose that we are able to evaluate the *exact* partition function $Q_N(V, T)$ of the given system. This function would possess certain properties which have been recognized and accepted for quite some time, though a rigorous proof of these was given by van Hove as late as in 1949. These properties can be expressed as follows:

(i) In the thermodynamic limit, i.e. when N and $V \to \infty$ while the ratio N/V stays constant, the quantity $N^{-1} \ln Q$ tends to be a function only of the specific volume $v(=V/N)$ and the temperature T; this limiting form may be denoted by the symbol $f(v, T)$. It is quite natural to identify $f(v, T)$ with the *intensive* variable $-A/NkT$, where A denotes the Helmholtz free energy of the system. The thermodynamic pressure P is then given by

$$P(v, T) = -\left(\frac{\partial A}{\partial V}\right)_{N,T} = +kT\left(\frac{\partial f}{\partial v}\right)_T, \tag{2}$$

which turns out to be a strictly *non-negative* quantity.

(ii) The slope $(\partial P/\partial v)_T$ of the (P, v)-curve is *never* positive. While at high temperatures the slope is negative for all values of v, at low temperatures there *can* exist a region (or regions) in which the slope is zero, whereby the system becomes infinitely compressible! The existence of such regions, in the (P, v)-diagram, corresponds to the coexistence of two or more phases in the given system; in other words, it constitutes a direct evidence of the onset of a phase transition in the system. In this connection we note that, so long as one uses the *exact* partition function of the system, isotherms of the van der Waals type, which possess unphysical regions of positive slope, would *never* appear. Of course, if the partition function is evaluated under certain approximations, as one does in the derivation of the van der Waals equation of state (see Sec. 9.3), isotherms with unphysical regions might as well appear. In that case the isotherms in question have got to be modified, by introducing a region of "flatness" ($\partial P/\partial v = 0$), with the help of the *Maxwell construction of equal areas*; see Fig. 12.2.* The real reason for the appearance of unphysical regions in the isotherms is that the approximate evaluations of the partition function introduce, almost invariably (though implicitly), the restraint of a *uniform* particle density throughout the system. This restraint eliminates the very possibility of the system passing through a state in which there exist, side by side, two phases of different densities; in other words, the existence of a region of "flatness" is automatically denied. On the other hand, a rigorous evaluation of the partition function must take into account *all* possible configurations of the system, including the ones characterized by a simultaneous existence of two or more phases of different densities.

* The physical basis of the Maxwell construction is provided by the "principle of *minimization of free energy*". For details, see Huang (1963), sec. 2.3.

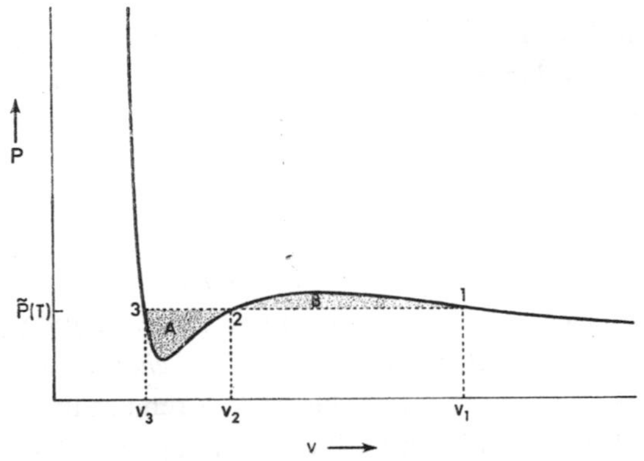

FIG. 12.2. An unphysical isotherm corrected with the help of the Maxwell construction; the horizontal line is such that the areas A and B are equal. The corrected isotherm corresponds to a phase transition, taking place at pressure $\tilde{P}(T)$, with densities v_1^{-1} and v_3^{-1} of the respective phases.

Under favorable conditions (for instance, when the temperature is sufficiently low), such a configuration might turn out to be the *equilibrium configuration* of the system, with the result that the system turns out to be in a multiphase, rather than a single-phase, state. We should remember in this connection that if in the evaluation of the partition function we introduce no other approximation except the implicit assumption of a uniform density throughout the system, then the resulting isotherms, corrected with the help of the Maxwell construction, would be the *exact* isotherms of the problem. Accordingly, the description of the multiphase state, as obtained from these "*corrected*" isotherms, would also be *exact*.

(iii) The presence of an absolutely flat portion in an isotherm ($\partial P/\partial v \equiv 0$), with mathematical singularities at its ends, is, strictly speaking, a consequence of the limiting process $N \to \infty$. If N were *finite* but large, and if the exact partition function were used, then the quantity P', defined by the relation

$$P' = kT\left(\frac{\partial \ln Q}{\partial V}\right)_{N,\,T}, \tag{3}$$

would be free from mathematical singularities. The ordinarily sharp corners in an isotherm would be rounded off; at the same time, the ordinarily flat portion of the isotherm would not be really flat—it would have instead a small, negative slope. In fact, for *finite* N, the quantity P' is not a function of v and T alone; it depends on the number N as well, though in a thermodynamically negligible manner. We further note that the absence of mathematical singularities in the thermodynamic functions of the system is closely related to the finiteness of N.

If we employ the grand partition function $\mathcal{Q}$, as obtained from the rigorous partition functions Q_N, viz.

$$\mathcal{Q}(z, V, T) = \sum_{N \geqq 0} Q_N(V, T) z^N, \tag{4}$$

a similar picture results. To see this, we first of all note that, for *real* molecules, with a given value of V, the variable N will be bounded by an upper limit, say N_m, which is the number of molecules that fill the volume V "tight-packed"; obviously, $N_m = \alpha V/\sigma^3$, where α is a number of the order of unity. For $N > N_m$, the potential energy of the system will be infinite; accordingly,

$$Q_N(N > N_m) \equiv 0. \tag{5}$$

Hence, for all practical purposes, the series (4) is a *polynomial* in z (which is $\geqslant 0$) and is of degree N_m. Since the coefficients Q_N of the series are all positive, and $Q_0 \equiv 1$, the sum $\mathscr{Q} \geqslant 1$. The thermodynamic potential $\ln \mathscr{Q}$ is, therefore, a *well-behaved* function of the parameters z, V and T. Consequently, so long as V (and hence N_m) remain finite, we do not expect that any mathematical singularities or discontinuities would appear in any of the functions derived from the thermodynamic potential. A non-analytic behavior could appear only in the limit $(V, N_m) \to \infty$.

We now define P' by the relation

$$P' = \frac{kT}{V} \ln \mathscr{Q} \qquad (V \text{ finite}); \tag{6}$$

since $\mathscr{Q} \geqslant 1$, $P' \geqslant 0$. The mean number of particles and the mean square deviation in this number are given by the formulae

$$\bar{N} = \left(\frac{\partial \ln \mathscr{Q}}{\partial \ln z}\right)_{V,T} \tag{7}$$

and

$$\overline{N^2} - \bar{N}^2 \equiv \overline{(N-\bar{N})^2} = \left(\frac{\partial \bar{N}}{\partial \ln z}\right)_{V,T}; \tag{8}$$

see eqns. (4.5.3) and (4.5.5). Accordingly,

$$\left(\frac{\partial \ln \mathscr{Q}}{\partial \bar{N}}\right)_{V,T} = \left(\frac{\partial \ln \mathscr{Q}}{\partial \ln z}\right)_{V,T} \Big/ \left(\frac{\partial \bar{N}}{\partial \ln z}\right)_{V,T} = \frac{\bar{N}}{\overline{N^2} - \bar{N}^2}. \tag{9}$$

On the other hand, writing $\bar{v}$ for $V/\bar{N}$ and using (6), we have

$$\left(\frac{\partial \ln \mathscr{Q}}{\partial \bar{N}}\right)_{V,T} = \frac{V}{kT}\left(\frac{\partial P'}{\partial \bar{N}}\right)_{V,T} = -\frac{\bar{v}^2}{kT}\left(\frac{\partial P'}{\partial \bar{v}}\right)_{V,T}. \tag{10}$$

Comparing (9) and (10), we obtain

$$\left(\frac{\partial P'}{\partial \bar{v}}\right)_{V,T} = -\frac{kT}{V^2} \frac{\bar{N}^3}{\overline{N^2} - \bar{N}^2}, \tag{11}$$

which is clearly non-positive.* For finite V, expression (11) will *never* vanish; accordingly, P' will *never* be strictly constant. Nevertheless, the slope $(\partial P'/\partial \bar{v})$ can, in a certain region, be extremely small, say $O(1/\bar{N})$; such a region would not be distinguishable from

* Compare eqn. (11), which has been derived *non-thermodynamically*, with a parallel result derived earlier, namely (4.5.10).

a phase transition because, on a macroscopic scale, the value of P' in such a region would be as good as a constant.*

If we now define the pressure of the system by the limiting relationship

$$P(\bar{v}, T) = \lim_{V\to\infty} P'(\bar{v}, T; V) = kT \lim_{V\to\infty}\left(\frac{1}{V}\ln \mathcal{Q}\right), \tag{12}$$

then we can expect, in a set of isotherms, an absolutely flat portion ($\partial P/\partial \bar{v} \equiv 0$), with sharp corners implying mathematical singularities. At the same time, the mean particle density would be given by

$$\bar{n} = \frac{\bar{N}}{V} = \lim_{V\to\infty}\left[\frac{1}{V}\frac{\partial \ln \mathcal{Q}(z, V, T)}{\partial \ln z}\right]. \tag{13}$$

It is important to note that the operation $V \to \infty$ and the operation $\partial/\partial \ln z$ cannot be interchanged freely.

We wish to mention in passing that the physical features associated with the grand partition function $\mathcal{Q}$, which is supposed to have been obtained from the *rigorous* partition functions Q_N, remain valid even if the quantity $\mathcal{Q}$ is obtained from a set of *approximate* Q_N's. This is due to the fact that the argument developed in the preceding paragraphs makes no use whatsoever of the precise form of the functions Q_N. Thus, if an approximate Q_N leads to the van der Waals type of loop in the canonical ensemble, the corresponding set of Q_N's, when employed in a grand canonical ensemble, would lead to isotherms free from such loops (Hill, 1953).

12.2. Mayer's theory of condensation

It is quite natural to expect that within the framework of statistical mechanics there exists a possibility of understanding phase transitions as a special consequence of the interparticle interactions. Accordingly, the statistical mechanics of interacting systems should be able to account for the occurrence of different kinds of transitions in various physical systems. Let us, first of all, see what Mayer's theory of cluster expansions, developed in Sec. 9.1, has to offer in this context.

According to Mayer's theory, the partition function of an N-particle system, obeying classical statistics, is given by eqn. (9.1.29):

$$Q_N(V, T) = \sum_{\{m_l\}}{}' \left[\prod_{l=1}^{N} \{(b_l V/\lambda^3)^{m_l}/m_l!\}\right], \tag{1}$$

where b_l's are the *cluster integrals*, as defined by eqns. (9.1.16)–(9.1.19), $\lambda[=h/\sqrt{(2\pi mkT)}]$ is the *mean thermal wavelength* of the particles, m_l's denote the numbers of clusters of

* The presence of such a region requires that $(\overline{N^2}-\bar{N}^2)$ be $O(\bar{N}^2)$. This implies that the thermodynamic fluctuations in the variable N be *macroscopically* large, which in turn implies equally large fluctuations in the variable $\bar{v}$ within the system and hence the coexistence of two or more phases with different values of $\bar{v}$. In a single-phase state, $(\overline{N^2}-\bar{N}^2)$ is $O(\bar{N})$; the slope $(\partial P'/\partial \bar{v})$ is then $O(\bar{N}^0)$, as an intensive quantity should be!

various sizes while the (restricted) summation Σ' goes over all the sets $\{m_l\}$ that conform to the condition

$$\sum_{l=1}^{N} lm_l = N, \quad m_l = 0, 1, 2, \ldots, N. \tag{2}$$

At high temperatures we do not expect that large-sized clusters ($l \gg 1$) will be present in macroscopically significant numbers. As temperature falls, a critical stage is reached when a sudden formation of large-sized clusters takes place. This stage may be identified with the *condensation point* of the system. To locate this point, we must examine the behavior of the numbers m_l for large values of l.

Heuristically, we apply the familiar method of picking out the *largest* term from the sum (1) and taking $\ln Q$ to be asymptotically equal to the logarithm of this term; of course, in this particular case, we will not be justified in regarding the *maximal* set $\{m_l^*\}$ as the "most probable" set of numbers, for the reason that the individual terms in the sum (1) do not appear in the nature of *weights* or *probabilities*. We thus have

$$\ln Q \simeq \sum_l m_l^* \{\ln (b_l V/\lambda^3) - \ln m_l^* + 1\}, \tag{3}$$

where the m_l^*'s are determined by the conditions

$$\sum_l \{\ln (b_l V/\lambda^3) - \ln m_l\} \, \delta m_l = 0 \tag{4}$$

and

$$\sum_l l \, \delta m_l = 0. \tag{5}$$

Using Lagrange's method of undetermined multipliers, we obtain from eqns. (4) and (5)

$$\ln (b_l V/\lambda^3) - \ln m_l^* - \alpha l = 0,$$

whence it follows that

$$m_l^* = \frac{V}{\lambda^3} b_l e^{-\alpha l}. \tag{6}$$

Substituting (6) into (3), we obtain

$$\ln Q \simeq \sum_l m_l^* (\alpha l + 1) = \alpha N + \frac{V}{\lambda^3} \sum_l b_l e^{-\alpha l}. \tag{7}$$

Comparing (7) with the equation of state (9.1.34, 35), we observe that the undetermined multiplier α should be identified with the quantity $-\ln z$. Equation (6) then becomes

$$m_l^* = \frac{V}{\lambda^3} b_l z^l, \tag{8}$$

which in turn leads to the highly suggestive result, namely $(PV/kT) = \Sigma_l m_l^*$—as if the imperfect gas composed of real particles could be looked upon as a "perfect-gas mixture" of different-sized clusters in equilibrium with one another! Clearly, this way of looking at the problem cannot be regarded as physically sound because the whole calculation breaks down *unless all the b_l's are non-negative*. If any of the b_l's happen to be negative, which in a real physical system is quite possible (see Sec. 9.3), then the corresponding m_l^* would also become negative, which is simply not acceptable. This also shows that the results following from these considerations, especially from eqn. (8), cannot be taken literally.

Now, in order to study the behavior of the numbers m_l^* for large values of l, we must consider the corresponding behavior of the cluster integrals b_l. For this purpose we recall eqn. (9.2.10), viz.

$$b_l = \frac{1}{l^2} \sum_{\{m_k\}}' \left[\prod_{k=1}^{l-1} \left\{ \frac{(l\beta_k)^{m_k}}{m_k!} \right\} \right], \tag{9}$$

where β_k's are the *irreducible cluster integrals*, defined by eqn. (9.2.8), while the summation Σ' goes over all the sets $\{m_k\}$ that conform to the condition

$$\sum_{k=1}^{l-1} km_k = l-1, \qquad m_k = 0, 1, 2, \ldots . \tag{10}$$

For large l, the number $\ln b_l$ would be very nearly equal to the logarithm of the largest term in the sum (9), i.e.

$$\ln b_l \simeq -2 \ln l + \sum_{k=1}^{l-1} m_k^* \{\ln (l\beta_k) - \ln m_k^* + 1\}, \tag{11}$$

where the m_k^*'s are determined by the conditions

$$\sum_k \{\ln (l\beta_k) - \ln m_k\} \, \delta m_k = 0 \tag{12}$$

and

$$\sum_k k \, \delta m_k = 0. \tag{13}$$

Again, using the method of undetermined multipliers, we obtain

$$\ln (l\beta_k) - \ln m_k^* - \gamma k = 0,$$

whence it follows that

$$m_k^* = l\beta_k e^{-\gamma k}. \tag{14}$$

Substituting (14) into (10), we obtain

$$\sum_k k\beta_k e^{-\gamma k} = \frac{l-1}{l} \simeq 1. \tag{15}$$

Equation (15) determines the undetermined multiplier γ. Next, we substitute (14) into (11) and obtain

$$\begin{aligned} \ln b_l &\simeq -2 \ln l + \sum_k m_k^* (\gamma k + 1) \\ &= -2 \ln l + (l-1)\gamma + l \sum_k \beta_k e^{-\gamma k} \\ &\simeq l \left(\gamma + \sum_k \beta_k e^{-\gamma k} \right); \end{aligned} \tag{16}$$

here use has been made of the fact that $l \gg 1$. We thus conclude that, for large values of l, the dependence of b_l on l is given by the formula

$$b_l = c_l b_0^l, \tag{17}$$

where

$$b_0(T) = \exp\left(\gamma + \sum_k \beta_k e^{-\gamma k}\right). \tag{18}$$

Of course, the factor c_l in eqn. (17) is also a function of the variable l; however, this dependence does not play any significant role in the present context.

Substituting (17) into (8), we obtain for the population of the large-sized clusters

$$m_l^* = \frac{V}{\lambda^3} c_l (b_0 z)^l. \tag{19}$$

Now, so long as $z < b_0^{-1}$, m_l^* is practically equal to zero ($\because$ $b_0 z < 1$ and $l \gg 1$); this means that in the *macroscopic* sense the population of large-sized clusters is not at all significant. On the other hand, if $z > b_0^{-1}$, then m_l^* becomes exceedingly large. Thus, as z increases and passes through the *critical value* $z_c = b_0^{-1}$, a sudden formation of large-sized clusters takes place. The corresponding value of v, at which this phenomenon sets in, can be obtained from the general relationship

$$\frac{1}{v} = \frac{1}{\lambda^3} \sum_{l \geqslant 1} l b_l z^l \tag{20}$$

by substituting $z = b_0^{-1}$; thus

$$\frac{1}{v_c} = \frac{1}{\lambda^3} \sum_{l \geqslant 1} l b_l b_0^{-l}. \tag{21}$$

Now, for large l, the term $l b_l z^l$ in eqn. (20) can be written as $l c_l (b_0 z)^l$. If z is increased (even slightly) above the critical value b_0^{-1}, these terms become exceedingly large, with the result that the sum (20) shoots up exceedingly fast and hence v falls down equally rapidly. In other words, v is an extremely sensitive function of z for $z > b_0^{-1}$; see Fig. 12.3.

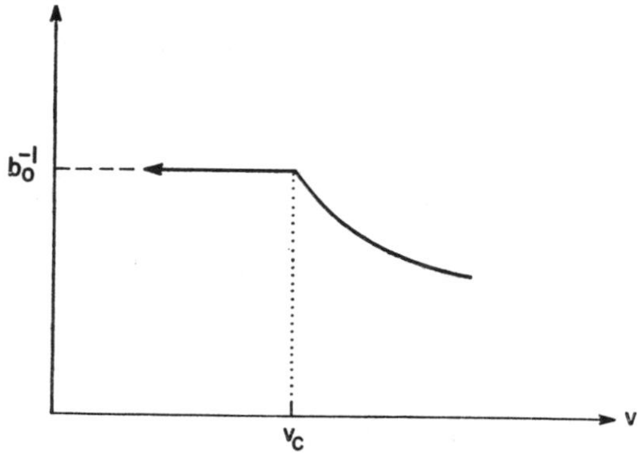

FIG. 12.3. Fugacity z, as a function of the specific volume v, in the region of a gas–liquid transition.

The horizontal portion in the figure represents a preferential formation of large-sized clusters, and hence the appearance of a two-phase state, in the system. Now, since the pressure of the system is an explicit function of z and T (and depends on V only in a thermodynamically negligible manner), see eqn. (12.1.6), a plot of P versus v would have the same qualitative appearance as the plot of z versus v. Mayer's theory, therefore, does provide a "mechanism" for the onset of a gas–liquid transition.

12.3. The theory of Yang and Lee

In the formalism of cluster expansions the interparticle interaction plays a very decisive role. The formalism, however, suffers from the disadvantage that if one wants to include in it the phenomenon of condensation and the theory of the liquid state in a rigorous fashion one has to go through stages beset with serious mathematical difficulties.* Yang and Lee (1952), on the other hand, have suggested an alternative approach that enables one to carry out a rigorous mathematical discussion of the phenomenon of condensation and of the liquid state in as natural a manner as that of the gaseous state; however, the connection with interparticle interactions is rather remote in this approach. The theory of Yang and Lee is indeed very general, in that the two-body interaction is supposed to be no more restricted than the one given by eqns. (12.1.1) which, as we already noted, is quite realistic.

In this theory one is primarily concerned with the *analytic behavior* of the quantities

$$P(z, T) = kT \operatorname*{Lim}_{V\to\infty} \left(\frac{1}{V} \ln \mathcal{Q}\right) \tag{1}$$

and

$$\frac{1}{v(z, T)} = \operatorname*{Lim}_{V\to\infty} \left\{\frac{1}{V}\left(\frac{\partial \ln \mathcal{Q}}{\partial \ln z}\right)_{V,T}\right\} = \operatorname*{Lim}_{V\to\infty} \left\{\frac{\partial}{\partial \ln z}\left(\frac{1}{V} \ln \mathcal{Q}\right)\right\}_{V,T}; \tag{2}$$

if the operation $V \to \infty$ and the operation $\partial/\partial \ln z$ are interchangeable, which is not always true, then eqn. (2) takes the form

$$\frac{1}{v(z, T)} = \frac{1}{kT}\left\{\frac{\partial P(z, T)}{\partial \ln z}\right\}_T. \tag{3}$$

To study the desired behavior, we write the grand partition function $\mathcal{Q}$, which is a polynomial (of degree N_m) in z, see eqns. (12.1.4) and (12.1.5), in terms of its *zeros in the complex z-plane*:

$$\mathcal{Q}(z, V, T) \equiv \sum_{N=0}^{N_m} Q_N(V, T) z^N = \prod_{k=1}^{N_m} \left(1 - \frac{z}{z_k}\right). \tag{4}^\dagger$$

The numbers z_k appearing here are the N_m roots of the algebraic equation

$$\sum_{N=0}^{N_m} Q_N(V, T) z^N = 0. \tag{5}$$

* It will be recalled that the relatively simple treatment developed in Sec. 12.2 suffers from a series of drawbacks, of which the most objectionable was the "possibility of some of the numbers m_i^* being negative".

† Note that, in writing this, we have kept in mind the fact that $\mathcal{Q}(z = 0) \equiv 1$.

It is obvious that the roots z_k are functions of the parameters V and T, and (through Q's) they depend upon the nature of the interparticle interaction as well; thus, we may formally write

$$z_k = z_k(V, T), \qquad k = 1, 2, \ldots, N_m. \tag{6}$$

The roots z_k are generally complex; if some of them happen to be real, they must be negative (because, in view of the fact that all the Q's are real and positive, eqn. (5) cannot be satisfied by any real, positive z). Moreover, since the complex roots of an algebraic equation with real coefficients always occur *in pairs*, i.e. if z_k is a root of the given equation, then z_k^* is also a root of that equation, the distribution of z_k's in the complex z-plane is symmetrical about the real axis.

The total number N_m of the z_k's is precisely equal to the *largest* number of particles that can be accommodated in the system; it is, therefore, directly proportional to the volume V of the system. As V increases, the number of zeros increases proportionately. At the same time, they continually move about in the z-plane; this is due to the fact that not only their number but their values also depend upon the parameter V, see eqn. (6). In the limit $V \to \infty$, the number N_m also tends to infinity; the distribution of the zeros may then be representable by a *distribution function* $g(z)$. The limiting distribution will still be a function of the temperature T, see again eqn. (6); of course, the precise nature of this dependence will be determined by the interparticle interactions operating in the system. If the limiting distribution is such that some of the zeros *converge* upon the real, positive axis at one or more points, z_c say, approaching it as close as we like (though never reaching it, because none of the zeros can itself be real, positive), then the analytic behavior of the function

$$F(z, T) = \lim_{V \to \infty} \left(\frac{1}{V} \ln \mathcal{Q} \right) \equiv P(z, T)/kT, \tag{7}^{\dagger}$$

and of any other functions derived from it, *along the real positive axis itself*, might turn out to be singular at the points $z = z_c$! The presence of such a singularity would imply the onset of a phase transition in the system.

We have already noticed that, for any given value of V and for any physical value of z (which means a real, positive z), the function $\mathcal{Q}(z)$ increases monotonically with z. Accordingly, the functions $\ln \mathcal{Q}$ and $V^{-1} \ln \mathcal{Q}$ also increase monotonically with z. We then have, by the *first theorem* of Yang and Lee,

(i) the limiting function $F(z, T)$, defined in (7), exists for all real, positive z,
(ii) it is a continuous, monotonically increasing function of z, and
(iii) it is independent of the shape of the container (unless the latter is so queer that its surface area increases faster than $V^{2/3}$).

Next, *for any finite value of* V, the function $\mathcal{Q}(z)$ is an analytic function of z throughout the complex z-plane, while the function $V^{-1} \ln \mathcal{Q}$ is analytic everywhere except at the points z_k (where $\mathcal{Q} = 0$). Remembering that none of the z_k's lie on the real positive axis, the function

† Note that the zeros of the grand partition function $\mathcal{Q}$ are automatically the poles of the thermodynamic potential $\ln \mathcal{Q}$.

$V^{-1} \ln \mathcal{Q}$ and its derivatives

$$\frac{\partial}{\partial \ln z}(V^{-1} \ln \mathcal{Q}), \quad \frac{\partial^2}{\partial(\ln z)^2}(V^{-1} \ln \mathcal{Q}), \ \ldots \tag{8}$$

are analytic *everywhere* on the real positive axis. Hence, for any physical value of z, no singularities in the thermodynamic functions can arise *so long as V is finite.*

Let us now consider a region R (of the complex plane), containing a segment of the real positive axis, such that R remains free of zeros as $V \to \infty$. Then, by the *second theorem* of Yang and Lee, the quantity $V^{-1} \ln \mathcal{Q}$, for all z in R, converges *uniformly* to a limit $F(z)$ as $V \to \infty$; consequently, the function $F(z)$ is analytic throughout R. In turn, the derivative $\partial/\partial \ln z$ of $F(z)$ would also be analytic throughout R. Further, in view of the uniform convergence of the quantity $V^{-1} \ln \mathcal{Q}$ in R, the operations $V \to \infty$ and $\partial/\partial \ln z$ become interchangeable. The equation of state of the system is then given by the *parametric equations*

$$\frac{P}{kT} = F(z) \quad \text{and} \quad \frac{1}{v} = \frac{\partial}{\partial \ln z} F(z), \tag{9}$$

for all z in R. The continued analyticity of these functions, in the region R, implies the persistence of a single-phase state throughout R. Singularities may occur if the variable z, proceeding along the real positive axis, tends to "pierce through" the boundary of R and in doing so passes too close to the zeros z_k that might be converging upon the axis at that point. For an explicit observation, we consider some of the interesting possibilities one by one.

Since the real positive axis is itself free of zeros, we might expect offhand that in a variety of cases a region R, free of zeros as $V \to \infty$, might exist and might even be unlimited in extent (thus enclosing the *entire* positive half of the real axis); see Fig. 12.4. Then, the functions P/kT and v would be analytic for all physical values of z; eliminating z, one obtains the isotherms, P versus v, whose slope is throughout negative. Clearly, a single-phase state persists in the system for all values of v. Experimentally, this sort of a situation obtains for all temperatures *above the critical temperature* T_c.

Although no zeros can lie on the real positive axis, it could be that, as $V \to \infty$, the zeros "close in" on this axis, in such a way that for a certain physical value of z, $z = t$ say, every neighborhood, however small, of the point t does contain some zeros. Then, our region R cannot include the point $z = t$. The second theorem of Yang and Lee then applies to the regions R_1 and R_2, respectively; see Fig. 12.5. As z increases along the real positive axis, the function $F(z)$ and its derivatives remain analytic everywhere in R_1 and R_2, *except at the point* $z = t$. We, therefore, have a single-phase system for all physical values of z in R_1 (which may be called Phase 1), a single-phase system for all physical values of z in R_2 (which may be called Phase 2) and a coexistence of the two phases at $z = t$. The function P/kT possesses a discontinuity of slope, and the function v a discontinuity of value, at $z = t$; in the (P, v)-diagram, the pressure P of the system stays constant (at value $\tilde{P}$ corresponding to $z = t$) while the specific volume v undergoes a finite change (from value v_1 corresponding to Phase 1 to value v_2 corresponding to Phase 2). Clearly, we are encountering a phase transition taking place in the system. Experimentally, this sort of a situation would obtain for temperatures *below* T_c.

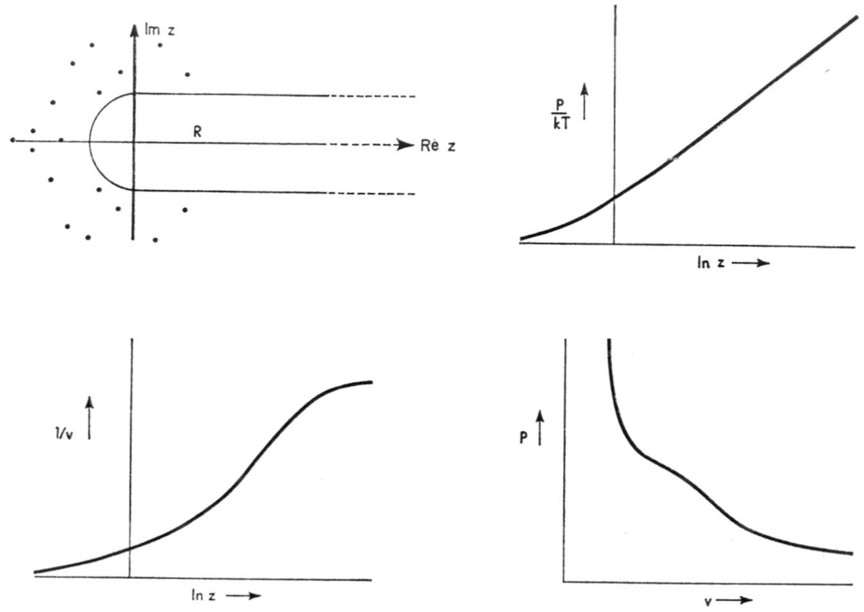

FIG. 12.4. The pressure P, the fugacity z and the specific volume v of a physical system at a temperature $T > T_c$.

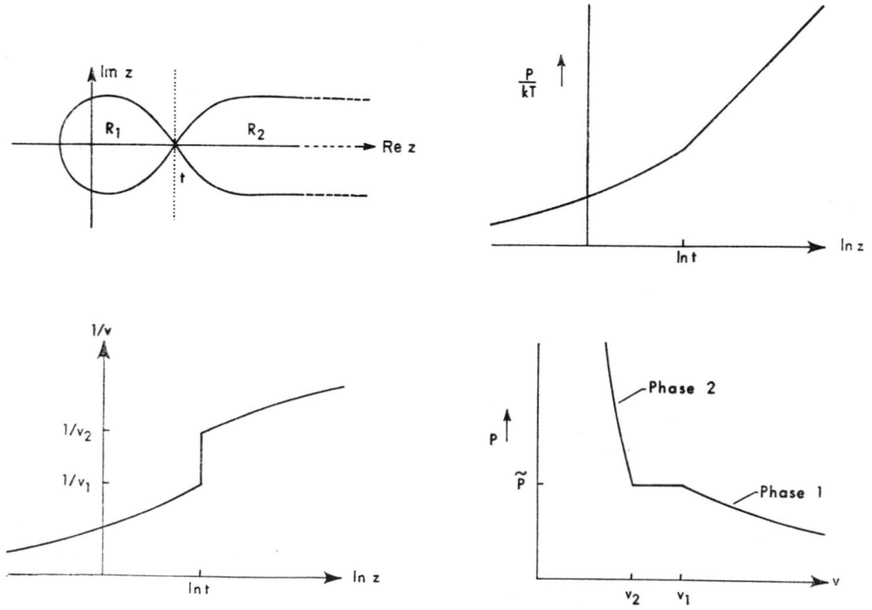

FIG. 12.5. The pressure P, the fugacity z and the specific volume v of a physical system at a temperature T at which a first-order phase transition occurs.

A more involved situation is shown in Fig. 12.6. Here, we encounter two phase transitions—one at $z = t_1$ with pressure staying constant at $\tilde{P}_1$ and volume undergoing a finite change from v_1 to v_2, and the other at $z = t_2$ with pressure staying constant at $\tilde{P}_2 (> \tilde{P}_1)$ and volume undergoing a finite change from v_3 to v_4. At $z = t_1$, Phases 1 and 2 coexist; at $z = t_2$, Phases 2 and 3 coexist.

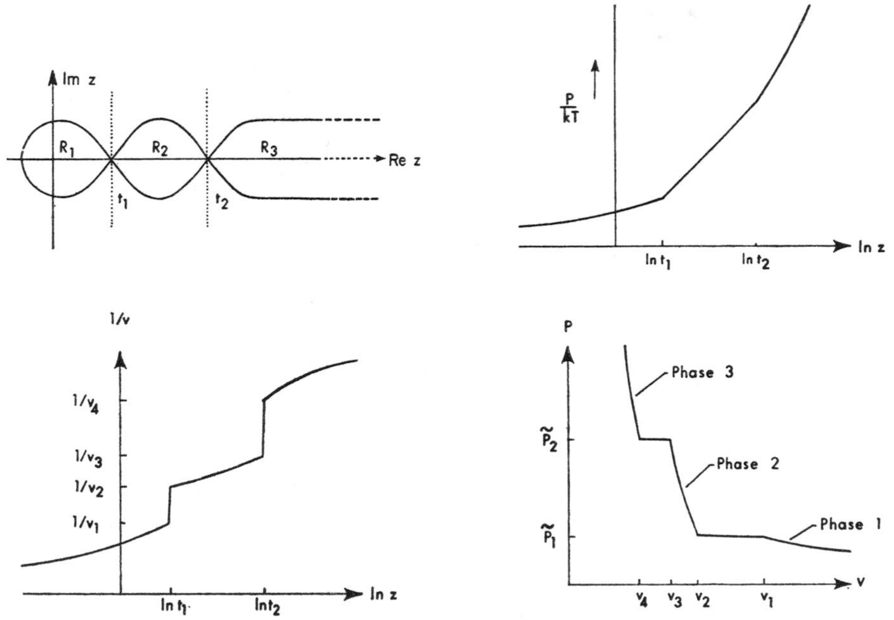

FIG. 12.6. The pressure P, the fugacity z and the specific volume v of a physical system at a temperature T at which two first-order phase transitions occur.

We shall now make a few remarks which might help to elucidate the role of the Yang–Lee theory in understanding phase transitions. In the first place, we consider what we mean by the *order* of a phase transition. In the case of transitions shown in Figs. 12.5 and 12.6, analytic discontinuities appear in the very first derivative of the function $F(z)$ $[\equiv P(z)/kT]$; consequently, the specific volume v undergoes a finite change during the transition. Such transitions are referred to as the *first-order* phase transitions. In certain cases, the singularity is of such a nature that no discontinuity appears in the first derivative of the function $F(z)$ but there does appear one in the second derivative; the specific volume then remains continuous at the transition point but it possesses a discontinuous slope there, see Fig. 12.7. Such a transition is referred to as a *second-order* phase transition. In general, if a discontinuity appears in the nth but in no lower derivative of $F(z)$, then the transition in question is said to be an *n-th order* phase transition.* Cases arise where a certain derivative of $F(z)$ diverges at some value of z, not because of the presence of a discontinuity in the previous derivative but because of a genuine, infinite slope thereof.

* For Ehrenfest's classification of phase transitions, see Epstein (1957) or Pippard (1957). Some examples of second-order phase transitions, of the type envisaged by Ehrenfest, have been discussed by Grindlay (1968).

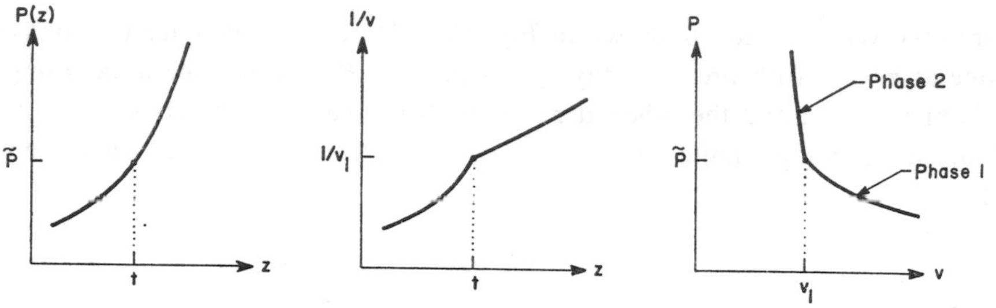

FIG. 12.7. The pressure P, the fugacity z and the specific volume v of a system at a temperature T at which a second-order phase transition occurs.

In such a case the order of the transition cannot be fixed in unambiguous terms—this situation bears some resemblance to the transition met with in liquid helium.

Next we wish to point out that in a first-order phase transition the volume $v(z)$ must *decrease* as z increases through the critical value z_c. To see this, we observe that, for any finite V,

$$\frac{\partial}{\partial \ln z}\left\{\frac{1}{v(z)}\right\} = \frac{\partial}{\partial \ln z}\left\{\frac{\partial}{\partial \ln z}\left(\frac{1}{V}\ln \mathcal{Q}\right)\right\}$$

$$= \frac{1}{V}(\overline{N^2}-\bar{N}^2) \geqslant 0; \tag{10}$$

see eqns. (12.1.7) and (12.1.8). Consequently

$$\frac{1}{v(z_c+0)} - \frac{1}{v(z_c-0)} \geqslant 0$$

and hence

$$v(z_c+0) - v(z_c-0) \leqslant 0. \tag{11}$$

Now, in a first-order phase transition v must change by a finite amount; therefore, we must have

$$v(z_c+0) - v(z_c-0) < 0. \tag{12}$$

Closely related to this result is the question of the slope of an isotherm. By eqns. (3) and (10), we have

$$\left(\frac{\partial P}{\partial v}\right)_T = \left(\frac{\partial P}{\partial \ln z}\right)_T \Big/ \left(\frac{\partial v}{\partial \ln z}\right)_T$$

$$= -\left(\frac{kT}{v^3}\right) \Big/ \left(\frac{\partial v^{-1}}{\partial \ln z}\right) \leqslant 0. \tag{13}$$

Accordingly, an isotherm can never have a positive slope; at best, the slope can be zero (which happens only when there appears a region of infinite slope in the curve v^{-1} versus $\ln z$; see Figs. 12.5 and 12.6).

Finally, we recall that the distribution of the zeros z_k, in the limit $V \to \infty$, is a function of the temperature T of the system. For temperatures below a particular value, say T_1, we expect z_k's to be "closing in" towards the real positive axis of the z-plane at a point, say $z = t_1(T)$. Suppose that, for *any* $T > T_1$, we do not encounter such a "closing in". The temperature T_1 must then be understood as the *critical temperature* of the system, namely the *highest* temperature at which a phase transition can set in. If, on the other hand, the system undergoes two transitions, one at $z = t_1(T)$ from Phase 1 to Phase 2 and the other at $z = t_2(T)$ from Phase 2 to Phase 3, the values t_1 and t_2, of the variable z, may vary with T in such a manner that for a certain temperature, say T_2, they become equal. We then have, at this common value of z, a coexistence of three phases; the temperature T_2 is naturally the *triple point* of the system. In certain other cases, the zeros of the grand partition function might approach the real positive axis along a finite segment of the axis rather than at an isolated point. In such a case we again expect to encounter a phase transition but the physical nature of such a transition is bound to be complicated.

All in all, the theory of Yang and Lee has been able to connect the problem of phase transitions with the "distribution of zeros of the grand partition function in the complex z-plane" in a very simple and satisfying manner. However, we cannot be sure that the sequence of the Q_N's pertaining to a typical interacting system met with in nature will at all lead to a "distribution of zeros" which, in the limit $V \to \infty$, will behave in the manner required by the theory. Though a general proof in this direction is still lacking, we have at least a few systems, such as the one- and two-dimensional Ising models, see Secs. 12.9 and 12.10, for which something definite can be said about the nature of the distribution of the zeros;* it is gratifying to note that the physical conclusions drawn on the basis of this theory turn out to be in complete agreement with the ones drawn on the basis of other, unrelated approaches.

12.4. Further comments on the theory of Yang and Lee

A. The gaseous phase and the cluster integrals

The gaseous phase of a system, if it exists, must be defined as "the phase corresponding to a region R which includes the origin of the complex z-plane". According to the second theorem of Yang and Lee, the function $F(z)$ is analytic in the neighborhood of the point $z = 0$; hence, for small physical values of z, this function may be represented by a power series in z. By analytic continuation, the validity of this series (and of any others derived from it) may be extended *throughout* the region R. Clearly, this continuation *cannot* be carried right into the liquid phase; in fact, we cannot even estimate the extent of the region R without a very critical study of the function $F(z)$.† Nevertheless, for the gaseous phase we can write, see eqns. (12.3.1) and (12.3.4),

$$\frac{P(z,T)}{kT} = \operatorname*{Lim}_{V\to\infty}\left(\frac{1}{V}\ln\mathcal{Q}\right) = \operatorname*{Lim}_{V\to\infty}\left\{\frac{1}{V}\sum_{k=1}^{N_m}\ln\left(1-\frac{z}{z_k}\right)\right\}$$

$$= \frac{1}{\lambda^3}\operatorname*{Lim}_{V\to\infty}\left\{\sum_{j\geq 1} b_j(V,T)z^j\right\}, \tag{1}$$

* For another case where the distribution of zeros has been studied in complete detail, see Hemmer, Hange and Aasen, *J. Math. Phys.* **7**, 35 (1966).

† See, in this connection, the review article by S. Katsura, *Adv. in Phys.* **12**, 391 (1963). Other relevant references are listed in this review.

where

$$b_j(V, T) = -\frac{\lambda^3}{jV} \sum_{k=1}^{N_m} \left\{ \frac{1}{z_k(V, T)} \right\}^j. \tag{2}$$

In the limit $V \to \infty$, $b_j(V, T) \to \bar{b}_j(T)$; accordingly, we obtain the *Mayer equations*

$$\frac{P(z, T)}{kT} = \frac{1}{\lambda^3} \sum_{j \geq 1} \bar{b}_j(T) z^j \tag{3}$$

and

$$\frac{1}{v(z, T)} = \frac{1}{\lambda^3} \sum_{j \geq 1} j \bar{b}_j(T) z^j. \tag{4}$$

As a sample calculation, we obtain from eqn. (2)

$$\begin{aligned} \bar{b}_1 &= -\lambda^3 \operatorname*{Lim}_{V \to \infty} \left\{ \frac{1}{V} \times (\text{the sum of the reciprocals of the zeros } z_k) \right\} \\ &= -\lambda^3 \operatorname*{Lim}_{V \to \infty} \left\{ \frac{1}{V} \times \frac{-Q_1(V, T)}{Q_0} \right\}; \end{aligned} \tag{5}$$

see the algebraic equation (12.3.5), of which z_k's are the roots. In view of the fact that Q_0 and Q_1 are identically equal to 1 and V/λ^3, respectively, eqn. (5) reduces to the trivial result: $\bar{b}_1 \equiv 1$. Calculation of higher-order $\bar{b}_j$'s would require a knowledge of the "sums of the correspondingly higher powers of the reciprocals of the z_k's", which in turn would require a knowledge of the correspondingly higher-order Q's. Ultimately, we obtain results identical with the ones contained in eqns. (9.4.5)–(9.4.8).

B. An electrostatic analogue

The dependence of the quantities P/kT and $1/v$ on the parameter z is determined by the form of the function $F(z)$; see eqn. (12.3.9). The function $F(z)$, on the other hand, can be determined from the (limiting) distribution of the zeros of the grand partition function:

$$F(z) \equiv \operatorname*{Lim}_{V \to \infty} \left(\frac{1}{V} \ln \mathcal{Q} \right) = \operatorname*{Lim}_{V \to \infty} \left\{ \frac{1}{V} \sum_{k=1}^{N_m} \ln \left[1 - \frac{z}{z_k(V, T)} \right] \right\}. \tag{6}$$

If we denote the *limiting values* of the zeros by the simpler symbols z_k, rather than $z_k(T)$, we can write (6) as

$$F(z) = \frac{1}{V} \sum_{k=1}^{N_m} \ln \left(1 - \frac{z}{z_k} \right). \tag{7}$$

In view of the fact that if z_k is a zero of the grand partition function then its complex conjugate z_k^* is also a zero of this function, we may rewrite (7) as

$$\begin{aligned} F(z) &= \frac{1}{2V} \sum_{k=1}^{N_m} \left\{ \ln \left(1 - \frac{z}{z_k} \right) + \ln \left(1 - \frac{z}{z_k^*} \right) \right\} \\ &= \frac{1}{V} \sum_{k=1}^{N_m} \ln \frac{\sqrt{(z^2 - 2zr_k \cos \theta_k + r_k^2)}}{r_k}, \end{aligned} \tag{8}$$

where we have substituted: $z_k = r_k(\cos\theta_k + i\sin\theta_k)$ and $z_k^* = r_k(\cos\theta_k - i\sin\theta_k)$. The first derivative of $F(z)$ follows readily from (8):

$$\frac{\partial F(z)}{\partial \ln z} = \frac{z}{V}\sum_{k=1}^{N_m}\frac{z - r_k\cos\theta_k}{z^2 - 2zr_k\cos\theta_k + r_k^2}. \tag{9}$$

Now, the analogy proceeds like this. We represent each of the zeros z_k, in the complex z-plane, by a uniform line charge (with charge density $c\lambda^3/V$, where c is a constant and hence the total charge density $N_m(c\lambda^3/V)$ is again a constant, but *independent of* V) perpendicular to this plane; then, the pressure P and the specific volume v, for any physical value of z, are straightforwardly related to the net electric potential φ and the net electric field E (due to all the N_m line charges) at the point z on the real positive axis! To see this, we refer to Fig. 12.8 in which P denotes the point of observation (actually, the physical state in question),

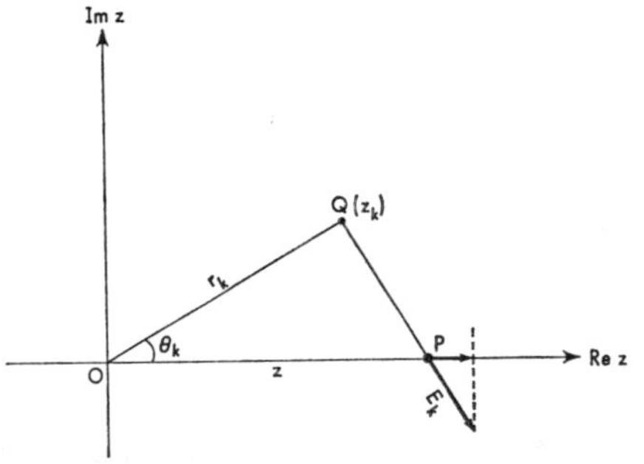

FIG. 12.8. The electric field at the point P due to a line charge passing through the point $Q(z_k)$.

while Q denotes the location of one of the zeros. The potential at the point P due to the line charge passing through the point Q would be given by†

$$-2\frac{c\lambda^3}{V}\ln(PQ) = -2\frac{c\lambda^3}{V}\ln\sqrt{(z^2 - 2zr_k\cos\theta_k + r_k^2)}; \tag{10}$$

accordingly, the net potential at P due to all the N_m line charges would be

$$\varphi(z) = -2\frac{c\lambda^3}{V}\sum_{k=1}^{N_m}\ln\sqrt{(z^2 - 2zr_k\cos\theta_k + r_k^2)}. \tag{11}$$

Next, the electric field at the point P due to the line charge passing through the point Q would be determined by the gradient of the scalar quantity (10), and the net electric field at the point P, due to all the N_m charges, would be determined by the gradient of (11). While the former will be directed along the straight line QP, the latter, in view of the *symmetrical* nature of the distribution, will be directed along the positive direction of the real axis. The magnitude of the net field will be given by

$$E(z) = -\frac{\partial\varphi(z)}{\partial z} = 2\frac{c\lambda^3}{V}\sum_{k=1}^{N_m}\frac{z - r_k\cos\theta_k}{z^2 - 2zr_k\cos\theta_k + r_k^2}. \tag{12}$$

† By convention, the electric potential at a unit distance from the line charge is taken to be zero.

Comparing (11) and (12) with (8) and (9), we obtain

$$\frac{P(z)}{kT} = -\frac{1}{2c\lambda^3}\{\varphi(z)-\varphi(0)\} \tag{13}$$

and

$$\frac{1}{v(z)} = \frac{z}{2c\lambda^3}E(z). \tag{14}$$

The foregoing results, coupled with a knowledge of the limiting distribution of the zeros in the complex z-plane, enable us to visualize the behavior of the quantities P/kT and $1/v$ as one proceeds along the real positive axis. Suppose that, as $V \to \infty$, the zeros converge upon this axis in such a way as to give a "continuous" distribution, with nonzero density $g(z)$, at the point $z = t$. Since each zero corresponds to a uniform line charge perpendicular to the z-plane, the supposed distribution would correspond to a sheet of charge, again perpendicular to the z-plane, intersecting the real positive axis at $z = t$. The surface charge density at $z = t$ could well be nonzero; let us denote it by the symbol σ. Then, as one moves along the real axis through the sheet of charge at the point $z = t$, the potential (and hence the quantity P/kT) would remain continuous while the electric field (and hence the quantity $1/v$) would show a discontinuity of magnitude $4\pi\sigma$. This is precisely the behavior associated with a ***first-order*** phase transition taking place at the fugacity value t.

12.5. A dynamical model for phase transitions

A number of physico-chemical systems which undergo phase transitions can be represented, to varying degrees of accuracy, by an "array of lattice sites, with a nearest-neighbor interaction that depends upon the manner of occupation of the neighboring sites". This simple-minded model turns out to be good enough to provide a unified theoretical basis for understanding a variety of phenomena such as ferromagnetism and antiferromagnetism, gas–liquid and liquid–solid transitions, order–disorder transitions in alloys, phase separation in binary solutions, etc. There is no doubt that this model considerably oversimplifies the actual physical systems it is supposed to represent; nevertheless, it does retain the essential statistical features of the problem—features that account for the propagation of *long-range order* in the systems under consideration. Accordingly, it does lead to the onset of a phase transition in the given system, which appears to arise in the nature of a *cooperative* phenomenon.

We find it convenient to formulate the problem in the language of ferromagnetism; later on, we shall establish the essential correspondence between this language and the languages appropriate to other physical phenomena. We regard each of the N lattice sites to be occupied by an atom possessing a magnetic moment μ, of magnitude $g\mu_B J$, which is capable of $(2J+1)$ discrete orientations in space. These orientations define different possible manners of occupation of a given lattice site; accordingly, the whole lattice is capable of $(2J+1)^N$ different configurations. Associated with each configuration is an amount of energy E that arises from mutual interactions between the neighboring atoms of the lattice (which in turn depends upon the *relative* states of orientation of their magnetic moments) and from the interaction of the whole lattice with an external field $\boldsymbol{B}$ (which in turn depends upon the *net* magnetic moment $\boldsymbol{M}$ of the system). A statistical analysis in the canonical ensemble should, therefore, enable us to determine the expectation value $\overline{M}(B, T)$ of the variable M. The presence of a *spontaneous magnetization* $\overline{M}(0, T)$ for $T < T_c$, and its absence for $T > T_c$, will then be interpreted as implying a ferromagnetic phase transition, with *Curie temperature* T_c.

Detailed studies, both theoretical and experimental, reveal that for *all* ferromagnetic materials data on the temperature dependence of spontaneous magnetization $\overline{M}(0, T)$ fit best with the value $J = \frac{1}{2}$; see Fig. 12.9. One is, therefore, tempted to conclude that the phenomenon of ferromagnetism is associated only with the spins of the electrons and not with

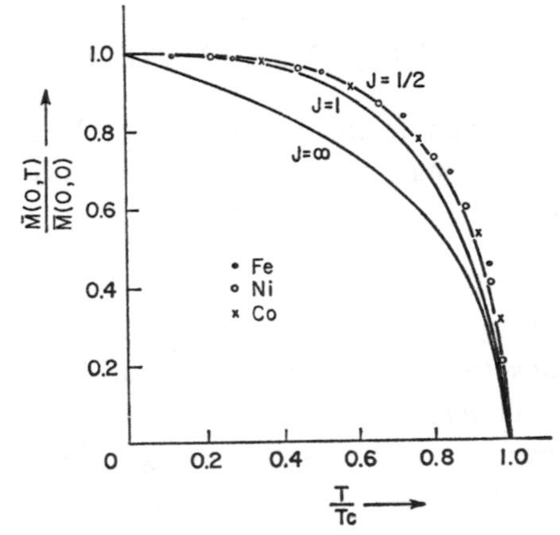

FIG. 12.9. Spontaneous magnetization of iron, nickel and cobalt as a function of temperature. Theoretical curves are based on the Weiss theory of ferromagnetism.

their orbital motion. This is further confirmed by the gyromagnetic experiments (Barnett, 1944; Scott, 1951, 1952), in which one *either* reverses the magnetization of a freely suspended specimen and observes the resulting rotation *or* imparts a rotation to the specimen and observes the resulting magnetization; the former is known as the *Einstein–de Haas method*, the latter the *Barnett method*. From these experiments one can derive the relevant g-value of the specimen which, in every case, has turned out to be very close to the value 2; this, however, pertains to the electron spin. Therefore, in discussing the problem of ferromagnetism, we may specifically take: $\boldsymbol{\mu} = 2\mu_B \boldsymbol{s}$, where $\boldsymbol{s}$ is the spin variable associated with a lattice site. With $s = \frac{1}{2}$, only two orientations are possible for each lattice site, namely $s_z = +\frac{1}{2}$ (with $\mu_z = +\mu_B$) and $s_z = -\frac{1}{2}$ (with $\mu_z = -\mu_B$). The whole lattice is then capable of 2^N configurations; one such configuration is shown in Fig. 12.10.

Let us now consider the question of the interaction energy between a pair of neighboring spins $\boldsymbol{s}_i$ and $\boldsymbol{s}_j$. According to quantum mechanics, this energy is of the form $K_{ij} \pm J_{ij}$; the upper sign holds for "antiparallel" spins ($S = 0$), the lower sign for "parallel" spins ($S = 1$). Here, K_{ij} is the *direct* or *Coulomb* energy between the two spins, while J_{ij} is the *exchange* energy between them; the latter is of purely quantum-mechanical origin:

$$K_{ij} = \int \psi_i^*(1)\, \psi_j^*(2)\, u_{ij} \psi_j(2)\, \psi_i(1)\, d\tau_1\, d\tau_2, \tag{1}$$

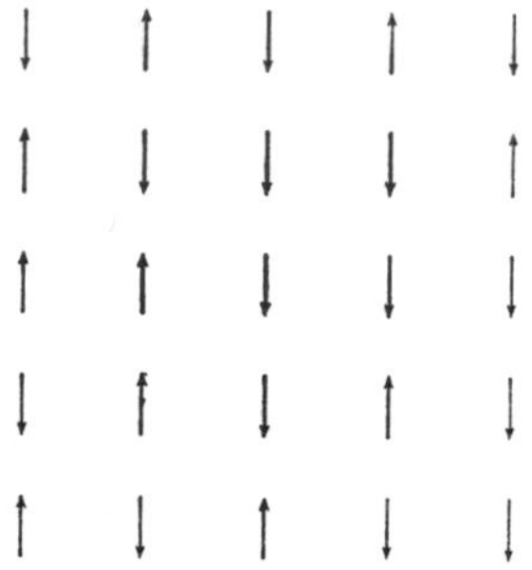

FIG. 12.10. One of the 2^N possible configurations of a system composed of N spins; here, $N=25$.

while

$$J_{ij} = \int \psi_j^*(1)\, \psi_i^*(2)\, u_{ij} \psi_j(2)\, \psi_i(1)\, d\tau_1\, d\tau_2. \tag{2}$$

The symbol u_{ij} denotes the relevant interaction potential. The energy difference between the state of "parallel" spins and the state of "antiparallel" spins is of magnitude $2J_{ij}$; algebraically,

$$\varepsilon_{\uparrow\uparrow} - \varepsilon_{\uparrow\downarrow} = -2J_{ij}. \tag{3}$$

If $J_{ij} > 0$, the state $\uparrow\uparrow$ is energetically favored against the state $\uparrow\downarrow$; we then look forward to the possibility of *ferromagnetism*. If, on the other hand, $J_{ij} < 0$, the situation is reversed; we then have the possibility of *antiferromagnetism*.

It might be useful to express the interaction energy of the two states, $\uparrow\uparrow$ and $\uparrow\downarrow$, by a single expression; to do this, we consider the scalar product

$$\begin{aligned} \boldsymbol{s}_i \cdot \boldsymbol{s}_j &= \tfrac{1}{2}\{(\boldsymbol{s}_i + \boldsymbol{s}_j)^2 - \boldsymbol{s}_i^2 - \boldsymbol{s}_j^2\} \\ &= \tfrac{1}{2} S(S+1) - s(s+1), \end{aligned} \tag{4}$$

which is equal to $+\frac{1}{4}$ if $S=1$ and $-\frac{3}{4}$ if $S=0$. We may, therefore, write for the interaction energy of the spins i and j

$$\varepsilon_{ij} = \text{const.} - 2J_{ij}(\boldsymbol{s}_i \cdot \boldsymbol{s}_j), \tag{5}$$

which is consistent with the energy difference (3). The drecise value of the constant appearing in (5) is immaterial because the potential energy is always arbitrary to the extent of an additive constant. Typically, the exchange interaction J_{ij} is of the same order of magnitude as the direct interaction K_{ij}; however, for increasing separation of the two spins it falls off much faster. To a first approximation, therefore, we may regard J_{ij} as negligible *except* for the nearest-neighbor pairs (in which case its value may be denoted by the simpler symbol J). The interaction energy of the whole lattice is then given by

$$E = \text{const.} - 2J \sum_{\text{n.n.}} (\boldsymbol{s}_i \cdot \boldsymbol{s}_j), \tag{6}$$

where the summation goes over all nearest-neighbor pairs in the lattice. The model based on expression (6) for the interaction energy of the lattice is known as the *Heisenberg model* of ferromagnetism (1928).

A simpler model results if we use, instead of (6), a *truncated* expression for the interaction energy of the lattice. This is obtained by replacing the product $(\boldsymbol{s}_i \cdot \boldsymbol{s}_j)$, which is equal to the sum $(s_{ix}s_{jx}+s_{iy}s_{jy}+s_{iz}s_{jz})$, by a single term $s_{iz}s_{jz}$; qualitatively, this appears plausible because, if the quantization takes place along the z-axis, then it is only the term $s_{iz}s_{jz}$ which will be diagonal—the expectation values of the other terms will be zero.* An additional reason for adopting the simpler model is that it does not necessarily require a quantum-mechanical treatment (because all the variables in the truncated expression for E are mutually commuting). The truncated expression for the interaction energy may now be written as

$$E = \text{const.} - J \sum_{\text{n.n.}} \sigma_i \sigma_j, \tag{7}$$

where the new symbol σ_i (or σ_j) $= +1$ for an "up" spin and -1 for a "down" spin; we note that, with the introduction of the new symbol, we still have: $\varepsilon_{\uparrow\uparrow} - \varepsilon_{\uparrow\downarrow} = -2J$. The model based on expression (7) is known as the *Ising model* of ferromagnetism; it originated with Lenz (1920) and was subsequently investigated by his student Ising (1925).†

To study the statistical mechanics of the Ising model, we may disregard the kinetic energy of the atoms associated with the lattice sites, for the phenomenon of phase transitions is essentially a consequence of the interaction energy between the atoms; in the interaction energy again, we may take into account only the nearest-neighbor contributions, in the hope that the remaining contributions would not affect the results qualitatively. Further, in the nearest-neighbor contributions too, we may retain only those terms that arise from the z-components of the spins; of course, the suppression of the x- and y-components of the spins does produce consequences which cannot be ignored while interpreting the physical results of the model.‡ To fix the z-direction, and to be able to study properties such as *magnetic susceptibility*, we subject the lattice to an external magnetic field $\boldsymbol{B}$, directed "upward"; the spins ↑ and ↓ then possess an additional potential energy which is given by the respective expressions $-\mu B$ and $+\mu B$ or by the general expression $-\mu B \sigma_i$.§ The Hamiltonian of the system in configuration $\{\sigma_1, \sigma_2, \ldots, \sigma_N\}$ is then given by

$$H\{\sigma_i\} = -J \sum_{\text{n.n.}} \sigma_i \sigma_j - \mu B \sum_i \sigma_i, \tag{8}$$

and the (configurational) partition function by

$$\begin{aligned} Q_N(B, T) &= \sum_{\sigma_1} \sum_{\sigma_2} \cdots \sum_{\sigma_N} \exp\,[-\beta H\{\sigma_i\}] \\ &= \sum_{\sigma_1} \sum_{\sigma_2} \cdots \sum_{\sigma_N} \exp\left[\beta J \sum_{\text{n.n.}} \sigma_i \sigma_j + \beta \mu B \sum_i \sigma_i\right]. \end{aligned} \tag{9}$$

* A different model results if we suppress the z-components of the spins and retain the x- and y-components. This model was originally introduced by Matsubara and Matsuda (1956) as a model of a quantum lattice gas, with possible relevance to the lambda-transition in helium. The critical behavior of this model has been investigated in detail by Betts and coworkers (1968–70) who have also emphasized the relevance of this model to the study of insulating ferromagnets.

† For an historical account of the origin and development of the Lenz–Ising model, reference may be made to an excellent review article by Brush (1967). This review gives a large number of other references as well.

‡ It may be noted that these consequences do not cause any serious harm to the model so long as $T \gg T_c$. At sufficiently low temperatures, however, they do produce serious inaccuracies.

§ Henceforth, we shall use the symbol μ instead of μ_B.

The Helmholtz free energy, the internal energy, the specific heat and the net magnetization of the system then follow from the formulae

$$A(B, T) = -kT \ln Q_N(B, T), \tag{10}$$

$$U(B, T) = -T^2 \frac{\partial}{\partial T}\left(\frac{A}{T}\right) = kT^2 \frac{\partial}{\partial T} \ln Q_N, \tag{11}$$

$$C(B, T) = \frac{\partial U}{\partial T} = -T\frac{\partial^2 A}{\partial T^2}, \tag{12}$$

and

$$\overline{M}(B, T) = \mu \overline{\left(\sum_i \sigma_i\right)} = \overline{\left(-\frac{\partial H}{\partial B}\right)} = \frac{1}{\beta}\left(\frac{\partial \ln Q_N}{\partial B}\right)_T = -\left(\frac{\partial A}{\partial B}\right)_T. \tag{13}$$

Obviously, the quantity $\overline{M}(0, T)$ gives the *spontaneous magnetization* of the system; if it is nonzero for temperatures below a critical temperature T_c, the system would be ferromagnetic for $T < T_c$ and paramagnetic for $T > T_c$. At the transition temperature, the specific heat of the system might show some sort of a singular behavior!

It is obvious that the energy levels of the system under study will be *degenerate*, in that the different configurations $\{\sigma_i\}$ will not all possess *distinct* energy values. In fact, the energy of a given configuration does not depend upon the detailed values of the variables σ_i; it only depends upon a few numbers, such as the total number N_+ of the "up" spins, the total number N_{++} of the "up–up" nearest-neighbor pairs, and so on. To see the precise nature of this dependence, we define other numbers as well, e.g. N_- as the total number of "down" spins, N_{--} as the total number of "down–down" nearest-neighbor pairs and N_{+-} as the total number of nearest-neighbor pairs with opposite spins. Of course, the numbers N_+ and N_- must satisfy the relationship:

$$N_+ + N_- = N. \tag{14}$$

Further, if q denotes the *coordination number* of the lattice, i.e. the number of the nearest neighbors for each lattice site, then we also have the relationships:

$$qN_+ = 2N_{++} + N_{+-}, \qquad (15)^*$$

$$qN_- = 2N_{--} + N_{+-}. \qquad (16)^*$$

With the help of these relationships, we can express all the numbers in terms of any two of them, say N_+ and N_{++}. We have

$$N_- = N - N_+; \qquad N_{+-} = qN_+ - 2N_{++}; \qquad N_{--} = \tfrac{1}{2}qN - qN_+ + N_{++}; \tag{17}$$

it will be noted that the total number of nearest-neighbor pairs in the lattice is given, quite expectedly, by the expression

$$N_{++} + N_{--} + N_{+-} = \tfrac{1}{2}qN. \tag{18}$$

Naturally, the Hamiltonian of the system can also be expressed in terms of N_+ and N_{++}; we

* The value of q for a linear chain is 2; for two-dimensional lattices, viz. honeycomb, square and triangular, it is 3, 4 and 6, respectively; for three-dimensional lattices, viz. simple cubic, body-centered cubic and face-centered cubic, it is 6, 8 and 12, respectively.

have from (8), with the help of the relationships established above,

$$\begin{aligned} H_N(N_+, N_{++}) &= -J(N_{++}+N_{--}-N_{+-})-\mu B(N_+-N_-) \\ &= -J(\tfrac{1}{2}qN-2qN_++4N_{++})-\mu B(2N_+-N). \end{aligned} \tag{19}$$

Now, "the number of lattice configurations which possess a desired amount of energy" is directly given by "the number of distinct ways in which the N spins of the lattice can be so arranged as to obtain the desired values of the numbers N_+ and N_{++}". Denoting this number by $g_N(N_+, N_{++})$, the partition function of the system can be written as

$$Q_N(B, T) = \sum_{N_+, N_{++}}' \exp\{-\beta H_N(N_+, N_{++})\},$$

that is,

$$e^{-\beta A} = e^{\beta N(\frac{1}{2}qJ-\mu B)} \sum_{N_+=0}^{N} e^{-2\beta(qJ-\mu B)N_+} \sum_{N_{++}}' g_N(N_+, N_{++}) e^{4\beta J N_{++}}, \tag{20}$$

where the primed summation goes over all values of N_{++} that are consistent with a *fixed* value of N_+ and is followed by a summation over all possible values of N_+, viz. from $N_+=0$ to $N_+ = N$. The central problem thus consists of determining the *combinatorial function* $g_N(N_+, N_{++})$ for the various lattices of interest; see, in particular, Sec. 12.9 and Problem 12.12.

12.6. The lattice gas and the binary alloy

Apart from the ferromagnet, Ising model can be readily adapted to simulate the behavior of certain other systems as well. More commonly discussed among these are (i) the lattice gas and (ii) the binary alloy.

(i) *The lattice gas.* Though it had already been recognized that the results derived for the Ising model could apply equally well to a system of "occupied" and "unoccupied" lattice sites, i.e. to a system of "atoms" and "holes" comprising a lattice, it was Yang and Lee (1952) who first used the term "lattice gas" to describe this system. By definition, a lattice gas is a collection of atoms, N_a in number, which can take up only discrete positions—positions that constitute a lattice structure with coordination number q. Each lattice site can be occupied by *at most* one atom, and the interaction energy between two occupied sites is nonzero, say $-\varepsilon_0$, only if the sites involved constitute a *nearest-neighbor pair*. The configurational energy of the gas is then given by

$$E = -\varepsilon_0 N_{aa}, \tag{1}$$

where N_{aa} is the total number of nearest-neighbor pairs (of occupied sites) in a given configuration of the system. Let $g_N(N_a, N_{aa})$ denote "the number of distinct ways in which the N_a atoms of the gas, assumed indistinguishable, can be distributed among the N sites of the lattice so as to obtain a desired value of the number N_{aa}". The partition function of the system, neglecting the kinetic energy of the atoms, is then given by

$$Q_{N_a}(N, T) = \sum_{N_{aa}}' g_N(N_a, N_{aa}) e^{\beta\varepsilon_0 N_{aa}}, \tag{2}$$

where the primed summation goes over all values of N_{aa} that are consistent with the given values of N_a and N; clearly, the number N here plays the role of the "total volume" available to the gas, the "volume of a primitive cell of the lattice" being the relevant unit for this purpose.

Going over to the grand canonical ensemble, we write for the grand partition function of the system

$$\mathcal{Q}(z, N, T) = \sum_{N_a=0}^{N} z^{N_a} Q_{N_a}(N, T). \tag{3}$$

The pressure P and the mean number $\bar{N}_a$ of the atoms in the gas are then given by

$$e^{\beta PN} = \sum_{N_a=0}^{N} z^{N_a} \sum_{N_{aa}}' g_N(N_a, N_{aa}) e^{\beta\varepsilon_0 N_{aa}} \tag{4}$$

and

$$\frac{\bar{N}_a}{N} = \frac{1}{v} = \frac{z}{kT}\left(\frac{\partial P}{\partial z}\right)_T; \tag{5}$$

here, v denotes the average volume per particle of the gas (again in the unit of the "volume of a primitive cell of the lattice").

To establish a formal correspondence between the lattice gas and the ferromagnet, we compare the present formulae with the ones established in the preceding section—in particular, the formula (4) with the formula (12.5.20). The first thing we observe here is that the canonical ensemble of the ferromagnet corresponds to the grand canonical ensemble of the lattice gas. The rest of the correspondence is summarized in the following chart:

The lattice gas		*The ferromagnet*
$N_a, N-N_a$	$\longleftrightarrow$	$N_+, N-N_+ (= N_-)$
ε_0	$\longleftrightarrow$	$4J$
z	$\longleftrightarrow$	$\exp\{-2\beta(qJ-\mu B)\}$
P	$\longleftrightarrow$	$-\left(\frac{A}{N}+\frac{1}{2}qJ-\mu B\right)$
$\frac{\bar{N}_a}{N}\left(=\frac{1}{v}\right)$	$\longleftrightarrow$	$\frac{\bar{N}_+}{N}\left(=\frac{1}{2}\left\{\frac{\bar{M}}{N\mu}+1\right\}\right)$,

where

$$\bar{M} = \mu(\bar{N}_+ - \bar{N}_-) = \mu(2\bar{N}_+ - N). \tag{6}$$

We note that the ferromagnetic analogue of formula (5) would be

$$\frac{\bar{N}_+}{N} = \frac{1}{kT}\left\{\frac{\partial(A/N+\frac{1}{2}qJ-\mu B)}{2\beta\,\partial(qJ-\mu B)}\right\}_T = \frac{1}{2}\left[-\frac{1}{N\mu}\left(\frac{\partial A}{\partial B}\right)_T+1\right] \tag{7}$$

which, by eqn. (12.5.13), assumes a more familiar form:

$$\frac{\bar{N}_+}{N} = \frac{1}{2}\left(\frac{\bar{M}}{N\mu}+1\right). \tag{8}$$

It is quite natural to ask: does lattice gas correspond to any real physical system in nature? The immediate answer seems to be that if we let the lattice constant tend to zero (thus going from a discrete structure to a continuous one) and also add, to the lattice-gas formulae, terms corresponding to an ideal gas (namely, the kinetic energy terms), then the model might simulate the behavior of a gas of real atoms interacting through a delta function potential. The investigation of the possibility of a phase transition in such a system may, therefore, be of value in understanding the onset of phase transitions in real gases. The case with $\varepsilon_0 > 0$, which implies an *attractive* interaction among the nearest-neighbors, has been frequently studied as a possible model for the gas–liquid transition and the critical point. On the other hand, if the interaction is *repulsive* ($\varepsilon_0 < 0$), so that configurations with alternating sites being "occupied" and "unoccupied" are the more favored ones, then we obtain a model which is of interest in connection with the theory of solidification; in such a study, however, the lattice constant has to be kept finite. Thus, several authors have pursued the study of the antiferromagnetic version of the Ising model, in the hope that this study might throw some light on the liquid–solid transition. For a bibliography of these pursuits, the reader is again referred to the review article by Brush (1967).

(ii) *The binary alloy*. Much of the early activity in the theoretical study of the Ising model has been due to developments in the field of order–disorder phenomena in alloys. In an alloy—to be specific, a *binary* alloy—we are concerned with a lattice structure consisting of two types of atoms, say 1 and 2, numbering N_1 and N_2, respectively. In a configuration characterized by the numbers N_{11}, N_{22} and N_{12} of the three types of nearest-neighbor pairs, the configurational energy of the alloy may be written as

$$E = \varepsilon_{11}N_{11}+\varepsilon_{22}N_{22}+\varepsilon_{12}N_{12}, \tag{9}$$

where ε_{11}, ε_{22} and ε_{12} denote the interaction energies of the different types of pairs. As in the case of the ferromagnet, the various numbers appearing in the expression for E may be expressed in terms of N, N_1 and N_{11} (of which only N_{11} is variable here). Equation (9) then takes the form

$$\begin{aligned} E &= \varepsilon_{11}N_{11}+\varepsilon_{22}(\tfrac{1}{2}qN-qN_1+N_{11})+\varepsilon_{12}(qN_1-2N_{11}) \\ &= \tfrac{1}{2}q\varepsilon_{22}N+q(\varepsilon_{12}-\varepsilon_{22})N_1+(\varepsilon_{11}+\varepsilon_{22}-2\varepsilon_{12})N_{11}. \end{aligned} \tag{10}$$

The correspondence between this system and the lattice gas, in the canonical ensemble, is then straightforward:

The lattice gas		*The binary alloy*
$N_a,\ N-N_a$	$\longleftrightarrow$	$N_1,\ N-N_1(=N_2)$
$-\varepsilon_0$	$\longleftrightarrow$	$(\varepsilon_{11}+\varepsilon_{22}-2\varepsilon_{12})$
A	$\longleftrightarrow$	$A-\frac{1}{2}q\varepsilon_{22}N-q(\varepsilon_{12}-\varepsilon_{22})N_1.$

The correspondence with a ferromagnet can be established likewise. In particular, this requires: $\varepsilon_{11} = \varepsilon_{22} = -J$ and $\varepsilon_{12} = +J$, which implies: $\varepsilon_0 \leftrightarrow 4J$, as before.

At absolute zero, the alloy must be in the state of minimum configurational energy, which would also be the state of maximum configurational order. The two types of atoms then occupy *mutually exclusive* sites, so that one might speak of the atoms 1 being only at the sites *a* and the atoms 2 being only at the sites *b*. As temperature rises, an exchange of sites results. And, in the face of thermal agitation, the order in the system gradually gives way. Ultimately, the two types of atoms get so "mixed up" that the very notion of the sites *a* being the "right" ones for the atoms 1 and the sites *b* being the "right" ones for the atoms 2 breaks down; the system then behaves, from the *configurational* point of view, as an assembly of N_1+N_2 atoms of essentially the same species!

12.7. Ising model in the zeroth approximation

In 1928 Gorsky attempted a statistical study of the order–disorder transitions in binary alloys on the basis of the assumption that the work expended in transferring an atom from na "ordered" position to a "disordered" position (or, in other words, from a "right" site to a "wrong" site) is directly proportional to the *degree of order* prevailing in the system. This idea was further developed by Bragg and Williams (1934, 1935) who, for the first time, introduced the concept of the *long-range order* in the sense we understand it now and, with relatively simple mathematics, obtained results that could explain the main qualitative features of the relevant experimental data. The main assumption in the Bragg–Williams approximation is that the energy of an individual atom in any given configuration of the system is determined by the *average* degree of order prevailing in the entire system rather than by the *fluctuating* configurations of the neighboring atoms. In this sense, the approximation is equivalent to the *mean molecular field* (or the *internal field*) theory of Weiss, which was put forward in 1907 to explain the magnetic behavior of ferromagnetic domains. It must be noted, however, that the real origin of the Weiss field could not be understood until Heisenberg (1928) developed the quantum-mechanical theory of atomic forces; of course, three years earlier Ising (1925) had introduced his own form of interatomic interaction on essentially empirical grounds.

It seems natural to call this approximation the *zeroth* approximation, for its features are totally insensitive to the detailed structure, or even to the dimensionality, of the lattice. We, therefore, expect the results following from this approximation to become more reliable as the number of neighbors interacting with a given atom increases (i.e. as $q \to \infty$), thus diminishing the importance of fluctuating influences.*

We define the long-range order parameter L by the very suggestive relationship

$$L = \frac{1}{N}\sum_i \sigma_i = \frac{N_+ - N_-}{N} = 2\frac{N_+}{N} - 1 \qquad (-1 \leqslant L \leqslant +1), \tag{1}$$

whence it follows that

$$N_+ = \frac{N}{2}(1+L) \quad \text{and} \quad N_- = \frac{N}{2}(1-L). \tag{2}$$

* In connection with this approximation, we may as well mention that the early attempts to construct a theory of binary solutions were based on the assumption that the atoms in the solution mix randomly. One finds that the results following from the *assumption of random mixing* are mathematically equivalent to the ones following from the *mean field approximation*.

The magnetization M is then given by

$$M = (N_+ - N_-)\mu = N\mu L \qquad (-N\mu \leqslant M \leqslant +N\mu); \tag{3}$$

the parameter L is, therefore, a direct measure of the magnetization obtaining in the system. For a *completely random* configuration, $\bar{N}_+ = \bar{N}_- = \frac{1}{2}N$; accordingly, the expectation values of both L and M would be identically zero.

Now, in the spirit of the present approximation, we replace the first part of the Hamiltonian (12.5.8) by the expression $-J\left(\frac{1}{2}q\bar{\sigma}\right)\Sigma_i\sigma_i$; note that, for a given σ_i, each of the $q\,\sigma_j$'s has been replaced by $\bar{\sigma}$ while the factor $\frac{1}{2}$ has been included to avoid duplication in the counting of the nearest-neighbor pairs. Making use of eqn. (1), and noting that $\bar{\sigma} \equiv \bar{L}$, we obtain for the total configurational energy of the system

$$E = -\tfrac{1}{2}(qJ\bar{L})NL - (\mu B)NL. \tag{4}$$

The expectation value of E is then given by

$$U = -\tfrac{1}{2}qJN\bar{L}^2 - \mu BN\bar{L}. \tag{5}$$

In the same approximation, the difference $\Delta\varepsilon$ between the configurational energy of an "up" spin and the configurational energy of a "down" spin (or the energy expended in changing an "up" spin into a "down" spin) is given by

$$\begin{aligned}\Delta\varepsilon &= -J(q\bar{\sigma})\{(-1)-(+1)\} - \mu B\{(-1)-(+1)\} \\ &= 2\mu\left(\frac{qJ}{\mu}\bar{\sigma} + B\right).\end{aligned} \tag{6}$$

The quantity $qJ\bar{\sigma}/\mu$ thus plays the role of the *internal molecular field* of Weiss; it is determined by (i) the mean value of the long-range order prevailing in the system and (ii) the strength of the coupling, qJ, between a given spin i and all the spins that are its nearest neighbors. The relative values of the numbers $\bar{N}_+$ and $\bar{N}_-$ then follow from the *Boltzmann principle*, viz.

$$\frac{\bar{N}_-}{\bar{N}_+} = \exp(-\Delta\varepsilon/kT) = \exp\{-2\mu(B'+B)/kT\}, \tag{7}$$

where B' denotes the internal molecular field:

$$B' = qJ\bar{\sigma}/\mu = qJ(\bar{M}/N\mu^2). \tag{8}$$

Substituting (2) into (7), and keeping in mind eqns. (3) and (8), we obtain for $\bar{L}$

$$\frac{1-\bar{L}}{1+\bar{L}} = \exp\left\{-2(qJ\bar{L}+\mu B)/kT\right\} \tag{9}$$

or equivalently

$$\frac{qJ\bar{L}+\mu B}{kT} = \frac{1}{2}\ln\frac{1+\bar{L}}{1-\bar{L}} = \tanh^{-1}\bar{L}. \tag{10}$$

To investigate the possibility of spontaneous magnetization, we set $B = 0$ and obtain the implicit relationship

$$\bar{L}_0 = \tanh\left(\frac{qJ\bar{L}_0}{kT}\right), \tag{11}$$

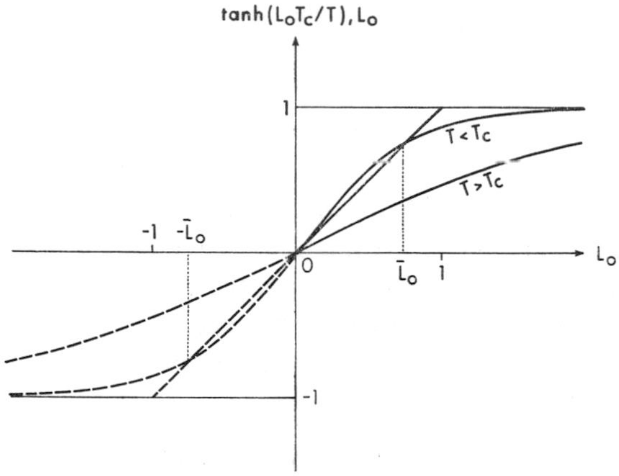

FIG. 12.11. The graphical solution of the implicit equation (12.7.11), with $T_c = qJ/k$.

which may be solved graphically; see Fig. 12.11. For any temperature T, the appropriate value of $\bar{L}_0(T)$ is determined by the point of intersection of (i) the straight line $y = L_0$ and (ii) the curve $y = \tanh(qJL_0/kT)$. Clearly, the solution $\bar{L}_0 = 0$ is always there; however, we are primarily interested in the *nonzero* solutions. For those, we note that since the slope of the curve (ii) varies from the initial value qJ/kT to the final value zero while the slope of the line (i) is constantly equal to unity, an intersection other than the one at the origin is possible if, and only if,

$$qJ/kT > 1, \tag{12}$$

that is,

$$T < qJ/k = T_c, \quad \text{say.} \tag{13}$$

We thus obtain a *critical temperature* T_c, determined solely by the strength of the interaction among neighboring spins, below which the system *can* acquire a nonzero spontaneous magnetization and above which it *cannot*. It is natural to identify this temperature with the *Curie temperature* of the system—the temperature that marks a transition from the ferromagnetic to the paramagnetic behavior.

It is clear from Fig. 12.11, as well as from eqn. (11), that if $\bar{L}_0$ is a solution of the problem, then $-\bar{L}_0$ is also a solution. The reason for this duplicity of solutions rests with the fact that, in the absence of an external field, there do not exist any means of assigning a "positive" direction, as opposed to a "negative" direction, to the alignment of the atomic spins! In view of this symmetry, we may confine our discussion to the positive solutions alone, viz. $0 \leqslant \bar{L}_0 \leqslant 1$.

The precise variation of $\bar{L}_0(T)$ with T can be obtained by solving eqn. (11) numerically; however, the general trend of this variation can be realized from Fig. 12.11. We note that for $T = qJ/k(= T_c)$ the straight line $y = L_0$ is tangential to the curve $y = \tanh(qJL_0/kT)$ at the origin; the relevant solution then is: $\bar{L}_0(T_c) = 0$. As T decreases, the initial slope of the curve becomes larger than that of the straight line and the relevant point of intersection rapidly moves away from the origin. Accordingly, $\bar{L}_0(T)$ rises very fast as T decreases below T_c, the approximate dependence being

$$\bar{L}_0(T) \simeq \{3(1-T/T_c)\}^{1/2} \quad \text{for } (1-T/T_c) \ll 1. \tag{14}$$

On the other hand, as $T \to 0$, $\bar{L}_0 \to 1$, in accordance with the asymptotic relationship

$$\bar{L}_0(T) \simeq 1-2\exp(-2T_c/T) \quad \text{for } (T/T_c) \ll 1. \tag{15}$$

Figure 12.12 shows a plot of $\bar{L}_0(T)$ versus T, along with the relevant experimental results for iron, nickel, cobalt and magnetite; we find that the agreement is not too bad.

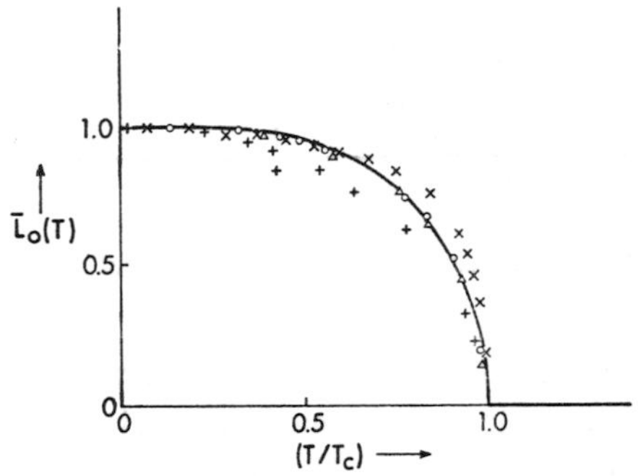

FIG. 12.12. The spontaneous magnetization of a Weiss ferromagnet as a function of temperature. The experimental points (after Becker) are for iron (×), nickel (○), cobalt (△) and magnetite (+).

The *field-free* configurational energy and the *field-free* specific heat of the system are given by, see eqn. (5),

$$U_0(T) = -\tfrac{1}{2}qJN\bar{L}_0^2 \tag{16}$$

and

$$C_0(T) = -qJN\bar{L}_0\frac{d\bar{L}_0}{dT} = \frac{Nk\bar{L}_0^2}{\dfrac{(T/T_c)^2}{1-\bar{L}_0^2}-\dfrac{T}{T_c}}, \tag{17}$$

where the reduction in the last step has been carried out with the help of the implicit relationship (11). Thus, for *all* $T > T_c$, both $U_0(T)$ and $C_0(T)$ are identically equal to zero. The value of the specific heat at the transition temperature T_c is given by, see eqns. (14) and (17),

$$C_0(T_c-0) = \lim_{x\to 0}\left\{\frac{Nk\cdot 3x}{\dfrac{(1-x)^2}{1-3x}-(1-x)}\right\} = \frac{3}{2}Nk. \tag{18}$$

However, $C_0(T_c+0) = 0$. The specific heat, therefore, displays a discontinuity at the transition point. On the other hand, as $T \to 0$, the specific heat vanishes, in accordance with the formula

$$C_0(T) \simeq 4Nk\left(\frac{T_c}{T}\right)^2 \exp(-2T_c/T). \tag{19}$$

The complete trend of the function $C_0(T)$ is shown in Fig. 12.13.

It is important to note that the vanishing of the configurational energy and the specific heat of the system for temperatures above T_c is directly related to the fact that, in the present approximation, the configurational order which prevails in the system at lower temperatures

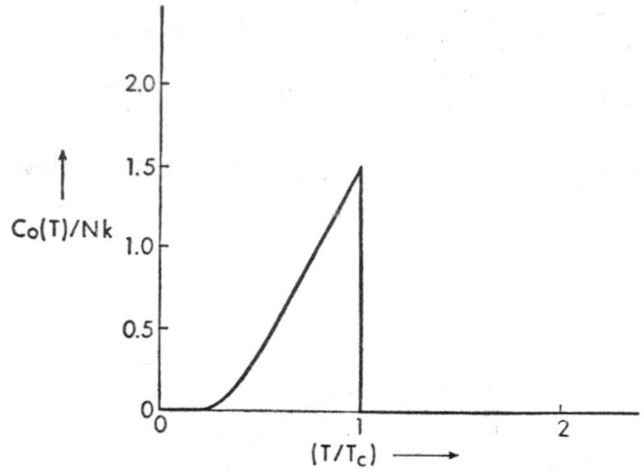

FIG. 12.13. The field-free specific heat of a Weiss ferromagnet as a function of temperature.

is *completely lost* as $T \to T_c$. Consequently, the configurational entropy and the configurational energy of the system attain their maximum values at $T = T_c$; beyond that stage the system is thermodynamically "inert". For instance, we may evaluate the configurational entropy of the system at $T = T_c$; we obtain, with the help of eqns. (11) and (17),

$$S_0(T_c) = \int_0^{T_c} \frac{C_0(T)\,dT}{T} = -qJN \int_1^0 \frac{\bar{L}_0}{T}\,d\bar{L}_0$$

$$= Nk \int_0^1 (\tanh^{-1} \bar{L}_0)\,d\bar{L}_0 = Nk \ln 2. \tag{20}$$

This is precisely what we expect for a system capable of 2^N *equally likely* microscopic complexions.* The fact that these (configurational) complexions are *equally likely* to occur is related to the fact that there does not exist any (configurational) order in the system.

We now proceed to study the magnetic susceptibility of the system at high temperatures ($\bar{L} \ll 1$); of course, now the magnetic field must be nonzero. Equation (10) then gives

$$\frac{qJ\bar{L}+\mu B}{kT} \simeq \bar{L},$$

whence we obtain for the susceptibility (per unit volume) of the system

$$\chi = \frac{\bar{M}}{VB} = \frac{N\mu\bar{L}}{VB} \simeq \frac{C}{T-T_c}, \quad \text{where} \quad C = \frac{N\mu^2}{Vk}. \tag{21}$$

This is the famous *Curie–Weiss law* for the paramagnetic susceptibility of a ferromagnetic material. It differs from the previously derived *Curie law*, in that T is replaced by $(T-T_c)$; cf. eqn. (3.8.15). Experimentally, one finds that the Curie–Weiss law is satisfied with considerable accuracy, except that the empirical value of T_c thus obtained is always somewhat

* Recall eqn. (3.3.21), whereby $S = k \ln \Omega$.

larger than the true transition temperature of the material; for instance, in the case of nickel, the empirical value of T_c obtained in this manner turns out to be about 650°K while the actual transition takes place at about 631°K.

We shall now proceed to demonstrate that the Bragg–Williams approximation, under which we have been working in this section, corresponds exactly to the *random mixing approximation* which was originally employed in the theory of regular solutions. According to eqn. (12.5.19), the configurational energy in the absence of the field is given by

$$U_0 = -J(\tfrac{1}{2}qN - 2q\bar{N}_+ + 4\bar{N}_{++}). \tag{22}$$

On the other hand, eqns. (2) and (16) give

$$\bar{N}_+ = N\left(\frac{1+\bar{L}_0}{2}\right) \quad \text{and} \quad U_0 = -\frac{1}{2}qJN\bar{L}_0^2. \tag{23}$$

Combining (22) and (23), we obtain

$$\bar{N}_{++} = \frac{1}{2}qN\left(\frac{1+\bar{L}_0}{2}\right)^2, \tag{24}$$

that is,

$$\frac{\bar{N}_{++}}{\frac{1}{2}qN} = \left(\frac{\bar{N}_+}{N}\right)^2. \tag{25}$$

Thus, the probability of obtaining an "up-up" nearest-neighbor pair in the lattice is precisely equal to the square of the probability of obtaining a single "up" spin; in other words, there does not exist, in spite of the presence of a spin–spin interaction characterized by the coupling constant J, any statistical correlation between the neighboring spins of the lattice. Put differently, there does not exist *any short-range order* in the system, apart from what follows statistically from the long-range order (characterized by the parameter $\bar{L}$). Therefore, in the present approximation, our system consists of a specified number of "up" spins, viz. $N(1+\bar{L})/2$, and a corresponding number of "down" spins, viz. $N(1-\bar{L})/2$, distributed over the N lattice sites *in a completely random manner*—similar to the mixing of $N(1+\bar{L})/2$ atoms of one kind with $N(1-\bar{L})/2$ atoms of another kind in a completely random manner to obtain a binary solution of N atoms. For this sort of a mixing, we must have

$$\left.\begin{aligned} \bar{N}_{++} &= \frac{1}{2}qN\left(\frac{1+\bar{L}}{2}\right)^2; \quad \bar{N}_{--} = \frac{1}{2}qN\left(\frac{1-\bar{L}}{2}\right)^2; \\ \bar{N}_{+-} &= 2\cdot\frac{1}{2}qN\left(\frac{1+\bar{L}}{2}\right)\left(\frac{1-\bar{L}}{2}\right), \end{aligned}\right\} \tag{26}$$

whence follows the important relationship

$$\frac{\bar{N}_{++}\bar{N}_{--}}{(\bar{N}_{+-})^2} = \frac{1}{4}. \tag{27}$$

It would be quite instructive to discuss our problem from the point of view of *random mixing*. With L as an unknown parameter, we have for the energy of the system

$$E(L) = -\tfrac{1}{2}qJNL^2 - \mu BNL \tag{28}$$

and for the corresponding "degree of degeneracy"

$$g(L) = \frac{N!}{\{(N/2)(1+L)\}!\,\{(N/2)(1-L)\}!}. \tag{29}$$

The partition function of the system is then given by

$$Q(B, T) = \sum_L \frac{N!}{\{(N/2)(1+L)\}!\,\{(N/2)(1-L)\}!} e^{\beta N(\frac{1}{2}qJL^2 + \mu BL)}. \tag{30}$$

In view of the fact that the numbers involved are large, the logarithm of the sum Σ_L may be approximated by the logarithm of the largest term in the sum; at the same time, we may apply Stirling's formula to the factorials involved. We thus obtain for the free energy of the system

$$\frac{1}{N}A(B, T) \equiv -kT\frac{1}{N}\ln Q(B, T) \simeq kT\left\{\frac{1+\bar{L}}{2}\ln\left(\frac{1+\bar{L}}{2}\right) + \frac{1-\bar{L}}{2}\ln\left(\frac{1-\bar{L}}{2}\right)\right\} - \left(\frac{1}{2}qJ\bar{L}^2 + \mu B\bar{L}\right), \tag{31}$$

where the maximal value $\bar{L}(T)$ is the root of the equation

$$\frac{kT}{2}\left\{\ln\left(\frac{1+\bar{L}}{2}\right) - \ln\left(\frac{1-\bar{L}}{2}\right)\right\} - (qJ\bar{L} + \mu B) = 0,$$

that is,

$$\tanh^{-1}\bar{L} = \frac{qJ\bar{L} + \mu B}{kT}, \tag{32}$$

which is precisely the same as the relationship obtained earlier; see eqn. (10). The random mixing approach, therefore, leads to the same expression for the order parameter of the problem as the mean field approach. The energy of the system follows readily from eqn. (31):

$$U(B, T) \equiv -T^2\frac{\partial}{\partial T}\left(\frac{A}{T}\right) = N\left(-\frac{1}{2}qJ\bar{L}^2 - \mu B\bar{L}\right), \tag{33}$$

which is again identical with the corresponding result obtained earlier. Combining (31) and (33), we obtain for the entropy of the system

$$S(B, T) \equiv \frac{U-A}{T} = -Nk\left\{\frac{1+\bar{L}}{2}\ln\left(\frac{1+\bar{L}}{2}\right) + \frac{1-\bar{L}}{2}\ln\left(\frac{1-\bar{L}}{2}\right)\right\}, \tag{34}$$

which, for $\bar{L} = 0$, reduces to the limiting value $Nk \ln 2$.* We also note that, with the help of

* It may be noted here that eqn. (34) is in formal agreement with the general result, $S/k = -\Sigma_r P_r \ln P_r$, obtained in Sec. 3.3. Here, it appears to have been applied to a single spin in the system, with the result: $S/Nk = -(p_1 \ln p_1 + p_2 \ln p_2)$, where $p_1 = \bar{N}_+/N = (1+\bar{L})/2$ and $p_2 = \bar{N}_-/N = (1-\bar{L})/2$.

the condition (32), eqn. (31) can be simplified to

$$\frac{1}{N}A(B,T) = \frac{kT}{2}\ln\left(\frac{1-\bar{L}^2}{4}\right)+\frac{1}{2}qJ\bar{L}^2, \tag{35}$$

which, for $\bar{L} = 0$, yields the expected result $-kT\ln 2 = -TS/N$ (since $U_{\bar{L}=0} = 0$).

As another illustration of this study, we consider the case of the *lattice gas*. In view of the correspondence chart drawn on p. 398, we obtain for the pressure and the volume (per particle) of the gas

$$P = \mu B - \frac{1}{8}q\varepsilon_0(1+\bar{L}^2) - \frac{kT}{2}\ln\left(\frac{1-\bar{L}^2}{4}\right) \tag{36}$$

and

$$\frac{1}{v} = \frac{1}{2}(1+\bar{L}). \tag{37}$$

The quantity μB is now a *free* parameter of the problem and is related to the fugacity z of the gas by the relation

$$z = \exp\left\{-2\beta(\tfrac{1}{4}q\varepsilon_0-\mu B)\right\}, \tag{38}$$

while the order parameter $\bar{L}(B, T)$ is determined by the equation

$$\bar{L} = \tanh\left(\frac{\frac{1}{4}q\varepsilon_0\bar{L}+\mu B}{kT}\right). \tag{39}$$

Equations (36) and (37) are *parametric equations* that determine the isotherms of the lattice gas in the mean-field approximation.

If we set $B = 0$, which means taking $z = \exp(-\frac{1}{2}\beta q\varepsilon_0)$, we naturally restrict ourselves to a rather limited region of the (P, v)-diagram. In this region, at any rate, we have the following situation: for temperatures less than a *critical temperature* T_c $(= q\varepsilon_0/4k)$, the parameter $\bar{L}_0$ has two equal and opposite values, which are determined by the equation

$$\bar{L}_0 = \tanh(\bar{L}_0 T_c/T), \tag{40}$$

while for temperatures greater than T_c, $\bar{L}_0 = 0$. Accordingly, for $T < T_c$, the volume v can have any of the two possible values

$$v_1(T) = \frac{2}{1+|\bar{L}_0|} \quad \text{and} \quad v_2(T) = \frac{2}{1-|\bar{L}_0|}, \tag{41}$$

while the pressure P has a common value, as given by the expression (36) with $B = 0$. On the other hand, for $T > T_c$, the volume v has a unique value, viz. 2, while the pressure is given by the formula

$$P(T) = -\tfrac{1}{8}q\varepsilon_0 + kT\ln 2 = -\tfrac{1}{2}kT_c + kT\ln 2. \tag{42}^*$$

* At $T = T_c$, the pressure $P_c = kT_c(\ln 2 - \frac{1}{2}) = 0.193kT_c$. This leads to the interesting result: $P_c v_c/kT_c = 0.386$, which may be compared with the van der Waals value of 0.375 and with the empirical value of about 0.293 obtained for several real gases.

The corresponding physical situation is described by the solid curve in Fig. 12.14. To obtain complete isotherms, we must include states with $B \neq 0$; this would supply the dotted parts of the curves. The qualitative similarity of these isotherms with the ones pertaining to a real gas is indeed remarkable; the specific volumes v_1 and v_2, as given by eqn. (41), thus correspond to the liquid phase and the gaseous phase, respectively.

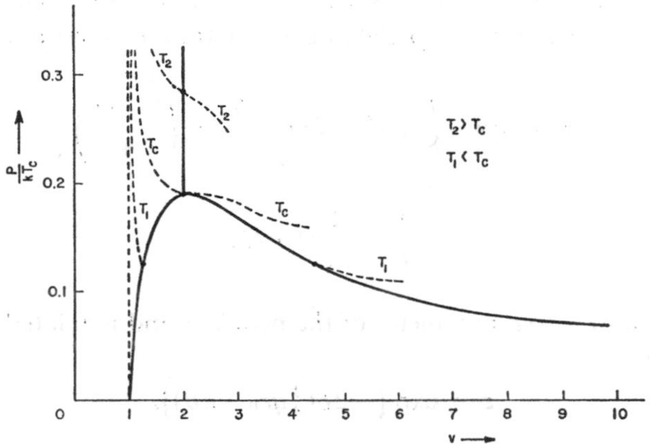

FIG. 12.14. The isotherms of a lattice gas in the mean field approximation.

In passing we note that as $T \to 0$, and hence $|\bar{L}_0| \to 1-2\exp(-2T_c/T)$, the pressure P, as given by eqn. (36) with $B = 0$, tends to vanish. At the same time, the volumes v_1 and v_2 approach their respective limiting values 1 and ∞; these values correspond to the special situation in which *each* lattice site in the denser phase of the system is occupied by an atom while *each* lattice site in the lighter phase is unoccupied.

12.8. Ising model in the first approximation

The approaches developed in Sec. 12.7 have their natural generalization towards an improved approximation. The mean field approach leads naturally to the *Bethe approximation* (Bethe, 1935; Rushbrooke, 1938), which treats somewhat more accurately the interaction of a given spin with its nearest neighbors. The random mixing approach, on the other hand, leads to the *quasi-chemical* approximation (Guggenheim, 1935; Fowler and Guggenheim, 1940), which takes into account the *specific* short-range order of the lattice—apart from the one that follows statistically from the long-range order. It was shown by Guggenheim (1938) and by Chang (1939) that the two methods yield identical results for the Ising model. It seems worth while to mention here that the extension of these approximations to higher orders, or their application to the Heisenberg model, does not produce identical results.

In the Bethe approximation, a given spin σ_0 is regarded as the central member of a group, which consists of this spin and its q nearest neighbors, and in writing down the Hamiltonian of this group the interaction between the central spin and the q neighbors is taken into account exactly while the interaction of these neighbors with other spins of the lattice is taken into account through a mean molecular field, B' say. Thus, one writes

$$H_{q+1} = -\mu B\sigma_0 - \mu(B+B')\sum_{j=1}^{q}\sigma_j - J\sum_{j=1}^{q}\sigma_0\sigma_j; \tag{1}$$

here, B denotes the external magnetic field acting on the lattice. The internal field B' is determined by the *condition of self-consistency*, which requires that the average value $\bar{\sigma}_0$ of the central spin be exactly the same as the average value $\bar{\sigma}_j$ of any of the q neighbors. The partition function Z of the group is now given by

$$Z = \sum_{\sigma_0,\,\sigma_j=\pm 1}\exp\left[\frac{1}{kT}\left\{\mu B\sigma_0 + \mu(B+B')\sum_{j=1}^{q}\sigma_j + J\sum_{j=1}^{q}\sigma_0\sigma_j\right\}\right]$$

$$= \sum_{\sigma_0,\,\sigma_j=\pm 1}\exp\left[\alpha\sigma_0 + (\alpha+\alpha')\sum_{j=1}^{q}\sigma_j + \gamma\sum_{j=1}^{q}\sigma_0\sigma_j\right], \tag{2}$$

where

$$\alpha = \frac{\mu B}{kT}, \quad \alpha' = \frac{\mu B'}{kT} \quad \text{and} \quad \gamma = \frac{J}{kT}. \tag{3}$$

Now, the right-hand side of eqn. (2) can be written as a sum of two terms, one pertaining to $\sigma_0 = +1$ and the other pertaining to $\sigma_0 = -1$:

$$Z = Z_+ + Z_-,$$

where

$$Z_\pm = \sum_{\sigma_j=\pm 1}\exp\left[\pm\alpha + (\alpha+\alpha'\pm\gamma)\sum_{j=1}^{q}\sigma_j\right]$$

$$= e^{\pm\alpha}[2\cosh(\alpha+\alpha'\pm\gamma)]^q. \tag{4}$$

The mean value of the central spin is then given by

$$\bar{\sigma}_0 = \frac{Z_+ - Z_-}{Z}, \tag{5}$$

while the mean value of a neighboring spin is given by

$$\bar{\sigma}_j = \frac{1}{q}\left(\overline{\sum_{j=1}^{q}\sigma_j}\right) = \frac{1}{q}\left(\frac{1}{Z}\frac{\partial Z}{\partial\alpha'}\right)$$

$$= \frac{1}{Z}\{Z_+\tanh(\alpha+\alpha'+\gamma) + Z_-\tanh(\alpha+\alpha'-\gamma)\}. \tag{6}$$

Equating (5) and (6), we obtain

$$Z_+\{1-\tanh(\alpha+\alpha'+\gamma)\} = Z_-\{1+\tanh(\alpha+\alpha'-\gamma)\}. \tag{7}$$

Substituting for Z_+ and Z_- from (4), we finally obtain

$$e^{2\alpha'} = \left\{\frac{\cosh(\alpha+\alpha'+\gamma)}{\cosh(\alpha+\alpha'-\gamma)}\right\}^{q-1}. \tag{8}$$

This is the condition that determines α' which in turn determines the magnetic behavior of the lattice.

To study the possibility of spontaneous magnetization, we set $\alpha\,(=\mu B/kT) = 0$. Our condition of self-consistency then reduces to

$$\alpha' = \frac{q-1}{2} \ln \left\{\frac{\cosh(\alpha'+\gamma)}{\cosh(\alpha'-\gamma)}\right\}. \tag{9}$$

In the absence of interactions ($\gamma = 0$), α' is necessarily equal to zero. In the presence of interactions ($\gamma \neq 0$), α' may still be zero unless γ exceeds a certain critical value, γ_c say. To determine this critical value, we expand the right-hand side of (9) as a Taylor series around $\alpha' = 0$; we obtain

$$\alpha' = (q-1) \tanh \gamma \left\{\alpha' - \frac{\alpha'^3}{3} \operatorname{sech}^2 \gamma + \ldots\right\}. \tag{10}$$

We note that, for all values of γ, $\alpha' = 0$ is one of the possible solutions of the problem; this, however, does not interest us. A nonzero solution clearly requires that

$$(q-1) \tanh \gamma > 1,$$

that is,

$$\gamma > \gamma_c = \tanh^{-1}\left(\frac{1}{q-1}\right) = \frac{1}{2} \ln\left(\frac{q}{q-2}\right). \tag{11}$$

In terms of temperature, this requires that

$$T < T_c = \frac{2J}{k} \Big/ \ln\left(\frac{q}{q-2}\right). \tag{12}$$

This defines the *Curie temperature* of the lattice. From eqn. (10), it also follows that, for temperatures close to the Curie temperature,

$$\begin{aligned}\alpha' &\simeq \{3 \cosh^2 \gamma_c[(q-1) \tanh \gamma - 1]\}^{1/2} \simeq \{3(q-1)(\gamma-\gamma_c)\}^{1/2} \\ &\simeq \left\{3(q-1)\frac{J}{kT_c}\left(1-\frac{T}{T_c}\right)\right\}^{1/2}.\end{aligned} \tag{13}$$

The parameter L of the long-range order is, by definition, equal to $\bar{\sigma}$. From eqns. (5) and (7), we obtain for this parameter

$$L = \frac{(Z_+/Z_-)-1}{(Z_+/Z_-)+1} = \frac{\sinh(2\alpha+2\alpha')}{\cosh(2\alpha+2\alpha')+\exp(-2\gamma)}. \tag{14}$$

In the absence of an external field ($\alpha = 0$) and at temperatures close to the Curie temperature ($\gamma \simeq \gamma_c$; $\alpha' \simeq 0$), we obtain

$$L_0 = \frac{\sinh(2\alpha')}{\cosh(2\alpha')+\exp(-2\gamma)} \simeq \frac{2\alpha'}{1+(q-2)/q} = \frac{q}{q-1}\alpha'. \tag{15}$$

Substituting from (12) and (13), this becomes

$$\bar{L}_0 \simeq \left[3\frac{q}{q-1}\left\{\frac{q}{2}\ln\left(\frac{q}{q-2}\right)\right\}\left(1-\frac{T}{T_c}\right)\right]^{1/2}. \tag{16}$$

We note that, for $q \gg 1$, eqns. (12) and (16) reduce to their zeroth-order counterparts, namely eqns. (12.7.13) and (12.7.14); in any case, $\bar{L}_0 \to 0$ as $T \to T_c$. On the other hand, as $T \to 0$, both γ and α' approach infinity and, by (14), $\bar{L}_0 \to 1$. In fact, the spontaneous magnetization curve in the present approximation has the same general shape as in the zeroth-order approximation; see Fig. 12.12. Of course, in the present case the curve depends explicitly on the coordination number q, being steepest for small values of q and becoming less steep as q increases, tending ultimately to the limiting form given by the mean field approximation.

We shall now study the *correlations* that might exist among the neighboring spins of the lattice. For this, we evaluate the numbers $\bar{N}_{++}$, $\bar{N}_{--}$ and $\bar{N}_{+-}$ in terms of the parameters α, α' and γ, and compare the resulting expressions with the ones obtained under the mean field approximation. Carrying out summations in expression (2) for the partition function Z, over all the spins (of the group) except σ_0 and σ_1, we obtain

$$Z = \sum_{\sigma_0,\,\sigma_1=\pm 1}\left[\exp\{\alpha\sigma_0+(\alpha+\alpha')\,\sigma_1+\gamma\sigma_0\sigma_1\}\{2\cosh(\alpha+\alpha'+\gamma\sigma_0)\}^{q-1}\right]. \tag{17}$$

Writing this as a sum of three parts, pertaining respectively to the values (i) $\sigma_0 = \sigma_1 = +1$, (ii) $\sigma_0 = \sigma_1 = -1$ and (iii) $\sigma_0 = -\sigma_1 = \pm 1$, we have

$$Z = Z_{++}+Z_{--}+Z_{+-}, \tag{18}$$

where, naturally enough,

$$\bar{N}_{++}:\bar{N}_{--}:\bar{N}_{+-}::Z_{++}:Z_{--}:Z_{+-}. \tag{19}$$

We thus obtain, using (8) as well,

$$\bar{N}_{++} \propto e^{(2\alpha+\alpha'+\gamma)}\{2\cosh(\alpha+\alpha'+\gamma)\}^{q-1}$$

$$\bar{N}_{--} \propto e^{(-2\alpha-\alpha'+\gamma)}\{2\cosh(\alpha+\alpha'-\gamma)\}^{q-1} = e^{(-2\alpha-3\alpha'+\gamma)}\{2\cosh(\alpha+\alpha'+\gamma)\}^{q-1}$$

and

$$\bar{N}_{+-} \propto e^{(-\alpha'-\gamma)}\{2\cosh(\alpha+\alpha'+\gamma)\}^{q-1}+e^{(\alpha'-\gamma)}\{2\cosh(\alpha+\alpha'-\gamma)\}^{q-1}$$
$$= 2e^{(-\alpha'-\gamma)}\{2\cosh(\alpha+\alpha'+\gamma)\}^{q-1}.$$

Normalizing with the help of the relationship

$$\bar{N}_{++}+\bar{N}_{--}+\bar{N}_{+-} = \tfrac{1}{2}qN, \tag{20}$$

we obtain the desired results:

$$(\bar{N}_{++},\bar{N}_{--},\bar{N}_{+-}) = \frac{1}{2}qN\frac{(e^{2\alpha+2\alpha'+\gamma},\ e^{-2\alpha-2\alpha'+\gamma},\ 2e^{-\gamma})}{2\{e^{\gamma}\cosh(2\alpha+2\alpha')+e^{-\gamma}\}}, \tag{21}$$

whence it follows that

$$\frac{\bar{N}_{++}\bar{N}_{--}}{(\bar{N}_{+-})^2} = \frac{1}{4}e^{4\gamma} = \frac{1}{4}e^{4J/kT}. \tag{22}$$

The last result is significantly different from the one, namely (12.7.27), that followed from the random mixing approximation. The difference lies in the appearance of the factor $\exp(4J/kT)$ which, for $J > 0$, favors the formation of parallel-spin pairs $\uparrow\uparrow$ and $\downarrow\downarrow$ as opposed to antiparallel-spin pairs $\uparrow\downarrow$ and $\downarrow\uparrow$. In fact, one may regard the elementary process

$$\uparrow\uparrow + \downarrow\downarrow \rightleftarrows 2\uparrow\downarrow, \tag{23}$$

which leaves the total number of "up" spins and of "down" spins unaltered, as some sort of a "chemical reaction" which, proceeding from left to right, is endothermic (requiring an amount of energy $4J$ to get through) and, proceeding from right to left, is exothermic (releasing an amount of energy $4J$). Equation (22) then constitutes the *law of mass action* for this reaction, the expression on the right-hand side being the *equilibrium constant* of the reaction. Historically, this equation was adopted by Guggenheim as the starting-point of his "quasi-chemical" treatment of the Ising model; only later on did he show that his treatment was equivalent to the Bethe approximation.

Equation (22) clearly tells us that, for $J > 0$, there exists among *like* neighbors ($\uparrow$ and $\uparrow$ or $\downarrow$ and $\downarrow$) a *positive* correlation and among *unlike* neighbors ($\uparrow$ and $\downarrow$) a *negative* correlation, and that these correlations are a direct consequence of the nearest-neighbor interactions. Accordingly, there must exist a *specific* short-range order in the system, quite apart from the one that follows statistically from the long-range order. To see it more explicitly, we note that even when the long-range order disappears ($\alpha+\alpha' = 0$), some short-range order still persists; from eqn. (21), we obtain

$$(\bar{N}_{++}, \bar{N}_{--}, \bar{N}_{+-})_{\bar{L}=0} = \frac{1}{2} qN \frac{(e^{\gamma}, e^{\gamma}, 2e^{-\gamma})}{4\cosh\gamma}, \tag{24}$$

which, only in the limit $\gamma \to 0$, goes over to the (correlation-free) random-mixing result

$$(\bar{N}_{++}, \bar{N}_{--}, \bar{N}_{+-})_{\bar{L}=0} = \frac{1}{2} qN \frac{(1, 1, 2)}{4}. \tag{25}$$

In the zeroth approximation, eqn. (25) is supposed to hold at *all* temperatures above T_c. We now find that a better approximation at these temperatures is provided by eqn. (24).

Finally, we evaluate the configurational energy U_0 and specific heat C_0 of the lattice in the absence of an external field ($\alpha = 0$). In view of eqn. (12.7.22), we have

$$U_0 = -J(\tfrac{1}{2}qN - 2q\bar{N}_+ + 4\bar{N}_{++})_{\alpha=0}. \tag{26}$$

The expression for $\bar{N}_{++}$ is contained in eqn. (21) while that for $\bar{N}_+$ can be obtained from eqn. (14):

$$(\bar{N}_+)_{\alpha=0} = \frac{1}{2} N(1+\bar{L}_0) = \frac{1}{2} N \frac{\exp(2\alpha')+\exp(-2\gamma)}{\cosh(2\alpha')+\exp(-2\gamma)}. \tag{27}$$

Equation (26) then gives

$$U_0 = -\frac{1}{2} qJN \frac{\cosh(2\alpha')-\exp(-2\gamma)}{\cosh(2\alpha')+\exp(-2\gamma)}, \tag{28}$$

where α' is determined by eqn. (9). For $T > T_c$, $\alpha' = 0$; eqn. (28) then takes the form

$$U_0 = -\frac{1}{2} qJN \frac{1-\exp(-2\gamma)}{1+\exp(-2\gamma)} = -\frac{1}{2} qJN \tanh\gamma. \tag{29}$$

It is obvious that the foregoing result arises solely because of the short-range order that persists in the system even above T_c. As for the specific heat, we obtain for $T > T_c$

$$C_0/Nk = \tfrac{1}{2}q\gamma^2 \operatorname{sech}^2 \gamma. \tag{30}$$

As $T \to \infty$, C_0 vanishes like T^{-2}. We note that a finite specific heat above the transition temperature is a welcome feature of the present approximation, because it brings our model considerably closer to real physical systems. In this connection, we recall that in the previous approximation the specific heat was zero for *all* $T > T_c$. Figure 12.15 gives the specific heat plot for a lattice, with coordination number 4, treated according to the Bethe approximation; for comparison, we have included the Bragg–Williams result as well.

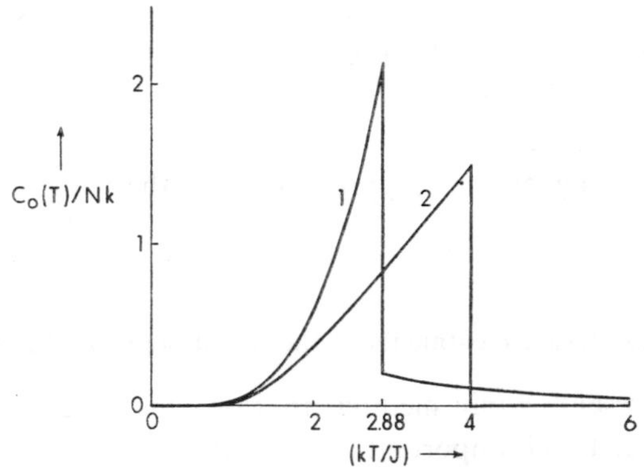

FIG. 12.15. The field-free specific heat of an Ising lattice with coordination number 4. Curve 1 obtains in the Bethe approximation, curve 2 in the Bragg–Williams approximation.

Finally, we study the specific heat discontinuity at $T = T_c$. The limiting value of the specific heat, as T approaches T_c from the *upper* side, can be obtained from eqn. (30) by letting $\gamma \to \gamma_c$. One obtains, with the help of (11),

$$\frac{1}{Nk} C_0(T_c+) = \frac{1}{2} q\gamma_c^2 \operatorname{sech}^2 \gamma_c = \frac{1}{8} \frac{q^2(q-2)}{(q-1)^2} \left\{ \ln \left(\frac{q}{q-2} \right) \right\}^2. \tag{31}$$

To obtain the limiting value of the specific heat, as T approaches T_c from the *lower* side, we must use the general expression (28) for U_0—of course, with $\gamma \to \gamma_c$ and $\alpha' \to 0$. Expanding (28) in powers of $(\gamma - \gamma_c)$ and α', and making use of the formula (13), we obtain for $(1 - T/T_c) \ll 1$

$$\begin{aligned} U_0 &= -\frac{1}{2} qJN \left[\frac{1}{(q-1)} + \frac{q(q-2)}{(q-1)^2} (\gamma - \gamma_c) + \frac{q(q-2)}{(q-1)^2} \alpha'^2 + \dots \right] \\ &= -\frac{1}{2} qJN \left[\frac{1}{(q-1)} + \frac{q(q-2)(3q-2)}{(q-1)^2} \frac{J}{kT_c} \left(1 - \frac{T}{T_c} \right) + \dots \right]. \end{aligned} \tag{32}$$

Differentiating (32) with respect to T, we obtain

$$\frac{1}{Nk} C_0(T_c-) = \frac{1}{8} \frac{q^2(q-2)(3q-2)}{(q-1)^2} \left\{ \ln \left(\frac{q}{q-2} \right) \right\}^2, \tag{33}$$

which is $(3q-2)$ times larger than the corresponding number at $T=T_c+$; cf. eqn. (31). The magnitude of the specific-heat discontinuity at the transition point is, therefore, given by

$$\frac{1}{Nk}\Delta C_0 = \frac{3}{8}\frac{q^2(q-2)}{(q-1)}\left\{\ln\left(\frac{q}{q-2}\right)\right\}^2. \tag{34}$$

One may check that, for $q \gg 1$, the foregoing results go over to the ones following from the mean field approximation.

In passing, we note that, according to eqn. (12), the transition temperature for a lattice with $q = 2$ turns out to be zero, which means that a one-dimensional Ising chain does not undergo phase transition. As will be seen in Sec. 12.9, this result is in complete agreement with the one following from an *exact* treatment of the one-dimensional lattice. In fact, for a lattice with $q = 2$, any results following from the Bethe approximation are completely identical with the corresponding exact results; on the other hand, the Bragg–Williams approximation is least reliable when $q = 2$.

For an account of the higher-order approximations, the reader is referred to the review article by Domb (1960).

12.9. Exact treatments of the one-dimensional lattice

In this section we present certain methods that provide an exact treatment of the Ising model in one dimension. This is important for several reasons. Firstly, there do exist certain phenomena that can be looked upon as one-dimensional nearest-neighbor problems, such as adsorption on a linear polymer or on a protein chain, the elastic properties of fibrous proteins, etc. Secondly, it would help us in evolving mathematical techniques for treating lattices in a larger number of dimensions, which is necessary for understanding the behavior of a variety of physical systems actually met with in nature. Thirdly, it would enable us to estimate the status of the Bethe approximation as a "possible" theory of the Ising model, for it will demonstrate mathematically that *at least* in the one-dimensional case this approximation leads to *exact* results.

A. The combinatorial method

In a short paper published in 1925, Ising himself gave an exact solution to the problem in one dimension. He employed a combinatorial approach which was essentially equivalent to the one being presented here. For this, we express the lattice Hamiltonian in terms of the numbers N_+ and N_{+-} rather than in terms of N_+ and N_{++}; cf. the expression (12.5.19). Making use of the relation (12.5.15), namely $2N_{++} = qN_+ - N_{+-}$, we obtain (for $q = 2$)

$$H_N(N_+, N_{+-}) = -J(N-2N_{+-}) - \mu B(2N^+ - N). \tag{1}$$

The partition function of the system is then given by

$$Q_N(B, T) = \sum_{N_+, N_{+-}}{}' \exp\{-\beta H_N(N_+, N_{+-})\},$$

that is,

$$e^{-\beta A} = e^{\beta N(J-\mu B)} \sum_{N_+=0}^{N} e^{2\beta\mu BN_+} \sum_{N_{+-}}' g_N(N_+, N_{+-}) e^{-2\beta JN_{+-}}, \tag{2}$$

where the primed summation Σ' goes over all values of N_{+-} that are consistent with a *fixed* value of N_+, and is followed by a summation over all possible values of N_+, viz. from $N_+ = 0$ to $N_+ = N$. The symbol $g_N(N_+, N_{+-})$ here denotes the "number of distinct ways in which the N spins of the (linear) chain can be so arranged as to obtain certain definite values of the numbers N_+ and N_{+-}". To determine $g_N(N_+, N_{+-})$, we proceed as follows.

As soon as we fix the number of "up" spins as N_+, the number of "down" spins is automatically fixed as $(N-N_+)$. The problem then reduces to determining the "number of distinct ways in which N_+ entities of one kind, say A, and $(N-N_+)$ entities of another kind, say B, can be distributed over a row of N sites, such that there occur N_{+-} links of the type AB or BA in the distribution". For instance, the arrangements shown below indicate two of the many different ways in which eight entities of the type A and seven entities of the type B can be distributed such that there exist nine links of the type AB or BA:

$$AAA\,|\,B\,|\,A\,|\,BB\,|\,A\,|\,B\,|\,A\,|\,BB\,|\,AA\,|\,B \tag{i}$$

and

$$B\,|\,AA\,|\,BB\,|\,A\,|\,B\,|\,A\,|\,BB\,|\,A\,|\,B\,|\,AAA. \tag{ii}$$

In arrangements such as (i), we have five links of the type AB and four links of the type BA, while in arrangements such as (ii), we have four links of the type AB and five links of the type BA; in this example, we have purposely chosen N_{+-} to be odd. The total number of arrangements in the categories (i) and (ii) will indeed be equal, and will be given by the "number of ways in which (a) the N_+ A's can be divided into $\frac{1}{2}(N_{+-}+1)$ groups, each group containing at least one A, and (b) the $(N-N_+)$ B's can be divided into $\frac{1}{2}(N_{+-}+1)$ groups, each group containing at least one B". We thus have, for *odd* values of N_{+-},

$$g_N(N_+, N_{+-}) = 2\cdot\frac{(N_+-1)!}{\{N_+-\frac{1}{2}(N_{+-}+1)\}!\,\{\frac{1}{2}(N_{+-}-1)\}!} \times\frac{(N-N_+-1)!}{\{(N-N_+)-\frac{1}{2}(N_{+-}+1)\}!\,\{\frac{1}{2}(N_{+-}-1)\}!}. \tag{3}^*$$

Using Stirling's formula, we obtain

$$\begin{aligned}\ln g_N(N_+, N_{+-}) \simeq{}& N_+ \ln N_+ + (N-N_+)\ln(N-N_+)\\ &-(N_+-\tfrac{1}{2}N_{+-})\ln(N_+-\tfrac{1}{2}N_{+-})\\ &-(N-N_+-\tfrac{1}{2}N_{+-})\ln(N-N_+-\tfrac{1}{2}N_{+-})\\ &-2(\tfrac{1}{2}N_{+-})\ln(\tfrac{1}{2}N_{+-}).\end{aligned} \tag{4}$$

A little reflection will show that *asymptotically* the same result holds for *even* values of N_{+-}.†

* We recall that the "number of ways of distributing n *indistinguishable* particles over g states, the number of particles in any one state being 0 or 1 or 2 or ..., is given by $(n+g-1)!/n!\,(g-1)!$". If, however, each state must have at least one particle, then the number of ways reduces to $(n-1)!/(n-g)!\,(g-1)!$.

† In this case, the number of links of the type AB, as well as the number of links of the type BA, is equal to $\frac{1}{2}N_{+-}$. The factor 2 will now arise from the fact that the linear array might be flanked by the A's or by the B's. At any rate, this factor is unimportant from the point of view of $\ln g_N(N_+, N_{+-})$.

From (2), we obtain for the Helmholtz free energy of the chain

$$A(B, T) = N(\mu B - J) - kT \ln \left\{ \sum_{N_+, N_{+-}}' g_N(N_+, N_{+-}) e^{2\beta(\mu B N_+ - J N_{+-})} \right\}. \tag{5}$$

Replacing the logarithm of the sum Σ' by the logarithm of the *largest* term in the sum, we obtain instead

$$\begin{aligned} A(B, T) &\simeq N(\mu B - J) - kT \ln g_N(\bar{N}_+, \bar{N}_{+-}) - 2(\mu B \bar{N}_+ - J \bar{N}_{+-}) \\ &= -J(N - 2\bar{N}_{+-}) - \mu B(2\bar{N}_+ - N) - kT \ln g_N(\bar{N}_+, \bar{N}_{+-}), \end{aligned} \tag{6}$$

where the equilibrium values $\bar{N}_+$ and $\bar{N}_{+-}$ of the numbers N_+ and N_{+-} are determined by the *maximizing* conditions [obtained by equating the relevant differential coefficients of the (logarithm of the) general term in the sum Σ' to zero]

$$\ln \left\{ \frac{\bar{N}_+(N - \bar{N}_+ - \frac{1}{2}\bar{N}_{+-})}{(N - \bar{N}_+)(\bar{N}_+ - \frac{1}{2}\bar{N}_{+-})} \right\} + 2\beta\mu B = 0 \tag{7}$$

and

$$\ln \left\{ \frac{(\bar{N}_+ - \frac{1}{2}\bar{N}_{+-})(N - \bar{N}_+ - \frac{1}{2}\bar{N}_{+-})}{(\frac{1}{2}\bar{N}_{+-})^2} \right\} - 4\beta J = 0. \tag{8}$$

The problem is essentially solved; however, it appears worth while to make a few remarks on the physical aspects of the solution.

Firstly, we note that the condition (8) is completely identical with the *quasi-chemical* condition (12.8.22), namely

$$\frac{\bar{N}_{++}\bar{N}_{--}}{(\bar{N}_{+-})^2} = \frac{1}{4} e^{4\beta J}. \tag{9}$$

Consequently, the quasi-chemical approximation, and hence the Bethe approximation, provide an *exact* treatment for the one-dimensional Ising problem!

Secondly, we note that, in the absence of an external field ($B = 0$), eqns. (7) and (8) give

$$\bar{N}_+ = \frac{1}{2} N \quad \text{and} \quad \bar{N}_{+-} = N \frac{1}{\exp(2\beta J) + 1}. \tag{10}$$

The first result rules out the possibility of spontaneous magnetization; we, therefore, conclude that the one-dimensional Ising lattice cannot be ferromagnetic! This is again in agreement with the result following from the Bethe approximation, viz. that the transition point for an Ising lattice with coordination number 2 coincides with the absolute zero of temperature. Substituting expressions (10) into formulae (4) and (6), we obtain for the entropy and the free energy of the system

$$\frac{S_0}{k} \simeq \ln g_N(\bar{N}_+, \bar{N}_{+-}) \simeq N \left[\ln (e^{2\beta J} + 1) - \frac{2\beta J}{1 + \exp(-2\beta J)} \right] \tag{11}$$

and

$$\begin{aligned} A_0 &\simeq N \left[-J \tanh(\beta J) + \left\{ \frac{2J}{1 + \exp(-2\beta J)} - kT \ln (e^{2\beta J} + 1) \right\} \right] \\ &= -NkT \ln \{2 \cosh(\beta J)\}. \end{aligned} \tag{12}$$

These results lead to the following expressions for the energy and the specific heat of the lattice:

$$U_0 = -NJ \tanh(\beta J) \tag{13}$$

and

$$C_0 = Nk(\beta J)^2 \operatorname{sech}^2(\beta J). \tag{14}$$

The agreement of these results with eqns. (12.8.29) and (12.8.30) shows once again that for a one-dimensional lattice ($q = 2$) the Bethe approximation yields *exact* results. Figure 12.16 depicts the variation of the field-free specific heat of the lattice with temperature;

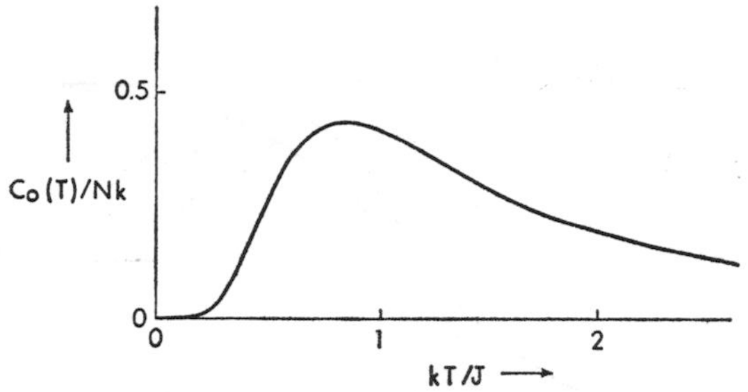

FIG. 12.16. The field-free specific heat of an Ising chain.

quite expectedly, the specific heat does not show any singular behavior in this case. Further, it follows from eqn. (12) that the field-free partition function of the lattice is given by

$$Q_N(0, T) = \{2 \cosh(J/kT)\}^N, \tag{15}$$

a result that follows straightforwardly from another combinatorial method which originated with van der Waerden (1941) and was developed systematically by Kac and Ward (1952); for an exposition of this method, reference may be made to the original papers or to the review articles by Newell and Montroll (1953) and by Domb (1960).

Finally, let us have a look at the situation in the presence of an external magnetic field ($B \neq 0$). Solving the simultaneous equations (7) and (8), we now obtain

$$\frac{\bar{N}_+}{N} = \frac{1}{2}\left[1 + \frac{\sinh(\beta\mu B)}{\{e^{-4\beta J} + \sinh^2(\beta\mu B)\}^{1/2}}\right] \tag{16}$$

and

$$\frac{\bar{N}_{+-}}{N} = \frac{e^{-2\beta J}}{[e^{\beta J}\cosh(\beta\mu B) + \{e^{-2\beta J} + e^{2\beta J}\sinh^2(\beta\mu B)\}^{1/2}]\,\{e^{-2\beta J} + e^{2\beta J}\sinh^2(\beta\mu B)\}^{1/2}}. \tag{17}$$

If $B \to 0$, these results reduce to the ones given by (10). On the other hand, if $J \to 0$, they reduce to

$$\frac{\bar{N}_+}{N} = \frac{e^{\beta\mu B}}{e^{\beta\mu B} + e^{-\beta\mu B}} = \frac{1}{2} e^{\beta\mu B} \operatorname{sech}(\beta\mu B) \tag{18}$$

and

$$\frac{\bar{N}_{+-}}{N} = \frac{1}{2} \operatorname{sech}^2(\beta\mu B); \tag{19}$$

these results are consistent with the fact that, in the absence of interactions (and hence correlations), $\bar{N}_+ : \bar{N}_- :: e^{\beta\mu B} : e^{-\beta\mu B}$. The resulting magnetization can be obtained from eqn. (16):

$$\frac{1}{N\mu}\bar{M}(B,T) = \frac{\bar{N}_+ - \bar{N}_-}{N} = 2\frac{\bar{N}_+}{N} - 1 = \frac{\sinh(\beta\mu B)}{\{e^{-4\beta J} + \sinh^2(\beta\mu B)\}^{1/2}}. \tag{20}$$

Figure 12.17 shows the degree of magnetization of the lattice as a function of the parameter $(\beta\mu B)$, the latter being a measure of the relative strengths of the magnetic and thermal

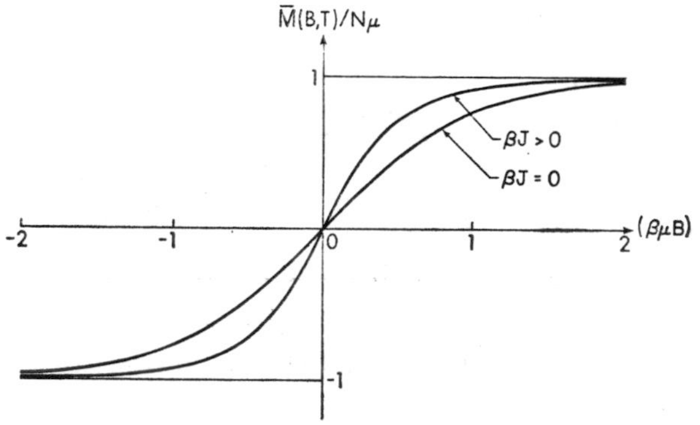

FIG. 12.17. The degree of magnetization of an Ising chain as a function of the parameter $(\beta\mu B)$.

influences experienced by the spins. As expected, the system reaches a state of magnetic saturation as the parameter $(\beta\mu B)$ becomes large in comparison with unity. The influence of spin–spin interactions is also clear; the coupling constant J (assumed positive) leads to comparatively larger values of magnetization and hence to a faster approach towards saturation. The low-field susceptibility of the lattice is given by the initial slope of the function (20); one obtains

$$\chi_0(T) = \frac{N\mu^2}{kT} e^{2J/kT}. \tag{21}$$

Before proceeding further, we observe, on the basis of eqn. (12.5.13), that by integrating (20) with respect to B we can obtain the free energy A, and hence the partition function Q, of the lattice:

$$A \equiv -kT \ln Q = -NkT \ln [\cosh(\beta\mu B) + \{e^{-4\beta J} + \sinh^2(\beta\mu B)\}^{1/2}] + C.$$

The constant of integration C may be chosen to meet the requirement that, for $B = 0$, the foregoing expression reduces to the former expression (12). This requires: $C = -NJ$. We thus obtain

$$A \equiv -kT \ln Q = -NkT \ln [e^{\beta J}\cosh(\beta\mu B) + \{e^{-2\beta J} + e^{2\beta J}\sinh^2(\beta\mu B)\}^{1/2}]. \tag{22}$$

This result should be obtainable directly by substituting the expressions (16) and (17) for $\bar{N}_+$ and $\bar{N}_{+-}$ into the formula (6) for A. The present derivation, however, is simpler.

B. The matrix method

In 1941 Kramers and Wannier introduced the so-called matrix method into the theory of the Ising model. In the one-dimensional case this method worked with immediate success. Three years later, in 1944, it became, due to Onsager's ingenuity, the first method to treat successfully the *field-free* Ising problem in *two* dimensions. To apply this method, we replace the actual lattice by one having the topology of a closed, endless structure; for instance, in the one-dimensional case we replace the straight chain by a curved one, such that the Nth spin becomes a neighbor of the first spin—see Fig. 12.18. This replacement eliminates

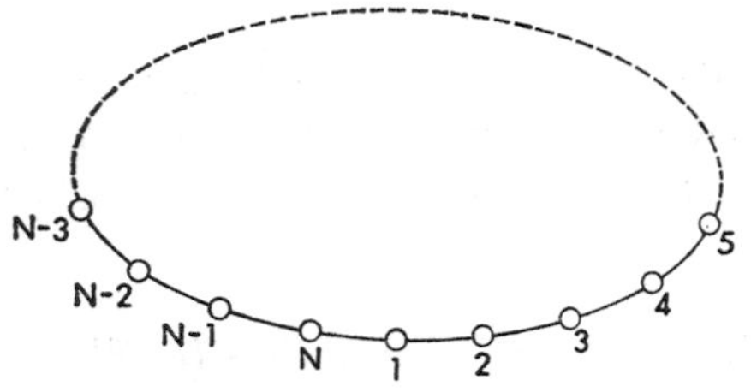

FIG. 12.18. An Ising chain with a closed, endless structure.

the inconvenient end effects; it does not, however, alter the thermodynamic properties of the (infinitely long) system. The important advantage of this replacement is that it enables us to write the Hamiltonian of the system,

$$H_N\{\sigma_i\} = -J\sum_{\text{n.n.}}\sigma_i\sigma_j - \mu B\sum_{i=1}^{N}\sigma_i, \tag{23}$$

in a symmetrical form, namely

$$H_N\{\sigma_i\} = -J\sum_{i=1}^{N}\sigma_i\sigma_{i+1} - \tfrac{1}{2}\mu B\sum_{i=1}^{N}(\sigma_i+\sigma_{i+1}), \tag{24}$$

because $\sigma_{N+1} \equiv \sigma_1$. The partition function of the system is then given by

$$\begin{aligned} Q_N(B,T) &= \sum_{\sigma_1=\pm1}\cdots\sum_{\sigma_N=\pm1}\exp\left[\beta\sum_{i=1}^{N}\{J\sigma_i\sigma_{i+1}+\tfrac{1}{2}\mu B(\sigma_i+\sigma_{i+1})\}\right] \\ &= \sum_{\sigma_1=\pm1}\cdots\sum_{\sigma_N=\pm1}\langle\sigma_1|\boldsymbol{P}|\sigma_2\rangle\langle\sigma_2|\boldsymbol{P}|\sigma_3\rangle\ldots\langle\sigma_{N-1}|\boldsymbol{P}|\sigma_N\rangle\langle\sigma_N|\boldsymbol{P}|\sigma_1\rangle, \end{aligned} \tag{25}$$

where $\boldsymbol{P}$ denotes an operator whose matrix elements are defined by

$$\langle\sigma_i|\boldsymbol{P}|\sigma_{i+1}\rangle = \exp\left[\beta\{J\sigma_i\sigma_{i+1}+\tfrac{1}{2}\mu B(\sigma_i+\sigma_{i+1})\}\right],$$

that is,

$$(\boldsymbol{P}) = \begin{pmatrix} e^{\beta(J+\mu B)} & e^{-\beta J} \\ e^{-\beta J} & e^{\beta(J-\mu B)} \end{pmatrix}. \tag{26}$$

According to the rules of the matrix algebra, the summations over the various σ's, in the expression (25), lead to the simple result

$$Q_N(B,T) = \sum_{\sigma_1=\pm1}\langle\sigma_1|\boldsymbol{P}^N|\sigma_1\rangle = \text{Trace}\,(\boldsymbol{P}^N) = \lambda_1^N+\lambda_2^N, \tag{27}$$

where λ_1 and λ_2 are the eigenvalues of the matrix $\boldsymbol{P}$. Again, by the rules of the matrix algebra, these eigenvalues are given by the *secular equation*

$$\begin{vmatrix} e^{\beta(J+\mu B)}-\lambda & e^{-\beta J} \\ e^{-\beta J} & e^{\beta(J-\mu B)}-\lambda \end{vmatrix} = 0, \tag{28}$$

that is,

$$\lambda^2 - 2\lambda e^{\beta J}\cosh(\beta\mu B) + 2\sinh(2\beta J) = 0. \tag{29}$$

One readily obtains the roots of this equation, viz.

$$\begin{pmatrix}\lambda_1\\ \lambda_2\end{pmatrix} = e^{\beta J}\cosh(\beta\mu B) \pm \{e^{-2\beta J} + e^{2\beta J}\sinh^2(\beta\mu B)\}^{1/2}. \tag{30}$$

For all physical values of the parameters involved, $\lambda_2 < \lambda_1$; consequently, $(\lambda_2/\lambda_1)^N \to 0$ as $N \to \infty$. Thus, it is only the *larger* eigenvalue, λ_1, that really determines the partition function of the system. Consequently, we obtain, from eqns. (27) and (30), *in the thermodynamic limit* ($N \to \infty$),

$$\frac{1}{N}\ln Q_N(B, T) \simeq \ln \lambda_1 = \ln\,[e^{\beta J}\cosh(\beta\mu B) + \{e^{-2\beta J} + e^{2\beta J}\sinh^2(\beta\mu B)\}^{1/2}], \tag{31}$$

which is in complete agreement with our previous result (22).* Needless to say, the complete thermodynamics of the system can be derived from the expression (31).

C. The zeros of the grand partition function

The foregoing results enable us to study the distribution of the zeros of the grand partition function of a one-dimensional lattice gas in the complex z-plane. In view of the correspondence established in Sec. 12.6, the grand partition function of the lattice gas is directly related to the partition function of the ferromagnet; accordingly, the vanishing of one of them implies the vanishing of the other. We also note that the fugacity z of the lattice gas is related to the parameters βJ and $\beta\mu B$ of the ferromagnet by the formula

$$z = \exp(-4\beta J + 2\beta\mu B). \tag{32}$$

So far as the location of the zeros is concerned, we have an important theorem due to Yang and Lee (1952) which states that "if the interaction energy between any two atoms of the lattice gas is $+\infty$ if the atoms occupy the same lattice site and is $\leqslant 0$ otherwise, then *all* the zeros of the grand partition function lie on a circle, in the complex z-plane, with its centre at the point $z = 0$". In the proof of the theorem, no assumptions are made about the range of the interatomic interaction or about the dimensionality, the size, the structure or the periodicity of the lattice. Further, if the interaction exists only between the *nearest* neighbors in the lattice, then the radius of the circle (of zeros) is given by

$$r = \exp\{\tfrac{1}{2}q\beta(\varepsilon_{++} - \varepsilon_{--})\}. \tag{33}$$

* In the special case $B = 0$, the eigenvalues λ_1 and λ_2 are, respectively, $2\cosh(\beta J)$ and $2\sinh(\beta J)$. The partition function $Q_N(0, T)$ is then given by $2^N\{\cosh^N(\beta J) + \sinh^N(\beta J)\}$ or essentially by $2^N\cosh^N(\beta J)$ because, in the limit $N \to \infty$, the first term completely dominates. This result is in complete accord with our previous finding, namely eqn. (15). For a derivation of this result, along the lines of the combinatorial method of van der Waerden, see Kubo (1965), Problem 5.16.

In the case of a ferromagnet, this gives $r = 1$, while in the case of a lattice gas (with $q = 2$), we get $r = \exp(\beta\varepsilon_{++}) = \exp(-4\beta J)$; see the correspondence chart of Sec. 12.6 whereby $\varepsilon_{++} = -\varepsilon_0 = -4J$. Therefore, while looking for the zeros of this problem, it seems natural to write

$$z = e^{-4\beta J + i\theta}, \tag{34}$$

where θ determines the location of the various zeros on the circle (of zeros). Comparing (34) with (32), we observe that in effect we are writing $(\beta\mu B) = i\theta/2$;* the equation $\lambda_1^N + \lambda_2^N = 0$, which determines the zeros of the problem, thus takes the form

$$\left[\cos\frac{\theta}{2} + \left(\eta^2 - \sin^2\frac{\theta}{2}\right)^{1/2}\right]^N + \left[\cos\frac{\theta}{2} - \left(\eta^2 - \sin^2\frac{\theta}{2}\right)^{1/2}\right]^N = 0, \tag{35}$$

where

$$\eta = \exp(-2\beta J). \tag{36}$$

The structure of this equation suggests that we introduce a new variable φ, such that

$$\cos\frac{\theta}{2} = (1-\eta^2)^{1/2}\cos\varphi; \tag{37}$$

the equation thereby reduces to

$$(1-\eta^2)^{N/2}\{(\cos\varphi + i\sin\varphi)^N + (\cos\varphi - i\sin\varphi)^N\} = 0, \tag{38}$$

that is,

$$\cos(N\varphi) = 0, \tag{39}$$

with the solutions

$$\varphi = \pm(k - \tfrac{1}{2})\pi/N; \qquad k = 1, 2, \ldots, N. \tag{40}$$

The corresponding values of θ are given by

$$\begin{aligned}\cos\theta &= -\eta^2 + (1-\eta^2)\cos 2\varphi \\ &= -\eta^2 + (1-\eta^2)\cos\{(2k-1)\pi/N\}; \qquad k = 1, 2, \ldots, N.\end{aligned} \tag{41}$$

Equations (34) and (41) determine the precise location of the N zeros of the problem on the circle (of zeros) in the complex z-plane.

As $N \to \infty$, the distribution of the zeros becomes continuous. For any given value of η, the distribution is confined to a "horse-shoe" segment of the circle, starting with the value $\theta_i = \cos^{-1}(1-2\eta^2) = 2\sin^{-1}\eta$ and ending with the value $\theta_f = 2\pi - \theta_i$; see Fig. 12.19. The distribution thus has an "opening", of angular width $2\cos^{-1}(1-2\eta^2) = 4\sin^{-1}\eta$ towards the real axis, which "closes up" *only if* $\eta = 0$, *i.e. if* $T = 0°$K. This confirms our earlier finding that the one-dimensional Ising lattice does not undergo phase transition at any finite temperature!

* We recall that the variable $(\beta\mu B)$, which plays the role of an *external* parameter in the problem of the ferromagnet, provides a variable fugacity in the problem of the lattice gas.

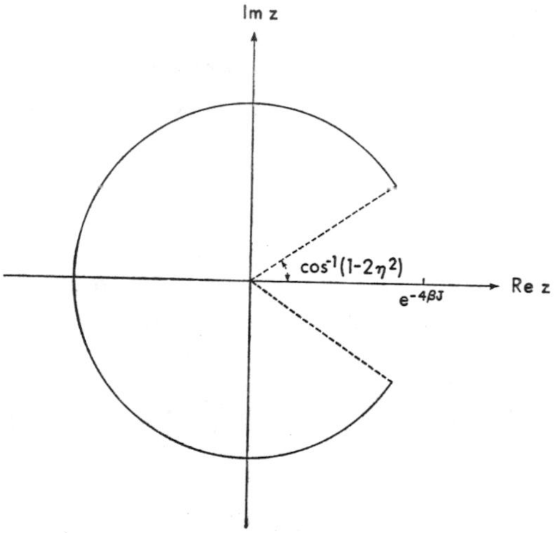

FIG. 12.19. The zeros of the grand partition function of a one-dimensional lattice gas spread over a circle in the complex z-plane.

In the case of a continuous distribution, we may introduce a *distribution function* $g(\theta)$; then, by definition, the product $g(\theta)\,d\theta$ determines the fraction of the zeros lying in the range $(\theta, \theta+d\theta)$. Clearly,

$$g(\theta) = \frac{1}{N}\left(\frac{dk}{d\theta}\right) = \begin{cases} \dfrac{1}{2\pi}\,\dfrac{\sin(\theta/2)}{\{\sin^2(\theta/2)-\eta^2\}^{1/2}} & \text{in the "horse-shoe" segment} \\ 0 & \text{elsewhere.} \end{cases} \tag{42}$$

It may be verified that the function $g(\theta)$ satisfies the normalization condition

$$\int_0^{2\pi} g(\theta)\,d\theta = 1. \tag{43}$$

In the absence of interatomic interactions ($J = 0$), or in the limit $T \to \infty$, the parameter η may be replaced by unity. The function $g(\theta)$ then vanishes everywhere except at the point $\theta = \pi$ where it has an infinitely large value; in fact, the function $g(\theta)$ in this case assumes the form of a delta function. As the interaction is "switched on", or the temperature lowered, the distribution of the zeros broadens out, the function $g(\theta)$ being minimum at the (central) point $\theta = \pi$ and becoming larger and larger towards the ends of the segment. Finally, as $T \to 0$ (and hence $\eta \to 0$), we obtain the *limiting* distribution which is uniformly spread over the entire circle, from $\theta = 0$ to $\theta = 2\pi$, with a constant value $1/(2\pi)$. The salient features of the "movement" of the zeros along the circle (of zeros), with the variation of the parameter η, are shown in the distribution curves of Fig. 12.20.

The physical properties of the lattice gas can be derived from the distribution function $g(\theta)$ as well. According to formulae (12.4.8) and (12.4.9), the pressure P and the volume per particle v of the gas, as functions of the parameters z and T, are given by

$$\frac{P}{kT} = \frac{1}{N}\sum_{k=1}^{N} \ln\left\{\left(\frac{z}{r_k}\right)^2 - 2\left(\frac{z}{r_k}\right)\cos\theta_k + 1\right\}^{1/2} \tag{44}$$

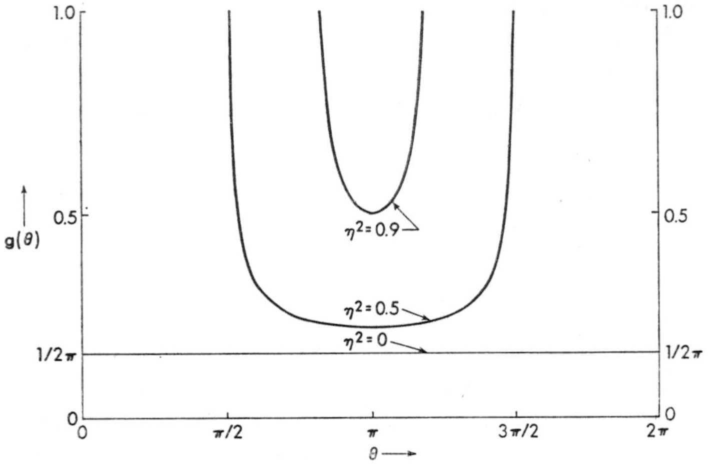

FIG. 12.20. The distribution function $g(\theta)$ for a lattice gas in one dimension; the parameter $\eta = \exp(-2\beta J)$.

and

$$\frac{1}{v} = \{\partial(P/kT)/\partial(\ln z)\}_T. \tag{45}$$

In these formulae, z is a *real, positive* parameter of the problem; so, it is different from *all* the zeros $z_k(= r_k e^{i\theta_k})$ of the grand partition function. Now, $r_k = \exp(-4\beta J)$ for all k; therefore, by the relation (32),

$$z/r_k = \exp(2\beta\mu B) = y, \quad \text{say.} \tag{46}$$

Of course, the parameter y will also be *real* and *positive*. Equations (44) and (45) now become

$$\frac{P}{kT} = \frac{1}{2N}\sum_{k=1}^{N} \ln\{y^2 - 2y\cos\theta_k + 1\} \tag{47}$$

and

$$\frac{1}{v} = \{\partial(P/kT)/\partial(\ln y)\}_T. \tag{48}$$

In the thermodynamic limit ($N \to \infty$), the summation over k may be replaced by an integration over θ; see eqn. (42) for $dk/d\theta$. We obtain

$$\frac{P}{kT} = \frac{1}{4\pi}\int_{\theta_i}^{\theta_f} \frac{\sin(\theta/2)}{\sqrt{(\sin^2(\theta/2) - \eta^2)}} \ln\{y^2 - 2y\cos\theta + 1\}\, d\theta$$

$$= \ln\left[\frac{(y+1) + \{(y-1)^2 + 4y\eta^2\}^{1/2}}{2}\right], \tag{49}$$

whence it follows that

$$\frac{1}{v} = \frac{1}{2}\left[1 + \frac{y-1}{\{(y-1)^2 + 4y\eta^2\}^{1/2}}\right]. \tag{50}$$

There is no doubt that these results could also be derived from those obtained earlier for the ferromagnet by making use of the correspondence between the two systems; thus, the expressions for P/kT and $1/v$ for the lattice gas should be the same as the expressions for $(\beta\mu B - \beta J - \beta A/N)$ and $\bar{N}_+/N$ for the ferromagnet. A reference to eqns. (16) and (22) shows that this is indeed true.

12.10. Study of the two- and three-dimensional lattices

As stated in Sec. 12.9, Ising (1925) himself carried out a combinatorial analysis of the one-dimensional problem and found that there was no phase transition in the system. This led him to conclude, erroneously though, that his model would not exhibit phase transition in higher dimensions either. In fact, it was this "supposed failure" of the Ising model that motivated Heisenberg to develop, in 1928, the theory of ferromagnetism based on a more sophisticated interaction between the spins; compare the Heisenberg interaction (12.5.6) with the Ising interaction (12.5.7). It was only after some exploitation of the sophisticated model of Heisenberg that people returned to investigate the properties of the simpler model of Ising.

Peierls (1936) was the first to demonstrate that at sufficiently low temperatures the Ising model in two or three dimensions must exhibit ferromagnetism. He considered the lattice as made up of two kinds of *regions*, of "up" spins and of "down" spins, separated by a set of *boundaries* between the neighboring regions, and argued that in a two- or three-dimensional lattice the long-range order that exists at 0°K would persist at finite temperatures. The same argument, when applied to a one-dimensional lattice, showed that the order that exists at 0°K would be destroyed at any finite temperature, however small it may be.* It was, therefore, natural to pursue the study of the Ising model in higher dimensions, in the firm hope that this pursuit would lead to useful information about the nature of the order–disorder transitions taking place in real physical systems. It was, however, soon realized that an exact treatment of the problem in higher dimensions involved formidable mathematical difficulties. Accordingly, Kirkwood (1938) initiated the development of a systematic method of carrying out a power expansion of the partition function of the lattice in terms of the quantity $\tanh(\beta J)$, or (βJ), which could be used for studying at least the *high-temperature* properties of the lattice. Continued by Bethe and Kirkwood (1939), Chang (1941) and Wakefield (1951), and considerably enriched by the Domb school at London, this method has played a vital role in making an *almost exact* study of a variety of two- and three-dimensional lattices. For an assessment of the status of this method, as regards its role in determining the critical point and in elucidating the properties of a system in the vicinity of the critical point, reference may be made to the review articles of Domb (1960, 1968) and Fisher (1963, 1965). It may be added that the foregoing expansions have been supplemented by other expansions which proceed in powers of the quantity $\eta\{=\exp(-2\beta J)\}$ and hence are useful for studying the *low-temperature* properties instead. At the critical point, the parameters (βJ) and η are comparable in value; consequently, the two types of expansions play "comple-

* This is essentially due to the fact that in the case of a one-dimensional lattice the "communication" between any two parts of the lattice can be completely disrupted by a single defect in between. This is not the case with two- or three-dimensional lattices.

mentary" roles in the study of the critical behavior. For the latter expansions too, reference may be made to the review articles of Domb and Fisher.

The first exact and quantitative result for the two-dimensional Ising model was obtained by Kramers and Wannier (1941) who, as mentioned earlier, were the first to introduce the *matrix method* into the theory of this model. Making use of the symmetry properties of the *simple quadratic lattice* ($q = 4$), Kramers and Wannier showed that, for this particular lattice, the field-free partition function $Q(0, T)$ at temperature T and the field-free partition function $Q(0, T^*)$ at temperature T^* are connected by the relationship

$$Q(0, T^*) = \left\{\frac{\cosh(2\gamma^*)}{\cosh(2\gamma)}\right\}^N Q(0, T), \tag{1}$$

where T^* and T are related as follows:

$$\tanh \gamma^* = \exp(-2\gamma) \quad \text{or} \quad \sinh(2\gamma^*) \sinh(2\gamma) = 1, \tag{2}$$

the parameters γ and γ^* being

$$\gamma = J/kT, \quad \gamma^* = J/kT^*. \tag{3}$$

It will be noted that as $T \to \infty$, $T^* \to 0$ and as $T \to 0$, $T^* \to \infty$; eqn. (1), therefore, establishes a one-to-one correspondence between the high-temperature and the low-temperature values of the partition function of the lattice. It then follows that if there exists a singularity in the partition function at a particular temperature T_c, there must exist an equivalent singularity at the corresponding temperature T_c^*. Now, if the thermodynamic functions of the system possess *only one* singularity, as indeed follows from one of the theorems of Yang and Lee (1952), it must exist at a temperature T_c such that $T_c^* = T_c$, i.e.

$$\tanh \gamma_c = \exp(-2\gamma_c) \quad \text{or} \quad \sinh(2\gamma_c) = 1, \tag{4}$$

whence one obtains

$$\eta_c \equiv \exp(-2\gamma_c) = \sqrt{2}-1 = 0.4142 \ldots \tag{5}$$

and

$$\gamma_c = \tfrac{1}{2} \ln(\sqrt{2}+1) = \tfrac{1}{2} \ln(\cot \pi/8) = \tfrac{1}{2} \sinh^{-1} 1 = 0.4407 \ldots. \tag{6}$$

For comparison, we note that for the same lattice the Bragg–Williams approximation and the Bethe approximation lead, respectively, to the values 0.6065 and 0.5000 for η_c.

Kramers and Wannier also showed that the partition function of the Ising lattice could be expressed in terms of the *largest* eigenvalue of a certain matrix and that the occurrence of a phase transition was related to the *degeneracy* of this eigenvalue; see also Montroll (1941, 1942) and Kubo (1943). Kramers and Wannier developed a method that yields the largest eigenvalue of a "sequence of *finite* matrices" which should, in principle, converge to the exact solution if sufficiently large matrices are analyzed. They could not obtain the exact solution in a closed form but they did develop variational techniques which were fairly accurate and which enabled them to make reliable estimates of some of the numbers connected with the *critical* behavior of the lattice.

The exact expression for the field-free partition function of the simple quadratic lattice was first obtained by Onsager (1944), who followed the same approach as that of Kramers and Wannier except that he emphasized "the abstract properties of relatively simple operators rather than their explicit representation by unwieldy matrices"; see also Wannier (1945). Onsager obtained for the partition function of this lattice*

$$\frac{1}{N}\ln Q(0, T) \simeq \ln(\lambda_{\max})$$

$$= \ln\{2\cosh(2\beta J)\}+\frac{1}{2\pi}\int_0^{\pi} d\varphi \ln\left\{\frac{1+\sqrt{(1-\varkappa^2\sin^2\varphi)}}{2}\right\}, \tag{7}$$

where

$$\varkappa = \frac{2\sinh(2\beta J)}{\cosh^2(2\beta J)} = \frac{4\eta(1-\eta^2)}{(1+\eta^2)^2}, \tag{8}$$

η being equal to $\exp(-2\beta J)$. Differentiating (7) with respect to $(-\beta)$, one obtains for the internal energy per spin

$$\frac{1}{N}U_0(T) = -2J\tanh(2\beta J)+\frac{1}{2\pi}\left(\varkappa\frac{d\varkappa}{d\beta}\right)\int_0^{\pi} d\varphi\frac{\sin^2\varphi}{\{1+\sqrt{(1-\varkappa^2\sin^2\varphi)}\}\sqrt{(1-\varkappa^2\sin^2\varphi)}}. \tag{9}$$

Rationalizing the integrand, the foregoing integral can be written as

$$\frac{1}{\varkappa^2}\{-\pi+2K(\varkappa)\}, \tag{10}$$

where $K(\varkappa)$ is the complete elliptic integral of the *first* kind, $\varkappa$ being the modulus of the integral:

$$K(\varkappa) = \int_0^{\pi/2}(1-\varkappa^2\sin^2\varphi)^{-1/2}\,d\varphi. \tag{11}$$

Now, a logarithmic differentiation of (8) with respect to β gives

$$\frac{1}{\varkappa}\frac{d\varkappa}{d\beta} = 2J\{\coth(2\beta J)-2\tanh(2\beta J)\}. \tag{12}$$

Substituting these results into (9), one obtains

$$\frac{1}{N}U_0(T) = -J\coth(2\beta J)\left\{1+\frac{2\varkappa'}{\pi}K(\varkappa)\right\}, \tag{13}$$

* Inspired by the Onsager solution, Kac and Ward (1952) developed a combinatorial method which provided a much simpler derivation of the partition function and other properties of the two-dimensional Ising model. This method throws considerable light on the "topological conditions" which give rise to an exact solution in two dimensions and which are clearly absent in three dimensions! In 1960 Hurst and Green introduced another approach to study the Ising problem; this involves the use of "triangular arrays of quantities directly related to antisymmetric determinants" and is known as the method of *Pfaffians*. This method applies very naturally to the study of "configurations of dimer molecules on a given lattice" which in turn is closely related to the Ising problem; for details, see Montroll (1964) and Green and Hurst (1964).

where $\varkappa'$ is the *complementary* modulus:

$$\varkappa' = 2\tanh^2(2\beta J)-1 \qquad (\varkappa^2+\varkappa'^2 = 1). \tag{14}$$

Figure 12.21 shows the variation of the moduli $\varkappa$ and $\varkappa'$ with the parameter $(\beta J)^{-1}$. We note that while $\varkappa$ is always positive $\varkappa'$ can be negative as well as positive; actually, $\varkappa$ lies between 0 and 1 while $\varkappa'$ lies between -1 and 1. At the *critical* point, where $(\beta_c J)^{-1} = \gamma_c^{-1} = 2.269$ and $\sinh(2\beta_c J) = 1$, the moduli $\varkappa$ and $\varkappa'$ are equal to 1 and 0, respectively.

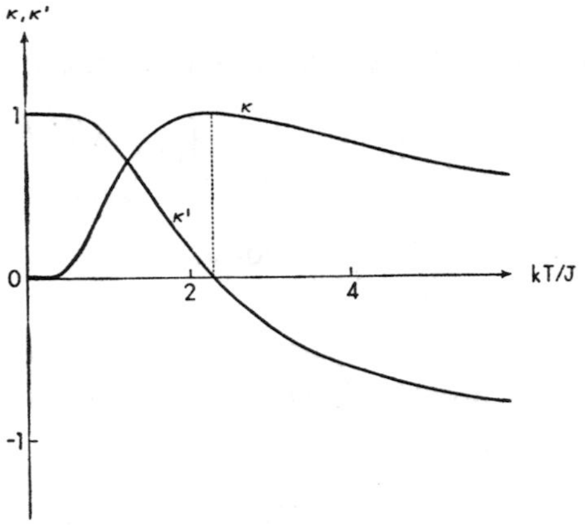

FIG. 12.21. The variation of the moduli $\varkappa$ and $\varkappa'$ with the parameter (kT/J).

To determine the specific heat of the lattice, we have to differentiate (13) with respect to temperature. In this process, we make use of the following results:

$$\frac{d\varkappa}{d\beta} = -\frac{\varkappa'}{\varkappa}\frac{d\varkappa'}{d\beta}, \quad \frac{d\varkappa'}{d\beta} = 8J\tanh(2\beta J)\{1-\tanh^2(2\beta J)\} \tag{15}$$

and

$$\frac{dK(\varkappa)}{d\varkappa} = \frac{1}{\varkappa'^2\varkappa}\{E(\varkappa)-\varkappa'^2K(\varkappa)\}, \tag{16}$$

where $E(\varkappa)$ is the complete elliptic integral of the *second* kind:

$$E(\varkappa) = \int_0^{\pi/2} (1-\varkappa^2\sin^2\varphi)^{1/2}\,d\varphi. \tag{17}$$

We finally obtain

$$\frac{1}{Nk}C_0(T) = \frac{2}{\pi}\{\beta J\coth(2\beta J)\}^2\left[2\{K(\varkappa)-E(\varkappa)\}-(1-\varkappa')\left\{\frac{\pi}{2}+\varkappa' K(\varkappa)\right\}\right]. \tag{18}$$

Now, the elliptic integral $K(\varkappa)$ has a singularity at $\varkappa = 1$ (or at $\varkappa' = 0$), in the neighborhood of which

$$K(\varkappa) \approx \ln\{4/|\varkappa'|\} \quad \text{and} \quad E(\varkappa) \approx 1. \tag{19}$$

As a result of this, the specific heat of the lattice displays a *logarithmic singularity* at temperature T_c, given by the condition $\varkappa_c = 1$ (or $\varkappa'_c = 0$) which is identical with the condition of Kramers and Wannier. In the vicinity of the critical point, eqn. (18) reduces to

$$\frac{1}{Nk}C_0(T) \simeq \frac{8}{\pi}\left(\frac{J}{kT_c}\right)^2\left[\ln\{4/|\varkappa'|\}-\left(1+\frac{\pi}{4}\right)\right]; \tag{20}$$

at the same time, the parameter $\varkappa'$ reduces to, see eqn. (15),

$$\varkappa' \simeq 2\sqrt{2}\left(\frac{J}{kT_c}\right)\left(1-\frac{T}{T_c}\right). \tag{21}$$

The exact nature of the specific heat singularity is, therefore, given by

$$\begin{aligned}\frac{1}{Nk}C_0(T) &\simeq \frac{8}{\pi}\left(\frac{J}{kT_c}\right)^2\left[-\ln\left|1-\frac{T}{T_c}\right|+\left\{\ln\left(\frac{\sqrt{2}kT_c}{J}\right)-\left(1+\frac{\pi}{4}\right)\right\}\right]\\ &\simeq -0.4945\ln\left|1-\frac{T}{T_c}\right|+\text{const.}\end{aligned} \tag{22}$$

Figures 12.22 and 12.23 depict the temperature dependence of the internal energy and the specific heat of the lattice, as given by the Onsager expressions (13) and (18); for compar-

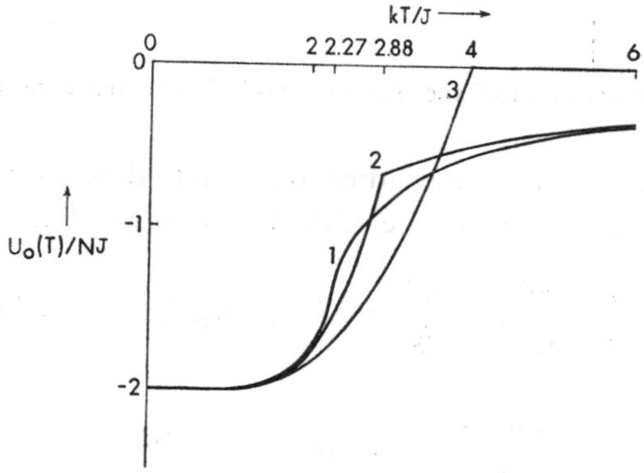

FIG. 12.22. The internal energy of a simple quadratic lattice ($q=4$) according to (1) the Onsager solution, (2) the Bethe approximation and (3) the Bragg–Williams approximation.

ison, the results of the Bragg–Williams approximation and the Bethe approximation (with $q = 4$) are also included. The physical nature of the specific-heat singularity, which is given by the Onsager expression (22) and is seen as a (logarithmic) peak in Fig. 12.23, bears close resemblance to the actual situation in several real systems, such as liquid He^4. We also note that the internal energy of the lattice is *continuous* at the critical point, having a value of $-2^{1/2}J$ per spin and an infinite, positive slope; needless to say, the continuity of the internal energy implies that the transition proceeds without any latent heat.

The next important property is the *spontaneous magnetization*, or the *long-range order*, obtaining in the lattice. The exact expression for this was first derived by Onsager (1949), though he never published the details of his derivation. The first printed derivation is due to Yang (1952), who studied the problem in the presence of a low magnetic field B and then proceeded to the limit $B \to 0$. The result can be expressed in a remarkably simple form, viz.

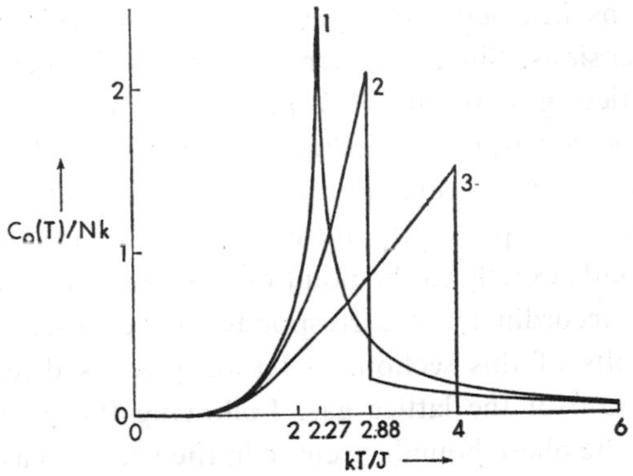

FIG. 12.23. The specific heat of a simple quadratic lattice ($q = 4$) according to (1) the Onsager solution, (2) the Bethe approximation and (3) the Bragg–Williams approximation.

$$\bar{L}_0(T) \equiv \frac{1}{N\mu}\bar{M}(0, T) = [1-\{\sinh(2\beta J)\}^{-4}]^{1/8} \quad \text{for} \quad T \leqslant T_c \tag{23}^*$$

$$= 0 \qquad \text{for} \quad T \geqslant T_c;$$

note that $\sinh(2\beta_c J) = 1$. In the limit $T \to 0$,

$$\bar{L}_0(T) \simeq 1-2\exp(-8\beta J), \tag{24}$$

which implies an *extremely slow* variation with temperature. On the other hand, in the limit $T \to T_c$,

$$\bar{L}_0(T) \simeq \left\{8\sqrt{2}\left(\frac{J}{kT_c}\right)\left(1-\frac{T}{T_c}\right)\right\}^{1/8} \simeq 1.2224\left(1-\frac{T}{T_c}\right)^{1/8}, \tag{25}$$

which indicates an *extremely fast* variation with temperature. The detailed dependence of $\bar{L}_0$ on T is shown in Fig. 12.24; for comparison, the results of the Bragg–Williams approximation and the Bethe approximation are also included.

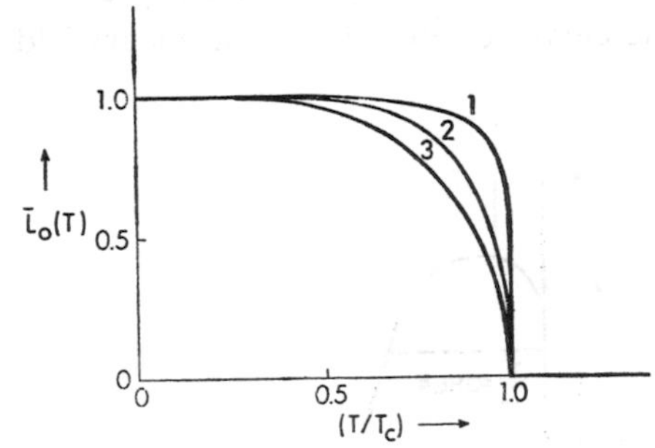

FIG. 12.24. The spontaneous magnetization of a simple quadratic lattice ($q = 4$) according to (1) the Onsager solution, (2) the Bethe approximation and (3) the Bragg–Williams approximation.

* An equivalent form may also be noted:

$$\bar{L}_0(T) = \left[\frac{(1+\eta^2)(1-6\eta^2+\eta^4)^{1/2}}{(1-\eta^2)^2}\right]^{1/4} \quad \text{for} \quad T \leqslant T_c$$

$$= 0 \qquad \text{for} \quad T \geqslant T_c;$$

note that

$$\eta_c = (\sqrt{2}-1).$$

At this stage, it seems instructive to apply the results of this section to the study of a *lattice gas* in two dimensions. Since our ferromagnet results pertain to the case $B = 0$, the corresponding lattice gas results will pertain to a *single* value of z, namely $z = \exp(-8\beta J)$; see the correspondence chart set up in Sec. 12.6. We also observe that the zeros of the grand partition function of the gas are all confined to a circle of radius $r = \exp\left(-\frac{1}{2}q\beta\varepsilon_0\right) = \exp(-8\beta J)$ in the complex z-plane; see eqn. (12.9.33). Thus, our value of z corresponds exactly to the point of intersection of the circle of zeros with the real, positive axis; accordingly, it corresponds to the onset of gas–liquid transition in the system! The results of this section, therefore, give us directly the *phase boundary curve* in the phase diagram of the lattice gas. Following the procedure adopted in Sec. 12.7, we conclude that the phase boundary curve in the present case is determined by the expressions

$$\frac{P}{kT} = \ln(1+\eta^2) + \frac{1}{2\pi}\int_0^\pi d\varphi \ln\left\{\frac{1+\sqrt{(1-\varkappa^2\sin^2\varphi)}}{2}\right\}, \tag{26}$$

$$v_l = 2\left[1+\left\{\frac{(1+\eta^2)(1-6\eta^2+\eta^4)^{1/2}}{(1-\eta^2)^2}\right\}^{1/4}\right]^{-1} \tag{27}$$

and

$$v_g = 2\left[1-\left\{\frac{(1+\eta^2)(1-6\eta^2+\eta^4)^{1/2}}{(1-\eta^2)^2}\right\}^{1/4}\right]^{-1}, \tag{28}$$

where v_l and v_g are the specific volumes in the two phases, $\eta = \exp(-2\beta J)$, $\varkappa = 4\eta(1-\eta^2)(1+\eta^2)^{-2}$. Of course, the expressions (27) and (28) hold only for $T \leqslant T_c$, i.e. for $\eta \geqslant \eta_c = (\sqrt{2}-1)$; the corresponding values of P are given by the expression (26). For $T > T_c$, we have only one value of v, namely $v_l = v_g = 2$; the pressure, of course, goes on increasing with T. The phase boundary curve resulting from these expressions is shown in Fig. 12.25; it may be compared with the curve resulting from the mean-field approximation, viz. Fig. 12.14.

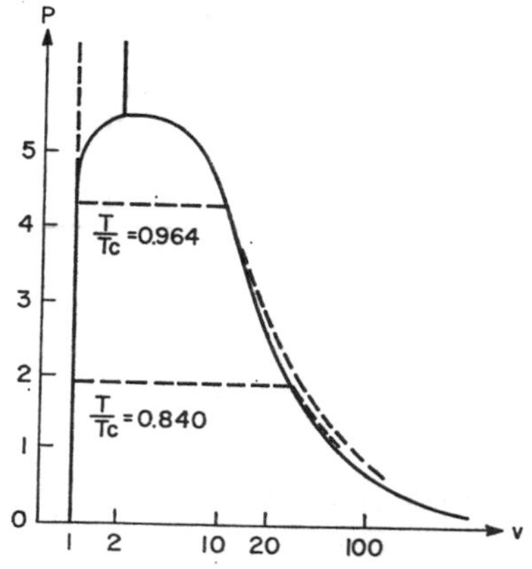

FIG. 12.25. The phase boundary curve and the isotherms of a lattice gas in two dimensions (after Lee and Yang, 1952).

There is one important difference between the two theories; this relates to the temperature dependence of the difference between the densities of the two phases in the vicinity of the critical point. We may write, for $(1-T/T_c) \ll 1$,

$$(\varrho_l - \varrho_g) \equiv \bar{L}_0 \propto (1-T/T_c)^{\beta}. \tag{29}$$

According to the exact solution in two dimensions, $\beta = \frac{1}{8}$; the mean-field approximation, on the other hand, gives: $\beta = \frac{1}{2}$. For actual systems, however, β turns out to be very close to $\frac{1}{3}$. In another respect, the two theories give identical results, viz.

$$(\varrho_l + \varrho_g) = 1 \quad \text{for all } T; \tag{30}$$

in fact, by the very definitions of ϱ_l and ϱ_g (namely $\bar{N}_+/N$ and $\bar{N}_-/N$, respectively), this result must hold for *all* lattice models. The point to note here is that this result resembles very closely the behavior of real gases, for which a so-called "law of *rectilinear* diameter" is known to hold. According to this law, the sum of the densities of the vapor and the liquid, in equilibrium, increases *linearly* as the temperature of the system decreases. This increment is, however, very slow, being not more than 10 per cent for a temperature variation over a factor of 2 for helium near its critical point. The lattice models, on the other hand, give what may be called a "law of *constant* diameter" and may be regarded as a first approximation to the actual law.

In passing, we remark that eqns. (26) and (28) for the pressure and the specific volume of the gaseous phase give, through eqns. (12.9.44) and (12.9.45), certain averages with respect to the distribution function, $g(\theta)$, of the zeros of the grand partition function of the lattice gas. The mathematical forms of these averages are quite suggestive; however, it has not been possible so far to construct the function $g(\theta)$ from them.

12.11. The critical indices

A basic problem in the theory of phase transitions is to study the behavior of a given physical system in the *neighborhood* of the *critical point*. We know that the behavior of a physical system in this region is marked by the fact that the various physical quantities pertaining to the system possess singularities at the critical point. It is customary to express the nature of these singularities in terms of a set of *critical indices*, which determine the temperature dependence of the various quantities in the neighborhood of the critical point. For instance, the *order parameter* $\bar{L}_0(T)$ which determines the intensity of spontaneous magnetization of a ferromagnet, or the quantity $(\varrho_l - \varrho_g)$ of a gas–liquid system, and goes to zero as $T \to T_c$ is written as

$$\bar{L}_0(T) \equiv \bar{L}(B, T)\Big|_{B=0} \propto (1-T/T_c)^{\beta}, \tag{1}$$

$\beta\ (>0)$ being the corresponding critical index; for $T > T_c$, $\bar{L}_0(T) \equiv 0$. The *low-field susceptibility* $\chi_0(T)$ $[\propto (\partial\bar{L}/\partial B)_T]_{B=0}$, however, diverges at $T \to T_c$; it may, therefore, be represented in a similar manner as the order parameter $\bar{L}_0(T)$ but with a *negative* exponent. Furthermore, the exponent may have different values for temperatures above and for temperatures below

T_c. Thus, we may write

$$\chi_0(T) \propto |(1-T/T_c)|^{-\gamma} \quad \text{for} \quad T > T_c \tag{2}$$
$$\propto (1-T/T_c)^{-\gamma'} \quad \text{for} \quad T < T_c; \tag{3}$$

the numbers γ and γ' (both > 0) denote the critical indices for the low-field susceptibility of the given system. The analogous quantity in the gas–liquid transition is the *isothermal compressibility* $K_T[\propto (\partial\varrho/\partial P)_T]$; this also diverges at the critical point and the nature of this divergence is again determined by the indices γ and γ'. Another relevant property of the gas–liquid system is the rate of variation of the quantity $(\varrho-\varrho_c)$ with the quantity $(P-P_c)$, where ϱ_c and P_c are the critical density and the critical pressure of the system. This variation may be expressed as

$$(\varrho-\varrho_c)\Big|_{T\to T_c} \propto (P-P_c)^{1/\delta}, \tag{4}$$

$\delta\,(> 0)$ being the corresponding critical index. The analogous relationship in the case of a ferromagnet is

$$\bar{L}(B, T_c) \propto B^{1/\delta}. \tag{5}$$

On physical grounds, one expects that

$$\gamma' = \beta(\delta-1). \tag{6}$$

Another quantity that diverges at the critical point is the *specific heat* $C_V(T)$, for which we may write, in the same manner as in eqns. (2) and (3),

$$C_V(T) \propto |(1-T/T_c)|^{-\alpha} \quad \text{for} \quad T > T_c \tag{7}$$
$$\propto (1-T/T_c)^{-\alpha'} \quad \text{for} \quad T < T_c. \tag{8}$$

The indices α and α' will indeed be positive; of course, a vanishingly small value of these indices might imply a *logarithmic* rather than an *algebraic* singularity!

Our theoretical knowledge regarding these indices is somewhat handicapped by the fact that we have not been able to find so far the *exact* solution of the Ising problem in three dimensions; all that we have been able to achieve in this direction is the Onsager solution of the two-dimensional problem—and that too in the absence of an external field ($B = 0$). On the other hand, we have certain approximate approaches, such as the mean field approximation, which do lead to results that possess some of the essential *qualitative* features of the actual situation but unfortunately they cannot be relied upon *quantitatively*; this has already been noticed in the case of the index β. The predictions of these approximations are essentially a consequence of the implicit, or explicit, assumption that the free energy A and the pressure P of the given system can be expanded as a *Taylor-type series*, in terms of the density and the temperature, right at the critical point itself! This clearly shows the need to pursue the study of the three-dimensional Ising problem in a manner that may not be exact in the mathematical sense but may be good enough to elucidate the nature of the singularities displayed by the various physical quantities.

The most promising approach in this respect is the one that employs *systematic, successive* approximations whose limiting behavior can be studied *numerically*. Expanding the partition function at *low* fields or at *low* temperatures or at *high* temperatures, one obtains, for quantities such as the intensity of (spontaneous) magnetization, the paramagnetic susceptibility, the specific heat, etc., power expansions of the type

$$A(w) = \sum_{n\geq 0} a_n w^n . \tag{9}$$

The evaluation of the coefficients a_n involves a lot of combinatorial work, systematic methods for which have been developed in considerable detail (see Domb, 1960); in practice, one can normally compute about a dozen of these coefficients *exactly*. To analyze these power expansions numerically, two independent techniques are available, viz. the *ratio method* of Domb and Sykes (1961) and the *Padé approximant procedure* of Baker, Gammel and Wills (1961).

The *ratio method* is applicable when the coefficients a_n of the given series (or of some transformed series, such as the logarithmic derivative of the given series) are of the *same* sign; the dominant singularity of $A(w)$ then lies on the real axis and (normally) corresponds to the transition point. In most cases one finds that the successive ratios $r_n(= a_n/a_{n-1})$ approach rapidly a *linear* behavior with respect to $1/n$. The gradient of this line determines the nature of the singularity while its intercept with the r_n-axis yields the transition point.

The *Padé approximant procedure* is basically a method of analytically continuing a series. The $[L, M]$-approximant to the function $A(w)$ is the ratio of two polynomials, $P(w)$ of degree L and $Q(w)$ of degree M, whose coefficients are so chosen that the expansion of the quotient $P(w)/Q(w)$ in powers of w agrees with the exact expansion of $A(w)$ up to the term in w^{L+M}. The distribution of the zeros and the poles of the successive approximants reveals the analytic behavior of $A(w)$ and enables one to make an accurate estimate of $A(w)$ well beyond the radius of convergence of the original series.

It is gratifying that the estimates of T_c, and of the various critical indices, as obtained by these methods, are in remarkable agreement with one another (and with the exact results, wherever available); accordingly, one hopes that the results for three-dimensional lattices, obtained in this manner, will also be reliable.

In Table 12.1, we have compiled both theoretical and experimental values of the critical indices β, γ, γ', δ, α and α'. On the experimental side, our data pertain to the gas–liquid as well as the ferromagnetic transition; the correspondence between the two sets of values is indeed striking. On the theoretical side, the results pertain to the mean field approximation as well as the exact (or almost exact) treatment of the Ising model (in two and three dimensions); the inadequacy of the mean field approximation is quite evident.

In the end, we would like to mention another critical index, μ, which is defined by the statement that, in the two-phase region, as $T \to T_c$ the *surface tension* of a liquid vanishes as $(1-T/T_c)^{\mu}$. An analysis of the experimental data on surface tension yields: $\mu = 1.27 \pm 0.02$.

Another useful piece of information is provided by a measurement of the *superfluid density* ϱ_s in liquid He^4 near the lambda point. The most accurate data for ϱ_s are provided by the experiments of Clow and Reppy (1966) and of Tyson and Douglass (1966); according to them,

$$\varrho_s \propto (1-T/T_\lambda)^{\zeta} , \tag{10}$$

where $\zeta = 0.666 \pm 0.006$. According to the theory of Ginzburg and Pitaevskii (1958), ϱ_s

TABLE 12.1. THE VALUES OF THE CRITICAL INDICES (compiled on the basis of a review article by Kadanoff *et al.* (1967), in which a detailed bibliography of the original works is also given)*

Critical index	Theoretical values			Experimental values	
	mean field approximation	Ising model		ferromagnetic transition	gas–liquid transition
		2-dim.	3-dim.		
β	$\frac{1}{2}$	$\frac{1}{8}$	0.313 ± 0.004	0.33 ± 0.03	0.346 ± 0.01
γ	1	$\frac{7}{4}$	1.250 ± 0.001 (a)	1.33 ± 0.03	$1.37\ \pm0.2$
γ'	1	$\frac{7}{4}$	$1.31\ \pm0.05$	(1.0 ± 0.1) (b)	$1.0\ \pm0.3$
δ	3	15	$5.2\ \pm0.15$	4.1 ± 0.1	$4.4\ \pm0.4$
α	$=0$ (c)	$\to 0$ (d)	$0.1\ \pm0.1$	$\lesssim 0.16$	$0.2\ \pm0.2$
α'	$=0$ (c)	$\to 0$ (d)	$0.07^{+0.16}_{-0.04}$	$\lesssim 0.16$	$0.12\ \pm0.12$

(a) For the three-dimensional Heisenberg model, $\gamma = 1.33\pm0.05$.

(b) Derived from the relation $\gamma' = \beta(\delta-1)$. The only ferromagnetic case reported in the review article of Kadanoff *et al.* is that of $YFeO_3$, for which $\gamma' = 0.7\pm0.1$.

(c) These values of α and α' correspond to a finite discontinuity, not a divergence, in the magnitude of C_V.

(d) These values of α and α' correspond to a logarithmic divergence of C_V.

is proportional to the *order parameter squared*:

$$\varrho_s \propto |\langle\psi\rangle|^2. \tag{11}$$

Now, if $\langle\psi\rangle$ goes to zero as $(1-T/T_\lambda)^\beta$, see eqn. (1), then we must have: $\zeta = 2\beta$, which does seem right.†

12.12. The law of corresponding states

To predict unknown properties of a given substance, about which only limited data are available, on the basis of a more complete knowledge of other "similar" substances, one can make use of a *law of corresponding states*. Such a law was first introduced into classical thermodynamics by van der Waals who showed that the equation of state of a gas, if it contains only two adjustable parameters, can be written in a *universal, dimensionless* form by expressing the pressure P, volume (per particle) v and temperature T as "reduced variables" in terms of the critical pressure P_c, the critical volume (per particle) v_c and the critical temperature T_e. Thus, if we start with the van der Waals equation of state

$$\left(P+\frac{a}{v^2}\right)(v-b) = kT, \tag{1}$$

* See, as well, the review articles of Fisher (1967), Heller (1967) and Smith (1969), and the monograph of Stanley (1971).

† In this context, we note that, in an *ideal* Bose gas, $\varrho_s \propto \{1-(T/T_\lambda)^{3/2}\}$ which, for $T \lesssim T_\lambda$, becomes: $\varrho_s \propto (1-T/T_\lambda)$. This means: $\zeta = 1$ and hence $\beta = \frac{1}{2}$ (which in turn corresponds to the assumption of a "mean field" approximation).

in which a and b are the two adjustable parameters, and transform the variables P, v and T to the "reduced" variables $\bar{P}$, $\bar{v}$ and $\bar{T}$ with the help of the critical pressure P_c ($= a/27b^2$), the critical volume (per particle) v_c ($= 3b$) and the critical temperature T_c ($= 8a/27bk$), so that

$$\bar{P} = \frac{P}{P_c}, \quad \bar{v} = \frac{v}{v_c} \quad \text{and} \quad \bar{T} = \frac{T}{T_c}, \tag{2}$$

then the equation of state reduces to the dimensionless form

$$\left(\bar{P} + \frac{3}{\bar{v}^2}\right)(3\bar{v} - 1) = 8\bar{T}. \tag{1a}$$

Clearly, the state $\bar{P} = 1$, $\bar{v} = 1$ and $\bar{T} = 1$ corresponds to the *critical point* of the gas; see

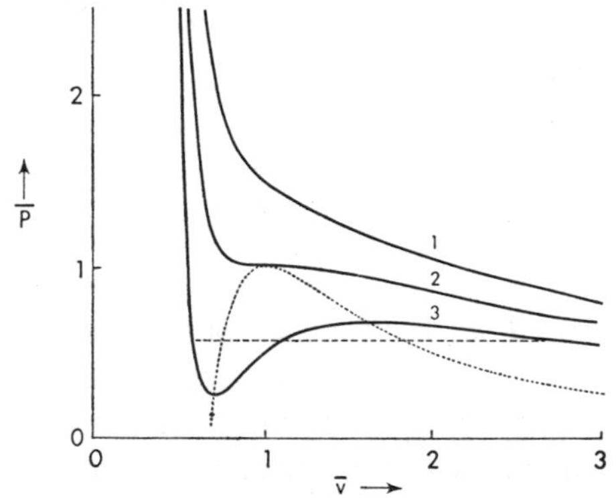

FIG. 12.26. The phase boundary curve and the isotherms of a van der Waals gas, in terms of the reduced variables $\bar{P}$, $\bar{v}$ and $\bar{T}$. Curve 1 is for $\bar{T} > 1$, curve 2 for $\bar{T} = 1$ and curve 3 for $\bar{T} < 1$; the horizontal line carries out the Maxwell construction on curve 3.

Fig. 12.26. At the same time, the relation

$$P_c v_c = \tfrac{3}{8} k T_c \tag{3}$$

reduces to the trivial identity

$$\bar{P}_c \bar{v}_c = \bar{T}_c. \tag{3a}$$

The *reduced equation of state* (1a) does not contain any adjustable parameters; it is, therefore, *universally* applicable to all gases that otherwise conform to the van der Waals equation of state (1). The usefulness of the reduced equation of state rests with the fact that if, for two "similar" systems, the values of two out of the three reduced variables are mutually the same then the values of the third reduced variable would also be the same; the two systems are then said to be in the *corresponding states*. The word "similar" here implies that the microscopic interactions in the two systems must be such as to lead to "similar" macroscopic behavior (or to "similar" equations of state).

The existence of a law of corresponding states can be understood, quite generally, by going back to the theory of the equation of state. To be specific, let us consider a class of gaseous systems whose molecules interact through a central potential. We assume that, for *all* systems in this class, the two-body interaction can be represented by a *common* mathematical function f which depends upon two *molecular parameters* σ and ε:

$$u(r) = \varepsilon f(r/\sigma). \tag{4}$$

For instance, $u(r)$ could be the Lennard-Jones six–twelve potential,

$$u(r) = 4\varepsilon\left\{\left(\frac{\sigma}{r}\right)^{12} - \left(\frac{\sigma}{r}\right)^{6}\right\}, \tag{5}$$

according to which $u(r) \geqslant 0$ for $r \leqslant \sigma$, has a minimum value of $-\varepsilon$ at $r = 2^{1/6}\sigma$ and tends to zero as $r \to \infty$; see Fig. 9.1. If the systems can be regarded as Boltzmannian, then their equation of state is determined by the formula

$$\frac{P}{kT} = \frac{\partial}{\partial V}\{\ln Q_N(V, T)\} = \frac{\partial}{\partial V}\{\ln Z_N(V, T)\}, \tag{6}$$

where $Z_N(V, T)$ is the *configuration integral*:

$$Z_N(V, T) = \int_V d^{3N}r \exp\left\{-\frac{1}{kT}\sum_{i<j} u(r_{ij})\right\}; \tag{7}$$

see Sec. 9.1. Thus, the quantity P/kT is a function of the quantities N/V, ε/kT and σ^3; the precise form of the function is determined by the nature of the two-body potential function $f(r/\sigma)$. Invoking dimensional considerations, we can express this result as follows: for *all* classical systems characterized by a *common* two-body potential function $u(r) = \varepsilon f(r/\sigma)$, the dimensionless quantity $P\sigma^3/\varepsilon$ is a "universal" function of the dimensionless quantities v/σ^3 and kT/ε. Taking these dimensionless quantities as our *reduced variables,* and denoting them by the symbols P^*, v^* and T^*, the *reduced equation of state* may be written as

$$P^* = P^*(v^*, T^*), \tag{8}$$

where the mathematical form of the function P^* is the same for all systems characterized by a common potential function f; here,

$$P^* = \frac{P\sigma^3}{\varepsilon}, \quad v^* = \frac{v}{\sigma^3} \quad \text{and} \quad T^* = \frac{kT}{\varepsilon}. \tag{9}$$

The main advantage of this formulation is that even if we do not have sufficient information regarding the functional form of eqn. (8), the very existence of a "universal" equation of this type enables us to learn a lot about a new member of a class of "similar" systems on the basis of the knowledge we already have about the existing members of the class.

The reduction of the variables P, v and T in terms of the molecular parameters σ and ε, and the subsequent derivation of the reduced equation of state in terms of the reduced variables P^*, v^* and T^*, is due to de Boer and Michels (1938–9). It is not difficult to see that the reduced equation of state due to van der Waals, viz. eqn. (1a), is only a specialized variation of the basic equation of state due to de Boer and Michels, and is applicable only to that class of systems whose intermolecular potential function f is supposed to lead, in the first place, to the original van der Waals equation of state (1).

A generalization of eqn. (8), which has proved to be of considerable physical interest, is the so-called *quantum-mechanical law of corresponding states.* This generalization was first considered by Byk (1921, 1922) in terms of the variables $\bar{P}$, $\bar{v}$ and $\bar{T}$; a systematic study in terms of the variables P^*, v^* and T^*, and a fruitful exploitation in the field of low-temperature physics, are due to de Boer and co-workers (1948, 1949). The driving force behind this work was the realization that at sufficiently low temperatures, when the mean thermal wavelength, $\lambda = h/\sqrt{(2\pi mkT)}$, of the molecules becomes comparable with the range σ of the two-body interactions, the *diffraction effects* arising from the quantum character of the molecular motion must become important. The deviations from classical behavior would become significant, and it seems natural that our equation of state would then have yet another parameter in it which takes care of the quantum aspect of the situation. The most appropriate parameter for this purpose turns out to be the dimensionless quantity Λ^*, defined by the formula

$$\Lambda^* = \frac{h}{(m\varepsilon)^{1/2}\sigma}; \tag{10}$$

we note that the parameter Λ^* correctly incorporates the fact that quantum effects are more important in the case of lighter molecules.† The new law of corresponding states, therefore, reads:

$$P^* = P^*(v^*, T^*, \Lambda^*); \tag{11}$$

cf. eqn. (8). In the limit $\Lambda^* \to 0$, the dependence of the function P^* on the parameter Λ^* disappears and the quantum-mechanical law (11) reduces to the classical version (8).

The actual values of Λ^* for various gases can be computed from formula (10), with the help of the interaction parameters ε and σ (which are available from the high-temperature data on the virial coefficients of these gases). For the special class of *inert* gases, the values of Λ^* turn out to be

Xe	Kr	A	Ne	He^4	He^3
0.064	0.102	0.187	0.591	2.64	3.05;

it may be mentioned here that in computing the value of Λ^* for He^3 the parameters ε and σ are taken to be the same as those for He^4, which is justified because the interatomic interaction for the two isotopes of helium must be essentially the same. The hydrogen-molecule isotopes H_2, HD, D_2, HT, DT and T_2 have Λ^*'s ranging from about 1.73 to about 1.00, while for gases such as N_2, O_2, CO_2 and CH_4, Λ^* falls in the range 0.20–0.25. We, therefore,

† The fact that the magnitude of the quantum effects is also governed by the temperature of the system need not be incorporated into the parameter Λ^* separately for the role of temperature in the equation of state of the system is going to come in anyway; see eqns. (8) and (11). We also note that, in the definition of the parameter Λ^*, the ratio of the "quantum length" $h/(m\varepsilon)^{1/2}$ is taken with respect to the range σ of the two-body interaction and not with respect to the mean interparticle distance $v^{1/3}$. This is related to the *implicit* assumption that we are working in that range of temperatures where the mean thermal wavelength of the molecules continues to be much smaller than the mean interparticle distance and, hence, the *symmetry effects*, that arise from the mutual interchange of indistinguishable particles, can still be neglected.

expect quantum effects to be especially important in the case of the hydrogen-molecule isotopes and the helium isotopes. We shall now give a brief account of the admirable manner in which the new law of corresponding states was employed by de Boer and co-workers (1948, 1949) for making a number of significant predictions about the low-temperature behavior of liquid He^3.

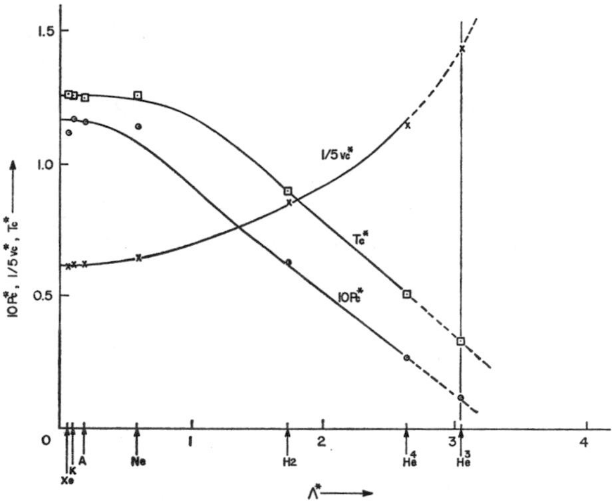

FIG. 12.27. The *reduced* critical data P_c^*, v_c^* and T_c^* for various gases, as a function of the quantum-mechanical parameter Λ^*. Extrapolations yield the theoretical estimates for He^3.

For this we plot, in Fig. 12.27, the *known* values of the quantities P_c^*, v_c^* and T_c^* for the inert gases Xe, Kr, A, Ne and He^4 as functions of the parameter Λ^*. The very fact that the resulting curves are reasonably smooth is, by itself, a proof that the law of corresponding states is applicable. The values of P_c^*, v_c^* and T_c^* for He^3 can then be obtained by simply extrapolating these curves to the relevant value of Λ^*. Proceeding in this manner, de Boer and Lunbeck (1948) obtained the following (extrapolated) values for He^3:

$$P_c^* = 0.0135, \quad v_c^* = 7.0 \quad \text{and} \quad T_c^* = 0.33$$

and, correspondingly,

$$P_c = 1.12 \text{ atm}, \quad v_c = 70 \text{ cm}^3 \text{ mole}^{-1} \quad \text{and} \quad T_c = 3.37°\text{K}.$$

In the following year, Sydoriak, Grilly and Hammel (1949) determined these data experimentally and obtained the following results:

$$P_c = 1.15 \text{ atm}, \quad v_c = 72 \text{ cm}^3 \text{ mole}^{-1} \quad \text{and} \quad T_c = 3.34°\text{K}.$$

Keeping in mind the fact, in the limit $\Lambda^* \to 0$, $P_c^* = 0.117$, $v_c^* = 3.10$ and $T_c^* = 1.26$, it is indeed remarkable that so large a deviation from the classical behavior, as shown by He^3, could be handled in such a simple manner by making use of the law of corresponding states.†

† In this connection, we would like to mention that the quantity $P_c^* v_c^* / T_c^*$ which, in the limit $\Lambda^* \to 0$, has the empirical value 0.288 (against the value 0.375 for a van der Waals gas) has very nearly the same value in the case of He^3. Clearly, this (composite) quantity is quite insensitive to the quantum-mechanical influences.

de Boer and Lunbeck also computed the vapor pressure curve for He^3 by analyzing the P^* versus T^* data for known cases as a function of the parameter Λ^*. Their results for the function $P(T)$ are tabulated in Table 12.2; for comparison, the experimental results of Abraham, Osborne and Weinstock (1950) are also included.

TABLE 12.2

T (in °K)		2.0	2.4	2.8	3.0	3.2	3.3
P (in cm of Hg)	theoretical estimates	$\simeq 14$	27	48	59	75	82
	experimental values	15.2	29.1	49.1	61.8	76.4	84.5

By a glance at the theoretical estimates one could say that the *normal* boiling point of He^3 (the point where the actual vapor pressure becomes equal to the atmospheric pressure) should be very nearly 3.2°K; the experimental result turned out to be 3.19°K. The accuracy of theoretical prediction is again remarkable.

Studies along these lines have also been carried out for the hydrogen-molecule isotopes; once again, one finds good agreement between the predicted and the observed values of the various physical quantities. These investigations have also been extended to other properties, such as the surface tension, the transport coefficients, etc. For details, see Hirschfelder, Curtiss and Bird (1954).

Problems

12.1. Assume that in the virial expansion

$$\frac{Pv}{kT} = 1 - \sum_{j=1}^{\infty} \left\{ \frac{j}{j+1} \beta_j \left(\frac{\lambda^3}{v} \right)^j \right\}, \qquad (9.4.22)$$

where β_j's are the *irreducible cluster integrals* of the system, only the terms $j = 1$ and $j = 2$ are appreciable at the critical temperature and density of the system. Determine the relationship between β_1 and β_2 at the critical point, and show that this leads to the result: $P_c v_c / kT_c = \frac{1}{3}$.

12.2. Assuming the *Dietrici equation of state*,

$$P(v-b) = kT \exp(-a/kTv),$$

evaluate the critical constants P_c, v_c and T_c in terms of the parameters a and b, and show that the quantity $(P_c v_c / kT_c) = 2/e^2 \simeq 0.271$. Further show that the following statements hold for the Dietrici equation of state:

(a) It yields the same expression for the second virial coefficient B_2 as the van der Waals equation does.
(b) For all values of P and for all $T \geqslant T_c$, it yields a *unique* value of v.
(c) For all $T < T_c$, there are *three* possible values of v for any given value of P and the critical volume v_c is always intermediate between the largest and the smallest of the three volumes.

12.3. Study the Heisenberg model of a ferromagnet, based on the interaction (12.5.6), in the *mean field approximation* and show that this also leads to a phase transition of the kind met with in the Ising model. Show, in particular, that the transition temperature T_c and the Curie–Weiss constant C are given by

$$T_c = \frac{qJ}{k} \frac{2S(S+1)}{3} \quad \text{and} \quad C = \frac{N(g\mu_B)^2}{Vk} \frac{S(S+1)}{3}.$$

Note that the ratio $T_c/CV = 2qJ/N(g\mu_B)^2$ is the *molecular field constant* of the problem; cf. eqn. (12.7.8).

12.4. Study the spontaneous magnetization of the Heisenberg model in the *mean field approximation* and examine the dependence of $\bar{L}_0$ on T (i) in the neighborhood of the critical temperature: $(1-T/T_c) \ll 1$, and (ii) at sufficiently low temperatures: $T/T_c \ll 1$. Compare these results with the corresponding ones for the Ising model.

[In this connection, it may be pointed out that, at very low temperatures, the experimental data do not agree with the theoretical formula derived here. We find instead a much better agreement with the formula $\bar{L}_0 = \{1-A(kT/J)^{3/2}\}$, where A is a numerical constant (which is equal to 0.1174 in the case of a simple cubic lattice). This formula is known as *Bloch's* $T^{3/2}$-*law* and is derivable from the spin-wave theory of ferromagnetism; see Wannier (1966), sec. 15.5.]

12.5. An antiferromagnet is characterized by the fact that the exchange integral J is negative, which naturally tends to align the neighboring spins *antiparallel* to one another. Assuming that a given crystal structure is such that the whole lattice can be divided into two interpenetrating sub-lattices, a and b say, then the spins belonging to each sub-lattice, a as well as b, will tend to align themselves in the same direction, while the directions of alignment in the two sub-lattices will be opposite to one another. Using the *Heisenberg* type of interaction and working in the *mean field* approximation, evaluate the paramagnetic susceptibility of the lattice at high temperatures.

12.6. The *Néel temperature* T_N of an antiferromagnet is defined as that temperature below which the sub-lattices a and b possess nonzero spontaneous magnetizations M_a and M_b. Determine T_N for the model considered in the previous problem.

12.7. Suppose that each atom of a crystal lattice can be in one of its r *internal* states, which may be denoted by the symbols σ, and that the interaction energy between an atom in the state σ' and its nearest neighbor in the state σ'' is given by $u(\sigma', \sigma'')$ $\{= u(\sigma'', \sigma')\}$. Let $f(\sigma)$ be the probability of an atom being in the particular state σ, *independently* of the states in which its nearest neighbors are. The interaction energy and the entropy of the lattice may then be written as

$$E = \tfrac{1}{2}qN \sum_{\sigma', \sigma''} u(\sigma', \sigma'') f(\sigma') f(\sigma'')$$

and

$$S/Nk = -\sum_{\sigma} f(\sigma) \ln f(\sigma),$$

respectively. Minimizing the free energy $(E-TS)$, show that the equation determining the equilibrium form of the function $f(\sigma)$ is $f(\sigma) = C \exp\{-(q/kT) \Sigma_{\sigma'} u(\sigma, \sigma') f(\sigma')\}$, where C is the constant of normalization. Further show that, for the special case $u(\sigma', \sigma'') = -J\sigma'\sigma''$, where the σ's can be either $+1$ or -1, this equation reduces to the Weiss equation (12.7.11), with $f(\sigma) = \tfrac{1}{2}(1+\bar{L}_0\sigma)$.

12.8. Consider a binary alloy containing N_A atoms of the kind A and N_B atoms of the kind B, so that the relative concentrations of the two components are given by $x_A = N_A/(N_A+N_B) \leqslant \tfrac{1}{2}$ and $x_B = N_B/(N_A+N_B) \geqslant \tfrac{1}{2}$. Define the degree of *long-range order*, X, by the formulae

$$\begin{bmatrix} A \\ a \end{bmatrix} = \frac{1}{2} N x_A (1+X), \qquad \begin{bmatrix} A \\ b \end{bmatrix} = \frac{1}{2} N x_A (1-X),$$

$$\begin{bmatrix} B \\ a \end{bmatrix} = \frac{1}{2} N (x_B - x_A X), \qquad \begin{bmatrix} B \\ b \end{bmatrix} = \frac{1}{2} N (x_B + x_A X),$$

where $N = N_A + N_B$, while the symbol $\begin{bmatrix} A \\ a \end{bmatrix}$ denotes the number of atoms of the kind A occupying sites of the sub-lattice a, and so on. In the *Bragg–Williams approximation*, the number of nearest-neighbor pairs of different kinds can be written down straightaway; for instance,

$$\begin{bmatrix} AA \\ ab \end{bmatrix} = \frac{1}{2} qN \cdot x_A(1+X) \cdot x_A(1-X),$$

and so on. The configurational energy of the lattice then follows from eqn. (12.6.9). In the same approximation, the entropy of the lattice is given by $S = k \ln W$, where

$$W = \frac{(\tfrac{1}{2}N)!}{\begin{bmatrix} A \\ a \end{bmatrix}! \begin{bmatrix} B \\ a \end{bmatrix}!} \cdot \frac{(\tfrac{1}{2}N)!}{\begin{bmatrix} A \\ b \end{bmatrix}! \begin{bmatrix} B \\ b \end{bmatrix}!}.$$

Minimizing the free energy of the lattice, show that the equilibrium value of X is determined by the equation

$$\frac{X}{x_B+x_A X^2} = \tanh\left(\frac{2qx_A\varepsilon}{kT}X\right); \qquad \varepsilon = \frac{\varepsilon_{11}+\varepsilon_{22}}{2}-\varepsilon_{12}.$$

Note that in the special case of equal concentrations ($x_A = x_B = \frac{1}{2}$) this equation reduces to the more familiar equation, viz.

$$X = \tanh\left(\frac{q\varepsilon}{2kT}X\right).$$

Further show that the *transition temperature* of the system is given by

$$T_c = 4x_A(1-x_A)T_c^0,$$

where T_c^0 ($= q\varepsilon/2k$) is the transition temperature in the special case of equal concentrations.

[Note. In the *Kirkwood approximation* (see Kubo (1965), problem 5.19), T_c^0 turns out to be $(\varepsilon/k)\{1-\sqrt{[1-(4/q)]}\}$, which may be written as $(q\varepsilon/2k)$ $(1-1/q+\ldots)$. To this order, the Bethe approximation also yields the same result.]

12.9. Consider a two-component solution of N_A atoms of the kind A and N_B atoms of the kind B, which are supposed to be randomly distributed over $N (= N_A+N_B)$ sites of a *single* lattice. Denoting the energies of the nearest-neighbor pairs AA, BB and AB by ε_{11}, ε_{22} and ε_{12}, respectively, write down the free energy of the system in the *Bragg–Williams* approximation and evaluate the chemical potentials μ_A and μ_B of the two components.

Next show that if $\varepsilon = \{\frac{1}{2}(\varepsilon_{11}+\varepsilon_{22})-\varepsilon_{12}\} < 0$, i.e. if atoms of the same species display greater affinity to be neighborly, then for all temperatures below a critical temperature T_c, which is given by the expression $q|\varepsilon|/2k$, the solution separates out into two phases of *unequal* relative concentrations.

[Note. For a study of the occurrence of phase separation in an isotopic mixture of hard-sphere bosons and fermions, and the relevance of this study to the actual behavior of He^3–He^4 solutions, see Cohen and van Leeuwen (1960, 1961).]

12.10. The concept of a *mean field* can also be applied to a system of particles interacting through Coulomb forces. Consider, for example, a *classical* system of charged particles. The *mean charge density* ϱ at point $\boldsymbol{r}$ would be given by

$$\varrho(\boldsymbol{r}) = \sum_s e_s n_{s0} e^{-e_s\varphi(\boldsymbol{r})/kT},$$

where s denotes the species of the particle, e_s the electric charge on it, $\varphi(\boldsymbol{r})$ the mean electric potential and n_{s0} the number density of the s-particles at the point where $\varphi(\boldsymbol{r}) = 0$. The potential, on the other hand, is determined by the charge density, through the *Poisson equation*

$$\nabla^2\varphi(\boldsymbol{r}) = -\frac{4\pi}{D}\varrho(\boldsymbol{r}),$$

D being the dielectric constant of the medium. Solving the two equations together, under appropriate boundary conditions, one obtains *self-consistent* solutions for the functions $\varrho(\boldsymbol{r})$ and $\varphi(\boldsymbol{r})$.

Assuming, for simplicity, that the electrostatic energy $e_s\varphi(\boldsymbol{r})$ in all regions of interest is much smaller than the thermal energy kT, the exponential factor in the first equation may be expanded and only the linear term retained. Moreover, if the system on the whole is neutral the main term of the expansion simply drops out. We are then left with the *linear equation*

$$\nabla^2\varphi(\boldsymbol{r}) = \varkappa^2\varphi(\boldsymbol{r}); \qquad \varkappa^2 = \frac{4\pi}{DkT}\sum_s e_s^2 n_{s0}.$$

Debye and Hückel applied this idea to study the equilibrium properties of ionic solutions. Taking an ion of species s as the origin, the potential around it turns out to be

$$\varphi(\boldsymbol{r}) = \frac{e_s}{D(1+\varkappa a)} e^{-\varkappa(r-a)}\frac{1}{r};$$

here, a denotes a finite distance such that the neighboring ions cannot be present within this distance from the central ion. The constant factor in the solution is so chosen that the electric field at $r = a$ is precisely e_s/Da^2. The quantity $1/\varkappa$ is known as the *Debye screening length* because at distances large compared to $1/\varkappa$ the field rapidly vanishes.

Verify the Debye–Hückel result, and evaluate the electrostatic free energy per unit volume of the solution.

12.11. Show that in the *Bethe approximation* the entropy of the Ising lattice at $T = T_c$ is given by the expression

$$\frac{S_c}{Nk} = \ln 2 + \frac{q}{2} \ln \left(1 - \frac{1}{q}\right) - \frac{q(q-2)}{4(q-1)} \ln \left(1 - \frac{2}{q}\right).$$

Compare this result with the one following from the *Bragg–Williams approximation.*

12.12. Using the *approximate* expression, see Fowler and Guggenheim (1940),

$$g_N(N_1, N_{12}) \simeq \frac{(\frac{1}{2}qN)!}{N_{11}!\, N_{22}!\, [(\frac{1}{2}N_{12})!]^2} \left(\frac{N_1!\, N_2!}{N!}\right)^{q-1},$$

for evaluating the partition function of the Ising lattice, show that one is led to the same results as the ones following from the Bethe approximation.

[Note that, for $q = 2$, the quantity $\ln g$ is *asymptotically* exact; cf. eqns. (12.9.3) and (12.9.4). No wonder, the Bethe approximation gives exact results in the case of a one-dimensional chain.]

12.13. Derive eqn. (12.9.22) by substituting the expressions (12.9.16) and (12.9.17) for $\bar{N}_+$ and $\bar{N}_{+-}$ in the formula (12.9.6) for the free energy of an Ising chain.

12.14. According to Onsager, the field-free partition function of a *rectangular* lattice (with different interaction constants, J and J', in the two perpendicular directions) is given by

$$\frac{1}{N} \ln Q(0, T) = \ln 2 + \frac{1}{2\pi^2} \int_0^{\pi} \int_0^{\pi} \ln \{\cosh (2\gamma) \cosh (2\gamma') - \sinh (2\gamma) \cos w - \sinh (2\gamma') \cos w'\}\, dw\, dw',$$

where $\gamma = J/kT$ and $\gamma' = J'/kT$. Show that if $J' = 0$, this leads to the expression (12.9.15) for the linear chain, while if $J' = J$, we are led to the expression (12.10.7) for the square net. Locate the *Curie point* of the rectangular lattice and study its thermodynamic behavior in the neighborhood of this point.

12.15. Prove *thermodynamically* the inequality (Rushbrooke, 1963)

$$(\alpha' + 2\beta + \gamma') \geqslant 2,$$

and check the extent to which the values of α', β and γ' tabulated in Sec. 12.11 conform to this inequality.

CHAPTER 13

FLUCTUATIONS

IN THIS course we have been mostly concerned with the evaluation of *statistical averages* of the various physical quantities; these averages represent, with a high degree of accuracy, the results expected to be obtained from relevant measurements on the given system *in equilibrium*. Nevertheless, there do occur *deviations* from, or *fluctuations* about, these mean values. Though they are generally small, a study of these fluctuations is of great physical interest for several reasons.

Firstly, it enables us to develop a mathematical scheme with the help of which the magnitude of the relevant fluctuations, under a variety of physical situations, can be estimated. We find that while in a single-phase system the fluctuations are thermodynamically negligible they can assume considerable importance in multi-phase systems, especially in the neighborhood of the critical points. In the latter case we obtain a rather high degree of *spatial correlation* among the molecules of the system which in turn gives rise to phenomena such as the *critical opalescence.*

Secondly, it provides a natural framework for understanding a class of physical phenomena which come under the common heading of "Brownian motion"; these phenomena relate properties such as the mobility of a fluid system, its coefficient of diffusion, etc., with temperature through the so-called *Einstein's relations.* The mechanism of the Brownian motion is vital in formulating, and in a certain sense solving, problems as to how "a given physical system, which is not in a state of equilibrium, finally approaches a state of equilibrium", while "a physical system, which is already in a state of equilibrium, persists to be in that state".

Thirdly, the study of fluctuations, as a function of time, leads to the concept of "correlation functions" which play an important role in relating the dissipative properties of a system, such as the viscous resistance of a fluid or the electrical resistance of a conductor, with the microscopic properties of the system in a state of equilibrium; this relationship (between irreversible processes on one hand and equilibrium properties on the other) manifests itself in the so-called *fluctuation-dissipation theorem.* At the same time, a study of the "frequency spectrum" of fluctuations, which is related to the time-dependent correlation function through the fundamental theorem of Wiener and Khintchine, is of considerable value in assessing the "noise" met with in electrical circuits as well as in the transmission of electromagnetic signals.

13.1. Thermodynamic fluctuations

We begin by deriving a *probability distribution law* for the fluctuations of the basic thermodynamic quantities pertaining to a given physical system; the *mean square fluctuations* can be evaluated, in a straightforward manner, with the help of this law. We assume that the given physical system, which may be referred to as 1, is embedded in a reservoir, which may be referred to as 2, such that a mutual exchange of energy, and of volume, can take place between 1 and 2; of course, the total energy E and the total volume V are supposed to remain fixed. For convenience, we do not envisage an exchange of particles; so the numbers N_1 and N_2 remain individually constant. The equilibrium division of E into $\bar{E}_1$ and $\bar{E}_2$, and of V into $\bar{V}_1$ and $\bar{V}_2$, must be such that the parts 1 and 2 of the composite system $(1+2)$ have a *common* temperature T^* and a *common* pressure P^*; see Secs. 1.2 and 1.3, especially eqn. (1.3.7). Of course, the entropy of the composite system will have its largest value in the equilibrium state; in any other state, such as the one characterized by a fluctuation, it must have a lower value. Let ΔS denote the deviation, in the actual value S of the entropy, from its equilibrium value S_0. Then

$$\Delta S = S - S_0 = k \ln \Omega_f - k \ln \Omega_0, \tag{1}$$

where Ω_f (or Ω_0) denotes the number of distinct, microscopic complexions accessible to the system $(1+2)$ in the presence (or in the absence) of a fluctuation from the equilibrium state; see eqn. (1.2.11). The probability that the proposed fluctuation may occur in the system is then given by

$$p \propto \Omega_f \propto \exp(\Delta S/k); \tag{2}$$

see Sec. 3.1, especially eqn. (3.1.3). In terms of other thermodynamic quantities, we may write

$$\Delta S = \Delta S_1 + \Delta S_2 = \Delta S_1 + \int_0^f \frac{dE_2 + P_2\, dV_2}{T_2}; \tag{3}$$

we note that the pressure P_2 and the temperature T_2 of the reservoir may, in principle, vary during the build-up of the fluctuation. Now, even if the fluctuations are sizable from the point of view of the system 1 they will be small from the point of view of 2. The "variables" P_2 and T_2 may, therefore, be replaced by the constant values P^* and T^*, respectively; at the same time, the increments dE_2 and dV_2 may be replaced by $-dE_1$ and $-dV_1$, respectively. Equation (3) then becomes

$$\Delta S = \Delta S_1 - (\Delta E_1 + P^*\, \Delta V_1)/T^*; \tag{4}$$

accordingly, formula (2) takes the form

$$p \propto \exp\{-(\Delta E_1 - T^*\, \Delta S_1 + P^*\, \Delta V_1)/kT^*\}. \tag{5}$$

In this form, the probability distribution law does not depend, *in any manner*, on the peculiarities of the reservoir in which the given system was supposed to be embedded. Formula (5), therefore, applies equally well to a system that "attained" equilibrium in a statistical ensemble (or, for that matter, to any macroscopic part of a given system). Consequently, we may drop the suffix 1 from the symbols ΔE_1, ΔS_1 and ΔV_1, and the star from the symbols P^* and T^*, and write

$$p \propto \exp\{-(\Delta E - T\,\Delta S + P\,\Delta V)/kT\}. \tag{6}$$

In most cases, the fluctuations are exceedingly small in magnitude; the quantity ΔE may therefore be expanded as a Taylor series about the equilibrium value $(\Delta E)_0 = 0$, with the result

$$\begin{aligned}\Delta E = {} & \left(\frac{\partial E}{\partial S}\right)_0 \Delta S + \left(\frac{\partial E}{\partial V}\right)_0 \Delta V \\ & + \frac{1}{2}\left[\left(\frac{\partial^2 E}{\partial S^2}\right)_0 (\Delta S)^2 + 2\left(\frac{\partial^2 E}{\partial S\,\partial V}\right)_0 \Delta S\,\Delta V + \left(\frac{\partial^2 E}{\partial V^2}\right)_0 (\Delta V)^2\right] + \dots .\end{aligned} \tag{7}$$

Substituting (7) into (6) and retaining terms up to second order only, we obtain

$$P \propto \exp\{-(\Delta T\,\Delta S - \Delta P\,\Delta V)/2kT\}; \tag{8}$$

here, use has been made of the relations

$$\left(\frac{\partial E}{\partial S}\right)_0 = T \quad \text{and} \quad \left(\frac{\partial E}{\partial V}\right)_0 = -P, \tag{9}$$

and of the fact that the expression within the long brackets in (7) is equivalent to

$$\Delta\left(\frac{\partial E}{\partial S}\right)_0 \Delta S + \Delta\left(\frac{\partial E}{\partial V}\right)_0 \Delta V = \Delta T\,\Delta S - \Delta P\,\Delta V. \tag{10}$$

With the help of (8), the mean square fluctuations of various physical quantities, and the statistical correlations among various fluctuations, can be readily calculated. We must, however, note that of the four Δ's appearing in this formula only two can be chosen independently; the remaining two must assume the role of "derived quantities". For instance, if we choose ΔT and ΔV as the *independent variables*, then ΔS and ΔP can be written as

$$\Delta S = \left(\frac{\partial S}{\partial T}\right)_V \Delta T + \left(\frac{\partial S}{\partial V}\right)_T \Delta V = \frac{C_V}{T}\Delta T + \left(\frac{\partial P}{\partial T}\right)_V \Delta V \tag{11}$$

and

$$\Delta P = \left(\frac{\partial P}{\partial T}\right)_V \Delta T + \left(\frac{\partial P}{\partial V}\right)_T \Delta V = \left(\frac{\partial P}{\partial T}\right)_V \Delta T - \frac{1}{\varkappa_T V}\Delta V, \tag{12}$$

where $\varkappa_T$ denotes the *isothermal compressibility* of the system. Substituting (11) and (12) into (8), we obtain

$$p \propto \exp\left\{-\frac{C_V}{2kT^2}(\Delta T)^2 - \frac{1}{2kT\varkappa_T V}(\Delta V)^2\right\}, \tag{13}$$

which shows that the fluctuations in T and V are *statistically independent, Gaussian variables*. A glance at (13) yields the results:

$$\overline{(\Delta T)^2} = \frac{kT^2}{C_V}, \quad \overline{(\Delta V)^2} = kT\varkappa_T V, \tag{14a}$$

and

$$\overline{(\Delta T\,\Delta V)} = 0. \tag{14b}$$

Similarly, if we choose the fluctuations ΔS and ΔP as our *independent variables*, we are led to the distribution law

$$p \propto \exp\left\{-\frac{1}{2kC_P}(\Delta S)^2 - \frac{\varkappa_S V}{2kT}(\Delta P)^2\right\}, \tag{15}$$

whence follow the results

$$\overline{(\Delta S)^2} = kC_P, \quad \overline{(\Delta P)^2} = \frac{kT}{\varkappa_S V}, \tag{16a}$$

and

$$\overline{(\Delta S\,\Delta P)} = 0; \tag{16b}$$

here, $\varkappa_S$ denotes the *adiabatic compressibility* of the system.

We note that, in general, the mean square fluctuation of an extensive quantity is directly proportional to the size of the system while that of an intensive quantity is inversely proportional to the same; in either case, the *relative* magnitude of the *root-mean-square fluctuation* of any quantity is inversely proportional to the square root of the size of the system. Thus, except for situations such as the ones met with in the critical region, normal fluctuations are thermodynamically negligible. This does not mean that the fluctuations are altogether irrelevant to the physical phenomena taking place in system; in fact, as will be seen in the sequel, the very existence of fluctuations at the *microscopic* level is of fundamental importance to several properties of the system displayed at the *macroscopic* level!

With the help of the foregoing results, we can evaluate the mean square fluctuation in the energy of the system. With T and V as independent variables, we have

$$\Delta E = \left(\frac{\partial E}{\partial T}\right)_V \Delta T + \left(\frac{\partial E}{\partial V}\right)_T \Delta V. \tag{17}$$

Squaring and taking averages, keeping in mind eqns. (14), we obtain

$$\begin{aligned}\overline{(\Delta E)^2} &= kT^2C_V + kT\varkappa_T V\left\{\left(\frac{\partial E}{\partial V}\right)_T\right\}^2 \\ &= kT^2C_V + kT\varkappa_T\left(\frac{N^2}{V}\right)\left\{\left(\frac{\partial E}{\partial N}\right)_T\right\}^2.\end{aligned} \tag{18}$$

The results derived in the preceding paragraphs determine the fluctuations of the various physical quantities pertaining to *any macroscopic subsystem* of a given physical system, provided that the number of particles in the subsystem remains fixed. The expression for $\overline{(\Delta V)^2}$ may, therefore, be used to derive an expression for the mean square fluctuation of the variable v (the volume per particle) and the variable n (the particle density) of the subsystem. We readily obtain

$$\overline{(\Delta v)^2} = kT\varkappa_T V/N^2; \quad \overline{(\Delta n)^2} = \frac{1}{v^4}\overline{(\Delta v)^2} = kT\varkappa_T N^2/V^3. \tag{19}$$

We note that the last result obtained here is in complete agreement with eqn. (4.5.10), which was derived on the basis of the grand canonical ensemble. A little reflection will show that this result applies equally well to a subsystem with a fixed volume V and a fluctuating number of particles N. The mean square fluctuation in the value of N is, therefore, given by

$$\overline{(\Delta N)^2} = V^2\overline{(\Delta n)^2} = kT\varkappa_T N^2/V. \tag{20}$$

Substituting (20) into (18), we obtain once again the grand canonical result for $\overline{(\Delta E)^2}$, namely

$$\overline{(\Delta E)^2} = kT^2C_V + \overline{(\Delta N)^2}\{(\partial E/\partial N)_T\}^2; \tag{21}$$

cf. eqn. (4.5.18).

In passing, we note that the first part of the expression (21) denotes the mean square fluctuation in the energy E of a subsystem for which both N and V are fixed, just as we have for a system in the canonical ensemble (N, V, T). Conversely, if we assume the energy to be fixed, then the temperature T of the subsystem will fluctuate and the mean square value of the quantity ΔT will be given by (kT^2C_V) divided by the square of the thermal capacity of the subsystem. The final result will, therefore, be (kT^2/C_V), which is the same as our earlier result in (14a).

13.2. Spatial correlations in a fluid

It is well known that the particles constituting a homogeneous, isotropic system, such as a liquid or a gas, are equally likely to be at any point $\boldsymbol{r}$ in the space available to them. This statement holds for each individual particle, under the condition that the positions of all the other particles in the system are completely *arbitrary*. For instance, if we consider two particles at a time, then, for a *given* position of one particle, different positions of the other particle may no longer be equally likely to obtain! In fact, because of the interparticle interactions and the symmetry properties of the wave functions, different values of the relative position $(\boldsymbol{r}_2 - \boldsymbol{r}_1)$ of any two particles in the system do not appear with equal likelihood. In other words, there exists, in general, a definite amount of *correlation* between the simultaneous positions $\boldsymbol{r}_1$ and $\boldsymbol{r}_2$ of the two particles. Interestingly enough, there exists a simple and straightforward relationship between the space integral of the relevant correlation function on one hand and the mean square fluctuation in the density of the system on the other.

To study the nature of these correlations, we introduce a few definitions. First of all, we have the *configurational distribution function* $F_N(\boldsymbol{r}_1, \ldots, \boldsymbol{r}_N)$ which appears in the nature of a *probability density* and satisfies the normalization condition

$$\int_V F_N(\boldsymbol{r}_1, \ldots, \boldsymbol{r}_N)\, d^3r_1 \ldots d^3r_N = 1. \tag{1}$$

Integrating $F_N(\boldsymbol{r}_1, \ldots, \boldsymbol{r}_N)$ over the coordinates $\boldsymbol{r}_2, \ldots, \boldsymbol{r}_N$ and multiplying the result by N, we obtain a *single-particle distribution function*

$$F_1(\boldsymbol{r}_1) = N\int_V F_N(\boldsymbol{r}_1, \ldots, \boldsymbol{r}_N)\, d^3r_2 \ldots d^3r_N = n(\boldsymbol{r}_1). \tag{2}$$

The function $F_1(\boldsymbol{r}_1)$ represents the particle density at the point $\boldsymbol{r}_1$, for $\int_V F_1(\boldsymbol{r}_1)\, d^3r_1 = N$; for a homogeneous system, $F_1(\boldsymbol{r}_1)$ is a constant (which may be denoted by the symbol n). The *two-particle distribution function* is now defined as

$$F_2(\boldsymbol{r}_1, \boldsymbol{r}_2) = N(N-1) \int_V F_N(\boldsymbol{r}_1, \ldots, \boldsymbol{r}_N)\, d^3r_3 \ldots d^3r_N = n^2 g(\boldsymbol{r}), \tag{3}$$

where $\boldsymbol{r} = (\boldsymbol{r}_2 - \boldsymbol{r}_1)$. Equation (3) then defines the *pair distribution function* $g(\boldsymbol{r})$ of the system; clearly, the product $g(\boldsymbol{r})\, d^3r$ determines the probability of finding a particle in the volume element d^3r around the point $\boldsymbol{r}$ when we already have a particle at the point $\boldsymbol{r} = 0$. In the absence of spatial correlations, which holds for a *classical* gas composed of *noninteracting* particles, the function $g(\boldsymbol{r})$ is identically equal to unity; for real systems, $g(\boldsymbol{r})$ is generally different from unity. It is then natural to introduce a function $\nu(\boldsymbol{r})$, defined by the formula

$$\nu(\boldsymbol{r}) = g(\boldsymbol{r}) - 1, \tag{4}$$

as a measure of the degree of spatial correlation in the system; the function $\nu(\boldsymbol{r})$ is generally referred to as the *pair correlation function*. In the absence of spatial correlations, $\nu(\boldsymbol{r})$ would be identically equal to zero.*

We now consider a *macroscopic* region V_A in the fluid and evaluate the mean square fluctuation in the number N_A of the particles occupying this region. For this, we introduce a mathematical function $\mu(\boldsymbol{r})$, such that $\mu(\boldsymbol{r}) = 1$ (or 0) according as the point $\boldsymbol{r}$ lies inside (or outside) the region V_A. The number N_A, which is indeed fluctuating, is then given by the formula

$$N_A = \sum_{i=1}^{N} \mu(\boldsymbol{r}_i), \tag{5}$$

where summation goes over *all* the N particles of the system. It then follows that

$$\begin{aligned} \bar{N}_A &= \sum_{i=1}^{N} \int_V \mu(\boldsymbol{r}_i)\, F_N(\boldsymbol{r}_1, \ldots, \boldsymbol{r}_N)\, d^{3N}r \\ &= N \int_V \mu(\boldsymbol{r}_1)\, d^3r_1\, F_N(\boldsymbol{r}_1, \ldots, \boldsymbol{r}_N)\, d^{3(N-1)}r \\ &= \int_V \mu(\boldsymbol{r}_1)\, F_1(\boldsymbol{r}_1)\, d^3r_1 = \int_{V_A} F_1(\boldsymbol{r})\, d^3r. \end{aligned} \tag{6}$$

* Some of the statements made here are true only for $\boldsymbol{r} \neq 0$; for $\boldsymbol{r} = 0$, they have to be modified. This is due to the fact that "the possibility of finding a particle at the point $\boldsymbol{r} \to 0$, when we already know that there is a particle at the point $\boldsymbol{r} = 0$, tends to be rather a certainty". This *singular* situation may be taken care of by including in the expression for $g(\boldsymbol{r})$, and hence in the expression for $\nu(\boldsymbol{r})$, a term with a delta function at $\boldsymbol{r}=0$. Now, eqn. (3) tells us that, since the integral $\int_V F_2(\boldsymbol{r}_1, \boldsymbol{r}_2)\, d^3r_1\, d^3r_2$ must be equal to $N(N-1)$, the integral $\int_V g(\boldsymbol{r})\, d^3r$ must be equal to $V(1-1/N)$. Therefore, in the absence of spatial correlations, we must have

$$g(\boldsymbol{r}) = 1 - n^{-1}\, \delta(\boldsymbol{r})$$

rather than $g(\boldsymbol{r}) = 1$; correspondingly, the correlation function $\nu(\boldsymbol{r})$ must be $-n^{-1}\delta(\boldsymbol{r})$ rather than zero.

It then follows that the statements made in the text become applicable to all values of $\boldsymbol{r}$ (including $\boldsymbol{r}=0$) if we add to the functions $g(\boldsymbol{r})$ and $\nu(\boldsymbol{r})$, as defined by eqns. (3) and (4), the *ad hoc* term $n^{-1}\delta(\boldsymbol{r})$.

Substituting from (2), we obtain for a homogeneous system

$$\bar{N}_A = nV_A. \tag{7}$$

Next, we have

$$\begin{aligned} N_A^2 &= \sum_{i=1}^{N} \mu(\boldsymbol{r}_i) \sum_{j=1}^{N} \mu(\boldsymbol{r}_j) \\ &= \sum_{i \neq j} \mu(\boldsymbol{r}_i)\, \mu(\boldsymbol{r}_j) + \sum_i \mu(\boldsymbol{r}_i); \end{aligned} \tag{8}$$

in writing the last term, we have made use of the fact that, for all i, $\mu(\boldsymbol{r}_i)\,\mu(\boldsymbol{r}_i) \equiv \mu(\boldsymbol{r}_i)$. Accordingly,

$$\begin{aligned} \overline{N_A^2} &= \bar{N}_A + \sum_{i \neq j} \int_V \mu(\boldsymbol{r}_i)\, \mu(\boldsymbol{r}_j)\, F_N(\boldsymbol{r}_1, \ldots, \boldsymbol{r}_N)\, d^{3N}r \\ &= \bar{N}_A + N(N-1) \int_V \mu(\boldsymbol{r}_1)\, \mu(\boldsymbol{r}_2)\, d^3r_1\, d^3r_2\, F_N(\boldsymbol{r}_1, \ldots, \boldsymbol{r}_N)\, d^{3(N-2)}r \\ &= \bar{N}_A + \int_V \mu(\boldsymbol{r}_1)\, \mu(\boldsymbol{r}_2)\, F_2(\boldsymbol{r}_1, \boldsymbol{r}_2)\, d^3r_1\, d^3r_2 \\ &= \bar{N}_A + n^2 V_A \int_{V_A} g(\boldsymbol{r})\, d^3r. \end{aligned} \tag{9}$$

Combining (7) and (9), we obtain the desired result

$$\frac{\overline{N_A^2} - \bar{N}_A^2}{\bar{N}_A} = 1 + n \int_{V_A} \{g(\boldsymbol{r}) - 1\}\, d^3r = 1 + n \int_{V_A} \nu(\boldsymbol{r})\, d^3r. \tag{10}$$

Equation (10) brings out the intimate connection between (i) the density fluctuations and (ii) the spatial correlations in any fluid system.*

Making use of eqn. (13.1.20), as applied to region V_A of the given system, eqn. (10) yields the important relationship:

$$\int \nu(\boldsymbol{r})\, d^3r = \frac{1}{n}\, [(nkT\varkappa_T) - 1]. \tag{11}$$

For a classical ideal gas, or for a real gas at high temperature and low density, $PV = NkT$; therefore

$$\varkappa_T \equiv -\frac{1}{V}\left(\frac{\partial V}{\partial P}\right)_T = \frac{1}{P} = \frac{1}{nkT}. \tag{12}$$

Accordingly, the space integral of the function $\nu(\boldsymbol{r})$ vanishes! This is in keeping with the fact that, in the absence of interparticle interactions and (quantum-mechanical) symmetry effects, there do not exist any spatial correlations among the particles of a given system. On the other hand, in a liquid (at temperatures not too close to the critical point) the compressibility $\varkappa_T$ is negligibly small in comparison with the quantity $(nkT)^{-1}$, so the integral of the function $\nu(\boldsymbol{r})$ turns out to be very close to $-n^{-1}$. This result is connected with the *impenetrability* of the particles, which makes the functions $g(\boldsymbol{r})$ and $\nu(\boldsymbol{r})$ tend to the respective values 0 and -1 as the interparticle distance r tends to a finite value, say σ.

† Compare eqn. (10) with eqns. (2.6.20) and (2.6.21), which connect the pressure P and the internal energy E of a system of interacting particles with the pair distribution function $g(\boldsymbol{r})$. See also Problem 13.6.

At this stage, a mention must be made of the intimate connection that exists between (i) the density fluctuations in a fluid and (ii) the scattering of electromagnetic waves by the fluid. It has been known for a long time that the scattering of electromagnetic waves results entirely from the density fluctuations in a given fluid; in fact, if fluctuations were totally absent, the waves scattered by the various molecules of the fluid would exactly cancel out by interference and no net scattering would result. For light waves, the scattered intensity is directly proportional to the expression (10); for x-rays, on the other hand, it is proportional to the more general expression

$$1+n\int\{g(\boldsymbol{r})-1\}\,\frac{\sin(sr)}{sr}\,d^3r. \tag{13}$$

Here, $\boldsymbol{s}\ (=\boldsymbol{k}-\boldsymbol{k}_0)$ is the vector difference between the scattered and the incident wave vectors; hence, $s=(4\pi\sin\theta)/\lambda$, where λ is the wavelength and θ the angle of scattering.† Denoting (13) by the symbol $i(s)$, we can write

$$s\{i(s)-1\} = 4\pi n\int_0^\infty\{g(\boldsymbol{r})-1\}\sin(sr)r\,dr, \tag{14}$$

whence it follows, by taking the inverse Fourier transform,

$$\{g(\boldsymbol{r})-1\} = \frac{1}{2\pi^2 nr}\int_0^\infty s\{i(s)-1\}\sin(sr)\,ds. \tag{15}$$

This formula enables us to derive the pair distribution function $g(r)$ from the experimental values of the quantity $i(s)$, i.e. from the intensity distribution of the scattered waves over various values of the variable s.

In passing, we note that in the region of phase transitions, especially at the critical points, the isothermal compressibility of a fluid and, with it, the degree of density fluctuations in the fluid become abnormally high. Consequently, the intensity of the scattered waves as well becomes abnormally large, which gives rise to the spectacular phenomenon of *critical opalescence*.‡ Analysis shows (see Landau and Lifshitz, 1958; sec. 116) that under these circumstances the correlation function $\nu(r)$, which is ordinarily *short-ranged* and is roughly given by the formula

$$\nu(r) \propto \frac{kT}{r}\exp(-r/r^*), \tag{16}$$

r^* being of the order of mean interparticle distance, becomes *long-ranged* and is roughly given by the formula

$$\nu(r) \propto \frac{kT}{r}. \tag{17}$$

That spatial correlations among the molecules of a fluid should extend over *macroscopic* distances is a feature that typifies the presence of a spatial order in the system.

† We note that for light waves, since λ is relatively large, for *all* values of r for which the pair distribution function $g(\boldsymbol{r})$ is significantly different from unity, the product $(sr) \ll 1$; consequently, $\sin(sr)/(sr) \simeq 1$. Expression (13) then reduces to the right-hand side of eqn. (10).

‡ For a discussion of the *critical fluctuations*, see Klein and Tisza (1949).

13.3. Einstein–Smoluchowski theory of the Brownian motion

The term "Brownian motion" derives its name from the botanist Robert Brown who, in 1828, made careful observations on the tiny pollen grains of a plant under a microscope. In his own words: "While examining the form of the particles immersed in water, I observed many of them very evidently in motion. These motions were such as to satisfy me ... that they arose neither from currents in the fluid nor from its gradual evaporation, but belonged to the particle itself." We now know that the real source of this motion lies in the incessant, and more or less random, bombardment of the *Brownian particles*, as these grains (or, for that matter, any colloidal suspensions) are usually referred to, by the *molecules* of the surrounding fluid. It was Einstein who, in a number of classic papers (beginning 1905), first provided a sound theoretical analysis of the Brownian motion on the basis of the so-called "random walk problem" and thereby established a far-reaching relationship between (i) the irreversible nature of this phenomenon (as well as of the phenomena of diffusion and viscosity) and (ii) the mechanism of molecular fluctuations.

To illustrate the essential theme of Einstein's approach, let us consider the problem in *one* dimension. Let $x(t)$ denote the position of the Brownian particle at time t, given that its position coincided with the point $x = 0$ at time $t = 0$. To simplify matters, we may assume that each molecular impact (which, on an average, takes place after a time τ^*) causes the particle to jump a (small) distance l—of *constant* magnitude—in either the positive or negative direction of the x-axis. It seems natural to regard the possibilities $\Delta x = +l$ and $\Delta x = -l$ as *equally likely*; though somewhat less natural, we may also regard the successive impacts on, and hence the successive jumps of, the Brownian particle to be *mutually uncorrelated*. The probability that the particle is found at the point x at time t is now equal to the probability that, in a series of $n(= t/\tau^*)$ successive jumps, the particle makes $m\,(=x/l)$ more jumps in the positive direction of the axis than in the negative direction, i.e. it makes $\frac{1}{2}(n+m)$ jumps in the positive direction and $\frac{1}{2}(n-m)$ jumps in the negative direction.† The desired probability is given by the binomial expression

$$p_n(m) = \left(\frac{1}{2}\right)^n \frac{n!}{\{\frac{1}{2}(n+m)\}!\{\frac{1}{2}(n-m)\}!}, \tag{1}$$

whence it follows that

$$\overline{m} = 0 \quad \text{and} \quad \overline{m^2} = n. \tag{2}$$

Thus, for $t \gg \tau^*$, we have for the net displacement of the particle

$$\overline{x(t)} = 0 \quad \text{and} \quad \overline{x^2(t)} = l^2 \frac{t}{\tau^*} \propto t^1. \tag{3}$$

Accordingly, the root-mean-square displacement of the particle is proportional to the *square root* of the time elapsed:

$$x_{\text{r.m.s.}} \equiv \sqrt{(\overline{x^2(t)})} = l\sqrt{(t/\tau^*)} \propto t^{1/2}. \tag{4}$$

† Since the quantities x and t are *macroscopic* in magnitude while the quantities l and τ^* are *microscopic*, the numbers n and m are much larger than unity; consequently, it is safe to assume that they are *integral* as well.

It should be noticed that the proportionality of the magnitude $x_{\text{r.m.s.}}$ of the *net* displacement of the Brownian particle to the *square root* of the total number of elementary steps is a typical consequence of the random nature of the steps and it manifests itself in a large variety of physical phenomena occurring in nature. In contrast, if the successive steps were fully coherent (or else if the motion were completely mechanical, predictable and reversible over the time interval t),* then the quantity $\overline{x^2}$ would have been proportional to t^2 and, accordingly, $x_{\text{r.m.s.}}$ would have been proportional to t^1.

Smoluchowski's approach to the problem of Brownian motion, which appeared in 1906, was essentially the same as that of Einstein; the difference lay primarily in the mathematical procedure. Smoluchowski introduced the *probability function* $p_n(x_0|x)$ which denotes the "probability that, in a series of n steps, the Brownian particle, initially at the point x_0, reaches the point x"; the number x here denotes the distance in terms of the length of the elementary step. Clearly,

$$p_n(x_0|x) = \sum_{z=-\infty}^{\infty} p_{n-1}(x_0|z)\, p_1(z|x) \qquad (n \geqslant 1); \tag{5}$$

moreover, since a single step is equally likely to take the particle to the right or to the left,

$$p_1(z|x) = \tfrac{1}{2}\delta_{z,\,x-1} + \tfrac{1}{2}\delta_{z,\,x+1}, \tag{6}$$

while

$$p_0(z|x) = \delta_{z,\,x}. \tag{7}$$

Equation (5) is known as the *Smoluchowski equation.* To solve it, we introduce a *generating function* $Q_n(\xi)$:

$$Q_n(\xi) = \sum_{x=-\infty}^{\infty} p_n(x_0|x)\xi^{x-x_0}, \tag{8}$$

whence it follows that

$$Q_0(\xi) = \sum_{x=-\infty}^{\infty} p_0(x_0|x)\xi^{x-x_0} = \sum_{x=-\infty}^{\infty} \delta_{x_0,\,x}\,\xi^{x-x_0} = 1. \tag{9}$$

Substituting (6) into (5), we obtain

$$p_n(x_0|x) = \tfrac{1}{2}p_{n-1}(x_0|x-1) + \tfrac{1}{2}p_{n-1}(x_0|x+1). \tag{10}$$

Multiplying (10) by ξ^{x-x_0} and adding over all values of x, we obtain the recurrence relation

$$Q_n(\xi) = \tfrac{1}{2}[\xi + (1/\xi)]\, Q_{n-1}(\xi), \tag{11}$$

* The term "reversible" is related to the fact that the Newtonian equations of motion, which govern this class of phenomena, preserve their form if the direction of time is reversed (i.e. if we change t to $-t$); alternatively, if at any instant of time we reverse the velocity of all the particles in a given mechanical system, the system would "retrace" its steps exactly. This is not true for the equations describing "irreversible" phenomena, e.g. the *diffusion equation* (19), with which the phenomenon of Brownian motion is intimately connected.

whence it follows that

$$Q_n(\xi) = \left\{\tfrac{1}{2}[\xi+(1/\xi)]\right\}^n Q_0(\xi) = (1/2)^n[\xi+(1/\xi)]^n. \tag{12}$$

Expanding this expression binomially and comparing it with (8), we get†

$$\begin{aligned} p_n(x_0|x) &= \frac{1}{2^n}\frac{n!}{\left\{\frac{1}{2}(n+x-x_0)\right\}!\left\{\frac{1}{2}(n-x+x_0)\right\}!} && \text{for } |x-x_0| \le n \\ &= 0 && \text{for } |x-x_0| > n. \end{aligned} \tag{13}$$

Identifying $(x-x_0)$ with m, we find this result to be in complete agreement with our previous result (1). Therefore, any conclusions drawn from the Smoluchowski approach will be the same as the corresponding conclusions drawn from the Einstein approach.

To obtain an *asymptotic* form of the probability function $p_n(m)$, we apply Stirling's formula, $n! \simeq (2\pi n)^{1/2}(n/e)^n$, to the factorials appearing in the expression (1), with the result

$$\begin{aligned} \ln p_n(m) \simeq{} & (n+\tfrac{1}{2})\ln n - \tfrac{1}{2}(n+m+1)\ln\left\{\tfrac{1}{2}n[1+(m/n)]\right\} \\ & -\tfrac{1}{2}(n-m+1)\ln\left\{\tfrac{1}{2}n[1-(m/n)]\right\} - n\ln 2 - \tfrac{1}{2}\ln(2\pi). \end{aligned}$$

For $m \ll n$ (which is generally true because $m_{\text{r.m.s.}} = n^{1/2}$ while $n \gg 1$), we obtain

$$p_n(m) \simeq \frac{2}{\sqrt{(2\pi n)}}\exp(-m^2/2n), \tag{14}$$

which is consistent with the mean values (2). Taking x to be a continuous variable (and remembering that $p_n(m) \equiv 0$ either for even values of m or for odd values of m, i.e. in the distribution (14), $\Delta m = 2$ and not 1), we can write this result in the *Gaussian* form:

$$p(x)\,dx = \frac{dx}{\sqrt{(4\pi Dt)}}\exp\left(-\frac{x^2}{4Dt}\right), \tag{15}$$

where

$$D = \frac{1}{2}\left(\frac{x}{m}\right)^2\Big/\left(\frac{t}{n}\right) = \frac{1}{2}(l^2/\tau^*). \tag{16}$$

Later on we shall discover that the quantity D introduced here is identical with the (macroscopic) *diffusion coefficient* of the given system; eqn. (16) connects it with the (microscopic) quantities l and τ^*. To appreciate this connection, one has simply to note that the problem of Brownian motion can also be looked upon as a problem of "diffusion" of the Brownian particles through the medium of the fluid; this point of view is again due to Einstein. However, before embarking upon these considerations, we would like to present here the results of an actual observation made on the Brownian motion of a spherical particle; see Lee, Sears and Turcotte (1963). It was found that the 403 values of the net displacement Δx of the particle, observed after successive intervals of 2 seconds each, were distributed as follows:

† It is easy to recognize the additional result that if n is even, then $p_n(m) \equiv 0$ for odd m, and if n is odd, then $p_n(m) \equiv 0$ for even m.

Displacement Δx, in units of μ (= 10^{-4} cm)	*Frequency of occurrence n*
less than -5.5	0
between -5.5 and -4.5	1
between -4.5 and -3.5	2
between -3.5 and -2.5	15
between -2.5 and -1.5	32
between -1.5 and -0.5	95
between -0.5 and $+0.5$	111
between $+0.5$ and $+1.5$	87
between $+1.5$ and $+2.5$	47
between $+2.5$ and $+3.5$	8
between $+3.5$ and $+4.5$	5
between $+4.5$ and $+5.5$	0
greater than $+5.5$	0

The mean square value of the displacement turns out to be: $\overline{(\Delta x)^2} = 2.09\times 10^{-8}\ \text{cm}^2$. The observed frequency distribution has been plotted as a "block diagram" in Fig. 13.1.

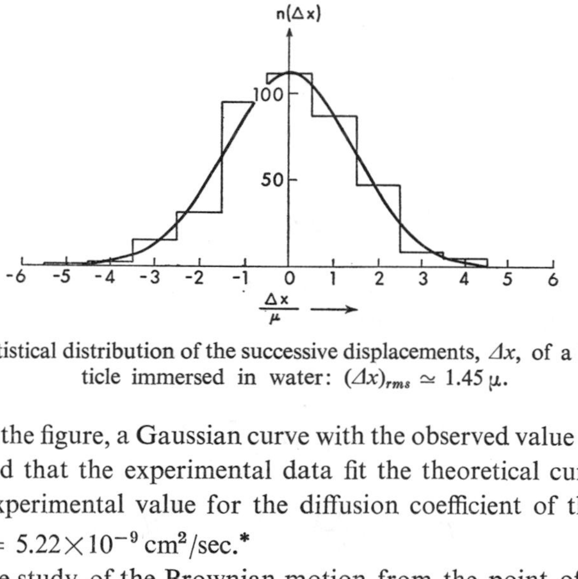

FIG. 13.1. The statistical distribution of the successive displacements, Δx, of a Brownian particle immersed in water: $(\Delta x)_{rms} \simeq 1.45\,\mu$.

We have included, in the figure, a Gaussian curve with the observed value of the mean square displacement; we find that the experimental data fit the theoretical curve fairly well. We can also derive an experimental value for the diffusion coefficient of the system; we obtain: $D = \overline{(\Delta x)^2}/2t = 5.22\times 10^{-9}\ \text{cm}^2/\text{sec}$.*

We now turn to the study of the Brownian motion from the point of view of diffusion. We denote the number density of the Brownian particles in the fluid by the symbol $n(\boldsymbol{r}, t)$ and their current density by the symbol $\boldsymbol{j}(\boldsymbol{r}, t)\,\{= n(\boldsymbol{r}, t)\,\boldsymbol{v}(\boldsymbol{r}, t)\}$; then, according to *Fick's law*,

$$\boldsymbol{j}(\boldsymbol{r}, t) = -D\,\nabla n(\boldsymbol{r}, t), \tag{17}$$

where D stands for the *diffusion coefficient* of the system. We also have the equation of continuity, viz.

$$\nabla\cdot\boldsymbol{j}(\boldsymbol{r}, t) + \frac{\partial n(\boldsymbol{r}, t)}{\partial t} = 0. \tag{18}$$

* In the next section, we shall see that, for a *spherical* particle, $D = kT/6\pi\eta a$, where η is the coefficient of viscosity of the medium and a the radius of the Brownian particle. In the case under discussion, $T \simeq 300°$K, $\eta \simeq 10^{-2}$ poise and $a \simeq 4\times 10^{-5}$ cm. Substituting these values, we obtain for the Boltzmann constant: $k \simeq 1.3\times 10^{-16}$ erg/°K.

Substituting (17) into (18), we obtain the *diffusion equation*:

$$\nabla^2 n(\boldsymbol{r}, t) - \frac{1}{D}\frac{\partial n(\boldsymbol{r}, t)}{\partial t} = 0. \tag{19}$$

Of the various possible solutions of this equation, the one relevant to the present problem is

$$n(\boldsymbol{r}, t) = \frac{N}{(4\pi D t)^{3/2}} \exp\left(-\frac{r^2}{4Dt}\right). \tag{20}$$

This is a spherically symmetric solution and is already normalized:

$$\int_0^\infty n(\boldsymbol{r}, t) 4\pi r^2\, dr = N, \tag{21}$$

N being the total number of (Brownian) particles immersed in the fluid. A comparison of the (three-dimensional) formula (20) with the (one-dimensional) formula (15) brings out most vividly the relationship between the random walk problem on one hand and the phenomenon of diffusion on the other.

It is clear that in the last approach we have considered the motion of an "ensemble" of N Brownian particles placed under "equivalent" physical conditions, rather than considering the motion of a single particle over a length of time (as was done in the random walk approach). Accordingly, the averages of the various physical quantities obtained here will be in the nature of "ensemble averages"; they must, of course, agree with the averages of the same quantities obtained earlier. Now, by virtue of the distribution function (20), we obtain

$$\langle \boldsymbol{r}(t) \rangle = 0; \qquad \langle r^2(t) \rangle = \frac{4\pi}{N} \int_0^\infty n(\boldsymbol{r}, t) r^4\, dr = 6Dt \propto t^1, \tag{22}$$

in complete agreement with our earlier results, namely

$$\overline{x(t)} = 0; \qquad \overline{x^2(t)} = l^2 t/\tau^* = 2Dt \propto t^1. \tag{23}$$

Thus, the "ensemble" of the Brownian particles, initially concentrated at the origin, "diffuses out" as time increases, the nature and the extent of its spread at any time t being given by eqns. (20) and (22), respectively. The diffusion process, which is clearly *irreversible*, gives us a fairly good picture of the statistical behavior of a single particle in the ensemble. However, the important thing to bear in mind is that, whether we focus our attention on a single particle in the ensemble or look at the ensemble as a whole, the ultimate source of the phenomenon lies in the incessant, and more or less random, impacts received by the Brownian particles from the molecules of the fluid. In other words, the irreversible character of the phenomenon ultimately arises from the random, fluctuating forces exerted by the fluid molecules on the Brownian particles. This then leads to another systematic and promising theory of the Brownian motion, viz. the theory of Langevin (1908). For a detailed analysis of the problem, see Uhlenbeck and Ornstein (1930), Chandrasekhar (1943, 1949), MacDonald (1948–9) and Wax (1954).

13.4. Langevin theory of the Brownian motion

We consider the simplest case of a "free" Brownian particle, surrounded by a fluid environment: the particle is assumed to be free in the sense that it is not acted upon by any other force except the one arising from the molecular bombardment. The equation of motion of the particle will be

$$M\frac{d\boldsymbol{v}}{dt} = \mathfrak{F}(t), \tag{1}$$

where M is the particle mass, $\boldsymbol{v}(t)$ the particle velocity and $\mathfrak{F}(t)$ the force acting upon the particle by virtue of the incessant impacts from the fluid molecules. Langevin suggested that the force $\mathfrak{F}(t)$ may be written as a sum of two parts: (i) an "averaged-out" part, which represents the *viscous drag*, $-\boldsymbol{v}/B$, experienced by the particle (accordingly, B is the *mobility* of the system, i.e. the drift velocity acquired by the particle by virtue of a unit "external" force)† and (ii) a "rapidly fluctuating" part $\boldsymbol{F}(t)$ which, over long intervals of time (as compared with the characteristic time τ^*), averages out to zero; thus, we may write

$$M\frac{d\boldsymbol{v}}{dt} = -\frac{\boldsymbol{v}}{B}+\boldsymbol{F}(t); \qquad \overline{\boldsymbol{F}(t)} = 0. \tag{2}$$

Taking the ensemble average of (2), we obtain‡

$$M\frac{d}{dt}\langle\boldsymbol{v}\rangle = -\frac{1}{B}\langle\boldsymbol{v}\rangle, \tag{3}$$

whence it follows that

$$\langle\boldsymbol{v}(t)\rangle = \boldsymbol{v}(0)\exp(-t/\tau); \qquad (\tau = MB). \tag{4}$$

Thus, the drift velocity of the particle decays, at a rate determined by the *relaxation time* τ, to the ultimate value zero. We note that this result is typical of the phenomena governed by *dissipative* properties such as the viscosity of the fluid; the *irreversible* nature of the result is also evident.

Dividing eqn. (2) by the mass of the particle, we obtain an equation for the *instantaneous* acceleration, viz.

$$\frac{d\boldsymbol{v}}{dt} = -\frac{\boldsymbol{v}}{\tau}+\boldsymbol{A}(t); \qquad \overline{\boldsymbol{A}(t)} = 0. \tag{5}$$

We construct the scalar product of (5) with the *instantaneous* position $\boldsymbol{r}$ of the particle and take the ensemble average of the product. In doing so, we make use of the facts that (i) $\boldsymbol{r}\cdot\boldsymbol{v} = \frac{1}{2}(dr^2/dt)$, (ii) $\boldsymbol{r}\cdot(d\boldsymbol{v}/dt) = \frac{1}{2}(d^2r^2/dt^2)-v^2$, and (iii) $\langle\boldsymbol{r}\cdot\boldsymbol{A}\rangle = 0$.§ We obtain

$$\frac{d^2}{dt^2}\langle r^2\rangle+\frac{1}{\tau}\frac{d}{dt}\langle r^2\rangle = 2\langle v^2\rangle. \tag{6}$$

† If Stokes' law is applicable, then $B = 1/(6\pi\eta a)$, where η is the coefficient of viscosity of the fluid and a the radius of the particle (assumed spherical).

‡ The process of "averaging over an ensemble" implies that we are imagining a large number of systems similar to the one originally under consideration and are taking an average over this collection at *any* instant of time t. By the very nature of the function $\boldsymbol{F}(t)$, the ensemble average $\langle\boldsymbol{F}(t)\rangle$ must be zero at all times!

§ This is so because we have no reason here to expect a statistical correlation between the position $\boldsymbol{r}(t)$ of the Brownian particle and the force $\boldsymbol{F}(t)$ exerted on it by the molecules of the fluid. Of course, we do expect a correlation between the variables $\boldsymbol{v}(t)$ and $\boldsymbol{F}(t)$; consequently, $\langle\boldsymbol{v}\cdot\boldsymbol{F}\rangle \neq 0$ (see Problem 13.10).

If the Brownian particle has already attained *thermal equilibrium* with the molecules of the fluid, the quantity $\langle v^2 \rangle$ in this equation may be replaced by its *equipartition value* $3kT/M$. The equation is then readily integrated, with the result

$$\langle r^2 \rangle = \frac{6kT\tau^2}{M}\left\{\frac{t}{\tau}-(1-e^{-t/\tau})\right\}, \tag{7}$$

where the constants of integration have been so chosen that at $t = 0$ both $\langle r^2 \rangle$ and its first time-derivative vanish. We observe that, for $t \ll \tau$,

$$\langle r^2 \rangle \simeq \frac{3kT}{M}t^2 = \langle v^2 \rangle t^2, \tag{8}^*$$

which is consistent with the *reversible* equations of motion (of Newton), whereby one should simply have

$$\boldsymbol{r} = \boldsymbol{v}t. \tag{9}$$

On the other hand, if $t \gg \tau$, then

$$\langle r^2 \rangle \simeq \frac{6kT\tau}{M}t = (6BkT)t, \tag{10}^\dagger$$

which is essentially the same as the Einstein–Smoluchowski result (13.3.22); incidentally, we obtain here a simple, but important, relationship between the coefficient of diffusion D and the mobility B, viz.

$$D = BkT, \tag{11}$$

which is generally referred to as the *Einstein relation*. The irreversible character of this result is self-evident; it is also clear that it arises essentially from the viscosity of the system. Moreover, the Einstein relation (11), which connects the coefficient of diffusion D with the mobility B of the system, tells us that the ultimate source of the viscosity of the medium (as well as that of diffusion) lies in the random, fluctuating forces arising from the incessant motion of the fluid molecules.

In this context, if we consider a particle of charge e and mass M moving in a viscous fluid under the influence of an external electric field of intensity $\boldsymbol{E}$, then the "coarse-grained" motion of the particle will be determined by the equation

$$M\frac{d}{dt}\langle \boldsymbol{v} \rangle = -\frac{1}{B}\langle \boldsymbol{v} \rangle + e\boldsymbol{E}; \tag{12}$$

cf. eqn. (3). The "terminal" drift velocity of the particle would be given by the expression $(eB)\boldsymbol{E}$, which prompts one to define (eB) as the "mobility" of the system and denote it by the symbol μ. Consequently, one obtains, instead of (11),

$$D = \frac{kT}{e}\mu, \tag{13}$$

* We note that the limiting solution (8) corresponds to "dropping out" the second term on the left-hand side of eqn. (6).

† We note that the limiting solution (10) corresponds to "dropping out" the first term on the left-hand side of eqn. (6).

which is, in fact, the original version of the Einstein relation; sometimes it is also referred to as the *Nernst relation.*

So far we have not felt any *direct* influence of the rapidly fluctuating term $\boldsymbol{A}(t)$ which is present in the equation of motion (5) of the Brownian particle. For this, let us try to evaluate the quantity $\langle v^2(t)\rangle$ which, in the preceding analysis, was assumed to have already attained its "limiting" value, viz. $3kT/M$. For this we replace the variable t in eqn. (5) by u, multiply both sides of the equation by $\exp(u/\tau)$, rearrange and integrate between the limits $u = 0$ and $u = t$; we thus obtain

$$\boldsymbol{v}(t) = \boldsymbol{v}(0)e^{-t/\tau}+e^{-t/\tau}\int_0^t e^{u/\tau}\boldsymbol{A}(u)\,du. \tag{14}$$

Thus, the drift velocity $\boldsymbol{v}(t)$ of the particle is also a fluctuating function of time; of course, since $\langle \boldsymbol{A}(u)\rangle = 0$ for all u, the *average* drift velocity is given by the first term alone, viz.

$$\langle \boldsymbol{v}(t)\rangle = \boldsymbol{v}(0)e^{-t/\tau}, \tag{15}$$

which is the same as our earlier result (4). On the other hand, things are very different if we consider instead the mean square velocity $\langle v^2(t)\rangle$; we obtain from (14)

$$\langle v^2(t)\rangle = v^2(0)e^{-2t/\tau}+2e^{-2t/\tau}\left[\boldsymbol{v}(0)\cdot\int_0^t e^{u/\tau}\langle \boldsymbol{A}(u)\rangle\,du\right]$$
$$+e^{-2t/\tau}\int_0^t\int_0^t e^{(u_1+u_2)/\tau}\langle \boldsymbol{A}(u_1)\cdot\boldsymbol{A}(u_2)\rangle\,du_1\,du_2. \tag{16}$$

The second term on the right-hand side of this equation is identically zero, because $\langle \boldsymbol{A}(u)\rangle$ vanishes for all u. In the third term, we have the important quantity $\langle \boldsymbol{A}(u_1)\cdot\boldsymbol{A}(u_2)\rangle$, which is a measure of the "statistical correlation between the value of the fluctuating variable $\boldsymbol{A}$ at time u_1 and its value at time u_2"; we call it the *autocorrelation function* of the variable $\boldsymbol{A}$ and denote it by the symbol $K_A(u_1, u_2)$ or simply by $K(u_1, u_2)$. Before proceeding with (16) any further, it seems desirable to place on record some of the important properties of the function $K(u_1, u_2)$.

(i) In a stationary ensemble (which means an ensemble whose macroscopic behavior does not, statistically speaking, change with time), the function $K(u_1, u_2)$ depends only on the time interval (u_2-u_1). Denoting this interval by the symbol s, we have

$$K(u_1, u_1+s) \equiv \langle \boldsymbol{A}(u_1)\cdot\boldsymbol{A}(u_1+s)\rangle = K(s), \quad \text{independently of } u_1. \tag{17}$$

(ii) The quantity $K(0)$, which is identically equal to the mean square value of the variable $\boldsymbol{A}$ at time u_1, must be *positive definite*. In a stationary ensemble, it would be a constant, independent of u_1:

$$K(0) = \text{const.} > 0. \tag{18}$$

(iii) For *any* value of s, the magnitude of the function $K(s)$ cannot exceed the value $K(0)$.

Proof. Since

$$\langle |A(u_1) \pm A(u_2)|^2 \rangle = \langle A^2(u_1) \rangle + \langle A^2(u_2) \rangle \pm 2\langle A(u_1) \cdot A(u_2) \rangle$$
$$= 2\{K(0) \pm K(s)\} \geqslant 0,$$

the function $K(s)$ cannot go outside the limits $-K(0)$ and $+K(0)$; consequently, we must have

$$|K(s)| \leqslant K(0) \quad \text{for } all\ s. \tag{19}$$

(iv) The function $K(s)$ is *symmetric* about the value $s = 0$, i.e.

$$K(-s) = K(s) = K(|s|). \tag{20}$$

Proof.

$$K(s) \equiv \langle A(u_1) \cdot A(u_1+s) \rangle = \langle A(u_1-s) \cdot A(u_1) \rangle^{\dagger}$$
$$= \langle A(u_1) \cdot A(u_1-s) \rangle \equiv K(-s).$$

(v) As s becomes large in comparison with the characteristic time τ^*, the values $A(u_1)$ and $A(u_1+s)$ become *uncorrelated*, that is

$$K(s) \equiv \langle A(u_1) \cdot A(u_1+s) \rangle \xrightarrow[s \gg \tau^*]{} \langle A(u_1) \rangle \cdot \langle A(u_1+s) \rangle = 0. \tag{21}$$

In other words, the "memory" of the molecular impacts received during a given interval of time, say between u_1 and u_1+du_1, is "completely lost" after a lapse of time large in comparison with τ^*; accordingly, the magnitude of the function $K(s)$ is significant only when the variable s is of the same order of magnitude as τ^*.

Figures 13.7–13.9 show the s-dependence of certain typical correlation functions $K(s)$; we note that they fully conform to the properties (ii)–(v) which are mathematically represented by eqns. (18)–(21), respectively.

We shall now evaluate the double integral appearing in (16), namely

$$I = \int_0^t \int_0^t e^{(u_1+u_2)/\tau} K(u_2-u_1)\, du_1\, du_2. \tag{22}$$

Changing over to the variables

$$S = \tfrac{1}{2}(u_1+u_2) \quad \text{and} \quad s = (u_2-u_1), \tag{23}$$

the integrand takes the form $\exp(2S/\tau)\, K(s)$, the element $(du_1\, du_2)$ gets replaced by the corresponding element $(dS\, ds)$ while the limits of integration, in terms of the variables S and s, can be read from Fig. 13.2; we find that, for $0 \leqslant S \leqslant t/2$, s goes from $-2S$ to $+2S$ while, for $t/2 \leqslant S \leqslant t$, it goes from $-2(t-S)$ to $+2(t-S)$. Accordingly, we have

$$I = \int_0^{t/2} e^{2S/\tau}\, dS \int_{-2S}^{+2S} K(s)\, ds + \int_{t/2}^{t} e^{2S/\tau}\, dS \int_{-2(t-S)}^{+2(t-S)} K(s)\, ds. \tag{24}$$

† This is the only crucial step in the proof. It involves a "shift", by an amount s, in both the instants of the measurement; the equality results from the fact that the ensemble has been supposed to be stationary!

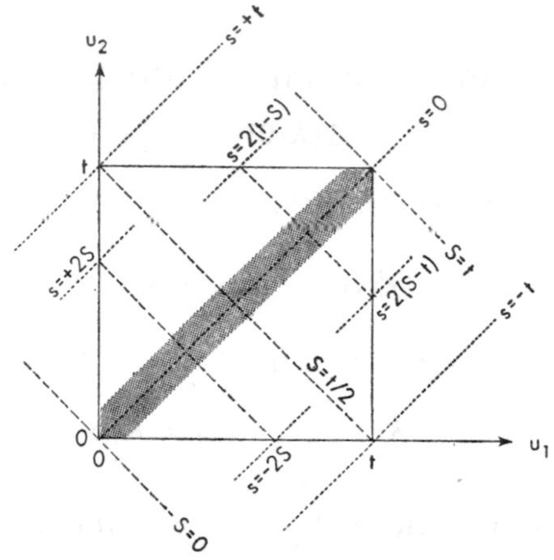

FIG. 13.2. Limits of integration, of the double integral I, in terms of the variables S and s.

In view of the property (v) of the function $K(s)$, see eqn. (21), the integrals over s draw significant contributions only from a very narrow region, of the order of τ^*, around the central value $s = 0$, i.e. from the shaded region in Fig. 13.2; contributions from regions with larger values of $|s|$ are negligible. Therefore, if $t \gg \tau^*$, the limits of integration for s may be replaced by $-\infty$ and $+\infty$, with the result

$$I \simeq C\int_0^t e^{2S/\tau}\,dS = C\frac{\tau}{2}(e^{2t/\tau}-1), \tag{25}$$

where

$$C = \int_{-\infty}^{\infty} K(s)\,ds; \tag{26}$$

clearly, the constant C stands for the "total area under the curve $K(s)$ versus s" and contains considerable information regarding the "molecular dynamics" of the system. Substituting (25) into (16), we obtain

$$\langle v^2(t)\rangle = v^2(0)e^{-2t/\tau}+C\frac{\tau}{2}(1-e^{-2t/\tau}). \tag{27}$$

Now, as $t \to \infty$, $\langle v^2(t)\rangle$ must tend to the equipartition value $3kT/M$; therefore, we must have

$$C = 6kT/M\tau \tag{28}$$

and, hence,

$$\langle v^2(t)\rangle = v^2(0)+\left\{\frac{3kT}{M}-v^2(0)\right\}(1-e^{-2t/\tau}). \tag{29}^\dagger$$

† One can verify that

$$\frac{d}{dt}\langle v^2(t)\rangle = \frac{2}{\tau}[v^2(\infty)-\langle v^2(t)\rangle] = -\frac{2}{\tau}\Delta\langle v^2(t)\rangle,$$

where $v^2(\infty) = 3kT/M$ and $\Delta\langle v^2(t)\rangle$ is the "deviation of the quantity concerned from its equilibrium value". In this form of the equation, we have a typical example of a "relaxation phenomenon", with a *relaxation time* $\tau/2$.

We note that if $v^2(0)$ were itself equal to the equipartition value $3kT/M$, then $\langle v^2(t)\rangle$ would always remain the same, which shows that statistical equilibrium, once attained, has a natural tendency to persist!

Substituting (29) into the right-hand side of (6), we obtain a more representative description of the manner in which the quantity $\langle r^2\rangle$ varies with t; we have

$$\frac{d^2}{dt^2}\langle r^2\rangle + \frac{1}{\tau}\frac{d}{dt}\langle r^2\rangle = 2v^2(0)e^{-2t/\tau} + \frac{6kT}{M}(1-e^{-2t/\tau}), \tag{30}$$

the relevant solution being

$$\langle r^2\rangle = v^2(0)\,\tau^2(1-e^{-t/\tau})^2 - \frac{3kT}{M}\tau^2(1-e^{-t/\tau})(3-e^{-t/\tau}) + \frac{6kT\tau}{M}t. \tag{31}$$

Solution (31) satisfies the initial conditions that both $\langle r^2\rangle$ and its first time-derivative vanish at $t = 0$; moreover, if we put $v^2(0) = 3kT/M$, it reduces to the solution (7) obtained earlier. Once again, we note the *reversible* nature of the motion for $t \ll \tau$, when $\langle r^2\rangle \simeq v^2(0)t^2$, and the *irreversible* nature for $t \gg \tau$, when $\langle r^2\rangle \simeq (6BkT)t$.

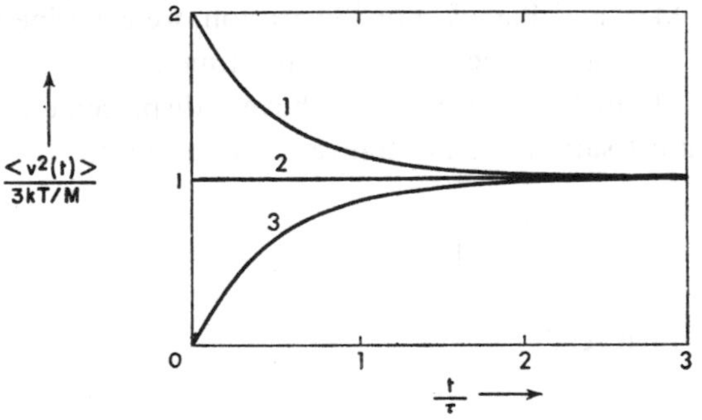

FIG. 13.3. The mean square velocity of a Brownian particle as a function of time. Curves 1, 2 and 3 correspond, respectively, to the initial conditions: $v^2(0) = 6kT/M$, $3kT/M$ and 0.

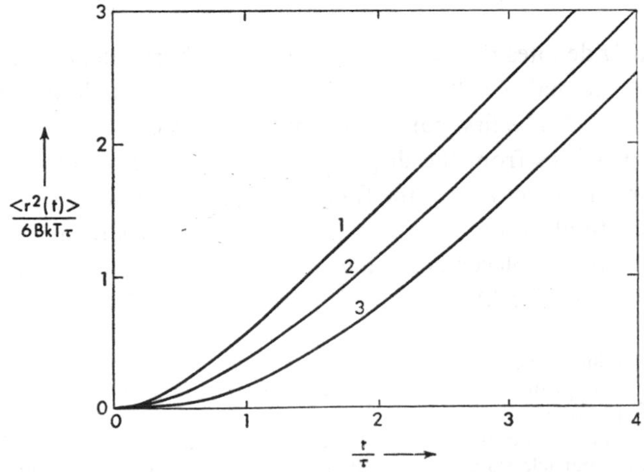

FIG. 13.4. The mean square displacement of a Brownian particle as a function of time. Curves 1, 2 and 3 correspond, respectively, to the initial conditions: $v^2(0) = 6kT/M$, $3kT/M$ and 0.

Figures 13.3 and 13.4 show the variation, with time, of the ensemble averages $\langle v^2(t)\rangle$ and $\langle r^2(t)\rangle$ of a Brownian particle, as given by eqns. (29) and (31), respectively. All the important features of Brownian motion studied so far are manifestly evident in these plots.

13.5. Approach to equilibrium: the Fokker–Planck equation

In our analysis of the Brownian motion we have considered the behavior of a dynamical variable, such as the position $\boldsymbol{r}(t)$ or the velocity $\boldsymbol{v}(t)$ of a Brownian particle, from the point of view of statistical fluctuations in the value of the variable. To determine the average behavior of such a variable, we sometimes invoked an "ensemble" of Brownian particles immersed in identical environments and undergoing diffusion. A treatment along these lines was carried out towards the end of Sec. 13.3, and the most important results of that treatment are summarized in eqn. (13.3.20) for the density function $n(\boldsymbol{r}, t)$ and in eqn. (13.3.22) for the mean square displacement $\langle r^2(t)\rangle$.

A more generalized way of looking at "the manner in which, and the rate at which, a given *arbitrary* distribution of Brownian particles approaches a state of thermal equilibrium" is provided by the so-called *Master Equation*, a simplified version of which goes after the names of Fokker and Planck. For illustration, we examine the displacement, $x(t)$, of the given set of particles along the x-axis. At any time t, let $f(x, t)\,dx$ be the probability that an arbitrary particle in the ensemble may have a displacement between x and $x+dx$. The function $f(x, t)$ must satisfy the normalization condition:

$$\int_{-\infty}^{\infty} f(x, t)\, dx = 1. \tag{1}$$

The Master Equation then reads:

$$\frac{\partial f(x, t)}{\partial t} = \int_{-\infty}^{\infty} \{-f(x, t)\, W(x, x')+f(x', t)\, W(x', x)\}\, dx', \tag{2}$$

where $W(x, x')\, dx'\, \delta t$ denotes the probability that in a short interval of time δt a particle having displacement x makes a "transition" to become a particle having a displacement between x' and $x'+dx'$.* The first part of the integral thus corresponds to all those transitions that remove particles from the displacement x at time t to some other displacement x' and, hence, represent a net *loss* to the function $f(x, t)$; similarly, the second part of the integral corresponds to all those transitions that bring particles from some other displacement x' at time t to the displacement x and, hence, represent a net *gain* to the function $f(x, t)$.† The structure of the Master Equation is, therefore, founded on very simple and

* We are tacitly assuming here a "Markovian" situation where the *transition probability function* $W(x, x')$ depends *only* on the present position x (and, of course, the next position x') of the particle, but *not* on the previous history of the particle.

† In the case of fermions, an account must be taken of the *Pauli exclusion principle* which controls the "occupation of the single-particle states in the system"; for instance, we cannot, in that case, consider a transition that tends to transfer a particle to a state which is already occupied! This calls for an appropriate modification of the Master Equation.

straightforward premises. Of course, under certain approximate conditions, this equation, or any generalization thereof (such as the one including velocity, or momentum, coordinates in the argument of f), can be reduced to the simpler form

$$\frac{\partial f}{\partial t} = -\frac{f-f_0}{\tau}, \tag{3}$$

which has proved to be a very useful first approximation for studying problems related to the so-called *transport phenomena*. Here, f_0 denotes the *equilibrium distribution function* (for $\partial f/\partial t = 0$ when $f = f_0$), while τ denotes the *relaxation time* that characterizes the average rate at which the fluctuations in the system tend to restore a state of equilibrium.

In studying Brownian motion on the basis of eqn. (2), we can safely assume that it is only the transitions between "closely neighboring" states x and x' that have an appreciable probability of occurring; in other words, the transition probability function $W(x, x')$ is sharply peaked around the value $x' = x$ and falls rapidly to zero elsewhere. Denoting the interval $(x'-x)$ by the symbol ξ, we may write

$$W(x, x') \rightarrow W(x; \xi), \tag{4}$$

where $W(x; \xi)$ is a function with a sharp peak around the value $\xi = 0$, falling rapidly to zero elsewhere.* This enables us to expand the right-hand side of eqn. (2) as a Taylor series around the value $\xi = 0$. Retaining terms up to second order, we obtain

$$\frac{\partial f(x, t)}{\partial t} = -\frac{\partial}{\partial x}\{\mu_1(x) f(x, t)\}+\frac{1}{2}\frac{\partial^2}{\partial x^2}\{\mu_2(x) f(x, t)\}, \tag{5}$$

where

$$\mu_1(x) = \int_{-\infty}^{\infty} \xi W(x; \xi)\, d\xi = \frac{\langle \delta x\rangle_{\delta t}}{\delta t} = \langle v_x\rangle, \tag{6}$$

and

$$\mu_2(x) = \int_{-\infty}^{\infty} \xi^2 W(x; \xi)\, d\xi = \frac{\langle (\delta x)^2\rangle_{\delta t}}{\delta t}. \tag{7}$$

Equation (5) is known as the *Fokker–Planck equation* which occupies a classic place in the field of "Brownian motion and fluctuations".

We now consider a specific system of Brownian particles (of *negligible* mass), each particle being acted upon by a linear (elastic) restoring force, $F_x = -\lambda x$, and having a mobility B in the surrounding medium. The mean viscous force, $-\langle v_x\rangle/B$, must then be

* Clearly, this assumption limits our analysis to what may be termed the "Brownian motion approximation", in which the body under consideration is presumed to be on a very different scale of magnitude from the molecules constituting the environment. It is obvious that if one tries to apply the subsequent analysis to "understand" the behavior of molecules *themselves*, one cannot hope for anything more than a "crude, semiquantitative" outcome.

balanced by the linear restoring force; consequently,

$$-\frac{\langle v_x \rangle}{B} + F_x = 0 \tag{8}$$

and, hence,

$$\langle v_x \rangle \equiv \mu_1(x) = -\lambda Bx. \tag{9}$$

Next, in view of eqn. (13.4.10), we have

$$\frac{\langle (\delta x)^2 \rangle}{\delta t} \equiv \mu_2(x) = 2BkT. \tag{10}$$

Substituting (9) and (10) into the Fokker–Planck equation (5), we obtain

$$\frac{\partial f}{\partial t} = \lambda B \frac{\partial}{\partial x}(xf) + BkT \frac{\partial^2 f}{\partial x^2}. \tag{11}$$

Let us apply this equation to an "ensemble" of Brownian particles, initially concentrated at the point $x = x_0$. In this context, we note that, in the absence of the restoring force ($\lambda = 0$), eqn. (11) reduces to the one-dimensional *diffusion equation*

$$\frac{\partial f}{\partial t} = D \frac{\partial^2 f}{\partial x^2} \qquad (D = BkT), \tag{12}$$

which conforms to our earlier results (13.3.19) and (13.4.11). The present derivation shows that the process of diffusion is neither more nor less than a "random walk, at the molecular level". In view of the solution (13.3.20), the function $f(x, t)$ would be

$$f(x, t) = \frac{1}{(4\pi Dt)^{1/2}} \exp\left\{-\frac{(x-x_0)^2}{4Dt}\right\}, \tag{13}$$

with

$$\bar{x} = x_0 \quad \text{and} \quad \overline{x^2} = x_0^2 + 2Dt; \tag{14}$$

the last result implies that the mean square distance traversed by the particle(s) increases *linearly* with time, without any upper limit on its value. A restoring force, on the other hand, puts a "check" on the diffusive tendency of the particles. For instance, in the presence of a restoring force ($\lambda \neq 0$), the *limiting* distribution f_∞ (for which $\partial f/\partial t = 0$) is determined by the equation

$$\frac{\partial}{\partial x}(xf_\infty) + \frac{kT}{\lambda}\frac{\partial^2 f_\infty}{\partial x^2} = 0, \tag{15}$$

whence it follows that

$$f_\infty(x) = \left(\frac{\lambda}{2\pi kT}\right)^{1/2} \exp\left(-\frac{\lambda x^2}{2kT}\right), \tag{16}$$

with

$$\bar{x} = 0 \quad \text{and} \quad \overline{x^2} = kT/\lambda. \tag{17}$$

The last result is in complete agreement with the fact that the mean square value of x

must ultimately comply with the *equipartition theorem*, viz. $\overline{(\frac{1}{2}\lambda x^2)}_\infty = \frac{1}{2}kT$. From the point of view of equilibrium statistical mechanics, if we regard the Brownian particles, with kinetic energy $p_x^2/2m$ and potential energy $\frac{1}{2}\lambda x^2$, as *loosely coupled* to a thermal environment at temperature T, then we can directly write

$$f_{eq}(x, p_x)\, dx\, dp_x \propto e^{-(p_x^2/2m+\lambda x^2/2)/kT}\, dx\, dp_x. \tag{18}$$

On integration over all possible values of p_x, the foregoing expression leads exactly to the distribution function (16).

The general solution of eqn. (11), relevant to the ensemble under consideration, is given by

$$f(x, t) = \left\{\frac{\lambda}{2\pi kT(1-e^{-2\lambda Bt})}\right\}^{1/2} \exp\left\{-\frac{\lambda(x-x_0\, e^{-\lambda Bt})^2}{2kT(1-e^{-2\lambda Bt})}\right\}, \tag{19}$$

with

$$\bar{x} = x_0\, e^{-\lambda Bt} \quad \text{and} \quad \overline{x^2} = x_0^2\, e^{-2\lambda Bt} + \frac{kT}{\lambda}(1-e^{-2\lambda Bt}); \tag{20}$$

in the limit $\lambda \to 0$, we recover the purely "diffusive" situation, as described by eqns. (13) and (14), while for $t \gg (\lambda B)^{-1}$, we approach the "limiting" situation, as described by eqns. (16) and (17). Figure 13.5 shows the manner in which an ensemble of Brownian particles approaches a state of equilibrium under the combined influence of the restoring force and the molecular bombardment.

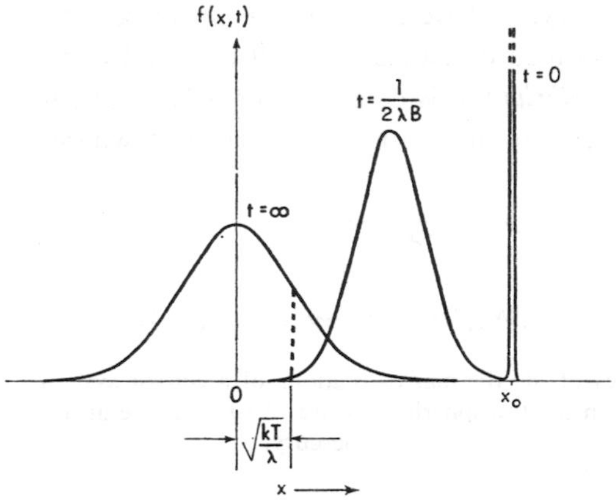

FIG. 13.5. The distribution function (13.5.19) at times $t = 0$, $t = 1/(2\lambda B)$ and $t = \infty$.

A physical system to which the foregoing theory is readily applicable is provided by the oscillating component of a moving-coil galvanometer. Here, we have a coil of wire and a mirror which are suspended by a fine fiber and are capable of rotation about a vertical axis. Random, incessant collisions of air molecules with the suspended system produce a succession of torques of *fluctuating* intensity; as a result, the angular position θ of the system continually fluctuates and the system exhibits an *unsteady* zero. This is indeed another example of the Brownian motion! The role of the viscous force in this case is played by the mechanism of air damping (or, else, electromagnetic damping) of the galvanometer, while the restoring torque, $F_\theta = -c\theta$, arises from the torsional properties of the fiber. In thermal equilibrium, we expect that

$$\overline{\left(\frac{1}{2}c\theta^2\right)} = \frac{1}{2}kT, \quad \text{i.e.} \quad \overline{\theta^2} = \frac{kT}{c}; \tag{21}$$

cf. eqn. (17). An experimental determination of the mean square deflection, $\overline{\theta^2}$, of a galvanometer system was made by Kappler (1931) who, in turn, applied his results to derive, with the help of eqn. (21), an empirical value for the Boltzmann constant k (or, for that matter, the Avogadro number N_A). The system used by Kappler had the moment of inertia $I = 4.552 \times 10^{-4}$ g cm² and the time period of oscillation $\tau = 1379$ sec; accordingly, the constant c of the restoring torque had a value given by the formula: $\tau = 2\pi(I/c)^{1/2}$, i.e.

$$c = 4\pi^2(I/\tau^2) = 9.443 \times 10^{-9} \text{ g cm}^2 \text{ sec}^{-2}/\text{rad}.$$

The observed value of $\overline{\theta^2}$, at a temperature of 287.1°K, was 4.178×10^{-6}. Substituting these numbers in (21), Kappler obtained: $k = 1.374 \times 10^{-16}$ erg °K^{-1}. And, since the gas constant R is equal to 8.31×10^7 erg °K^{-1} mole^{-1}, Kappler obtained for the Avogadro number: $N_A \equiv R/k = 6.06 \times 10^{23}$ mole^{-1}.

One might expect that by suspending the mirror system in an "evacuated" casing the fluctuations caused by the collisions of the air molecules could be *severely* minimized. This is not true because even at the lowest possible pressures there still remain a tremendously large number of molecules in the system which keep the Brownian motion "alive". The interesting part of the story, however, is that the mean square deflection of the system, caused by the molecular bombardment, is not at all affected by the density of the molecules; for a given system *in equilibrium*, it is determined solely by the temperature of the system. The situation is depicted, rather dramatically, in Fig. 13.6 where we have two traces of the

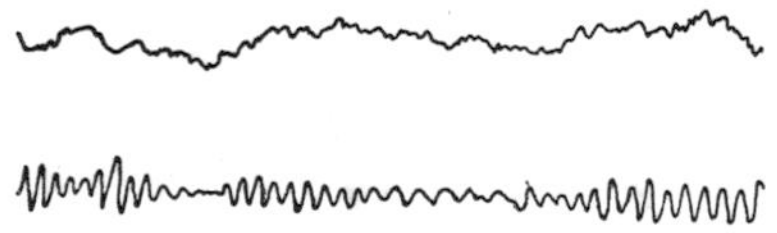

FIG. 13.6. Two traces of the thermal oscillations of a mirror system suspended in air; the upper trace was taken at atmospheric pressure, the lower one at a pressure of 10^{-4} mm of mercury.

oscillations of the mirror system, the upper one having been taken at the atmospheric pressure and the lower one at a pressure of 10^{-4} mm of mercury. The root-mean-square deviation is very nearly the same in the two cases! Nevertheless, one does note a difference of "quality" between the two traces, which relates to the "frequency spectrum" of the fluctuations and arises from the following situation. When the density of the surrounding gas is relatively high, the molecular impulses come in rapid succession, with the result that the *individual* deflections of the system are large in number but weak in magnitude. As the pressure is lowered, the intervals between the impulses become longer, making the *individual* deflections smaller in number but stronger in magnitude. However, the overall deflection, observed over a large interval of time, remains practically the same.

13.6. Spectral analysis of fluctuations: the Wiener–Khintchine theorem

We have already made a reference to the (spectral) quality of a fluctuation pattern. Referring once again to the two patterns shown in Fig. 13.6, we note that, even though the mean square fluctuation of the variable θ is the same in the two cases, the second pattern is far more "jagged" than the first; in other words, the high frequency components are far more prominent in the second pattern. At the same time, we note that there is a lot more "predictability" in the first pattern (insofar as it is represented by a much smoother curve); in other words, the *correlation function*, or the *memory function*, $K(s)$ of the first pattern extends over much larger values of s. In fact, these two aspects of a fluctuation process, viz. its time-dependence and its frequency spectrum, are very closely related to one another. And the most natural course for studying this relationship is to carry out a Fourier analysis of the given process.

For the proposed analysis we consider only those variables, $y(t)$, whose *mean square value*, $\langle y^2(t)\rangle$, has already attained an equilibrium, or stationary, value:

$$\langle y^2(t)\rangle = \text{const.} \tag{1}$$

Such a variable is said to be *statistically stationary*. As an example of a statistically stationary variable, we recall the velocity $v(t)$ of a Brownian particle at times t much larger than the relaxation time τ, see eqn. (13.4.29), or the displacement $x(t)$ of a Brownian particle moving under the influence of a restoring force ($F_x = -\lambda x$) at times t much larger than $(\lambda B)^{-1}$, see eqn. (13.5.20). Now, if the variable $y(t)$ were *strictly periodic* (and hence *completely predictable*), with a time period $T = 1/f_0$, then we could write, in terms of the fundamental frequency f_0 and its harmonics nf_0,

$$y(t) = a_0 + \sum_n a_n \cos(2\pi n f_0 t) + \sum_n b_n \sin(2\pi n f_0 t), \tag{2}$$

where

$$a_0 = \frac{1}{T}\int_0^T y(t)\,dt, \tag{3}$$

$$a_n = \frac{2}{T}\int_0^T y(t)\cos(2\pi n f_0 t)\,dt \tag{4}$$

and

$$b_n = \frac{2}{T}\int_0^T y(t)\sin(2\pi n f_0 t)\,dt; \tag{5}$$

in this special case, the coefficients a's and b's would be completely known numbers and would define, with no uncertainty, the frequency spectrum of the variable $y(t)$. If, on the other hand, the given variable is a *more or less* random function of time, then the coefficients a's and b's would themselves be statistical in nature. To apply the concept of periodicity to such a function, we must take the "time interval of repetition" as infinitely large, i.e. we must let $f_0 \to 0$.

Under the aforementioned circumstances, eqn. (3) would read

$$a_0 = \lim_{T\to\infty} \frac{1}{T}\int_0^T y(t)\,dt \equiv \langle y(t)\rangle; \tag{6}$$

thus, the coefficient a_0, which represents the mean (or d.c.) value of the variable y, may be determined *either* by taking a time average (over a sufficiently long duration) of the variable *or* by taking an ensemble average (at any instant of time t). For convenience, and without loss of generality, we take: $a_0 = 0$. In other words, we assume that from the actual values of the variable $y(t)$ its mean value, $\langle y(t)\rangle$, has already been removed.* Taking the ensemble average of eqns. (4) and (5), we conclude that, for all n,

$$\langle a_n\rangle = \langle b_n\rangle = 0. \tag{7}$$

Next, by taking the ensemble average of eqn. (2) *squared*, we obtain

$$\begin{aligned}\langle y^2(t)\rangle &= \sum_n \tfrac{1}{2}\langle a_n^2\rangle + \sum_n \tfrac{1}{2}\langle b_n^2\rangle \\ &= \sum_n \tfrac{1}{2}\{\langle a_n^2\rangle + \langle b_n^2\rangle\} = \text{const.}\end{aligned} \tag{8}$$

The term $\frac{1}{2}\{\langle a_n^2\rangle + \langle b_n^2\rangle\}$ represents the respective "share" of the component frequency nf_0 in the total, time-independent magnitude of the quantity $\langle y^2(t)\rangle$. Now, in view of the randomness of the phases of the various components, we have, for all n, $\langle a_n^2\rangle = \langle b_n^2\rangle$; consequently, eqn. (8) may be written as

$$\langle y^2\rangle = \sum_n \langle a_n^2\rangle \simeq \int_0^\infty w(f)\,df, \tag{9}$$

where

$$\langle a_n^2\rangle = w(nf_0)\,\Delta(nf_0), \quad \text{i.e.} \quad w(nf_0) = \frac{1}{f_0}\langle a_n^2\rangle; \tag{10}$$

the function $w(f)$ defines the *power spectrum* of the variable $y(t)$.

We shall now show that the power spectrum $w(f)$ of the fluctuating variable $y(t)$ is completely determined by the auto-correlation function $K(s)$ of the same variable. For this, we make use of eqn. (4) whereby

$$\langle a_n^2\rangle = 4f_0^2 \int_0^{1/f_0}\int_0^{1/f_0} \langle y(t_1)\,y(t_2)\rangle \cos(2\pi nf_0t_1)\cos(2\pi nf_0t_2)\,dt_1\,dt_2. \tag{11}$$

Changing over to the variables

$$S = \tfrac{1}{2}(t_1+t_2) \quad \text{and} \quad s = (t_2-t_1),$$

and remembering that the interval T over which the integrations extend is much larger than

* Obviously, this does not affect the spectral quality of the fluctuations, except that now we have no component belonging to the frequency $f = 0$! To represent the actual situation, one has to add, to the resulting spectrum, a suitably weighted $\delta(f)$-term.

the duration over which the "memory" of the variable lasts, we obtain

$$\langle a_n^2 \rangle \simeq 2f_0^2 \int_{S=0}^{1/f_0} \int_{s=-\infty}^{\infty} K(s)\{\cos(2\pi n f_0 s)+\cos(4\pi n f_0 S)\}\, dS\, ds; \tag{12}$$

cf. the steps leading from eqn. (13.4.22) to (13.4.25). The second part of the integral is identically equal to zero; the first part gives

$$\langle a_n^2 \rangle = 4f_0 \int_0^{\infty} K(s) \cos(2\pi n f_0 s)\, ds. \tag{13}$$

Comparing (13) with (10), we obtain the desired formula

$$w(f) = 4 \int_0^{\infty} K(s) \cos(2\pi f s)\, ds. \tag{14}$$

Taking the inverse of (14), we obtain the complementary formula

$$K(s) = \int_0^{\infty} w(f) \cos(2\pi f s)\, df. \tag{15}$$

For $s = 0$, formula (15) yields the important relationship, viz.

$$K(0) = \int_0^{\infty} w(f)\, df = \langle y^2 \rangle; \tag{16}$$

see eqn. (9) as well as the definition of the autocorrelation function: $K(s) = \langle y(t_1)\, y(t_1+s)\rangle$. Equations (14) and (15), which connect the complementary functions $w(f)$ and $K(s)$, constitute the theorem that goes after the names of Wiener (1930) and Khintchine (1934).

We shall now examine some special cases to illustrate the use of the *Wiener–Khintchine theorem*:

(i) If the given variable $y(t)$ is extremely irregular, and hence unpredictable, then the correlation function $K(s)$ would extend over a negligibly small range of the time interval s.* Under these circumstances, we may like to write:

$$K(s) = c\delta(s). \tag{17a}$$

Equation (14) then gives:

$$w(f) = 2c \qquad \text{for } \textit{all } f. \tag{17b}$$

A spectrum in which the distribution (of power) over the various frequencies f is uniform is known as a "flat" or a "white" spectrum. We must, however, note that if the uniformity of distribution were literally true for *all* frequencies, from $f = 0$ to $f = \infty$, then the integral in (16), which is identically equal to $\langle y^2 \rangle$, would be diverging! This is hardly acceptable. We, therefore, expect that, in any real situation, the correlation function $K(s)$ will not be so

* This is essentially true of the rapidly fluctuating force $F(t)$ experienced by a Brownian particle due to the incessant molecular impulses received by it.

sharply peaked as the expression (17a) suggests. Realistically, $K(s)$ will extend over a *small* range, $0(\sigma)$, of the variable s, which will introduce a "frequency zone", with $f = 0(1/\sigma)$, such that the power distribution function $w(f)$ undergoes a significant change of character as the variable f passes through this zone; towards lower frequencies $w(f) \to \text{const.} \neq 0$, while towards higher frequencies $w(f) \to \text{const.} = 0$. An interesting representation of this

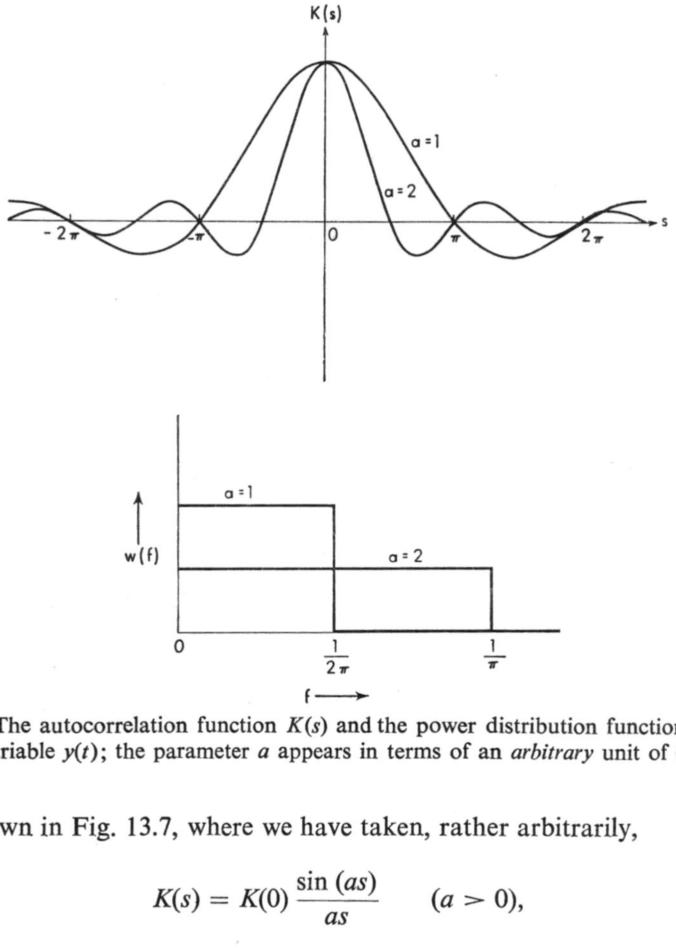

FIG. 13.7. The autocorrelation function $K(s)$ and the power distribution function $w(f)$ of a given variable $y(t)$; the parameter a appears in terms of an *arbitrary* unit of (time)$^{-1}$.

situation is shown in Fig. 13.7, where we have taken, rather arbitrarily,

$$K(s) = K(0)\frac{\sin(as)}{as} \qquad (a > 0), \tag{18a}$$

for which

$$\begin{aligned} w(f) &= \frac{2\pi}{a}K(0) \quad \text{for} \quad f < \frac{a}{2\pi} \\ &= 0 \qquad\qquad \text{for} \quad f > \frac{a}{2\pi}. \end{aligned} \tag{18b}$$

In the limit $a \to \infty$, eqns. (18) reduce to the idealized ones, viz. (17), with $c = \pi a^{-1} K(0)$.

(ii) On the other hand, if the variable $y(t)$ is extremely regular, and hence predictable, then its correlation function would extend over considerably large values of s and its power spectrum would appear in the form of "peaks", located at the "characteristic frequencies" of the variable. In the simplest case of a *monochromatic* variable, with characteristic frequency f^*, the correlation function would be given by

$$K(s) = K(0) \cos (2\pi f^* s), \tag{19a}$$

for which

$$w(f) = K(0)\, \delta(f - f^*); \tag{19b}$$

scc Fig. 13.8. A special case arises when $f^* = 0$; then, both $y(t)$ and $K(s)$ are constant in value, and the function $w(f)$ is peaked at the d.c. frequency ($f = 0$).

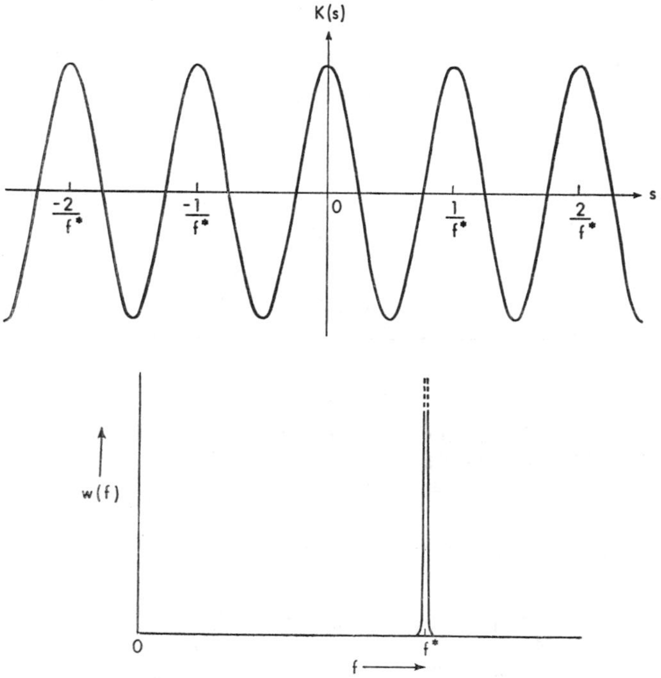

FIG. 13.8. The autocorrelation function $K(s)$ and the power distribution function $w(f)$ of a monochromatic variable $y(t)$, with characteristic frequency f^*.

(iii) If the variable $y(t)$ represents a signal that arises from, or has been filtered through, a lightly damped tuned circuit (a "narrow-band" filter), then its power would be distributed in the form of a "hump", around the *mean* frequency f^*. The function $K(s)$ will then appear in the nature of an "attenuated" function whose time scale, σ, is determined by the width, Δf, of the hump in the power spectrum. A situation of this kind is shown in Fig. 13.9.

The relevance of the *spectral analysis* of a fluctuating variable to the problem of its *physical observation* is best brought out by examining the power spectrum of the velocity $v(t)$ of a Brownian particle. Considering the x-component alone, the autocorrelation function $K_{v_x}(s)$, or simply $K(s)$, is given by

$$K(s) = \frac{kT}{M} e^{-|s|/\tau} \qquad (\tau = MB); \tag{20}$$

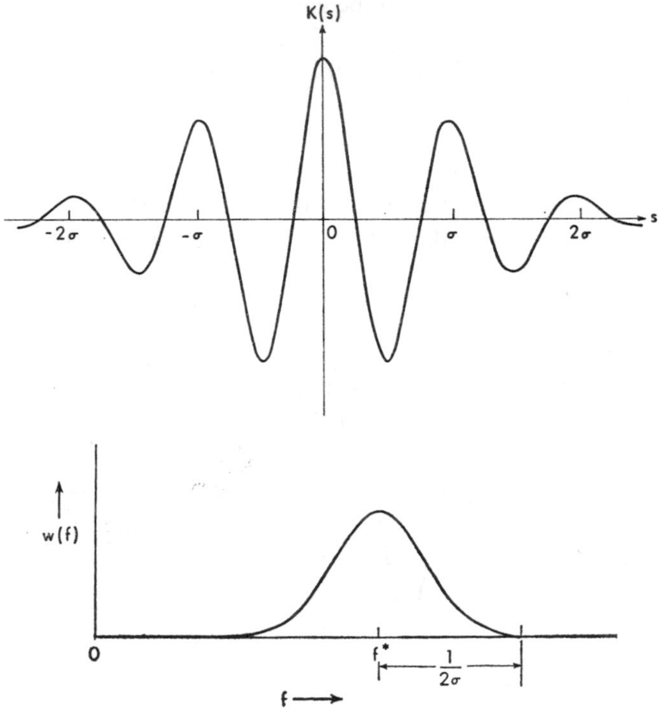

FIG. 13.9. The autocorrelation function $K(s)$ and the power distribution function $w(f)$ of a variable that has been filtered through a lightly damped tuned circuit, with mean frequency f^* and width $\Delta f \approx (1/\sigma)$.

see eqn. (13.7.10). The power spectrum $w(f)$ is then given by

$$w(f) = \frac{4kT}{M}\int_0^\infty e^{-s/\tau}\cos(2\pi fs)\,ds = \frac{4kT\tau}{M}\,\frac{1}{1+(2\pi f\tau)^2}, \tag{21}$$

which indeed satisfies the condition

$$\int_0^\infty w(f)\,df = \frac{2kT}{\pi M}\tan^{-1}(2\pi f\tau)\Big|_0^\infty = \frac{2kT}{\pi M}\,\frac{\pi}{2}$$

$$= \frac{kT}{M} = \langle v_x^2\rangle, \tag{22}$$

in agreement with the equipartition theorem (as applied to a single component of the velocity $\boldsymbol{v}$). For $f \ll \tau^{-1}$, the power distribution is practically independent of f; thus, for relatively low frequencies, we have a practically "white" spectrum:

$$w(f) \simeq \frac{4kT\tau}{M} = 4BkT. \tag{23}$$

Accordingly, we can write for the velocity fluctuations in the frequency range $(f, f+\Delta f)$, with $f \ll \tau^{-1}$,

$$\langle \Delta v_x^2 \rangle_{(f, f+\Delta f)} \simeq w(f)\,\Delta f \simeq (4BkT)\,\Delta f. \tag{24}$$

In general, our measuring instrument (or the eye, in the case of a visual examination of the particle) has a *finite* response time τ_0, as a consequence of which it is unable to respond to frequencies larger than, say, τ_0^{-1}. The *observed* fluctuation is then given by the "pruned" expression

$$\langle v_x^2 \rangle_{\text{obs}} = \int_0^{1/\tau_0} w(f)\,df = \frac{2kT}{\pi M} \tan^{-1}\left(2\pi \frac{\tau}{\tau_0}\right), \tag{25}$$

instead of the "full" expression (22). In a typical case, the mass of the Brownian particle $M \approx 10^{-12}$ g, its diameter $2a \approx 10^{-4}$ cm and the coefficient of viscosity of the fluid $\eta \approx 10^{-2}$ poise, whence it follows that the relaxation time $\tau = M/(6\pi\eta a) \approx 10^{-7}$ sec. However, the response time τ_0, in the case of visual observation, is of the order of 10^{-1} sec; accordingly, the ratio $\tau/\tau_0 \approx 10^{-6} \ll 1$. Equation (25), therefore, reduces to

$$\langle v_x^2 \rangle_{\text{obs}} \simeq \frac{4kT\tau}{M\tau_0} \ll \frac{kT}{M}; \tag{26*}$$

thus, in view of the *finiteness* of the response time τ_0, the observed value of the root-mean-square velocity of the Brownian particle goes down by a factor of $2(\tau/\tau_0)^{1/2}$, which is of the order of 10^{-3}; numerically, it takes us down from the equipartition value $(kT/M)^{1/2}$, which at room temperatures should be of the order of 10^{-1} cm/sec, to the reduced value $2(kT\tau/M\tau_0)^{1/2}$, which is of the order of 10^{-4} cm/sec instead. It is gratifying to note that the outcome of actual observations of the Brownian particles is in complete agreement with the latter result, and not with the dictates of the equipartition theorem; for a more detailed analysis of this question, see MacDonald (1950). The foregoing discussion clearly shows that, in the process of observing a fluctuating variable, our measuring instrument "picks up" signals over only a *limited* range of frequencies (as determined by the response time of the instrument); signals belonging to higher frequencies are simply left out!

The theory of this section can be readily applied to the spontaneous fluctuations in the motion of electrons in an (L, R) circuit. Corresponding to eqns. (21)–(24), we have for the fluctuations in the electric current I

$$w(f) = \frac{4kT\tau'}{L} \frac{1}{1+(2\pi f\tau')^2} \qquad \left(\tau' = \frac{L}{R}\right), \tag{27}$$

so that

$$\int_0^\infty w(f)\,df = \frac{kT}{L} = \langle I^2 \rangle, \tag{28}$$

in agreement with the equipartition theorem: $\langle \frac{1}{2}LI^2 \rangle = \frac{1}{2}kT$. For $f \ll 1/\tau'$, eqn. (27) reduces to

$$w(f) \simeq \frac{4kT}{R}, \tag{29}$$

* It will be noted that the fluctuations constituting the net result (26) belong *entirely* to the region of the "white" noise, with $\Delta f = 1/\tau_0$; see eqn. (24), with $B = \tau/M$.

which again corresponds to the region of the "white" noise; accordingly, for frequencies belonging to this region,

$$\langle \Delta I^2 \rangle_{(f, f+\Delta f)} \simeq w(f)\,\Delta f \simeq \frac{4kT}{R}\,\Delta f. \tag{30}$$

Equivalently, we obtain for the voltage fluctuations

$$\langle \Delta V^2 \rangle_{(f, f+\Delta f)} \simeq (4RkT)\,\Delta f. \tag{31}$$

Equation (31) constitutes the so-called *Nyquist theorem*, which was first discovered empirically by Johnson (1927–8) and was later derived by Nyquist (1927–8) on the basis of an argument involving the second law of thermodynamics and the exchange of energy between two resistances in thermal equilibrium.*

13.7. The fluctuation–dissipation theorem

In Sec. 13.4 we obtained a result of considerable importance, namely

$$\frac{1}{B} \equiv \frac{M}{\tau} = \frac{M^2}{6kT}C = \frac{M^2}{6kT}\int_{-\infty}^{\infty} K_A(s)\,ds$$

$$= \frac{1}{6kT}\int_{-\infty}^{\infty} K_F(s)\,ds; \tag{1}$$

see eqns. (13.4.26) and (13.4.28). Here, $K_A(s)$ and $K_F(s)$ are, respectively, the autocorrelation functions of the fluctuating acceleration $A(t)$ of the Brownian particle and of the fluctuating force $F(t)$ experienced by it:

$$K_A(s) = \langle A(0)\cdot A(s)\rangle = \frac{1}{M^2}\langle F(0)\cdot F(s)\rangle = \frac{1}{M^2}K_F(s). \tag{2}^{\dagger}$$

Equation (1) establishes a fundamental relationship between (i) the coefficient $1/B$ of the "averaged-out" part of the total force $\mathfrak{F}(t)$ experienced by the Brownian particle due to the impacts of the fluid molecules and (ii) the statistical character of the "fluctuating" part $F(t)$ of that force; see *Langevin's equation* (13.4.2). In other words, it connects the coefficient of viscosity of the fluid, which represents the *dissipative* forces operating in the system,

* It is interesting to note that the foregoing results are essentially equivalent to Einstein's original result for the charge fluctuations in a conductor, viz.

$$\langle \delta q^2 \rangle_t = \frac{2kT}{R}\,t;$$

compare, as well, the Brownian-particle result: $\langle x^2 \rangle_t = 2BkTt$.

† We notice that the functions $K_A(s)$ and $K_F(s)$, which are nonzero only if $s = 0(\tau^*)$, may, for certain purposes, be written as

$$K_A(s) = \frac{6kT}{M^2B}\,\delta(s) \quad \text{and} \quad K_F(s) = \frac{6kT}{B}\,\delta(s),$$

where $\delta(s)$ is the *Dirac delta function*. In this form, the functions are nonzero only for $s = 0$.

with the temporal character of the molecular *fluctuations*; the content of eqn. (1) is, therefore, referred to as a *fluctuation–dissipation theorem*. The most striking feature of this theorem is that it relates, in a most fundamental manner, the fluctuations of a physical quantity pertaining to the *equilibrium state* of the given system to a dissipative process which, in practice, is realized only when the system is subject to an external force that drives it *away from equilibrium*. Consequently, this theorem enables us to determine the *nonequilibrium properties* of a given statistical system on the basis of a knowledge of the thermal fluctuations occurring in the system when the system is in one of its *equilibrium states*! Thus, the statistical mechanics of *irreversible processes* is, in a sense, reduced to the statistical mechanics of equilibrium states (provided that we are in a position to handle the time-dependent fluctuation processes occurring in the system). For an expository account of the fluctuation–dissipation theorem, the reader may refer to Kubo (1966).

At this stage we recall that in eqn. (13.4.11) we obtained a relationship between the *diffusion coefficient* D and the *mobility* B, viz. $D = BkT$. Combining this relationship with eqn. (1), we obtain

$$\frac{1}{D} = \frac{1}{6(kT)^2} \int_{-\infty}^{\infty} K_F(s)\, ds. \tag{3}$$

Now, the diffusion coefficient D can be directly related to the autocorrelation function $K_v(s)$ of the fluctuating variable $\boldsymbol{v}(t)$. For this, one starts with the observation that, by definition,

$$\boldsymbol{r}(t) = \int_0^t \boldsymbol{v}(u)\, du, \tag{4}$$

whence it follows that

$$\langle r^2(t)\rangle = \int_0^t \int_0^t \langle \boldsymbol{v}(u_1)\cdot\boldsymbol{v}(u_2)\rangle\, du_1\, du_2 . \tag{5}$$

Proceeding in the same manner as for the integral in eqn. (13.4.22), one obtains

$$\langle r^2(t)\rangle = \int_0^{t/2} dS \int_{-2S}^{+2S} K_v(s)\, ds + \int_{t/2}^{t} dS \int_{-2(t-S)}^{+2(t-S)} K_v(s)\, ds; \tag{6}$$

cf. eqn. (13.4.24). The function $K_v(s)$ can be determined by making use of the expression (13.4.14) for $\boldsymbol{v}(t)$ and following exactly the same procedure as was followed for determining the quantity $\langle v^2(t)\rangle$, which is nothing but the maximal value, $K_v(0)$, of the desired function. One obtains

$$K_v(s) = \begin{cases} v^2(0)e^{-(2t+s)/\tau} + \dfrac{3kT}{M} e^{-s/\tau}(1 - e^{-2t/\tau}) & \text{for } s > 0 \quad (7) \\ v^2(0)e^{-(2t+s)/\tau} + \dfrac{3kT}{M} e^{s/\tau}(1 - e^{-2(t+s)/\tau}) & \text{for } s < 0; \quad (8) \end{cases}$$

cf. eqn. (13.4.27). It is easily seen that formulae (7) and (8) can be combined into a single formula, viz.

$$K_v(s) = v^2(0)e^{-|s|/\tau} + \left\{\frac{3kT}{M} - v^2(0)\right\}(e^{-|s|/\tau} - e^{-(2t+s)/\tau}) \quad \text{for all } s; \tag{9}$$

cf. eqn. (13.4.29). In the case of a "stationary ensemble", the function $K_v(s)$ would be

$$K_v(s) = \frac{3kT}{M} e^{-|s|/\tau}, \tag{10}$$

which is consistent with the property (13.4.20). It will be noted that the time scale for the correlation function $K_v(s)$ is provided by the *relaxation time* τ of the (Brownian) motion, which is many orders of magnitude larger than the *characteristic time* τ^* that provides the time scale for correlation functions $K_A(s)$ and $K_F(s)$. It is instructive to verify that the substitution of expression (10) into eqn. (6) leads to the formula (13.4.7) for $\langle r^2 \rangle$. Similarly, the substitution of the more general expression (9) into eqn. (6) leads to the equally general formula (13.4.31) for $\langle r^2 \rangle$; see Problem 13.20. In either case,

$$\langle r^2 \rangle \xrightarrow[t \gg \tau]{} 6Dt. \tag{11}$$

In the same limit, eqn. (6) reduces to

$$\langle r^2 \rangle \simeq \int_0^t dS \int_{-\infty}^{\infty} K_v(s)\, ds = t \int_{-\infty}^{\infty} K_v(s)\, ds. \tag{12}$$

Comparing the two results, we obtain the desired relationship, viz.

$$D = \tfrac{1}{6} \int_{-\infty}^{\infty} K_v(s)\, ds. \tag{13}$$

In passing, we note, from eqns. (3) and (13),

$$\int_{-\infty}^{\infty} K_v(s)\, ds \int_{-\infty}^{\infty} K_F(s)\, ds = (6kT)^2; \tag{14}$$

see also Problem 13.10.

It is obvious that the equations which contain a fluctuation–dissipation theorem can be adapted to any situation that involves a dissipative mechanism. For instance, the fluctuations in the motion of electrons in an electrical resistor give rise to a "spontaneous" thermal e.m.f., which may be denoted as $\mathfrak{V}(t)$. In the spirit of the Langevin theory, this e.m.f. may be split into two parts: (i) an "averaged-out" part, $-RI(t)$, which represents the resistive (or dissipative) aspect of the situation, and (ii) a "fast-fluctuating" part $V(t)$ which, over long intervals of time, averages out to zero. The "spontaneous" current in the resistor is then given by the equation

$$L\frac{dI}{dt} = -RI + V(t); \qquad \langle V(t) \rangle = 0. \tag{15}$$

Comparing this with the *Langevin equation* (13.4.2), and pushing the analogy further, we conclude that there exists a direct relationship between the resistance R and the temporal character of the fluctuations in the variable $V(t)$. According to eqn. (1), this relationship would be

$$R = \frac{1}{6kT} \int_{-\infty}^{\infty} \langle V(0) \cdot V(s) \rangle\, ds \tag{16}$$

or, equivalently,

$$\frac{1}{R} = \frac{1}{6kT}\int_{-\infty}^{\infty}\langle \boldsymbol{I}(0)\cdot\boldsymbol{I}(s)\rangle\,ds. \tag{17}$$

A generalization of the foregoing result has been given by Kubo (1957, 1959); see, for instance, Kubo (1965), problem 6.19, and Wannier (1966), sec. 23.2. On generalization, the *electric current density* $\boldsymbol{j}(t)$ is given by the expression

$$j_i(t) = \sum_l \int_{-\infty}^{t} E_l(t')\,\Phi_{li}(t-t')\,dt' \qquad (i, l = x, y, z); \tag{18}$$

here, $\boldsymbol{E}(t)$ denotes the *applied electric field* while

$$\Phi_{li}(s) = \frac{1}{kT}\langle j_l(0)\,j_i(s)\rangle. \tag{19}$$

Clearly, the quantities $kT\Phi_{li}(s)$ are the components of the *autocorrelation tensor* of the fluctuating vector $\boldsymbol{j}(t)$. In particular, if we consider the static case $\boldsymbol{E} = (E, 0, 0)$, we obtain for the (static) *conductivity* of the system

$$\sigma_{xx} = \frac{j_x}{E} = \int_{-\infty}^{t}\Phi_{xx}(t-t')\,dt' = \int_{0}^{\infty}\Phi_{xx}(s)\,ds$$

$$= \frac{1}{2kT}\int_{-\infty}^{\infty}\langle j_x(0)\,j_x(s)\rangle\,ds, \tag{20}$$

which may be compared with eqn. (17). If, on the other hand, we take $\boldsymbol{E} = (E\cos\omega t, 0, 0)$, we obtain instead

$$\sigma_{xx}(\omega) = \frac{1}{2kT}\int_{-\infty}^{\infty}\langle j_x(0)\,j_x(s)\rangle e^{-i\omega s}\,ds. \tag{21}$$

Taking the inverse of (21), we get

$$\langle j_x(0)\,j_x(s)\rangle = \frac{kT}{\pi}\int_{-\infty}^{\infty}\sigma_{xx}(\omega)e^{i\omega s}\,d\omega. \tag{22}$$

If we now assume that $\sigma_{xx}(\omega)$ does not depend on ω (and may, therefore, be denoted by the simpler symbol σ), we obtain

$$\langle j_x(0)\,j_x(s)\rangle = (2kT\sigma)\,\delta(s); \tag{23}$$

cf. the footnote to eqn. (2). A reference to eqns. (13.6.17) shows that, in the present approximation, thermal fluctuations in the electric current correspond to a system displaying "white" noise.

13.8. The Onsager relations

Most physical phenomena exhibit a kind of symmetry, sometimes referred to as *reciprocity*, which arises from certain basic properties of the microscopic processes that operate behind the observable macroscopic situations. A notable example of this is met with in the *thermodynamics of irreversible processes* where one deals with a variety of flow processes, such as heat flow, electric current, mass transfer, etc. These flows (or "currents") are driven by "forces", such as a temperature difference, a potential difference, a pressure difference, etc., which come into play because of a natural tendency among physical systems, which happen to be out of equilibrium, to approach a state of equilibrium. If the given state (of the system) is not too far removed from the state of equilibrium, then one might assume a *linear* relationship between the forces X_i and the currents $\dot{x}_i$:

$$\dot{x}_i = \gamma_{ij} X_j, \tag{1}$$

where γ_{ij}'s are referred to as the *kinetic coefficients* of the system.* Straightforward examples of the kinetic coefficients are thermal conductivity, electrical conductivity, diffusion coefficient, etc. There are, however, *nondiagonal elements* γ_{ij} $(i \neq j)$ as well, which may or may not vanish; they are responsible for the so-called *cross effects*. It is the symmetry properties of the matrix (γ_{ij}) that form the subject matter of this section.

The most appropriate way to approach this problem is to consider the entropy, $S(x_i)$, of the system in the *disturbed state* in relation to its (maximal) value, $S(\tilde{x}_i)$, in the *state of equilibrium*. It is the natural tendency of the function $S(x_i)$ to approach its maximal value $S(\tilde{x}_i)$ that brings into play the driving forces X_i; these forces give rise to currents $\dot{x}_i$ (which take the "coordinates" x_i of the system towards their equilibrium values $\tilde{x}_i$). If the deviations $(x_i - \tilde{x}_i)$ are small, then the function $S(x_i)$ may be expressed as a Taylor series around the values $x_i = \tilde{x}_i$; retaining terms up to the second order, we have

$$S(x_i) = S(\tilde{x}_i) + \left(\frac{\partial S}{\partial x_i}\right)_{x_i = \tilde{x}_i} (x_i - \tilde{x}_i) + \frac{1}{2}\left(\frac{\partial^2 S}{\partial x_i \, \partial x_j}\right)_{x_{i,j} = \tilde{x}_{i,j}} (x_i - \tilde{x}_i)(x_j - \tilde{x}_j). \tag{2}$$

In view of the fact that the function $S(x_i)$ is maximum at $x_i = \tilde{x}_i$, the first order terms identically vanish; we may, therefore, write

$$\Delta S = S(x_i) - S(\tilde{x}_i) = -\tfrac{1}{2}\beta_{ij}(x_i - \tilde{x}_i)(x_j - \tilde{x}_j), \tag{3}$$

where

$$\beta_{ij} = -\left(\frac{\partial^2 S}{\partial x_i \, \partial x_j}\right)_{x_{i,j} = \tilde{x}_{i,j}} = \beta_{ji}. \tag{4}$$

The driving forces X_i may be defined in the spirit of the *first law of thermodynamics*, i.e.

$$X_i = \left(\frac{\partial S}{\partial x_i}\right) = -\beta_{ij}(x_j - \tilde{x}_j). \tag{5}$$

We note that the forces X_i depend linearly on the displacements $(x_i - \tilde{x}_i)$; in the state of

* In writing eqn. (1), and other subsequent equations, we have followed the *summation convention* which implies *an automatic summation over a repeated index*.

equilibrium, they identically vanish. Now, in view of eqn. (13.1.2), the *ensemble average* of the quantity $(x_i - \tilde{x}_i)X_j$ is given by

$$\langle (x_i - \tilde{x}_i)X_j \rangle = \langle x_i X_j \rangle = \frac{\int_{-\infty}^{\infty} x_i X_j \exp\left\{-\frac{1}{2k}\beta_{ij}(x_i - \tilde{x}_i)(x_j - \tilde{x}_j)\right\} \prod_i dx_i}{\int_{-\infty}^{\infty} \exp\left\{-\frac{1}{2k}\beta_{ij}(x_i - \tilde{x}_i)(x_j - \tilde{x}_j)\right\} \prod_i dx_i}; \tag{6}$$

the limits of integration in (6) have been extended to $-\infty$ and $+\infty$ because the integrals here do not draw any significant contribution from large values of the variables. In the same way,

$$\langle x_i \rangle = \frac{\int_{-\infty}^{\infty} x_i \exp\left\{-\frac{1}{2k}\beta_{ij}(x_i - \tilde{x}_i)(x_j - \tilde{x}_j)\right\} \prod_i dx_i}{\int_{-\infty}^{\infty} \exp\left\{-\frac{1}{2k}\beta_{ij}(x_i - \tilde{x}_i)(x_j - \tilde{x}_j)\right\} \prod_i dx_i} = \tilde{x}_i. \tag{7}$$

Differentiating (7) with respect to $\tilde{x}_j$ (and remembering that the integral in the denominator is a constant, independent of the actual values of the quantities $\tilde{x}_i$), and comparing the result with (6), we obtain

$$\langle x_i X_j \rangle = -k\delta_{ij}. \tag{8}$$

We now proceed towards the key point of the argument. First of all we note that although the phenomenological equations (1) are concerned with *irreversible* phenomena, the microscopic processes underlying these phenomena do obey *time reversal*, which means that the temporal correlations of the relevant variables are the same whether measured forward or backward. Thus, we have

$$\langle x_i(0)\, x_j(s) \rangle = \langle x_i(0)\, x_j(-s) \rangle; \tag{9}$$

we also have, by a shift in the scale of time,

$$\langle x_i(0)\, x_j(-s) \rangle = \langle x_i(s)\, x_j(0) \rangle. \tag{10}$$

Combining (9) and (10), we obtain

$$\langle x_i(0)\, x_j(s) \rangle = \langle x_i(s)\, x_j(0) \rangle. \tag{11}$$

If we now subtract, from both sides of this equation, the quantity $\langle x_i(0)\, x_j(0) \rangle$, divide the resulting equation by s and let $s \to 0$, we obtain

$$\langle x_i(0)\, \dot{x}_j(0) \rangle = \langle \dot{x}_i(0)\, x_j(0) \rangle. \tag{12}$$

Substituting from (1) and making use of (8), we obtain on the left-hand side

$$\langle x_i(0)\, \gamma_{jl} X_l(0) \rangle = -k\gamma_{jl}\delta_{il} = -k\gamma_{ji}$$

and on the right-hand side

$$\langle \gamma_{il} X_l(0)\, x_j(0) \rangle = -k\gamma_{il}\delta_{jl} = -k\gamma_{ij}.$$

Hence, the desired relations

$$\gamma_{ij} = \gamma_{ji}. \tag{13}$$

Equations (13) constitute the *Onsager reciprocity relations*; they were first derived by Onsager in 1931 and have formed an essential basis for the thermodynamics of irreversible phenomena.

In view of eqns. (1) and (13), the currents $\dot{x}_i$ may be written as

$$\dot{x}_i = \frac{\partial f}{\partial X_i}, \tag{14}$$

where the *generating function* f is a quadratic function of the forces X_i:

$$f = \tfrac{1}{2}\gamma_{ij} X_i X_j. \tag{15}$$

The function f is of considerable importance in that it determines directly the rate at which the entropy of the system changes with time:

$$\dot{S} = \frac{\partial S}{\partial x_i}\dot{x}_i = X_i \dot{x}_i = X_i \frac{\partial f}{\partial X_i} = 2f. \tag{16}$$

As the system approaches the state of equilibrium, its entropy must *increase* towards the equilibrium value $S(\tilde{x}_i)$. The function f must, therefore, be *positive definite*; this clearly places certain restrictions on the values of the coefficients γ_{ij}.

Just as we have eqn. (1), we could also have

$$\dot{X}_i = \zeta_{ij}(x_j - \tilde{x}_j); \tag{17}$$

the quantities ζ_{ij} are another set of coefficients pertaining to the system. From eqns. (1) and (5), on the other hand, we obtain

$$\begin{aligned} \dot{X} &= -\beta_{ij}\dot{x}_j = -\beta_{ij}(\gamma_{jl} X_l) = -\beta_{ij}\gamma_{jl}\{-\beta_{lm}(x_m - \tilde{x}_m)\} \\ &= \beta_{ij}\gamma_{jl}\beta_{lm}(x_m - \tilde{x}_m). \end{aligned} \tag{18}$$

Comparing (17) and (18), we obtain a relationship between the set of coefficients ζ_{ij} and the set of (kinetic) coefficients γ_{ij}:

$$\zeta_{im} = \beta_{ij}\gamma_{jl}\beta_{lm}. \tag{19}$$

Further, in view of the symmetry properties of the matrices β and γ, we obtain

$$\zeta_{im} = \zeta_{mi}; \tag{20}$$

thus, the coefficients ζ_{ij} introduced through the phenomenological equations (17) also obey the reciprocity relations. It then follows that the quantities $\dot{X}_i$ may be written as

$$\dot{X}_i = \frac{\partial f'}{\partial x_i}, \tag{21}$$

where

$$f' = \tfrac{1}{2}\zeta_{ij}(x_i - \tilde{x}_i)(x_j - \tilde{x}_j). \tag{22}$$

The entropy change dS may now be written as

$$dS = \frac{\partial S}{\partial x_j}\,dx_j = X_j\,dx_j = -\beta_{ji}(x_i-\tilde{x}_i)\,dx_j$$
$$= (x_i-\tilde{x}_i)\,d\{-\beta_{ij}(x_j-\tilde{x}_j)\} = (x_i-\tilde{x}_i)\,dX_i, \tag{23}$$

whence it follows that

$$\frac{\partial S}{\partial X_i} = (x_i-\tilde{x}_i); \tag{24}$$

of course, the entropy S is now assumed to be an *explicit* function of the forces X_i (rather than of the coordinates x_i). The time derivative of S now takes the form

$$\dot{S} = \frac{\partial S}{\partial X_i}\dot{X}_i = (x_i-\tilde{x}_i)\,\frac{\partial f'}{\partial x_i} = 2f'. \tag{25}$$

Comparing (16) and (25), we conclude that the functions f and f' represent one and the same thing; they are only expressed in terms of two different sets of variables.

It seems important to mention here that Onsager's reciprocity relations have a deep connection with the fluctuation–dissipation theorem of the preceding section. Following eqns. (13.7.18) and (13.7.19), and adopting the summation convention, we have in the present context

$$\dot{x}_i(t) = \frac{1}{kT}\int_{-\infty}^{t} E_l(t')\,\langle\dot{x}_l(t')\,\dot{x}_i(t)\rangle\,dt' \tag{26}$$

or, putting $(t-t') = s$,

$$\dot{x}_i(t) = \frac{1}{kT}\int_{0}^{\infty} E_l(t-s)\,\langle\dot{x}_l(t-s)\,\dot{x}_i(t)\rangle\,ds; \tag{27}$$

cf. eqn. (1). Interchanging the indices i and l, we obtain

$$\dot{x}_l(t) = \frac{1}{kT}\int_{0}^{\infty} E_i(t-s)\,\langle\dot{x}_i(t-s)\,\dot{x}_l(t)\rangle\,ds. \tag{28}$$

The crucial point now is that the correlation coefficients appearing within the integrals (27) and (28) are identical in value:

$$\langle\dot{x}_l(t-s)\,\dot{x}_i(t)\rangle = \langle\dot{x}_l(0)\,\dot{x}_i(s)\rangle = \langle\dot{x}_l(0)\,\dot{x}_i(-s)\rangle = \langle\dot{x}_l(t)\,\dot{x}_i(t-s)\rangle; \tag{29}$$

in establishing (29), the first and third steps followed from "a shift in the time scale" while the second step followed from the "principle of *dynamical reversibility* of microscopic processes". The equivalence (29) is, in essence, the content of Onsager's reciprocity relations. In particular, if the correlation functions appearing in (27) and (28) are sharply peaked at the value $s = 0$, then these equations reduce to the phenomenological equations (1) and the equivalence (29) becomes literally identical with the Onsager relations (13).

In the end, we propose to make some remarks concerning the relations (13). We recall that, for arriving at these relations, we had to appeal to the invariance of the microscopic processes under time reversal. The situation is somewhat different in the case of a "system in rotation" (or a "system in an external magnetic field"), for the invariance under time reversal holds only if there is a simultaneous change of sign of the angular velocity $\boldsymbol{\Omega}$ (or of the magnetic field $\boldsymbol{B}$). The kinetic coefficients, which might now depend upon the parameter $\boldsymbol{\Omega}$ (or $\boldsymbol{B}$), will satisfy the relations

$$\gamma_{ij}(\boldsymbol{\Omega}) = \gamma_{ji}(-\boldsymbol{\Omega}) \tag{13a}$$

and

$$\gamma_{ij}(\boldsymbol{B}) = \gamma_{ji}(-\boldsymbol{B}). \tag{13b}$$

Secondly, our proof contained the *implicit* assumption that the quantities x_i themselves do not change under time reversal. If, however, these quantities are proportional to the velocities of a certain macroscopic motion, then they will also change their sign with that of time. Now, if both x_i and x_j belong to this category, then the result (12), which is crucial to our proof, remains unaltered; consequently, the coefficients γ_{ij} and γ_{ji} continue to be equal. However, if only one of them belongs to this category while the other does not, then eqn. (12) changes to

$$\langle x_i(0)\,\dot{x}_j(0)\rangle = -\langle \dot{x}_i(0)\,x_j(0)\rangle; \tag{12'}$$

the coefficients γ_{ij} and γ_{ji} then satisfy the relations

$$\gamma_{ij} = -\gamma_{ji}. \tag{13'}$$

For the application of Onsager's relations to various physical problems, reference may be made to the monographs by de Groot (1951), de Groot and Mazur (1962) and Prigogine (1967).

Problems

13.1. Making use of the expressions (13.1.11) and (13.1.12) for ΔS and ΔP, and the expressions (13.1.14) for $\overline{(\Delta T)^2}$, $\overline{(\Delta V)^2}$ and $\overline{(\Delta T\,\Delta V)}$, derive the following results:

(i) $\overline{(\Delta T\,\Delta S)} = kT$; (ii) $\overline{(\Delta P\,\Delta V)} = -kT$;

(iii) $\overline{(\Delta S\,\Delta V)} = kT(\partial V/\partial T)_P$; (iv) $\overline{(\Delta P\,\Delta T)} = kT^2 C_V^{-1}(\partial P/\partial T)_V$.

[Note that the results (i) and (ii) give: $\overline{(\Delta T\,\Delta S - \Delta P\,\Delta V)} = 2kT$, which follows directly from the probability distribution function (13.1.8), for then

$$\bar{x} = \int_0^\infty e^{-x/\xi} x\,dx \Big/ \int_0^\infty e^{-x/\xi}\,dx = \xi.]$$

13.2. Establish the probability distribution formula (13.1.15), which leads to the results (13.1.16) for $\overline{(\Delta S)^2}$, $\overline{(\Delta P)^2}$ and $\overline{(\Delta S\,\Delta P)}$. Show that these results can also be obtained by following the procedure of the preceding problem.

13.3. If we choose the quantities E and V as "independent" variables, then the probability distribution function (13.1.8) does not reduce to a form as simple as that of the formulae (13.1.13) and (13.1.15); it is marked by the presence, in the exponent, of a cross term proportional to the product $\Delta E\,\Delta V$. Consequently, the variables E and V are not *statistically independent*; they are rather correlated, in the sense that $\overline{(\Delta E\,\Delta V)} \neq 0$.

Show that

$$\overline{(\Delta E\, \Delta V)} = kT\left\{T\left(\frac{\partial V}{\partial T}\right)_P + P\left(\frac{\partial V}{\partial P}\right)_T\right\};$$

verify as well the expressions (13.1.14) and (13.1.18) for $\overline{(\Delta V)^2}$ and $\overline{(\Delta E)^2}$.

[Note that in the case of a two-dimensional *normal* distribution, viz.

$$p(x, y) \propto \exp\{-\tfrac{1}{2}(ax^2+2bxy+cy^2)\},$$

the mean values $\langle x^2\rangle$, $\langle xy\rangle$ and $\langle y^2\rangle$ can be obtained in a straightforward manner by carrying out a logarithmic differentiation of the formula

$$\int_{-\infty}^{\infty}\int_{-\infty}^{\infty} \exp\left\{-\frac{1}{2}(ax^2+2bxy+cy^2)\right\} dx\, dy = \frac{2\pi}{\sqrt{(ac-b^2)}}$$

with respect to the parameters a, b and c. This leads to the *covariance matrix* of the distribution, namely

$$\begin{pmatrix} \langle x^2\rangle & \langle xy\rangle \\ \langle yx\rangle & \langle y^2\rangle \end{pmatrix} = \frac{1}{(ac-b^2)}\begin{pmatrix} c & -b \\ -b & a \end{pmatrix}.$$

If $b = 0$, we obtain the simple-minded results:

$$\langle x^2\rangle = 1/a, \quad \langle xy\rangle = 0, \quad \langle y^2\rangle = 1/c.]^*$$

13.4. A string of length l is stretched, under a constant tension F, between two fixed points A and B. Show that the mean square value of the fluctuational displacement $y(x)$ at the point P, distant x from A, is given by

$$\overline{\{y(x)\}^2} = \frac{kT}{Fl}x(l-x).$$

Further show that, for $x_2 \geqslant x_1$,

$$\overline{y(x_1)\, y(x_2)} = \frac{kT}{Fl}x_1(l-x_2).$$

[Hint. Evaluate the energy Φ associated with the fluctuation in question; the relevant probability distribution is then given by $p \propto \exp(-\Phi/kT)$, from which the desired averages can be computed.]

13.5. How small must the volume V_A of a gaseous subsystem (at n.t.p.) be, so that the root-mean-square deviation in the number N_A of the particles occupying this volume is 1 per cent of the mean value $\bar{N}_A$ of this number?

13.6. Consider a gas, of infinite extent, divided into regions A and B by an imaginary sheet running through the system. The molecules of the system are supposed to be interacting through a potential energy function $u(r)$. Show that the net force F experienced by all the molecules on the A-side of the sheet due to all the molecules on the B-side of the sheet is *perpendicular* to the plane of the sheet, and its magnitude (per unit area) is given by

$$\frac{F}{A} = -\frac{2\pi n^2}{3}\int_0^{\infty}\left(\frac{du}{dr}\right) g(r) r^3\, dr,$$

where the quantity $n^2 g(r)$ has been defined in eqn. (13.2.3). Compare this result with eqn. (2.6.20), which was obtained with the help of the *virial theorem*.

13.7. Show that for a gas of *noninteracting* bosons, or fermions, the *correlation function* $\nu(r)$ is given by the formula

$$\nu(r) = \pm\frac{g}{n^2h^6}\left|\int_{-\infty}^{\infty}\frac{e^{i(\boldsymbol{p}\cdot\boldsymbol{r})/\hbar}\, d^3p}{e^{(p^2/2m-\mu)/kT}\mp 1}\right|^2,$$

where $g(= 2S+1)$ is the spin multiplicity factor while the other symbols have their usual meaning; the upper sign holds for the bosons, the lower one for the fermions.†

---

* For the covariance matrix of an n-dimensional *normal* distribution, see Landau and Lifshitz (1958), sec. 110.

† It may be noted that, in the classical limit ($\hbar \to 0$), the infinitely rapid oscillations of the factor $\exp\{i(\boldsymbol{p}\cdot\boldsymbol{r})/\hbar\}$ make the integral vanish. Consequently, for an ideal classical gas, the function $\nu(\boldsymbol{r})$ is identically equal to zero. *(Continued on p. 484.)*

[Hint. To solve this problem, it is preferable to follow the method of second quantization, as developed in Chapter 11. The *particle density operator* $\hat{n}$ is then given by the sum, see eqn. (11.1.25),

$$\sum_{\alpha,\beta} a_\alpha^\dagger a_\beta u_\alpha^*(r) u_\beta(r),$$

whose diagonal terms are directly related to the mean particle density n in the system. The non-diagonal terms give the *density fluctuation operator* $(\hat{n}-n)$. And so on....]

13.8. Show that, in the case of a degenerate gas of fermions ($T \ll T_F$), the correlation function $\nu(r)$, for $r \gg \hbar/p_F$, reduces to the expression

$$\nu(r) = -\frac{3(mkT)^2}{4p_F^3\hbar r^2}\left\{\sinh\left(\frac{\pi mkTr}{p_F\hbar}\right)\right\}^{-2}.$$

Note that, as $T \to 0$, this expression tends to the limiting form

$$\nu(r) = -\frac{3\hbar}{4\pi^2 p_F r^4} \propto \frac{1}{r^4}.$$

13.9. Pospišil (1927) observed the Brownian motion of soot particles, of radii 0.4×10^{-4} cm, immersed in a water–glycerine solution, of viscosity 0.0278 poise, at a temperature of 18.8°C. The observed value of $\overline{x^2}$, in a 10-second time interval, was 3.3×10^{-8} cm². Making use of these data, determine the Boltzmann constant k.

13.10. Prove that the cross-correlation function of the variables $\boldsymbol{v}(t)$ and $\boldsymbol{F}(t)$ of a Brownian particle is given by

$$\langle \boldsymbol{v}(t)\cdot\boldsymbol{F}(t)\rangle = 3kT/\tau.$$

13.11. Integrate eqn. (13.4.14) to obtain

$$\boldsymbol{r}(t) = \boldsymbol{v}(0)\,\tau(1-e^{-t/\tau}) + \tau\int_0^t \{1-e^{-(u-t)/\tau}\}\,\boldsymbol{A}(u)\,du,$$

so that $\boldsymbol{r}(0) = 0$. Taking the square of this result and making use of an appropriate expression for the autocorrelation function $K_A(s)$, derive the general formula (13.4.31) for $\langle r^2(t)\rangle$.

13.12. While detecting a very feeble current with the help of a moving-coil galvanometer, one must ensure that an observed deflection is not just a stray kick arising from the Brownian motion of the suspended system. If we agree that a deflection θ, whose magnitude exceeds $4\theta_{\text{r.m.s.}} = 4(kT/c)^{1/2}$, is highly unlikely to be due to the Brownian motion, we obtain a *lower* limit to the magnitude of the current than can be safely recorded with the given galvanometer. Express this limiting value of the current in terms of the time period τ and the critical damping resistance R_c of the galvanometer.

13.13. (a) Integrate the Langevin equation (13.4.5), for the velocity component v_x, over a *small* interval of time δt, and show that

$$\frac{\langle \delta v_x\rangle}{\delta t} = -\frac{v_x}{\tau} \quad\text{and}\quad \frac{\langle(\delta v_x)^2\rangle}{\delta t} = \frac{2kT}{M\tau}.$$

(b) Now, set up the Fokker–Planck equation for the *distribution function* $f(v_x, t)$ and, making use of the foregoing results for $\mu_1(v_x)$ and $\mu_2(v_x)$, derive an explicit expression for this function. Study the different cases of interest, especially the situation pertaining to $t \gg \tau$.

13.14. The Langevin equation for a particle, of mass M, executing Brownian motion, under the constraint of a restoring force $-\lambda x$, would be (Kappler, 1938)

$$M\frac{d^2x}{dt^2} + \frac{1}{B}\frac{dx}{dt} + \lambda x = F(t);$$

cf. eqn. (13.4.2) which pertains to the case $\lambda = 0$. Derive, on the basis of this equation, the relevant expressions for the quantities $\langle x^2(t)\rangle$ and $\langle v_x^2(t)\rangle$, and show that, in the limit $\lambda \to 0$, these expressions correspond to eqns. (13.4.29) and (13.4.31) while, in the limit $M \to 0$, they lead to the relevant results of Sec. 13.5.

(Footnote † continued from p. 483.)
It is not difficult to see that, for $n\lambda^3 \ll 1$ (where $\lambda = h/\sqrt{(2\pi mkT)}$),

$$\nu(r) \simeq \pm\frac{1}{g}\exp(-2\pi r^2/\lambda^2);$$

cf. eqn. (5.5.27). We also note that $\nu(0)$ is *identically* equal to $\pm 1/g$ (or ± 1 if $s = 0$).

13.15. Generalize the Fokker–Planck equation to a particle executing Brownian motion in *three* dimensions. Determine the general solution of this equation and study its important features.

13.16. The autocorrelation function $K(s)$ of a *statistically stationary* variable $y(t)$ is given by

$$\text{(i)}\quad K(s) = K(0)e^{-\alpha^2 s^2} \cos(2\pi f^* s)$$

or by

$$\text{(ii)}\quad K(s) = K(0)e^{-\alpha|s|} \cos(2\pi f^* s),$$

where $\alpha > 0$. Determine, and discuss the nature of, the power spectrum $w(f)$ in both the cases (i) and (ii), and investigate its behavior in the limits (a) $\alpha \to 0$, (b) $f^* \to 0$, and (c) both α and $f^* \to 0$.

13.17. Show that if the autocorrelation function $K(s)$ of a *statistically stationary* variable $y(t)$ is given by

$$K(s) = K(0)\frac{\sin(as)}{as}\frac{\sin(bs)}{bs} \qquad (a > b > 0),$$

then the power spectrum $w(f)$ of the variable is given by

$$\begin{aligned} w(f) &= \frac{2\pi}{a}K(0) && \text{for} \quad 0 < f \leqslant \frac{a-b}{2\pi} \\ &= \frac{2\pi}{ab}K(0)\left\{\frac{a+b}{2} - \pi f\right\} && \text{for} \quad \frac{a-b}{2\pi} \leqslant f \leqslant \frac{a+b}{2\pi} \\ &= 0 && \text{for} \quad \frac{a+b}{2\pi} \leqslant f < \infty. \end{aligned}$$

Verify that the function $w(f)$ satisfies the requirement (13.6.16).

[Note that, in the limit $b \to 0$, we recover the situation pertaining to eqns. (13.6.18).]

13.18. (a) Show that the mean square value of the variable $Y(t)$, defined by the formula

$$Y(t) = \int_u^{u+t} y(u)\,du,$$

where $y(u)$ is a *statistically stationary* variable with power spectrum $w(f)$, is given by

$$\langle Y^2(t)\rangle = \frac{1}{2\pi^2}\int_0^\infty \frac{w(f)}{f^2}\{1 - \cos(2\pi ft)\}\,df$$

and, accordingly,

$$w(f) = 4\pi f \int_0^\infty \frac{\partial}{\partial t}\langle Y^2(t)\rangle \sin(2\pi ft)\,dt$$

$$= 2\int_0^\infty \frac{\partial^2}{\partial t^2}\langle Y^2(t)\rangle \cos(2\pi ft)\,dt;$$

see MacDonald (1962), sec. 2.2.1. A comparison of the last result with eqn. (13.6.14) suggests that

$$K_y(s) = \frac{1}{2}\frac{\partial^2}{\partial s^2}\langle Y^2(s)\rangle.$$

(b) Apply the foregoing analysis to the fluctuating motion of a Brownian particle, taking y to be the velocity of the particle and Y its displacement.

13.19. Show that the power spectra $w_v(f)$ and $w\ (f)$ of the fluctuating variables $v(t)$ and $A(t)$, which appear in the Langevin equation (13.4.5), are connected by the relation

$$w_v(f) = w_A(f)\frac{\tau^2}{1 + (2\pi f\tau)^2},$$

where τ denotes the relaxation time of the problem. Hence, $w_A(f) = 12kT/M\tau$.

13.20. (a) Verify eqns. (13.7.7)–(13.7.9).

(b) Substituting the expression (13.7.9) for $K_v(s)$ into eqn. (13.7.6), derive the formula (13.4.31) for $\langle r^2(t)\rangle$.

[Note that the expression for $K_v(s)$ can also be written as

$$K_v(s) = \left(v^2(0) - \frac{3kT}{M}\right)e^{-(u_1+u_2)/\tau} + \frac{3kT}{M}\,e^{-|u_2-u_1|/\tau}.$$

The first part, substituted into (13.7.5), gives

$$\left(v^2(0) - \frac{3kT}{M}\right)\{\tau(1-e^{-t/\tau})\}^2,$$

while the second part, substituted into (13.7.6), gives

$$\frac{6kT\tau^2}{M}\left\{\frac{t}{\tau} - (1-e^{-t/\tau})\right\}. \tag{13.4.7}$$

Combining the two, we obtain the desired result.]

APPENDIXES

A. Influence of Boundary Conditions on the Distribution of Quantum States

IN THIS appendix we propose to examine, under different boundary conditions, the asymptotic distribution of the single-particle states in a bounded continuum. The main purpose of this study is to examine the manner in which the geometry of the bounding surface affects the final result. For simplicity, we consider a cuboidal enclosure of sides a, b and c. The admissible solutions of the free-particle Schrödinger equation

$$\nabla^2\psi + k^2\psi = 0 \qquad (\boldsymbol{k} = \boldsymbol{p}\hbar^{-1}), \tag{1}$$

which satisfy the Dirichlet boundary conditions (namely, $\psi=0$ everywhere at the boundary), are then given by the expression

$$\psi_{lmn}(\boldsymbol{r}) \propto \sin\left(\frac{l\pi x}{a}\right)\sin\left(\frac{m\pi y}{b}\right)\sin\left(\frac{n\pi z}{c}\right), \tag{2}$$

with

$$k = \pi\left(\frac{l^2}{a^2}+\frac{m^2}{b^2}+\frac{n^2}{c^2}\right)^{1/2}; \qquad l, m, n = 1, 2, 3, \ldots. \tag{3}$$

It is important to note that in this case none of the quantum numbers l, m or n can be zero, for otherwise our wave function would identically vanish. If, on the other hand, we impose the Neumann boundary conditions (namely, $\partial\psi/\partial n = 0$ everywhere at the boundary), the desired solutions turn out to be

$$\psi_{lmn}(\boldsymbol{r}) \propto \cos\left(\frac{l\pi x}{a}\right)\cos\left(\frac{m\pi y}{b}\right)\cos\left(\frac{n\pi z}{c}\right), \tag{4}$$

with

$$l, m, n = 0, 1, 2, \ldots; \tag{5}$$

the value zero of the quantum numbers is no longer disallowed! In each case, however, the negative-integral values of the quantum numbers do not lead to any new wave functions.

The total number $g(K)$ of the distinct wave functions ψ, with wave number k not exceeding a given value K, can be written as

$$g(K) = \sum_{l,m,n}{}' f(l, m, n), \tag{6}$$

where $f(l, m, n) = 1$ for (l, m, n) corresponding to the values (3) or (5), as the case may be;

the summation Σ' in each case is restricted by the condition

$$\left(\frac{l^2}{a^2}+\frac{m^2}{b^2}+\frac{n^2}{c^2}\right) \leqslant \frac{K^2}{\pi^2}. \tag{7}$$

We now define a mathematical sum

$$G(K) = \sum_{l,m,n}{}' f^*(l, m, n), \tag{8}$$

where $f^*(l, m, n) = 1$ for *all* integral values of l, m and n (negative, positive or zero), the restriction on the sum being formally the same as (7). One can then show, by setting up correspondence of terms, that

$$\begin{aligned}\sum_{l,m,n}{}' f(l, m, n) = \tfrac{1}{8}\Bigg[&\sum_{l,m,n}{}' f^*(l, m, n)\\ &\mp\left\{\sum_{l,m}{}' f^*(l, m, 0)+\sum_{l,n}{}' f^*(l, 0, n)+\sum_{m,n}{}' f^*(0, m, n)\right\}\\ &+\left\{\sum_{l}{}' f^*(l, 0, 0) +\sum_{m}{}' f^*(0, m, 0)+\sum_{n}{}' f^*(0, 0, n)\right\}\mp 1\Bigg];\end{aligned} \tag{9}$$

the upper and the lower signs correspond, respectively, to the Dirichlet and the Neumann boundary conditions.

Clearly, the first sum on the right-hand side of eqn. (9) denotes the number of lattice points in the ellipsoid $(X^2/a^2+Y^2/b^2+Z^2/c^2) = K^2/\pi^2$,* the next three sums denote the numbers of lattice points in the ellipses which are cross-sections of the aforementioned ellipsoid with the Z-, Y- and X-planes, while the last three sums denote the numbers of lattice points on the principal axes of the ellipsoid. Now, if a, b and c are sufficiently larger than π/K, one may replace these numbers by the corresponding volume, areas and lengths, with the result

$$g(K) = \frac{K^3}{6\pi^2}(abc)\mp\frac{K^2}{8\pi}(ab+ca+bc)+\frac{K}{4\pi}(a+b+c)\mp\frac{1}{8}+E(K); \tag{10}$$

the term $E(K)$ denotes the *net* error committed in making the (approximate) replacements.† We thus find that the main term of our result is directly proportional to the volume of the enclosure while the first correction term is proportional to its surface area (and, hence, represents the "surface effect"); moreover, the next order term appears in the nature of an "edge effect".

Now, a reference to the mathematical literature on the subject dealing with the determination of the "number of lattice points in a given domain" reveals that the error term $E(K)$ in eqn. (10) is $0(K^\alpha)$ where $1 < \alpha < 1.4$; hence, expression (10) for $g(K)$ is safe only up to the surface term (inclusive). In view of this, we better write

$$g(K) = \frac{K^3}{6\pi^2}V\mp\frac{K^2}{16\pi}S+\text{a lower-order remainder}; \tag{11}$$

* By the term "in the ellipsoid" we mean "not external to the ellipsoid", i.e. the lattice points "*on* the ellipsoid" are also included. Other such expressions in the sequel carry a similar meaning.

† For different derivations of (10), see Maa (1939) and Husimi (1939).

in terms of ε^*, where

$$\varepsilon^* = \frac{8mL^2}{h^2}\varepsilon = \frac{4L^2}{h^2}P^2 = \frac{L^2}{\pi^2}K^2, \tag{12}$$

eqn. (11) reduces to eqns. (1.4.15) and (1.4.16).

In the case of *periodic* boundary conditions, namely

$$\psi(x, y, z) = \psi(x+a, y+b, z+c), \tag{13}$$

the appropriate wave functions are

$$\psi_{lnm}(\boldsymbol{r}) \propto \exp\{i(\boldsymbol{k}\cdot\boldsymbol{r})\}, \tag{14}$$

with

$$\boldsymbol{k} = 2\pi\left(\frac{l}{a}, \frac{m}{b}, \frac{n}{c}\right); \quad l, m, n = 0, \pm 1, \pm 2, \ldots. \tag{15}$$

The number of free-particle states $g(K)$ is then given by

$$g(K) = \sum_{l, m, n}{}' f^*(l, m, n), \tag{16}$$

such that

$$(l^2/a^2 + m^2/b^2 + n^2/c^2) \leqslant K^2/(4\pi^2). \tag{17}$$

This is precisely the number of lattice points in the ellipsoid with semiaxes $Ka/2\pi$, $Kb/2\pi$ and $Kc/2\pi$, which, allowing for the approximation made in the earlier cases, is just equal to the volume term in (11). Thus, in the case of periodic boundary conditions, we do not obtain a surface term in the expression for the density of states. This result is physically consistent with the spirit in which one imposes periodic boundary conditions on a system, namely that in the thermodynamic limit the box is *indeed* going to be extended infinitely in *each* direction, with the result that surface effects, if any, must eventually disappear.

Fedosov (1963, 1964) has studied this very problem for enclosures of *arbitrary* shape. He finds that, quite generally,

$$g(K) = \frac{K^3}{6\pi^2}V \mp \frac{K^2}{16\pi}S + 0(K^2). \tag{18}$$

In the general case, the error term seems rather comparable to the surface term. Of course, in the special case of a cuboidal enclosure, the error term was decidedly inferior to the one here. Though no corresponding demonstration is available in the general case, one could possibly expect the same to be true for enclosures of arbitrary shape. If so, the error term in (18) would better be written as $o(K^2)$, rather than $0(K^2)$, i.e.

$$g(K) = \frac{K^3}{6\pi^2}V \mp \frac{K^2}{16\pi}S + o(K^2); \tag{19}$$

this implies that, as $K \to \infty$, the ratio of the remainder term to K^2 would tend to vanish.

Intuitively, the foregoing statement was inspired by the feeling that, like the volume term, the surface term too should be independent of the geometry of the enclosure. This would be possible only if the rate of increase of the error term with K is slower than that of the surface term. By computing the *actual* values of the numbers $g(K)$ for spherical and cylindrical enclosures, under varying boundary conditions, Pathria (1966b) has found that, quite generally, this does appear to be the case.

B. Certain Mathematical Functions

In this appendix we propose to outline the main properties of certain mathematical functions which are of importance to the subject matter of this text.

We start with the *Gamma function* $\Gamma(n+1)$, which is identical with the *factorial function* $n!$ and is defined by the equation

$$\Gamma(n+1) \equiv n! = \int_0^\infty e^{-x} x^n \, dx; \qquad n > -1. \tag{1}$$

First of all we note that

$$\Gamma(1) \equiv 0! = 1. \tag{2}$$

Next, integrating by parts, we obtain the reduction formula

$$\Gamma(n+1) = n\Gamma(n), \tag{3}$$

whence it follows that

$$\Gamma(n+1) = n(n-1) \dots (1+p)\, p\, \Gamma(p), \qquad 1 \geqslant p > 0, \tag{4}$$

p being the fractional part of n. For integral values of n, we have the most familiar representation, viz.

$$\Gamma(n+1) \equiv n! = n(n-1) \dots 2 \cdot 1; \tag{5}$$

on the other hand, if n is half-odd integral, say $n = m - \frac{1}{2}$, then

$$\begin{aligned}\Gamma\left(m+\frac{1}{2}\right) \equiv \left(m-\frac{1}{2}\right)! &= \left(m-\frac{1}{2}\right)\left(m-\frac{3}{2}\right) \cdots \frac{3}{2} \cdot \left(\frac{1}{2}\right) \Gamma\left(\frac{1}{2}\right) \\ &= \frac{(2m-1)(2m-3) \dots 3 \cdot 1}{2^m} \pi^{1/2},\end{aligned} \tag{6}$$

where use has been made of the fact that

$$\Gamma(\tfrac{1}{2}) \equiv (-\tfrac{1}{2})! = \pi^{1/2}, \tag{7}$$

which will be proved subsequently; see eqn. (21).

Replacing x by αy^2 and n by $(n-1)$, we obtain from (1)

$$\Gamma(n) = 2\alpha^n \int_0^\infty e^{-\alpha y^2} y^{2n-1} \, dy, \qquad n > 0. \tag{8}$$

Thus, we obtain another closely related integral, namely

$$I_{2n-1} \equiv \int_0^\infty e^{-\alpha y^2} y^{2n-1} \, dy = \frac{1}{2\alpha^n} \Gamma(n), \qquad n > 0; \tag{9}$$

by a change of notation, this can be written as

$$I_n \equiv \int_0^\infty e^{-\alpha y^2} y^n \, dy = \frac{1}{2\alpha^{(n+1)/2}} \Gamma\left(\frac{n+1}{2}\right), \qquad n > -1. \tag{10}$$

One can casily see that these integrals satisfy the following relationship:

$$I_{n+2} = -\frac{\partial}{\partial \alpha}(I_n). \tag{11}$$

The integrals I_n appear so frequently in our study that write down the values of some of them rather explicitly. Thus, we have

$$I_0 = \frac{1}{2}\left(\frac{\pi}{\alpha}\right)^{1/2}, \quad I_2 = \frac{1}{4}\left(\frac{\pi}{\alpha^3}\right)^{1/2}, \quad I_4 = \frac{3}{8}\left(\frac{\pi}{\alpha^5}\right)^{1/2}, \ldots, \tag{12}$$

while

$$I_1 = \frac{1}{2\alpha}, \quad I_3 = \frac{1}{2\alpha^2}, \quad I_5 = \frac{1}{\alpha^3}, \ldots. \tag{13}$$

In connection with these integrals, it may also be noted that

$$\begin{aligned} \int_{-\infty}^{\infty} e^{-\alpha y^2} y^n \, dy &= 0 \quad \text{if } n \text{ is an odd integer} \\ &= 2I_n \quad \text{if } n \text{ is an even integer.} \end{aligned} \tag{14}$$

Next we consider the product of two gamma functions, say $\Gamma(m)$ and $\Gamma(n)$. Following the representation (8), with $\alpha = 1$, we obtain

$$\Gamma(m)\,\Gamma(n) = 4 \int_0^\infty \int_0^\infty e^{-(x^2+y^2)} x^{2m-1} y^{2n-1} \, dx \, dy, \qquad m > 0, \ n > 0. \tag{15}$$

Changing over to the polar coordinates (r, θ), eqn. (15) becomes

$$\begin{aligned} \Gamma(m)\,\Gamma(n) &= 4 \int_0^\infty e^{-r^2} r^{2(m+n)-1} \, dr \int_0^{\pi/2} \cos^{2m-1}\theta \sin^{2n-1}\theta \, d\theta \\ &= 2\Gamma(m+n) \int_0^{\pi/2} \cos^{2m-1}\theta \sin^{2n-1}\theta \, d\theta, \end{aligned} \tag{16}$$

where we have made use of (9), again with $\alpha = 1$. Defining the *beta function* $B(m, n)$ by the equation

$$B(m, n) = 2 \int_0^{\pi/2} \cos^{2m-1}\theta \sin^{2n-1}\theta \, d\theta, \qquad m > 0, \ n > 0, \tag{17}$$

we obtain the important relationship

$$B(m, n) = \frac{\Gamma(m)\,\Gamma(n)}{\Gamma(m+n)} = B(n, m). \tag{18}$$

Substituting $\cos^2\theta = \eta$, eqn. (17) takes the form

$$B(m, n) = \int_0^1 \eta^{m-1}(1-\eta)^{n-1}\, d\eta, \quad m > 0, n > 0, \tag{19}$$

while in the special case $m = n = \frac{1}{2}$, it gives

$$B(\tfrac{1}{2}, \tfrac{1}{2}) = 2\int_0^{\pi/2} d\theta = \pi; \tag{20}$$

coupled with (2) and (18), this yields the important result (7), viz.

$$\Gamma(\tfrac{1}{2}) = \pi^{1/2}. \tag{21}$$

The Gaussian distribution function and the error function

For $n = 0$, eqn. (10) becomes

$$\int_{-\infty}^{\infty} e^{-\alpha y^2}\, dy = 2I_0 = \left(\frac{\pi}{\alpha}\right)^{1/2}, \tag{22}$$

which, on substituting $y = x - x_0$ and $\alpha = (2\sigma^2)^{-1}$, takes the form

$$\frac{1}{\sqrt{(2\pi)}\sigma} \int_{-\infty}^{\infty} e^{-(x-x_0)^2/(2\sigma^2)}\, dx = 1. \tag{23}$$

We now define the Gaussian probability distribution function $p(x)$ as

$$p(x)\, dx = \frac{1}{\sqrt{(2\pi)}\sigma} e^{-(x-x_0)^2/(2\sigma^2)}\, dx, \quad -\infty < x < \infty, \tag{24}$$

so that, by virtue of (23),

$$\int_{-\infty}^{\infty} p(x)\, dx = 1, \tag{25}$$

which means that the distribution function $p(x)$ is properly normalized. Clearly, x_0 represents the mean value of the variable x while σ represents its root-mean-square deviation from the mean value.

Of many physical situations governed by the Gaussian distribution, a typical one relates to the distribution of *random errors* involved in a sequence of measurements. According to this, the probability that the magnitude of the error $(x - x_0)$ in the measurement of the quantity x does not exceed a pre-assigned value t is given by

$$\begin{aligned} P(t) &= \frac{1}{\sqrt{(2\pi)}\sigma} \int_{-t}^{t} e^{-(x-x_0)^2/(2\sigma^2)}\, dx \\ &= \frac{2}{\sqrt{\pi}} \int_0^{t/\sqrt{(2)}\sigma} e^{-z^2}\, dz = \operatorname{erf}\left(\frac{t}{\sqrt{(2)}\sigma}\right). \end{aligned} \tag{26}$$

This equation defines the so-called *error function*; obviously,

$$\text{erf}\,(0) = 0 \quad \text{and} \quad \text{erf}\,(\infty) = 1. \tag{27}$$

In particular, the probability that $|x - x_0| \leqslant n\sigma$ is given by $\text{erf}\,(n/\sqrt{2})$. Some special values of this probability are

$$\text{erf}\,(n/\sqrt{2}) = 0.6827,\ 0.9545 \quad \text{and} \quad 0.9973$$

for $n = 1, 2$ and 3, respectively. Alternatively,

$$\text{erf}\,(n/\sqrt{2}) = 0.5,\ 0.9 \quad \text{and} \quad 0.99$$

for $n = 0.6745$, 1.6449 and 2.5758, respectively; the first value here implies that, for errors distributed in the Gaussian manner, the magnitude of any particular error is *as likely as not* to exceed the value 0.6745σ. This special value is generally referred to as the *probable error* of the measurement concerned and is often the one stated along with the mean value of an experimentally observed quantity; the value σ itself is usually referred to as the *standard error*.

Detailed numerical tables of the error function are available in the mathematical literature; see, for example, Abramovitz and Stegun (1964); Beyer (1966). Nevertheless, a few analytical expansions are useful for approximate evaluations: for instance,

$$\text{erf}\,(z) = \frac{2}{\sqrt{\pi}}\left[z - \frac{z^3}{3\cdot 1!} + \frac{z^5}{5\cdot 2!} - \frac{z^7}{7\cdot 3!} + \ldots\right], \tag{28}$$

which is convergent for all values of z but is particularly useful for small z;

$$\text{erf}\,(z) = \frac{2}{\sqrt{\pi}}\,e^{-z^2}\left[z + \frac{2z^3}{1\cdot 3} + \frac{4z^5}{1\cdot 3\cdot 5} + \frac{8z^7}{1\cdot 3\cdot 5\cdot 7} + \ldots\right], \tag{29}$$

which is again useful for small z; finally,

$$\text{erf}\,(z) = 1 - \frac{e^{-z^2}}{\sqrt{(\pi)}z}\left[1 - \frac{1}{2z^2} + \frac{1\cdot 3}{(2z^2)^2} - \frac{1\cdot 3\cdot 5}{(2z^2)^3} + \ldots\right]. \tag{30}$$

The last expression, strictly speaking, does not converge for any value of z; however, with an appropriate number of terms used, it provides a good approximation for large z.

Stirling's formula for $n!$

We now derive an *asymptotic* expression for the factorial function (1), namely

$$n! = \int_0^\infty e^{-x}x^n\,dx. \tag{1}$$

It is not difficult to see that, for $n \gg 1$, the major contribution to this integral comes from that region of x which lies around the value $x = n$ and extends over a range of the order of $\sqrt{n}$. It is thus natural to invoke a substitution such as

$$x = n+\sqrt{(n)}u, \tag{31}$$

whereby eqn. (1) takes the form

$$n! = \sqrt{n}\left(\frac{n}{e}\right)^n \int_{-\sqrt{n}}^{\infty} e^{-\sqrt{(n)}u}\left(1+\frac{u}{\sqrt{n}}\right)^n du. \tag{32}$$

The integrand of (32) attains its maximum value, which is unity, at $u = 0$ and on both sides of the maximum it falls to zero, which strongly suggests that the integrand might be approximated by a Gaussian function in the variable u. We may, therefore, expand the logarithm of the integrand around the value $u = 0$ and then reconstruct the integrand by taking the exponential of the resulting expression; this gives

$$n! = \sqrt{n}\left(\frac{n}{e}\right)^n \int_{-\sqrt{n}}^{\infty} \exp\left\{-\frac{u^2}{2}+\frac{u^3}{3\sqrt{n}}-\frac{u^4}{4n}+\ldots\right\} du. \tag{33}$$

If n is sufficiently large, the integrand here might be replaced by the single factor $\exp(-u^2/2)$; moreover, since the major contribution to the integral would be coming only from that part of the range for which $|u|$ is of the order of unity, the lower limit of integration might be replaced by $-\infty$. Then, making use of the formula (22), with $\alpha = \frac{1}{2}$, we obtain the desired result:

$$n! \simeq \sqrt{(2\pi n)}\,(n/e)^n, \quad n \gg 1. \tag{34}$$

This important result is known as the *Stirling formula.*

It is not difficult to determine the first correction to the formula (34). Since n is a large number, the factor

$$\exp\left\{\frac{u^3}{3\sqrt{n}}-\frac{u^4}{4n}+\ldots\right\}$$

in the integrand of (33) may be expanded as a power series in $1/n$. The lower limit of integration may still be taken as $-\infty$, for that only causes an error $0(e^{-n/2})$ which is far too weak in comparison with terms of the order of $1/n$, $1/n^2$, etc. Equation (33) then becomes

$$\begin{aligned} n! &\simeq \sqrt{n}\left(\frac{n}{e}\right)^n \int_{-\infty}^{\infty} e^{-u^2/2}\left\{1+\frac{u^3}{3\sqrt{n}}+\left(\frac{u^6}{18n}-\frac{u^4}{4n}\right)+\ldots\right\} du \\ &= \sqrt{(2\pi n)}\left(\frac{n}{e}\right)^n\left\{1+\frac{1}{12n}+\ldots\right\}. \end{aligned} \tag{35}$$

Thus, even when $n = 10$ (which can hardly be called a large number) formula (34) is already accurate to better than 1 per cent. In passing, it may be noted that a more detailed analysis leads to the following asymptotic expansion for the factorial function:

$$n! = \sqrt{(2\pi n)}\left(\frac{n}{e}\right)^n\left[1+\frac{1}{12n}+\frac{1}{288n^2}-\frac{139}{51840n^3}-\frac{571}{2488320n^4}+\ldots\right]. \tag{36}$$

Finally, we consider the function ln $(n!)$. Corresponding to formula (36) we have, for large n,

$$\ln(n!) = \left(n+\frac{1}{2}\right)\ln n - n + \frac{1}{2}\ln(2\pi) + \left[\frac{1}{12n} - \frac{1}{360n^3} + \frac{1}{1260n^5} - \frac{1}{1680n^7} + \ldots\right]. \quad (37)$$

Now, the ratio $(\ln n)/n$, in the limit $n \to \infty$, tends to zero; therefore, for most purposes, we can write

$$\ln(n!) \simeq (n \ln n - n). \quad (38)$$

It might be remarked here that formula (38) can be obtained very simply by a crude application of the Euler–Maclaurin formula. Since n is very large, we may consider the integral values alone; then, by definition,

$$\ln(n!) = \sum_{i=1}^{n} (\ln i).$$

Replacing summation by integration, we obtain

$$\ln(n!) \simeq \int_1^n (\ln x)\,dx = (x \ln x - x)\Big|_{x=1}^{x=n}$$
$$\simeq (n \ln n - n),$$

which is identical with (38).

We must, however, be warned that whereas approximation (38) is all right in its own place, it would be wrong to invert it and adopt $n! \simeq (n/e)^n$, for that would affect the evaluation of $n!$ by a *factor* which is $0(n^{1/2})$ and can be considerably large. In the expression for ln $(n!)$, the corresponding *term* is indeed negligible.

Numerical accuracy obtainable by using the two versions of Stirling's formula is depicted in Table B.1. Comparative features of the computations are self-evident.

TABLE B.1. STIRLING'S FORMULA AT WORK

n	$n!$			ln $(n!)$		
	From (38)	From (34)	Actual	From (38)	From (34)	Actual
5	21.06	118.0	120	3.0472	4.7709	4.7875
10	4.540×10^{5}	3.599×10^{6}	3.629×10^{6}	13.0258	15.0961	15.1044
20	2.161×10^{17}	2.423×10^{18}	2.433×10^{18}	39.9146	42.3314	42.3356
50	1.713×10^{63}	3.036×10^{64}	3.041×10^{64}	145.6012	148.4761	148.4778
100	3.720×10^{156}	9.325×10^{157}	9.333×10^{157}	360.5170	363.7386	363.7394

The Dirac δ-function

We had in eqn. (24) the Gaussian distribution function, viz.

$$p(x; x_0, \sigma) = \frac{1}{\sqrt{(2\pi)}\sigma} e^{-(x-x_0)^2/(2\sigma^2)}, \quad (39)$$

which satisfies the normalization condition

$$\int_{-\infty}^{\infty} p(x; x_0, \sigma)\, dx = 1. \tag{40}$$

The function $p(x)$ is symmetric about the value x_0 where it shows a maximum; the height of the maximum is inversely proportional to the parameter σ while its width is directly proportional to σ, the total area under the curve being a constant. As σ becomes smaller and smaller, the function becomes narrower and narrower in width and grows higher and higher at the central point, condition (40) being satisfied for all values of σ; see Fig. B.1.

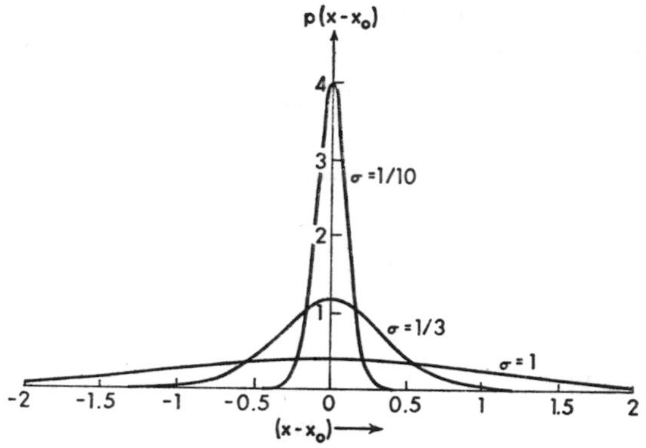

FIG. B.1.

In the limit $\sigma \to 0$, we are confronted with a function whose value at $x = x_0$ is infinitely large while at $x \neq x_0$ it is vanishingly small, the area under the curve being still equal to unity. This limiting form of the function $p(x; x_0, \sigma)$ defines the *δ-function* of Dirac. We can define this function anew as the one satisfying the following conditions:

$$\text{(i)} \quad \delta(x-x_0) = 0 \quad \text{for all } x \neq x_0 \tag{41}$$

and

$$\text{(ii)} \quad \int_{-\infty}^{\infty} \delta(x-x_0)\, dx = 1. \tag{42}$$

These conditions inherently imply that, at $x = x_0$, $\delta(x-x_0) = \infty$ and that the range of integration in (42) need not extend all the way from $-\infty$ to $+\infty$; actually, any range that includes the point $x = x_0$ would suffice. Thus, we may rewrite (42) as

$$\int_{A}^{B} \delta(x-x_0)\, dx = 1 \quad \text{if } A < x_0 < B, \tag{42a}$$

whence it follows that for *any* well-behaved function $f(x)$

$$\int_{A}^{B} f(x)\, \delta(x-x_0)\, dx = f(x_0) \quad \text{if } A < x_0 < B. \tag{43}$$

There is another limiting process which is frequently employed to represent the δ-function, viz.

$$\delta(x-x_0) = \lim_{\gamma \to 0} \frac{\gamma}{\pi\{(x-x_0)^2+\gamma^2\}} \cdot \tag{44}$$

To see the appropriateness of this representation, we note that, for $x \neq x_0$, this function vanishes (like γ) while, for $x = x_0$, it diverges (like γ^{-1}); moreover, for all γ,

$$\int_{-\infty}^{\infty} \frac{\gamma}{\pi\{(x-x_0)^2+\gamma^2\}}\, dx = \frac{1}{\pi}\left[\tan^{-1}\frac{(x-x_0)}{\gamma}\right]_{-\infty}^{\infty} = 1.$$

The above representations of the δ-function are indeed very good; still it is the integral representation that has proved most useful. In this, the δ-function is not defined as a limiting form of some other mathematical function but rather as the "Fourier transform of a constant number":

$$\delta(x-x_0) = \frac{1}{2\pi}\int_{-\infty}^{\infty} e^{ik(x-x_0)}\, dk. \tag{45}$$

We see here that, for $x = x_0$, the integrand is throughout equal to unity whereby the function diverges. On the other hand, for $x \neq x_0$, the oscillatory character of the integrand makes the whole thing vanish. And, finally, the integration of this function, over a range of x that includes the value $x = x_0$, gives

$$\frac{1}{2\pi}\int_{-\infty}^{\infty}\left[\int_{x_0-L}^{x_0+L} e^{ik(x-x_0)}\, dx\right] dk = \int_{-\infty}^{\infty} \frac{e^{ikL}-e^{-ikL}}{2\pi(ik)}\, dk$$

$$= \frac{1}{\pi}\int_{-\infty}^{\infty} \frac{\sin(kL)}{k}\, dk = 1, \tag{46}$$

independently of the choice of L. Thus, one feels convinced that the representation (45) is perfectly sound.

It is quite instructive to see how the integral representation of the δ-function is related to its previous representations. To see this, we introduce into the integrand of (45) a factor $\exp(-\gamma k^2)$, where γ is an infinitesimally small, positive number. The resulting function, in the limit $\gamma \to 0$, must reduce to the δ-function; accordingly, we must have

$$\delta(x-x_0) = \frac{1}{2\pi}\lim_{\gamma \to 0}\int_{-\infty}^{\infty} e^{ik(x-x_0)-\gamma k^2}\, dk. \tag{47}$$

The real advantage of introducing the new factor here is that it avoids any convergence ambiguities that might otherwise arise from the limiting values of k, viz. $k = \pm\infty$. The integral in (47) can be evaluated very easily if we recall that

$$\int_{-\infty}^{\infty} \cos(kx)e^{-\gamma k^2}\,dk = 2\int_{0}^{\infty} \cos(kx)e^{-\gamma k^2}\,dk = \sqrt{\left(\frac{\pi}{\gamma}\right)}e^{-x^2/(4\gamma)} \tag{48}$$

and

$$\int_{-\infty}^{\infty} \sin(kx)e^{-\gamma k^2}\,dk = 0. \tag{49}$$

Accordingly, (47) becomes

$$\delta(x-x_0) = \lim_{\gamma\to 0} \frac{1}{\sqrt{(4\pi\gamma)}}e^{-(x-x_0)^2/(4\gamma)}. \tag{50}$$

This is precisely the representation we started with, except that there we had $\frac{1}{2}\sigma^2$ instead of γ.*

The last steps leading to eqn. (46) suggest that we may take a somewhat closer look at the behavior of the function $\sin(gx)/(\pi x)$ for varying values of g. First of all, we have

$$\int_{-\infty}^{\infty} \frac{\sin(gx)}{\pi x}\,dx = 1, \tag{51}$$

irrespective of the value of g. Secondly, as $x \to 0$, this function approaches the limiting value g/π; accordingly, in the limit $g \to \infty$, the function diverges at the origin. Its integral from $-\infty$ to $+\infty$, however, remains the same. The heart of the matter is that, as g becomes larger and larger, the *nodal separation* π/g of the function becomes smaller and smaller, so that, in the limit $g \to \infty$, any finite interval of x contains an infinitely large number of cycles, with the result that any integral with this limiting function in its integrand would receive practically no contribution from any other region of x except from the extreme neighborhood of $x = 0$. Thus, we have

$$\lim_{g\to\infty} \int_{-\infty}^{\infty} \frac{\sin(gx)}{\pi x} f(x) = f(0), \tag{52}$$

a result that can be readily verified for functions such as $\exp(-\gamma x^2)$, $\exp(-\gamma|x|)$, etc. Comparing (52) with (43), we conclude that the limiting form of the function $\sin(gx)/(\pi x)$ serves *practically* the same purpose as the Dirac function $\delta(x)$. Of course, we must not overlook an important difference, namely that the function $\sin(gx)/(\pi x)$, even in the limit $g \to \infty$, *does not vanish for all* $x \neq 0$; it is only its integral over *any* finite interval of x (not including $x = 0$) that vanishes!

We observe that representation (45) is no different from the "closure" property of the plane-wave solutions of the free-particle Schrödinger equation, namely

$$\int_{-\infty}^{\infty} \psi_k^*(x_1)\,\psi_k(x_2)\,dk = \delta(x_2-x_1), \tag{53}$$

* The reader may like to check that the introduction of a factor $\exp(-\gamma|k|)$, instead of the factor $\exp(-\gamma k^2)$, leads to the representation (44).

where

$$\psi_k(x) = \frac{1}{\sqrt{(2\pi)}} e^{ikx}. \tag{54}$$

Analogously, the "orthogonality" of the wave functions may be expressed as

$$\int_{-\infty}^{\infty} \psi_{k_1}^*(x)\, \psi_{k_2}(x)\, dx = \delta(k_2-k_1). \tag{55}$$

An important and frequently occurring situation, which inevitably requires the use of the δ-function, is the one arising from the differentiation of a step function $\varepsilon(x-x_0)$; here,

$$\begin{aligned} \varepsilon(x-x_0) &= 0 \quad \text{for } x < x_0, \\ &= 1 \quad \text{for } x > x_0. \end{aligned} \tag{56}$$

Clearly, the derivative of $\varepsilon(x)$ with respect to x satisfies all the properties of the δ-function. Accordingly, we have

$$\varepsilon'(x-x_0) \equiv \frac{d}{dx} \varepsilon(x-x_0) = \delta(x-x_0). \tag{57}$$

One can verify that for all A and B, such that $A < x_0 < B$,

$$\int_A^B f(x)\, \varepsilon'(x-x_0)\, dx = f(x_0), \tag{58}$$

which is the same as (43).

For illustration, we observe that the function $\ln x$, where x is real, possesses (for all values of x) the property

$$\ln(-x) = \ln(x) + \ln(-1),$$

which, on taking principal values, becomes

$$\ln(-x) = \ln(x) + i\pi. \tag{59}$$

Thus, by virtue of the real part alone, the function $\ln x$ is symmetric about $x = 0$; however, at any point on the negative side of the x-axis the function contains an extra imaginary part $i\pi$ which it does not contain at the corresponding point on the positive side of the axis. In other words, the function $\ln x$ possesses a built-in step function, of value $i\pi\{1-\varepsilon(x)\}$; accordingly, its derivative with respect to x must contain a δ-function, i.e.

$$\frac{d}{dx}(\ln x) = \frac{1}{x} - i\pi\, \delta(x), \tag{60}$$

and not simply $1/x$. By (60), the function $1/x$ gets well defined (in the sense of an improper function) in the neighborhood of the point $x = 0$, in that its integral from $-a$ to $+a$ ($a > 0$) is now exactly equal to zero (as it should be for any odd function of the variable).

Finally, we note that the notation of the δ-function can be readily extended to variables in more than one dimension. For instance, in three dimensions, we have

$$\delta(\boldsymbol{r}) \equiv \delta(x_1)\,\delta(x_2)\,\delta(x_3), \tag{61}$$

so that (i)

$$\delta(\boldsymbol{r}) = 0 \qquad (\text{unless } x = y = z = 0) \tag{62}$$

and (ii)

$$\iiint_{-\infty}^{\infty} \delta(\boldsymbol{r})\, dx_1\, dx_2\, dx_3 = 1. \tag{63}$$

The integral representation of $\delta(\boldsymbol{r})$ is

$$\delta(\boldsymbol{r}) = \frac{1}{(2\pi)^3} \int_{-\infty}^{\infty} e^{i(\boldsymbol{k}\cdot\boldsymbol{r})}\, d\boldsymbol{k}. \tag{64}$$

Here again one would like to obtain a representation for $\delta(\boldsymbol{r})$ as a limiting form of a certain analytic function of $\boldsymbol{r}$. For this, we have, after (47),

$$\delta(\boldsymbol{r}) = \frac{1}{(2\pi)^3} \operatorname*{Lim}_{\gamma\to 0} \int_{-\infty}^{\infty} e^{i(\boldsymbol{k}\cdot\boldsymbol{r})-\gamma k^2}\, d\boldsymbol{k}. \tag{65}$$

Integrating over k_1, k_2 and k_3, we obtain

$$\delta(\boldsymbol{r}) = \operatorname*{Lim}_{\gamma\to 0} \left(\frac{1}{4\pi\gamma}\right)^{3/2} e^{-r^2/(4\gamma)}. \tag{66}$$

It is quite straightforward to verify that

$$\int_{-\infty}^{\infty} \delta(\boldsymbol{r})\, d\boldsymbol{r} = \int_{0}^{\infty} \left(\frac{1}{4\pi\gamma}\right)^{3/2} e^{-r^2/(4\gamma)} 4\pi r^2\, dr = 1, \tag{67}$$

as is required by the defining property (63).

The use of δ-function in physical problems is rather commonplace. In electrostatics, for instance, the charge density $\varrho(\boldsymbol{r})$ is written as

$$\varrho(\boldsymbol{r}) = \sum_i q_i\, \delta(\boldsymbol{r}-\boldsymbol{r}_i), \tag{68}$$

the summation going over all the points $\boldsymbol{r}_i$ which denote the positions of the charges q_i. Clearly, $\varrho(\boldsymbol{r}) = 0$ for all $\boldsymbol{r} \neq \boldsymbol{r}_i$ $(i = 1, 2, 3, \ldots)$, while

$$\int \varrho(\boldsymbol{r})\, d\boldsymbol{r} = \sum_i q_i = Q, \tag{69}$$

where Q denotes the total electric charge present in the relevant region of the physical space. The electrostatic potential at $\boldsymbol{r}$ will obviously be given by

$$\varphi(\boldsymbol{r}) = \sum_i q_i \Big/ |\boldsymbol{r}-\boldsymbol{r}_i|. \tag{70}$$

Applying Laplacian operator to (70), we have

$$\nabla^2\varphi(\boldsymbol{r}) = \sum_i q_i \nabla^2 \frac{1}{|\boldsymbol{r}-\boldsymbol{r}_i|}. \tag{71}$$

Now, it is quite easy to see that

$$\nabla^2\left(\frac{1}{r}\right) = 0 \quad \text{for all } r \neq 0; \tag{72}$$

however, at the origin ($r = 0$), $\nabla^2(1/r)$ must diverge. Thus, we clearly see here the need of a δ-function. To be explicit, let us consider the integration of the function $\nabla^2(1/r)$ over a spherical region of radius R centered at the origin. Making use of the divergence theorem, we obtain

$$\int\limits_{0\leq r\leq R} \left\{\nabla\cdot\nabla\left(\frac{1}{r}\right)\right\} dV \equiv \int\limits_{r=R} \left\{\nabla\left(\frac{1}{r}\right)\right\}\cdot d\boldsymbol{S}$$

$$= -\frac{1}{R^2}(4\pi R^2) = -4\pi. \tag{73}$$

Thus, we better replace (72) by

$$\nabla^2\left(\frac{1}{r}\right) = -4\pi\,\delta(\boldsymbol{r}) \quad \text{for } all\ \boldsymbol{r}, \tag{74}$$

a result we required in setting up the method of pseudopotentials to treat the many-body problem at low temperatures; see Chapter 10. By (68), (71) and (74), we finally obtain

$$\nabla^2\varphi(\boldsymbol{r}) = \sum_i q_i[-4\pi\,\delta(\boldsymbol{r}-\boldsymbol{r}_i)] = -4\pi\varrho(\boldsymbol{r}), \tag{75}$$

which is the well-known *Poisson equation.*

C. "Volume" and "Surface Area" of an n-Dimensional Sphere of Radius R

Consider an n-dimensional space in which the position of a point is denoted by the vector $\boldsymbol{r}$, with rectangular components $(x_1, x_2, \ldots, x_n)$. The volume element dV_n in this space would be

$$d\boldsymbol{r} \equiv \prod_{i=1}^{n} (dx_i);$$

accordingly, the "volume" V_n of a sphere of radius R would be given by

$$V_n(R) = \int \cdots \int\limits_{0\leq \sum_{i=1}^{n} x_i^2 \leq R^2} \prod_{i=1}^{n} (dx_i). \tag{1}$$

Obviously, V_n will be proportional to R^n, so let it be written as

$$V_n(R) = C_n R^n; \tag{2}$$

here, C_n is a constant which depends only on the dimensionality of the space. Clearly, the volume element dV_n can also be written as

$$dV_n \equiv S_n(R)\, dR = nC_nR^{n-1}\, dR, \tag{3}$$

where $S_n(R)$ denotes the "surface area" of the sphere.

To evaluate C_n, we make use of formula (B.22), with $\alpha = 1$, viz.

$$\int_{-\infty}^{\infty} \exp(-x^2)\, dx = \pi^{1/2}. \tag{4}$$

Multiplying n such integrals, one for each of variables x_i, we obtain

$$\begin{aligned}\pi^{n/2} &= \int_{x_i=-\infty}^{\infty} \cdots \int \exp\left(-\sum_{i=1}^{n} x_i^2\right) \prod_{i=1}^{n} (dx_i) \\ &= \int_0^\infty \exp(-R^2) nC_nR^{n-1}\, dR \\ &= nC_n \cdot \frac{1}{2}\Gamma\left(\frac{n}{2}\right) = \left(\frac{n}{2}\right)!\; C_n;\end{aligned} \tag{5}$$

here, use has been made of expression (3) and of formula (B.10), again with $\alpha = 1$. Thus

$$C_n = \pi^{n/2} \Big/ \left(\frac{n}{2}\right)!, \tag{6}$$

whence it follows that

$$V_n(R) = \frac{\pi^{n/2}}{(n/2)!} R^n \quad \text{and} \quad S_n(R) = \frac{2\pi^{n/2}}{((n/2)-1)!} R^{n-1}, \tag{7}$$

which are the desired results.

In the three-dimensional case, (7) yields the well-known results: $V_3 = 4\pi R^3/3$ and $S_3 = 4\pi R^2$. Moreover, in the two-dimensional case, we obtain: $V_2 = \pi R^2$ and $S_2 = 2\pi R$, and in the one-dimensional case, $V_1 = 2R$ and $S_1 = 2$. The last case is especially interesting, for V_1 now represents the length of a straight line extending between the points $-R$ and $+R$, while S_1 represents the "count" of its ends!

D. On the Bose–Einstein Integrals

In the theory of Bose–Einstein systems we come across integrals of the type

$$G_n(z) = \int_0^\infty \frac{x^{n-1}\, dx}{z^{-1}e^x - 1}, \qquad 0 \leqslant z \leqslant 1. \tag{1}$$

In this appendix we propose to study the behavior of $G_n(z)$ over the stated range of the parameter z.* First of all, we note that

$$\lim_{z\to 0} G_n(z) = \int_0^\infty ze^{-x}x^{n-1}\,dx = z\Gamma(n). \tag{2}$$

Hence, it appears useful to introduce another function, say $g_n(z)$, such that, for all relevant values of z,

$$G_n(z) \equiv \Gamma(n)\, g_n(z), \tag{3}$$

that is,

$$g_n(z) = \frac{1}{\Gamma(n)}\int_0^\infty \frac{x^{n-1}\,dx}{z^{-1}e^x - 1}. \tag{4}$$

For small z, the integrand may be expanded in powers of z, with the result

$$\begin{aligned} g_n(z) &= \frac{1}{\Gamma(n)}\int_0^\infty x^{n-1}\sum_{l=1}^\infty (ze^{-x})^l\,dx \\ &= \sum_{l=1}^\infty \frac{z^l}{l^n} = z + \frac{z^2}{2^n} + \frac{z^3}{3^n} + \ldots; \end{aligned} \tag{5}$$

thus, for $z \ll 1$, the function $g_n(z)$, *for all* n, behaves like z itself. Moreover, $g_n(z)$ is a monotonically increasing function of z. Its largest value in the physical range of interest obtains when $z = 1$; then, for all $n > 1$, it becomes identical with the Riemann *zeta function* $\zeta(n)$:

$$g_n(1) = \sum_{l=1}^\infty \frac{1}{l^n} = \zeta(n) \qquad (n > 1). \tag{6}$$

For $n \leqslant 1$, $g_n(1)$ diverges!

The numerical values of some of the $\zeta(n)$ are

$$\zeta(2) = \frac{\pi^2}{6} \simeq 1.645, \qquad \zeta(4) = \frac{\pi^4}{90} \simeq 1.082,$$

$$\zeta(6) = \frac{\pi^6}{945} \simeq 1.017, \qquad \zeta(8) = \frac{\pi^8}{9450} \simeq 1.004,$$

$$\zeta(\tfrac{3}{2}) \simeq 2.612, \quad \zeta(\tfrac{5}{2}) \simeq 1.341, \quad \zeta(\tfrac{7}{2}) \simeq 1.127,$$

and, finally,

$$\zeta(3) \simeq 1.202, \qquad \zeta(5) \simeq 1.037, \qquad \zeta(7) \simeq 1.008.$$

A simple process of differentiation brings out an important relationship between $g_n(z)$ and $g_{n-1}(z)$:

$$g_{n-1}(z) = z\frac{\partial}{\partial z}[g_n(z)] = \frac{\partial}{\partial(\ln z)}\, g_n(z). \tag{7}$$

* The behavior of $G_n(z)$ for $z > 1$ has been discussed by Clunie (1954).

This relationship follows quite readily from the series expansion (5), but can also be derived from the defining integral (4). We thus have

$$z\frac{\partial}{\partial z}[g_n(z)] = \frac{z}{\Gamma(n)}\int_0^\infty \frac{e^x x^{n-1}\,dx}{(e^x - z)^2}.$$

Integrating by parts, we get

$$z\frac{\partial}{\partial z}[g_n(z)] = \frac{z}{\Gamma(n)}\left[-\frac{x^{n-1}}{e^x - z}\Bigg|_0^\infty + (n-1)\int_0^\infty \frac{x^{n-2}\,dx}{e^x - z}\right].$$

The integrated part vanishes at both the limits (of course, for $n > 1$), while the part yet to be integrated yields precisely $g_{n-1}(z)$. Indeed, by analytic continuation of the function $g_n(z)$, as defined by (4), to complex n, the validity of relationship (7) can be established for *all* n; see Robinson (1951).

Robinson has also derived a power series expansion for the function $g_n(z)$ which is very useful for $z \simeq 1$. Obviously, it would be better in this case to work with the parameter α ($\equiv -\ln z$), which will be a small, positive number. The desired expansion, which is patently an *analytic* function of n if $n \leqslant 0$ and for all non-integral n, is

$$g_n(\alpha) = \Gamma(1-n)\alpha^{n-1} + \sum_{i=0}^{\infty}\frac{(-1)^i}{i!}\zeta(n-i)\alpha^i; \tag{8}$$

here, $\zeta(n-i)$ is again the Riemann zeta function, of order $(n-i)$, but analytically continued so as to be appropriate for all values of the argument. It can be shown by analytic continuation that series (8) actually holds for *all n*.

In terms of α, eqn. (7) becomes

$$g_{n-1}(\alpha) = -\frac{\partial}{\partial\alpha}[g_n(\alpha)], \tag{9}$$

which is indeed satisfied by the expansion (8).

For the special cases of $n = \frac{5}{2}, \frac{3}{2}$ and $\frac{1}{2}$, we obtain from (8)

$$\begin{aligned}
g_{5/2}(\alpha) &= 2.36\alpha^{3/2} + 1.34 - 2.61\alpha - 0.730\alpha^2 + 0.0347\alpha^3 + \ldots,\\
g_{3/2}(\alpha) &= -3.54\alpha^{1/2} + 2.61 + 1.46\alpha - 0.104\alpha^2 + 0.00425\alpha^3 + \ldots,\\
g_{1/2}(\alpha) &= 1.77\alpha^{-1/2} - 1.46 + 0.208\alpha - 0.0128\alpha^2 + \ldots.
\end{aligned}$$

The terms quoted here are sufficient to yield a better-than-1-per-cent accuracy for all $\alpha \leqslant 1$. The numerical values of these functions have been tabulated by London (1954) over the range $0 \leqslant \alpha \leqslant 2$.

Finally, we note from (8) that in the limit $\alpha \to 0$ (that is, in the limit $z \to 1$), the function $g_n(z)$, for all $n < 1$, diverges like $\alpha^{-(1-n)}$. For $n = 1$, the integral in (4) can be evaluated *exactly*, with the result

$$g_1(z) = -\ln(1-z). \tag{10}$$

As $z \to 1$, $g_1(z)$ diverges logarithmically or, since $z = e^{-\alpha}$, like $\ln(1/\alpha)$. For $n > 1$, $g_n(z)$ tends to a finite limit given by $\zeta(n)$, as was noted in (6) above.

E. On the Fermi–Dirac Integrals

In the theory of Fermi–Dirac systems we come across integrals of the type

$$F_n(z) = \int_0^\infty \frac{x^{n-1}\,dx}{z^{-1}e^x+1}, \qquad 0 \leqslant z < \infty. \tag{1}$$

In this appendix we propose to study the behavior of $F_n(z)$ over the entire range of the parameter z. For the same reason as in the case of Bose–Einstein integrals, we introduce here another function, $f_n(z)$, such that for all z

$$F_n(z) = \Gamma(n)\, f_n(z), \tag{2}$$

that is,

$$f_n(z) = \frac{1}{\Gamma(n)} \int_0^\infty \frac{x^{n-1}\,dx}{z^{-1}e^x+1}. \tag{3}$$

For small z, the integrand may be expanded in powers of z, with the result

$$\begin{aligned} f_n(z) &= \frac{1}{\Gamma(n)} \int_0^\infty x^{n-1} \sum_{l=1}^\infty (-1)^{l-1} (ze^{-x})^l\, dx \\ &= \sum_{l=1}^\infty (-1)^{l-1} \frac{z^l}{l^n} = z - \frac{z^2}{2^n} + \frac{z^3}{3^n} - \ldots; \end{aligned} \tag{4}$$

thus, for $z \ll 1$, the function $f_n(z)$, *for all n*, behaves like z itself.

The functions $f_n(z)$ and $f_{n-1}(z)$ are connected through the relationship

$$f_{n-1}(z) = z\frac{\partial}{\partial z}[f_n(z)] \equiv \frac{\partial}{\partial(\ln z)} f_n(z); \tag{5}$$

this result follows quite readily from the series expansion (4), but can also be derived from the defining integral (3).

To study the behavior of Fermi–Dirac integrals for large values of z, it is advisable to introduce the variable

$$\xi \equiv \ln z, \tag{6}$$

so that

$$F_n(\xi) = \Gamma(n)\, f_n(\xi) = \int_0^\infty \frac{x^{n-1}\,dx}{e^{x-\xi}+1}. \tag{7}$$

For large values of ξ, the situation is primarily controlled by the factor $(e^{x-\xi}+1)^{-1}$, whose departure from its limiting values, namely zero (as $x \to \infty$) and *almost* unity (as $x \to 0$), is significant only in the neighborhood of the point $x = \xi$; see Fig. E.1. The width of this "region of significance" is $O(1)$ and, hence, *much smaller* than the total, effective range of integration, which is $O(\xi)$. Therefore, in the lowest approximation, one may replace the

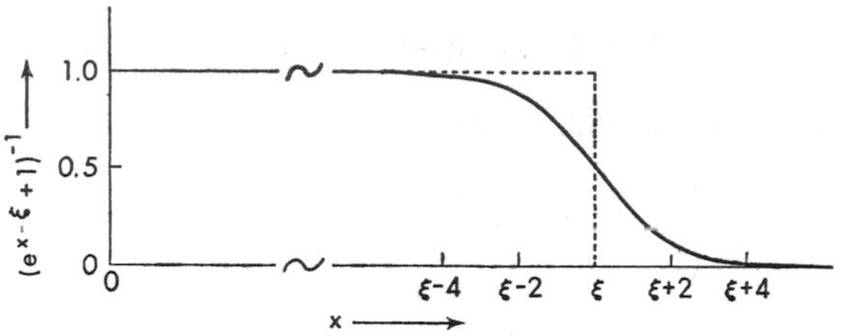

FIG. E.1.

actual curve of Fig. E.1 by a step function, as shown by the dotted line in the diagram. Equation (7) then reduces to

$$F_n(\xi) \simeq \int_0^{\xi} x^{n-1}\,dx = \frac{\xi^n}{n} \tag{8}$$

and, accordingly,

$$f_n(\xi) \simeq \frac{\xi^n}{\Gamma(n+1)}. \tag{9}$$

To evolve a better approximation, we rewrite (7) as

$$F_n(\xi) = \int_0^{\xi} x^{n-1}\left[1-\frac{1}{e^{\xi-x}+1}\right]dx + \int_{\xi}^{\infty} x^{n-1}\frac{1}{e^{x-\xi}+1}\,dx$$

and substitute in the respective integrals

$$x = \xi-\eta_1 \quad \text{and} \quad x = \xi+\eta_2,$$

with the result

$$F_n(\xi) = \frac{\xi^n}{n} + \int_{\xi}^{0}\frac{(\xi-\eta_1)^{n-1}\,d\eta_1}{(e^{\eta_1}+1)} + \int_0^{\infty}\frac{(\xi+\eta_2)^{n-1}\,d\eta_2}{e^{\eta_2}+1}. \tag{10}$$

Since $\xi \gg 1$ while our integrands are significant only for small values of the respective variables, the lower limit in the first integral may be replaced by ∞. Moreover, one might use the same variable η in both the integrals, with the result

$$F_n(\xi) = \frac{\xi^n}{n} + \int_0^{\infty}\frac{(\xi+\eta)^{n-1}-(\xi-\eta)^{n-1}}{e^{\eta}+1}\,d\eta$$

$$= \frac{\xi^n}{n} + 2\sum_{j=1,3,5,\ldots}\binom{n-1}{j}\left[\xi^{n-1-j}\int_0^{\infty}\frac{\eta^j}{e^{\eta}+1}\,d\eta\right]; \tag{11}$$

in the last step the numerator in the integrand has been expanded in powers of η. Now,

$$\frac{1}{\Gamma(j+1)}\int_0^\infty \frac{\eta^j}{e^\eta+1}\,d\eta = 1-\frac{1}{2^{j+1}}+\frac{1}{3^{j+1}}-\dots$$

$$= \left(1-\frac{1}{2^j}\right)\zeta(j+1); \tag{12}$$

cf. the corresponding Bose–Einstein integral (D.4, 5), with $n = j+1$ and $z = 1$. Substituting (12) into (11), we finally obtain

$$f_n(\xi) = \frac{\xi^n}{\Gamma(n+1)}\left[1+\sum_{j=2,4,6,\dots}\left\{2n(n-1)\dots(n+1-j)\left(1-\frac{1}{2^{j-1}}\right)\frac{\zeta(j)}{\xi^j}\right\}\right]$$

$$= \frac{\xi^n}{\Gamma(n+1)}\left[1+n(n-1)\frac{\pi^2}{6}\frac{1}{\xi^2}+n(n-1)(n-2)(n-3)\frac{7\pi^4}{360}\frac{1}{\xi^4}+\dots\right], \tag{13}$$

which is the desired asymptotic formula—commonly known as Sommerfeld's lemma (see Sommerfeld, 1928).*

By the same procedure, one can derive the following asymptotic result, which is clearly more general than (13):

$$\int_0^\infty \frac{\varphi(x)\,dx}{e^{x-\xi}+1} = \int_0^\xi \varphi(x)\,dx+\frac{\pi^2}{6}\left(\frac{d\varphi}{dx}\right)_{x=\xi}+\frac{7\pi^4}{360}\left(\frac{d^3\varphi}{dx^3}\right)_{x=\xi}+\frac{31\pi^6}{15120}\left(\frac{d^5\varphi}{dx^5}\right)_{x=\xi}+\dots; \tag{14}$$

here, $\varphi(x)$ is any well-behaved function of x. It may be noted that the numerical coefficients in this expansion approach a limiting value of 2.

Blakemore (1962) has tabulated the numerical values of the function $f_n(\xi)$ for the range $-4 \leq \xi \leq +10$; his tables cover all integral orders from $n = 0$ to $+5$ and all half-odd integral orders from $n = -\frac{1}{2}$ to $+\frac{9}{2}$.

* A more careful analysis carried out by Rhodes (1950), and followed up by Dingle (1956), shows that the expansion (13) omits a term which, for large ξ, is of the order $e^{-\xi}$. This term turns out to be $\cos\{(n-1)\pi\}$ $f_n(-\xi) \equiv \cos\{(n-1)\pi\} f_n^{(1/z)}$. For large z, this term would be very nearly equal to $\cos\{(n-1)\pi\}/z$ and would be negligible in comparison with any of the terms in the expansion (13). Of course, for $n = \frac{1}{2}, \frac{3}{2}, \frac{5}{2}, \dots$, which are the values occurring in most of the important applications of Fermi–Dirac statistics, the missing term is identically equal to zero.

F. General Physical Constants

Constant	*Symbol*	*Value*	*System of units* (MKS A)	(CGS)
Speed of light (in vacuum)	c	2.998	$\times 10^{8}$ m s^{-1}	$\times 10^{10}$ cm s^{-1}
Planck constant	h	6.626	10^{-34} J s	10^{-27} erg s
	$\hbar(=h/2\pi)$	1.055	10^{-34} J s	10^{-27} erg s
Elementary charge	e	1.602	10^{-19} C	10^{-20} cm$^{1/2}$ g$^{1/2}$*
		4.803		10^{-10} cm$^{3/2}$ g$^{1/2}$ s^{-1}†
Fine structure constant ($e^2/\hbar c$)	α	7.297	10^{-3}	10^{-3}
	$1/\alpha$	1.370	10^{2}	10^{2}
Faraday constant	F	9.649	10^{4} C mol^{-1}	10^{3} cm$^{1/2}$ g$^{1/2}$ mol^{-1}*
		2.893		10^{14} cm$^{3/2}$ g$^{1/2}$ s^{-1} mol^{-1}†
Avogadro number (F/e)	N_A	6.023	10^{23} mol^{-1}	10^{23} mol^{-1}
Mass unit	u	1.660	10^{-27} kg	10^{-24} g
Normal volume (perfect gas)	V_0	2.241	10^{-2} m^{3} mol^{-1}	10^{4} cm^{3} mol^{-1}
Gas constant	R	8.314	10^{0} J °K^{-1} mol^{-1}	10^{7} erg°K^{-1} mol^{-1}
Boltzmann constant (R/N_A)	k	1.381	10^{-23} J °K^{-1}	10^{-16} erg°K^{-1}
First radiation constant ($2\pi hc^2$)	c_1	3.742	10^{-16} W m^{2}	10^{-5} erg cm^{2} s^{-1}
Second radiation constant (hc/k)	c_2	1.439	10^{-2} m°K	10^{0} cm°K
Wien displacement constant ($c_2/4.96511\ldots$)	b	2.898	10^{-3} m°K	10^{-1} cm°K
Stefan–Boltzmann constant ($2\pi^5k^4/15c^2h^3$)	σ	5.670	10^{-8} W m^{-2}°K^{-4}	10^{-5} erg cm^{-2} s^{-1}°K^{-4}
Electron rest mass	m_e	9.109	10^{-31} kg	10^{-28} g
		5.486	10^{-4} u	10^{-4} u
Proton rest mass	m_p	1.673	10^{-27} kg	10^{-24} g
		1.007277	10^{0} u	10^{0} u
Neutron rest mass	m_n	1.675	10^{-27} kg	10^{-24} g
		1.008665	10^{0} u	10^{0} u
Charge-to-mass ratio for electron	e/m_e	1.759	10^{11} C kg^{-1}	10^{7} cm$^{1/2}$ g$^{-1/2}$*
		5.273		10^{17} cm$^{3/2}$ g$^{-1/2}$ s^{-1}†
Rydberg constant ($m_e e^4/4\pi\hbar^3 c$)	R_∞	1.097	10^{7} m^{-1}	10^{5} cm^{-1}
Bohr radius ($\hbar^2/m_e e^2$)	a_B	5.292	10^{-11} m	10^{-9} cm
Bohr magneton ($e\hbar/2m_e c$)	μ_B	9.273	10^{-24} J T^{-1}	10^{-21} erg G^{-1}*
Nuclear magneton ($e\hbar/2m_p c$)	μ_N	5.051	10^{-27} J T^{-1}	10^{-24} erg G^{-1}*
Gravitational constant	G	6.67	10^{-11} N m^{2} kg^{-2}	10^{-8} dyn cm^{2} g^{-2}

* Electromagnetic system. † Electrostatic system.

C — coulomb J — joule W — watt N — newton T — tesla G — gauss

u — atomic mass unit (1 u = 931.5 MeV, while 1 eV = 1.602×10^{-12} erg).

G. Defined Values and Equivalents*

Meter (m)	1650763.73 wavelengths of the unperturbed transition $2p_{10}-5d_5$ in ^{86}Kr
Kilogram (kg)	Mass of the international kilogram
Second (s)	(i) Astronomical 1/31556925.9747 of the tropical year at 12^h ET, 0 January 1900 (yr = $365^d5^h48^m45^s.9747$) (ii) Physical 9192631770 cycles of the hyperfine transition (4, 0 → 3, 0) of the ground state of ^{133}Cs unperturbed by external fields
Degree Kelvin (°K)	On the thermodynamic scale, 273.16°K = triple-point of water t (°C) = T (°K)−273.15. (F.P. of water: 0.0000±0.0002°C)
Unified atomic mass unit (u)	1/12 the mass of an atom of the ^{12}C nuclide
Standard acceleration of free fall (g_n)	9.80665 m s^{-2} 980.665 cm s^{-2}
Normal atmosphere (atm)	101325 N m^{-2} 1013250 dyn cm^{-2}
Thermochemical calorie (cal_{th})	4.184 J 4.184×10^7 erg
Int. Steam Table calorie (cal_{IT})	4.1868 J 4.1868×10^7 erg
Liter (l.)	0.001000028 m^3 1000.028 cm^3
Inch (in.)	0.0254 m 2.54 cm
Pound (lb)	0.45359237 kg 453.59237 g

* The *defined values and equivalents* tabulated here are the ones adopted by the 11th General Conference on Weights and Measures, held in October 1960. For a detailed story of the circumstances leading to this adoption, see Pathria (1964), *School Science* (NCERT, New Delhi), vol. 3, pp. 5, 120, 219, 327.

H. General Mathematical Constants

	Value	Common logarithm	Natural logarithm
π	3.14159	0.49715	1.14473
π^2	9.86960	0.99430	2.28946
π^3	31.00628	1.49145	3.43419
$\pi^{1/2}$	1.77245	0.24857	0.57236
$\pi^{1/3}$	1.46459	0.16572	0.38158
π^{-1}	0.31831	$\bar{1}.50285$	−1.14473
$\pi^{-1/2}$	0.56419	$\bar{1}.75143$	−0.57236
e	2.71828	0.43429	1.00000
e^{-1}	0.36788	$\bar{1}.56571$	−1.00000
	2.00000	0.30103	0.69315
	3.00000	0.47712	1.09861
	10.00000	1.00000	2.30259

BIBLIOGRAPHY

ABRAHAM, B. M., OSBORNE, D. W. and WEINSTOCK, B. (1950) *Phys. Rev.* **80,** 366.

ABRAMOVITZ, M. and STEGUN, I. A. (eds.) (1964) *Handbook of Mathematical Functions* (National Bureau of Standards, Washington).

ABRIKOSOV, A. A. and KHALATNIKOV, I. M. (1957) *J. Exptl. Theoret. Phys. USSR* **33**, 1154; English transl. *Soviet Phys. JETP* **6,** 888 (1958).

ABRIKOSOV, A. A., GOR'KOV, L. P. and DZYALOSHINSKII, I. Y. (1965) *Quantum Field Theoretical Metho in Statistical Physics* (Pergamon Press, Oxford).

ANDREWS, F. C. (1963) *Equilibrium Statistical Mechanics* (John Wiley, New York).

ARZELIÈS, H. (1965) *Nuovo Cim.* **35,** 792.

ATKINS, K. R. (1953) *Can. J. Phys.* **31,** 1165.

ATKINS, K. R. (1957) *Physica* **23,** 1143.

ATKINS, K. R. (1959) *Liquid Helium* (Cambridge University Press).

AULUCK, F. C. and KOTHARI, D. S. (1946) *Proc. Camb. Phil. Soc.* **42,** 272.

AULUCK, F. C. and KOTHARI, D. S. (1946–7) *Proc. Roy. Irish Academy*, p. 13.

BAKER, Jr., G. A., GAMMEL, J. L. and WILLS, J. G. (1961) *J. Math. Anal. Appl.* **2,** 405; also BAKER, Jr., G. A. (1965) in *Advances in Theoretical Physics*, ed. K. A. BRUECKNER (Academic Press, New York).

BAND, W. (1955) *An Introduction to Quantum Statistics* (Van Nostrand, Princeton).

BARDEEN, J., COOPER, L. N. and SCHRIEFFER, J. R. (1957) *Phys. Rev.* **108,** 1175.

BARNETT, S. J. (1944) *Proc. Am. Acad. Arts & Sci.* **75,** 109.

BAZAROV, I. P. (1964) *Thermodynamics* (Pergamon Press, New York).

BELINFANTE, F. J. (1939) *Physica* **6,** 849, 870.

BERNOULLI, D. (1738) *Hydrodynamica* (Argentorati).

BETH, E. and UHLENBECK, G. E. (1937) *Physica* **4,** 915; see also (1936) *Physica* **3,** 729.

BETHE, H. A. (1935) *Proc. Roy. Soc. London* A **150,** 552.

BETHE, H. A. and KIRKWOOD, J. G. (1939) *J. Chem. Phys.* **7,** 578.

BETTS, D. D. *et al.* (1968) *Phys. Rev. Letters* **20,** 1507.

BETTS, D. D. *et al.* (1969) *Physics Letters* **29A,** 150.

BETTS, D. D. *et al.* (1970) *Can. J. Phys.* **48,** 1566.

BETTS, D. D. *et al.* (1970) *Physics Letters* **32A,** 152.

BETTS, D. S. (1969) *Contemporary Physics* **10,** 241.

BEYER, W. H. (ed.) (1966) *Handbook of Tables for Probability and Statistics* (Chemical Rubber Company, Cleveland).

BLAKEMORE, J. S. (1962) *Semiconductor Statistics* (Pergamon Press, Oxford).

DE BOER, J. (1948) *Physica* **14,** 139.

DE BOER, J. (1949) *Rep. Prog. Phys.* **12,** 305.

DE BOER, J. and BLAISSE, B. S. (1948) *Physica* **14,** 149.

DE BOER, J. and LUNBECK, R. J. (1948) *Physica* **14,** 510, 520.

DE BOER, J. and MICHELS, A. (1938) *Physica* **5,** 945.

DE BOER, J. and MICHELS, A. (1939) *Physica* **6,** 97, 409.

DE BOER, J. and UHLENBECK, G. E. (eds.) (1962, 1964, 1965, 1969) *Studies in Statistical Mechanics*, Vols. I, II, III and IV (North-Holland Publishing Company, Amsterdam).

BOGOLIUBOV, N. N. (1947) *J. Phys. USSR* **11,** 23.

BOGOLIUBOV, N. N. (1963) *Lectures on Quantum Statistics* (Gordon & Breach, New York).

BOLTZMANN, L. (1868) *Wien. Ber.* **58,** 517.

BOLTZMANN, L. (1871) *Wien. Ber.* **63,** 397, 679, 712.

BOLTZMANN, L. (1872) *Wien. Ber.* **66,** 275.
BOLTZMANN, L. (1875) *Wien. Ber.* **72,** 427.
BOLTZMANN, L. (1876) *Wien. Ber.* **74,** 503.
BOLTZMANN, L. (1877) *Wien. Ber.* **76,** 373.
BOLTZMANN, L. (1879) *Wien. Ber.* **78,** 7.
BOLTZMANN, L. (1884) *Ann. Phys.* **22,** 31, 291, 616.
BOLTZMANN, L. (1896, 1898) *Vorlesungen über Gastheorie*, 2 vols. (J. A. Barth, Leipzig). English transl. (1964) *Lectures on Gas Theory* (Berkeley, California).
BOLTZMANN, L. (1899) *Amsterdam Ber.* 477; see also (1909) *Wiss. Abhandl. von L. Boltzmann* III, 658.
BONCH-BRUEVICH, V. L. and TYABLIKOV, S. V. (1962) *The Green Function Method in Statistical Mechanics* (Interscience Publishers, New York).
BOSE, S. N. (1924) *Z. Physik* **26,** 178.
BRAGG, W. L. and WILLIAMS, E. J. (1934) *Proc. Roy. Soc. London* A **145,** 699.
BRAGG, W. L. and WILLIAMS, E. J. (1935) *Proc. Roy. Soc. London* A **151,** 540; A **152,** 231.
BROUT, R. (1965) *Phase Transitions* (W. A. Benjamin, New York).
BROUWER, W. and PATHRIA, R. K. (1967) *Phys. Rev.* **163,** 200.
BRUECKNER, K. A. and GAMMEL, J. L. (1958) *Phys. Rev.* **109,** 1040.
BRUECKNER, K. A. and SAWADA, K. (1957) *Phys. Rev.* **106,** 1117, 1128.
BRUSH, S. G. (1957) *Annals of Sc.* **13,** 188, 273.
BRUSH, S. G. (1958) *Annals of Sc.* **14,** 185, 243.
BRUSH, S. G. (1961) *Am. J. Phys.* **29,** 593.
BRUSH, S. G. (1961) *Am. Scientist* **49,** 202.
BRUSH, S. G. (1965–6) *Kinetic Theory*, 3 vols. (Pergamon Press, Oxford).
BRUSH, S. G. (1967) *Rev. Mod. Phys.* **39,** 883.
BUSH, V. and CALDWELL, S. H. (1931) *Phys. Rev.* **38,** 1898.
BYK, A. (1921) *Ann. Phys.* **66,** 157.
BYK, A. (1922) *Ann. Phys.* **69,** 161.
CHANDRASEKHAR, S. (1939) *Introduction to the Study of Stellar Structure* (University of Chicago Press, Chicago); now available in a Dover edition.
CHANDRASEKHAR, S. (1943) *Rev. Mod. Phys.* **15,** 1.
CHANDRASEKHAR, S. (1949) *Rev. Mod. Phys.* **21,** 383.
CHANG, T. S. (1939) *Proc. Camb. Phil. Soc.* **35,** 265.
CHANG, T. S. (1941) *J. Chem. Phys.* **9,** 169.
CHAPMAN, S. (1916) *Trans. Roy. Soc. London* A **216,** 279.
CHAPMAN, S. (1917) *Trans. Roy. Soc. London* A **217,** 115.
CHAPMAN, S. and COWLING, T. G. (1939) *The Mathematical Theory of Non-uniform Gases* (Cambridge University Press).
CHISHOLM, J. S. R. and BORDE, A. H. (1958) *An Introduction to Statistical Mechanics* (Pergamon Press, New York).
CHRÈTIEN, M., GROSS, E. P. and DESER, S. (eds.) (1968) *Statistical Physics, Phase Transitions and Superfluidity*, 2 vols. (Gordon & Breach, New York).
CLAUSIUS, R. (1857) *Ann. Phys.* **100,** 353 (= (1857) *Phil. Mag.* **14,** 108).
CLAUSIUS, R. (1859) *Ann. Phys.* **105,** 239 (= (1859) *Phil. Mag.* **17,** 81).
CLAUSIUS, R. (1870) *Ann. Phys.* **141,** 124 (= (1870) *Phil. Mag.* **40,** 122).
CLOW, J. R. and REPPY, J. D. (1966) *Phys. Rev. Letters* **16,** 887.
CLUNIE, J. (1954) *Proc. Phys. Soc.* **67,** 632.
COHEN, E. G. D. and VAN LEEUWEN, J. M. J. (1960) *Physica* **26,** 1171.
COHEN, E. G. D. and VAN LEEUWEN, J. M. J. (1961) *Physica* **27,** 1157.
COHEN, E. G. D. (ed.) (1962) *Fundamental Problems in Statistical Mechanics* (John Wiley, New York).
COHEN, E. H. and DUMOND, J. W. M. (1965) *Rev. Mod. Phys.* **37,** 537.
COHEN, M. and FEYNMAN, R. P. (1957) *Phys. Rev.* **107,** 13.
COMPTON, A. H. (1923) *Phys. Rev.* **21,** 207, 483.
COMPTON, A. H. (1923) *Phil. Mag.* **46,** 897.
COOPER, L. N. (1956) *Phys. Rev.* **104,** 1189.
COOPER, L. N. (1960) *Am. J. Phys.* **28,** 91.
CORAK, GARFUNKEL, SATTERTHWAITE and WEXLER (1955) *Phys. Rev.* **98,** 1699.
CRAWFORD, F. H. (1963) *Heat, Thermodynamics and Statistical Physics* (Harcourt-Brace and World, New York).
DARWIN, C. G. and FOWLER, R. H. (1922) *Phil. Mag.* **44,** 450, 823.

DARWIN, C. G. and FOWLER, R. H. (1922) *Proc. Cambridge Phil. Soc.* **21,** 262.
DARWIN, C. G. and FOWLER, R. H. (1923) *Proc. Cambridge Phil. Soc.* **21,** 391, 730; see also FOWLER, R. H. (1923) *Phil. Mag.* **45,** 1, 497; (1925) *Proc. Cambridge Phil. Soc.* **22,** 861; (1926) *Phil. Mag.* **1,** 845; (1926) *Proc. Roy. Soc. London* A **113,** 432.
DENNISON, D. M. (1927) *Proc. Roy. Soc. London* A **115,** 483.
DESLOGE, E. A. (1966) *Statistical Physics* (Holt, Rinehart & Winston, New York).
DINGLE, R. B. (1956) *J. App. Res.* B **6,** 225.
DIRAC, P. A. M. (1926) *Proc. Roy. Soc. London* A **112,** 661, 671.
DIRAC, P. A. M. (1929) *Proc. Cambridge Phil. Soc.* **25,** 62.
DIRAC, P. A. M. (1930) *Proc. Cambridge Phil. Soc.* **26,** 361, 376.
DIRAC, P. A. M. (1931) *Proc. Cambridge Phil. Soc.* **27,** 240.
DOMB, C. (1960) *Advances in Physics* **9,** 150.
DOMB, C. (1968) *Bull. Inst. Phys. London* **19,** 36.
DOMB, C. and SYKES, M. F. (1961) *J. Math. Phys.* **2,** 63.
DRUDE, P. (1900) *Ann. Phys.* **1,** 566; **3,** 369.
EHRENFEST, P. (1905) *Wiener Ber.* **114,** 1301.
EHRENFEST, P. (1906) *Phys. Zeits.* **7,** 528.
EHRENFEST, P. and T. (1912) *Enzyklopädie der mathematischen Wissenschaften,* vol. IV (Teubner, Leipzig). English transl. (1959) *The Conceptual Foundations of the Statistical Approach in Mechanics* (Cornell University Press, Ithaca).
EINSTEIN, A. (1902) *Ann. Phys.* **9,** 417.
EINSTEIN, A. (1903) *Ann. Phys.* **11,** 170.
EINSTEIN, A. (1905a) *Ann. Phys.* **17,** 132; see also (1906b) *ibid.* **20,** 199.
EINSTEIN, A. (1905b) *Ann. Phys.* **17,** 549; see also (1906a) *ibid.* **19,** 289, 371.
EINSTEIN, A. (1905c) *Ann. Phys.* **17,** 891; see also (1905d) *ibid.* **18,** 639 and (1906c) *ibid.* **20,** 627.
EINSTEIN, A. (1907) *Jb. Radioakt.* **4,** 411.
EINSTEIN, A. (1909) *Physik Z.* **10,** 185.
EINSTEIN, A. (1910) *Ann. Phys.* **33,** 1275.
EINSTEIN, A. (1924) *Berliner Ber.* **22,** 261.
EINSTEIN, A. (1925) *Berliner Ber.* **1,** 3.
EISENSCHITZ, R. (1958) *Statistical Theory of Irreversible Processes* (Oxford University Press).
ELCOCK, E. W. and LANDSBERG, P. T. (1957) *Proc. Phys. Soc. London* **70,** 161.
EMERY, V. J. and SESSLER, A. M. (1960) *Phys. Rev.* **119,** 43.
ENSKOG, D. (1917) *Kinetische Theorie der Vorgänge in mässig verdünnten Gasen,* dissertation (Almquist & Wiksell, Uppsala).
EPSTEIN, P. S. (1957) *Textbook of Thermodynamics* (John Wiley, New York).
EYRING, H., HENDERSON, D., STOVER, B. J. and EYRING, E. M. (1963) *Statistical Mechanics and Dynamics* (John Wiley, New York).
FARKAS, A. (1935) *Orthohydrogen, Parahydrogen and Heavy Hydrogen* (Cambridge University Press).
FARQUHAR, I. E. (1964) *Ergodic Theory in Statistical Mechanics* (Interscience Publishers, New York).
FAY, J. A. (1965) *Molecular Thermodynamics* (Addison-Wesley, Reading, Mass.).
FEDOSOV, B. V. (1963) *Dokl. Akad. Nauk SSSR* **151,** 786; English transl. *Sov. Math. (Doklady)* **4,** 1092.
FEDOSOV, B. V. (1964) *Dokl. Akad. Nauk SSSR* **157,** 536; English transl. *Sov. Math. (Doklady)* **5,** 988.
FERMI, E. (1926) *Z. Physik* **36,** 902.
FERMI, E. (1928) *Zeit. für Phys.* **48,** 73; see also (1927) *Acc. Lencei* **6,** 602.
FERMI, E. (1936) *Ricerca Sci.* **7,** 13.
FERMI, E. (1957) *Thermodynamics* (Dover Publications, New York).
FETTER, A. L. (1963) *Phys. Rev. Letters* **10,** 507.
FETTER, A. L. (1965) *Phys. Rev.* **138,** A 429.
FEYNMAN, R. P. (1953) *Phys. Rev.* **91,** 1291, 1301.
FEYNMAN, R. P. (1954) *Phys. Rev.* **94,** 262.
FEYNMAN, R. P. (1955) *Progress in Low Temperature Physics,* ed. C. J. GORTER (North-Holland, Amsterdam) **1,** 17.
FEYNMAN, R. P. and COHEN, M. (1956) *Phys. Rev.* **102,** 1189.
FISHER, M. E. (1963) *J. Math. Phys.* **4,** 278.
FISHER, M. E. (1965) *Lectures in Theoretical Physics* (University of Colorado Press), Vol. VII C, p. 1.
FISHER, M. E. (1967) *Rep. Prog. Phys.* **30,** 615.
FOKKER, A. D. (1914) *Ann. Phys.* **43,** 812.
FOWLER, R. H. (1926) *Monthly Not. R.A.S.* **87,** 114.

Fowler, R. H. (1955) *Statistical Mechanics*, 2nd ed. (Cambridge University Press).
Fowler, R. H. and Guggenheim, E. A. (1940) *Proc. Roy. Soc. London* A **174,** 189.
Fowler, R. H. and Guggenheim, E. A. (1960) *Statistical Thermodynamics* (Cambridge University Press).
Fowler, R. H. and Nordheim, L. (1928) *Proc. Roy. Soc. London* A **119,** 173.
Frisch, H. L. and Lebowitz, J. L. (1964) *The Equilibrium Theory of Classical Fluids* (W. A. Benjamin, New York).
Fujiwara, ter Haar and Wergeland (1970) *J. Stat. Phys.* **2,** 329.
Galitskii, V. M. (1958) *J. Exptl. Theor. Phys. USSR* **34,** 151; English transl. *Soviet Physics JETP* **7,** 104.
Gibbs, J. W. (1902) *Elementary Principles in Statistical Mechanics* (Yale University Press, New Haven); reprinted by Dover Publications, New York (1960). See also *A Commentary on the Scientific Writings of J. Willard Gibbs*, ed. A. Haas, Yale University Press, New Haven (1936), Vol. II.
Ginzburg, V. L. and Pitaevskii, L. P. (1958) *Zh. Eksperim. i Teor. Fiz.* **34,** 1240; English transl. (1958) *Soviet Phys. JETP* **7,** 858.
Gol'dman, Krivchenkov, Kogan and Galitskii (1960) *Problems in Quantum Mechanics*, translated by D. ter Haar (Academic Press, New York).
Gombás, P. (1949) *Die statistische Theorie des Atoms und ihre Anwendungen* (Vienna, Springer).
Gombás, P. (1952) *Ann. Phys.* **10,** 253.
Green, H. S. and Hurst, C. A. (1964) *Order–Disorder Phenomena* (Interscience Publishers, New York).
Grindlay, J. (1968) *Can. J. Phys.* **46,** 2253.
de Groot, S. R. (1951) *Thermodynamics of Irreversible Processes* (North-Holland, Amsterdam).
de Groot, S. R. and Mazur, P. (1962) *Non-equilibrium Thermodynamics* (North-Holland, Amsterdam).
Gross, E. P. (1958) *Annals of Phys.* **4,** 57.
Gross, E. P. (1960) *Annals of Phys.* **9,** 292.
Gross, E. P. (1961) *Nuovo Cim.* **20,** 454.
Gross, E. P. (1963) *J. Math. Phys.* **4,** 195.
Guggenheim, E. A. (1935) *Proc. Roy. Soc. London* A **148,** 304.
Guggenheim, E. A. (1938) *Proc. Roy. Soc. London* A **169,** 134.
Guggenheim, E. A. (1959) *Boltzmann's Distribution Law* (North-Holland, Amsterdam).
Guggenheim, E. A. (1960) *Elements of the Kinetic Theory of Gases* (Pergamon Press, Oxford).
Guggenheim, E. A. (1960) *Thermodynamics*, 4th ed. (Interscience Publishers, New York).
Gupta, H. (1947) *Proc. Nat. Ins. Sci. India* **13,** 35.
ter Haar, D. (1954) *Elements of Statistical Mechanics* (Rinehart, New York).
ter Haar, D. (1955) *Rev. Mod. Phys.* **27,** 289.
ter Haar, D. (1961) *Rep. Prog. Phys.* **24,** 304.
ter Haar, D. (1966) *Elements of Thermostatistics* (Holt, Rinehart & Winston, New York).
ter Haar, D. (1967) *The Old Quantum Theory* (Pergamon Press, Oxford).
ter Haar, D. (1968) *On the History of Photon Statistics*, Course 42 of the International School of Physics "Enrico Fermi" (Academic Press, New York).
ter Haar, D. and Wergeland, H. (1966) *Elements of Thermodynamics* (Addison-Wesley, Reading, Mass.).
de Haas, W. J. and van Alphen, P. M. (1930) *Leiden Comm.* 208d, 212a.
Hall, H. E. (1960) *Adv. in Phys.* **9,** 89.
Hanbury Brown, R. and Twiss, R. Q. (1956) *Nature* **177,** 27; **178,** 1447.
Hanbury Brown, R. and Twiss, R. Q. (1957) *Proc. Roy. Soc. London* A **242,** 300; A **243,** 291.
Hanbury Brown, R. and Twiss, R. Q. (1958) *Proc. Roy. Soc. London* A **248,** 199, 222.
Heisenberg, W. (1928) *Z. Physik* **49,** 619.
Heller, P. (1967) *Rep. Prog. Phys.* **30,** 731.
Henderson, D. (1964) *Am. J. Phys.* **32,** 795.
Henry, W. E. (1952) *Phys. Rev.* **88,** 561.
Henshaw, D. G. (1960) *Phys. Rev.* **119,** 9.
Henshaw, D. G. and Woods, A. D. B. (1961) *Phys. Rev.* **121,** 1266.
Herapath, J. (1821) *Ann. Philos.* **1,** 273, 340, 401.
Hildebrand, J. H. (1963) *An Introduction to Molecular Kinetic Theory* (Reinhold, New York).
Hill, T. L. (1953) *J. Phys. Chem.* **57,** 324.
Hill, T. L. (1956) *Statistical Mechanics* (McGraw-Hill, New York).
Hill, T. L. (1960) *Introduction to Statistical Thermodynamics* (Addison-Wesley, Reading, Mass.).
Hirschfelder, J. O., Curtiss, C. F. and Bird, R. B. (1954) *Molecular Theory of Gases and Liquids* (John Wiley, New York).
Holliday, D. and Sage, M. L. (1964) *Annals of Physics* **29,** 125.
van Hove, L. (1949) *Physica,* **15,** 951.

VAN HOVE, L., HUGENHOLTZ, N. M. and HOWLAND, L. P. (1961) *Problems in Quantum Theory of Many-particle Systems* (W. A. Benjamin, New York).
HUANG, K. (1959) *Phys. Rev.* **115,** 765.
HUANG, K. (1960) *Phys. Rev.* **119,** 1129.
HUANG, K. (1963) *Statistical Mechanics* (John Wiley, New York).
HUANG, K. and KLEIN, A. (1964) *Annals of Physics* **30,** 203.
HUANG, K. and YANG, C. N. (1957) *Phys. Rev.* **105,** 767.
HUANG, K., YANG, C. N. and LUTTINGER, J. M. (1957) *Phys. Rev.* **105,** 776.
HUPSE, J. C. (1942) *Physica* **9,** 633.
HURST, C. A. and GREEN, H. S. (1960) *J. Chem. Phys.* **33,** 1059.
HUSIMI, K. (1939) *Proc. Phys.-Math. Soc. Japan* **21,** 759.
HUSIMI, K. (1940) *Proc. Phys.-Math. Soc. Japan* **22,** 264.
ISING, E. (1925) *Z. Physik* **31,** 253.
JACKSON, H. W. and FEENBERG, E. (1962) *Rev. Mod. Phys.* **34,** 686.
JEANS, J. H. (1905) *Phil. Mag.* **10,** 91.
JEANS, J. H. (1925) *The Dynamical Theory of Gases*, 4th ed. (Cambridge University Press); reprinted by Dover Publications, New York (1954).
JEANS, J. H. (1959) *An Introduction to the Kinetic Theory of Gases* (Cambridge University Press).
JOHNSON, J. B. (1927) *Nature* **119,** 50.
JOHNSON, J. B. (1927) *Phys. Rev.* **29,** 367.
JOHNSON, J. B. (1928) *Phys. Rev.* **32,** 97.
JOULE, J. P. (1851) *Mem. and Proc. Manchester Lit. and Phil. Soc.* **9,** 107 (= (1857) *Phil. Mag.* **14,** 211).
KAC, M. and WARD, J. C. (1952) *Phys. Rev.* **88,** 1332.
KADANOFF, L. P. and BAYM, G. (1962) *Quantum Statistical Mechanics* (W. A. Benjamin, New York).
KADANOFF, L. P. *et al.* (1967) *Rev. Mod. Phys.* **39,** 395.
KAHN, B. (1938) *On the Theory of the Equation of State*, Thesis, Utrecht; English translation appears in DE BOER and UHLENBECK, eds. (1965), *Studies in Statistical Mechanics*, Vol. III (North-Holland, Amsterdam).
KAHN, B. and UHLENBECK, G. E. (1938) *Physica* **5,** 399.
VAN KAMPEN, N. G. (1969) *J. Phys. Soc. Japan* **26** (suppl.), 316.
KAPPLER, E. (1931) *Ann. Phys.* **11,** 233.
KAPPLER, E. (1938) *Ann. Phys.* **31,** 377, 619.
KATSURA, S. (1959) *Phys. Rev.* **115,** 1417.
KATSURA, S. (1963) *Advances in Physics* **12,** 391.
KAWATRA, M. P. and PATHRIA, R. K. (1966) *Phys. Rev.* **151,** 132.
KEESOM, P. H. and PEARLMAN, N. (1953) *Phys. Rev.* **91,** 1354.
KHALATNIKOV, I. M. (1965) *Introduction to the Theory of Superfluidity* (W. A. Benjamin, New York).
KHINCHIN, A. I. (1949) *Mathematical Foundations of Statistical Mechanics* (Dover Publications, New York).
KHINTCHINE, A. (1934) *Math. Ann.* **109,** 604.
KILPATRICK, J. E. (1953) *J. Chem. Phys.* **21,** 274.
KILPATRICK, J. E. and FORD, D. I. (1969) *Am. J. Phys.* **37,** 881.
KIRKWOOD, J. G. (1938) *J. Chem. Phys.* **6,** 70.
KIRKWOOD, J. G. (1965) *Quantum Statistics and Cooperative Phenomena* (Gordon & Breach, New York).
KITTEL, C. (1958) *Elementary Statistical Physics* (John Wiley, New York).
KITTEL, C. (1969) *Thermal Physics* (John Wiley, New York).
KLEIN, M. J. and TISZA, L. (1949) *Phys. Rev.* **76,** 1861.
DE KLERK, D., HUDSON, R. P. and PELLAM, J. R. (1953) *Phys. Rev.* **89,** 326, 662.
KOTHARI, D. S. and AULUCK, F. C. (1957) *Curr. Sci.* **26,** 169.
KOTHARI, D. S. and SINGH, B. N. (1941) *Proc. Roy. Soc. London* A **178,** 135.
KOTHARI, D. S. and SINGH, B. N. (1942) *Proc. Roy. Soc. London* A **180,** 414.
KRAMERS, H. A. (1938) *Proc. Kon. Ned. Akad. Wet. (Amsterdam)* **41,** 10.
KRAMERS, H. A. and WANNIER, G. H. (1941) *Phys. Rev.* **60,** 252, 263.
KRÖNIG, A. (1856) *Ann. Phys.* **99,** 315.
KUBO, R. (1943) *Busseiron-kenkyu* **1,** 1 (English transl. UCRL-trans. 1030 (L), available from the Lawrence Radiation Laboratory, Berkeley).
KUBO, R. (1956) *Can. J. Phys.* **34,** 1274.
KUBO, R. (1957) *Proc. Phys. Soc. Japan* **12,** 570.
KUBO, R. (1959) *Some Aspects of the Statistical Mechanical Theory of Irreversible Processes*, University of Colorado *Lectures in Theoretical Physics* (Interscience Publishers, New York), Vol. 1, p. 120.

KUBO, R. (1965) *Statistical Mechanics* (Interscience Publishers, New York).
KUBO, R. (1966) *Rep. Prog. Phys.* **29,** 255.
KUPER, C. G. (1956) *Phys.* **22,** 1291.
LANDAU, L. D. (1927) *Z. Phys.* **45,** 430; see also (1965) *Collected Papers*, Oxford, p. 8.
LANDAU, L. D. (1930) *Z. Phys.* **64,** 629.
LANDAU, L. D. (1941) *J. Phys. USSR* **5,** 71.
LANDAU, L. D. (1947) *J. Phys. USSR* **11,** 91.
LANDAU, L. D. (1956) *J. Exptl. Theoret. Phys. USSR* **30,** 1058; English transl. *Soviet Physics JETP* **3,** 920 (1957).
LANDAU, L. D. and LIFSHITZ, E. M. (1958) *Statistical Physics* (Pergamon Press, Oxford).
LANDSBERG, P. T. (1954a) *Proc. Nat. Acad. Sciences (U.S.A.)* **40,** 149.
LANDSBERG, P. T. (1954b) *Proc. Camb. Phil. Soc.* **50,** 65.
LANDSBERG, P. T. (1961) *Thermodynamics with Quantum Statistical Illustrations* (Interscience Publishers, New York).
LANDSBERG, P. T. and DUNNING-DAVIES, J. (1965) *Phys. Rev.* **138,** A1049; see also their article in the *Statistical Mechanics of Equilibrium and Non-equilibrium* (1965), ed. J. MEIXNER (North-Holland Publishing Co., Amsterdam).
LANE, C. T. (1962) *Superfluid Physics* (McGraw-Hill, New York).
LANGEVIN, P. (1905) *J. Phys.* **4,** 678.
LANGEVIN, P. (1905) *Ann. Chim. et Phys.* **5,** 70.
LANGEVIN, P. (1908) *Comptes Rend. Acad. Sci. Paris* **146,** 530.
VON LAUE, M. (1919) *Ann. Phys.* **58,** 695.
LEE, J. F., SEARS, F. W. and TURCOTTE, D. L. (1963) *Statistical Thermodynamics* (Addison-Wesley, Reading, Mass.).
LEE, T. D., HUANG, K. and YANG, C. N. (1957) *Phys. Rev.* **106,** 1135.
LEE, T. D. and YANG, C. N. (1952) *Phys. Rev.* **87,** 410; see also *ibid.*, p. 404.
LEE, T. D. and YANG, C. N. (1957) *Phys. Rev.* **105,** 1119.
LEE, T. D. and YANG, C. N. (1958) *Phys. Rev.* **112,** 1419.
LEE, T. D. and YANG, C. N. (1959 a, b) *Phys. Rev.* **113,** 1165; **116,** 25.
LEE, T. D. and YANG, C. N. (1959c) *Phys. Rev.* **113,** 1406.
LEE, T. D. and YANG, C. N. (1960 a, b, c) *Phys. Rev.* **117,** 12, 22, 897.
LENNARD-JONES, J. E. (1924) *Proc. Roy. Soc. London* A **106,** 463.
LENZ, W. (1929) *Z. Physik* **56,** 778.
LINDSAY, R. B. (1941) *Introduction to Physical Statistics* (John Wiley, New York).
LIOUVILLE, J. (1838) *J. de Math.* **3,** 348.
LIU, L., LIU, L. S. and WONG, K. W. (1964) *Phys. Rev.* **135,** A1166.
LONDON, F. (1938) *Nature* **141,** 643.
LONDON, F. (1938) *Phys. Rev.* **54,** 947.
LONDON, F. (1954) *Superfluids*, Vols. 1 and 2 (John Wiley, New York); reprinted by Dover Publications, New York (1964).
LORENTZ, H. A. (1904–5) *Proc. Kon. Ned. Akad. Wet. Amsterdam* **7,** 438, 585, 684. See also (1909) *The Theory of Electrons* (Teubner, Leipzig), secs. 47–50; reprinted by Dover Publications, New York (1952).
LORENTZ, H. A. (1905) *Arch. Néerland. Sci.* **10,** 336 (= (1936) *Collected Papers*, The Hague; Vol. III).
LOSCHMIDT, J. (1876) *Wien. Ber.* **73,** 139.
LOSCHMIDT, J. (1877) *Wien. Ber.* **75,** 67.
LÜDERS, G. and ZUMINO, B. (1958) *Phys. Rev.* **110,** 1450.
MAA, D. Y. (1939) *J. Acous. Soc. America* **10,** 235.
MACDONALD, D. K. C. (1948–9) *Rep. Prog. Phys.* **12,** 56.
MACDONALD, D. K. C. (1950) *Phil. Mag.* **41,** 814.
MACDONALD, D. K. C. (1962) *Noise and Fluctuations: An Introduction* (John Wiley, New York).
MACDONALD, D. K. C. (1963) *Introductory Statistical Mechanics for Physicists* (John Wiley, New York).
MAJUMDAR, R. (1929) *Bull. Calcutta Math. Soc.* **21,** 107.
MANDEL, L., SUDARSHAN, E. C. G. and WOLF, E. (1964) *Proc. Phys. Soc. London* **84,** 435.
MARCH, N. H. (1957) *Adv. in Phys.* **6,** 1.
MARTIN, P. C. and DE DOMINICIS, C. (1957) *Phys. Rev.* **105,** 1417.
MATSUBARA, T. and MATSUDA, H. (1956) *Prog. Theor. Phys. Japan* **16,** 416.
MAXWELL, J. C. (1860) *Phil. Mag.* **19,** 19; **20,** 1.
MAXWELL, J. C. (1867) *Trans. Roy. Soc. London* **157,** 49 (= (1868) *Phil. Mag.* **35,** 129, 185).

MAXWELL, J. C. (1879) *Camb. Phil. Soc. Trans.* **12,** 547.
MAYER, J. E. *et al.* (1937) *J. Chem. Phys.* **5,** 67, 74.
MAYER, J. E. *et al.* (1938) *J. Chem. Phys.* **6,** 87, 101.
MAYER, J. E. *et al.* (1942) *J. Chem. Phys.* **10,** 629.
MAYER, J. E. *et al.* (1951) *J. Chem. Phys.* **19,** 1024.
MAYER, J. E. and MAYER, M. G. (1940) *Statistical Mechanics* (John Wiley, New York).
MEERON, E. (1965–6) *Physics of Many-particle Systems* (Gordon & Breach, New York).
MENDELSSOHN, K. (1956) "Liquid Helium" in *Handbuch der Physik* (Springer-Verlag, Berlin), Vol. 15, p. 370.
MENDELSSOHN, K. (1960) *Cryophysics* (Interscience Publishers, New York).
MILNE, E. (1927) *Proc. Camb. Phil. Soc.* **23,** 794.
MOHLING, F. (1961) *Phys. Rev.* **122,** 1062.
MOHLING, F. (1964) *Phys. Rev.* **135,** A 831, A 855, A 876; **136,** A 938.
MOHLING, F. (1965) *Phys. Rev.* **139,** A 664.
MOHLING, F. and GRANDY, Jr., W. T. (1965) *J. Math. Phys.* **6,** 348.
MONTROLL, E. (1941) *J. Chem. Phys.* **9,** 706.
MONTROLL, E. (1942) *J. Chem. Phys.* **10,** 61.
MONTROLL, E. (1963) *The Many-body Problem*, ed. J. K. PERCUS (Interscience Publishers, New York).
MONTROLL, E. (1964) *Applied Combinatorial Mathematics*, ed. E. F. BECKENBACH (John Wiley, New York).
MORSE, P. M. (1969) *Thermal Physics*, 2nd ed. (W. A. Benjamin, New York).
MOSENGEIL, K. v. (1907) *Ann. Phys.* **22,** 867.
MÜNSTER, A. (1959) *Statistische Thermodynamik* (Springer-Verlag, Berlin).
MÜNSTER, A. (1959) "Prinzipien der statistischen Mechanik" in *Handbuch der Physik*, Vol. 3_2, p. 176.
NANDA, V. S. (1953) *Proc. Nat. Inst. Sci. (India)* **19,** 681.
NERNST, W. (1906) *Göttinger Nachr.* 1.
VON NEUMANN, J. (1927) *Göttinger Nachr.* 1, 24, 273.
NEWELL, G. F. and MONTROLL, E. W. (1953) *Rev. Mod. Phys.* **25,** 353.
NOZIÈRES, P. (1964) *Theory of Interacting Fermi Systems* (W. A. Benjamin, New York).
NYQUIST, H. (1927) *Phys. Rev.* **29,** 614.
NYQUIST, H. (1928) *Phys. Rev.* **32,** 110.
ONO, S. (1951) *J. Chem. Phys.* **19,** 504.
ONSAGER, L. (1931) *Phys. Rev.* **37,** 405; **38,** 2265.
ONSAGER, L. (1944) *Phys. Rev.* **65,** 117.
ONSAGER, L. (1949) *Nuovo Cim.* **6** (Suppl. 2), 249, 261.
OSBORNE, D. V. (1950) *Proc. Phys. Soc. London* A **63,** 909.
OSBORNE, D. V. (1963) *Can. J. Phys.* **41,** 820; with Turtington and Brown.
OTT, H. (1963) *Z. Phys.* **175,** 70.
PAIS, A. and UHLENBECK, G. E. (1959) *Phys. Rev.* **116,** 250.
PATHRIA, R. K. (1955) *Proc. Nat. Inst. Sci. India* **21,** 331.
PATHRIA, R. K. (1957) *Proc. Nat. Inst. Sci. India* **23,** 168.
PATHRIA, R. K. (1963) *The Theory of Relativity* (Hindustan Publishing Corporation, Delhi).
PATHRIA, R. K. (1966a) *Proc. Phys. Soc. London* **88,** 791.
PATHRIA, R. K. (1966b) *Nuovo Cim. (Supp.), Ser. I,* **4,** 276.
PATHRIA, R. K. (1967) *Proc. Phys. Soc. London* **91,** 1.
PATHRIA, R. K. (1970) *A Critical Review of Thermodynamics*, eds. STUART, GAL-OR and BRAINARD (Mono Book Corp., Baltimore).
PATHRIA, R. K. and KAWATRA, M. P. (1962) *Prog. Theor. Phys. Japan* **27,** 638, 1085.
PATHRIA, R. K. and KAWATRA, M. P. (1963) *Phys. Rev.* **129,** 944.
PATHRIA, R. K. and SINGH, A. D. (1960) *Proc. Nat. Ins. Sci. India* **26,** 520.
PAULI, W. (1925) *Z. Physik* **31,** 776.
PAULI, W. (1927) *Z. Physik* **41,** 81.
PAULI, W. (1940) *Phys. Rev.* **58,** 716.
PAULI, W. and BELINFANTE, F. J. (1940) *Physica* **7,** 177.
PEIERLS, R. E. (1933) *Z. Phys.* **81,** 186.
PEIERLS, R. E. (1936) *Proc. Camb. Phil. Soc.* **32,** 471, 477.
PENROSE, O. and ONSAGER, L. (1956) *Phys. Rev.* **104,** 576.
PERCUS, J. K. (1963) *The Many-body Problem* (Interscience Publishers, New York).
PINES, D. (1962) *The Many-body Problem* (W. A. Benjamin, New York).
PINES, D. and NOZIÈRES, P. (1966) *The Theory of Quantum Liquids*, 2 vols. (W. A. Benjamin, New York).
PIPPARD, A. B. (1957) *The Elements of Classical Thermodynamics* (Cambridge University Press).

PITAEVSKII, L. P. (1959) *Soviet Phys. JETP* **9**, 830.
PITAEVSKII, L. P. (1961) *Soviet Phys. JETP* **13**, 451.
PLANCK, M. (1900) *Verhandl. Deut. Physik Ges.* **2**, 202, 237.
PLANCK, M. (1907) *S. B. Preuss. Akad. Wiss.*, p. 542.
PLANCK, M. (1908) *Ann. Phys.* **26**, 1.
PLANCK, M. (1917) *Sitz. der Preuss. Akad.*, p. 324.
POSPIŠIL, W. (1927) *Ann. Phys.* **83**, 735.
PRESENT, R. D. (1958) *Introduction to the Kinetic Theory of Gases* (McGraw-Hill, New York).
PRIGOGINE, I. (1967) *Introduction to Thermodynamics of Irreversible Processes*, 3rd ed. (John Wiley, New York).
PURCELL, E. M. (1956) *Nature* **178**, 1449.
PURCELL, E. M. and POUND, R. V. (1951) *Phys. Rev.* **81**, 279.
RAMSEY, N. F. (1956) *Phys. Rev.* **103**, 20.
RAYFIELD, G. W. and REIF, F. (1964) *Phys. Rev.* **136**, A 1194; see also (1963) *Phys. Rev. Letters* **11**, 305.
RAYLEIGH, LORD (1900) *Phil. Mag.* **49**, 539.
REE, F. H. and HOOVER, W. G. (1964) *J. Chem. Phys.* **40**, 939.
REIF, F. (1965) *Fundamentals of Statistical and Thermal Physics* (McGraw-Hill, New York).
RHODES, P. (1950) *Proc. Roy. Soc. London* A **204**, 396.
RIECKE, E. (1898) *Ann. Phys.* **66**, 353, 545.
RIECKE, E. (1900) *Ann. Phys.* **2**, 835.
ROBERTS, T. R. and SYDORIAK, S. G. (1955) *Phys. Rev.* **98**, 1672.
ROBINSON, J. E. (1951) *Phys. Rev.* **83**, 678.
RUSHBROOKE, G. S. (1938) *Proc. Roy. Soc. London* A **166**, 296.
RUSHBROOKE, G. S. (1955) *Introduction to Statistical Mechanics* (Clarendon Press, Oxford).
RUSHBROOKE, G. S. (1963) *J. Chem. Phys.* **39**, 842.
SACKUR, O. (1911) *Ann. Phys.* **36**, 958.
SACKUR, O. (1912) *Ann. Phys.* **40**, 67.
SCHIFF, L. I. (1968) *Quantum Mechanics*, 3rd ed. (McGraw-Hill, New York).
SCHRIEFFER, J. R. (1964) *Theory of Superconductivity* (W. A. Benjamin, New York).
SCHRÖDINGER, E. (1960) *Statistical Thermodynamics* (Cambridge University Press).
SCHULTZ, T. D. (1963) *Quantum Field Theory and the Many-body Problem* (Gordon & Breach, New York).
SCOTT, G. G. (1951) *Phys. Rev.* **82**, 542.
SCOTT, G. G. (1952) *Phys. Rev.* **87**, 697.
SHANNON, C. E. (1948) *Bell Syst. Tech. J.* **27**, 379, 623.
SHANNON, C. E. (1949) *The Mathematical Theory of Communication* (University of Illinois Press, Urbana).
SHOENBERG, D. (1957) *Progress in Low Temperature Physics*, ed. C. J. GORTER, Vol. II, p. 226.
SIMON, F. (1930) *Ergeb. Exakt. Naturwiss.* **9**, 222.
SINGH, A. D. (1968) *Can. J. Phys.* **46**, 1801, 2607.
SMITH, B. L. (1969) *Contemporary Physics* **10**, 305.
SMOLUCHOWSKI, M. v. (1906) *Ann. Phys.* **21**, 756.
SMOLUCHOWSKI, M. v. (1908) *Ann. Phys.* **25**, 205.
SOMMERFELD, A. (1928) *Z. Physik* **47**, 1; see also SOMMERFELD and FRANK (1931) *Rev. Mod. Phys.* **3**, 1.
SOMMERFELD, A. (1932) *Z. Physik* **78**, 283.
SOMMERFELD, A. (1956) *Thermodynamics and Statistical Mechanics* (Academic Press, New York).
STANLEY, H. E. (1971) *Introduction to Phase Transitions and Critical Phenomena* (Oxford University Press).
STEFAN, J. (1879) *Wien. Ber.* **79**, 391.
STUART, E. B., GAL-OR, B. and BRAINARD, A. J. (eds.) (1970) *A Critical Review of Thermodynamics* (Mono Book Corp., Baltimore).
SYDORIAK, S. G., GRILLY, E. R. and HAMMEL, E. F. (1949) *Phys. Rev.* **75**, 303.
TEMPERLEY, H. N. V. (1949) *Proc. Roy. Soc. London* A **199**, 361.
TEMPERLEY, H. N. V. (1956) *Changes of State* (Interscience Publishers, New York).
TETRODE, H. (1912) *Ann. Phys.* **38**, 434 (corrections to this paper in (1912) *ibid.* **39**, 255).
TETRODE, H. (1915) *Proc. Kon. Ned. Akad. Amsterdam* **17**, 1167.
THOMAS, L. H. (1927) *Proc. Camb. Phil. Soc.* **23**, 542.
TISZA, L. (1938) *Nature* **141**, 913.
TISZA, L. (1938) *Compt. Rend.* **207**, 1035, 1186.
TOLMAN, R. C. (1934) *Relativity, Thermodynamics and Cosmology* (Clarendon Press, Oxford).
TOLMAN, R. C. (1938) *The Principles of Statistical Mechanics* (Oxford University Press).
TUTTLE, E. R. and MOHLING, F. (1966) *Annals of Physics* **38**, 510.

TYSON, J. A. and DOUGLASS, Jr., D. H. (1966) *Phys. Rev. Letters* **17,** 472, 622.
UHLENBECK, G. E. and BETH, E. (1936) *Physica* **3,** 729.
UHLENBECK, G. E. and BETH, E. (1937) *Physica* **4,** 915.
UHLENBECK, G. E. and FORD, G. W. (1963) *Lectures in Statistical Mechanics* (American Mathematical Society, Providence).
UHLENBECK, G. E. and GROPPER, L. (1932) *Phys. Rev.* **41,** 79.
UHLENBECK, G. E. and ORNSTEIN, L. S. (1930) *Phys. Rev.* **36,** 823.
URSELL, H. D. (1927) *Proc. Camb. Phil. Soc.* **23,** 685.
VANDERSLICE, J. T., SCHAMP, Jr., H. W. and MASON, E. A. (1966) *Thermodynamics* (Prentice-Hall, Englewood Cliffs, N.J.).
VINEN, W. F. (1961) *Proc. Roy. Soc. London* A **260,** 218; see also *Progress in Low Temperature Physics*, Vol. III (1961), ed. C. J. GORTER (North-Holland Publishing Co., Amsterdam).
VINEN, W. F. (1968) *Rep. Prog. Phys.* **31,** 61.
VAN DER WAALS, J. D. (1873) *Over de Continuïteit van den Gas- en Vloeistoftoestand*, Thesis, Leiden.
VAN DER WAERDAN, B. L. (1941) *Z. Physik* **118,** 473.
WAKEFIELD, A. J. (1951) *Proc. Camb. Phil. Soc.* **47,** 419, 799.
WALKER, C. B. (1956) *Phys. Rev.* **103,** 547.
WALTON, A. J. (1969) *Contemporary Physics* **10,** 181.
WANNIER, G. H. (1945) *Rev. Mod. Phys.* **17,** 50.
WANNIER, G. H. (1966) *Statistical Physics* (John Wiley, New York).
WATERSTON, J. J. (1892) *Phil. Trans. Roy. Soc. London* A **183,** 5, 79; reprinted in his collected papers (1928), ed. J. S. HALDANE (Oliver & Boyd, Edinburgh), pp. 207, 318. An abstract of Waterston's work did appear earlier; see *Proc. Roy. Soc. London* **5,** 604 (1846).
WAX, N. (ed.) (1954) *Selected Papers on Noise and Stochastic Processes* (Dover Publications, New York).
WEINSTOCK, R. (1969) *Am. J. Phys.* **37,** 1273.
WELLER, W. (1963) *Z. Naturforsch.* **18***A*, 79.
WERGELAND, H. (1969) *Lettere al Nuovo Cimento* **1,** 49.
WHITMORE, S. C. and ZIMMERMANN, W. (1965) *Phys. Rev. Letters* **15,** 389.
WIEBES, NIELS-HAKKENBERG and KRAMERS (1957) *Physica* **23,** 625.
WIEN, W. (1896) *Ann. Phys.* **58,** 662.
WIENER, N. (1930) *Act. Math. Stockholm* **55,** 117.
WILKS, J. (1961) *The Third Law of Thermodynamics* (Oxford University Press).
WILSON, A. H. (1960) *Thermodynamics and Statistical Mechanics* (Cambridge University Press).
WOODS, A. D. B. (1966) *Quantum Fluids*, ed. D. F. BREWER (North-Holland Publishing Co., Amsterdam), p. 239.
WU, T. T. (1959) *Phys. Rev.* **115,** 1390.
WU, T. T. (1961) *J. Math. Phys.* **2,** 105.
YANG, C. N. (1952) *Phys. Rev.* **85,** 808.
YANG, C. N. and LEE, T. D. (1952) *Phys. Rev.* **87**, 404; see also *ibid.*, p. 410.
YANG, C. P. (1962) *J. Math. Phys.* **3,** 797.
YARNELL, J. L. *et al.* (1959) *Phys. Rev.* **113,** 1379, 1386.
YUEN, C. K. (1970) *Am. J. Phys.* **38,** 246.
ZERMELO, E. (1896) *Ann. Phys.* **57,** 485; **59,** 793.
VAN DER ZIEL, A. (1956) *Noise* (Prentice-Hall, Englewood Cliffs, N.J.).

INDEX